Calcium Antagonists in Clinical Medicine

SECOND EDITION

Murray Epstein, MD, FACP
Professor of Medicine
University of Miami School of Medicine
Attending Physician
Veterans Affairs Medical Center
Jackson Memorial Medical Center
Miami, Florida

HANLEY & BELFUS, INC. / Philadelphia

Publisher: HANLEY & BELFUS, INC.
Medical Publishers
210 South 13th Street
Philadelphia, PA 19107
215-546-7293; 800-962-1892
FAX 215-790-9330
Web site: http://www.hanleyandbelfus.com

Disclaimer: Although the information in this book has been carefully reviewed for correctness of dosage and indications, neither the authors nor the editor nor the publisher can accept any legal responsibility for any errors or omissions that may be made. Neither the publisher nor the editor makes any warranty, expressed or implied, with respect to the material contained herein. Experimental compounds and off-label uses of approved products are discussed. Before prescribing any drug, the reader must review the manufacturer's current product information (package inserts) for accepted indications, absolute dosage recommendations, and other information pertinent to the safe and effective use of the product described. This is especially important when drugs are given in combination or as an adjunct to other forms of therapy.

Library of Congress Cataloging-in-Publication Data

Calcium antagonists in clinical medicine / edited by Murray Epstein.—2nd ed.
p. cm.
Includes bibliographical references and index.
ISBN 1-56053-223-8 (alk. paper)
1. Calcium—Antagonists—Therapeutic use. I. Epstein, Murray, 1937– .
[DNLM: 1. Calcium Channel Blockers—metabolism. 2. Calcium Channel Blockers—therapeutic use. 3. Heart Diseases—drug therapy.
QV 150 C14368 1997]
RC684.C34C337 1997
616.1'061—dc21
DNLM/DLC
for Library of Congress 97-25775
CIP

CALCIUM ANTAGONISTS IN CLINICAL MEDICINE, 2nd ed. ISBN 1-56053-223-8

© 1998 by Hanley & Belfus, Inc. All rights reserved. No part of this book may be reproduced, reused, republished, or transmitted in any form, or stored in a data base or retrieval system, without written permission of the publisher.

To Nina,

and David, Susanna, and Jonathan

Contents

Emerging Perspectives

Contributors

AYUB AKBARI, MD
Fellow, Nephrology Section
Department of Medicine
Pritzker School of Medicine
University of Chicago
Chicago, Illinois

GEORGE L. BAKRIS, MD, FACP
Associate Professor of Preventive and Internal Medicine and Vice Chairman, Department of Preventive Medicine
Director, Rush University Hypertension Training Center
Rush Medical College of Rush University
Rush–Presbyterian–St. Luke's Medical Center
Chicago, Illinois

ARTHUR L. BASSETT, PhD
Professor of Cellular and Molecular Pharmacology
Department of Cellular and Molecular Pharmacology
University of Miami School of Medicine
Miami, Florida

GEORGE E. BILLMAN, PhD
Professor of Physiology
Department of Physiology
Ohio State University College of Medicine
Columbus, Ohio

BRYAN J. BURNS, MD
Department of Medicine
Cornell Medical Center
New York Hospital
New York, New York

SIMON CHAKKO, MD
Professor of Medicine, Division of Cardiology
University of Miami School of Medicine
Chief, Cardiology Section
Veterans Affairs Medical Center
Miami, Florida

MARK E. COOPER, MD, FRACP
Associate Professor
Department of Medicine
University of Melbourne
Austin and Repatriation Medical Centre
Victoria, Australia

DALILA B. CORRY, MD
Associate Professor of Medicine
Department of Medicine
UCLA School of Medicine
Chief, Renal Division
Olive View Medical Center
Los Angeles, California

AMY J. DAVIDOFF, PhD
Assistant Professor of Medicine and Physiology
Department of Medicine, Division of Cardiology, and Department of Physiology
Wayne State University School of Medicine
Detroit, Michigan

JESUS EGIDO, MD
Professor of Medicine
Department of Medicine
Universidad Autónoma
Laboratory of Nephrology
Fundación Jiménez Díaz
Madrid, Spain

HENRY L. ELLIOTT, MD, FRCP
Senior Lecturer
Department of Medicine and Therapeutics
University of Glasgow
Western Infirmary
Glasgow, Scotland

WILLIAM J. ELLIOTT, MD, PhD
Associate Professor of Preventive Medicine
Rush Medical College of Rush University
Director, Section of Clinical Research
Department of Preventive Medicine
Rush–Presbyterian–St. Luke's Medical Center
Chicago, Illinois

MURRAY EPSTEIN, MD, FACP
Professor of Medicine
Department of Medicine
University of Miami School of Medicine
Attending Physician
Veterans Affairs Medical Center
Jackson Memorial Medical Center
Miami, Florida

JOHN M. FLACK, MD, MPH
Professor of Medicine and Community Medicine and Associate Chairman of Medicine
Director, Cardiovascular Epidemiology and Clinical Applications
Department of Internal Medicine and Community Medicine
Wayne State University Medical School
Detroit Medical Center
Detroit, Michigan

HERMES FLOREZ, MD
Research Associate
Diabetes Research Institute
Department of Medicine
University of Miami School of Medicine
Miami, Florida

WILLIAM H. FRISHMAN, MD
Professor of Epidemiology and Social Medicine and Associate Chairman of Medicine
Department of Medicine
Montefiore Medical Center
Albert Einstein College of Medicine
Bronx, New York

RONALD B. GOLDBERG, MD
Professor of Medicine
Diabetes Research Institute
Department of Medicine
University of Miami School of Medicine
Miami, Florida

KOICHI HAYASHI, MD, PhD
Assistant Professor of Medicine
Department of Internal Medicine
School of Medicine
Keio University
Tokyo, Japan

CATHERINE HURAUX, MD
Research Fellow
Department of Anesthesiology
Emory University Hospital
Atlanta, Georgia

FRANS H. H. LEENEN, MD, PHD, FRCPC
Professor of Medicine and Pharmacology
Department of Medicine
University of Ottawa
Director, Hypertension Unit
University of Ottawa Heart Institute
Ottawa Civic Hospital
Ottawa, Ontario, Canada

JERROLD H. LEVY, MD
Professor of Anesthesiology
Division of Cardiothoracic Anesthesia and Critical Care
Department of Anesthesiology
Emory University School of Medicine
Atlanta, Georgia

MARSHALL D. LINDHEIMER, MD, FACP, FRCOG
Professor, Departments of Medicine, Obstetrics and Gynecology, and Clinical Pharmacology
Pritzker School of Medicine
University of Chicago
Chicago, Illinois

GERARD M. LONDON, MD
Chief, Department of Nephrology
F. H. Mahnes Hospital
Fleury-Mérogis
Ste Geneviéve des Bois, France

PER LUND-JOHANSEN, MD, PhD
Professor of Medicine
Department of Cardiology
University of Bergen Medical School
Bergen, Norway

THOMAS F. LÜSCHER, MD
Professor and Head of Cardiology
University Hospital Zurich
Zurich, Switzerland

KONSTANTINOS MAKRILAKIS, MD, PhD
Fellow in Hypertension
Department of Preventive Medicine
Rush Medical College of Rush University
Rush–Presbyterian–St. Luke's Medical Center
Chicago, Illinois

JAMES D. MARSH, MD
Professor, Department of Internal Medicine
Center for Molecular Medicine and Genetics
Wayne State University School of Medicine
Harper Hospital and Detroit Medical Center
Detroit, Michigan

PETER A. MEREDITH, PHD
Reader
Department of Medicine and Therapeutics
University of Glasgow
Western Infirmary
Glasgow, Scotland

FRANZ H. MESSERLI, MD
Section on Hypertensive Diseases
Department of Internal Medicine
Ochsner Clinic
Alton Ochsner Medical Foundation
New Orleans, Louisiana

LESZEK MICHALEWICZ, MD
Ochsner Clinic
Alton Ochsner Medical Foundation
New Orleans, Louisiana

GEORG NOLL, MD
Assistant Professor, Division of Cardiology
University Hospital Zurich
Zurich, Switzerland

MARGARETA NORDLANDER, PhD
Associate Professor, Department of Physiology
University of Gothenburg
Department of Cardiovascular Pharmacology
Astra Hässle
Mölndal, Sweden

JOHN L. REID, DM, FRCP
Regius Professor of Medicine and Therapeutics
Department of Medicine and Therapeutics
University of Glasgow
Western Infirmary
Glasgow, Scotland

MICHEL E. SAFAR, MD
Professor of Medicine
Department of Internal Medicine
Broussais Hospital
Paris, France

TAKAO SARUTA, MD, PhD
Professor of Medicine
Department of Internal Medicine
School of Medicine
Keio University
Tokyo, Japan

ALEXANDER SCRIABINE, MD
Lecturer, Department of Pharmacology
Yale University School of Medicine
New Haven, Connecticut

JAMES R. SOWERS, MD
Professor of Medicine and Physiology
Director, Division of Endocrinology, Metabolism, and Hypertension
Department of Internal Medicine
Wayne State University School of Medicine
Detroit, Michigan

NIEVES TARIN, MD
Specialist in Cardiology, Department of Cardiology
Fundación Jiménez Díaz
Madrid, Spain

DAVID J. TRIGGLE, PhD
University Distinguished Professor
Vice Provost for Graduate Education and Research
Dean, The Graduate School
State University of New York at Buffalo
Buffalo, New York

MICHAEL L. TUCK, MD
Professor of Medicine, Department of Medicine
UCLA School of Medicine
Los Angeles, California
Chief, Endocrinology and Metabolism
Veterans Affairs Medical Center
Sepulveda, California

JOSE TUÑON, MD
Specialist in Cardiology, Department of Cardiology
Fundación Jiménez Díaz
Madrid, Spain

GEORGE W. VETROVEC, MD
Professor of Medicine
Chairman, Division of Cardiology
Department of Internal Medicine
Medical College of Virginia
Virginia Commonwealth University
Richmond, Virginia

BERNARD WAEBER, MD
Professor, Division of Hypertension
University Hospital
Lausanne, Switzerland

MARY F. WALSH, PHD
Research Scientist
Veterans Affairs Medical Center
Detroit, Michigan

MATTHEW R. WEIR, MD
Professor and Director, Division of Nephrology and Clinical Research Unit
Department of Medicine
University of Maryland School of Medicine
Associate Director, Organ Transplant Services
University of Maryland Hospital
Baltimore, Maryland

RENE R. WENZEL, MD
Fellow, Departments of Internal Medicine and Nephrology
University of Essen
Essen, Germany

Preface to the First Edition

In the 25 years since Fleckenstein introduced the concept of "calcium antagonism," we have witnessed an exponential advance in our knowledge of the pharmacology and clinical application of calcium antagonists. Initially, attention focused on the mechanisms of action of these agents in cardiac and vascular smooth muscle, and on their use in the management of cardiovascular disorders. Subsequently, investigations expanded into the effects of calcium antagonists on other organ systems, exploring a wider range of potential therapeutic applications. This book surveys the progress and current status of these diverse applications.

Invited experts review the current state of knowledge of the effects of calcium antagonists on a number of organ systems. In the first chapter, Dr. Triggle provides an overview of the mechanisms of action of calcium antagonists. He considers how calcium antagonists modify excitation-contraction coupling in vascular smooth muscle.

The next contribution by Drs. Ram, Standley, and Sowers provides a framework for the succeeding chapters by examining the normal function of Ca^{2+}-dependent mechanisms in vascular smooth muscle and the changes that occur in hypertension.

In chapter 3, Drs. Morris, Meredith, and Reid critically consider the pharmacokinetic properties and concentration-effect relationship of calcium antagonists and discuss how these properties of calcium antagonists may influence their pharmacologic response and therapeutic usefulness.

In the next seven chapters, experts in cardiovascular physiology and medicine consider the important role of calcium antagonists in the management of cardiac disorders. Specific topics include the pharmacotherapy of myocardial infarction and of congestive heart failure, the hemodynamic effects of calcium antagonists, and experimental data suggesting their vasoprotective role in experimental models of ischemia. A number of timely controversies are reviewed. Although the negative ionotropic and chronotropic effects of calcium antagonists have dampened some of the enthusiasm regarding their utilization in patients with congestive heart failure, it is becoming clear that these agents may acutely increase cardiac output and heart rate in some patients. In addition, calcium antagonists tend to reduce left ventricular mass to a variable extent when given for a sustained period. The use of calcium antagonists in patients with cardiovascular disease is complicated by the realization that the effects on afterload, left ventricular hypertrophy, coronary blood flow, left ventricular filling, conduction, and contractility vary from one calcium antagonist to another. This section concludes with a comprehensive discussion of the antiarrhythmic effects of calcium antagonists. The role of calcium ions in the development of ventricular arrhythmias associated with myocardial ischemia and reperfusion is explored, followed by a review of experimental and clinical studies that illustrate the therapeutic potential of individual calcium channel antagonists in the management of various arrhythmias.

In chapter 11, I built on the foundation provided by colleagues and review practical aspects of the use of calcium antagonists in the management of hypertension, including dosing, combined therapy with other antihypertensive agents, side effects, metabolic effects (e.g., homeostasis), and effects on serum electrolytes and the lipid profile.

The final section of the book is devoted to the use of calcium antagonists for the treatment of noncardiovascular disorders. Drs. Katzka and Castell examine the role of calcium antagonists in the management of gastrointestinal disorders, including achalasia. Next, Drs. Ahmed and Wanner consider the potential role of calcium antagonists in managing pulmonary disorders. Because many pathophysiologic events that underlie airway disease—disorders of pulmonary circulation and acute lung injury—are calcium-dependent, it is possible that appropriate calcium antagonists may prove useful in both the therapy and prophylaxis of these conditions.

In chapter 14 Dr. Gelmers reviews the efficacy of calcium antagonists in the management of patients with disorders of the cerebral circulation. Although calcium antagonists are not yet approved for use in the treatment of acute ischemic stroke, preliminary trials have suggested a therapeutic potential for calcium antagonists in the treatment of selected stroke patients. Furthermore, Dr. Gelmers interprets the available clinical evidence of suggesting that calcium antagonists such as nimodipine reduce the occurrence of delayed ischemic deficit and hence improve clinical outcome in patients with subarachnoid hemorrhage.

Drs. Corry and Tuck address the alterations of glucoregulatory hormones, including insulin, in hypertension. They suggest that antihypertensive agents vary with respect to their effects on insulin resistance and its metabolic consequences. Although insulin resistance might represent only an epiphenomenon in hypertension, the authors suggest that metabolic neutrality should constitute an additional factor in selecting agents for initial monotherapy.

The final five chapters are devoted to an examination of the effects of calcium antagonists on renal physiology and function. Chapter 16 constitutes an extensive overview of the renal effects of calcium antagonists and the current and emerging therapeutic applications for these agents in diverse renal disorders. In the next chapter, I review the available data on the natriuretic effects of calcium antagonists. Data are marshalled to suggest that calcium antagonists may act by potentiating the natriuretic response to volume-expansive maneuvers including saline administration and water immersion. Dr. Bakris considers the alterations of calcium in diabetes and highlights a growing controversy about the selection of angiotensin-converting enzyme (ACE) inhibitors vs. calcium antagonists in the management of diabetic hypertension. In the penultimate chapter, Dr. Weir reviews the emerging role of calcium antagonists in the setting of renal transplantation. In the final chapter Drs. Sweeney and Raij elucidate the role of the endothelium and mesangium in glomerular injury, suggesting that future investigations should explore the role of ACE inhibitors and calcium antagonists in attenuating renal injury by interrupting endothelial-mesangial cell communication and/or modulating the response to injurious stimuli.

I wish to express my gratitude to all the contributors for their excellent and willing cooperation. I have considered it good fortune indeed to be associated with them in this undertaking. Finally, I would like to express my deep gratitude to my wife for her ongoing support and encouragement.

Murray Epstein, M.D.
Miami, Florida

Preface to the Second Edition

The past 30 years have witnessed a major increase in our knowledge of the pharmacology of a diverse group of agents known as calcium antagonists or calcium channel blockers (CCBs). This has been accompanied by an exponential advance in the development of newer calcium antagonists and their clinical applications. In this book invited experts review the current state of our knowledge of calcium antagonists.

The first section entitled **Basic Principles** deals with the fundamental biochemical and mechanistic concepts of calcium antagonists. In the opening chapter, Dr. Triggle provides the overview of the biochemistry of calcium antagonists. He discusses the molecular structure of this large and heterogeneous group of agents and explains the molecular mechanisms underlying their inhibitory actions upon calcium channels. He emphasizes that calcium antagonists are substantially selective in their actions and demonstrates how this selectivity of action has important implications for both the therapeutic profile of the calcium antagonists, and their contraindications and safety profile. Next chapter by Davidoff and colleagues considers abnormalities in Ca^{2+} metabolism that are generally associated with the diabetic state and may account for many pathophysiological findings. They specify that increased $[Ca^{2+}]_i$ in vascular smooth muscle cells (VSMCs) and attendant responses to vasoactive calcium-mobilizing hormones such as vasopressin and angiotensin II probably contribute to the characteristically enhanced vascular reactivity seen in people with type I and type II diabetes. The following chapter by Drs. Elliott, Meredith, and Reid critically considers the pharmacokinetic properties of calcium antagonists and how they may influence their pharmacologic response and therapeutic usefulness. They emphasize the emerging therapeutic possibilities brought about by novel or modified release formulations of existing agents and the appearance of new drugs on the market. In chapter 4, Wenzel and colleagues survey a burgeoning subject of intense investigative interest—endothelial regulation of vascular tone and growth and their modulation by calcium antagonists, and also review endothelial dysfunction in diverse diseases including hypertension. They conclude with an overview of the effects of calcium antagonists on vascular growth and sympathetic nervous system activity.

The section entitled **The Heart** comprises five chapters contributed by experts in the field of cardiovascular medicine who explore the potential of calcium antagonists for management of cardiac disorders. The chapter by Messerli and Michalewicz focuses on their cardiac effects in hypertension, while the two chapters by Vetrovec and Billman consider the role of calcium antagonists in the management and targeting of myocardial ischemia. The full appreciation of the actions of calcium antagonists in cardiovascular diseases is complicated by the fact that they are a heterogeneous group of drugs that vary in structure, elimination half-life, blocking activity of metabolites, relative affinities for myocardium and smooth muscles, cardiac electrophysiologic effects, formulations, and the mode of administration. Analysis of the cardiac effects of calcium antagonists becomes even more difficult because normal and diseased myocardium may respond differently. This section continues with a comprehensive discussion by Chakko and Bassett of the antiarrhythmic effects of calcium antagonists and their role in the development and management of various arrhythmias. The final chapter in this section by Bassett et al. critically reviews the available data regarding the arrhythmogenic potential of calcium antagonists. Calcium overload in myocardial cells is an important factor in the genesis of various serious arrhythmias. Verapamil, diltiazem, and the dihydropyridine (DHP) calcium antagonists have been shown to prevent ventricular ischemic and reperfusion arrhythmias in the laboratory. Despite extensive data indicating that calcium antagonists are antiarrhythmic, a recent controversy has focused on the possibility that certain calcium antagonists are unsafe, especially in patients with coronary heart disease (see chapter 29).

In the next four chapters that constitute the section entitled **Hypertension**, the important role of calcium antagonists in the management of hypertension is reviewed. In chapter 10, I review practical aspects of the use of calcium antagonists in the management of hypertension, including dosing and combined therapy with other antihypertensive agents. I consider a number of crucial issues including differences among calcium antagonists, and provide a review of the newer slow-release formulations that have profoundly altered the pharmacokinetic and pharmacodynamic profile of these agents. I also briefly consider the metabolic effects and side effects of these agents. In the next chapter, Dr. Lund-Johansen provides a comprehensive review of the hemodynamic effects of calcium antagonists in hypertension. This is complemented by an in-depth review by London and Safar of the effects of calcium antagonists on the capacitative properties of central and peripheral large arteries. Drs. Corry and Tuck provide an elegant review of alterations in insulin and glucoregulatory hormones associated with hypertension. Based on these considerations, they present a thoughtful approach for selecting specific classes of antihypertensive agents depending on their metabolic effects and their ability to counteract the consequences of such derangements in hypertension.

The next section deals with the unique aspects of calcium antagonist therapy in **special populations**. The initial two chapters in this section by Dr. Cooper and I and by Drs. Makrilakis and Bakris provide a comprehensive review of the role of calcium antagonists in the management of the diabetic patient. We address the pivotal question of whether calcium antagonists should be used despite the fact that they do not normalize glomerular capillary pressure (P_{GC}). These two chapters present somewhat differing viewpoints and hopefully provide the reader with a balanced overview of this major controversy, that should be resolved by ongoing randomized clinical trials. The next two chapters deal with the unique attributes of calcium antagonists in two special populations of hypertensive patients—African-Americans, and the elderly. Flack and Weir review the pathophysiological profile of the African-American hypertensive patient and the unique pharmacological attributes of calcium antagonists that reverse many of the hallmark physiological aberrations of this high-risk demographic group. In the next chapter, Burns and Frishman review the pathophysiology of hypertension in the elderly and the rationale for drug treatment. Based on the pharmacologic properties of calcium antagonists, the authors suggest that this antihypertensive class is well-suited for use in elderly patients whose hypertensive profile is based on increasing arterial stiffness, decreased vascular compliance, and diastolic dysfunction secondary to atrial and ventricular stiffness. Akbari and Lindheimer review a nettlesome problem of management of hypertension in pregnancy. Hypertension is the most common medical complication of gestation, and the approach to this challenging clinical problem differs considerably from that employed in nonpregnant populations. The text provides a comprehensive overview of the pathophysiology of hypertension in pregnancy and available therapeutic options. Levy and colleagues provide an extensive look into the management of perioperative hypertension, which differs considerably from that for chronic hypertension. Furthermore, because oral therapy is not possible, patients require parenteral therapy. They summarize the potential approaches with parenteral therapy and discuss the applications of current formulations of calcium antagonists for intravenous use. Clevidipine, a new ultrashort-acting vascular-selective dihydropyridine currently undergoing clinical investigations is also discussed.

The remaining chapters discuss the potentials of calcium antagonists for the treatment and management of **noncardiovascular disorders**. Goldberg and Florez consider the effects of calcium antagonists on lipids. Increasing attention has been devoted to the concept that hypertension is one component of a multifactorial syndrome of increased cardiovascular risk, and lipid disorders have attracted particular interest. This chapter reviews the effects of the available calcium antagonists on plasma lipids and lipoproteins and their potential for favorably influencing atherogenesis. Dr. Scriabine reviews the efficacy of calcium antagonists in the management of patients with disorders of the cerebral circulation and central nervous system (CNS). Although binding sites for calcium antagonists in the brain were discovered soon after their introduction for the therapy of angina pectoris and hypertension, even today the only application in CNS disorders approved by the Food and Drug Administration (FDA) is the use of nimodipine for prevention of neurologic deficits after subarachnoid hemorrhage. This chapter summarizes the available information about the effects of calcium antagonists in animal and human models of CNS diseases and assesses their potential applications in neurology and psychiatry.

Four chapters are devoted to the effects of calcium antagonists on renal physiology and function. Hayashi, Saruta and I review the renal microcirculatory effects of calcium antagonists. Calcium antagonists preferentially attenuate afferent arteriolar vasoconstriction, with a concomitant reduction in systemic blood pressure. Since their net effects depend on the balance between these hemodynamic factors, the ability of calcium antagonists to alter glomerular hemodynamics may vary depending on the underlying basal vascular tone observed in a variety of renal diseases. Finally, we consider the recent development of newer calcium antagonists that appear to have a different microcirculatory profile and possess certain efferent vasodilatory effects. Tuñon et al. provide a comprehensive overview of the effects of calcium antagonists on vasoactive hormones with a focus on the renin-angiotensin system (RAS) and endothelins. Calcium is an intracellular messenger that plays a central role in angiotensin II and endothelin signaling. Thus, drugs that block calcium entry into the cells can interfere with these systems. A discussion of the effects of calcium antagonists on the levels of the different RAS components is followed by a review of their interactions with angiotensin II and endothelins in cardiovascular and renal diseases that may explain their pharmacologic actions. Chapter 24 constitutes an extensive overview of the renal effects of calcium antagonists and the current and emerging therapeutic applications for these agents in various renal disorders. Although the renal microcirculatory effects of calcium antagonists should not favor an attenuation of glomerular hypertension, these agents have additional properties that may contribute to their ability to provide renal protection. Several recent prospective, randomized trials have suggested that calcium antagonists and ACE inhibitors may be equally renoprotective. Collectively, the data available thus far suggest that

ACE inhibitors and probably calcium antagonists attenuate the progression of chronic renal failure. Dr. Weir reviews the emerging role of calcium antagonists in the setting of renal transplantation as agents that could potentially attenuate the nephrotoxic effects of cyclosporine. In addition to their established efficacy in managing renal hypertension, Weir reviews the controversial notion that calcium antagonists may act to diminish allograft dysfunction and enhance the effects of conventional immunosuppression.

The final section entitled **Emerging Perspectives** comprises four chapters on topics that are timely, controversial, and are the focus of current clinical and investigative efforts. Dr. Waeber and I summarize an exciting new initiative in the management of hypertension—the resurgence of interest in fixed-dose combination agents. The rationale for combination therapy relates to the concept that antihypertensive efficacy may be enhanced when two classes of agents are combined. In addition, combination therapy enhances tolerability—the notion that one drug of a fixed combination can antagonize some of the adverse effects of the second drug. The recent approval by the FDA of four fixed-dose ACE inhibitor/calcium antagonist combinations has once again focused attention and prompted reexamination of this issue. In the next chapter Dr. Elliott provides a thoughtful and probing discourse on the costs of treating hypertension and its impact on clinical management. In light of the exponential growth of health-related expenditures, policy-makers have focused on hypertension as the first target for restriction of further growth. Outpatient prescription drugs for hypertension are currently favored targets for cost-containment, with recommendations from many authorities to use more diuretics or beta-blockers. Elliott concludes by suggesting physician-dependent initiatives for improving the cost-effectiveness of treating hypertension. In the penultimate chapter Leenen provides a thoughtful review of disparate effects of various calcium antagonists on sympathetic activity, focusing on dihydropyridines (DHPs). He clearly demonstrates that although DHPs decrease sympathetic activity, rapid lowering of blood pressure induced by short-acting DHPs may override this effect, resulting in intermittent increases in sympathetic activity. In contrast, long-acting DHPs that have sustained hemodynamic effects do not seem to cause sympathetic hyperactivity (and may even decrease it). As a consequence, long-acting DHPs are associated with an improved outcome and decreased incidence of adverse outcomes compared with the short-acting agents. In the final chapter I review the controversy surrounding the long-term safety of calcium antagonists that has been a subject of intense debate since 1995. I focus on a number of key issues that are pivotal to this discussion. First, I review the markedly disparate effects of different calcium antagonist formulations and their clinical implications. Second, I also consider the putative mechanisms proposed by Furberg to account for the enhanced morbidity. Finally, I present data obtained from recent prospective studies that are inconsistent with Furberg and Psaty's findings that calcium antagonists, as a group, are dangerous and argue that such allegations are not relevant to the calcium antagonists in their current usage.

Hereby I wish to thank all of the contributors and once again express my appreciation of their excellent cooperation. Finally, I would like to express my deep gratitude to my wife for her ongoing support and encouragement.

Murray Epstein, M.D.
Miami, Florida

DAVID J. TRIGGLE Ph.D.

1

Mechanisms of Action of Calcium Channel Antagonists

The calcium channel antagonists are a chemically, pharmacologically, and therapeutically heterogeneous group. First-generation agents, which include verapamil, nifedipine, and diltiazem, owe their effectiveness to interactions at specific sites associated with a major protein of the L class of voltage-gated calcium channel. The L channel is the best characterized of several major classes of voltage-gated channel that are distinguishable by electrophysiologic and pharmacologic characteristics. Calcium channels may be considered as pharmacologic receptors, with specific sites for activator and antagonist ligands that are linked to the functional machinery of the channel. These sites are subject to homologous and heterologous regulation and are altered in expression in a number of disease states, both experimental and clinical. The calcium channel antagonists are substantially selective in their actions and are clearly not interchangeable. This distinctiveness of action exists among the first-generation agents and among the collection of second-generation 1,4-dihydropyridines. The calcium channel antagonists are quantitatively and qualitatively distinct species. Their selectivity stems from several mechanisms, including state-dependent interactions and class and subclass of channel, with attendant pharmacologic discrimination. This selectivity of action has important implications for both the therapeutic profile of the calcium antagonists and their contraindications and safety profile.

CALCIUM ANTAGONISTS

Calcium plays critical roles in cellular communication and regulation. It is a ubiquitous intracellular messenger serving to couple membrane-mediated stimuli with cellular responses.[1,2] The cell maintains a low resting intracellular concentration of ionized calcium in the face of large and inwardly directed concentration and electrochemical gradients; intracellular calcium-binding proteins, including the ubiquitous calmodulin, couple calcium entry to cellular response.[3–6] It may be anticipated that the cellular movements and storage of calcium are subject to various regulatory processes (Fig. 1). These processes are not of equal importance in every cell type or during every stimulus mode.

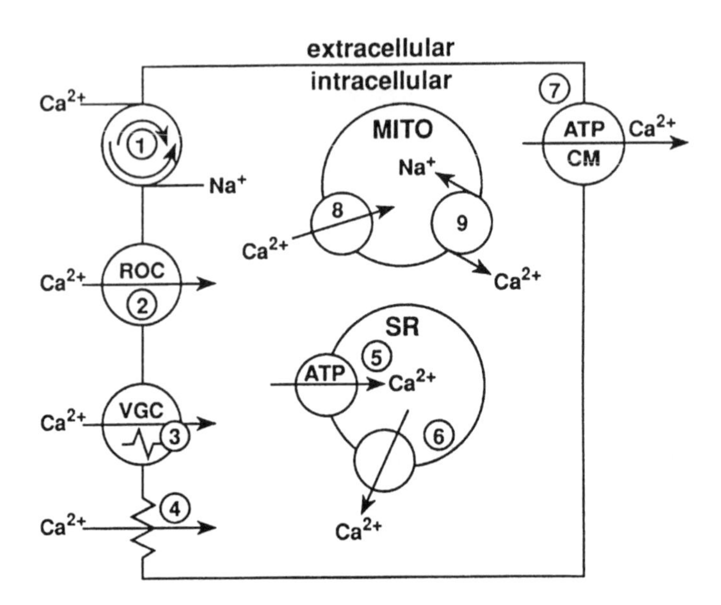

FIGURE 1. Pathways of cellular calcium regulation:. (1) Na^+/Ca^{2+} exchanger; (2) receptor-operated channels (ROC); (3) voltage-gated channels; (4) leak pathway; (5 and 6) Ca^{2+} pump and release channels in sarcoplasmic reticulum; (7) plasmalemmal Ca^{2+} pump; (8 and 9) Ca^{2+} uptake and release processes in mitochondria.

However, calcium in excess is a lethal cation, and its uncontrolled movements after cellular injury may lead ultimately to irreversible cell destruction and death.[7,8] Part of the impetus in the search for pharmacologic agents that control calcium storage and movements is control of diseases and pathologic states in which excessive calcium mobilization is a critical component.

Whole body calcium metabolism, whereby plasma calcium levels are maintained at approximately 2.5 millimolar, is controlled by a triumvirate of hormones—vitamin D, parathyroid hormone, and calcitonin.[2] Important links between intra- and extracellular calcium metabolism have received increasing attention. Parathyroid hormone and vitamin D mobilize calcium from bone, the major storage depot; increase calcium absorption; and reduce calcium excretion. The concentrations of these hormones stand in inverse relationship to serum calcium levels. In contrast, calcitonin serves as a calcium-conserving agent that promotes calcium deposition. Vitamin D and parathyroid hormone may be linked to intracellular calcium metabolism in at least two ways: (1) vitamin D exerts acute stimulatory effects on calcium influx into a number of cell types, including vascular smooth muscle,[9] and (2) parathyroid hormone (1–84) and its fragments increase Ca^{2+} current in vascular and cardiac cells.[10] In addition, various observations suggest that some essential hypertension may be associated with a serum calcium deficiency in which serum calcium levels are low and urinary calcium excretion is elevated with compensating changes in parathyroid, vitamin D, and calcitonin levels.[11–13] These changes may well accord with the long-held view that hypertension, associated vascular smooth muscle hyperreactivity, and clinical use of calcium antagonists as antihypertensive agents relate to elevated levels of intracellular calcium and an enhanced calcium entry process.[14,15]

Parathyroid hypertensive factor is proposed to be a modulator of cellular Ca^{2+} influx in the cardiovascular system and to activate L-type channels.[13,16,17] The release of parathyroid hypertensive factor is linked to extracellular Ca^{2+} levels and is increased with serum Ca^{2+} deficiency. The observations are not incompatible: low extracellular calcium may serve to promote a leakiness of plasma membranes, including those of vascular smooth muscle, and to activate the release of parathyroid hypertensive factor. Both factors contribute to an augmented calcium influx. In addition, the alteration in effective electric field across the membrane, caused by a reduction in calcium interaction with the external negative membrane charge, alters the sensitivity of vascular smooth muscle to drugs, including calcium antagonists, that function through voltage-sensitive interactions[18,19]

In principle, specific drugs should interact discretely with the sites depicted in Figure 1: at the cell membrane, receptor-operated channels, voltage-gated channels, Na^+:Ca^{2+} antiporter, calcium-binding sites, and Ca^{2+}-activated adenosine triphosphatase (ATPase), and at the intracellular level, calcium-binding proteins, calcium-release channels (sarcoplasmic reticulum [SR]), mitochondrial uptake and release sites, and Ca^{2+}-ATPase (SR). In practice, such agents have assumed both specificity of action and pharmacologic and therapeutic significance at only one site: the L-type voltage-gated calcium channel. This group of drugs, the calcium channel antagonists, includes verapamil, nifedipine, and diltiazem as the major cardiovascular agents.[20–24] Given the multiplicity of calcium mobilization and control processes in the cell, the available agents exhibit an automatic selectivity of action because they block only one of several mobilization processes.

Verapamil, nifedipine, and diltiazem, members of a chemically heterogeneous group of agents (Fig. 2), are widely used in cardiovascular therapy (Table 1) and are first-line antihypertensive agents. Second-generation derivatives,

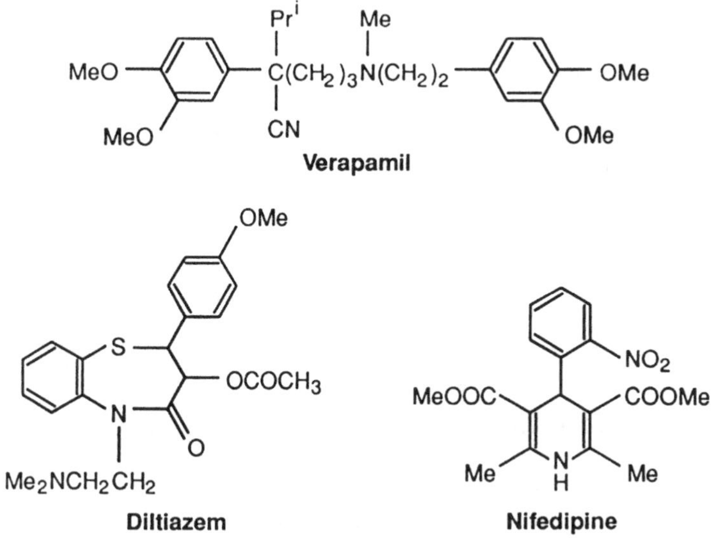

FIGURE 2. Structural formulas of calcium channel antagonists.

TABLE 1. Therapeutic Uses of Calcium Channel Blockers

	Blocking Agents		
Use	**Verapamil**	**Nifedipine**	**Diltiazem**
Angina			
Exertional	+++*	+++	+++
Prinzmetal's	+++	++	+++
Variant	+++	+	+++
Arrhythmias			
Paroxysmal	+++	–	++
Supraventricular			
Tachyarrhythmias			
Atrial fibrillation and flutter	++	–	++
Hypertension	++	+++	++
Hypertrophic cardiomyopathy	+	–	–
Raynaud's phenomenon	++	++	++
Cerebral vasospasm (post-hemorrhage)	–	+†	–

* Number of plus signs indicates extent of use: +++ = very common; – = not used.
† Nimodipine is a nifedipine analog.

particularly in the 1,4-dihydropyridine family (Fig. 3), have modified pharmacologic profiles. In addition, the calcium channel antagonists are being examined in an increasing number of experimental and clinical situations (Table 2), some of which may be unrelated to the process of calcium channel blockade. Thus, verapamil and several 1,4-dihydropyridines are blockers of the P-glycoprotein

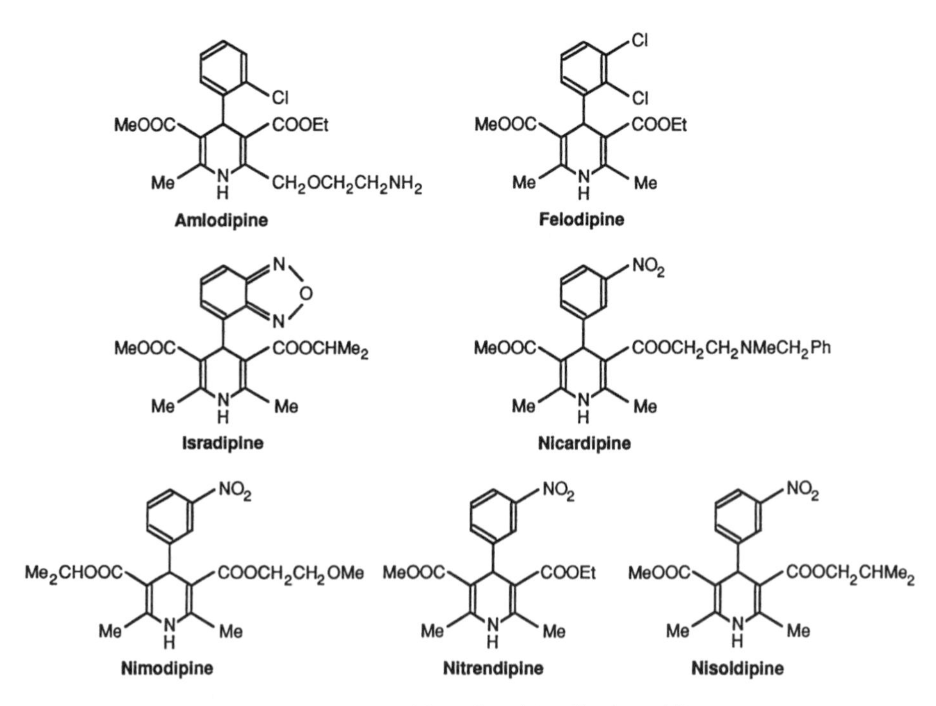

FIGURE 3. Structural formulas of 1,4-dihydropyridines.

TABLE 2. Additional and Potential Uses for Calcium Blocking Agents

Cardiovascular

- Aortic and mitral valvular insufficiency
- Atherosclerosis
- Cardiac arrest, prevention of neurologic damage
- Cardiac surgery, cardioplegia
- Cerebral ischemia and vasospasm
- Cerebral surgery
- Heart failure, congestive
- Headache, cluster and migraine
- Hypertension, associated with pregnancy and pulmonary hypertension
- Intermittent claudication
- Percutaneous transluminal angioplasty
- Peripheral vascular disease
- Provocative test for sick sinus syndrome (verapamil)
- Renal failure, inflammatory response to injuries
- Stroke
- Sudden hearing loss
- Ventricular tachycardia, exercise-induced
- Vestibular nystagmus (vertebrobasilar insufficiency)

Nonvascular

- Achalasia, esophageal motor disorders
- Asthma
- Diarrhea due to carcinoid
- Dysmenorrhea (primary)
- Myometrial hyperactivity
- Premature labor
- Urinary incontinence; ureteral spasms

Other

- Age-associated memory loss; Alzheimer's disease
- Aldosteronism, primary
- Allergic reactions
- Anticalcinotic, retardation of age-dependent calcinosis
- Cancer: enhancement of anticancer drug effect or decreased drug resistance
- Cirrhosis of the liver
- Cocaine intoxication
- Epilepsy
- External muscle pain syndrome (benign myalgia)
- Glaucoma
- Immunosuppressive therapy, adjunct
- Manic and schizomanic syndrome
- Panic disorders
- Psychosis, phencyclidine-induced
- Schistosomal infection
- Schizophrenia, chronic (as adjunct therapy)
- Seafood (ciquatera) poisoning
- Shock, *Escherichia coli*-induced, endotoxic
- Sickle cell anemia
- Spinal cord injury
- Tinnitus
- Tourette's disorder (tics)
- Ulcers
- Vertigo
- Withdrawal symptoms (e.g., ethanol, morphine, phencyclidine)

responsible for multidrug resistance in cancer chemotherapy. These actions, however, can be distinguished from actions at the L-type channel by different structure-activity relationships, including the absence of stereoselectivity.[25] Calcium channel antagonists clearly differ both qualitatively and quantitatively in their therapeutic roles: verapamil and diltiazem have significant class

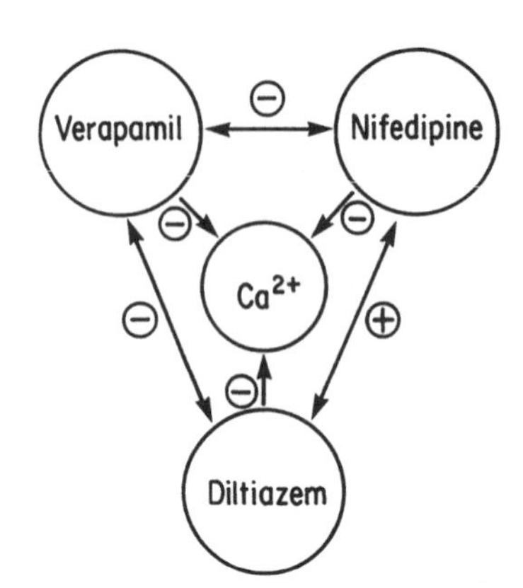

FIGURE 4. Schematic arrangement of binding sites at the L-type calcium channel. These discrete binding sites are depicted allosterically, linked one to the other and to the functional machinery of the channel. The + and – signs depict the nature of the allosteric linkage.

IV antiarrhythmic activity, which is not observed in nifedipine or other 1,4-dihydropyridines. Despite such differences in pharmacologic activities, it is generally agreed that calcium channel antagonists interact at a set of allosterically linked sites on a major protein—the $alpha_1$ subunit—of the voltage-gated L-type calcium channel[21,26–30] (Fig. 4). These binding sites have been characterized as discrete entities. Their specific interactions with the ligand species doubtless contribute to the selectivity of actions of the agents as revealed by the differing therapeutic (see Table 1) and cardiovascular (Table 3) profiles. Furthermore, the L-type channel enjoys widespread distribution in excitable cells outside the cardiovascular axis, including the central nervous system.[31,32] Despite the widespread distribution of the L-type calcium channel, the therapeutic actions of the calcium antagonists are directed almost exclusively at the cardiovascular system.

The above considerations underlie the critical issue of selectivity of action of the calcium channel antagonists. This issue is most usefully discussed if the calcium channel is considered as a pharmacologic receptor or, more properly, as a group of pharmacologic receptors. Molecular biology studies indicate the existence of isoforms of the $alpha_1$-subunit for heart and skeletal muscle and for vascular smooth muscle, derived from distinct genes or alternative splicing, respectively.[33,34]

TABLE 3. Cardiovascular Profile of Calcium Antagonists

	Nifedipine	Diltiazem	Verapamil
Coronary vessels			
Tone	– – –	– –	– –
Flow	+++	++	++
Peripheral vasodilation	+++	+	++
Heart rate	++	–	–
Contractility	0, +	0, –	0, –
A-V node conduction	0	–	–
A-V node ERP	0	–	–

+ = increase, – = decrease, 0 = no effect.

TABLE 4. Classification of Voltage-gated Calcium Channels

Property	T	L	N	P	R
Conductance, pS	7–10	11–25	10–25	10–20	15
Activation threshold, mV	LVA >–170	HVA >–30	HVA >–30	HVA >–40	HVA >–40
Inactivation rate	Fast	Slow	Moderate	Moderate to slow	Fast
Conductance	$Ba^{2+} = Ca^{2+}$	$Ba^{2+} > Ca^{2+}$	$Ba^{2+} > Ca^{2+}$	?	$Ba^{2+} > Ca^{2+}$
Localization	Widespread	Cardiovascular, neurons endocrine	Neurons	Neurons	Neurons
Function	Pacemaking	E-coupling: action potentials	Neuro-transmitter release	Neuro-transmitter release	Neuro-transmitter release
Pharmacology:					
Nifedipine	–	✔	–	–	–
Verapamil	–	✔	–	–	–
Diltiazem	–	✔	–	–	–
Conotoxin GVIA	–	–	✔	–	–
Conotoxin MVIIC	–	–	✔	✔	–
Agatoxin IVA	–	–	–	✔	–
Agatoxin IIIA	–	✔	✔	✔	–
Cd^{2+}	–	✔	✔	–	✔
Ni^{2+}	✔	–	–	–	–

– = no activity, ✔ = activity.

CLASSIFICATION AND STRUCTURE OF CALCIUM CHANNELS

There are at least four major subtypes of voltage-gated calcium channels: T, L, N, and P. They may be distinguished by localization and function, electrophysiologic and biophysical characteristics, and structure[33–36] (Table 4). Voltage-gated calcium channels can be divided into two primary classes:

1. The low voltage-activated channels (T class) are activated and inactivated at low membrane potentials. They are found in a wide variety of cells and have an inadequate pharmacologic classification. The T-type channel is probably responsible for pacemaking activities in excitable cells.
2. The high voltage-activated channels (L, N, and P types) are distinct in localization and pharmacology. The L-type channel dominates the functional properties of the cardiovascular system, and the N and P types are confined to neurons. It is likely that at least one other major class of high voltage-activated channel exists because a significant fraction of neuronal high voltage-activated current is insensitive to the known channel antagonists.

The high voltage-activated channels are heteromeric associations of four subunits: $alpha_1$, $alpha_2$-δ, beta, and gamma. The gamma subunit appears to be associated only with the calcium channel of skeletal muscle. The $alpha_1$ subunit is the dominant component; it expresses pore structure and drug-binding

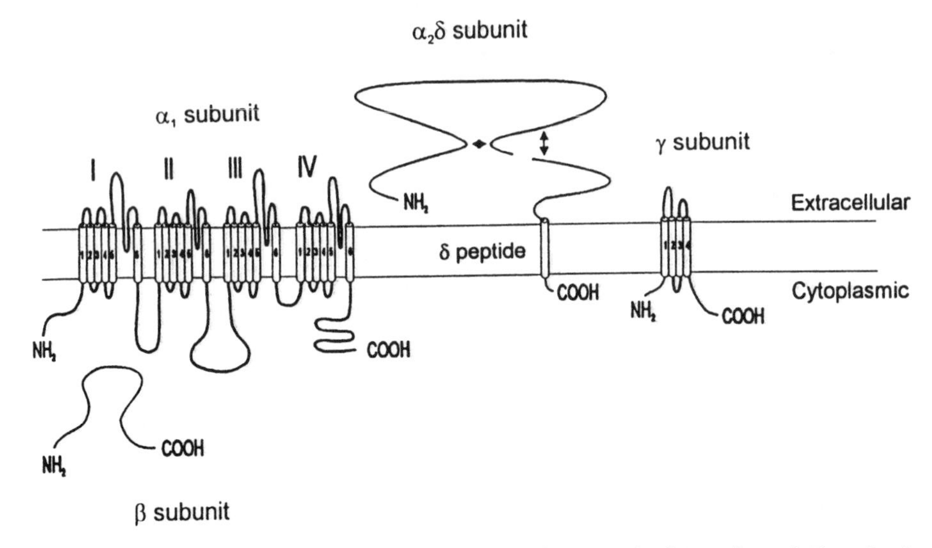

FIGURE 5. Arrangement of the subunits of the L-type voltage-gated calcium channel. The subunits may be arranged in various ways, but it is clear that the alpha$_1$ subunit expresses the major functional properties of the channel. (From DeWaard M, Gurnett CA, Campbell KP: Structural and functional diversity of voltage-activated calcium channels. In Narahashi T (ed): Ion Channels 4. New York, Plenum, 1995, pp. 41–87, with permission).

sites.[36,37] Mammalian alpha$_1$ subunits are coded by at least six different genes (S, A, B, C, D, and E), each of which expresses several splice variants. The membrane topology of the alpha$_1$ subunit reveals four homologous domains, each of which possesses six transmembrane segments, S1 to S6, and a region between S5 and S6 that, together with S5 and S6, is believed to comprise the pore of the channel. The S4 segment contains a positively charged residue at every third or fourth position followed by several hydrophobic residues. This characteristic sequence is believed to make up the voltage sensor region of the channel (Fig. 5). The A and C types of alpha$_1$ subunit are expressed in the cardiovascular system (Table 5). The beta subunits are coded by four different genes and, like the alpha$_1$ subunits, they are differentially spliced. The beta

TABLE 5. Properties of Alpha$_1$ Subunits

	S	A	B	C	D	E
Functional type	HVA L	HVA P	HVA N	HVA L	HVA L	HVA R
Pharmacologic sensitivity	Nifedipine	Agatoxin IVA	Conotoxin GVIA	Nifedipine	Nifedipine	?
Tissue location	Skeletal muscle	Neurons Cardiac	Neurons	Neurons Cardiac, smooth muscle	Neurons Endocrine	Neurons
Cell function	E-C coupling	NT release	NT release	Other	Other	NT release

subunit is not a membrane component but rather associates with the alpha$_1$ subunit through a specific sequence located on the cytoplasmic loop joining the first and second domains of the alpha$_1$ subunit.[38,39] The membrane topology of the alpha$_2$ subunit is less settled, but it probably contains at least one transmembrane segment.

The subunit organization of the calcium channel depicted in Figure 5 is both a functional and a structural association. In particular, the beta subunit significantly modifies the biophysical and pharmacologic properties of the expressed alpha$_1$ subunit, increasing current density and shifting the voltage dependence of channel activation in the hyperpolarizing direction to increase its efficiency.

CALCIUM CHANNELS AS PHARMACOLOGIC RECEPTORS

Ion channels, including voltage-gated calcium channels, are conveniently conceptualized as pharmacologic receptors. As pharmacologic receptors the channels are expected to possess the following general properties:

1. Specific binding sites for both activator and antagonist ligands
2. Coupling of the binding sites to the permeation and gating machinery of the channel
3. Regulation by homologous and heterologous influences
4. Alteration of expression and function in disease states

These expectations have been largely realized, and the several classes of calcium channels may be regarded as a homologous group of receptors defined by localization, function, and specific electrophysiologic and pharmacologic criteria (see Table 4). In particular, each channel class is characterized by greater or lesser specificity of action by pharmacologic agents. For N- and P-types, which are localized in neuronal tissues, the only available specific agents are complex peptide toxins, but for the L-type channel several structurally distinct groups of synthetic molecules are available, several of which are available as therapeutic drugs (see Figs. 2 and 6). Examples include the phenylalkylamines (e.g., verapamil), benzothiazepines (e.g., diltiazem), and 1,4-dihydropyridines (e.g., nifedipine).[21–24] However, other structurally distinct ligands interact at additional and probably discrete receptor sites on the L-type channel.[40,41] Examples include the diphenylbutylpiperidines, pimozide and fluspiriline; the indolizine, SR 33557; and the benzolactam, HOE 166. Mibefradil, a newly introduced agent, appears to share existing binding sites, including the phenylalkylamine receptor.[42,43] In addition, mibefradil shows T channel-blocking activity, although this contribution to the total pharmacologic profile remains unclear. There are few new activator structures, but the benzoylpyrrole FPL 64176 defines a separate site distinct from that occupied by the 1,4-dihydropyridine activators, including Bay K 8644.[44,45]

Structure-activity relationships have been described for the major drug classes, particularly for the 1,4-dihydropyridines, which include both potent activators and antagonists.[9,21,30,46] However, activator properties have also been indicated under certain conditions for verapamil, D600, and diltiazem. Thus, it is possible that activator characteristics are a general feature of the three major ligand classes active at the L channel.[47] There are no existing therapeutic applications for calcium channel activators.

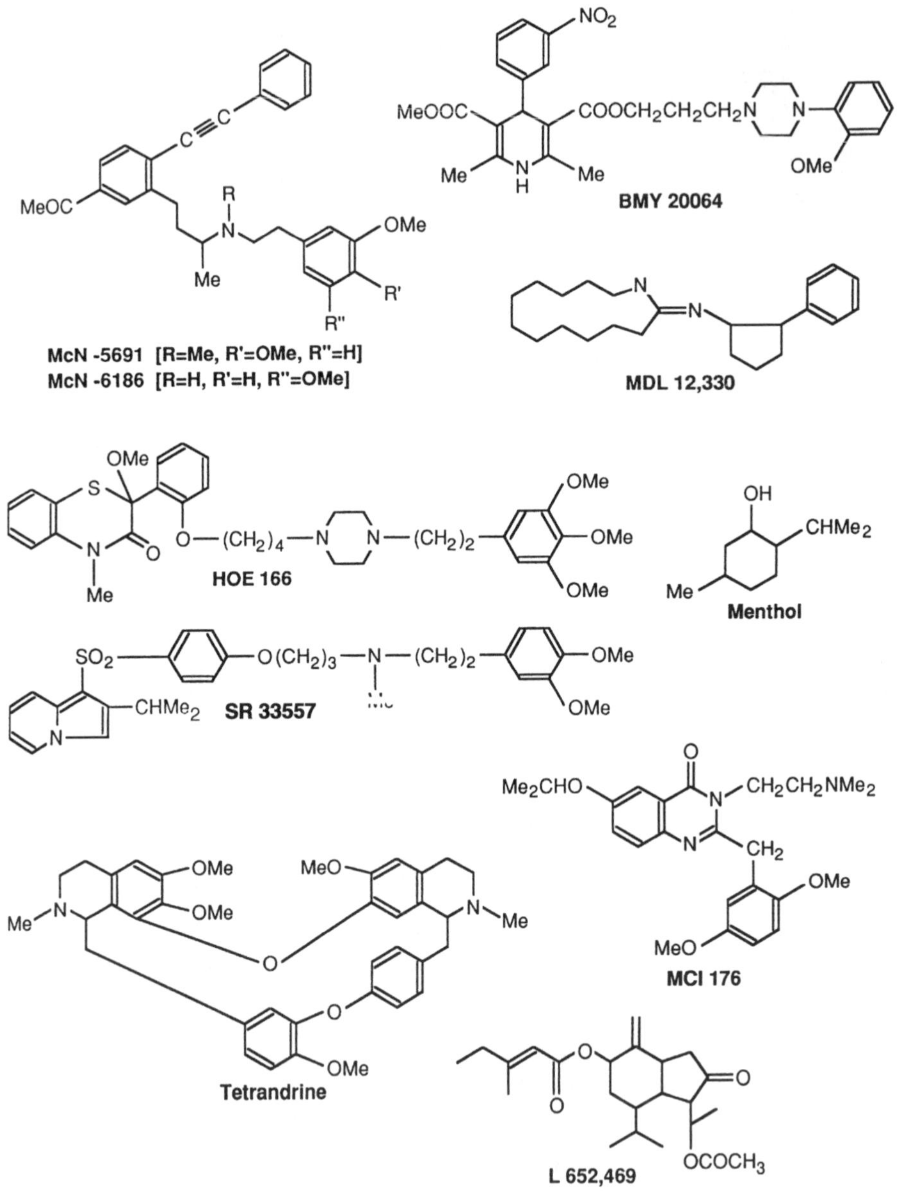

FIGURE 6. Structures of other ligands active at the L-type calcium channel.

Each of the three major structural classes of L-type calcium channel ligand interacts at a discrete site on the channel to define a specific structure-function relationship, including stereoselectivity of interaction.[21] The structure-function relationship for the 1,4-dihydropyridines has been particularly well investigated, and the general structural requirements are outlined in Figure 7.[19,21,30,46,48] Several features are of particular interest. Ester substituents at C-3 and C-5 of the 1,4-dihydropyridine ring provide optimal antagonist activity, but considerable

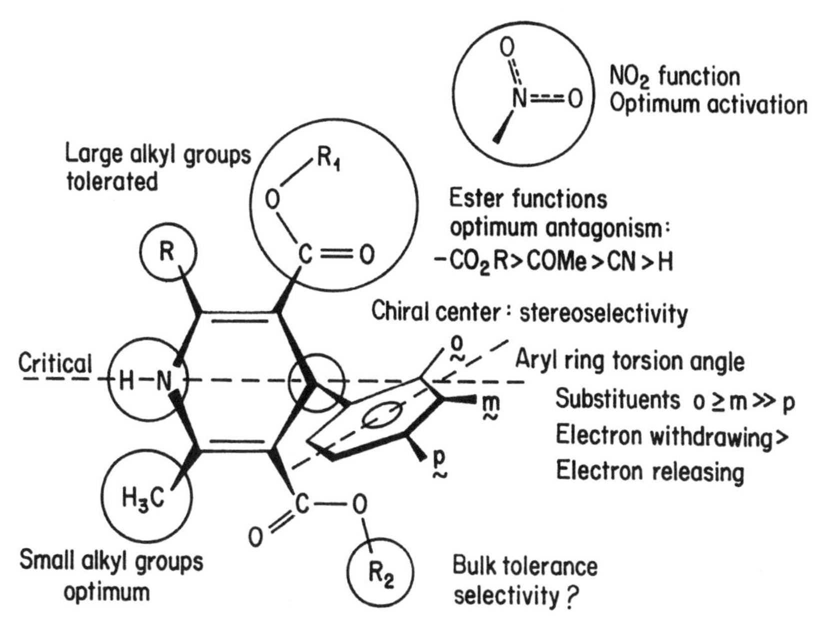

FIGURE 7. Structural requirements for antagonist and activator properties in 1,4-dihydropyridines.

variation in the nature and size of the groups is possible (see Fig. 3). The vascular selectivity differences of 1,4-dihydropyridines with different ester substituents may reflect interactions with differentially expressed channel subtypes. In addition, the presence of basic substituents in the 2-position, as in amlodipine (see Fig. 3) produces agents with slow onset, long duration of action, and an extremely high stereoselectivity of interaction that contrasts markedly with the neutral 1,4-dihydropyridines.[49,50] The presence of a C-5 nitro group in the 1,4-dihydropyridine ring is important for activator properties, but stereochemistry is also determinant. In enantiomeric pairs of C-3 ester, C-5-nitro substituted 1,4-dihydropyridines the S and R enantiomers are dominantly activators and antagonists, respectively (Fig. 8).

Consistent with the heterogeneity of chemical structure, pharmacologic activities, and therapeutic applications, the three principal classes of calcium antagonist interact at separate receptors on the L-type voltage-gated calcium channel. The discrete nature of these receptors has been demonstrated by pharmacologic, biochemical, and molecular biologic techniques. The discrete characters of the

FIGURE 8. An enantiomeric pair of 1,4-dihydropyridines that shows activator and antagonist properties. This property is shared by other C-5 nitro-substituted 1,4-dihydropyridines, including Bay K 8644.

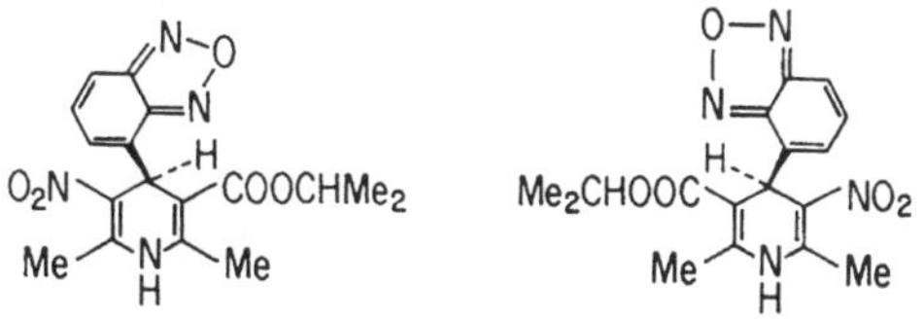

(+)S-202-791 Activator

(−)R-202-791 Antagonist

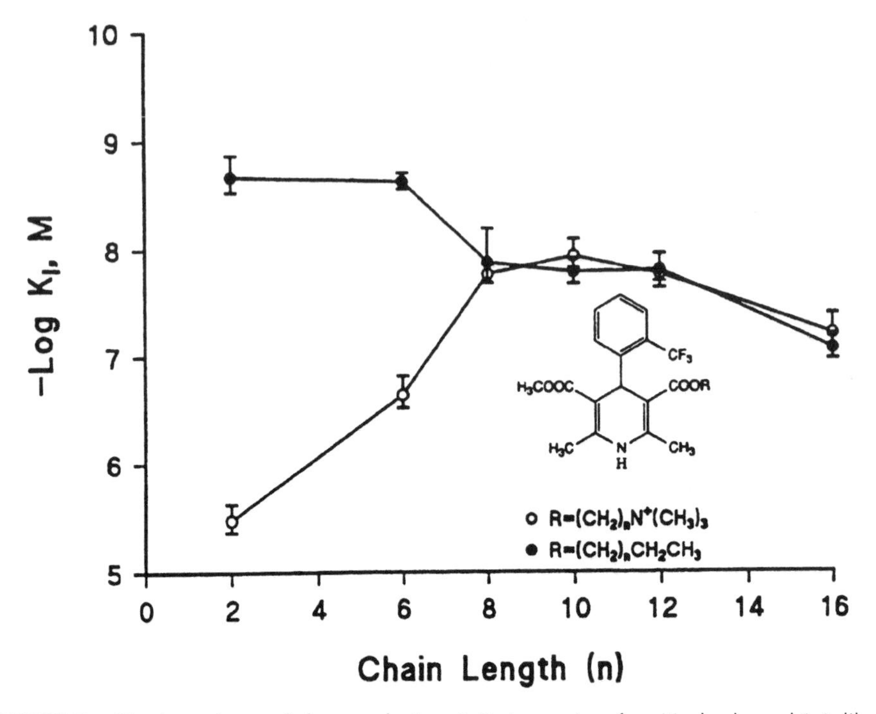

FIGURE 9. The dependence of pharmacologic activity in a series of positively charged 1,4-dihydropyridines on the chain length in the charged group. Optimal activity in the charged series occurs at n = 8 – 10, suggesting a corresponding insertion of the dihydropyridine nucleus into the cell membrane. In contrast, there is no peak of dependence of activity on chain length in the corresponding neutral series. (From Baindur N, Rutledge A, Triggle DJ: A homologous series of permanently charged 1,4-dihydropyridines: Novel probes designed to localize drug binding sites on ion channels. J Med Chem 36:3743, 1993, with permission).

binding sites represented in Figure 4 were demonstrated originally by radioligand binding techniques.[21,29,30] However, these studies neither localized nor characterized the topography of the sites. The use of impermeant analogs of diltiazem, verapamil, and nifedipine has revealed the extracellular localization of the dihydropyridine and benzothiazepine sites and the intracellular localization of the phenylalkylamine site.[51–55] That the 1,4-dihydropyridine site is extracellular but located some 10–14 A in the interior, is shown by the activity of a series of quaternary ammonium probes of nifedipine, in which the activity was steeply dependent on chain length[54] (Fig. 9). This location accords with the voltage-dependent interactions of the 1,4-dihydropyridines that are important in the definition of their selectivity of action (see below).

The drug-binding sites on the alpha$_1$ subunit of the L-type channel were localized more specifically through biochemical and molecular biologic techniques.[33,34,37,56] The verapamil binding site is localized to segment S6 of domain IV and a short segment of the carboxy terminal intracellular region[57,58] (Fig. 10). Of interest, this region corresponds to the local anesthetic-binding region of the sodium channel, and verapamil is a potent local anesthetic.[59] Amino acid residues apparently critical to phenylalkylamine binding have been identified in segment IVS6. These include Tyr$_{1463}$, Ala$_{1467}$, and Ile$_{1470}$, which can be aligned

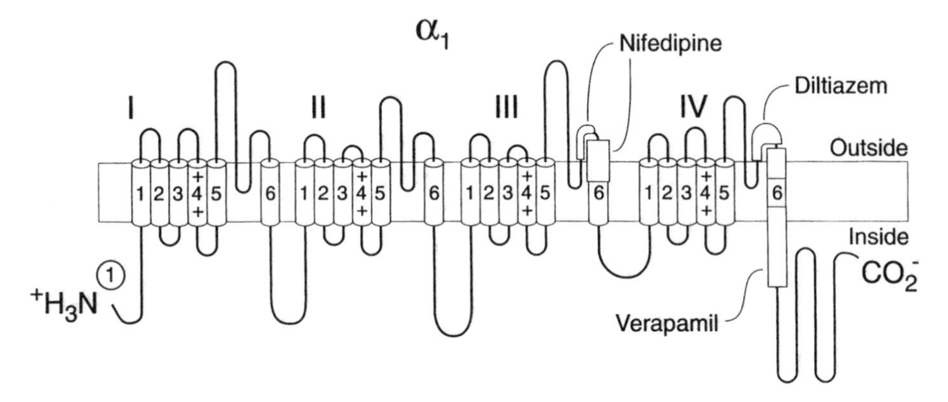

FIGURE 10. Location of the drug binding sites on the alpha$_1$ subunit.

on the same face of an alpha helix (Fig. 11).[60] In contrast, the 1,4-dihydropyridine site has been localized to segments IIIS6 and IVS6.[61–63] Mutagenesis studies have identified specific amino acid residues in these segments that are critical

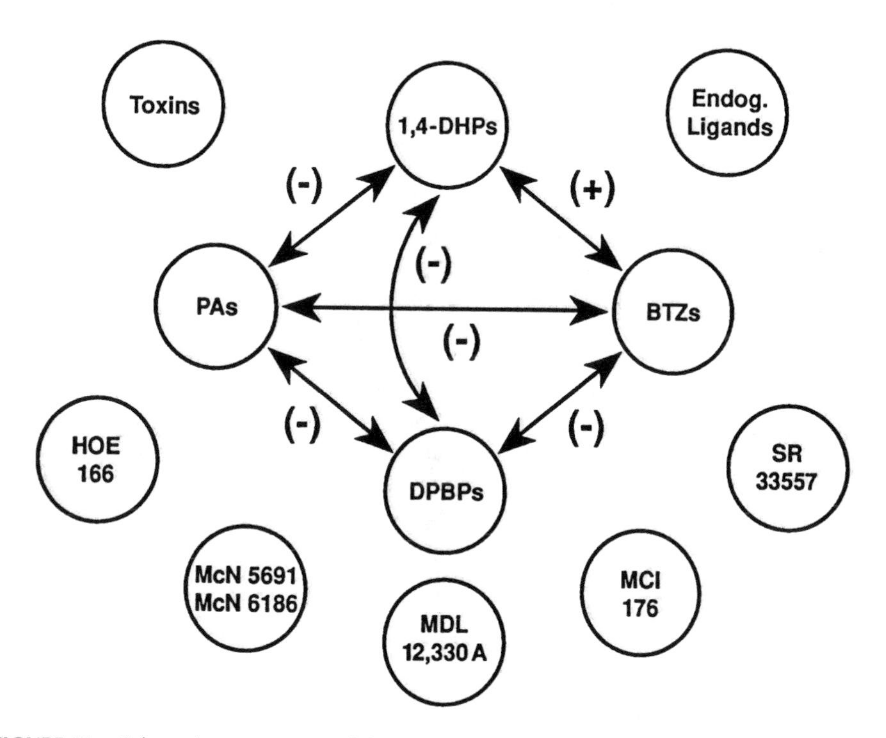

FIGURE 11. Schematic arrangement of drug binding sites at the L-type calcium channel. Depicted are the binding sites for 1,4-dihydropyridines (1,4-DHPs), phenylalkylamines (PAs), and benzothiazepines (BTZs) in the arrangement depicted in Figure 4. Also present is a site for diphenylbutylpiperidines (DPBPs), linked allosterically to the three primary binding sites. Additional and likely discrete sites are shown for HOE 166 and SR 33557 as well as sites for the ethynylbenzenealkanamines (McN 5691 and 6186), the lactamimide MDL 12,330A, and the quinazolinone MCI 176. Possible additional sites are depicted for potential endogenous ligands as well as toxin molecules that interact at the channel.

for 1,4-dihydropyridine interaction, including Tyr_{1048} of IIIS6.[62,63] These studies suggest that 1,4-dihydropyridines bind to a receptor at the interface of domains III and IV. This finding provides the observed allosteric interactions between nifedipine and verapamil. The high affinity 1,4-dihydropyridine interaction is linked to Ca^{2+} binding in the pore region formed by the $alpha_1$ subunit.[64] Because permeation of Ca^{2+} is believed to require the pore binding of two Ca^{2+} ions, it is possible that 1,4-dihydropyridines effectively block the channel by promoting high affinity pore binding of Ca^{2+}, thus effectively rendering the channel divalent cation-impermeant.[65] The binding site for the benzothiazepines has been less well characterized, but a region partially overlapping the binding site of phenylalkylamines has been suggested.[66] This suggestion is consistent with the similar pharmacologic profile of verapamil and diltiazem.

REGULATION OF CALCIUM CHANNELS

A further analogy between pharmacologic receptors and calcium channels is revealed by the regulatory behavior of calcium channels. Voltage-gated calcium channels are regulated by homologous and heterologous factors, including pathology, and an increasing number of genetic defects are known to alter channel function.[67–70] Thus, persistent activation of neuronal and neurosecretory, but not cardiac, cells results in downregulation of 1,4-dihydropyridine receptors and channel function.[67,71–74] Both beta-adrenoceptors and calcium channels are regulated, but in opposing directions, in cardiac tissue in hyper- and hypothyroid states.[75,76] Uterine calcium channels are altered in expression with hormonal status during pregnancy and parturition; an increase in these channels appears to be associated with labor.[77,78]

Human cardiomyopathy is associated with an increased density of cardiac channels,[79] and in experimental congestive heart failure downregulation of 1,4-dihydropyridine binding sites is accompanied by corresponding down regulation of ATP-sensitive potassium channels.[68,80] Similarly, downregulation occurs in porcine myocardium during ischemia.[81] Changes in channel number and/or function have been reported in other experimental conditions, including hypertension, cardiac hypertrophy, aortic insufficiency, and truncus arteriosus.[67] The causal relationship between these changes and the initiation and progression of disease remains to be established.

Increasing numbers of functional disorders—termed "channelopathies"—are now known to be associated with mutations in channel sequences.[69,70,82] Murine muscular dysgenesis is an autosomal recessive defect of skeletal muscle in which the $alpha_1$ subunit is not expressed. Hypokalemic periodic paralysis is a single amino acid defect in the skeletal muscle calcium channel in which paralysis is precipitated under conditions of low serum potassium.

It is clear that the voltage-gated calcium channel is a regulated species. However, the tachyphylaxis or rebound that might be expected after chronic administration or withdrawal of calcium antagonists is difficult to demonstrate clinically,[83] despite experimental documentation.[67,84–89] Few reports suggest either tachyphylaxis[90] or withdrawal phenomena,[91–96] and no human data are available for calcium channel number or function after chronic drug administration.

SELECTIVITY OF ACTION OF CALCIUM ANTAGONISTS

Despite the widespread presence of voltage-gated calcium channels in excitable tissues, the calcium antagonists are highly specific and selective agents. Verapamil, nifedipine, diltiazem, and their second-generation derivatives exhibit therapeutic activities primarily in the cardiovascular system. Major actions of existing agents are not routinely seen in the central nervous system (CNS), although calcium channels are widespread in CNS tissue.[20–24] Furthermore, differences in their activities are not only quantitative: verapamil and diltiazem exhibit significant cardiac depressant properties, whereas nifedipine is predominantly a vasodilator (see Table 3). In addition, drugs within a single structural group may have significant differences in profile. For example, felodipine, a second-generation 1,4-dihydropyridine, is more vascular selective than nifedipine; the vascular:cardiac selectivities of the two agents are approximately 100:1 and 10:1, respectively.[97] Nimodipine, on the other hand, exhibits cerebral vascular and neuronal selectivity.[98] A further expression of selectivity of action in the 1,4-dihydropyridine series is seen in correlations between pharmacologic and radioligand-binding activities in smooth, cardiac, and neuronal tissue.[21,99,100] Similar high-affinity binding sites are accompanied by different expressions of pharmacologic activity, with correspondingly high-affinity pharmacology in depolarized smooth muscle, medium-affinity pharmacology in cardiac preparations, and a frequent absence of obvious pharmacology in neuronal preparations.[19,21,22,99,100]

In principle, such selectivity of action stems from a number of sources:

1. Pharmacokinetic factors, including distribution
2. Mode of calcium mobilization (intracellular and extracellular sources)
3. Class and subclass of calcium channel activation
4. State-dependent interactions (voltage- and frequency-dependent behavior)
5. Pathologic state of tissue (channel regulation)

All of these factors may contribute to the profile of a particular agent.

An important and subtle process for selectivity of action is derived from state-dependent interactions between the channel and the drug. According to the modulated receptor hypothesis, drugs may exhibit preferential affinity for or access to different states of the channel—resting, open, or inactivated (state-dependent interactions)[101,102] (Fig. 12). Transitions between these states are determined by membrane potential and biochemical status, including phosphorylation. Drugs with equal affinity for all states do not exhibit selective interactions, but for drugs that selectively interact with the inactivated state, stimuli that increase the channel fraction in the inactivated state enhance the apparent affinity. In a simple two-state model in which the drug binds to state A or B with dissociation constants K_A and K_B, the apparent affinity is expressed as follows:

$$K_{app} = \frac{1}{h/K_A - (1\text{-}h)/K_B}$$

where h and 1-h are the channel fractions in states A and B, respectively. Furthermore, drug structure and physicochemical properties may influence the access of drug to a preferred site of interaction. Hydrophilic or charged species may access through the open channel state in a pathway of stabilization of the inactivated state. The scheme depicted in Figure 12 may be subject to further, subtle discrimination, by shifts of channel voltage-activation curves under

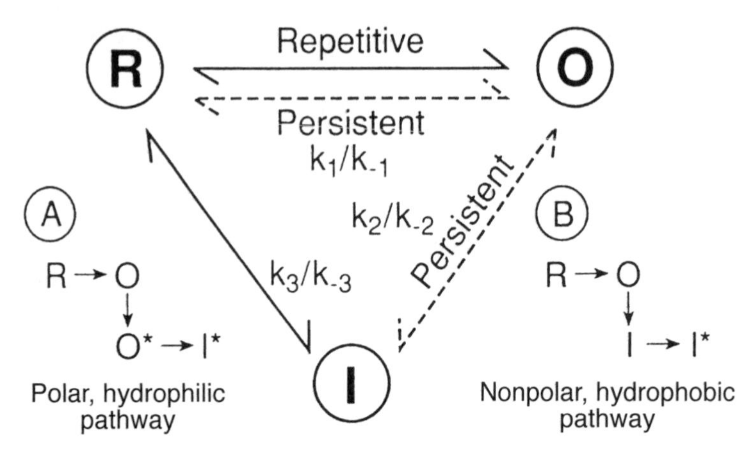

FIGURE 12. Calcium channel drug action at voltage-gated calcium channels according to the modulated receptor hypothesis. The channel is shown existing in three interconvertible states: resting, open, and inactivated. Interconversion among the states is governed by a set of time- and voltage-dependent rate constants. Repetitive firing favors an equilibrium between the resting and open states, and persistent depolarization shifts the equilibrium to the open and inactivated states. According to the physicochemical properties of the drug, its pathway to a favored binding site may differ. Thus, in pathway A for polar, hydrophilic species, binding to the inactivated state may access this state preferentially or exclusively when the channel is in an open state, whereas nonpolar, hydrophobic agents can access the same state through the membrane phase as depicted in pathway B.

modulator influence that may alter phosphorylation or other biochemical status, including G protein association, of the channel[103] (Fig. 13).

Verapamil and diltiazem exhibit prominent frequency-dependent interactions, whereas nifedipine exhibits prominent voltage-dependent antagonism, according to which activity increases with increasing frequency of repetitive stimulation or with increasing maintained depolarization.[101,102,104–109] These observations unquestionably underlie, at least in significant part, the antiarrhythmic properties of verapamil and diltiazem and the general vascular selectivity of nifedipine and other 1,4-dihydropyridines, which may be attributed to selective interactions through the open and inactivated channel states, respectively. Verapamil and diltiazem are charged species at physiologic pH, whereas the 1,4-dihydropyridines are predominantly hydrophobic species with high membrane:buffer partition coefficients. It has been proposed that 1,4-dihydropyridines approach their channel binding sites through diffusional

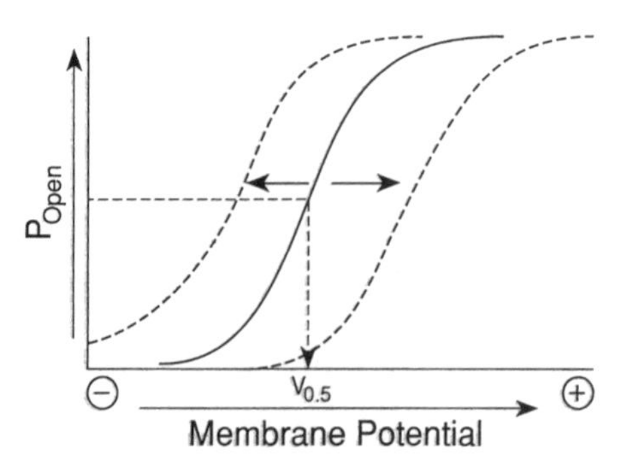

FIGURE 13. The probability that a given channel may open is depicted as dependent on membrane potential. Increasing levels of membrane depolarization favor channel opening. Biochemical modulation of the channel may shift this curve in depolarizing or hyperpolarizing directions, as shown. The voltage-dependent binding of calcium antagonists is correspondingly shifted along the voltage axis by this channel modulation.

TABLE 6. State-dependent Interactions of Calcium Channel Ligand at Calcium Channels

		K^D, M			
Tissue	**Ligand**	**Resting**	**Open**	**Inactivated**	**Reference**
Cardiac Purkinje	Nisoldipine	1.3×10^{-6}		1×10^{-9}	108
Ventricle	Nitrendipine	7.3×10^{-7}		2.5×10^{-9}	109
Ventricle	Amiodarone	5.8×10^{-6}		3.6×10^{-7}	112
Dorsal root ganglion	Nimodipine	4.0×10^{-8}		1.3×10^{-9}	113
Dorsal root ganglion (F11 cells)	Nimodipine	2.0×10^{-6}		3×10^{-9}	114
GH3 cells	(+) R Bay K 8644	$> 1.0 \times 10^{-8}$		4×10^{-9}	115
CH_4C_1 cells	Nimodipine	7.0×10^{-6}	5×10^{-10}	—	116
Skeletal muscle	(+) Isradipine	1.3×10^{-8}		1.5×10^{-9}	117
Smooth muscle (A7r5)	Nimodipine	1×10^{-6}	2×10^{-11}		118
Smooth muscle (A7r5)	S12968 (amlodipine analog)	4×10^{-7}		5×10^{-8}	119
Saphenous vein	Nitrendipine	$\sim 10^{-8}$M		2.5×10^{-10}	120
Mesenteric artery	Nisoldpine	1.2×10^{-8}		7×10^{-11}	121
Mesenteric artery	Nitrendipine	2.2×10		4.6×10^{-10}	123
Uterus	(+) Isradipine	2.3×10^{-9}M		1×10^{-10}	124

access, whereby an initial and structured partitioning of the ligand into the membrane lipid is followed by two-dimensional diffusion to the specific binding site.[110,111]

Direct evidence for the state-dependent interactions of the calcium channel antagonists derives from electrophysiologic, radioligand binding, and pharmacologic studies. Electrophysiologic data from cardiac, smooth muscle, neuronal, and secretory cells indicate that calcium antagonists exhibit higher affinity for the inactivated state or, less commonly, the open state than for the resting state of the channel (Table 6). Data have been obtained under various experimental conditions and are not quantitatively comparable. However, it is clear that large (up to a thousand-fold) differences in affinity between the resting and open or inactivated states are exhibited by the calcium channel antagonists. Thus, even small differences in membrane potential between different vascular beds are sufficient to generate the modest levels of vascular selectivity exhibited by the 1,4-dihydropyridines.

Radioligand binding of calcium channel antagonists to cells or tissues under polarized and depolarized conditions also reveals voltage-dependent interactions (Table 7) that, in general, confirm the electrophysiologic data. Further, but less direct, support is derived from pharmacologic studies demonstrating the enhanced effectiveness of verapamil and diltiazem with increased frequency of stimulation and of nifedipine with increased depolarization and duration of the

TABLE 7. Calcium Channel Ligand Binding in Polarized and Depolarized Preparations

Tissue	Ligand	K_D, M Polarized	K_D, M Depolarized	Ratio	Reference
Rat cardiomyocytes	(+) [^{3}H] Isradipine	0.73×10^{-9}	5.9×10^{-11}	~12	125
	(+) [^{3}H] Isradipine	3.5×10^{-9}	5.3×10^{-11}	~55	126
Rat cardiomyocytes	(–) [^{3}H] Bay K 8644	5.2×10^{-9}	5.5×10^{-8}	~1	127
Rat aorta	(+) [^{3}H] Isradipine	2.5×10^{-10}	6.8×10^{-11}	~4	128
Horse portal vein	(+) [^{3}H] Isradipine	1.4×10^{-10}	4×10^{-11}	~4	129
Bovine coronary	[^{3}H] Nifedipine	2.0×10^{-9}	1.4×10^{-10}	~1	130
Rat mesenteric	(+) [^{3}H] Isradipine	2.5×10^{-11}	4.1×10^{-10}	~6	131,132
Nifedipine		3.2×10^{-9}	1.2×10^{-9}	~2	
Brain Blood	(+) [^{3}H] Isradipine	8.8×10^{-10}	3.5×10^{-10}	2.5	133
	(–) Nimodipine	7.7×10^{-10}	2.3×10^{-10}	~3	
	(–) Nimodipine	5.9×10^{-9}	2.0×10^{-10}	~3	
	Nifedipine	2.8×10^{-9}	2.9×10^{-19}	~1	
PC 12 cells	(+) [^{3}H} Nifedipine	2.7×10^{-10}	5.5×10^{-11}	~6	134

depolarizing stimulus. The preferential inhibition of the slow phase of potassium depolarization-induced contractions and the increased potency with increasing time of challenge in both vascular and nonvascular smooth muscle are quite consistent with a preferential interaction of the calcium channel antagonists with a channel state favored by depolarization.[126–128] Figure 14 shows that in intestinal smooth muscle preincubation of the calcium antagonist in depolarizing media before inhibition of the tension response augments potency.[135] Observations related to functional smooth muscle preparations confirm the importance, realized from electrophysiologic studies,[101,102] of the kinetics of drug–channel interaction in the determination of selectivity of action. When channel opening is brief, the time available for drug-channel equilibrium at therapeutically significant concentrations may be inadequate. Thus, in cardiac tissue the fractional availability of the high-affinity inactivated state is insufficient to permit significant interaction with 1,4-dihydropyridines at plasma concentrations that are effective in vascular smooth muscle.[136–138] In contrast, the interactions of verapamil and diltiazem through the open channel state are governed by the open channel

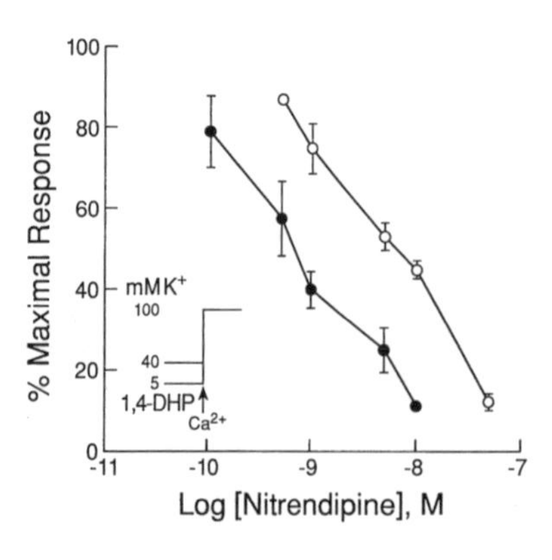

FIGURE 14. The influence of membrane potential (predepolarization) on the inhibitory activity of nitrendipine against the potassium depolarization-induced tension response in guinea pig ileal longitudinal smooth muscle. The specific protocol is depicted in the inset. Nitrendipine was incubated at 5 (O) or 40 mM K+ (O) in the absence of calcium, and calcium was then reintroduced, together with potassium, to bring the total concentration to 100 mM. Dose–response relationships for the inhibitory activity of nitrendipine against the initial phasic component of response were then determined. Nitrendipine is 5–10 times more active in the depolarized preparation. (From Triggle DJ, Hawthorn M, Zheng W: Potential-dependent interactions of nitrendipine and related 1,4-dihydropyridines in functional smooth muscle preparations. J Cardiovasc Pharmacol 12(Suppl 4):91–93, 1988, with permission).

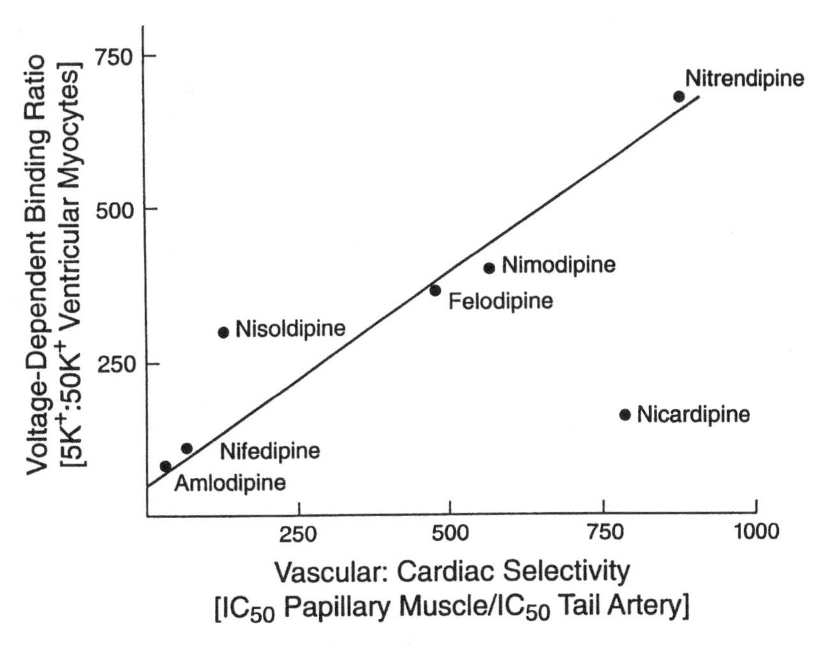

FIGURE 15 Relationship between voltage-dependent binding (ratio of binding constants of 1,4-dihydropyridines in polarized and depolarized cardiac cells) and vascular:cardiac selectivity determined pharmacologically for a series of clinically available 1,4-dihydropyridines. (From Sun J, Triggle DJ: Calcium channel antagonists—cardiovascular selectivity of action. J Pharmacol Exp Ther 274:419, 1995, with permission).

distribution time as well as the relative rates of association to and dissociation from the channel. An adequately slow channel dissociation rate permits the accumulation of blockade in a frequency-dependent manner.

Figure 15 demonstrates that the voltage dependence of the interactions of the 1,4-dihydropyridines is associated with vascular:cardiac selectivity.[139] The approximately linear relationship for the series of clinically available 1,4-dihydropyridines between voltage-dependent binding in cardiac cells and the ratio of the antagonist activities in vascular and cardiac muscle preparations is quite suggestive that tissue selectivity is determined significantly by structure-dependent and voltage-dependent interactions. It is clear that voltage dependence is not the only factor determining tissue selectivity and that the differential pharmacology of subtypes of the L-type channel is also important. Thus, inhibition by nisoldipine of recombinant channels derived from cardiac and vascular smooth muscle alpha$_1$ subunits revealed greater resting block and faster development of block in vascular smooth muscle channels.[140]

A further determinant of tissue selectivity, intimately related to state-dependent processes, is the activator-antagonist spectrum of pharmacologic properties. Activator properties are potently expressed in the 1,4-dihydropyridines but probably contribute also to verapamil and diltiazem structures. A number of 1,4-dihydropyridines, even as single chiral species, exhibit mixed activator-antagonist properties and thus are clearly partial agonists. Activator properties are present at polarized levels of membrane potential, and a transition to antagonist properties occurs with increasing depolarization.[141–144] 1,4-Dihydropyridine activators, including Bay K 8644 and related structures (see Fig. 9), exhibit, in contrast to the corresponding 1,4-dihydropyridine antagonists, little voltage dependency of interaction and presumably have comparable affinities for the several channel states. Accordingly, their observed pharmacologic activity is set by the level of membrane potential.

TISSUE SELECTIVITY AND CONTRAINDICATIONS

The factors that contribute to the determination of the pharmacologic and therapeutic profile of the calcium channel antagonists also determine their contraindication profile. The calcium antagonists have been advanced as cardioprotective agents, a theoretically attractive use. However, there is a clear difference between use of nifedipine and use of verapamil or diltiazem. Several trials with nifedipine, primarily the original short-acting formulation, revealed its ineffectiveness or trend to harm in postinfarct patients. In contrast, verapamil and, to a lesser extent, diltiazem revealed benefits, particularly in patients with good left ventricular function.[145–147] More recently, the safety of calcium antagonists, most specifically nifedipine, in the treatment of essential hypertension has been questioned in case control studies.[148–152] It is likely that pharmacokinetic and tissue selectivity considerations may contribute to any deleterious effects of calcium antagonists. Rapidly acting vasodilators may be pro-ischemic and, through reflex sympathetic activation, initiate excessive activation of the neurohormonal system with undesirable consequences.[153,154] The relative cardiac:vascular selectivity of these agents may contribute to their detrimental effects in patients with compromised left ventricular function. Enhanced vascular selectivity and slower onset and offset of action, as seen in second-generation 1,4-dihydropyridines, including amlodipine and felodipine, will probably reduce or eliminate such detrimental effects.[155,156]

CONCLUSION

The calcium antagonists are a remarkable groups of drugs—as therapeutic agents, pharmacologic tools, and molecular probes. They owe their selectivity of action to specific interactions with one class of voltage-gated calcium channel. From such specific interactions, dependent on both subclass of channel and subtle modulation of drug affinity through potential and biochemical influences, the calcium antagonists derive their therapeutic and molecular utility. Ongoing studies with available agents and development of new agents active at the L-type and other classes of calcium channels will increase interest in the benefits of calcium antagonists to patient and scientist alike. Continued study of the relationships between channel structure and function will facilitate therapeutic advances.[157]

References

1. Campbell AK: Intracellular Calcium: Its Universal Role as a Regulator. New York, John Wiley and Sons, 1983.
2. Triggle DJ: Drugs affecting calcium regulation and actions. In Smith CM, Reynard A (eds): Textbook of Pharmacology. Philadelphia, W.B. Saunders, 1991.

3. Gill DL, Waldron RT, Rys-Sikora KE, et al: Calcium pools, calcium entry and cell growth. Biosci Rep 16:139–157, 1996.
4. Barran T: Calcium and cell cycle progression: Possible effects of external perturbations on cell proliferation. Biophys J 70:1198–1213, 1996.
5. Finkbeiner S, Greenberg ME: Ca^{2+}-dependent routes to *ras*: mechanisms for neuronal survival, differentiation and plasticity. Neuron 16:233–236, 1996.
6. Hughes AD: Calcium channels in vascular smooth muscle cells. J Vasc Res 32:353–370, 1995.
7. Cheung JY, Bonventre JV, Malis CD, Leaf A: Calcium and ischemic injury. N Engl J Med 14:1670–1676, 1986.
8. Harman AW, Maxwell MJ: An evaluation of the role of calcium in cell injury. Annu Rev Pharmacol Toxicol 35:129–144, 1995.
9. Inoue T, Kawashima H: 1,25-Dihydroxyvitamin D3 stimulates $^{45}Ca^{2+}$ uptake by cultured vascular smooth muscle cells derived from rat aorta. Biochem Biophys Res Commun 152:1388–1394, 1988.
10. Wang R, Karpinski E, Pang PKT: Effect of three fragments of parathyroid hormone on calcium channel currents in neonatal rat ventricular cells. Regul Pept 54:445–456, 1994.
11. Resnick LM: Uniformity and diversity of calcium in hypertension. A conceptual framework. Am J Med 82(Suppl 1B):16–26, 1987.
12. Luft FC, McCarron DA: Heterogeneity of hypertension: The diverse role of electrolyte intake. Annu Rev Med 42:347–355, 1991.
13. Resnick LM: Calciotropic hormones in salt-sensitive essential hypertension: 1,25-dihydroxyvitamin D and parathyroid hypertensive factor. J Hypertens 12(Suppl 1):S3–S9, 1994.
14. Erne P, Bolli P, Burgisser E, Buhler FR: Correlation of platelet calcium with blood pressure. Effect of antihypertensive therapy. N Engl J Med 310:1084–1088, 1984.
15. Lau K, Eby B: The role of calcium in genetic hypertension. Hypertension 7:657–667, 1985.
16. Lewanczuk RZ, Resnick LM, Ho M-S, et al: Clinical aspects of parathyroid hypertensive factor. J Hypertens 12(Suppl 1):S11–S16, 1994.
17. Pang PKT, Shan JJ, Lewanczuk RZ, Benishin CG: Parathyroid hypertensive factor and intracellular calcium regulation. J Hypertens 14:1053–1060, 1996.
18. Kass RS, Krafte DS: Negative surface charge near heart calcium channels. Relevance to block by 1,4-dihydropyridines. J Gen Physiol 89:629–644, 1987.
19. Triggle, DJ: Structure-function correlations of 1,4-dihydropyridine calcium channel antagonists and activators, In Hondeghem LM (ed): Molecular and Cellular Mechanisms of Antiarrhythmic Agents. Mt. Kiscoe, NY, 1989, pp 269–291.
20. Fleckenstein A: Calcium Antagonism in Heart and Smooth Muscle. New York, John Wiley and Sons, 1983.
21. Janis RA, Silver P, Triggle DJ: Drug action and cellular calcium regulation. Adv Drug Res 16:309–591, 1987.
22. Triggle DJ: Calcium Antagonists. In Antonaccio M (ed): Cardiovascular Pharmacology, 3rd ed. New York, Raven, 1990, pp 107–160.
23. Opie LH: Calcium channel antagonists. I, II, III. Cardiovasc Drugs Ther 1:411–430, 460–491, 625–666, 1987.
24. Glossmann H, Striessnig J: Calcium channels. Vitam Horm 44:155–328, 1988.
25. Hollt V, Kouba M, Dietel M, Vogt G: Stereoisomers of calcium antagonists which differ markedly in their potencies as calcium blockers are equally effective in modulating drug transport by P-glycoprotein. Biochem Pharmacol 43:2601–2608, 1992.
26. Godfraind T: Calcium antagonists and vasodilatation. Pharmacol Ther 64:37–75, 1994.
27. Kuriyama H, Kitamura K, Nabata H: Pharmacodynamic and physiological significance of ion channels and factors that modulate them in vascular tissues. Pharmacol Rev 47:387–573, 1995.
28. Triggle DJ: Calcium antagonists. History and perspective. Stroke 21(Suppl IV):IV49–IV58, 1990.
29. Godfraind T, Miller RC, Wibo M: Calcium antagonism and calcium entry blockade. Pharmacol Rev 38:321–416, 1986.
30. Triggle DJ, Langs DA, Janis RA: Ca^{2+} channel ligands: Structure-function relationships of the 1,4-dihydropyridines. Med Res Rev 9:123–180, 1989.
31. Bean BP: Classes of calcium channels in vertebrate cells. Annu Rev Physiol 51:367–384, 1989.
32. Miljanich G, Ramachandran J: Antagonists of neuronal calcium channels. Structure, function and therapeutic implications. Annu Rev Pharmacol Toxicol 35:707–734, 1995.
33. Catterall, WA: Structure and function of voltage-gated ion channels. Annu Rev Biochem 64:493–531, 1995.
34. De Waard M, Gurnett CA, Campbell KP: Structure and functional diversity of voltage-activated calcium channels. In Narahashi T (ed): Ion Channels. New York, Plenum, 1996, pp 41–87.

35. Hofmann F, Biel M, Flockerzi V: Molecular basis for Ca^{2+} channel diversity. Annu Rev Neurosci 17:399–418, 1994.
36. Mori Y, Mikala G, Varadi G, et al: Molecular pharmacology of voltage-dependent calcium channels. Jpn J Pharmacol 72:83–109, 1996.
37. Catterall WA, Striessnig J: Receptor sites for Ca^{2+} channel antagonists. Trends Pharmacol Sci 13:256–262, 1992.
38. De Waard M, Pragnell M, Campbell,KP: Ca^{2+} channel regulation by a conserved beta-subunit domain. Neuron 13:495–503, 1994.
39. Witcher DR, De Waard M, Liu H, et al: Association of native Ca^{2+} channel beta-subunits with the $alpha_1$ subunit interaction domain. J Biol Chem 270:18088–18093, 1995.
40. Rampe D, Triggle DJ: New ligands for L-type Ca^{2+} channels. Trends Pharmacol Sci 11:112–116, 1990.
41. Spedding M, Kenny B, Chatelain P: New drug binding sites in Ca^{2+} channels. Trends Pharmacol Sci 16:139–142, 1995.
42. Rutledge A, Triggle DJ: The binding interactions of Ro 40-5967 at the L-type Ca^{2+} channel in cardiac tissue. Eur J Pharmacol 280:155–158, 1995.
43. Bernink PJL, Prager G, Schelling A, Kobrin, I: Antihypertensive properties of the novel calcium antagonist mibefradil (Ro 40-5967). A new generation of calcium antagonist. Hypertension 27:426–432, 1996.
44. Zheng W, Rampe D, Triggle DJ: Pharmacological, radioligand binding and electrophysiological characteristics of FPL 64176. Mol Pharmacol 40:734–742, 1991.
45. Rampe D, Lacerda AE: A new site for the activation of cardiac calcium channels defined by the nondihydropyridine FPL 64176. J Pharmacol Exp Ther 259:982–987, 1991.
46. Goldmann S, Stoltefuss J: 1,4-Dihydropyridines: effects of chirality and conformation on the calcium antagonist and calcium agonist activities. Angew Chemie (Int Ed) 30:1559–1578, 1991.
47. Scott RH, Dolphin AC: Activation of G protein promotes agonist response to calcium channel ligands. Nature 330:760–762, 1987.
48. Bossert F, Vater W: 1,4-Dihydropyridines—a basis for developing new drugs. Med Res Rev 9:291–324, 1989.
49. Arrowsmith JE, Campbell SF, Cross PE, et al: Long acting dihydropyridine calcium antagonists. I. 2-Alkoxymethyl derivatives incorporating basic substituents. J Med Chem 29:1696–1702, 1986.
50. Kwan Y-W, Zhong Q, Wei X-Y, et al: The interactions of 1,4-dihydropyridines bearing a 2-(2-aminoethylthio)methyl substituent at voltage-dependent Ca^{2+} channels of smooth muscle, cardiac muscle and neuronal tissues. Naunyn Schmiedebergs Arch Pharmacol 341:128–136, 1990.
51. Hescheler J, Pelzer D, Trube G, Trautwein W: Does the organic calcium channel blocker D600 act from inside or outside on the cardiac cell membrane? Pflugers Arch 393:287–291, 1982.
52. Leblanc N, Hume, JR: D600 block of L-type Ca^{2+} channels in vascular smooth muscle cells: Comparison with permanently charged derivative, D890. Am J Physiol 257:C689–C695, 1989.
53. Kass RS, Arena JP, Chin S: Block of L-type calcium channels by charged dihydropyridines. Sensitivity to side of application. J Gen Physiol 98: 63-75, 1991.
54. Bangelore R, Baindur N, Rutledge A, et al: L-type calcium channels. Asymmetrical intramembrane binding domain revealed by variable length, permanently charged 1,4-dihydropyridines. Mol Pharmacol 46:660–666, 1994.
55. Seydl K, Kimball D, Schindler H, Romanin C: The benzazepine/benzothiazepine binding domain of the cardiac L-type Ca^{2+} channel is accessible only from the extracellular side. Pflugers Arch 424:552–554, 1993.
56. Varadi G, Mori Y, Mikala G, Schwartz A: Molecular determinants of Ca^{2+} channel function and drug action. Trends Pharmacol Sci 16:43–49, 1995.
57. Striessnig J, Glossmann H, Catterall WA: Identification of a phenylalkylamine binding region within the $alpha_1$ subunit of skeletal muscle Ca^{2+} channels. Proc Natl Acad Sci U S A 87:9108–9112, 1990.
58. Doring F, Degtiar VE, Grabner M, et al: Transfer of L-type calcium channel IVS6 segment increases phenylalkylamine sensitivity of $alpha_{1A}$. J Biol Chem 271:11745–11749, 1996.
59. Ragsdale DS, McPhee JC, Scheuer T, Catterall WA: Molecular determinants of state-dependent block of Na^+ channels by local anesthetics. Science 265:1724–1728, 1994.
60. Hockerman GH, Johnson BD, Scheuer T, Catterall WA: Molecular determinants of high affinity phenylalkylamine block of L-type calcium channels. J Biol Chem 270:22119–22122, 1995.
61. Nakayama H, Taki M, Striessnig J, et al: Identification of 1,4-dihydropyridine binding regions within the $alpha_1$ subunit of skeletal muscle Ca^{2+} channels by photoaffinity labeling with diazipine. Proc Natl Acad Sci U S A 88:9203–9207, 1996.

62. Peterson BZ, Tanada TM, Catterall WA: Molecular determinants of high affinity dihydropyridine binding in L-type calcium channels. J Biol Chem 271:5293–5296, 1996.
63. Grabner M, Wang Z, Hering S, et al: Transfer of 1,4-dihydropyridine sensitivity from L-type to class A(B1) calcium channels. Neuron 16:207–218, 1996.
64. Peterson BZ, Catterall WA: Calcium binding in the pore of L-type calcium channels modulates high affinity 1,4-dihydropyridine binding. J Biol Chem 270:18201–18204, 1995.
65. Yang J, Ellinor PT, Sather WA, Zet al: Molecular determinants of Ca^{2+} selectivity and ion permeation in L-type Ca^{2+} channels. Nature 336:158–161, 1993.
66. Watanabe T, Kalasz H, Yabana H, et al: Azidobutyrylclentiazem, a new photoactivatable diltiazem analogs, labels benzothiazepine binding sites in the $alpha_1$ subunit of the skeletal muscle calcium channel. FEBS Lett 334:261–264, 1993.
67. Ferrante J, Triggle DJ: Drug- and disease-induced regulation of voltage-dependent Ca^{2+} channels. Pharmacol Rev 42:29–44, 1990.
68. Gopalakrishnan M, Triggle DJ: The regulation of receptors, ion channels and G proteins in congestive heart failure. Cardiovasc Drug Rev 8:255–302, 1990.
69. Hoffmann EP, Lehmann-Horn F, Rudel R: Overexcited or inactive: Ion channels in muscle disease. Cell 80:681–686, 1995.
70. Hess EJ: Migraines in mice? Cell 87:1149–1151, 1996.
71. DeLorme EM, Rabe CS, McGee R: Regulation of the number of functional voltage-sensitive Ca^{2+} channels in PC12 cells by chronic changes in membrane potential. J Pharmacol Exp Ther 44:838–843, 1988.
72. Ferrante J, Triggle DJ: The effects of chronic depolarization on L-type 1,4-dihydropyridine-sensitive Ca^{2+} channels in chick neural retina and rat cardiac cells. Can J Physiol Pharmacol 69:914–920, 1991.
73. Skattebol A, Triggle DJ, Brown AM: Homologous regulation of voltage-dependent Ca^{2+} channels by 1,4-dihydropyridines. Biochem Biophys Res Commun 60:929–936, 1989.
74. Liu J, Bangalore R, Rutledge A, Triggle DJ: Modulation of L-type Ca^{2+} channels in clonal rat pituitary cells by membrane depolarization. Mol Pharmacol 45:1198–1206, 1994.
75. Hawthorn M, Gengo P, Wei X-Y, et al: Effect of thyroid status on *beta*-adrenoceptors and calcium channels in rat cardiac and vascular tissue. Naunyn-Schmiedebergs Arch Pharmacol 337:539–544, 1988.
76. Wibo M, Kolar F, Zheng L, Godfraind T: Influence of thyroid status on postnatal maturation of calcium channels, *beta*-adrenoceptors and cation transport ATPases in rat ventricular tissue. J Mol Cell Cardiol 27:1731–1743, 1995.
77. Mershon JL, Mikala G, Schwartz A: Changes in the expression of the L-type voltage-dependent calcium channel during pregnancy and parturition in the rat. Biol Reprod 51:993–999, 1994.
78. Tezuka N, Ali M, Chwalisz K, Garfield RE: Changes in transcripts encoding calcium channel subunits or rat myometrium during pregnancy. Am J Physiol 269:C1008–C1017, 1995.
79. Wagner JA, Sax FL, Weisman HF, et al: Calcium antagonist receptors in the atrial tissue of patients with hypertrophic cardiomyopathy. N Engl J Med 320:755–761, 1989.
80. Gopalakrishnan M, Triggle DJ, Rutledge A, et al: Regulation of ATP-sensitive K^+ channels and 1,4-dihydropyridine sensitive Ca^{2+} channels in experimental heart failure. Am J Physiol 261:1979–1987, 1991.
81. Stokke M, Kirkeboen KA, Ness PA, et al: Equal changes in L-type calcium channel density after 60 min of ischemia in normal and ischemically preconditioned porcine myocardium. Acta Physiol Scand 157:147–155, 1996.
82. Hoffmann EP: Voltage-gated ion channelopathies: Inherited disorders caused by abnormal sodium, chloride, and calcium regulation in skeletal muscle. Annu Rev Med 46:432–441, 1995.
83. Raftery EB: Cardiovascular drug withdrawal syndromes. A potential problem with calcium antagonists? Drugs 28:371–374, 1984.
84. Gengo P, Bowling N, Wyss VL, Hayes JS: Effects of prolonged phenylephrine infusion on cardiac adrenoceptors and calcium channels. J Pharmacol Exp Ther 244:100–105, 1988.
85. Gengo P, Skattebol A, Moran JF, et al: Regulation by chronic drug administration of neuronal and cardiac calcium channels, *beta*-adrenoceptor and muscarinic receptor levels. Biochem Pharmacol 37:627–633, 1988.
86. Dolin S, Little H, Hudspith M, et al: Increased dihydropyridine-sensitive calcium channels in rat brain may underlie ethanol physical dependence. Neuropharmacology 26:275–279, 1987.
87. Garthoff B, Bellemann P: Effects of salt loading and nitrendipine on dihydropyridine receptors in hypertensive rats. J Cardiovasc Pharmacol 10(Suppl 10): S36–S38, 1987.
88. Galletti F, Rutledge A, Triggle DJ: Dietary sodium intake: Influence on calcium channels and urinary calcium excretion in spontaneously hypertensive rats. Biochem Pharmacol 41:893–896, 1991.

89. Chiappe de Cingolani GE, Mosca SM, Moreyra AE, Cingolani HE: Chronic nifedipine treatment diminishes cardiac inotropic response to nifedipine: Functional upregulation of dihydropyridine receptors. J Cardiovasc Pharmacol 27:240–246, 1996.
90. Dixon IMC, Lee S-L, Dhalla NS: Nitrendipine binding in congestive heart failure due to myocardial infarction. Circ Res 66:782–788, 1990.
91. Gopalakrishnan M, Triggle DJ, Rutledge A, et al: Regulation of ATP-sensitive K^+ channels and 1,4- dihydropyridine-sensitive Ca^{2+} channels in rat cardiac failure secondary to myocardial infarction. Am J Physiol 261:H1979–H1987, 1991.
92. Le Grand B, Hatem S, Deroubaix E, et al: Calcium current depression in isolated human atrial myocytes after cessation of chronic treatment with calcium antagonists. Circ Res 69:292–300, 1991.
93. Aderka D, Levy A, Pinkhas J: Tachyphylaxis to verapamil. Arch Intern Med 146:207, 1986.
94. Gottlieb SO, Gerstenblith G: Safety of acute calcium antagonist withdrawal: Status in patients with unstable angina withdrawn from nifedipine. Am J Cardiol 55:27E–30E, 1985.
95. Mehta J, Lopez LM: Calcium-blocker withdrawal phenomena: increase in affinity of $alpha_2$ adrenoceptors for agonist as a potential mechanism. Am J Cardiol 58:242–246, 1986.
96. Schroeder JS, Walker DS, Skallard ML, Hemberger JA: Absence of rebound from diltiazem therapy in Prinzmetal's variant angina. J Am Coll Cardiol 6:174–178, 1985.
97. Elvelin L, Elmfeldt D: Felodipine—a review of its pharmacological and clinical properties. Drugs Today 25:589–596, 1989.
98. Scriabine A, Schuurman T, Traber J: Pharmacological basis for the use of nimodipine in the central nervous system. FASEB J 3:1799–1806, 1989.
99. Janis RA, Triggle DJ: 1,4-Dihydropyridine Ca^{2+} channel antagonists and activators: Comparison of binding characteristics with pharmacology. Drug Dev Res 4:254–274, 1984.
100. Janis RA, Triggle DJ: Calcium channel antagonists: New perspectives from the radioligand binding assay. In Spector S, Back N (eds): Modern Methods in Pharmacology, vol 2, New York, A. Liss, 1984, pp 1–28.
101. Hondeghem LM, Katzung BG: Antiarrhythmic agents: The modulated receptor mechanism of action of sodium and calcium channel blocking drugs. Annu Rev Pharmacol Toxicol 24:387–423, 1985.
102. Hille B: Local anesthetics: Hydrophilic and hydrophobic pathways for the drug-receptor reaction. J Gen Physiol 69:475–496, 1977.
103. Nelson MT, Potlack B, Worley JF, Standen NB: Calcium channels, potassium channels and voltage-dependence of arterial smooth muscle tone. Am J Physiol 259:C3–C18, 1990.
104. Bayer R, Hennckes R, Kaufmann R, Mannhold R: Inotropic and electrophysiologic actions of verapamil and D600 in mammalian myocardium. Naunyn-Schmiedebergs Arch Pharmacol 290:49–68, 1975.
105. Peltzer D, Trautwein W, McDonald TF: Calcium channel block and recovery from block in mammalian ventricle muscle treated with organic channel inhibitors. Pflugers Arch 394:97–105, 1982.
106. McDonald TF, Pelzer D, Trautwein W: Cat ventricle muscle treated with D600: Characteristics of calcium channel block and unblock. J Physiol 352:217–244, 1984.
107. Talajic M, Nayebpour M, Jing W, Nattel S: Frequency-dependent effects of diltiazem on the atrioventricular node during experimental atrial fibrillation. Circulation 80:380–389, 1989.
108. Sanguinetti MC, Kass RS: Voltage-dependent block of calcium channel current in calf cardiac Purkinje fibers by dihydropyridine calcium channel antagonists. Circ Res 55:336–348, 1984.
109. Bean BP: Nitrendipine block of cardiac calcium channels: High affinity binding to the inactivated state. Proc Natl Acad Sci U S A 81:6388–6392, 1984.
110. Chester DW, Herbette LG, Mason RP, et al: Diffusion of dihydropyridine calcium channel antagonists in cardiac sarcolemmal lipid multilayers. Biophys J 52:1021–1030, 1987.
111. Mason RP, Rhodes DG, Herbette LG: Reevaluating equilibrium and kinetic binding parameters for lipophilic drugs based on a structural model for drug interaction with biological membranes. J Med Chem 34:869–877, 1991.
112. Nishimura M, Follmer CM, Singer DH: Amiodarone blocks calcium current in single guinea pig ventricular myocytes. J Pharmacol Exp Ther 251:650–657, 1989.
113. McCarthy RT: Nimodipine block of L-type calcium channels in dorsal root ganglion cells. In Traber J, Gispen W (eds): Nimodipine and Central Nervous System Function: New Vistas. Stuttgart, Schattauer, 1989, pp 35–51.
114. Boland LM, Dingledine RW: Multiple components of both transient and sustained barium currents in a rat dorsal root ganglion cell line. J Physiol 420:223–236, 1990.
115. McCarthy RT, Cohen CJ: The enantiomers of Bay K 8644 have differential effects on Ca^{2+} channel gating in rat anterior pituitary cells. Biophys J 49:432A, 1986.

116. Cohen CJ, McCarthy RT: Nimodipine block of calcium channels in rat anterior pituitary cells. J Physiol 387:195–225, 1987.
117. Cognard C, Romey G, Galizzi J-P, et al: Dihydropyridine-sensitive Ca^{2+} channels in mammalian muscle cells in culture: Electrophysiological properties and interactions with Ca^{2+} channel activator [Bay K 8644] and inhibitor [PN 200 110]. Proc Natl Acad Sci U S A 83:1518–1522, 1986.
118. McCarthy RT, Cohen CJ: Nimodipine block of calcium channels in vascular smooth muscle cell lines. J Gen Physiol 94:669–692 1989.
119. Randle JCR, Lombet A, Nagel N, et al: Ca^{2+} channel inhibition by a new dihydropyridine derivative, S11568, and its enantiomers S12967 and S12968. Eur J Pharmacol 190:85–96, 1990.
120. Yatani A, Seidel Cl, Allen JC, Brown AM: Whole-cell and single channel calcium currents of isolated smooth muscle cells from saphenous vein. Circ Res 60:523–533, 1987.
121. Nelson MT, Worley JF: Dihydropyridine inhibition of single calcium channels and contraction in rabbit mesenteric artery depends on voltage. J Physiol 412:65–91, 1989.
122. Bean BP, Sturek M, Puga A, Hersmeyer K: Calcium channels in muscle cells isolated from rat mesenteric arteries: modulation by dihydropyridine drugs. Circ Res 59:229–235, 1986.
123. Lang RJ, Paul RJ: Effects of 2,3-butanedione monoxime on whole cell Ca^{2+} channel current in single cells of the guinea pig taenia caeci. J Physiol 433:1–24, 1991.
124. Honore E, Amedee T, Martin C, et al: Calcium channel current and its sensitivity to (+)isradipine in cultured pregnant rat myometrial cells. Pflugers Arch 414:477–483, 1989.
125. Kokubun S, Prod'hom B, Becker C, et al: Studies on Ca^{2+} channels in intact cardiac cells: Voltage-dependent effects and cooperative interactions of dihydropyridine enantiomers. Mol Pharmacol 30:571–584, 1987.
126. Wei X-Y, Rutledge A, Triggle DJ: Voltage-dependent binding of 1,4-dihydropyridine Ca^{2+} antagonists and activators in cultured neonatal rat ventricular myocytes. Mol Pharmacol 35:541–552, 1989.
127. Ferrante J, Luchowski EM, Rutledge A, Triggle DJ: Binding of a 1,4-dihydropyridine calcium channel activator, (-)S Bay K 8644, to cardiac preparations. Biochem Biophys Res Commun 158:149–154, 1989.
128. Morel N, Godfraind T: Characterization in rat aorta of the binding sites responsible for blockade of noradrenaline-evoked calcium entry by nisoldipine. Br J Pharmacol 102:467–477, 1991.
129. Bacquet C, Loirand G, Rakotoarisoa L, et al: (+)[^{3}H]PN 200 110 binding to cell membranes and intact strips of portal veins smooth muscle: characterization and modulation by membrane potential and divalent cations. Br J Pharmacol 97:256–262, 1989.
130. Sumimoto K, Hiraka M, Kuriyama H: Characterization of [^{3}H]nifedipine binding to intact vascular smooth muscle cells. Am J Physiol 254:C45–C52, 1988.
131. Morel N, Godfrand T: Selective modulation by membrane potential of the interaction of some calcium entry blockers with calcium channels in mesenteric artery. Br J Pharmacol 95:252–258, 1988.
132. Morel N, Godfraind T: Prolonged depolarization increases the pharmacological effect of dihydropyridines and their binding affinity for calcium channels of vascular smooth muscle. J Pharmacol Exp Ther 243:711–715, 1987.
133. Morel N, Godfraind T: Pharmacological properties of voltage-dependent channels in functional microvessels isolated from rat brain. Naunyn-Schmiedebergs Arch Pharmacol 340:442–451, 1989.
134. Greenberg DA, Carpenter CA, Messing RO: Depolarization- dependent binding of the calcium channel antagonist (+)[^{3}H]PN 200 110, to intact cultured PC 12 cells. J Pharmacol Exp Ther 238:1021–1027, 1986.
135. Triggle DJ, Hawthorn M, Zheng W: Potential-dependent interactions of nitrendipine and related 1,4-dihydropyridines in functional smooth muscle preparations. J Cardiovasc Pharmacol 12(Suppl 4):S91–S93, 1988.
136. Godfraind T, Morel N, Wibo M: Tissue specificity of dihydropyridine-type calcium antagonists in human isolated tissues. Trends Pharmacol Sci 9:37–39, 1988.
137. Wibo M: Mode of action of calcium antagonists: Voltage-dependence and kinetics of drug-receptor interactions. Pharmacol Toxicol 65:1–8, 1989.
138. Godfraind T, Morel N, Wibo M: Modulation of the action of calcium antagonists in arteries. Blood Vessels 27:184–196, 1990.
139. Sun J, Triggle DJ: Calcium channel antagonists: Cardiovascular selectivity of action. J Pharmacol Exp Ther 274:419–426, 1995.
140. Welling A, Kwan YW, Bosse E, et al: Subunit-dependent modulation of recombinant L-type calcium channels. Molecular basis for dihydropyridine tissue selectivity. Circ Res 73:974–980, 1994.

141. Wei, X-Y, Luchowski EM, Rutledge A, et al: Pharmacologic and radioligand analysis of the actions of 1,4-dihydropyridine activator-antagonist pairs in smooth muscle. J Pharmacol Exp Ther 239:144–153, 1986.
142. Kass RS: Voltage-dependent modulation of cardiac calcium channel current by optical isomers of Bay K 8644: Implications for channel gating. Circ Res 61(Suppl I):I1–I5, 1987.
143. Kamp TJ, Sanguinetti MC, Miller RJ: Voltage- and use-dependent modulation of cardiac calcium channels by the dihydropyridine (+) 202 791. Circ Res 64:338–351, 1989.
144. Nakaya H, Hattori Y, Tohse N, Kanno M: Voltage-dependent effects of YC-170, a dihydropyridine calcium channel modulator, in cardiovascular tissues. Naunyn-Schmiedebergs Arch Pharmacol 333:421–430, 1986.
145. Held PH, Yusuf S, Furberg CD: Calcium channel blockers in acute myocardial infarction and unstable angina: An overview. BMJ 299:1187–1192, 1989.
146. Messerli FH: "Cardioprotection": Not all calcium antagonists are created equal. Am J Cardiol 66:855–856, 1989.
147. Opie, LH: Should calcium antagonists be used after myocardial infarction? Ischemia selectivity versus vascular selectivity. Cardiovasc Drugs Ther 6:19–24, 1992.
148. Sleight P: Calcium antagonists during and after myocardial infarction. Drugs 51:216–225, 1996.
149. Pasty BM, Heckbert SR, Koepsell TD, et al: The risk of myocardial infarction associated with anti-hypertensive drug therapies. JAMA 274:620–625, 1995.
150. Furberg CD, Patsy BM, Meyer, JV: Dose-related increase in mortality in patients with coronary artery disease. Circulation 92:1326–1331, 1995.
151. Opie LH: Calcium channel antagonists should be among the first-line drugs in the management of cardiovascular disease. Cardiovasc Drugs Ther 10:455–461, 1996.
152. Opie LH, Messerli FH: Nifedipine and mortality. Grave defects in the dossier. Circulation 92:1068–1073, 1995.
153. Packer M: Pathophysiological mechanisms underlying the adverse effects of calcium channel-blocking drugs in patients with chronic heart failure. Circulation 80 (Suppl IV):59–67, 1989.
154. Reicher-Reiss H, Barasch E: Calcium antagonists in patients with heart failure. A review. Drugs 42:343–364, 1991.
155. Little WC, Cheng C-P, Elvelin L, Nordlander M: Vascular selective calcium entry blockers in the treatment of cardiovascular disorders: Focus on felodipine. Cardiovasc Drugs Ther 9:657–663, 1995.
156. Piepho RW: Calcium antagonist use in congestive heart failure: Still a bridge too far? J Clin Pharmacol 35:443–453, 1995.
157. Hosey MM, Chien AJ, Puri TS: Calcium channels: A current assessment of the properties and roles of channel subunits. Trends Cardiovasc Med 6:265–273, 1996.

AMY DAVIDOFF Ph.D. / MARY F. WALSH Ph.D.
JAMES MARSH M.D. / JAMES R. SOWERS M.D.

2

Abnormalities of Cardiovascular Calcium Metabolism Associated with Diabetes Mellitus and Metabolic Hypertension

CARDIOVASCULAR DISEASE ASSOCIATED WITH DIABETES MELLITUS AND METABOLIC HYPERTENSION: AN OVERVIEW

Ischemic heart disease (IHD) and atherosclerotic vascular disease affect over 12 million persons in the United States and remain the greatest cause of mortality and morbidity necessitating hospitalization.[1–12] The propensity for IHD greatly increases with multiple risk factors, particularly if diabetes is among them.[4,5,8–12] Diabetes is often accompanied by hypertension that, along with lipid abnormalities, cigarette smoking, and physical inactivity and obesity, remains an eminently modifiable risk factor for IHD.[1–7] A cluster of metabolic derangements often exists even in hypertensive persons who do not have clinical diabetes mellitus.[4,5,8–12] Examples include insulin resistance, dyslipidemia, abnormalities of the coagulation-fibrinolytic system predisposing to a procoagulant state, and visceral obesity.[4,5] The dyslipidemia often present in both hypertension and diabetes mellitus consists of low levels of high-density lipoprotein (HDL) cholesterol and elevated levels of triglycerides and the more atherogenic low-density lipoprotein (LDL) cholesterol particles. A multiplicative interaction of hemodynamic factors (i.e., hypertension, endothelial dysfunction, and associated cardiomyopathy) and metabolic factors (i.e., dyslipidemia, increased platelet aggregation, and increased levels of procoagulant factors such as fibrinogen, plasminogen activator inhibitor-1, and endothelin) potentiates cardiovascular and renal diseases.[4,5,12]

Abnormal cardiovascular handling of $[Ca^{2+}]_i$ is often seen in both type I and type II diabetes mellitus and in persons with visceral obesity and associated metabolic derangements.[21] It appears that both hyperglycemia and impairment of insulin action (insulin deficiency or resistance) contribute to the $[Ca^{2+}]_i$ abnormalities.[10–12,22] On the other hand, it has been postulated that a primary defect in cellular calcium (Ca^{2+}) homeostasis contributes to impaired insulin action in cardiovascular as well as other tissues.[10–12,21] Abnormalities in Ca^{2+} metabolism are generalized in the diabetic state, as indicated by the observation of increased $[Ca^{2+}]_i$ in various tissues in experimental diabetes, including the heart, vascular smooth muscle cells (VSMCs), and adipocytes.[5,8,21–25] Increased $[Ca^{2+}]_i$ in VSMCs and attendant responses to vasoactive calcium-mobilizing hormones such as vasopressin[26] and angiotensin II[26] probably contribute to the characteristically enhanced vascular reactivity seen in people with type I and type II diabetes.[8–12,23,24]

Furthermore, abnormalities of $[Ca^{2+}]_i$ also may help to explain development of diabetic cardiomyopathy[27] and diminished insulin action.[25]

ENDOTHELIUM

Hypertension and its metabolic accompaniments, including diabetes mellitus, are characterized by diminished vascular and myocardial relaxation and enhanced vascular constriction.[5,9–13] These hemodynamic abnormalities are currently attributed to functional abnormalities of cardiovascular endothelium.[9–16] Both hyperglycemia and dyslipidemia exert toxic effects on the endothelium.[14,15,17] Indeed, vascular strips from diabetic rats demonstrate impairment of endothelium-dependent (nitric oxide [NO]-mediated) relaxation, which also can be produced by incubating normal vessels in high glucose media.[14] Hyperglycemia increases endothelial cell synthesis of collagen IV and fibronectin[15] and may affect renal as well as cardiovascular endothelial cell function.[12,15] Elevated glucose also delays replication and increases death of endothelial cells, in part by promoting glycooxidation.[13,18] Hypercholesterolemia and perhaps hypertriglyceridemia impair endothelium-dependent cardiovascular relaxation.[17] It appears that the major abnormalities of endothelial function in hypertension are related to associated metabolic abnormalities.[13]

Often the initial response of the endothelial cell to such diverse signals as shear stress and vasodilatory factors (e.g., bradykin and other biochemical and hormonal factors) involves an elevation of endothelial cell cytosolic calcium $[Ca^{2+}]_i$ and Ca^{2+}-dependent enzymes, including constitutive nitric oxide synthase (cNOS).[19,20] In endothelial cells the most important trigger for increases in $[Ca^{2+}]_i$ is inositol 1,4,5-triphosphate, which is generated through the action of phospholipase C, a plasmalemmal enzyme activated in many cases by G-protein-coupled receptors. Endothelial cell Ca^{2+} influx is related to the activity of receptor–G-protein enzyme complex and to the degree of fullness of the endoplasmic reticulum but does not appear to involve voltage-gated calcium channels.[19,20] The degree of Ca^{2+} influx is related to the electrochemical gradient for the ions, which is regulated by membrane stimuli (e.g., acetylcholine) that hyperpolarize the cells, thus increasing the electrochemical gradient for Ca^{2+}. This process is modulated by activation of Ca^{2+}-dependent potassium (K+) and chloride (Cl–) currents. Nevertheless, the lack of potent and specific blockers for these channels has hindered the study of endothelial cell Ca^{2+} handling in diabetes mellitus, hypertension, and associated metabolic conditions. Although acknowledging the importance of putative impairment of endothelial cell Ca^{2+} metabolism, this chapter focuses on Ca^{2+} homeostasis metabolism in VSMCs and cardiomyocytes. Much more is known about these cell types.

NORMAL CALCIUM METABOLISM IN VASCULAR SMOOTH MUSCLE CELLS

Normal calcium handling in VSMCs was extensively reviewed in an earlier volume in this series (Fig. 1).[69] In brief, an increase in $[Ca^{2+}]_i$ and thus in contraction may be brought about by influx across the plasma membrane and/or

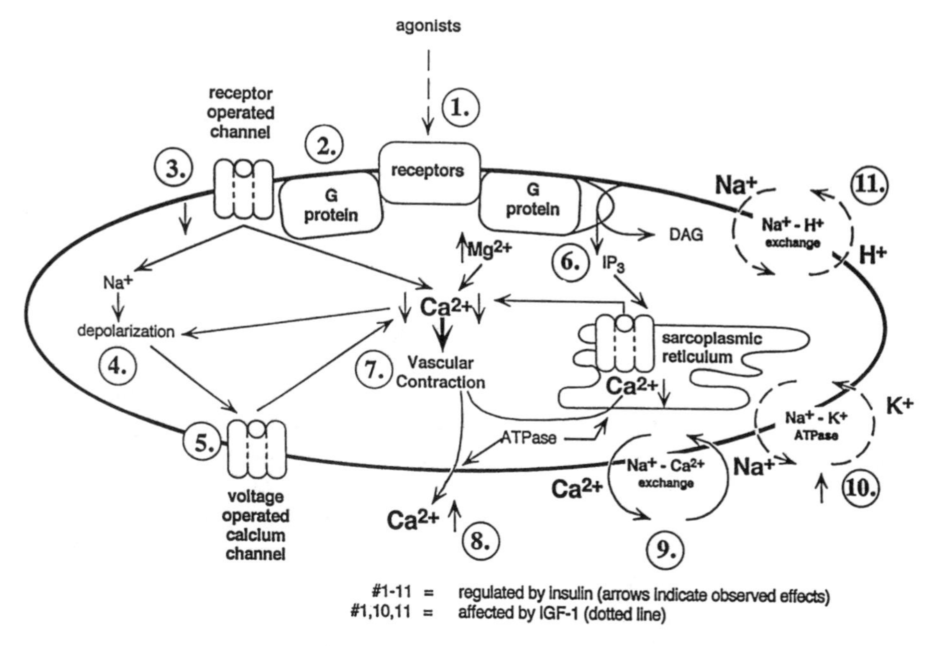

FIGURE 1. Normal calcium handling in vascular smooth muscle cells.

release from intracellular stores. Both voltage-gated and receptor-operated calcium channels have been documented in VSMCs.[80,81] Dihydropyridines have been shown to inhibit both to variable degrees.[80] Release from internal stores, in particular the sarcoplasmic reticulum (SR), follows agonist activation of phospholipase C and subsequent generation of inositol triphosphate (IP_3).[82] Normalization to resting $[Ca^{2+}]_i$ and relaxation accompany sequestration to intracellular storage sites, such as the SR, and efflux from the cell. These events require activation of two distinct adenosine triphosphate (ATP)-dependent Ca^{2+} transporters,[83] which may be variably stimulated by insulin[22] and inhibited by vanadate.[84] In concert, these mechanisms maintain VSMC intracellular free calcium concentration $[Ca^{2+}]_i$ in a range of 75–150 nM.[44,85] Perturbations of any or all of these mechanisms may result in altered $[Ca^{2+}]_i$ homeostasis in the diabetic state.

ROLE OF INSULIN AND INSULIN-LIKE GROWTH FACTOR-1

Both insulin and insulin-like growth factor-1 (IGF-1) normally attenuate vascular contractility and increase skeletal muscle blood flow (SMBF).[10–12,16,21–24,27–38] Unlike insulin, IGF-1 is produced in cardiovascular tissue,[39–41] where it probably exerts autocrine/paracrine effects.[8,12,16,22] Although both insulin and IGF-1 diminish vascular contractility, in part through effects on vascular NO production,[8,12,22,32–34,38,42,43] both hormones also directly lower VSMC contractility[8,29,37] through effects on VSMC $[Ca^{2+}]_i$ metabolism[8,35,42–47] (see Fig. 1). Both hormones reduce Ca^{2+} influx into VSMCs through inhibition of voltage- and receptor-operated calcium channels.[28,29,35,37,44–47] Insulin and IGF-1 also may modulate VSMC $[Ca^{2+}]_i$ and thus vascular resistance through stimulation of the Na^+,K^+-ATPase pump, thus reducing $[Ca^{2+}]_i$ by secondary activation of sodium-calcium exchange (see Fig 1). Insulin and IGF-1 enhance Na^+,K^+-ATPase activity through several mechanisms: (1) by increasing expression of new catalytic pump subunits mRNA,[22,48] (2) by stimulation of sodium-hydrogen exchange and consequent increases in Na^+,[8,37] (3) by modification of enzyme affinity for Na^+, K^+, and/or ATP, or (4) by increasing translocation of preformed Na^+,K^+-ATPase units to the plasma membrane.[8,22,49]

Recent reports that insulin and IGF-1 increase glucose transport in VSMCs[22] and cardiomyocytes[50] indicate that their effects on VSMC intermediary metabolism may play a role in the regulation of VSMC $[Ca^{2+}]_i$. Elevated cytosolic glucose would be anticipated to increase aerobic glycolysis and lactic acid production and thus stimulate sodium-hydrogen exchange activity by lowering pH.[37,51] Furthermore, ATP generated by aerobic glycolysis is an important energy source for the Na^+,K^+-ATPase pump that transports Na^+ and K^+ ions against a concentration gradient.[51] Recently it has also been demonstrated that IGF-1 and insulin activation of tyrosine kinase pathways reduces angiotensin II-elicited VSMC $[Ca^{2+}]_i$ transients, whereas tyrosine kinase inhibition prevents $[Ca^{2+}]_i$ recovery after agonist stimulation.[52] These observations imply an interaction between insulin- and IGF-1-induced tyrosine kinase phosphorylation and receptor-mediated, G-protein–coupled pathways.[8,51] Furthermore, Ca^{2+}-ATPase, which appears to be stimulated by insulin and IGF-1[8,22] (see Fig. 1) is linked to a tyrosine kinase receptor pathway; it is possible that insulin and IGF-1 may regulate the sarcoplasmic reticulum and plasmalemma Ca^{2+}-ATPase and thus $[Ca^{2+}]_i$ through this mechanism.

Insulin and IGF-1 also may modify VSMC divalent cation metabolism by several other mechanisms. Insulin and IGF-1 stimulation of Na^+,K^+-ATPase activity is secondary, in part, to increases in VSMC and cardiomyocyte cytosolic magnesium $[Mg^{2+}]_i$, a process that is defective in insulin resistance and deficiency states[8,22–24] (see Fig. 1). NO produced by endothelial cells and VSMCs also regulates the Na^+ pump.[53] Alternatively, insulin- and IGF-1-mediated increases in VSMC NO and cyclic guanosine monophosphate (cGMP) may directly activate Ca^{2+}-dependent potassium channels in VSMC,[54] thus indirectly decreasing Ca^{2+} influx via voltage-operated channels[12,22,37,54] (see Fig. 1).

VASCULAR REACTIVITY AND VASCULAR SMOOTH MUSCLE CELL $[CA^{2+}]_i$ IN DIABETES

Abnormal cardiovascular divalent cation metabolism may partially explain the relationship between diabetes and hypertension, metabolic hypertension, and associated vascular and renal disease.[8–12,22] In both insulinopenic and insulin-resistant states there are parallel increases in vascular contractile and $[Ca^{2+}]_i$ responses to various vasoconstrictors.[8–12] For example, both agonist-induced VSMC $[Ca^{2+}]_i$ and vascular reactivity are accentuated in insulin-deficient streptozotocin rats[23,24] and insulin-resistant Zucker rats.[30] These heightened VSMC $[Ca^{2+}]_i$ responses to various agonists may be due, in part, to diminished activity of the Na^+,K^+-ATPase pump.[22,37] Reduced activity of the pump has been observed in both insulin-deficient[55] and insulin-resistant[56] states. In tissues in which glucose transport and metabolism are regulated by insulin and IGF-1 (i.e., cardiac, VSM, and skeletal tissue), a reduction in pump activity with increased $[Ca^{2+}]_i$ and vascular contractility[57] has been seen in insulin-resistant rats. Exaggerated VSMC $[Ca^{2+}]_i$ is associated with attenuated insulin-stimulated glucose transport in various insulin-sensitive tissues.[58] Thus, abnormal VSMC $[Ca^{2+}]_i$ regulatory mechanisms may represent a fundamental abnormality associated with both impaired VSMC insulin and IGF-1 action, increased cardiovascular $[Ca^{2+}]_i$, and enhanced vascular resistance.[8,22,58]

Enhanced vascular contractility to vasoconstrictive factors in metabolic hypertension and other states of decreased insulin/IGF-1 vascular action also may be related to abnormalities of $[Mg^{2+}]_i$ metabolism.[8,22,59] Recently, using nuclear magnetic resonance techniques, our group has shown that IGF-1 increases cellular uptake of Mg^{2+}, as previously observed for insulin.[59] A reduction in $[Mg^{2+}]_i$ is often seen in conditions of decreased cellular insulin action (e.g., type I and type II diabetes mellitus).[59,60] Furthermore, depletion of tissue $[Mg^{2+}]_i$ contributes to decreased insulin-stimulated glucose uptake.[1,60] Although the mechanism by which $[Mg^{2+}]_i$ depletion leads to insulin resistance is unclear, decreases in $[Mg^{2+}]_i$ may lead to increases in $[Ca^{2+}]_i$, a known contributor to insulin resistance.[1,60] Thus, elevations in $[Ca^{2+}]_i$ and depletions in $[Mg^{2+}]_i$ probably contribute to resistance to the vasodilatory action of insulin and IGF-1, promoting enhanced peripheral vascular resistance[8,22,61–68] (see Fig. 1).

ROLE OF HYPERGLYCEMIA

Hyperglycemia is a major complication of all forms of diabetes mellitus. The cardiovascular effects of high glucose have recently become the focus of intense interest. Most of the research has defined the effects of hyperglycemia on the vasculature, and little is known about its impact on the myocardium. In vitro, high glucose is known to elevate VSMC $[Ca^{2+}]_i$, which in turn may alter protein function and gene expression. The mechanisms by which glucose alters $[Ca^{2+}]_i$ remain unknown, although activation of protein kinase C has been documented. High glucose also may affect the glycosylation state of various proteins in VSMCs. The effects of glycosylation on ion channels and other proteins have been studied in a number of other cell systems and shown to alter protein content and function (e.g., K^+ channel-gating properties and Na^+ channel receptor sites are altered with glycosylation). In red blood cells, high glucose inhibits sarcolemmal Ca^{2+}-ATPase through a glycosylation-dependent mechanism. Whether this is also the case in VSMCs remains to be determined.

References

1. Lipid Research Clinics Program: The Lipid Research Clinics Primary Prevention Trials: Reduction in incidence of coronary heart disease. JAMA 251:351–364, 1984.
2. Frick MH, Elo O, Haapa K, et al: Helsinki Heart Study: Primary prevention trial with gemfibrozil in middle-aged men with dyslipidemia. N Engl J Med 317:1237–1245, 1987.
3. Abbott RD, Wilson PWF, Kannel WB, et al: High density lipoprotein cholesterol, total cholesterol, and myocardial infarction: The Framingham Study. Arteriosclerosis 8:207–211, 1988.
4. Flack JM, Sowers JR: Epidemiologic and clinical aspects of insulin resistance and hyperinsulinemia. Am J Med 91:1S–11S, 1991.

5. Walsh MF, Dominguez LJ, Sowers JR: Metabolic abnormalities in cardiac ischemia. Cardiol Clin 13:529–538, 1995.
6. Pooling Project Research Group: Relationship of blood pressure, serum cholesterol, relative weight and ECG abnormalities to incidence of major coronary events: Final report of the Pooling Project. J Chronic Dis 31:201–306, 1978.
7. Stamler J, Neaton JD, Wentworth DN: Blood pressure (systolic and diastolic) and risk of fatal coronary heart disease. Hypertension 13(Suppl 1):2–12, 1993.
8. Sowers JR: Insulin and insulin-like growth factor in normal and pathological cardiovascular physiology. Hypertension 29:691–699, 1997.
9. The National High Blood Pressure Education Program Working Group: National High Blood Pressure Education Program Working Group report on hypertension in diabetes. Hypertension 23:145–158, 1994.
10. Sowers JR, Epstein M: Diabetes mellitus and associated hypertension, vascular disease, and nephropathy: An update. Hypertension 26(Pt 1):869–879, 1995.
11. Sowers JR, Epstein M: Diabetes mellitus and hypertension, emerging therapeutic perspectives. Cardiovasc Drug Rev 113:149–210, 1995.
12. Bakris GL, Standley PR, Palant CE, et al: Analogy between endothelial/mesangial cell and endothelial/vascular smooth muscle cell interactions. In Sowers JR (ed): Contemporary Endocrinology: Endocrinology of the Vasculature. Totowa, NJ, Humana Press, 1996, pp 341–355.
13. Hsueh WA, Anderson PW: Hypertension, the endothelial cell, and the vascular complications of diabetes mellitus. Hypertension 20:235–253, 1992.
14. Tesfamariam B, Brown ML, Cohen RA: Elevated glucose impairs endothelium-dependent relaxation by activating protein kinase C. J Clin Invest 87:1643–1648, 1991.
15. Cagliero E, Roth T, Roy S, Lorenzi M: Characteristics and mechanisms of high-glucose-induced overexpression of basement membrane components in cultured human endothelial cells. Diabetes 40:102–110, 1991.
16. Walsh MF, Barazi M, Pete G, et al: Insulin-like growth factor 1 diminishes in vivo and in vitro vascular contractility: role of vascular nitric oxide. Endocrinology 1137:2798–1803, 1996.
17. Creager MA, Cooke JP, Mendelsohn M, et al: Impaired vasodilation of forearm resistance vessels in hypercholesterolemic humans. J Clin Invest 86:228–234, 1990.
18. Vlassara H: Recent progress on the biological and clinical significance of advanced glycosylation end products. J Lab Clin Med 124:19–30, 1994.
19. Himmel HM, Whorton RA, Strauss HC: Intracellular calcium, currents, and stimulus-response coupling in endothelial cells. Hypertension 21:112–127, 1993.
20. Song J, Ram JL: Ionic mechanisms of peptide-induced responses in vascular endothelial cells. In Sowers JR (ed): Endocrinology of the Vasculature. Totowa, NJ, Humana Press, 1996, pp 21–37.
21. Levy J, Gavin JR, Sowers JR: Diabetes mellitus: A disease of abnormal cellular calcium metabolism. Am J Med 96:260–273, 1994.
22. Sowers JR: Effects of insulin and IGF-1 on vascular smooth muscle glucose and cation metabolism. Diabetes 45(3):S47–S51, 1996.
23. Sowers JR, Standley PR, Ram JL, Jacober S: Hyperinsulinemia, insulin resistance, and hyperglycemia: Contributing factors in the pathogenes of hypertension and atherosclerosis. Am J Hypertens 6:260S–270S, 1993.
24. Sowers JR, Sowers PS, Peuler JD: Role of insulin resistance and hyperinsulinemia in development of hypertension and atherosclerosis. J Lab Clin Metab 123:647–652, 1994.
25. Draznin B: Cytosolic calcium and insulin resistance. Am J Kidney Dis 21:32–38, 1993.
26. Johnson EM, Theler JM, Capponi AM, Wafloton MB: Characterization of oscillations in cytosolic free Ca^{2+} concentration and measurement of cytosolic Na^{+} concentration changes evoked by angiotensin II and vasopressin in individual rat aortic smooth muscle cells. Use of microfluorimetry and digital imaging. J Biol Chem 266:12618–12626, 1991.
27. Anderson EA, Hoffman RP, Balon TW, et al: Hyperinsulinemia produces both sympathetic neural activation and vasodilation in normal humans. J Clin Invest 87:2246–2252, 1991.
28. Kahn AM, Seidel CL, Allen JC, et al: Insulin reduces contraction and intracellular calcium concentration in vascular smooth muscle. Hypertension 22:735–742, 1993.
29. Ram JL, Fares MA, Standley PR, et al: Insulin inhibits vasopressin-elicited contraction of vascular smooth muscle cells. J Vasc Med Biol 4(5–6): 250–254, 1993.
30. Standley PR, Ram JL, Sowers JR: Insulin attenuation of vasopressin-induced calcium responses in arterial smooth muscle from Zucker rats. Endocrinology 133:1693–1699, 1993.
31. Lembo G, Iaccarino G, Rendina V, et al: Insulin blunts sympathetic vasoconstriction through the α_2-adrenergic pathway in humans. Hypertension 24:429–438, 1994.

32. Steinberg HO, Brechtel G, Johnson A, et al: Insulin-mediated skeletal muscle vasodilation is nitric oxide dependent. A novel action of insulin to increase nitric oxide release. J Clin Invest 94:1172–1179, 1994.
33. Scherrer U, Randin D, Vollenweider P, et al: Nitric oxide release accounts form insulin's vascular effects in humans. J Clin Invest 94:2511–2515, 1994.
34. Wu H, Jeng YY, Yue C, et al: Endothelial-dependent vascular effects of insulin and insulin-like growth factor 1 in the perfused rat mesenteric artery and aortic ring. Diabetes 43:1027–1032, 1994.
35. Han SZ, Ouchi Y, Karaki H, Orimo H: Inhibitory effects of insulin on cytosolic Ca^{2+} level and contraction in the rat aorta. Circ Res 77:673–678, 1995.
36. Copeland KC, Streekuran K: Recombinant human insulin-like growth factor-1 increases forearm blood flow. J Clin Endocrinol Metab 79:230–232, 1994.
37. Kahn AM, Song T: Effects of insulin on vascular smooth muscle contraction. In Sowers JR (ed): Endocrinology of the Vasculature. Totowa, NJ, Humana Press, 1996, pp 215–223.
38. Baron AD, Steinberg HO: Vascular actions of insulin in health and disease. In Sowers JR (ed): Endocrinology of the Vasculature. Totowa, NJ, Humana Press, 1996, pp 95–107.
39. Bornfeldt KE, Arnqvist HJ, Norstedt G: Regulation of insulin-like growth factor-1 gene expression by growth factors in cultured vascular smooth muscle cells. J Endocrinol 125:381–386, 1990.
40. Khorsandi MJ, Fagin JA, Giannella-Neto D, et al: Regulation of insulin-like growth factor-1 and its receptor in rat aorta after balloon denudation. Evidence for local bioactivity. J Clin Invest 90:1926–1931, 1992.
41. Delafontaine P, Lou H: Angiotensin II regulates insulin-like growth factor 1 gene expression in vascular smooth muscle cells. J Biol Chem 268:16866–16870, 1993.
42. Tsukahara H, Gordienko DV, Toushoff B, et al: Direct demonstration of insulin-like growth factor-1-induced nitric oxide production by endothelial cells. Kidney Int 45:598–604, 1994.
43. Zeng G, Quon MJ: Insulin-stimulated production of nitric oxide is inhibited by wortmannin direct measurement in vascular endothelial cells that are also involved with glucose metabolism. J Clin Invest 98:894–898, 1996.
44. Standley PR, Zhang F, Ram JL, Zemel MB, Sowers JR. Insulin attenuates vasopressin-induced calcium transient and voltage-dependent calcium response in rat vascular smooth muscle cells. J Clin Invest 88:1230-1236, 1991.
45. Saito F, Hori MT, Fittingott M, et al: Insulin attenuates agonist-mediated calcium mobilization in cultured rat vascular smooth muscle cells. J Clin Invest 92:1161–1168, 1993.
46. Kahn AM, Allen JC, Seidel CL, Song T: Insulin inhibits serotonin-induced Ca^{2+} influx in vascular smooth muscle. Circulation 90:384–390, 1994.
47. Touyz RM, Tolloczko B, Schiffrin EL: Insulin attenuates agonist-evoked calcium transients in vascular smooth muscle cells. Hypertension 23(I):I23–I28, 1994.
48. Tirupattur PR, Ram JL, Standley PR, Sowers JR: Regulation of Na^+,K^+-ATPase gene expression by insulin in vascular smooth muscle cells. Am J Hypertens 6:626–629, 1993.
49. Hundal HS, Marette A, Mitsumoto Y, et al: Insulin induces translocation of the α_2 and β_1 subunits of the Na^+/K^+-ATPase from intracellular compartments to the plasma membranes in mammalian skeletal muscle. J Biol Chem 267:5040–5043, 1992.
50. Eckel J, Gerlach-Eskuchen E, Reinauer H: G-protein-mediated regulation of the insulin-responsive glucose transporter in isolated cardiac myocytes. Biochem J 272:691–696, 1990.
51. Dominguez LJ, Peuler JD, Sowers JR: Endocrine regulation of vascular smooth muscle intermediary metabolism. In Sowers JR (ed): Endocrinology of the Vasculature. Totowa, NJ, Humana Press, 1996, pp 325–339.
52. Touyz RM, Schiffrin EL: Tyrosine kinase signaling pathways modulate angiotensin II-induced calcium $[Ca^{2+}]_i$ transients in vascular smooth muscle cells. Hypertension 27:1097–1103, 1996.
53. Gupta S, McArthur C, Grady C, Ruderman NB: Stimulation of vascular Na^+-K^+-ATPase activity by nitric oxide: A cGMP-independent effect. Am J Physiol 266:H2146–H2151, 1994.
54. Bolotina VM, Najibl S, Palacino JJ, et al: Nitric oxide directly activates calcium dependent potassium channels in vascular smooth muscle. Nature 368:850–853, 1994.
55. O'Hare T, Sussman KE, Draznin B: Effect of diabetes on cytosolic free Ca^{2+} and Na^+,K^+-ATPase in rat aorta. Diabetes 40:1560–1563, 1991.
56. Sowers JR, Whitfield L, Beck FWJ, et al: Role of enchanced sympathetic nervous sysem activity and reduced Na^+,K^+-dependent adenosine triphosphatase activity in the maintenance of elevated blood pressure in obesity: Effects of weight loss. Clin Sci 63:121S–124S, 1982.
57. Sada T, Koike H, Ikeda M, Sato K, Ozaki H, Karaki H. Cytosolic free calcium of aorta in hypertensive rats. Hypertension 16:245-251, 1990.
58. Begum N, Leitner W, Reusch JE, et al: GLUT-4 phosphorylation and its intrinsic activity: Mechanism of Ca^{2+}-induced inhibition of insulin-stimulated glucose transport. J Biol Chem 268:3352–3356, 1993.

59. Hwang DL, Yen CF, Nadler JL: Insulin increases intracellular magnesium transport in human platelets. J Clin Endocrinol Metab 76:549–553, 1993.
60. Nadler JL, Buchanan T, Natarajan R, et al: Magnesium deficiency produces insulin resistance and increased thromboxane synthesis. Hypertension 21:1024–1029, 1993.
61. Zhang F, Sowers JR, Ram JL, et al: Effects of pioglitazone on calcium channels in vascular smooth muscle. Hypertension 24:170–175, 1994.
62. Nadler JL, Scott S: Evidence that pioglitazone increases intracellulr free magnesium in freshly isolated rat adipocytes. Biochem Biophys Res Commun 202:416–421, 1995.
63. Oclov SN, Tumblay J, Hamet P: cAMP signaling inhibits dihydropyridine-sensitive Ca^{2+} influx in vascular smooth muscle cells. Hypertension 27:774–780, 1996.
64. Rivera AA, White CR, Elton TS, Marchase RB: Hyperglycemia alters cytoplasmic Ca^{2+} responses to capacitative Ca^{2+} influx in rat aortic smooth muscle cells. Am J Physiol 269: C1482–C1488, 1995.
65. Kishi Y, Watanabe T, Makita T, et al: Effect of nifedipine on cGMP turnover in cultured coronary smooth muscle cells. J Cardiovasc Pharmacol 26:590–595, 1995.
66. Chulia T, Gonzalez P, Del Rio M, Tijerina T: Comparative study of elgodipine and nisoldipine on the contractile responses of various isolated blood vessels. Eur J Pharmacol 285:115–122, 1995.
67. Magnon M, Gallix P, Cavero I: Intervessel (arteries and veins) and heart/vessel selectivities of therapeutically used calcium entry blockers: Variable, vessel-dependent indexes. J Pharmacol Exp Ther 275:1157, 1995.
68. Wilson TW, Quest DW: Comparative pharmacology of calcium antagonists. Can J Cardiol 11:243–249, 1995.
69. Ram JL, Standley PR, Sowers JR: Calcium function in vascular smooth muscle and its relationship to hypertension. In Epstein M (ed): Calcium Antagonists in Clinical Medicine. Philadelphia, Hanley & Belfus, 1992, pp 29–48.

HENRY L. ELLIOTT M.D., F.R.C.P. / PETER A. MEREDITH Ph.D.
JOHN L. REID D.M., F.R.C.P.

3

Pharmacokinetics of Calcium Antagonists: Implications for Therapy

Since the introduction of verapamil for the treatment of angina pectoris in the 1970s, many other new drugs with calcium antagonist properties have emerged, along with a number of modified-release formulations. Perhaps as a consequence of the plethora of available agents—different drugs and different formulations—there is a nondiscriminatory perception that all calcium antagonists are approximately equivalent in therapeutic terms. Thus, although the chemical and pharmacologic differences are well recognized, there remains a tendency to view calcium antagonists as a single therapeutic class. This is not the case, for clinical pharmacologic differences translate directly to therapeutic differences, with advantages and disadvantages for different drugs and different formulations.

BASIC PHARMACOLOGIC DIFFERENCES

The most appropriate terminology for subclassifying the heterogeneous agents capable of interfering with the entry of calcium ions into cells continues to be debated.[1] The term *calcium antagonist* is disliked by authors who argue that the agents do not antagonize the cellular effects of calcium ions but instead have the common characteristic of interfering with the availability of calcium ions necessary for the maintenance or induction of specific cellular activities. Hence, alternative terms have been proposed, including calcium slow channel blockers, calcium channel blockers, calcium channel modulators, and calcium modulators. In this chapter, the term *calcium antagonist* is used to encompass compounds that interfere with transmembrane calcium influx.

Although often considered by nonspecialist clinicians to constitute a broadly similar and notionally homogeneous group, calcium antagonists clearly are not equipotent in their vascular, myocardial, and electrophysiologic effects. In contrast to their pharmacodynamic differences, however, their pharmacokinetic characteristics show many common features, although there are important exceptions. The principal purpose of this chapter is to highlight the pharmacokinetic similarities and differences (with emphasis on the recently developed agents) and to discuss how the pharmacokinetic profiles influence the overall pharmacologic response and clinical usefulness of different calcium antagonists. The reader is directed to several more comprehensive texts that review the pharmacokinetic characteristics in more detail.[2–4]

DIFFERENT CLASSIFICATIONS: CONFUSION OR CLARIFICATION?

The importance of a rational classification system as a guide to appropriate prescribing is clear; unfortunately, most classification systems have been created on an ad hoc basis to cope with new developments. To date, at least eight different schemes have been proposed to classify calcium antagonists. These schemes tend to group agents according to (1) chemical structure, (2) specificity of slow current inhibition, (3) tissue selectivity, or (4) receptor specificity. An alternative specification, proposed by the World Health Organization (WHO), essentially splits the calcium antagonists into "selective" agents, which preferentially act on L-type slow calcium channels, and "nonselective" agents, which act on L-, T-, N-, and P-type channels.

Pharmacologic Classification

The most widely used classification is based on the structural differences among the three prototype agents: diltiazem, nifedipine, and verapamil. Nifedipine is the prototype for the most numerous group—the dihydropyridine derivatives. Verapamil is a phenylalkylamine derivative, and diltiazem is a benzothiazepine derivative. Mibefradil, a recently described agent, has been classified separately as a benzimidazolyl-substituted tetraline. Other drugs with calcium antagonist properties have different structures and a range of additional pharmacologic properties, but they are not widely used in current clinical practice (Table 1).

Physicochemical properties of calcium antagonists have been used by other authors as criteria for classification (Table 2). For example, Spedding[5] classified calcium antagonists according to lipophilicity and difference in surface charge with an experimental salicylate preparation. Alternatively, Fleckenstein[6] identified three groups, A, B, and C. Groups A and B are characterized as potent, specific inhibitors of calcium-dependent excitation-contraction coupling. Group A agents are more specific (e.g., nifedipine, verapamil, and diltiazem), whereas

TABLE 1. Classification of Calcium Antagonists by Structure

Phenylalkylamines	Dihydropyridines	Benzothiazepines	Benzimidazolyl-substituted Tetraline	Others
Verapamil	Nifedipine	Diltiazem	Mibefradil	Flunarizine
Gallopamil	Nitrendipine			Cinnarizine
Tiapamil	Nisoldipine			Lidoflazine
Anipamil	Nimodipine			Perhexiline
	Felodipine			Bepridil
	Nicardipine			
	Amlodipine			
	Isradipine			
	Nivaldipine			
	Niludipine			
	Ryosidine			
	Lacidipine			

TABLE 2. Physicochemical Classification of Calcium Antagonists

	Verapamil	Diltiazem	Dihydropyridines	Cinnarizine/ Flunarizine
Glossman and Ferry[7] (1983)	II	III	IA	IB
Spedding[5] (1984)	B	B	A	C
Fleckenstein[6] (1983)	A	A	A	B

group B agents are correspondingly potent but less specific (e.g. bepridil). Group C contains organic and inorganic compounds that have nonspecific, less potent effects on excitation-contraction coupling.

The usefulness of such classification schemes is limited in several respects, but most importantly because grouping according to physicochemical properties takes no account of tissue selectivity. Furthermore, although classification according to receptor affinity highlights potential pharmacodynamic differences among different agents,[7] it is not yet possible to define clearly the molecular pharmacology of calcium antagonist binding sites, which remain incompletely understood. To address the problem of tissue selectivity and overall pharmacodynamic profile, several authors have devised classifications based on the principal pharmacodynamic effects.

Clinical Pharmacologic Classification

Singh[8] classified calcium antagonists on the basis of their cardiac and peripheral activity; arguably, this is the most useful clinical pharmacologic classification (Table 3).

A further attempt at a practical classification was recently suggested by Toyooka and Naylor,[9] who propose three groups or "generations" (Table 4). This scheme focuses on the clinical effects of the drugs, as determined by receptor binding properties, tissue selectivity, and pharmacokinetic profiles, but also takes account of clinical criteria. According to the selected criteria, the original (instant-release) formulations of verapamil, diltiazem, and nifedipine are first-generation drugs. The second-generation calcium antagonists include compounds developed with an improved pharmacokinetic profile and/or increased

TABLE 3. Classification of Calcium Antagonists According to Cardiac and Peripheral Effects

Type 1	Typified by verapamil and diltiazem and leading to prolonged atrioventricular (AV) nodal conduction and refractoriness, with little effect on ventricular or atrial refractory periods.
Type 2	Nifedipine and other dihydropyridines that have no overt cardiac electrophysiologic effects in vivo but instead are potent peripheral vasodilators. Their net hemodynamic effects are peripheral vasodilatation and reflex activation.
Type 3	Piperazine derivatives (e.g., flunarizine, cinnarizine) that in vitro and in vivo are potent peripheral vasodilators with no effects on the heart.
Type 4	Agents with a broader pharmacologic profile, including blockade of calcium fluxes in the heart, peripheral blood vessels, or both.

TABLE 4. Clinical Considerations Selected by Toyo-oka and Naylor*

1. Duration of action and dosing frequency, as determined by receptor binding rate, plasma half-life, and volume of distribution.
2. Frequency and severity of unwanted side effects, particularly those related to rapid vasodilatation and dependent on time to peak plasma level, peak-to-trough variability, and receptor binding characteristics.
3. Negative chronotropic and inotropic cardiac effects, particularly slowing of atrioventricular conduction as determined by tissue selectivity, which may predispose to heart failure, bradycardia, and heart block.
4. Predictability of the pharmacokinetic profile (and of the response)—effects of distribution and peak-to-trough variability.

* From Toyo-oka T, Naylor WG: Third generation calcium entry blockers. Blood Pressure 5:206–208, 1986.

vascular selectivity. They are subdivided into slow-release formulations (IIa) and agents with a novel chemical structure (IIb). The third generation is exemplified by amlodipine, which can be differentiated from both first- and second-generation agents by its more predictable efficacy in relation to its small peak-to-trough variations in plasma levels at steady state and by its intrinsically long duration of action, which is characterized by both slow onset and slow offset.

In summary, to accommodate the ever increasing number of agents with calcium antagonist properties, classification schemes that attempt to subgroup the different agents have proliferated. The most recent attempt to create a rational classification system for calcium antagonists has yet to gain widespread acceptance, and it may well be that a definitive classification system may not emerge until the molecular pharmacology of specific binding sites has been more clearly defined. Because none of these classifications is entirely suitable or generally applicable, compromise is required. The most logical classification appears to depend firstly on dividing the calcium antagonists into groups according to chemical structure and then, for clinical purposes, discriminating among the different drugs and formulations within each group on the basis of overall pharmacodynamic profile and duration of action.

PHARMACOKINETIC CHARACTERISTICS OF FIRST-GENERATION DRUGS

The pharmacokinetic profiles of the three prototype calcium antagonists have been studied in great detail. Knowledge of pharmacokinetic parameters is important for the development of rational dosage regimens and for the appropriate dose adjustments in the presence of concomitant diseases or drugs that may significantly alter drug disposition. Increasingly, however, the development of modified-release formulations with more desirable pharmacokinetic profiles requires new, detailed pharmacokinetic information.

In contrast to their structural and pharmacodynamic differences, the galenic or instant-release formulations of the three prototype agents have remarkably similar pharmacokinetic characteristics[10] (Table 5).

TABLE 5. Pharmacokinetic Characteristics of Prototype and Recent Drugs (Instant-release Formulations)

	Prototype Drugs			Recent Drugs	
	Nifedipine	Verapamil	Diltiazem	Amlodipine	Mibefradil
Oral absorption (%)	> 90	> 90	> 90	> 90	> 90
Bioavailability (%)	30–60	10–30	30–60	52–88	70–90
Protein binding (%)	> 90	> 90	> 90	> 90	> 90
Elimination half-life (hr)	3.4 ± 10.4	3–7*	3–6*	35–50	17–25
Hepatic metabolism	+++	+++	+++	+++	+++
Active metabolites	? Yes	Yes	Yes	No	No

* Single dose.

Verapamil

After oral administration, the absorption of verapamil exceeds 90%, but because of extensive first-pass metabolism, verapamil has a relatively low bioavailability (about 10–30%).[11] The consequences of low bioavailability are plasma concentrations that exhibit wide inter- and intrapatient variability across steady-state dosing intervals. Verapamil is extensively metabolized in the liver, primarily by N-dealkylation and O-demethylation,[12] but it has at least 12 metabolites, of which norverapamil retains about 10–20% of the pharmacologic activity of the parent compound in animal studies. Furthermore, verapamil is characterized by an optically active center, and the two isomers (D- and L-verapamil) have been shown to have different pharmacologic activity as well as different disposition and clearance characteristics.[13]

Verapamil is widely distributed with a high volume of distribution (3–6 L/kg), and it is about 90% protein-bound. The pharmacokinetics of verapamil after chronic administration shows a significant increase in bioavailability relative to single-dose administration; steady-state concentrations are about 2–3 times higher than the concentrations predicted from the single-dose date. The mechanism of these changes reflects saturation of hepatic enzymes and changes in hepatic blood flow, which lead to a reduced clearance rate and increased half-life.[14–17] An elimination half-life of about 4–5 hours is identified after single doses, but during long-term administration, with reduction in clearance, the elimination half-life is prolonged to about 8–12 hours. In practice, therefore, a longer dose interval is appropriate during chronic administration.

Diltiazem

Diltiazem also is well absorbed after oral administration and undergoes extensive first-pass hepatic metabolism; the major metabolic pathways involve O-deacetylation and N-demethylation, followed by O-demethylation.[18] Less than 4% of the parent drug is excreted unchanged in the urine, and pharmacologically active metabolites are not thought to be clinically important in humans. For example, the major metabolite, deacetyl-diltiazem, contributes only about 15–35% of the measured diltiazem concentration and possesses only 40–50% of the pharmacologic activity of the parent drug. Diltiazem also has a large volume of distribution (about 3–8 L/kg), and the elimination half-life is about 4 hours after single dosing. Diltiazem, however, is similar to verapamil in that its clearance declines during repeated oral dosing, presumably because of saturation of hepatic biotransformation pathways, and the elimination half-life increases to about 8 hours.

Dihydropyridine Derivatives (Nifedipine)

The dihydropyridine derivatives, exemplified by nifedipine, constitute the largest group (see Table 1), and most are almost totally absorbed from the gastrointestinal tract but then undergo extensive first-pass hepatic metabolism.[2,3] The elimination half-lives are short, typically 2–6 hours. For this reason there have been many attempts to develop modified-release preparations to extend the effective half-life, and only a few of the early dihydropyridines are still available in their original, instant-release forms. It is beyond the scope of this chapter to review the characteristics of the early dihydropyridines (described in detail in earlier editions).

NEW DRUGS AND NEW FORMULATIONS

In therapeutic terms it is relatively straightforward to recognize the basic distinction between the so-called "rate-limiting" agents (verapamil and diltiazem) and the more numerous dihydropyridine derivatives (e.g., nifedipine, nisoldipine, amlodipine, lacidipine). However, the important differences within the dihydropyridine group are seldom appreciated, although clear clinical pharmacologic and therapeutic differences distinguish the intrinsically long-acting drugs (e.g., amlodipine) and formulations (e.g., nifedipine GITS) from the prototype nifedipine in its instant-release (capsule) formulation.

New Drugs

Amlodipine

Although it is a dihydropyridine derivative, amlodipine has a distinctly different pharmacokinetic profile characterized by a relatively high bioavailability of 52–88% and a slow absorption phase after oral administration. Maximal plasma concentrations are not achieved until about 6 hours after dosing.[19] Thereafter, although amlodipine is extensively metabolized in the liver, it does not undergo any significant presystemic metabolism and instead has a clearance rate of 4–8 ml/min/kg. It has a high volume of distribution (about 20 L). As a consequence of these pharmacokinetic features, amlodipine has a long terminal elimination half-life of 35–50 hours.[20]

Lacidipine

Lacidipine is subject to extensive presystemic hepatic metabolism and thus has a relatively low bioavailability.[21] The quoted values for the half-life of lacidipine vary from 3 to 15 hours. However, the lack of pharmacokinetic accumulation with once-daily maintenance therapy is compatible with a reported mean half-life of 7 hours.[22] The high volume of distribution[23] has been attributed to the drug's high membrane partition coefficient and a unique membrane-binding characteristic.[24]

Mibefradil

Mibefradil is a novel calcium antagonist from the new chemical structural class of the benzimidazolyl-substituted tetraline derivatives. At the molecular level mibefradil differs in two ways from existing calcium antagonists. Firstly, other calcium antagonists block only L-type channels, whereas mibefradil blocks both L- and T-type channels, with preferential selectivity for T-type channels. Secondly, it binds to its own unique receptor site.[25]

Pharmacokinetic studies have demonstrated that mibefradil is rapidly and completely absorbed after oral administration, with absolute bioavailability in the range of 70–90%. Food intake has no influence on the rate or extent of absorption of mibefradil, and peak plasma concentrations are achieved 1–2 hours after dosing. The area under the plasma concentration time curve increases disproportionately with increasing single doses, probably because of saturation of first-pass metabolism.[26] After chronic dosing the area under the curve is proportional to the administered dose, and the elimination half-life is 17–25 hours, suggesting suitability for once-daily administration. Steady-state plasma levels are achieved after 3–4 days of treatment, and peak plasma concentrations at steady state are about two-fold higher than after single-dose administration. These features are entirely compatible with the reported elimination half-life.[27] Mibefradil and its major metabolites are predominantly excreted in the feces; less than 3% is excreted as unchanged mibefradil.

Modified-release Formulations

With the dihydropyridine derivatives in particular (but also with verapamil and diltiazem), many attempts have been made to prolong the drug concentration-time profile with the development of a wide range of formulations. Perhaps the most sophisticated formulation is the gastrointestinal therapeutic system (GITS) developed for nifedipine. The bilayer nifedipine GITS tablet has been designed to release drug into the gastrointestinal lumen at a steady (zero order) rate for 16–18 hours to permit once-daily dosing with virtually uniform plasma concentrations across 24 hours.[28] This formulation not only extends the duration of action but also avoids the need for multiple daily dosing and wide fluctuations in the peak and trough drug concentrations.

CONCENTRATION–EFFECT RELATIONSHIPS

The development of long-acting calcium antagonists and long-acting formulations is a consequence of the direct relationship between the concentration–time profile and the response–time profile. Although some early studies with calcium antagonists failed to identify a clear relationship between plasma levels and measured effects, more recent reports have repeatedly confirmed a direct relationship between the two.[29] This relationship is particularly apparent when individual patients are studied and when the measured response takes account of placebo or background (circadian) changes. Most studies have assessed blood pressure changes, and, in general, a direct linear relationship has been identified between plasma drug concentration and measured blood pressure response. In simplistic terms, the plasma concentration–time profile is an appropriate predictor of the antihypertensive response–time profile and, correspondingly, an indicator of the duration and consistency of action of different calcium antagonists.

First-generation Drugs

Studies using a pharmacokinetic-pharmacodynamic (PK-PD) modeling approach have shown that the response to nifedipine in an individual patient can be expressed in terms of decrease in blood pressure per unit change in plasma drug concentration (mmHg per ng/ml). Furthermore, it has been demonstrated that the magnitude of the first-dose response correlates well with the response during steady-state treatment in terms of mmHg per ng/ml. This correlation creates the possibility of predicting the response to long-term treatment in each individual patient.[29,30]

A similar PK-PD modeling approach has been used to investigate the relationship between the plasma concentrations and blood pressure-lowering effect of verapamil. A direct linear relationship again can be identified. Despite the significant change in drug clearance during steady-state treatment, the magnitude of the blood pressure response was consistent during both single and multiple dosing in terms of decrease in blood pressure per unit drug concentration.[31,32]

In both studies, factors influencing the magnitude of the blood pressure response to treatment with a calcium antagonist were also investigated. The pretreatment level of blood pressure and the achieved drug concentration were the major determinants, whereas age, sex, and plasma renin activity did not significantly predict the responsiveness to either verapamil or nifedipine.[30–32]

Amlodipine and Felodipine

Of the newer dihydropyridines, the concentration–effect relationship for amlodipine has been explored in detail.[33,34] A recent comparative study investigated the relationship between amlodipine disposition and antihypertensive effect in young and elderly patients. Initially, when plasma drug concentrations after intravenous dosing were correlated with changes in mean blood pressure, it appeared that elderly patients had a greater antihypertensive response for any given drug concentration.[32] However, long-term oral treatment resulted in comparable decreases in blood pressure for given amlodipine concentrations in both young and elderly hypertensive patients. A possible explanation for the initial intravenous finding may lie with the less vigorous acute counterregulatory responses in elderly patients, such as lesser degrees of sympathetic activation and increased cardiac output in response to the acute vasodilator effect. Thus, there may have been a more pronounced initial reduction in blood pressure.

Identification of the PK-PD relationship for the first-dose response to oral amlodipine is complicated by the gradual absorption rate; effective concentrations are attained only gradually and with repeated dosing. Nevertheless, an integrated PK-PD modeling technique has made it possible to identify clearly a linear concentration–effect relationship after first-dose administration and 4 weeks of administration.[34] This finding further confirms that the antihypertensive response to amlodipine reflects the plasma concentration–time profile with a gradual onset and gradual increase in effect during the initial dosing period when the drug accumulates to establish its effective concentration range (typically within the first 1–2 weeks of continuous dosing).

Several studies in healthy subjects and hypertensive patients have confirmed similar relationships between the fall in diastolic blood pressure and plasma concentrations of felodipine after both acute and chronic dosing.[35–37] Despite differences in the details of the mathematical models (linear or E_{max}), close correlations again are observed between maximal fall after the first dose of felodipine and maximal fall around the time of maximal drug concentrations after 4 weeks of treatment.

Mibefradil

For mibefradil, concentration–effect relationships have been investigated in detail for supine blood pressure, electrocardiographic PQ interval, and heart rate.[38,39] For all of the measured effects, a graded concentration–response relationship was established, and the E_{max} model (Langmuir) proved superior to all other models tested.

Whereas concentration–effect relationships were demonstrated for each of the measured hemodynamic indices, concentration-dependent differences were clearly identified in the relationship between blood pressure response and cardiac effects. Thus, the EC_{50} value (i.e, the concentration required to produce 50% of the maximal effect) was two-fold higher for the effect on PQ interval compared with the effect on blood pressure. In contrast, the EC_{50} was lowest for the effect on heart rate. Thus, over most of the concentration range achieved during steady-state therapy, the heart rate response was at the flat or plateau portion at the top of the concentration–response curve.

The mibefradil analysis was extended by the nonlinear mixed effects model program, NONMEM, which is a population-based technique. NONMEM not only evaluates the PK-PD relationship but also seeks to identify the influence of demographic factors or covariates such as age, gender, and race on both pharmacokinetic and pharmacodynamic characteristics. For the PQ interval, none of the tested covariates was found to contribute significantly. However, for both blood pressure and heart rate, the baseline hemodynamic characteristics of the patients contributed significantly to the overall concentration–response relationship. For example, at 90 mmHg, the E_{max} for diastolic blood pressure was –14 mmHg, whereas the corresponding figure at 110 mmHg was –22.6 mmHg. Similarly, when the baseline heart rate was low (for example, 50 bpm), the E_{max} characterized by the model was –2.1 bpm, whereas when the baseline heart rate was 90 bpm, the corresponding E_{max} value was –14.3 bpm. None of the other factors, such as age, race, and sex, contributed significantly to the concentration–effect relationship.[39]

In summary, concentration–effect relationships have been established for most calcium antagonists, and overall a considerable volume of evidence indicates that the blood pressure response (and other pharmacologic responses) correlates directly with the concentration–time profile in terms of both magnitude and duration. This finding may be of value in individualizing and optimizing dose and dosage interval for antihypertensive therapy with calcium antagonists.[40]

LONG-ACTING VS. SHORT-ACTING AGENTS

The practical shortcomings of the prototype short-acting drugs, particularly the dihydropyridine derivatives, have encouraged the development of longer-acting drugs and formulations. Although initially developed mainly for reasons of dosage simplicity and patient compliance, long-acting drugs and formulations have significantly improved the overall therapeutic profile of the calcium antagonist group.[41]

Compliance

Dosage Simplicity

Evidence clearly indicates a modestly but significantly higher rate of compliance with once-daily regimens compared with twice-daily regimens.[42] However, it is also important to appreciate that particular pharmacologic characteristics have important therapeutic consequences; simplistic focus on the number of doses taken or omitted may lead to false conclusions about the therapeutic superiority of one drug relative to another. In fact, the crucial consideration is the persistence of therapeutic action, which is determined only in part by the percentage of prescribed doses taken. Drugs differ in their ability to forgive errors in compliance because they have differing abilities to maintain therapeutic action in the face of missed doses.[42] Thus, some agents demand punctual dosing to maintain an uninterrupted therapeutic effect, whereas others have greater leeway that in some cases may be sufficient to maintain the therapeutic effect even when a dose has been omitted. The concept of therapeutic coverage takes these factors into account by combining the drug's duration of action with the profile of dosing history (i.e., compliance) to calculate the percentage of time that the drug's action is maintained within a therapeutic target range.

Therapeutic Coverage

Continuity of therapeutic action or maintenance of therapeutic coverage is best ensured when the interval between doses is shorter than the duration of drug action; ideally, the dosage interval should be less than half of the duration of the action.[43] This principle applies regardless of the frequency of dosing.

Inferences about the ability of a given drug to provide therapeutic coverage can be drawn by relating the antihypertensive response to the rate of compliance. Figure 1 shows the correlation between the blood pressure responses to once-daily nitrendipine and the rates of compliance assessed by electronic monitoring (using the MEMS monitor).[44] The authors of this study noted a negative linear correlation between blood pressure response and rates of compliance. However, it is more important to note that levels of compliance must be sustained above 90% to achieve any significant fall in blood pressure. It thus may be concluded that nitrendipine has a relatively modest duration of action and can achieve blood pressure control only when adherence to the treatment regimen is relatively high. A further interpretation is that nitrendipine has a

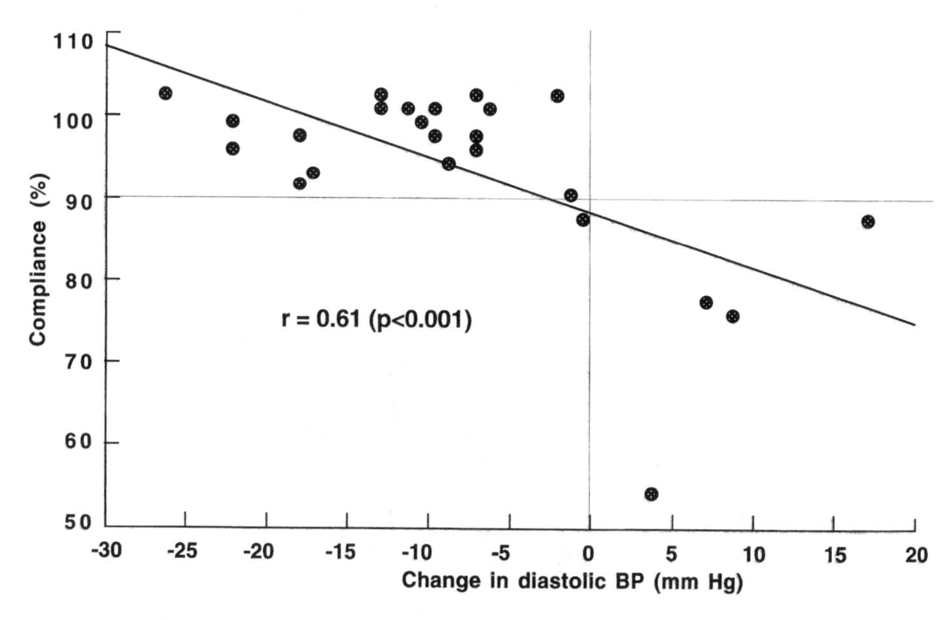

FIGURE 1. Change in diastolic blood pressure in relation to the measured compliance with once daily nitrendipine treatment. (Redrawn from the data of Mallion JM, Meilhac B, Tremel F, et al: Use of a microprocessor-equipped tablet in monitoring compliance with antihypertensive treatment. J Cardiovasc Pharmacol 19(Suppl 2):S41–S48, 1992.)

limited ability to maintain antihypertensive efficacy in the setting of poor compliance; it has only a limited therapeutic reserve and provides relatively limited therapeutic coverage.

Dose Omissions

The most direct and almost certainly the most appropriate method for testing therapeutic coverage with antihypertensive drugs is to create dosage omissions or belated dosing within a steady-state regimen. Such studies make clear that a high level of therapeutic coverage is achieved most consistently by drugs with an intrinsically long duration of action. For example, studies have clearly identified relatively little diminution in the antihypertensive response to amlodipine 48 hours after dosing compared with the response 24 hours after dosing.[34] This finding, which suggests that amlodipine has sufficient therapeutic coverage to compensate for at least a single dosage omission, was confirmed in a study that compared the antihypertensive responses to amlodipine and enalapril after a single dosage omission at steady state.[45] With perfect compliance, the overall profile of antihypertensive effect did not differ significantly between the drugs. However, after a dosage omission, amlodipine sustained much of its effect over a missed dose interval, whereas enalapril showed a significant loss of antihypertensive efficacy (Fig. 2).

The therapeutic coverage of amlodipine also has been assessed across more than a single missed dose interval in a comparison of once-daily amlodipine and twice-daily diltiazem (modified-release formulation).[46] Once again, the blood pressure-lowering effect of the two drugs was broadly comparable

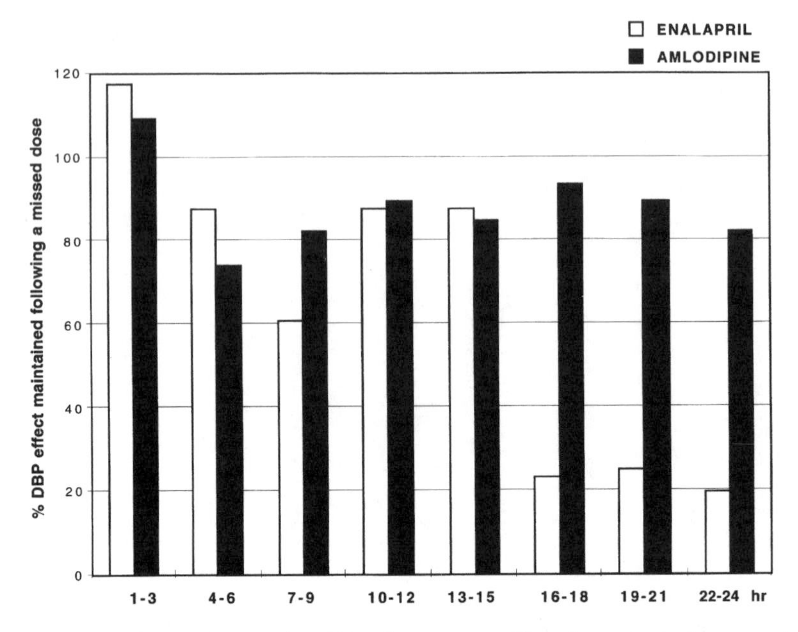

FIGURE 2. Comparison of the 24-hour blood pressure profiles after a single dosage omission at steady state treatment with enalapril (white bars) and amlodipine (black bars). (Redrawn from the data of Hernandez-Hernandez R, de Hernandez MJ, Armas-Padilla MC, et al: The effects of missing a dose of enalapril versus amlodipine on ambulatory blood pressure. Blood Pressure Monit 1:121–126, 1996.)

during active maintenance treatment. However, after the steady-state treatment was interrupted for 2 days by placebo insertion (using a double-blind, randomized design), it was clear that the antihypertensive effects of amlodipine persisted, whereas only a small antihypertensive effect was retained with diltiazem over the same period of dosage omission.

These observations about therapeutic response, frequency of dosing, and dosage omissions are based on the assumption that the intrinsic duration of action of any given drug is independent of the frequency of dosing. In general, this principle is valid. However, it is possible by pharmaceutical manipulation to extend the duration of action of a drug and, at the same time, to create a treatment regimen that significantly improves compliance. Despite potential limitations in using pharmaceutical manipulation to extend therapeutic coverage, in specific instances this strategy leads to potential therapeutic benefits. Table 6 summarizes the results of a randomized, parallel group comparison of nifedipine administered either once daily as the GITS formulation or twice daily as a sustained-release formulation.[47] The results clearly indicate that the once-daily GITS regimen offers significant benefit in three different measures of compliance and also improves blood pressure control.

Adverse-effect Profiles

Symptomatic Side Effects

Headache, flushing, dizziness, and peripheral edema are the well-recognized vasodilator adverse effects of the dihydropyridine derivatives in particular.

TABLE 6. Compliance, Once- vs. Twice-Daily Dosing, and Blood Pressure Control

	Nifedipine GITS (Once Daily)	Nifedipine SR (Twice Daily)
Pill count (%)	104 + 4	93 + 2
MEMS compliance		
Prescribed doses taken (%)	93*	84
Correct number of doses taken (%)	87*	72
Doses taken on time (%)	70*	50
Patients with > 80% compliance and DBP < 90 mmHg or a fall by > 10 mmHg		
Blood pressure responders (%)	82*	59

From Toal CB, Laplante L: Is there a difference in hypertensive patient compliance between once a day or twice a day nifedipine? Am J Hypertens 8(4 Pt 2):D17, 1995.
* Statistically significant difference ($p < 0.05$) between nifedipine GITS and nifedipine SR.

These vasodilator effects (and the accompanying reflex cardioacceleration) are particularly troublesome with rapid-onset, short-acting dihydropyridines. It may be anticipated that the long-acting agents should have a superior side-effect profile, and the available evidence tends to confirm this assumption. For example, a direct comparison of amlodipine with a modified-release formulation of nifedipine (nifedipine SR, which has an intermediate duration of action) demonstrated a lower incidence of adverse early vasodilator effects, particularly headache, flushing, and tachycardia, with the long-acting agent.[48] To date, however, peripheral edema remains a well-recognized adverse effect for which no definitive evidence indicates that the long-acting agents are superior to the short- and intermediate-acting dihydropyridine calcium antagonists.

Rate of Drug Delivery and Neurohumoral Activation

Increasing evidence suggests that fluctuations in plasma drug concentrations and corresponding fluctuations in vasodilator action and blood pressure response, particularly with rapid onset and offset, may be deleterious. The significant increases in sympathetic activity associated with short-acting dihydropyridine calcium antagonists are well recognized, but it is widely assumed that such activation diminishes and disappears during chronic treatment because of resetting of the arterial baroreflex. It is now clear, however, that this assumption is not invariably valid. Baroreceptor mechanisms remain active, and sympathetic activity continues to increase in response to a rapid decrease in blood pressure, as may be seen repeatedly after each dose of a rapidly absorbed short-acting agent. Thus, even during chronic treatment, significant increases in sympathetic activity are apparent in the first few hours after drug administration as drug concentrations rapidly rise during the absorption phase.[49] Such increased sympathetic activity is detectable from the circulating levels of noradrenaline, which are significantly elevated around the time of peak blood pressure response (1–2 hours after dosing), although they may be comparable to placebo at the time of trough.[50] Long-acting dihydropyridine calcium antagonists, such as amlodipine and nifedipine GITS, produce a sustained vasodilator antihypertensive effect with minimal fluctuations over each dosage interval; thus, dosing-related increases in plasma norepinephrine are minimal or absent during chronic treatment.[51,52]

Increased Sympathetic Activity and Cardiovascular Outcomes

Increased sympathetic activity is recognized as one of the non–pressure-related coronary risk factors in hypertensive patients. Increased sympathetic tone may directly and indirectly influence the development or maintenance of left ventricular hypertrophy and also enhance the development and progression of atherosclerosis.[49] Therefore, repeated increases in sympathetic activity, as associated with short-acting dihydropyridine calcium antagonists, may be implicated in reports that they cause less regression of left ventricular hypertrophy than anticipated from the decrease in blood pressure. Such increases also may contribute to the apparent failure of translating the antiatherosclerotic properties identified in animals and humans into outcome benefits in humans.

The hypothesis that drugs with long duration of action should produce more significant regression of left ventricular hypertrophy than short-acting dihydropyridine calcium antagonists is supported by the findings of study with felodipine in two different formulations.[50] Felodipine in its instant-release formulation, administered twice daily for 12 months, significantly reduced both systolic and diastolic blood pressure, but this reduction was accompanied by only modest decreases of left ventricular mass index and significant increases in plasma norepinephrine around the time of peak response. Furthermore, the correlation between systolic blood pressure and left ventricular mass during the 12 months of treatment was significantly improved when average plasma norepinephrine was incorporated into the correlation. On this basis, it may be anticipated that long-acting agents should be more effective than short-acting agents. A direct comparison of felodipine ER (administered once daily) and short-acting felodipine (administered twice daily) supports this contention. The long-acting formulation provides more consistent blood pressure control and a greater reduction in left ventricular mass over 8 weeks of treatment. Further support for the concept that a truly long-acting calcium antagonist causes regression of left ventricular hypertrophy to the extent predicted from the decrease in blood pressure is found in a comparison of amlodipine and enalapril.[53] Thus, over a 2-year period, amlodipine produced a reduction in left ventricular mass similar to that achieved by enalapril despite the popular perception based on metaanalyses that angiotensin-converting enzyme (ACE) inhibitors are the most effective antihypertensive agents in reversing left ventricular hypertrophy.

BLOOD PRESSURE CONTROL AND TROUGH-TO-PEAK RATIO

Blood Pressure Throughout 24 Hours

Although no direct prospective evidence indicates that sustained 24-hour blood pressure control is superior to intermittent and fluctuating blood pressure control, a number of factors suggest that this is the case.[54] Epidemiologic evidence includes the following:

1. Measures of 24-hour blood pressure more closely predict cardiovascular target organ damage than clinic or casual measurements.
2. The incidence of cardiovascular complications is higher when both daytime and nighttime blood pressure remain elevated.

3. High levels of blood pressure variability (i.e., wide fluctuations in achieved blood pressure) are an additional and independent determinant of target organ damage.
4. The incidence of cardiovascular events is highest in the morning when blood pressure increases rapidly from the low levels of sleep. This normally corresponds to the period of 24 hours after dosing in a once-daily regimen.

All of these factors suggest that optimal blood pressure control requires strategies that lower blood pressure consistently and fully throughout 24 hours while maintaining the normal circadian pattern of blood pressure without increasing blood pressure variability. The problem has been to identify agents that consistently provide these characteristics.

Trough-to-Peak Ratio

Trough-to-peak ratio is a relatively recent addition to the list of parameters characterizing the therapeutic profile of an antihypertensive drug. The concept arose from the deliberations of the Cardio-Renal Drug Advisory Committee of the Food and Drug Administration (FDA), which suggested that any new antihypertensive agent should retain most of its peak effect at trough and that the effect at trough should be no less than one-half to two-thirds of the peak effect.[55,56] This particular guideline was formulated largely because of the concern that excessive doses of drug may be used simply to control blood pressure at the end of the dosage interval (trough) without due regard to the possibility of an exaggerated fall in blood pressure at peak response. Nonetheless, it has proved to be a useful index of the duration of the antihypertensive response.[56]

Table 7 summarizes our interpretation of the published data about trough-to-peak ratios with the newer calcium antagonists. The value of these data is limited, however, because different analytical approaches have been adopted for calculating trough-to-peak ratio. For example, the original published data for felodipine ER were compromised by failure to incorporate a placebo correction. Furthermore, for both felodipine and lacidipine, the quoted values are based on estimates of trough-to-peak ratio, either derived from mean data or reported as a mean value with no evidence of the range or variability around the mean values. In contrast, the values for amlodipine, nifedipine GITS, and mibefradil are based on studies in which the estimates were calculated from the response in individual patients. At first sight, the data in Table 7 suggest that all of the agents are suitable for once-daily administration in the treatment

TABLE 7. Trough-to-Peak Ratios of Calcium Antagonists

	Mean Trough-to-Peak Ratio (%)	Range (%)
Amlodipine	68	53–75
Nifedipine GITS	65	60–100
Verapamil SR	64	40–70
Felodipine ER	52	29–56
Lacidipine	65	55–70
Mibefradil	70	60–78

of hypertension. However, caution must be exercised in interpreting the single-point estimates derived with lacidipine and felodipine ER. In addition, with felodipine ER the 50% threshold trough-to-peak ratio is not consistently achieved with lower doses.

SAFETY ISSUES

Controversy surrounds the use of short-acting calcium antagonists.[57–60] In brief, findings in case control and cohort studies suggest that antihypertensive treatment with short-acting calcium antagonists may be associated with an increased rate of myocardial infarction compared with beta blockers and diuretics.[57] Although such reports of adverse outcomes have provoked considerable interest, it is a reflection of the weakness of such retrospective studies that a less publicized but corresponding study from another research group found no differences in the relative risks of myocardial infarction with calcium antagonists, beta blockers, and diuretics. In fact, this analysis suggested that both calcium antagonists and ACE inhibitors were superior to diuretics and beta blockers during 5 years of treatment.[61] Similar case control reports have identified the possibility of differential rates of gastrointestinal hemorrhage and development of cancer.[62–64] Like earlier studies, these evaluations have limitations because the study populations were relatively small and they were not prospective or randomized. With respect to cancer, for example, a much larger database revealed no increased incidence of cancer in patients maintained on nifedipine for a period of not less than 4 years.[65] Similarly, examination of an alternative large database revealed that the incidence of cancer is comparable in patients treated with calcium antagonists, patients treated with other antihypertensives, and a matched population in a general cancer register database.[66]

EFFECTS OF CONCOMITANT CONDITIONS AND DRUGS

Age

As a group, calcium antagonists are widely prescribed for the treatment of cardiovascular disease in elderly patients. As a consequence, many studies have investigated their pharmacokinetics in elderly populations. For example, peak plasma concentrations of verapamil have been reported to be increased by about 25%. This increase has been attributed to a reduced rate of metabolism, presumably because of decreases in hepatic blood flow and drug biotransformation.[67] Similar results have been reported for diltiazem, nifedipine, and amlodipine.[68–71] This finding is consistent among calcium antagonists and independent of the formulation, although increases in steady-state concentrations in elderly populations do not attain statistical significance in all studies.

Hepatic Disease

Because the liver is the principal site of biotransformation for calcium antagonists, there has been considerable research in patients with hepatic disease, particularly with the older agents. In general, the changes in the pharmacokinetic characteristics are significant in patients with advanced hepatic impairment (e.g., cirrhosis) as a consequence of the reduced mass of the functioning liver and also of intra- and extrahepatic shunting of the portal blood supply.

For example, the pharmacokinetics of verapamil are grossly altered in patients with liver cirrhosis, with 2–5-fold decreases in oral and systemic clearances and volumes of distribution that are almost twice normal.[72,73] The terminal elimination half-life after intravenous administration of verapamil, for example, increases from 3.7 to 14.2 hours. After oral administration the bioavailability of verapamil increases from 22 to 52%. Furthermore, because of reduced protein binding in patients with liver cirrhosis, there is a 60% increase in free drug concentration.

In summary, because the liver is the principal site of biotransformation for all calcium antagonists, hepatic disease results in a consistent pattern of reduced drug clearance, higher drug concentrations, and augmented pharmacodynamic response. Although detailed data exist only for some of these agents, it seems prudent to reduce the dose of all calcium antagonists in patients with advanced liver impairment.

Renal Impairment

Because there have been fewer studies of the effects of renal impairment on the disposition of calcium antagonists, the pattern of results is less consistent than for hepatic impairment. Nonetheless, studies show that advanced renal failure may alter the pharmacokinetic characteristics. Overall, the pharmacokinetics of verapamil and diltiazem[74–76] does not appear to change to a clinically significant extent, and dosage adjustment is probably unnecessary. In contrast, evidence suggests that with several dihydropyridine derivatives statistically and clinically significant changes in disposition occur. For example, the pharmacokinetics of intravenous nifedipine has been shown to change significantly in patients with renal disease in relation to the degree of renal impairment. The volume of distribution and elimination half-life increases, whereas the extent of protein binding decreases.[77]

In summary, the pharmacokinetic parameters of the dihydropyridine derivatives are more variably affected by renal disease, and no general dosage recommendation is possible.

Drug Interactions

Because calcium antagonists are most likely to be prescribed for patients with cardiovascular disorders, interactions of clinical relevance tend to occur with other cardiovascular drugs (Table 8). Coadministration of verapamil and digoxin increases the plasma concentration of digoxin by 40–90% as a result of changes in digoxin clearance (both renal and nonrenal), although the exact mechanism remains unclear.[78,79] A similar interaction has been reported with nitrendipine,[80] but it has not been seen consistently with other dihydropyridine derivatives or diltiazem.[81,82]

With respect to beneficial interactions, the additive effect of combining a dihydropyridine calcium antagonist and a beta blocker in the treatment of both hypertension and angina is well recognized. This beneficial interaction includes

TABLE 8. Pharmacokinetic Interactions of Calcium Antagonists

Interacting Drug	Calcium Antagonist	Result
Other cardiovascular drugs		
Digoxin	Verapamil Nitrendipine	Digoxin levels increased by 40–90%
Alpha blockers	Verapamil Diltiazem Dihydropyridines	Excessive hypotension
Quinidine	Verapamil	Excessive hypotension
Propranolol	Dihydropyridines	Increased levels of propranolol
Noncardiovascular drugs		
Carbamazepine	Verapamil	Increased levels of carbamazepine
Enzyme inducers (e.g., rifampicin, phenytoin)	Verapamil	Reduced bioavailability of verapamil
Cimetidine	Verapamil Diltiazem Dihydropyridines	Increased area under the curve and plasma levels of calcium antagonist
Cyclosporine	Nicardipine Diltiazem	Increased levels of cyclosporine

the additive antihypertensive and antiischemic properties of each component and a tendency for each drug to ameliorate some of the other's potentially adverse effects.[83] For example, the tendency for reflex cardioacceleration due to vasodilation is countered by the beta blocker, and the potential for peripheral circulatory impairment with the beta blocker is attenuated by the vasodilator activity of the calcium antagonist. The combination of a beta blocker and a dihydropyridine calcium antagonist, therefore, is safe and effective. Because of the negative chronotropic and inotropic effects of both verapamil and diltiazem, it is not routinely recommended that either be combined with a beta blocker because of the potentially additive cardiodepressant effects,[84] although with appropriate specialist supervision these combinations can be safe and effective.

The antihypertensive consequences of combined alpha blockade and calcium antagonism can be construed as either particularly effective or potentially harmful.[85,86] It has been suggested that this combination has an exaggerated first-dose response that places patients at risk for hypotension, whereas the combination appears to be safe and effective once steady-state conditions are established.

Perhaps the interaction of greatest clinical interest and relevance, at least to the specialist, is between cyclosporine and either verapamil or diltiazem.[87,88] As a consequence of the effects on hepatic enzyme activity, clearance of cyclosporine tends to be reduced, and cyclosporine concentrations tend to increase. In addition, experimental evidence suggests that the potentially nephrotoxic effect of cyclosporine can be attenuated by administration of a calcium antagonist. This interaction is relevant to the treatment of patients after renal and cardiac transplantation.

There are several other interactions of lesser clinical significance—for example, between verapamil and the anticonvulsant drug carbamazepine[89] and between verapamil and enzyme-inhibiting and enzyme-inducing drugs that obviously affect the clearance of the calcium antagonist.[90] Finally, multidrug-resistant cancers may be rendered less resistant to cytotoxic treatment if verapamil is administered as adjuvant therapy.[91] An interesting interaction, albeit of limited clinical significance, is reported with grapefruit juice and some of the calcium antagonists, particularly felodipine.[92] This effect is attributable to specific inhibition of hepatic cytochrome P4503A4.

CONCLUSION

Calcium antagonists are well established in the treatment of cardiovascular disorders, particularly hypertension and angina. Although some recent publications have cast doubt on their safety (of the short-acting agents, in particular), interpretation of the studies is controversial and the adverse conclusions are neither definitive nor widely accepted. This controversy, however, has highlighted the increasing necessity for discriminating among the different calcium antagonists and for not assuming that they constitute a homogeneous group. In the past, this distinction has relied heavily on experimental pharmacologic differences, but there is now a clear need to identify pharmacodynamic and clinically relevant differences. For example, the short-acting, rapid-onset agents have obvious shortcomings because of pronounced fluctuations in their vasodilatory effects, which provoke neurohumoral activation and cardioacceleration. Pharmacokinetic and pharmacodynamic studies have demonstrated that the clinical response to a calcium antagonist is a direct consequence of the pharmacokinetic features and the drug concentration–time profile. The development of gradual-onset, genuinely long-acting agents has significantly improved the overall therapeutic profile via a more appropriate pharmacokinetic profile that translates to slow onset of action and protracted, more consistent duration of effect.

References

1. Fleckenstein A: A history of calcium antagonists. Circ Res 52(Suppl 11):1–16, 1983.
2. Schlanz KD, Myre SA, Bottorff B: Pharmacokinetic interactions with calcium channel antagonists. Part I. Clin Pharmacokinet 21:344–356, 1991.
3. Schlanz KD, Myre SA, Bottorff MB: Pharmacokinetic interactions with calcium channel antagonists. Part II. Clin Pharmacokinet 21:448–460, 1991.
4. Kelly JG, O'Malley K: Clinical pharmacokinetics of calcium antagonists: An update. Clin Pharmacokinet 22:416–433, 1992.
5. Spedding M: Changing surface charge with salicylate differentiates between sub-groups of calcium antagonists. Br J Pharmacol 83:211–220, 1984.
6. Fleckenstein A: Calcium Antagonism in Heart and Smooth Muscle. New York, Wiley & Sons, 1983.
7. Glossman H, Ferry DR: Solubilisation and partial purification of putative calcium channels labelled with `H-nimodipine. Naunyn Schemiedebergs Arch Pharmacol 323:279–291, 1983.
8. Singh B: The mechanism of action of calcium antagonists relative to their clinical applications. Br J Clin Pharmacol 21(Suppl):109–121, 1986.
9. Toyo-oka T, Naylor WG: Third generation calcium entry blockers. Blood Pressure 5:206–208, 1996.
10. Echizen H, Vogelgesang B, Eichelbaum M: Clinical pharmacokinetics of verapamil, nifedipine, and diltiazem. Clin Pharmacol 2:425–449, 1986.
11. Woodcock BG, Schulz W, Kober G, et al: Direct determination of hepatic extraction of verapamil in cardiac patients. Clin Pharmacol Ther 30:52–56, 1981.
12. Eichelbaum M, Eude M, Reinberg G, et al: The metabolism of ^{14}C-d-1 verapamil in man. Drug Metab Dispos 7:145–148, 1979.
13. Echizen H, Vogelgesang B, Eichelbaum M: Effects of d-1-verapamil on atrioventricular conduction in relation to its stereoselective first-pass metabolism. Clin Pharmacol Ther 38:71–76, 1985.

14. Meredith PA, Elliott HL, Pasanisi F, et al: Verapamil pharmacokinetics and hepatic and renal blood flow. Br J Clin Pharmacol 20:101–106, 1985.
15. Eichelbaum M, Somagyi A: Inter- and intra-subject variations in the first-pass elimination of highly cleared drugs during chronic dosing: Studies with deuterated verapamil. Eur J Clin Pharmacol 26:47–53, 1984.
16. Schwartz JB, Abernethy DR, Taylor AA, et al: An investigation of the cause of accumulation of verapamil during regular dosing in patients. Br J Clin Pharmacol 19:512–516, 1985.
17. Shand DG, Hammill SC, Aanousen L, et al: Reduced verapamil clearance during long-term oral administration. Clin Pharmacol Ther 30:701–703, 1981.
18. Sugihara J, Sugawara Y, Audo H, et al: Studies on the metabolism of diltiazem in man. J Pharmacobiodyn 7:24–32, 1984.
19. Faulkner JK, McGibney D, Chasseaud LF, et al: The pharmacokinetics of amlodipine in healthy volunteers after single intravenous and oral doses and after 14 repeated oral doses given once daily. Br J Clin Pharmacol 22:21–26, 1986.
20. Meredith PA, Elliott HL: Amlodipine: Clinical relevance of a unique pharmacokinetic profile. J Cardiovasc Pharmacol 22(Suppl A):S6–S8, 1993.
21. Squassante L, Caveggion E, Braggio S, et al: A study of plasma disposition kinetics of lacidipine after single oral ascending doses. J Cardiovasc Pharmacol 23(Suppl 5):S94–S97, 1994.
22. Meredith PA: The pharmacokinetics of lacidipine. Rev Contemp Pharmacother 6:9–15, 1995.
23. Hall ST, Harding SM, Evans GL, et al: Clinical pharmacology of lacidipine. J Cardiovasc Pharmacol 17(Suppl 4):S9–S13, 1991.
24. Herbette LG, Haviraghi G, Tulenko T, et al: Molecular interaction between lacidipine and biological membranes. J Hypertens 11(Suppl 1):S13–S19, 1993.
25. Rutledge A, Triggle DJ: The binding interactions of RO40-5967 at the L type Ca^{2+} channel in cardiac tissue. Eur J Pharmacol 290:155–158, 1995.
26. Abernethy D: Pharmacologic and pharmacokinetic profile of mibefradil, a T and L type channel antagonist. Am J Cardiol (in press).
27. Petrie JR, Glen SK, MacMahon M, et al: Haemodynamics, cardiac conduction and pharmacokinetics of mibefradil (RO40-5967), a novel calcium antagonist. J Hypertens 13:1842–1846, 1995.
28. Grundy JS, Foster RT: The nifedipine gastrointestinal and therapeutic system (GITS): Evaluation of pharmaceutical, pharmacokinetic and pharmacological properties. J Clin Pharmacokinet 30:28–51, 1996.
29. Donnelly R, Elliott HL, Meredith PA: Concentration-effect analysis of antihypertensive drug response: Focus on calcium antagonists. Clin Pharmacokinet 26:472–485, 1994.
30. Meredith PA, Donnelly R, Elliott HL: Prediction and optimisation of the antihypertensive response to nifedipine. Blood Pressure 3:303–308, 1994.
31. Meredith PA, Elliott HL, Ahmed JH, et al: Age and the antihypertensive effect of verapamil: An integrated pharmacokinetic-pharmacodynamic approach. J Hypertens 5(Suppl 5):S125–S221, 1987.
32. Meredith PA, Reid JL: The use of pharmacodynamic and pharmacokinetic profiles in drug development for planning individual therapy. In Laragh JH, Brenner BM (eds): Hypertension: Pathophysiology, Diagnosis and Management, 2nd ed. New York, Raven Press, 1995, pp 2771–2783.
33. Abernethy DR, Gutkowska J, Winterbottom LM: Effects of amlodipine, a long-acting dihydropyridine calcium antagonist, in aging hypertension: Pharmacodynamics in relation to disposition. Clin Pharmacol Ther 48:76–86, 1990.
34. Donnelly R, Meredith PA, Miller SHK, et al: Pharmacodynamic modelling of the antihypertensive response to amlodipine. Clin Pharmacol Ther 54:303–310, 1993.
35. Blychert E, Edgar B, Elmfeldt D, Hedner T: Plasma concentration–effect relationships for felodipine: A meta-analysis. Clin Pharmacol Ther 52:80–89, 1992.
36. Edgar B, Lundborg P, Regardh C: Clinical pharmacokinetics of felodipine—a summary. Drugs 34(Suppl 3):16–27, 1987.
37. Edgar B, Elmfeldt D: Relation between plasma concentration of felodipine and effect on diastolic blood pressure. Cardiovasc Drugs Ther 1:232, 1987.
38. Schmitt R, Kleinbloesem CH, Belz GG, et al: Hemodynamic and humoral effects of the novel calcium antagonist, RO 40-5967 in patients with hypertension. Clin Pharmacol Ther 52:314–323, 1992.
39. Meredith PA: Clinical relevance of optimal pharmacokinetics in the treatment of hypertension. J Cardiovasc Pharmacol 1997 (in press).
40. Donnelly R, Meredith PA, Elliott HL: The description and prediction of antihypertensive drug response: An individualized approach. Br J Clin Pharmacol 31:627–634, 1991.
41. Van Zwieten PA, Hansson L, Epstein M: Slowly acting calcium antagonists and their merits. Blood Pressure 6:78–80, 1997.

42. Cramer JA, Mattson RH, Prevey ML, et al: How often is medication taken as prescribed? A novel assessment technique. JAMA 261:3273–3277, 1989.
43. Meredith PA: Therapeutic implications of drug "holidays." Eur Heart J 17(Suppl A):21–24, 1996.
44. Mallion JM, Meilhac B, Tremel F, et al: Use of a microprocessor-equipped tablet in monitoring compliance with antihypertensive treatment. J Cardiovasc Pharmacol 19(Suppl 2):S41–S48, 1992.
45. Hernandez-Hernandez R, de Hernandez MJ, Armas-Padilla MC, et al: The effects of missing a dose of enalapril versus amlodipine on ambulatory blood pressure. Blood Pressure Monit 1:121–126, 1996.
46. Leenen FHH, Fourney A, Notman G, Tanner J: Persistence of antihypertensive effect after "missed doses" of calcium antagonist with long (amlodipine) vs short (diltiazem) elimination half-life. Br J Clin Pharmacol 41:83–88, 1996.
47. Toal CB, Laplante L: Is there a difference in hypertensive patient compliance between once a day or twice a day nifedipine? Am J Hypertens 8(Pt 2):D17, 1995.
48. Hosie J, Bremner AD, Fell PJ, et al: Side effects of dihydropyridine therapy: Comparison of amlodipine and nifedipine retard. J Cardiovasc Pharmacol 22(Suppl A):S99–S112, 1993.
49. Ruzicka M, Leenen FHH: Relevance of intermittent increases in sympathetic activity for adverse outcome on short-acting calcium antagonists. In Laragh JH, Brenner BM (eds): Hypertension: Pathophysiology, Diagnosis and Management, 2nd ed. New York, Raven Press, 1995, pp 2815–2825.
50. Leenen FHH, Holliwell DL: Antihypertensive effect of felodipine associated with persistent sympathetic activation and minimal regression of left ventricular hypertrophy. Am J Cardiol 69:639–645, 1992.
51. Packer M, O'Connor CM, Ghali JK, et al, for the PRAISE Study Group: Effect of amlodipine on morbidity and mortality in severe chronic heart failure. N Engl J Med 335:1107–1114, 1996.
52. Frohlich ED, McLoughlin MJ, Losem CJ, et al: Hemodynamic comparison of two nifedipine formulations in patients with essential hypertension. Am J Cardiol 68:1346–1350, 1991.
53. Picca M, Pelosi GC: Effects of enalapril and amlodipine on left ventricular hypertrophy and function in hypertension. Eur Heart J 13(Suppl):177, 1992.
54. Meredith PA, Perloff D, Mancia G, Pickering T: Blood pressure variability and its implications for antihypertensive therapy. Blood Pressure 4:5–11, 1995.
55. Rose M, McMahon FG: Some problems with antihypertensive drug studies in the context of the new guidelines. Am J Hypertens 3:151–155, 1990.
56. Elliott HL, Meredith PA: Trough:peak ratio: Clinically useful or practically irrelevant? J Hypertens 13:279–283, 1995.
57. Psaty BM, Heckbert SR, Koepsell TD, et al: The risk of myocardial infarction associated with antihypertensive drug therapies. JAMA 274:620–625, 1995.
58. Furberg CD, Psaty BM, Meyer JV: Nifedipine: Dose-related increase in mortality in patients with coronary heart disease. Circulation 92:1326–1331, 1995.
59. Buring JE, Glynn RJ, Hennekens CH: Calcium channel blockers and myocardial infarction: A hypothesis formulated but not yet tested. JAMA 274:654–655, 1995.
60. Opie LH, Messerli FH: Nifedipine and mortality: Grave defects in the dossier. Circulation 92:1068–1073, 1995.
61. Aursnes I, Litleskare I, Froyland H, Abdelnoor M: Association between various drugs used for hypertension and risk of acute myocardial infarction. Blood Pressure 4:157–163, 1995.
62. Pahor M, Guralnik JM, Corti MC, et al: Long-term survival and use of antihypertensive medications in older persons. J Am Geriatr Soc 43:1–7, 1995.
63. Pahor M, Guralnik JM, Furberg CD, et al: Risk of gastrointestinal haemorrhage with calcium antagonists in hypertensive persons over 67 years old. Lancet 347:1061–1065, 1996.
64. Pahor M, Guralnik JM, Salive ME, et al: Do calcium channel blockers increase the risk of cancer? Am J Hypertens 9:698–699, 1996.
65. Jick H, Jick S, Derby LE, et al: Calcium channel blockers and risk of cancer. Lancet 349:525–528, 1997.
66. Hole DJ, Gillis CR, McCallum IR, et al: Cancer risk in hypertensive patients taking calcium channel drugs. J Hypertens (in press).
67. Cox JP, O'Boyle CA, Mee F, et al: The antihypertensive efficacy of verapamil in the elderly evaluated by ambulatory blood pressure measurements. J Hum Hypertens 2:41–47, 1988.
68. Storstein L, Larsen A, Midrbo K, et al: Pharmacokinetics of calcium blockers in patients wth renal insufficiency and in geriatric patients. Acta Med Scand 681:25–30, 1983.
69. Robertson DRC, Waller DG, Renwick AG, et al: Age-related changes in the pharmacokinetics and pharmacodynamics of nifedipine. Br J Clin Pharmacol 25:297–305, 1988.

70. Elliott HL, Meredith PA, Reid JL, et al: A comparison of the deposition of single oral doses of amlodipine in young and elderly subjects. J Cardiovasc Pharmacol 12(Suppl 7):S64–S66, 1988.
71. Abernethy DR, Gutkowska J, Lambert MD: Amlodipine in elderly hypertensive patients: Pharmacokinetics and pharmacodynamics. J Cardiovasc Pharmacol 12(Suppl 7):S67–S71, 1988.
72. Giacomini KM, Massoud N, Wong FM, et al: Decreased binding of verapamil to plasma proteins in patients with liver disease. J Cardiovasc Pharmacol 6:924–928, 1984.
73. Sornogyi A, Albrecht M, Kliems G, et al: Pharmacokinetics, bioavailability and ECG response of verapamil in patients with cirrhosis. Br J Clin Pharmacol 12:51–60, 1981.
74. Mooy J, Shols M, van Baak M, et al: Pharmacokinetics of verapamil in patients with renal failure. Eur J Clin Pharmacol 28:405–410, 1985.
75. Shah GM, Winer RL: Verapamil kinetics during maintenance haemodialysis. Am J Nephrol 5:338–341, 1985.
76. Pozet N, Brazier JL, Aissa AH, et al: Pharmacokinetics of diltiazem in severe renal failure. Eur J Clin Pharmacol 24:635–638, 1983.
77. Kleinbloesem CH, van Brummelen P, van Harten J, et al: Nifedipine: Influence of renal function on pharmacokinetic/haemodynamic relationship. Clin Pharmacol Ther 37:563–574, 1985.
78. Pedersen KE: Digoxin interactions: The influence of quinidine and verapamil on the pharmacokinetics and receptor binding of digitalis glycosides. Acta Med Scand 697:12–40, 1985.
79. Belz GG, Doering W, Munkes R, et al: Interactions between digoxin and calcium antagonists and antiarrhythmic drugs. Clin Pharmacol Ther 33:410–417, 1983.
80. Kirch W, Hutt HJ, Heidemann H, et al: Drug interactions with nitrendipine. J Cardiovasc Pharmacol 6(Suppl 7):S982–S985, 1984.
81. Roth A, Harrison E, Mitani G, et al: Efficacy and safety of medium and high dose diltiazem alone and in combination with digoxin for control of heart rate at rest and during exercise in patients with chronic atrial fibrillation. Circulation 73:316–324, 1986.
82. Kuhlmann J: Effects of nifedipine and diltiazem on plasma levels and renal excretion of beta-acetyl digoxin. Clin Pharmacol Ther 37:150–156, 1985.
83. Lederballe-Pedersen O, Christensen CK, Mikkelsen E, et al: Relationship between the antihypertensive effect and steady-state plasma of nifedipine given alone or in combination with a beta-adrenoceptor blocking agent. Eur J Clin Pharmacol 18:287–293, 1980.
84. Kieval J, Kirstein EB, Kessler KM, et al: The effect of intravenous verapamil on haemodynamic status of patients with coronary artery disease receiving propranolol. Circulation 65:653–656, 1982.
85. Jee LD, Opie LH: Acute hypotensive response to nifedipine added to prazosin. BMJ 288:238, 1984.
86. Pasanisi F, Elliott HL, Meredith PA, et al: Combined alpha-adrenoceptor antagonists and calcium channel blockade in normal subjects. Clin Pharmacol Ther 36:716–723, 1984.
87. Neumayer HH, Wagner K: Diltiazem and the economic use of cyclosporin. Lancet 2:523, 1986.
88. Sabate I, Grino J, Castelao AM, et al: Evaluation of cyclosporin-verapamil interaction with observations on parent cyclosporin and metabolites. Clin Chem 34:2151, 1988.
89. MacPhee GIA, McInnes GT, Thompson GG: Verapamil potentiates carbamazepine neurotoxicity: A clinically important inhibitor interaction. Lancet 1:700–703, 1986.
90. Piepho RW, Culbertson VL, Rhodes RS: Drug interactions with the calcium entry blockers. Circulation 75(Suppl V):181–194, 1987.
91. Eichelbaum M, Hirth H-P, Schumacher K, Traugott U (eds): Dexverapamil: Circumventor of Multidrug Resistance. A Satellite Symposium held during Deutscher Krebskongress, Berlin March 16–20, 1992. Dordrecht/Boston/London, Kluwer Academic Publishers, 1993.
92. Lown KS, Bailey DG, Fontana RG, et al: Grapefruit juice increases felodipine oral availability in humans by decreasing intestinal CYP3A protein expression. J Clin Invest 99:2545–2553, 1997.

RENE R. WENZEL M.D. / GEORG NOLL M.D.
THOMAS F. LÜSCHER M.D.

4

Endothelial Regulation of Vascular Tone and Growth: Role of Calcium Antagonists

REGULATION OF VASCULAR TONE AND GROWTH

Physiology

Endothelium-derived Relaxing Factors

The endothelium can elicit relaxation when stimulated by neurotransmitters, hormones, substances derived from platelets, and the coagulation system[1,2] (Fig. 1). Furthermore, shear forces exerted by the circulating blood induce endothelium-dependent vasodilation, an important adaptive response of the vasculature during exercise. The mediator of these responses is a diffusible substance with a half-life of a few seconds, the so-called endothelium-derived relaxing factor[2] (EDRF), which has recently been identified as the free radical nitric oxide (NO). Nitric oxide is formed from L-arginine by oxidation of its guanidine-nitrogen terminal.[3] The catalyzing enzyme NO synthase is constitutively expressed and exists in several isoforms in endothelial cells, platelets, macrophages, and vascular smooth muscle cells as well as in the brain.[4] In endothelial cells, nitric oxide synthase gene expression—although constitutively activated—is upregulated by shear stress and estrogens. The activity of the enzyme is inhibited by the circulating amino acid, asymmetrical dimethyl-arginine (ADMA), which accumulates in patients with renal failure.[5]

In addition, an inducible enzyme exists in vascular smooth muscle, endothelium, and macrophages.[6] The enzyme is calcium-independent and produces large amounts of nitric oxide; it is induced by cytokines such as endotoxin, interleukin 1b and tumor necrosis factor; hence it is activated in inflammatory processes and endotoxin shock.

Endothelium-dependent relaxation can be pharmacologically inhibited by analogs of L-arginine, such as L-N^G-monomethyl arginine (L-NMMA) or L-nitroarginine methylester (L-NAME), that compete with the natural precursor L-arginine at the catalytic site of the enzyme[7,8] (see Fig. 1). In isolated arteries, such inhibitors cause endothelium-dependent contractions. In perfused hearts, inhibition of nitric oxide formation markedly decreases coronary flow. Local infusion of L-NMMA in the human forearm circulation induces an increase in peripheral vascular resistance. When infused intravenously, L-NMMA induces long-lasting increases in blood pressure.[9] This demonstrates that the vasculature is in a constant state of vasodilation due to the continuous basal release of nitric oxide by the endothelium. The intracellular mechanism by which nitric

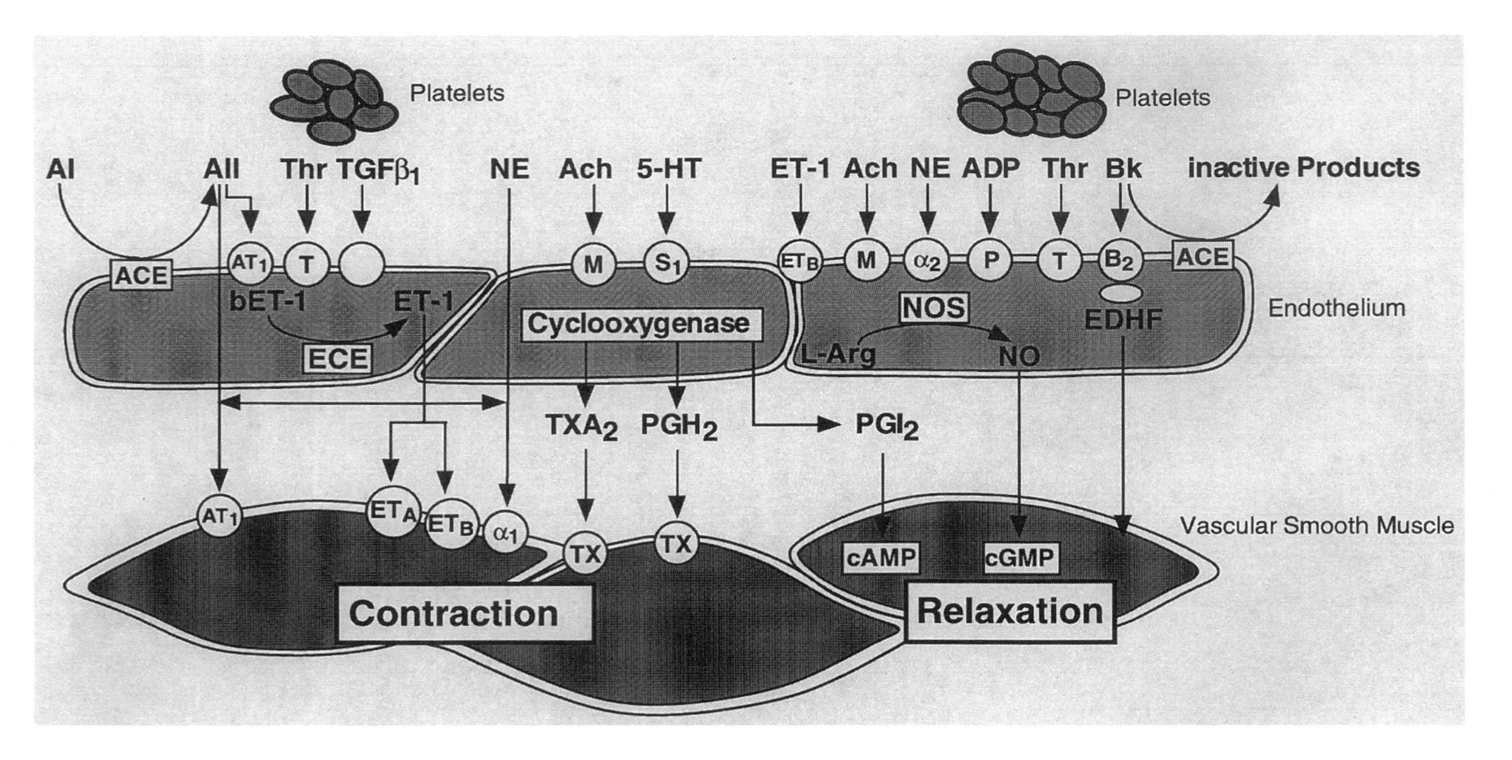

FIGURE 1. Endothelium-derived vasoactive substances. The endothelium is a source of relaxing *(right)* and contracting *(left)* factors. α_{12} = adrenoceptor subtypes, A = angiotensin, ACE = angiotensin-converting enzyme, Ach = acetylcholine, ADP = adenosine diphosphate, Bk = bradykinin, cAMP/cGMP = cyclic adenosine/guanosine monophosphate, ECE = endothelin-converting enzyme, EDHF = endothelium-derived hyperpolarizing factor, ET = endothelin-1, 5HT = 5-hydroxytryptamine (serotonin), L-Arg = L-arginine, NE = norepinephrine, NO = nitric oxide, O_2– = superoxide, PGH_2 = prostaglandin H_2, PGI_2 = prostacyclin, TGFβ1 = transforming growth factor β_1, Thr = thrombin, TXA_2 = thromboxane A_2. Circles represent receptors (AT = angiotensinergic, B = bradykinergic, M = muscarinic, P = purinergic, T = thrombin receptor).

oxide causes relaxation in vascular smooth muscle cells involves formation of cyclic 3′5′-guanosine monophosphate (cGMP) via the enzyme soluble guanylyl cyclase (see Fig. 1).[10]

Nitric oxide is released abluminally as well as luminally and interacts with circulating blood cells and proteins (see Fig. 1). Certain plasma proteins, such as albumin, become nitrosylated and may act as circulating reservoirs of nitric oxide. In platelets, an increase of intracellular cyclic 3′5′-guanosine monophosphate is associated with a reduced adhesion and aggregation. Platelets themselves possess an L-arginine/nitric oxide pathway that regulates their aggregability. Of great interest, platelets release substances such as adenosine diphosphate and triphosphate as well as serotonin that activate the release of nitric oxide and prostacyclin from the endothelium (see Fig. 1).[11,12] Furthermore, thrombin, the major enzyme of the coagulation cascade, stimulates the formation of nitric oxide by the endothelium.[13] Hence, at sites where platelets and the coagulation cascade are activated, intact endothelial cells immediately release nitric oxide and in turn cause vasodilation and platelet inhibition, thereby preventing vasoconstriction and thrombus formation.

In addition to nitric oxide, prostacyclin is released by endothelial cells in response to shear stress, hypoxia, and several substances (see above) that also release nitric oxide. Prostacyclin increases cyclic 3′,5′-adenosine monophosphate (cAMP) in smooth muscle and platelets.[14] Its platelet inhibitory effects are probably more important than its contribution to endothelium-dependent relaxation. In platelets nitric oxide and prostacyclin synergistically inhibit platelet aggregation, suggesting that the activity of both mediators is required to exert full antiplatelet activity.[15]

In the coronary circulation, not all endothelium-dependent relaxation is prevented by inhibitors of the L-arginine pathway.[16] Nitric oxide-independent relaxation is even more prominent in intramyocardial vessels.[17] Because vascular smooth muscle cells become hyperpolarized under these conditions, an endothelium-dependent hyperpolarizing factor of unknown chemical structure has been proposed (see Fig. 1).[18] Indirect evidence suggests that it may be a product of the cytochrome P450 pathway, but C-type natriuretic peptide (CNP) also may be a candidate.

Endothelium-derived Contracting Factors

Soon after the discovery of endothelium-derived relaxing factors, it became clear that endothelial cells also can mediate contraction, at least under certain conditions[1,19] (see Fig. 1). Endothelium-derived contracting factors include the 21-amino acid peptide endothelin, vasoconstrictor prostanoids (such as thromboxane A_2 and prostaglandin H_2), and components of the renin-angiotensin system.

Endothelin exists in three isoforms: endothelin-1, endothelin-2, and endothelin-3. Endothelial cells produce exclusively endothelin-1.[20] Translation of messenger RNA generates preproendothelin, which is converted to big endothelin; its conversion to the mature peptide endothelin-1 by the endothelin converting enzyme (ECE) is necessary for the development of full vascular activity. Two isoforms of ECE have recently been cloned.[21,22] The expression of messenger RNA and the release of the peptide is stimulated by thrombin, transforming growth factor β_1, interleukin-1, epinephrine, angiotensin II, arginine vasopressin, calcium ionophore, and phorbol ester[20,23] (see Fig. 1).

Endothelin-1 causes vasodilation at lower concentrations and marked, sustained contractions at higher concentrations.[20,24] In the heart such contractions eventually lead to ischemia, arrhythmias, and death. Intramyocardial vessels are more sensitive to the vasoconstrictor effects of endothelin-1 than epicardial coronary arteries, suggesting that the peptide is particularly important in the regulation of flow. In general, veins are more sensitive to endothelin-1 than arteries.

The circulating levels of endothelin-1 are quite low, suggesting that little of the peptide is formed physiologically. This may be due to the absence of stimuli for endothelin production, the presence of potent inhibitory mechanisms, or the preferentially abluminal release of endothelin toward smooth muscle cells.[25] Three inhibitory mechanisms regulating endothelin production have been delineated: (1) cGMP-dependent inhibition,[23] (2) cAMP-dependent inhibition,[26] and (3) an inhibitory factor produced by vascular smooth muscle cells.[27] After inhibition of the endothelial L-arginine pathway, the thrombin- or angiotensin-induced endothelin production is augmented;[23] on the other hand, nitrates and atrial natriuretic peptide (which activates particulate guanylyl cyclase) prevent thrombin-induced endothelin release via a cGMP-dependent mechanism. Endothelin also can release nitric oxide and prostacyclin from endothelial cells, which as a negative feedback mechanism reduce endothelin production in the endothelium and its vasoconstrictor action in smooth muscle.

Two distinct endothelin receptors exist, the ET_A- and ET_B-receptor.[28,29] Both are G-protein–coupled receptors with seven transmembrane domains and are linked to phospholipase C and protein kinase C. Endothelial cells express ET_B-receptors linked to the formation of nitric oxide and prostacyclin, which explains the transient vasodilator effects of endothelin when infused in intact organs or organisms. In vascular smooth muscle ET_A- and in part ET_B-receptors mediate contraction and proliferation. Several endothelin receptor antagonists have been developed and currently are under evaluation in normal subjects and patients.[30] Particularly in veins but also in the cerebral and ophthalmic circulation, agonists such as arachidonic acid, acetylcholine, histamine, and serotonin can evoke endothelium-dependent contractions that are mediated by thromboxane A_2 or prostaglandin H_2[1] (see Fig. 1). Thromboxane A_2 and prostaglandin H_2 activate the thromboxane receptor in vascular smooth muscle cells and platelets and hence counteract the effects of nitric oxide and prostacyclin in both cells. In addition, the cyclooxygenase pathway is a source of superoxide anions that inactivate nitric oxide (see Fig. 1). Besides this indirect effect, superoxide anions can also directly cause vasoconstriction.

Finally, the endothelium regulates the activity of the renin-angiotensin system; the angiotensin-converting enzyme (ACE), which activates angiotensin I into angiotensin II, is expressed on the endothelial cell membrane (see Fig. 1).[31] ACE is identical to kinase II, which breaks down bradykinin. Whether or not other components of the renin angiotensin system are produced in endothelial cells is controversial. Angiotensin II can activate endothelial angiotensin receptors; these receptors stimulate production of endothelin and possibly of other mediators, such as plasminogen activator inhibitor.

Endothelium-dependent Control of Vascular Structure

Removal of the endothelium—for example, mechanically by a balloon catheter—invariably leads to immediate deposition of platelets and white blood cells and, after days to weeks, to intimal hyperplasia at the site of injury.[32] This suggests that the endothelium also regulates vascular structure and that its presence ensures quiescence of vascular smooth muscle cells (Fig. 2). Endothelial dysfunction, on the other hand, may be an important factor in atherosclerosis, restenosis, and hypertensive vascular disease. Vascular structure is determined mainly by vascular smooth muscle cells and, in disease states, by white blood cells invading the intima. Endothelial cells have indirect and direct effects on vascular structure. Nitric oxide and prostacyclin inhibit the adhesion of platelets to the vessel wall.[33] If at sites of endothelial dysfunction or denudation platelets do adhere to the blood vessel wall, they cause contraction (through the release of thromboxane A_2 and serotonin)[12] and stimulate proliferation and migration of vascular smooth muscle cells (via the release of platelet-derived growth factor).[34] In addition, nitric oxide inhibits the adhesion of monocytes, which are an important component of atherosclerotic plaque and also are capable of releasing growth factors and cytokines.

Furthermore, endothelial cells are a source of growth promoters and inhibitors. It is thought that under physiologic conditions growth inhibitors prevail; this may explain why the blood vessel wall normally is quiescent and does not exhibit proliferative responses (see Fig. 2). Heparane sulfates, nitric oxide, and transforming growth factor β_1 are potent inhibitors of vascular smooth muscle migration and proliferation.[35–37] On the other hand, at least under certain conditions, endothelial cells can produce various growth factors, in particular platelet-derived growth factor and epidermal growth factor[38,39] (see Fig. 2). This may be important in disease states, such as atherosclerosis, in which the endothelium remains intact but may contribute to proliferative responses.

Effects of the Sympathetic Nervous System

The sympathetic nervous system (SNS) has important effects on the regulation of vascular tone and growth. Via stimulation of α_1-adrenoceptors on vascular smooth muscle cells, mediators of the SNS (i.e., norepinephrine and epinephrine) induce vasoconstriction (Figs. 1 and 3). On the other hand, stimulation of β_2-adrenoceptors induces vasodilation (see Figs. 1 and 3). The distribution and density of adrenoceptor subtypes depend on the circulatory area involved and determine which effect—vasoconstriction or vasodilation—predominates.

Of importance, the SNS interacts on various levels with other systems—including the vascular endothelium, vascular smooth muscle cells, and the renin-angiotensin system—to regulate vascular tone, blood pressure, and cardiac performance (see Fig. 3). On the level of the vascular endothelium, it stimulates the release of nitric oxide; stimulation of α_2-adrenoceptors induces concentration-dependent relaxation only in the presence of an intact endothelium[40] (see Figs. 1 and 3). Furthermore, inhibition of NO synthesis by L-NMMA enhances norepinephrine-induced vasoconstriction[40] (see Figs. 1 and 3). On the other hand, the endothelium-derived vasocontrictor endothelin potentiates the vasoconstrictor effects of norepinephrine[41] (see Figs. 1 and 3). Angiotensin II centrally activates the SNS; this effect can be inhibited by ACE inhibitors in humans under in vivo conditions[42] (see Fig. 3).

Volume and pressor receptors in the carotic bulb and lung control SNS activity; in case of a drop in blood pressure, SNS activation counteracts decreases in peripheral vascular resistance (see Fig. 3). This explains the fact that pure vasodilators such as nitrates and calcium antagonists (see below) activate the SNS.[42,43]

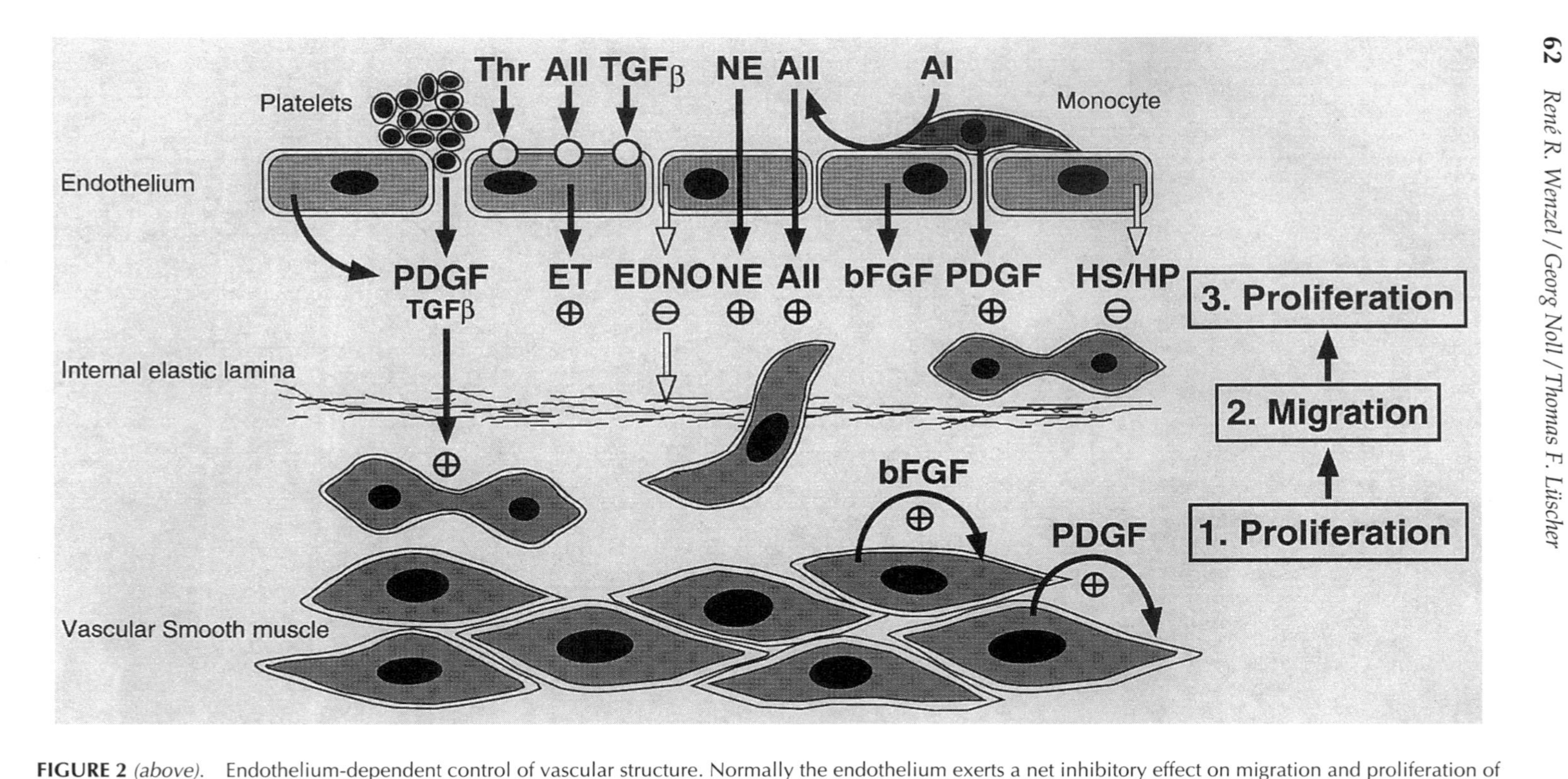

FIGURE 2 *(above).* Endothelium-dependent control of vascular structure. Normally the endothelium exerts a net inhibitory effect on migration and proliferation of vascular smooth muscle cells. With endothelium dysfunction, platelets and monocytes adhere to the vessel wall; growth factors are released by these cells as well as the endothelium. A = angiotensin; bFGF = basic fibroblast growth factor; ET = endothelin-1; NE = norepinephrine, NO = nitric oxide; HP/HS = heparan sulfates; PDGF = platelet-derived growth factor; $TGF\beta_1$ = transforming growth factor β_1; Thr = thrombin.

FIGURE 3 *(facing page).* Effects of the sympathetic nervous system (SNS) on vascular tone and interactions with other systems regulating vascular tone, i.e., the renin-angiotensin system (RAS), vascular endothelium, and vascular smooth muscle cells. $\alpha/\beta_{1/2}$ = adrenoceptor subtypes, AT II = angiotensin II, Ach = acetylcholine, ET = endothelin, PGI2 = prostacyclin; M = muscarinic receptor; NE = norepinephrine; NO = nitric oxide.

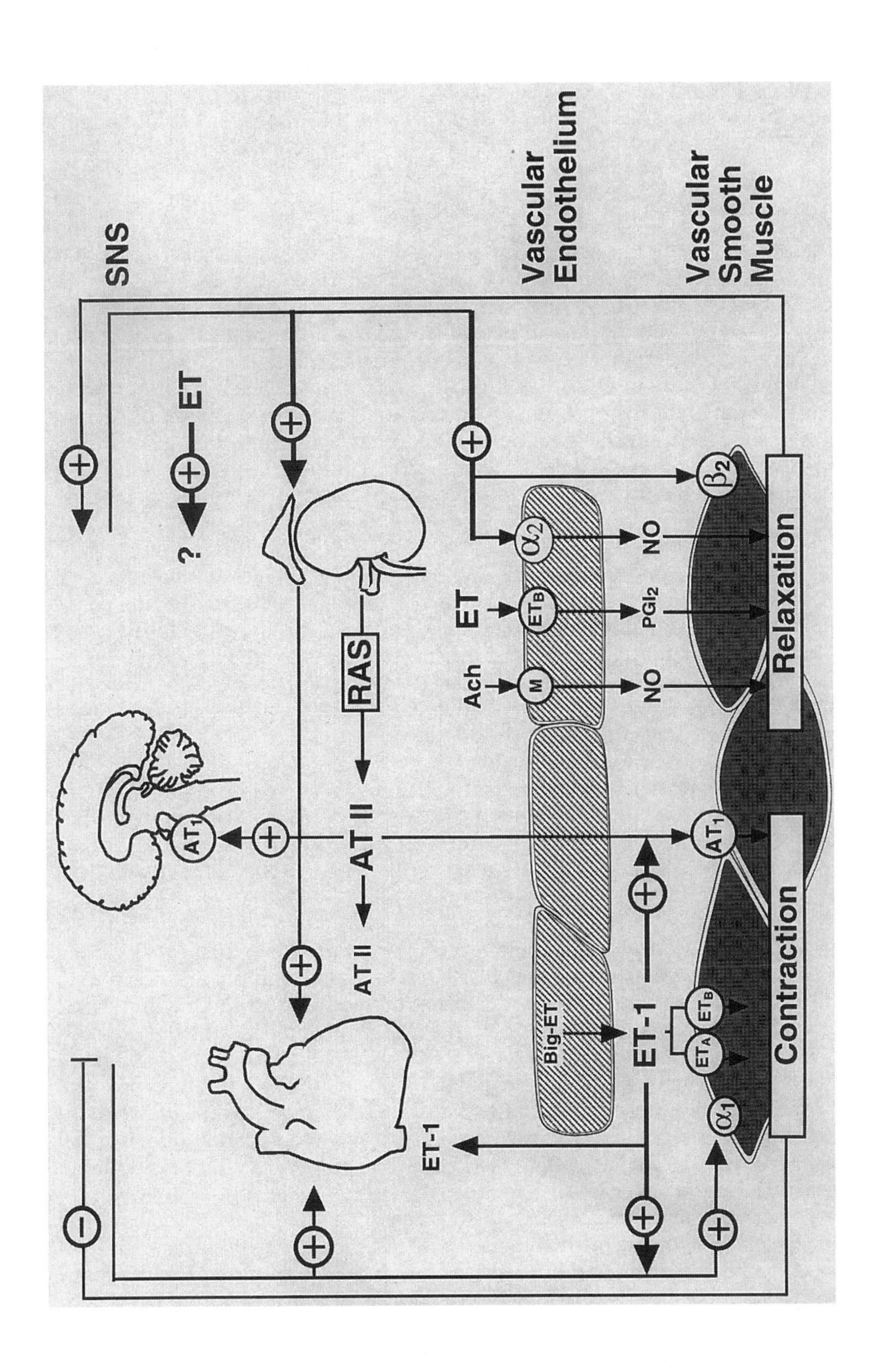
SNS
Vascular Endothelium
Vascular Smooth Muscle
?
ET
RAS
AT II
AT II
AT₁
Ach
M
ET
ET_B
α₂
NO
PGI₂
NO
β₂
Relaxation
AT₁
Big-ET
ET-1
ET_A
ET_B
α₁
Contraction
ET-1

Finally, mediators of the SNS are potent stimulators of vascular smooth muscle cell proliferation and cardiac hypertrophy[44] (see Fig. 2). This effect seems to be mediated via α_1-adrenoceptors on vascular smooth muscle cells, which trigger phosphoinositide hydrolysis and activate the mitogen-activating protein (MAP) kinase pathway, thus leading to DNA synthesis and cell proliferation.[44]

Pathophysiology

The endothelium is the structure of the blood vessel wall that is most exposed to the mechanical forces of blood and hormones and noxious substances therein. Morphologic studies have demonstrated changes in endothelial cell morphology with aging and disease, in particular increased endothelial cell turnover and density, a marked heterogeneity in endothelial cell size, and bulging of the cells into the lumen. Endothelial cell denudation, however, does not occur except in very late stages of atherosclerosis and plaque rupture. Almost invariably associated with these changes in endothelial cell morphology are functional alterations and intimal thickening with accumulation of white blood cells, vascular smooth muscle cells and fibroblasts, and matrix deposition.

Aging

All forms of of cardiovascular disease increase in frequency with age even in the absence of known cardiovascular risk factors, suggesting that aging per se alters vascular function. In most studies, endothelium-dependent relaxation decreased with aging. In humans, the increase in coronary flow induced by acetylcholine infusion decreases with age.[45] Direct measurement of nitric oxide has demonstrated a selective attenuation in the arterial but not the pulmonary circulation with age, most likely due to downregulation of nitric oxide synthase and increased superoxide production.[46]

Endothelin production was found to increase with age in some but not all studies.[47] The response to endothelin, however, decreases with age, presumably because of receptor downregulation. Little is known about the role of the antithrombotic and antiproliferative properties of the endothelium in aging.

Hypertension

Endothelial dysfunction in hypertension may contribute to the increase in peripheral vascular resistance (particularly if it occurs in resistant arteries) and to the vascular complications of the disease (if present in large and medium-sized conduit arteries). In most models of hypertension, high blood pressure is associated with reduced endothelium-dependent relaxation.[1] This defect is more dominant in some than other blood vessels and appears to occur as blood pressure rises; hence it is a consequence rather than a cause of hypertension. In hypertensive patients, acetylcholine causes paradoxical vasoconstriction of epicardial coronary arteries. The increase in blood flow due to acetylcholine was found to be reduced in the forearm and coronary circulation in all but one study.[48–50]

The mechanism of endothelial dysfunction differs in different models of hypertension. In spontaneously hypertensive rats, the activity of nitric oxide synthase is markedly increased but inefficacious, probably because of an increased deactivation of nitric oxide (Fig. 4).[51] In addition, the endothelium of spontaneously hypertensive and ren-2 transgenic rats produces increased amounts of

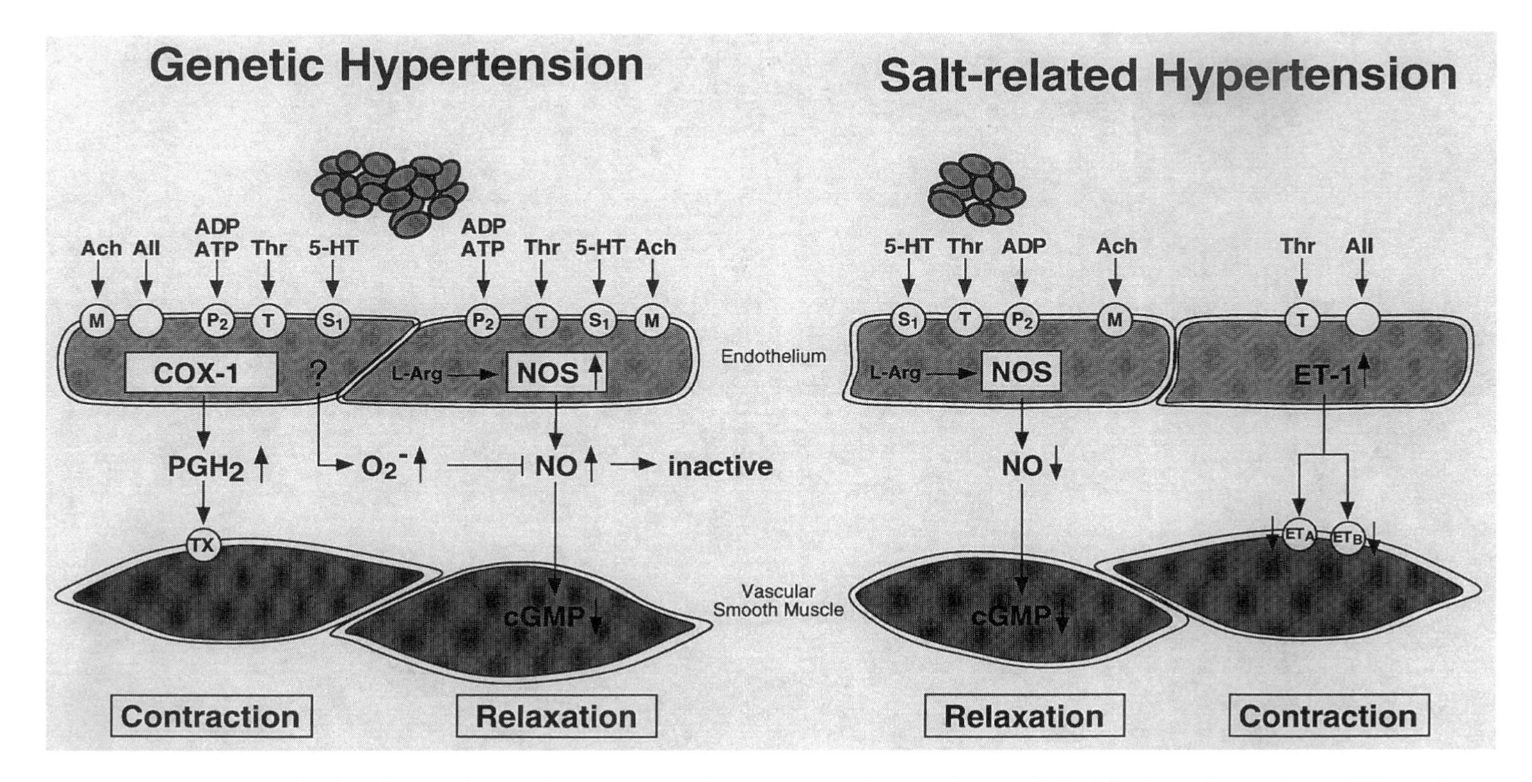

FIGURE 4. Heterogeneity of endothelium dysfunction in hypertension. In spontaneously hypertensive rats (SHR; *left*) nitric oxide synthase (NOS) activity is increased, but the biologic activity of nitric oxide (NO) is reduced, possibly because of inactivation. In addition, the production of thromboxane A_2 (TXA_2) and prostaglandin H_2 (PGH_2) via cyclooxygenase (COX-1) is increased. In contrast, in salt-related hypertension (Dahl rats, Sabra rat, DOCA-salt hypertension; *right*), NO production is reduced, but no TXA_2 or PGH_2 is produced. Endothelin (ET-1) production is increased in DOCA-salt hypertension but reduced in SHR. For key to abbreviations, see Figure 1.

prostaglandin H_2, which offset the effects of nitric oxide in vascular smooth muscle and platelets.[1,52] Whether this occurs in humans is uncertain; however, in the forearm circulation of patients with essential hypertension infusion of a cyclooxygenase inhibitor such as indomethacin enhances the vasodilation due to acetylcholine.[53] In rats with salt-induced hypertension, nitric oxide production is likely to be reduced[54] (see Fig. 4).

Endothelin plasma levels are normal in most patients with hypertension except in the presence of renal failure and atherosclerosis. Increased local vascular production of endothelin, however, remains a possibility because most of the peptide is released abluminally.[25] Hence plasma levels do not necessarily reflect local tissue levels of endothelin. Vascular endothelin production is reduced in spontaneously hypertensive rats but increased in DOCA-salt hypertensive rats,[55] Dahl salt-sensitive rats, and angiotensin II-induced hypertension.[56]

SNS activity, as assessed by microneurographic measurement of muscle sympathetic activity, correlates with age in subjects with evidence of no cardiovascular disease except hypertension.[57] However, the role of the SNS in hypertension is controversial. Most likely, the SNS plays an important role in the development of hypertension, because in early stages of hypertension (e.g., normotensive offspring of hypertensive parents and borderline hypertension) disregulation of SNS activity occurs.[58,59] This increased SNS activity may promote increases in vascular stretch and possibly enhance vascular proliferation.

Hyperlipidemia and Atherosclerosis

Endothelium-dependent relaxation is reduced in hyperlipidemia and atherosclerosis.[60] It is likely that low-density lipoproteins are a major determinant of this phenomenon. Indeed, incubation of isolated coronary arteries with oxidized but nonnative low-density lipoproteins selectively inhibits endothelium-dependent relaxation due to serotonin, aggregating platelets, and thrombin, whereas the response to bradykinin is unaffected.[61] A similar reduction of the response can be achieved by pertussis toxin or an inhibitor of nitric oxide formation, suggesting that activation of the L-arginine pathway by G_i protein-coupled receptors becomes defective[61,62] (Fig. 5). Because exogenous L-arginine improves or restores the reduced endothelium-dependent relaxation in the presence of oxidized low-density lipoproteins, it is possible that reduced intracellular availability of L-arginine also contributes to the effect. The active component of low-density lipoproteins appears to be lysolecithine, which mimicks most of the effects of low-density lipoproteins. In line with in vitro experiments, selective dysfunction of endothelium-dependent relaxation due to serotonin, aggregating platelets, and thrombin also is observed in the coronary artery of pigs with hypercholesteremia. In more advanced stages of atherosclerosis, endothelial dysfunction is more generalized. Experiments in the hypercholesteremic rabbit aorta suggest that the overall production of nitric oxide is not reduced but markedly augmented; however, nitric oxide is inactivated by superoxide radicals produced within the endothelium[63] (see Fig. 5). Similar observations have been made in rabbits with fully developed atherosclerosis. Under both conditions, the biologically active nitric oxide is markedly reduced, a fact also supported by bioassay experiments with coronary arteries of hypercholesteremic pigs.[64]

Endothelin may be important in atherosclerotic vascular disease. In hyperlipidemia and atherosclerosis vascular endothelin production is increased[65]

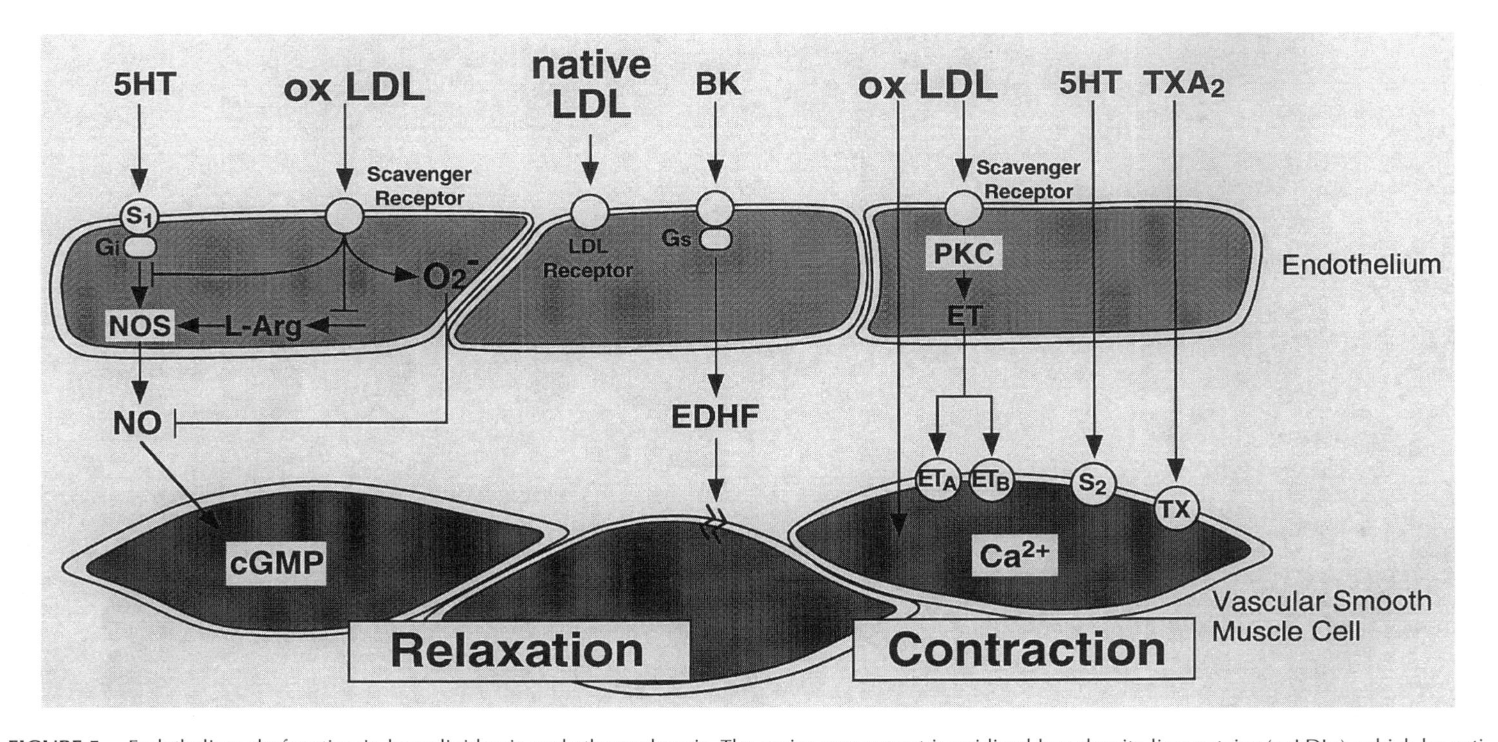

FIGURE 5. Endothelium dysfunction in hyperlipidemia and atherosclerosis. The major component is oxidized low-density lipoproteins (oxLDLs), which by activating scavenger receptors impair the activity of the L-arginine-nitric oxide (NO) pathway. The mechanism may involve inactivation of G_i proteins (G_i), decreased intracellular availability of L-arginine (L-Arg), and increased breakdown of NO by superoxide (O_2^-). OxLDLs further activate endothelin (ET) gene expression and production via protein kinase C (PKC). For key to abbreviatons, see Figure 1.

(see Fig. 5), whereas the expression of endothelin receptors is downregulated.[66] A most likely stimulus for the increased endothelin production is low-density lipoproteins, which increase endothelin gene expression and release of the peptide from porcine and human aortic endothelial cells[67] (see Fig. 5). Because these effects are specific for oxidized low-density lipoproteins and not shared by native low-density lipoproteins, it appears that the oxidation process in atherosclerotic human arteries is crucial. In addition to endothelial cells, vascular smooth muscle cells, particularly cells that migrated into the intima during the atherosclerotic process, also produce endothelin. In cultured vascular smooth muscle cells, endothelin can be released by growth factors, such as platelet-derived growth factor and transforming growth factor β_1, as well as by vasoconstrictors such as arginine vasopressin.[68] Hence, several mediators involved in atherosclerosis stimulate vascular endothelin production. This may explain why plasma endothelin levels are increased and positively correlated with the extent of the atherosclerotic process.[65] Furthermore, particularly unstable lesions removed from coronary arteries by atherectomy exhibit marked staining for endothelin-1.[69] It is possible that local vascular endothelin contributes to abnormal coronary vasomotion in patients with unstable angina. Triggers of endothelin production in patients with acute coronary syndromes may include ischemia and thrombin. In atherosclerosis endothelin may contribute to hypervasoconstriction and proliferation of vascular smooth muscle cells.

EFFECTS OF CALCIUM ANTAGONISTS ON VASCULAR TONE AND GROWTH

Because of the importance of the endothelium and the vascular smooth muscle cells in cardiovascular disease, endothelial effects of drugs currently used in clinical practice or under development are of great interest. Drugs can directly affect endothelial function, prevent the action of endothelial mediators, substitute for deficient endothelial factors, or indirectly exert protective effects by interfering with cardiovascular risk factors.

Mechanisms of Action of Calcium Antagonists

Calcium antagonists inhibit inflow of extracellular calcium through voltage-operated channels into the smooth muscle cells and thereby cause vasodilation. Three main classes of calcium antagonists have been delineated: (1) phenylalkylamines (prototype: verapamil), (2) dihydropyridines (prototype: nifedipine), and (3) benzothiazepines (prototype: diltiazem). All substances interfere with different parts of the voltage-operated calcium channels, thereby decreasing influx of extracellular calcium into cells—in particular, vascular smooth muscle cells. Increases in intracellular calcium are a basic biologic signal mechanism involved in numerous intracellular processes. In the context of coronary artery disease, increases in intracellular calcium are particularly important in platelets (leading to platelet activation), vascular smooth muscle cells (leading to vasoconstriction), and endothelial cells (leading to the release of vasoactive substances).

Effects of Calcium Antagonists on Endothelium-dependent Vasodilation

The formation of nitric oxide from L-arginine via the enzyme nitric oxide synthase[70] is associated with increases in intracellular calcium.[3,71,72] Indeed, under most conditions the calcium ionophore A23187 is a potent activator of the release of this relaxing factor. Although increases in intracellular calcium are most likely important for the release of nitric oxide,[71] calcium antagonists have not yet convincingly been shown to stimulate NO release. This is not surprising, as endothelial cells lack L-type calcium channels. Nevertheless, other mechanisms may be activated by the drugs. Indeed, mibefradil, which blocks both L-type and T-type calcium channels, seems to elicit endothelium-dependent relaxation.[73,74]

However, calcium antagonists facilitate the effects of nitric oxide at the level of vascular smooth muscle. Indeed, nitric oxide exerts its effects via guanylyl cyclase, leading to increases in intracellular concentrations of cGMP.[75] Cyclic GMP lowers intracellular calcium (most likely by increasing calcium efflux and reuptake of calcium into intracellular stores) and dephosphorylates myosin light chains. In vessels treated with calcium antagonists, endothelium-dependent relaxation and relaxation due to the endothelium-independent vasodilator sodium nitroprusside (which, like nitric oxide, causes relaxation via cGMP) are augmented in line with the previous interpretation of the effects of calcium antagonists.[73,76]

Chronic treatment with the calcium antagonist verapamil restores endothelium-dependent relaxation in experimental (i.e., L-NAME–induced) hypertension[77] (Fig. 6). Furthermore, in spontaneously hypertensive rats, endothelium-dependent relaxation is improved not only by a dihydropyridine calcium antagonist but also by other antihypertensive drugs, such as angiotensin II receptor antagonists or ACE inhibitors[78] (Fig. 7). In patients with essential hypertension, chronic treatment with nifedipine for 6 months markedly augments endothelium-dependent vasodilation to acetylcholine, whereas the response to sodium nitroprusside remains unaffected.[79]

Effects of Calcium Antagonists on Endothelium-dependent Vasoconstriction

In certain blood vessels, the endothelin receptors on vascular smooth muscle cells are linked to voltage-operated calcium channels via G_i-proteins.[80] This linkage may explain why calcium antagonists reduce endothelin-induced vasoconstriction in some vessels, such as porcine coronary arteries. Calcium antagonists are similarly active in the mammary artery, suggesting that the same mechanism is operative.[81] In other blood vessels, such as the human internal mammary artery,[82] most of the contractile effects induced by endothelin are mediated by release of intracellular calcium through activation of phospholipase C and formation of inositol trisphosphate and diacylglycerol.[83,84] This process may explain why calcium antagonists of all classes are unable to prevent endothelin-induced contractions in the internal mammary artery.[82] On the other hand, in most blood vessels, calcium antagonists are able to reverse endothelin-induced contractions.[82] The most likely explanation is the fact that endothelin lowers membrane potential of vascular smooth muscle cells[85] and

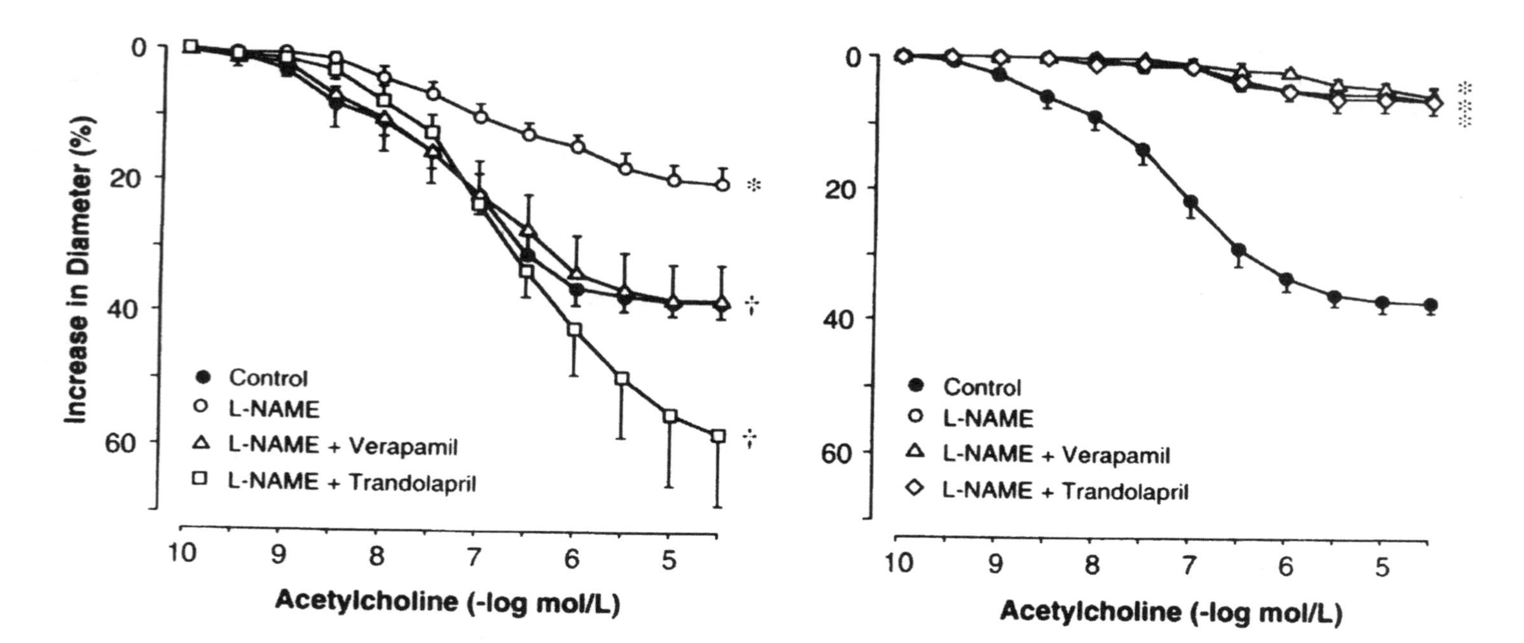

FIGURE 6. Effects of chronic antihypertensive treatment on endothelium-dependent relaxation as assessed by concentration-dependent relaxation due to acetylcholine in experimental hypertension (i.e., L-NAME-induced hypertension). Note that chronic *(left panel)* but not acute *(right panel)* treatment with a calcium antagonist or ACE-inhibitor restored endothelium-dependent relaxation. (Modified from Takase H, Moreau P, Küng CF, et al: Antihypertensive therapy prevents endothelial dysfunction in chronic nitric oxide deficiency. Effect of verapamil and trandolapril. Hypertension 27:25–31, 1996.)

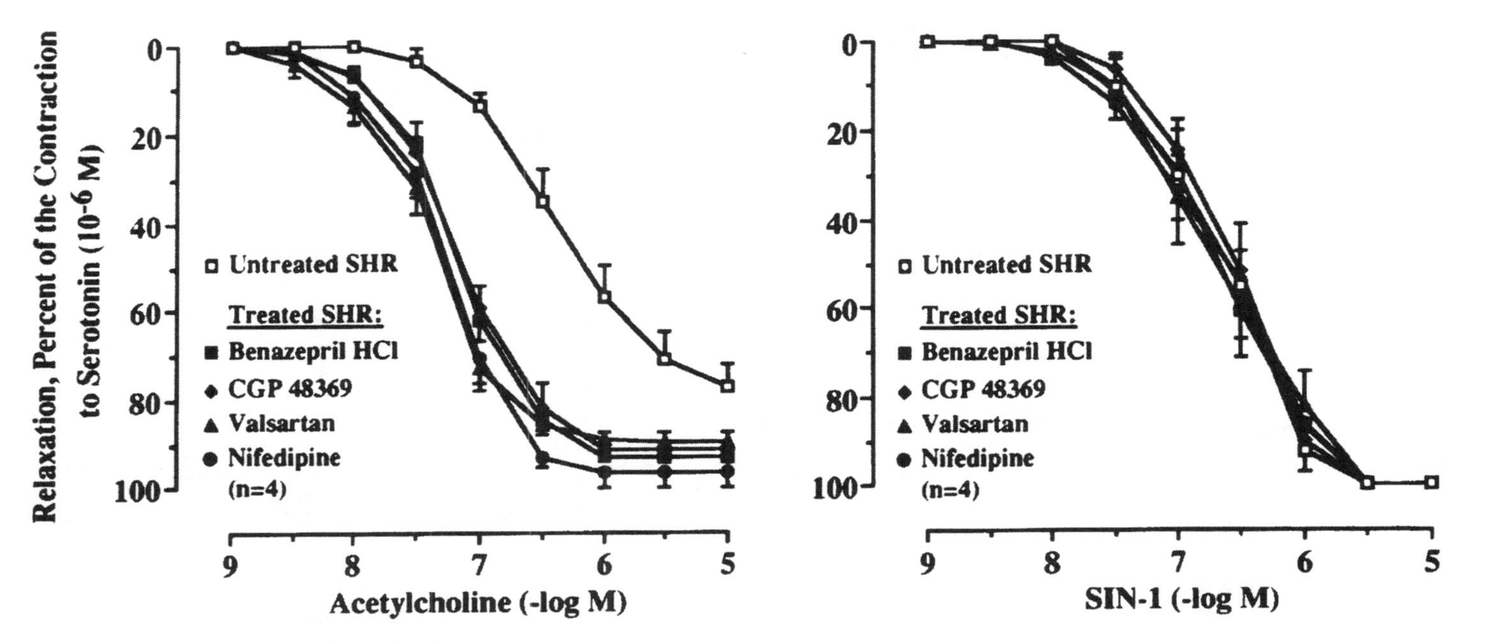

FIGURE 7. Effects of chronic antihypertensive treatment on endothelium-dependent relaxation as assessed by concentration-dependent relaxation due to acetylcholine in spontaneously hypertensive rats. Antihypertensive treatment with the calcium antagonist nifedipine and with other antihypertensive drugs (e.g., the ACE inhibitor benazepril, the angiotensin-II receptor antagonists CGP 48369 and valsartan) restored both sensitivity and potency of the endothelium-dependent vasodilator acetylcholine; in contrast, endothelium-independent relaxation with SIN-1 remains unaffected (Modified from Tschudi MR, Criscione L, Novosel D, et al: Antihypertensive therapy augments endothelium-dependent relaxations in coronary arteries of spontaneously hypertensive rats. Circulation 89:2212–2218, 1994.)

thereby opens voltage-operated calcium channels. Hence, once contractions have developed, calcium antagonists exert an inhibitory effect.

In vivo, calcium antagonists such as verapamil or nifedipine inhibit endothelin-1–induced contractions in the forearm circulation of healthy volunteers[24] (Fig. 8). In patients with coronary artery disease, high oral doses of the calcium antagonist diltiazem partially prevent endothelin-1–induced vasodilation in the skin microcirculation[86] (Fig. 9). However, local application of the calcium antagonist and, to a greater degree, specific inhibition of endothelin with an endothelin antagonist, are much more potent.[86] Chronic therapy with nifedipine for 6 months blunts contractions due to endothelin-1 in patients with essential hypertension.[87]

On the other hand, at least the slow-release formulation of the calcium antagonist nifedipine increases plasma endothelin levels.[43] Although the mechanism is unknown, under pathophysiologic conditions this effect may induce unwanted side effects. Future studies must evaluate the effects of calcium antagonists on plasma endothelin and the mechanism(s) involved.

Effects of Calcium Antagonists on Vascular Growth

Blood vessels with endothelial damage or functionally altered endothelial cells show enhanced platelet adhesion. Because platelets are an important source of platelet-derived growth factor (PDGF), which is a potent mitogen,[34] this effect may promote vascular smooth muscle cell proliferation. Indeed, in cultured smooth muscle cells derived from human coronary arteries, PDGF induces

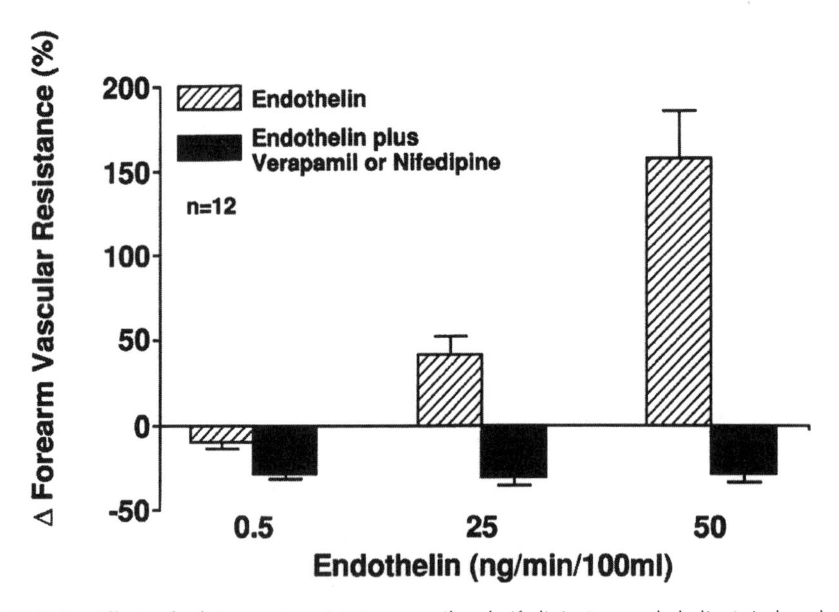

FIGURE 8. Effects of calcium antagonists (verapamil and nifedipine) on endothelin-1–induced vasoconstriction in the human forearm circulation. Note that endothelin-1 induces a marked increase in vascular resistance, which can be fully prevented by either of the calcium antagonists. (Modified from Kiowski W, Lüscher TF, Linder L, Bühler FR: Endothelin-1–induced vasoconstriction in humans. Reversal by calcium channel blockade but not by nitrovasodilators or endothelium-derived relaxing factor. Circulation 83:469–475, 1991.)

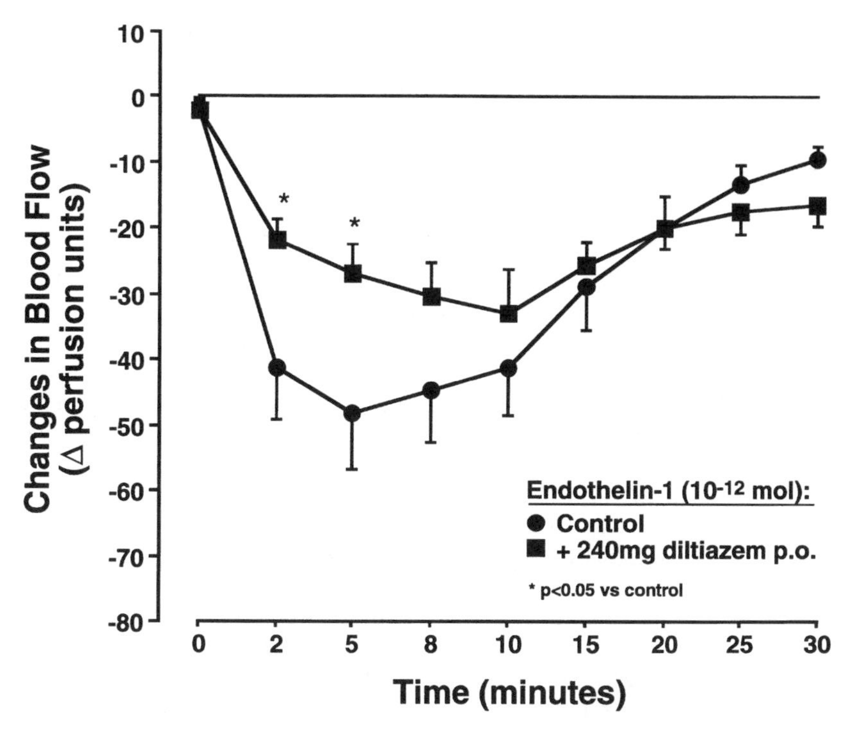

FIGURE 9. Effects of systemic application of the calcium antagonist diltiazem (240 mg orally) on endothelin-1–induced vasoconstriction (10^{-12} mol endothelin-1 intradermally). Note that diltiazem moderately inhibits endothelin-1–induced vasoconstriction. (Modified from Wenzel RR, Duthiers N, Noll G, et al: Endothelin and calcium antagonists in the skin microcirculation of patients with coronary artery disease. Circulation 94:316–322, 1996.)

marked proliferative responses with a half maximal concentration of 0.1–0.5 ng/ml.[88] The effect of PDGF involves activation of tyrosine kinase and autophosphorylation of PDGF receptors.[34,89] Of particular interest is the fact that a calcium antagonist such as verapamil in a concentration that may be achieved with therapeutic doses in vivo inhibits PDGF-induced proliferative responses in human coronary vascular smooth muscle cells.[88] In contrast, proliferative responses induced by other mitogen stimuli, such as pulsatile stretch, are unaffected by either verapamil or nifedipine. This contrast indicates that mechanical forces such as pulsatile stretch and receptor-operated growth factors such as PDGF operate via distinct mechanisms. The fact that neither the activity of tyrosine kinase nor activation of S6 kinase or MAP kinase is affected by calcium antagonists suggests that they interact at more distal intracellular signal transduction mechanisms.[90]

In cholesterol-fed rabbits several classes of calcium antagonists inhibit the atherosclerotic process.[91] Because in this model atherosclerosis is mainly related to monocyte invasion and foam cell formation, calcium antagonists also may interfere with this important process of atherosclerosis. Three large angiographic trials using nifedipine or nicardipine have demonstrated that chronic therapy with a calcium antagonist reduces the development of new lesions in the coronary circulation without affecting already existing stenoses.[92–94] These clinical studies have confirmed that calcium antagonists have an antiatherosclerotic action, although it is less pronounced compared with the angiographic results of lipid-lowering therapy.[95–97] The clinical relevance of these findings in terms of the incidence and mortality rate of myocardial infarction remains to be established.

Effects of Calcium Antagonists on the Sympathetic Nervous System

Calcium antagonists are potent vasodilators and thus may stimulate the baroreflex. Indeed, acute administration of a dihydropyridine calcium antagonist strongly activates muscle sympathetic nerve activity and increases heart rate in healthy young volunteers[43] (Fig. 10). The activation of peripheral sympathetic activity is independent of drug release formulation; the slow-release formulation of the same drug induces similar increases in muscle sympathetic nerve activity.[43] However, heart rate does not increase with the slow–release formulation, indicating a differential activation of cardiac and peripheral sympathetic activity[43] (see Fig. 10). Thus, heart rate cannot be used as an index of SNS activity, because peripheral sympathetic activation may occur despite an increase in heart rate.

The effects of other calcium antagonists on muscle sympathetic activity are not yet studied. However, studies assessing SNS activity indirectly (e.g., by measuring plasma catecholamines, forearm blood flow, or insulin sensitivity) show marked differences in the behavior of the various classes. Verapamil-type calcium antagonists do not increase plasma norepinephrine levels and reduce heart rate, whereas other dihydropyridine calcium antagonists, such as felodipine and amlodipine, seem to increase SNS activity, although this finding is controversial.[98,99] Thus, prospective chronic studies assessing the differential effects of different classes of calcium antagonist as the primary endpoint in patients with cardiovascular diseases are necessary. In diseases with already enhanced SNS activity, such as coronary artery disease, heart failure, and possibly hypertension, calcium antagonists with neutral or possibly inhibitory effects on SNS should be preferred.

Acknowledgment

The original research of the authors reported in this chapter was supported by grants from the Swiss National Research Foundation (32-32541.91, 32-32655.91, SCORE 32-35591.92) and the Swiss Heart Foundation. R.R. Wenzel is the recipient of a fellowship from the German Research Association (DFG, WE 1772/1-1).

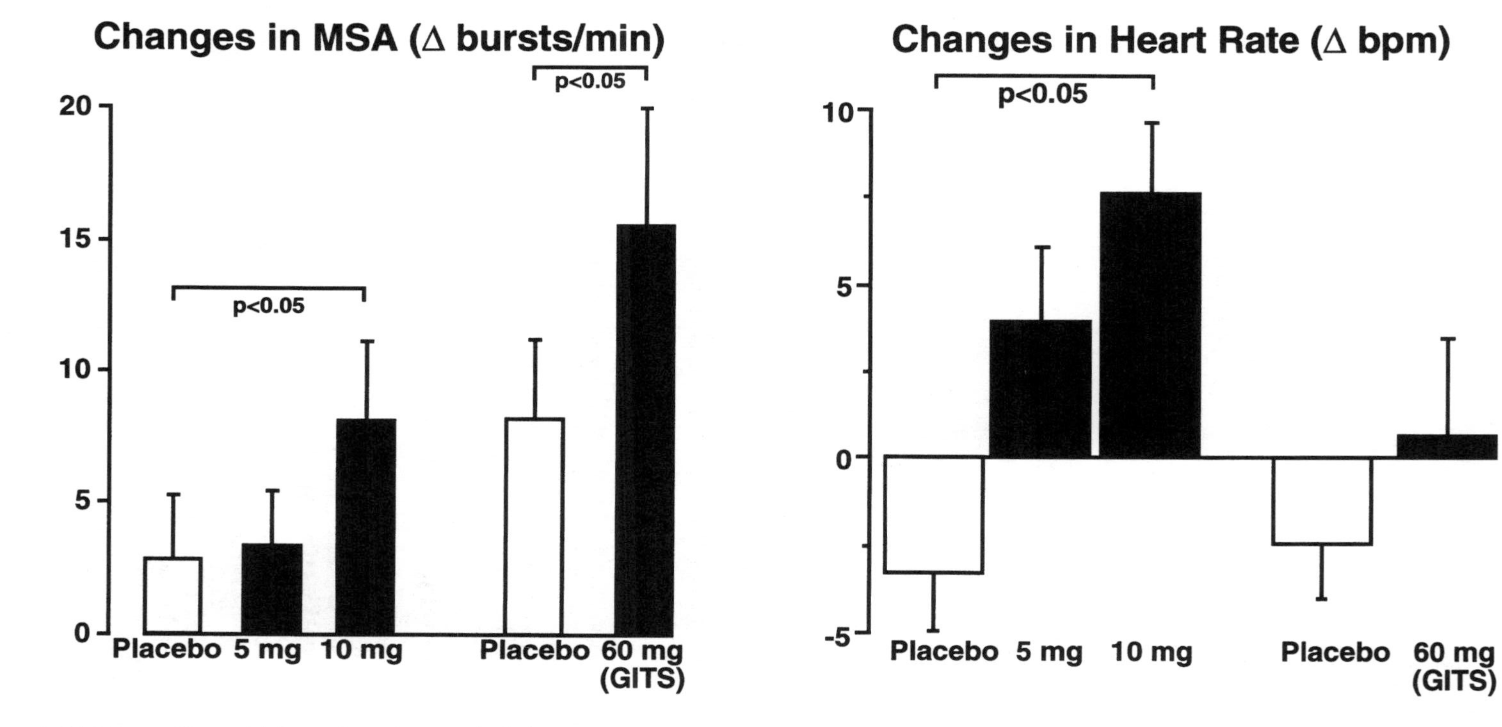

FIGURE 10. Effects of different formulations of nifedipine on muscle sympathetic nerve activity (*left panel,* bursts/min) and heart rate (*right panel,* beats/min) in young, healthy volunteers. Nifedipine activates peripheral sympathetic nerve activity independently of drug release formulation, whereas heart rate increases only with the short-acting compound. (Modified from Wenzel RR, Allegranza G, Binggeli C, et al: Differential activation of cardiac and peripheral sympathetic nervous system by nifedipine: Role of pharmacokinetics. J Am Coll Cardiol 29:1607–1614, 1997.)

References

1. Lüscher TF, Vanhoutte PM: The Endothelium: Modulator of Cardiovascular Function. Boca Raton, FL, CRC Press, 1990, pp 1–215.
2. Furchgott RF, Zawadzki JV: The obligatory role of endothelial cells in the relaxation of arterial smooth mucle by acetylcholine. Nature 299:373–376, 1980.
3. Palmer RM, Ashton DS, Moncada S: Vascular endothelial cells synthesize nitric oxide from L-arginine. Nature 333:664–666, 1988.
4. Bredt DS, Hwang PM, Snyder SH: Localization of nitric oxide synthase indicting a neural role for nitric oxide. Nature 347:768–770, 1990.
5. Vallance P, Leone A, Calver A, et al: Accumulation of an endogenous inhibitor of nitric oxide synthesis in chronic renal failure. Lancet 339:572–575, 1992.
6. Wright CE, Rees DD, Moncada S: Protective and pathological roles of nitric oxide in endotoxin shock. Cardiovasc Res 26:48–57, 1992.
7. Rees DD, Palmer RM, Schulz R, et al: Characterization of three inhibitors of endothelial nitric oxide synthase in vitro and in vivo. Br J Pharmacol 101:746–752, 1990.
8. Yang ZH, von Segesser L, Bauer E, et al: Different activation of the endothelial L-arginine and cyclooxygenase pathway in the human internal mammary artery and saphenous vein. Circ Res 68:52–60, 1991.
9. Rees DD, Palmer RMJ, Moncada S: Role of endothelium-derived nitric oxide in the regulation of blood pressure. Proc Natl Acad Sci USA 86:3375–3378, 1989.
10. Rapoport RM, Draznin MB, Murad F: Endothelium-dependent relaxation in rat aorta may be mediated through cyclic GMP-dependent protein phosphorylation. Nature 306:174–176, 1983.
11. Cohen RA, Shepherd JT, Vanhoutte PM: Inhibitory role of the endothelium in the response of isolated coronary arteries to platelets. Science 221:273–274, 1983.
12. Yang Z, Stulz P, Von SL, et al: Different interactions of platelets with arterial and venous coronary bypass vessels. Lancet 337:939–943, 1991.
13. Lüscher TF, Diedrich D, Siebenmann R, et al: Difference between endothelium-dependent relaxation in arterial and in venous coronary bypass grafts. N Engl J Med 319:462–467, 1988.
14. Moncada S, Vane VR: Pharmacology and endogenous roles of prostaglandin endoperoxides, thromboxane A2, and prostacyclin. Pharmacol Rev 30:293–331, 1979.
15. Radomski MW, Palmer RM, Moncada S: Comparative pharmacology of endothelium-derived relaxing factor, nitric oxide and prostacyclin in platelets. Br J Pharmacol 92:181–187, 1987.
16. Richard V, Tanner FC, Tschudi M, Lüscher TF: Different activation of L-arginine pathway by bradykinin, serotonin, and clonidine in coronary arteries. Am J Physiol 259:H1433–H1439, 1990.
17. Tschudi M, Richard V, Bühler FR, Lüscher TF: Importance of endothelium-derived nitric oxide in porcine coronary resistance arteries. Am J Physiol 260:H13–H20, 1991.
18. Vanhoutte PM: Vascular physiology: The end of the quest? [news]. Nature 327:459–460, 1987.
19. De Mey JG, Vanhoutte PM: Heterogeneous behavior of the canine arterial and venous wall. Importance of the endothelium. Circ Res 51:439–447, 1982.
20. Yanagisawa M, Kurihara H, Kimura S, et al: A novel potent vasoconstrictor peptide produced by vascular endothelial cells. Nature 332:411–415, 1988.
21. Ohnaka K, Takayanagi R, Nishikawa M, et al: Purification and characterization of a phosphoramidon-sensitive endothelin-converting enzyme in porcine aortic endothelium. J Biol Chem 268:26759–26766, 1993.
22. Xu D, Emoto N, Giaid A, et al: ECE-1: A membrane-bound metalloprotease that catalyzes the proteolytic activation of big endothelin-1. Cell 78:473–485, 1994.
23. Boulanger C, Lüscher TF: Release of endothelin from the porcine aorta. Inhibition of endothelium-derived nitric oxide. J Clin Invest 85:587–590, 1990.
24. Kiowski W, Lüscher TF, Linder L, Bühler FR: Endothelin-1-induced vasoconstriction in humans. Reversal by calcium channel blockade but not by nitrovasodilators or endothelium-derived relaxing factor. Circulation 83:469–475, 1991.
25. Wagner OF, Christ G, Wotja J, et al: Polar secretion of endothelin-1 by cultured endothelial cells. J Biol Chem 267:16066–16068, 1992.
26. Yokokawa K, Kohno M, Yasunari K, et al: Endothelin-3 regulates endothelin-1 production in cultured human endothelial cells. Hypertension 18:304–315, 1991.
27. Stewart DJ, Langleben D, Cernacek P, Cianflone K: Endothelin release is inhibited by coculture of endothelial cells with cells of vascular media. Am J Physiol 259:H1928–H1932, 1990.
28. Arai H, Hori S, Aramori I, et al: Cloning and expression of a cDNA encoding an endothelin receptor. Nature 348:730–732, 1990.

29. Sakurai T, Yanagisawa M, Takuwa Y, et al: Cloning of a cDNA encoding a non-isopeptide-selective subtype of the endothelin receptor. Nature 348:732–735, 1990.
30. Lüscher TF: Do we need endothelin antagonists? Cardiovasc Res 27:2089–2093, 1993.
31. Ng KK, Vane JR: Conversion of angiotensin I to angiotensin II. Nature 216:762–766, 1967.
32. Baumgartner HR, Studer A: Gezielte Ueberdehnung der Aorta abdominalis am normo- und hypercholesterinaemischen Kaninchen. Pathol Microbiol (Basel) 26:129–148, 1963.
33. Radomski MW, Palmer RM, Moncada S: The anti-aggregating properties of vascular endothelium: Interactions between prostacyclin and nitric oxide. Br J Pharmacol 92:639–646, 1987.
34. Ross R: The pathogenesis of atherosclerosis: A perspective for the 1990s. Nature 362:801–809, 1993.
35. Garg UC, Hassid A: Nitric oxide-generating vasodilators and 8-bromo-cyclic guanosine monophosphate inhibit mitogenesis and proliferation of cultured rat vascular smooth muscle cells. J Clin Invest 83:1774–1777, 1989.
36. Battegay EJ, Raines EW, Seifert RA, et al: TGF-beta induces bimodal proliferation of connective tissue cells via complex control of an autocrine PDGF loop. Cell 63:515–524, 1990.
37. Castellot JJ Jr, Beeler DL, Rosenberg RD, Karnovsky MJ: Structural determinants of the capacity of heparin to inhibit the proliferation of vascular smooth muscle cells. J Cell Physiol 120:315–320, 1984.
38. Hannan RL, Kourembanas S, Flanders KC, et al: Endothelial cells synthesize basic fibroblast growth factor and transforming growth factor beta. Growth Factors 1:7–17, 1988.
39. DiCorleto PE, Hassid A: Growth factor production by endothelial cells. In Ryan U (ed): Endothelial Cells, vol. II. Boca Raton, FL, CRC Press, 1990, pp 51–62.
40. Liu SF, Crawley DE, Evans TW, Barnes PF: Endogenous nitric oxide modulates adrenergic neural vasoconstriction in guinea-pig pulmonary artery. Br J Pharmacol 104:565–569, 1991.
41. Yang Z, Richard V, non Segesser L, et al: Threshold concentrations of endothelin-1 potentiate contractions to norepinephrine and serotonin in human arteries. A new mechanism of vasospasm? Circulation 82:188–195, 1990.
42. Noll G, Wenzel RR, de Marchi S, et al: Differential effects of captopril and nitrates on muscle sympathetic nerve activity in volunteers. Circulation 95:2286–2292, 1997.
43. Wenzel RR, Allegranza G, Binggeli C, et al: Differential activation of cardiac and peripheral sympathetic nervous system by nifedipine: Role of pharmacokinetics. J Am Coll Cardiol 29: 1607–1614, 1997.
44. Yu SM, Tsai SY, Guh JH, et al: Mechanism of catecholamine-induced proliferation of vascular smooth muscle cells. Circulation 94:547–554, 1996.
45. Zeiher AM, Drexler H, Saubrier B, Just H: Endothelium-mediated coronary blood flow modulation in humans. Effects of age, atherosclerosis, hypercholesterolemia, and hypertension. J Clin Invest 92:652–662, 1993.
46. Tschudi MR, Barton M, Bersinger NA, et al: Effect of age on kinetics of nitric oxide release in rat aorta and pulmonary artery. J Clin Invest 98:899–905, 1996.
47. Lüscher TF, Wenzel RR, Noll G: Local regulation of the coronary circulation in health and disease: Role of nitric oxide and endothelin. Eur Heart J 16:51–58, 1995.
48. Linder L, Kiowsky W, Bühler FR, Lüscher TF: Indirect evidence for release of endothelium-derived relaxing factor in human forearm circulation in vivo. Blunted response in essential hypertension. Circulation 81:1762–1767, 1990.
49. Panza JA, Quyyumi AA, Brush JJ, Esptein SE: Abnormal endothelium-dependent vascular relaxation in patients with essential hypertension. N Engl J Med 323:22–27, 1990.
50. Cockcroft JR, Chowienczyk PJ, Benjamin N, Ritter JM: Preserved endothelium-dependent vasodilation in patients with essential hypertension. N Engl J Med 330:1036–1040, 1994.
51. Nava E, Leone AM, Wiklund NP, Moncada S: Detection of release of nitric oxide by vasoactive substances in the anesthesized rat. In Feelisch M, Busse R, Moncada S (eds): The Biology of Nitric Oxide. London, Portland Press, 1994, pp 179–181.
52. Küng CF, Lüscher TF: Different mechanisms of endothelial dysfunction with aging and hypertension in rat aorta. Hypertension 25:194–200, 1995.
53. Taddei S, Virdis A, Mattei P, Salvetti A: Vasodilation to acetylcholine in primary and secondary forms of human hypertension. Hypertension 21:929–933, 1993.
54. Lüscher TF, Vanhoutte PM, Raij L: Antihypertensive treatment normalizes decreased endothelium-dependent relaxations in rats with salt-induced hypertension. Hypertension 9(Suppl III):193–197, 1987.
55. Larivière R, Thibault G, Schiffrin EL: Increased endothelin-1 content in blood vessels of deoxycorticosterone acetate-salt hypertensive but not in spontaneously hypertensive rats. Hypertension 21:294–300, 1993.

56. d'Uscio LV, Moreau P, Shaw S, et al: Effects of chronic ETA-receptor blockade in angiotensin-II–induced hypertension. Hypertension 29:435–441, 1997.
57. Yamada Y, Miyajiama E, Tochikubo O: Age-related changes in muscle sympathetic activity in essential hypertension. Hypertension 13:870–877, 1989.
58. Noll G, Wenzel RR, Schneider M, et al: Increased activation of sympathetic nervous system and endothelin by mental stress in normotensive offspring of hypertensive parents. Circulation 93:866–869, 1996.
59. Anderson EA, Sinkey CA, Lawton WJ, Mark AL: Elevated sympathetic nerve activity in borderline hypertensive humans. Evidence from direct intraneural recordings. Hypertension 14:177–183, 1989.
60. Harrison DG, Freiman PC, Armstrong ML, et al: Alterations of vascular reactivity in atherosclerosis. Circ Res 61:II74–II80, 1987.
61. Tanner FC, Noll G, Boulanger CM, Lüscher TF: Oxidized low density lipoproteins inhibit relaxations of porcine coronary arteries. Role of scavenger receptor and endothelium-derived nitric oxide. Circulation 83:2012–2020, 1991.
62. Shimokawa H, Flavahan NA, Vanhoutte PM: Natural course of the impairment of endothelium-dependent relaxations after balloon endothelium removal in porcine coronary arteries. Possible dysfunction of a pertussis toxin-sensitive G protein. Circ Res 65:740–753, 1989.
63. Minor R Jr, Myers PR, Guerra R Jr, et al: Diet-induced atherosclerosis increases the release of nitrogen oxides from rabbit aorta. J Clin Invest 86:2109–2116, 1990.
64. Shimokawa H, Vanhoutte PM: Impaired endothelium-dependent relaxation to aggregating platelets and related vasoactive substances in porcine coronary arteries in hypercholesterolemia and atherosclerosis. Circ Res 64:900–914, 1989.
65. Lerman A, Edwards BS, Hallett JW, et al: Circulating and tissue endothelin immunoreactivity in advanced atherosclerosis. N Engl J Med 325:997–1001, 1991.
66. Winkles JA, Alberts GF, Brogi E, Libby P: Endothelin-1 and endothelin receptor mRNA expression in normal and atherosclerotic human arteries. Biochem Biophys Res Commun 191: 1081–1088, 1993.
67. Boulanger CM, Tanner FC, Bea ML, et al: Oxidized low density lipoproteins induce mRNA expression and release of endothelin from human and porcine endothelium. Circ Res 70:1191–1197, 1992.
68. Hahn AW, Resink TJ, Scott-Burden T, et al: Stimulation of endothelin mRNA and secretion in rat vascular smooth muscle cells: A novel autocrine function. Cell Regul 1:649–659, 1990.
69. Zeiher AM, Ihling C, Pistorius K, et al: Increased tissue endothelin immunoreactivity in atherosclerotic lesions associated with acute coronary syndromes. Lancet 344:1405–1406, 1994.
70. Furchgott RF, Zawadzki JV: The obligatory role of endothelial cells in the relaxation of arterial smooth muscle by acetylcholine. Nature 288:373–376, 1980.
71. Palmer RM, Ferrige AG, Moncada S: Nitric oxide release accounts for the biological activity of endothelium-derived relaxing factor. Nature 327:524–526, 1987.
72. Busse R, Fichtner H, Luckhoff A, Kohlhardt M: Hyperpolarization and increased free calcium in acetylcholine-stimulated endothelial cells. Am J Physiol 255:H965–H969, 1988.
73. Yang ZH, von SL, Bauer E, et al: Different activation of the endothelial L-arginine and cyclooxygenase pathway in the human internal mammary artery and saphenous vein. Circ Res 68:52–60, 1991.
74. Karila-Cohen D, Dubois-Rande JL, Giudicelli JF, Berdeaux A: Effects of mibefradil on large and small coronary arteries in conscious dogs: Role of vascular endothelium. J Cardiovasc Pharmacol 28:271–277, 1996.
75. Lüscher TF, Diederich D, Siebenmann R, et al: Difference between endothelium-dependent relaxation in arterial and in venous coronary bypass grafts. N Engl J Med 319:462–467, 1988.
76. Richard V, Tanner PC, Tschudi M, Lüscher TF: Different activation of L-arginine pathway by bradykinin, serotonin, and clonidine in coronary arteries. Am J Physiol 259:H1433–H1439, 1990.
77. Takase H, Moreau P, Küng CF, et al: Antihypertensive therapy prevents endothelial dysfunction in chronic nitric oxide deficiency. Effect of verapamil and trandolapril. Hypertension 27:25–31, 1996.
78. Tschudi MR, Criscione L, Novosel D, et al: Antihypertensive therapy augments endothelium-dependent relaxations in coronary arteries of spontaneously hypertensive rats. Circulation 89:2212–2218, 1994.
79. Taddei S, Noll G, Salvetti A, Lüscher TF: Unpublished observation, 1997.
80. Goto K, Kasuya Y, Matsuki N, et al: Endothelin activates the dihydropyridine-sensitive, voltage-dependent Ca(2+) channel in vascular smooth muscle. Proc Natl Acad Sci USA 86:3915–3918, 1989.

81. Godfraind T, Mennig D, Morel N, Wibo M: Effect of endothelin-1 on calcium channel gating by agonists in vascular smooth muscle. J Cardiovasc Pharmacol 13:S112–S117, 1989.
82. Yang Z, Bauer E, von SL, et al: Different mobilization of calcium in endothelin-1-induced contractions in human arteries and veins: Effects of calcium antagonists. J Cardiovasc Pharmacol 16:654–660, 1990.
83. Resink TJ, Scott-Burden T, Bühler FR: Endothelin stimulates phospholipase C in cultured vascular smooth muscle cells. Biochem Biophys Res Commun 157:1360–1368, 1988.
84. Wallnöfer A, Weir S, Ruegg U, Cauvin C: The mechanism of action of endothelin-1 as compared with other agonists in vascular smooth muscle. J Cardiovasc Pharmacol 13(Suppl 5): S23–S31, 1989.
85. Miller VM, Komori K, Burnett JC Jr, Vanhoutte PM: Differential sensitivity to endothelin in canine arteries and veins. Am J Physiol 257:H1127–H1131, 1989.
86. Wenzel RR, Duthiers N, Noll G, et al: Endothelin and calcium antagonists in the skin microcirculation of patietns with coronary artery disease. Circulation 94:316–322, 1996.
87. Salvetti A, Virdis A, Taddei S, et al: Trough:peak ratio of nifedipine gastrointestinal therapeutic systems and nifedipine retard in essential hypertensive patients: An Italian multicentre study. J Hypertens 14:661–667, 1996.
88. Yang Z, Noll G, Lüscher TF: Calcium antagonists differently inhibit proliferation of human coronary smooth muscle cells in response to pulsatile stretch and platelet-derived growth factor. Circulation 88:832–836, 1993.
89. Ulrich A, Schlessinger J: Signal transduction by receptors with tyrosine kinase activity. Cell 61:203–212, 1990.
90. Yang Z, Oemar BS, Lüscher TF: Unpublished observation, 1994.
91. Henry PD, Bentley KI: Suppression of atherogenesis in cholesterol-fed rabbit treated with nifedipine. J Clin Invest 68:1366–1369, 1981.
92. Loaldi A, Polese A, Montorsi P, et al: Comparison of nifedipine, propranolol and isosorbide dinitrate on angiographic progression and regression of coronary arterial narrowings in angina pectoris. Am J Cardiol 64:433–439, 1989.
93. Waters D, Lesperance J, Francetich M, et al: A controlled clinical trial to assess the effect of a calcium channel blocker on the progression of coronary atherosclerosis. Circulation 82:1940–1953, 1990.
94. Lichtlen PR, Hugenholtz P, Rafflenbeul W, et al: Retardation of angiographic progression of coronary artery disease by nifedipine. Lancet 335:1109–1113, 1990.
95. Brown BG, Zhao XQ, Sacco DE, Albers JJ: Arteriographic view of treatment to achieve regression of coronary atherosclerosis and to prevent plaque disruption and clinical cardiovascular events. Br Heart J 69:S48–S53, 1993.
96. Blankenhorn DH, Azen SP, Kramsch DM, et al: Coronary angiographic changes with lovastatin therapy. The Monitored Atherosclerosis Regression Study (MARS). The MARS Research Group. Ann Intern Med 119:969–976, 1993.
97. Dumont JM: Effect of cholesterol reduction by simvastatin on progression of coronary atherosclerosis: Design, baseline characteristics, and progress of the Multicenter Anti-Atheroma Study (MAAS). Control Clin Trials 14:209–228, 1993.
98. Kailasam MT, Parmer RJ, Cervenka JH, et al: Divergent effects of dihydropyridine and phenylalkylamine calcium channel antagonist classes on autonomic function in human hypertension. Hypertension 26:143–149, 1995.
99. de Courten M, Ferrari P, Schneider M, et al: Lack of effect of long-term amlodipine on insulin sensitivity and plasma insulin in obese patients wtih essential hypertension. Eur J Clin Pharmacol 44:457–462, 1993.
100. Kiowski W, Lüscher TF, Linder L, Bühler FR: Endothelin-1–induced vasoconstriction in humans. Reversal by calcium channel blockade but not by nitrovasodilators or endothelium-derived relaxing factor. Circulation 83:469–475, 1991.

FRANZ H. MESSERLI M.D. / LESZEK MICHALEWICZ M.D.

5

Cardiac Effects of Calcium Antagonists in Hypertension

Since Fleckenstein's pioneering observations that calcium antagonists have a cardioprotective effect against isoproterenol-induced myocardial necrosis,[1] various other effects of calcium antagonists on the heart have been identified. It is important to differentiate between the direct and indirect cardiac effects of calcium antagonists.[2,3] Most calcium antagonists exert negative inotropic and chronotropic effects in isolated myocyte preparations. In hypertensive patients, however, calcium antagonists, because of their powerful unloading effect, may acutely increase cardiac output and heart rate and thereby be perceived as positive inotropic and chronotropic agents.[4–8] Even verapamil has been shown to increase cardiac output and heart rate when given intravenously.[6] Analysis of the cardiac effects of calcium antagonists becomes even more difficult because normal and diseased myocardium may respond differently. In addition, the effects on afterload, left ventricular hypertrophy (LVH), coronary blood flow, left ventricular filling, contractility, and conduction vary from one calcium antagonist to another.[4–9] Calcium antagonists are a heterogenous group, and even different galenic formulations of the same molecule may have significantly different pharmacologic effects. Thus, every formulation should be evaluated separately in carefully designed clinical trials.[10] The following discussion is an attempt to analyze the effects of calcium antagonists on cardiac function and structure in patients with hypertension.

EFFECTS ON ARTERIAL PRESSURE

All calcium antagonists lower arterial pressure when given acutely and with prolonged administration.[4–9] Head-to-head comparisons of five different calcium antagonists reveal little, if any, difference in blood pressure-lowering potency among the various agents, provided that adequate doses are given.[4–9] The effect on arterial pressure seems to be somewhat more powerful in elderly or black patients, who are characterized by low activity of the renin-angiotensin system, than in young or white patients,[10,11] although distinct age and race dependency of the antihypertensive effects have not been documented by all studies.[10,11]

Because calcium antagonists predominantly affect arteriolar resistance vessels, orthostatic hypotension is rare.[4–9] In a recent study Oren et al. found the vasodilatory effect of calcium antagonists to be most pronounced with verapamil and less pronounced with isradipine and diltiazem.[12] Conceivably the greater efficacy of calcium antagonists in the elderly may be related to higher rates of pretreatment for blood pressure and decreased metabolic activity. In at least one study, much of the correlation between age and antihypertensive response was lost when corrections for pretreatment pressures and plasma levels were made.[11]

EFFECTS ON HEART RATE

Calcium antagonists lower arterial pressure by decreasing total peripheral resistance. As a consequence, reflexive tachycardia and an increase in cardiac output are commonly seen, particularly with the first dose. An increase in plasma catecholamine and plasma renin activity usually accompanies this positive chronotropic effect, indicating a reflexive increase in activity of the sympathetic nervous system.[4–9] We recently analyzed 63 clinical studies reporting the effects of calcium antagonists on heart rate and plasma norepinephrine levels.[13] Acutely, after single dosing, short-acting calcium antagonists increased heart rate and plasma norepinephrine levels. Change in norepinephrine levels correlated positively with change in heart rate and inversely with change in arterial pressure in patients taking dihydropiridine calcium antagonists acutely; change in norepinephrine levels was less pronounced with nondihydropyridines. With sustained therapy both classes of single-acting calcium antagonists increased norepinephrine levels; heart rate remained increased with dihydropyridines, whereas it decreased with nondihydropyridines. Heart rate and norepinephrine levels remained slightly elevated with long-acting dihydropyridines; both heart rate and norepinephrine levels decreased with long-acting nondihydropyridines. Similarly, long-term administration of isradipine, felodipine, and amlodipine appears to have little if any effect on heart rate and appears not to activate the sympathetic nervous system or renin-angiotensin system[5,8,9,14,15] (Table 1). Conceivably, the direct effect of these drugs on the sinus nodes prevents the reflexive tachycardia seen with other dihydropyridine antagonists.[16,17]

EFFECTS ON CARDIAC CONDUCTION

In pharmacologic doses, most calcium antagonists diminish automaticity of the sinus node, slow conduction in the atrioventricular (A-V) node, and have little if any effect on the automaticity of the myocytes.[18] However, these electrophysiologic properties vary a great deal from one drug to another and also appear to depend on the form of application.[18–20] Verapamil, when given intravenously, has a marked effect on the A-V node that can be used therapeutically in patients with supraventricular tachycardia.[21–24] In fact, transient A-V block is not an uncommon occurrence with intravenous verapamil.[18] In contrast, complete A-V block is exceedingly rare with oral verapamil. A slight prolongation of the PR interval to about the same extent can be seen with both verapamil and diltiazem, provided that equipotent oral antihypertensive

TABLE 1. Acute and Long-term Effects of Calcium Antagonists on Heart Rate, Contractility, and Plasma Levels of Catecholamines

	Acute Effects		Long-term Effects		
Drug	**Heart Rate**	**Contractility**	**Heart Rate**	**Contractility**	**Plasma Levels of Catecholamines**
Verapamil	↑	↓↓	↓↓	↓↓	↔
Diltiazem	↑	↓↓	↓	↓↓	↔
Gallopamil	↑	↓↓	↓	↓↓	↔
Mibefradil	↓	↔	↓	↔	↔↓
Nifedipine	↑↑	↓	↑	↓	↑ (?)
Nitrendipine	↑↑	↓	↑	↓	↑ (?)
Nicardipine	↑↑	↔	↑	↓	↑ (?)
Isradipine	↑↑	↔	↔	↔	↔
Felodipine	↑	↔	↔	↔	↔
Amlodipine	↑	↔	↔	↔	↔
Nisoldipine	↑	↔	↔	↔	↔

↑ = increased, ↓ = decreased, ↔ = unchanged.
From Messerli FH, Aepfelbacher FC: Cardiac effects of calcium antagonists in hypertension. In Messerli FH (ed): Cardiovascular Drug Therapy, 2nd ed. Philadelphia, W.B. Saunders, 1996, pp 908–915, with permission.

doses are used.[18] The dihydropyridine calcium antagonists generally have less effect on cardiac conduction than the nondihydropyridine calcium antagonists.[18]

EFFECTS ON CARDIAC CONTRACTILITY

By and large, calcium antagonists are negative inotropic agents and therefore are likely to impair cardiac pump function to some extent[2,3,24–34] (see Table 1). The most profound negative inotropic effect is seen with verapamil and diltiazem.[19,21–25] This direct effect is partially overridden by afterload reduction and its reflexive sympathetic drive, which are elicited by most dihydropyridine derivates. Some of the differences in the negative inotropic effect of various calcium antagonists may be due to differences in binding affinity ratios of smooth muscle to cardiac muscle.[34,35] The dihydropiridines, particularly newer agents, are more vascular-selective than the nondihydropyridines, such as verapamil and diltiazem.[36] Some of the newer agents, such as isradipine, felodipine, amlodipine, nisoldipine, and mibefradil, seem to have little if any negative inotropic effect. In fact, some of these drugs have been used successfully in patients with congestive heart failure caused by predominantly systolic dysfunction. Thus, favorable hemodynamic responses such as decreases in pulmonary wedge pressure and end-diastolic ventricular pressure were documented when isradipine, felodipine, and amlodipine were given over a short period.[37] In contrast to other dihydropyridine calcium antagonists, little or no stimulation of the sympathetic nervous system or renin-angiotensin system occurs with isradipine, felodipine, or amlodipine. Improvement of hemodynamics in patients with congestive heart failure, however, does not necessarily parallel increased survival rate. In the PRAISE study, amlodipine comparably reduced both sudden death and death from pump failure in patients with dilated but not ischemic cardiomyopathy.[38] In contrast, felodipine in the V-Heft study had no effect on morbidity and mortality.[39]

EFFECTS ON LEFT VENTRICULAR FILLING

In simplistic terms, left ventricular filling consists of an early diastolic active relaxation phase and a late diastolic passive distensibility phase. Calcium antagonists have been documented to have a favorable effect on both.[40,41] When given intravenously, they improve early diastolic relaxation,[41] which is impaired early in the course of essential hypertension.[41] This effect may be heart rate-dependent and appears to be most pronounced with verapamil, less pronounced with diltiazem, and even less so with dihydropyridine calcium antagonists. With prolonged administration of most calcium antagonists, late diastolic distensibility is improved as well. This effect may be related to a decrease in left ventricular wall thickness resulting from prolonged administration.[42,43] The clinical significance of decreased left ventricular filling in patients with hypertension and of its improvement by calcium channel blockade is unclear.[44] However, certain patients with long-standing hypertension have been documented to have latent or overt congestive heart failure, primarily as the result of impaired filling. In such patients, the nondihydropyridine calcium antagonists that lower heart rate, especially verapamil, are the drugs of choice because they not only lower arterial pressure but also improve ventricular filling.[45]

EFFECTS ON LEFT VENTRICULAR MASS AND MYOCARDIAL FIBROSIS

LVH has been identified as one of the strongest pressure-independent risk factors for sudden death, acute myocardial infarction, congestive heart failure, and other cardiovascular morbidity and mortality.[46–55] In fact, recently LVH was identified as the most powerful cardiovascular risk factor in humans. These epidemiologic findings clearly negate the concept that LVH is a benign adaptive process to compensate for the increased hemodynamic burden. Although it is unclear whether a reduction in LVH confers benefit over and above the benefit of lowering blood pressure, the effects of various antihypertensive agents on LVH have come under scrutiny.

Not all antihypertensive agents are equally effective in regressing LVH. In addition to lowering raised blood pressure, the effects on the sympathetic system, renin-angiotensin system, cardiac cell growth factors, arterial compliance, and the high-velocity reflected wave, which raises central systolic blood pressure, may play important roles.[56] Most calcium antagonists tend to reduce left ventricular mass when given for a sustained period[57–95] (Table 2). In a recent metaanalysis of 39 randomized, double-blind trials, calcium antagonists decreased left ventricular mass by 9% compared with a 7% decrease by diuretics,

TABLE 2. Reduction of Left Ventricular Mass with Calcium Antagonists

Author	Year	Drug	Duration (months)	Reduction (%)	Number of Patients
Muiesan et al.[58]	1986	Verapamil	3	17	22
Schmeider et al.[6]	1987	Verapamil	3	6	7
Schulman et al.[59]	1990	Verapamil	6	18	15
Granier et al.[60]	1990	Verapamil	3	26	13
Amodeo et al.[4]	1986	Diltiazem	1	10	10
Weiss and Bent[61]	1987	Diltiazem	6	19	17
Szlachic et al.[43]	1989	Diltiazem	4	15	15
Senda et al.[62]	1990	Diltiazem	6	2*	9
Gottdiener et al.[63]	1991	Diltiazem	2	5	77
van Leeuwen et al.[64]	1995	Diltiazem	6	2*	16
Leenen and Fourney[65]	1996	Diltiazem	6	5	13
Ferrara et al.[66]	1984	Nifedipine	2	0*	10
Muiesan et al.[58]	1986	Nifedipine	3	13	19
Phillips et al.[67]	1991	Nifedipine	12	19	16
Totteri et al.[68]	1993	Nifedipine	12	19	10
Yamakado et al.[69]	1994	Nifedipine	50	15	9
Kirpizidis et al.[70]	1995	Nifedipine	6	9	31
Myers et al.[71]	1995	Nifedipine	2	2*	52
Ferrara et al.[72]	1985	Nitrendipine	2	3*	20
Drayer et al.[73]	1986	Nitrendipine	12	3*	30
Giles et al.[74]	1987	Nitrendipine	2	10	7
Grossman et al.[9]	1988	Nitrendipine	3	13	17
Machnig et al.[75]	1994	Nitrendipine	8	15	43
Costantino et al.[76]	1988	Nicardipine	2	5	12
Gökçe et al.[77]	1990	Nicardipine	1	7	18
Sumimoto et al.[78]	1994	Nicardipine	20	13	10
Langan et al.[79]	1997	Nisoldipine	4	14	60
Kloner et al.[80]	1995	Amlodipine	10	24	13
Bignotti et al.[81]	1995	Amlodipine	12	4	10
Skoulargis et al.[82]	1995	Amlodipine	3	20	54
Leenen and Fourney[65]	1996	Amlodipine	6	9	17
Carr and Prisnant[83]	1990	Isradipine	1	9	9†
Grossman et al.[5]	1991	Isradipine	3	10	10
Saragoca et al.[84]	1991	Isradipine	3	14	17
Vyssoulis et al.[85]	1993	Isradipine	6	14	45
Manolis et al.[86]	1993	Isradipine	6	17	15

Table continued on following page.

TABLE 2. Reduction of Left Ventricular Mass with Calcium Antagonists *(Continued)*

Author	Year	Drug	Duration (months)	Reduction (%)	Number of Patients
Modena et al.[87]	1994	Isradipine	3	30	13
Galderisi et al.[88]	1994	Isradipine	3	11	10
Grandi et al.[89]	1995	Isradipine	6	14	18
Pringle et al.[90]	1989	Felodipine	6	24	14
Cerasola et al.[91]	1990	Felodipine	12	22	10
Wetzchewald et al.[92]	1992	Felodipine	9	20	36
Leenen and Holliwell[93]	1992	Felodipine	12	9	20
Myers et al.[71]	1995	Felodipine	2	6	56
Nalbantgil et al.[94]	1996	Felodipine	12	32	50

* = nonsignificant, † = responders only.
From Michalewicz L, Messerli FH: Cardiac effects of calcium antagonists in hypertension. Am J Cardiol (in press), with permission.

6% decrease by beta blockers, and 13% decrease by angiotensin-converting enzyme (ACE) inhibitors[96] (Fig. 1). A metaanalysis by Cruickshank et al.[56] indicated that the dihydropyridine calcium antagonists have a less powerful effect on left ventricular mass than the nondihydropyridine calcium antagonists. Thus, for any given decrease in arterial pressure, verapamil and diltiazem reduce left ventricular mass more than nifedipine, nitrendipine, or nicardipine. The benefits of short-acting dihydropyridines conceivably may be counteracted by the stimulatory effects in the sympathetic and renin-angiotensin systems. A limited number of studies of felodipine, isradipine, and amlodipine are

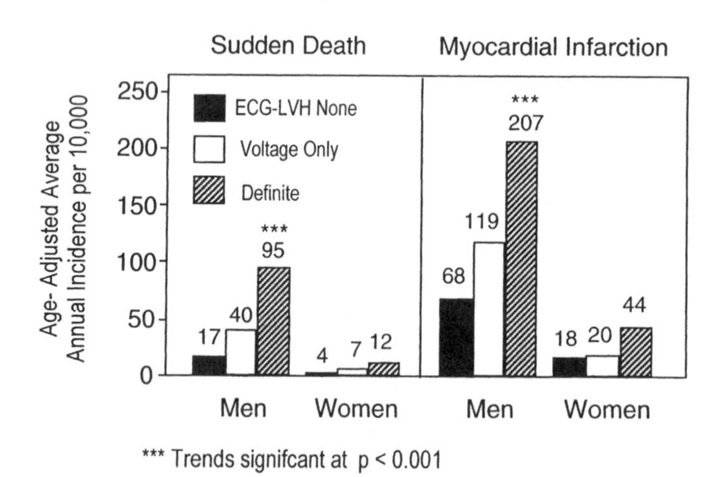

FIGURE 1. Risk of clinical manifestations of coronary disease according to electrocardiographic left ventricular hypertrophy status: 20-year follow-up, Framingham Study, subjects 54–74 years old. (From Kannel WB, Gordon T, Castelli WP, Margolis JR: Electrocardiographic left ventricuar hypertrophy and risk of coronary heart disease. The Framingham Study. Ann Intern Med 72:813–822, 1970, with permission.)

available, but each has been shown to decrease left ventricular mass (see Table 2). In a recent study by Langan et al., a new vascular-selective dihydropyridine calcium antagonist, nisoldipine, reduced left ventricular mass index with no deleterious effects on left ventricular systolic performance.[79] The decrease in left ventricular mass documented with calcium antagonists is associated with improved ventricular filling, diminished ventricular ectopy, and preserved contractility.[4–10,58–78, 80–95] In addition to myocyte growth, LVH is characterized by activation of nonmyocytic cells. Proliferation of fibroblasts that generate collagen, augmentation in local fibroblastic activity, and reduced local collagenolysis contribute to interstitial and perivascular fibrosis and coronary microangiopathy and macroangiopathy. Evidence from experimental studies indicates that nifedipine, nisoldipine, and amlodipine lead to regression of interstitial and perivascular myocardial fibrosis, which may contribute to the improvement of diastolic function and coronary reserve seen with calcium antagonists.[97–99] Thus, calcium antagonists reduce or prevent all pathophysiologic sequelae of LVH. For this reason, it seems a logical (although unproven) extrapolation that the melting away of a powerful risk factor, such as LVH, would be beneficial for hypertensive patients.

EFFECTS ON CORONARY BLOOD FLOW

All calcium antagonists are vasodilators and therefore increase coronary blood flow.[95,100–103] Coronary blood flow, however, is but one determinant of the myocardial oxygen supply-and-demand equilibrium.[104] Other determinants, such as heart rate, contractility, and arterial pressure, are also profoundly and variably affected by calcium antagonists. Thus, their overall effects on myocardial oxygenation depend on interplay of these mechanisms.[104–108] It should not be surprising, therefore, that in certain clinical situations some calcium antagonists may even have a detrimental effect on myocardial oxygenation. Thus, acute exacerbation of angina and even acute myocardial infarction were observed when arterial pressure was excessively lowered by short-acting nifedipine.[109–114] Because most coronary perfusion occurs in diastole, any increase in heart rate may reduce the oxygen supply of the heart. In addition, sudden dilation of nonischemic vessels may lead to steal from ischemic myocardium. With increased neurohumoral activity the steal phenomenon may result in enhanced platelet aggregation and lead to thrombus formation and occlusion of vessels. The present consensus is that short-acting dihydropyridines should not be used for treatment of coronary disease and hypertension.[115,116] Conversely, several studies attested to the safety and efficacy of longer-acting calcium antagonists.[37,38,117–123]

Calcium antagonists generally improve myocardial oxygenation by unloading the heart, increasing coronary blood flow, and reducing myocardial energy consumption.[100–103] Effects of calcium antagonists on LVH and left ventricular filling, which become manifest only after prolonged administration, also tend to improve the equilibrium between myocardial oxygen supply and demand.[4–9,35] These properties have made them useful in certain forms of angina pectoris, particularly when it is caused by vasospasm.[101,124] Thus, calcium antagonists control symptoms of vasospastic angina pectoris,[125,126] although no data confirm that they affect mortality rates. In chronic stable angina numerous small randomized clinical evaluations have attested to the efficacy of calcium antagonists in reducing diary-recorded anginal frequency and enhancing treadmill exercise performance.[127] The APSIS study, which compared verapamil with metoprolol in patients with stable angina pectoris, found no difference in morbidity and mortality rates between the two treatment groups.[118] Similarly, in comparing atenolol with nifedipine SR in patients with stable angina, Dargie et al. found no difference in morbidity and mortality.[123] It was hypothesized that calcium antagonists may improve systolic function after stunning or hibernation. The Doppler Flow and Echocardiography in Functional Cardiac Insufficiency Assessment of Nisoldipine Therapy (DEFIANT-II) study reported benefits in morbidity with nisoldipine in survivors of myocardial infarction.[117,119] Thus, in patients with chronic, stable angina pectoris long-acting calcium antagonists may be useful as an adjunctive treatment but not as initial monotherapy, unless beta blockers and nitrates are not tolerated.

REDUCTION IN REINFARCTION RATE (CARDIOPROTECTION)

After an acute myocardial infarction, the intact myocardium is the Achilles heel for further ischemia or necrosis. Whether this Achilles heel is amenable to medical therapy becomes a question of utmost importance. A distinct therapeutic benefit in patients after myocardial infarction has been documented with aspirin, beta blockers, ACE inhibitors, and, to a lesser extent, warfarin.

Initial studies with calcium antagonists yielded disappointing results in these patients.[128] The second Danish Verapamil Infarction Trial (DAVIT II) showed for the first time that verapamil reduced the reinfarction rate when therapy was started 1 week after the acute event.[129,130] A pooling of all patients in the DAVIT I and DAVIT II studies who survived day 7 after acute myocardial infarction showed significant decreases in mortality, major events, and reinfarction rate in verapamil-treated patients compared with patients receiving placebo.[131] The order of magnitude of these benefits was similar to that achieved in numerous beta-blocker trials. A subanalysis of the DAVIT II data showed that verapamil had little effect on ventricular arrhythmias but distinctly diminished ischemic ST-T changes. This finding indicates that the cardioprotective effect of verapamil is conferred by its antiischemic properties rather than its electrophysiologic effects.

In the recent study of Rengo et al., when verapamil was given to patients 1–3 weeks after myocardial infarction, a favorable trend for reduced rates of reinfarction and development of angina was observed.[132] A decrease in the reinfarction rate was previously documented with diltiazem in patients with non–Q-wave infarction only.[133]

Findings from the Multicenter Diltiazem Postinfarction Trial (MDIPT)[134] and both DAVIT trials clearly show that after myocardial infarction patients with congestive heart failure do not benefit from calcium channel blockade. In fact, in the MDIPT study, such patients had a higher mortality rate when treated with diltiazem than when receiving placebo. In contrast, in a recent pilot study of Fischer Hansen et al., verapamil reduced cardiac event rates when added to trandolapril and diuretic 3–10 days after the acute myocardial infarction in patients with congestive heart failure.[135]

Subanalysis of the MDIPT study showed that hypertension may enhance the ability of diltiazem to reduce reinfarction rate in postinfarction patients.[136] Subanalysis of DAVIT II data showed that angina pectoris and ST depression during exercise tests after myocardial infarction enhance the benefits of treatment with verapamil.[137–139]

Of note, a correlation between a decrease in heart rate and the decrease in reinfarction rate has been demonstrated for various beta blockers. Thus, the more negative the chronotropic effect of the beta blocker, the greater the benefit to the patient. Although such findings cannot be extrapolated directly to calcium antagonists, it is interesting that the calcium antagonists for which benefits have been documented are the heart rate-lowering agents.

At present, heart rate-lowering calcium antagonists should probably be reserved (1) for primary treatment for survivors of myocardial infarction without congestive heart failure who cannot tolerate a beta blocker or in whom a beta blocker is contraindicated and (2), as recent data suggest, for adjunctive treatment for survivors of myocardial infarction with congestive heart failure whose primary treatment includes an ACE inhibitor and diuretic.

ANTIATHEROMATOUS EFFECTS

Various experimental studies have documented that calcium antagonists exert antiatheromatous effects in certain animal models, such as cholesterol-fed rabbits.[140–147] Not all calcium antagonists have been shown to be equipotent in this regard.[140–150] Although these findings are provocative and have tremendous clinical appeal, they cannot be extrapolated directly to the clinical setting.[151] Nevertheless, some recent studies have shown that calcium antagonists may have a favorable influence on the progression of vascular disease in the coronary and carotid circulation[152–160] (Table 3).

HYPERTENSIVE HEART DISEASE

Hypertensive heart disease is characterized by LVH and its pathophysiologic sequelae: ventricular ectopy, myocardial ischemia, and impaired left ventricular systolic and diastolic function.[51,161,162] Numerous studies have shown that the classic adaptation to a sustained increase in arterial pressure consists of concentric LVH, that is, thickening of the wall at the expense of chamber volume. Eccentric LVH, defined as thickening of the chamber wall with concomitant chamber dilatation, occurs in the late phase of hypertensive heart disease and is a precursor of congestive heart failure. The observation of a poor correlation between arterial pressure and left ventricular mass indicates that the increased hemodynamic burden on the heart is not the sole determinant of left ventricular structure. Indeed, other factors (age, gender, race, obesity, salt intake, alcohol intake, job strain, neurohumoral factors) may independently influence the development of LVH.[163] Even before an increase in left ventricular mass can be identified, left ventricular filling (predominantly in the early relaxation phase) becomes impaired.[164] With established LVH, late diastolic filling also diminishes. The ventricle becomes progressively stiffer and therefore requires a

TABLE 3. Antiatheromatous Effects of Calcium Antagonists in Clinical Trials

Year	Study	Drug	Target Circulation	Duration (months)	Results
1990	Montreal Heart Institute Trial[152]	Nicardipine	Coronary	24	No significant effect
1990	INTACT[153]	Nifedipine	Coronary	36	Decrease in new lesions, no change in established lesions
1990	VAS[154]	Verapamil	Coronary	24	Decrease in restenosis rate after PTCA
1993	Heart transplant recipients[155]	Diltiazem	Coronary	24	Prevention of reduction of coronary artery diameters
1994	MIDAS[156]	Isradipine	Carotid	36	Slowed progression after 6 mos; no difference at end of study
1996	FIPS[157]	Verapamil	Coronary	36	Pending
1996	VHAS[158]	Verapamil	Carotid	36	Pending
1996	ELSA[159]	Lacidipine	Carotid	48	Pending

INTACT = International Nifedipine Trial on Antiatherosclerotic Therapy
VAS = Verapamil Angioplasty Study
MIDAS = Multicenter Isradipine/Diuretic Atherosclerosis Study (ongoing trial)
FIPS = Frankfurt Isoptin Progression Study
VHAS = Verapamil in Hypertension Atheroscerosis Study (ongoing trial)
ELSA = European Lacidipine Study on Atherosclerosis
PTCA = percutaneous transluminal coronary angioplasty

higher filling pressure to maintain a similar ejection fraction. Impaired filling usually is well compensated under resting conditions but may lead to symptoms and signs of congestive heart failure during exercise. Left ventricular contractility is usually well preserved in patients with LVH, although it has been shown to decline progressively as LVH progresses.[161,165–168]

LVH also unfavorably influences the myocardial oxygen supply-and-demand equilibrium. Various pathogenic mechanisms may be responsible. First, hypertension has been directly implicated in the pathogenesis of coronary atherosclerosis, which further impedes myocardial oxygen supply. Second, in hypertensive patients who have not yet developed LVH, exaggerated reactivity of cardiac arterioles has been documented, leading to the clinical picture of microvascular angina. Third, both an increase in arterial pressure, resulting from increased hemodynamic burden of the heart, and an increase in left ventricular mass require more oxygen for tissue perfusion. Finally, the growth of capillary beds in the hypertrophying myocardium does not keep pace with the increasing left ventricular mass.[54,55,161,162,164,168–170] Ventricular ectopy is common in patients with LVH and has been shown to occur in patients with concentric, eccentric, and isolated septal hypertrophy as well.[161,162,164,168–176] Although it is still controversial whether these findings can explain, at least in part, the higher incidence of sudden cardiac death in such patients, a study from the Framingham cohort indicates that in patients with LVH asymptomatic ventricular arrhythmias are indeed associated with a nearly twofold increase in mortality.[52,55,177,178] Calcium antagonists are particularly beneficial in patients with hypertensive heart disease. They not only diminish the hemodynamic burden and decrease left ventricular mass but also seem to confer specific benefit by diminishing ventricular ectopy[179] and improving myocardial oxygenation and left ventricular filling while conserving or improving contractile function. In regard to reduction of LVH, improvement of left ventricular filling, and reduction of ventricular dysrhythmias, verapamil and perhaps diltiazem seem to be somewhat preferable to other calcium antagonists. However, no head-to-head comparison between two different calcium antagonists has been done to document such clinical differences.

CONCLUSION

By definition, all antihypertensive drugs, including calcium antagonists, lower arterial pressure. In addition to lowering arterial pressure, however, calcium antagonists have a variety of beneficial effects in patients with hypertensive heart disease. They reduce LVH and improve its sequelae, such as ventricular dysrhythmias, impaired filling and contractility, and myocardial ischemia. Certain calcium antagonists reduce the reinfarction rate and have the potential for decreasing atherogenesis. Although efficacy in regard to some of these properties clearly varies from one calcium antagonist to the other, they are attractive first-choice agents for therapy in patients with essential hypertension, particularly those with cardiac involvement. Ongoing studies such as ALLHAT, HOT, STOP-2, and SYST-EUR[180–183] should give definite answers to questions about safety and efficacy of calcium antagonists and other drugs (such as alpha blockers, and ACE inhibitors) for which efficacy and safety have not yet been documented.

References

1. Fleckenstein A: Specific inhibitors and promoters of calcium action in the excitation-contraction coupling of heart muscle and their role in the prevention of myocardial lesions. In Harris P, Opie L (eds): Calcium and the Heart. Proceedings of the Meeting of the European Section of the International Study Group for Research in Cardiac Metabolism. New York, Academic Press, 1970, pp 135–188.
2. Kohlhardt M, Fleckenstein A: Inhibition of the slow inward current by nifedipine in mammalian ventricular myocardium. Naunyn Schmeidebergs Arch Pharmacol 298:267–272, 1977.
3. Nayler WG, Szeto J: Effect of verapamil on contractility, oxygen utilization, and calcium exchangeability in mammalian heart muscle. Cardiovasc Res 6:120–128, 1972.
4. Amodeo C, Kobrin I, Ventura HO, et al: Immediate and short-term hemodynamic effects of diltiazem in patients with hypertension. Circulation 73:108–113, 1986.
5. Grossman E, Messerli FH, Oren S, et al: Cardiovascular effects of isradipine in essential hypertension. Am J Cardiol 68:65–70, 1991.
6. Schmieder RE, Messerli FH, Garavaglia GE, Nunez BD: Cardiovascular effects of verapamil in patients with essential hypertension. Circulation 75:1030–1036, 1987.
7. Ventura HO, Messerli FH, Oigman W, et al: Immediate hemodynamic effects of a new calcium-channel blocking agent (nitrendipine) in essential hypertension. Am J Cardiol 51:783–786, 1983.
8. Little WC, Cheng CP, Elvelin L, Nordlander M: Vascular selective calcium entry blockers in the treatment of cardiovascular disorders: Focus on felodipine. Cardiovasc Drugs Ther 9:657–663, 1995.
9. Grossman E, Oren S, Garavaglia GE, Messerli FH, Frohlich ED: Systemic and regional hemodynamic and humoral effects of nitrendipine in essential hypertension. Circulation 78:1394–1400, 1988.

10. Wilson TW, Quest DW: Comparative pharmacology of calcium antagonists. Can J Cardiol 11:243–249, 1995.
11. Erne P, Bolli P, Bertel O, et al: Factors influencing the hypertensive effects of calcium antagonists. Hypertension 5(Suppl 4, Pt 2)II97–II102, 1983.
12. Oren S, Gossman E, Frohlich ED: Effects of calcium entry blockers on distribution of blood volume. J Hypertens 9:628–632, 1996.
13. Grossman E, Messerli FH: Calcium antagonists and sympathetic activity. Am J Cardiol, submitted.
14. Leonetti G, Gradnik R, Terzoli L, et al: Effects of single and repeated doses of the calcium antagonist felodipine in blood pressure, renal function, electrolytes and water balance, and renin-angiotensin-aldosterone system in hypertensive patients. J Cardiovasc Pharmacol 8:1243–1248, 1986.
15. Messerli FH, Aepfelbacher FC: Cardiac effects of calcium antagonists in hypertension. In Messerli FH (ed): Cardiovascular Drug Therapy, 2nd ed. Philadelphia, W.B. Saunders, 1996, pp 908–915.
16. Hossack KF: Conduction abnormalities due to diltiazem. N Engl J Med 307:953–954, 1982.
17. Taira N: Differences in cardiovascular profile among calcium antagonists. Am J Cardiol 59: 24B–29B, 1987.
18. Singh BN, Nademanee K: Use of calcium antagonists for cardiac arrhythmias. Am J Cardiol 59:153B–162B, 1987.
19. Angus JA, Richmond DR, Dhumma-Upakorn P, et al: Cardiovascular action of verapamil in the dog with particular reference to myocardial contractility and atrioventricular conduction. Cardiovasc Res 10:623–632, 1976.
20. Nakaya H, Schwartz A, Millard RW: Reflex chronotropic and inotropic effects of calcium channel-blocking agents in conscious dogs. Diltiazem, verapamil, and nifedipine compared. Circ Res 52:302–311, 1983.
21. Heng MK, Singh BN, Roche AHG, et al: Effects of intravenous verapamil on cardiac arrhythmias and on the electrocardiogram. Am Heart J 90:487–498, 1975.
22. Rinkenberger RL, Prystowsky EN, Heger JJ, et al: Effects of intravenous and chronic oral verapamil administration in patients with supraventricular tachyarrhythmias. Circulation 62:996–1010, 1980.
23. Singh BN, Nademanee K, Baky SH: Calcium antagonists: Clinical use in the treatment of arrhythmias. Drugs 25:125–153, 1983.
24. Sung RJ, Elser B, McAllister RG Jr: Intravenous verapamil for termination of re-entrant supraventricular tachycardias: Intracardiac studies correlated with plasma verapamil concentrations. Ann Intern Med 93:682–689, 1980.
25. Bonow RO, Leon MB, Rosing DR, et al: Effects of verapamil and propanolol on left ventricular systolic function and diastolic filling in patients with coronary artery disease: Radionuclide angiographic studies at rest and during exercise. Circulation 65:1337–1350, 1981.
26. Brooks N, Cattell M, Pidgeon J, Balcon R: Unpredictable response to nifedipine in severe cardiac failure. BMJ 281:1324, 1980.
27. Chew Cy, Hecht HS, Collett HT, et al: Influence of severity of ventricular dysfunction on hemodynamic responses to intravenously administered verapamil in ischemic heart disease. Am J Cardiol 47:917–922, 1981.
28. De Buitleir M, Rowland E, Krikler DM: Hemodynamic effects of nifedipine given alone or in combination with atenolol in patients with impaired left ventricular function. Am J Cardiol 55:15E–20E, 1985.
29. Elkayam U, Weber L, McKay C, Rahimtoola S: Spectrum of acute hemodynamic effects of nifedipine in severe congestive heart failure. Am J Cardiol 56:560–566, 1985.
30. Ferlinz J, Easthope JL, Aronow WS: Effects of verapamil on myocardial performance in coronary disease. Circulation 59:313–319, 1979.
31. Klein HO, Ninio R, Oren V, et al: The acute hemodynamic effects of intravenous verapamil in coronary artery disease. Assessment by equilibrium-gated radionuclide ventriculography. Circulation 67:101–110, 1983.
32. Lamping KA, Gross GJ: Differential effects of intravenous vs intracoronary nifedipine on myocardial segment function in ischemic canine hearts. J Pharmacol Exp Ther 228:28–32, 1984.
33. Serruys PW, Brower RW, ten Katen HJ, et al: Regional wall motion from radiopaque markers after intravenous and intracoronary injections of nifedipine. Circulation 63:584–591, 1981.
34. Thomas P, Sheridan DJ: Vascular selectivity of felodipine: Clinical experience. J Cardiovasc Pharmacol 15(Suppl 4):S17–S20, 1990.
35. Koenig W on behalf of the multicenter study group: Efficacy and tolerability of felodipine and amlodipine in the treatment of mild to moderate hypertension. Drug Invest 5:200–205, 1993.

36. Opie LH: Calcium channel antagonists in the treatment of coronary artery disease: Fundamental pharmacological properties relevant to clinical use. Prog Cardiovasc Dis 38:273–290, 1996.
37. Greenberg B, Siemienczuk D, Broudy D: Hemodynamic effects of PN 200-110 (isradipine) in congestive heart failure. Am J Cardiol 59:70B–74B, 1987.
38. O'Connor CM, Belkin RM, Carson PE, et al, for the PRAISE Investigators: Effect of amlodipine on mode of death in severe chronic heart failure: The PRAISE Trial. J Am Coll Cardiol 92(Suppl):143, 1995.
39. Cohn JN, Ziesche SM, Loss LE, et al, and the V-HeFT Study Group: Effect of felodipine on short-term exercise and neurohormones and long-term mortality in heart failure: Results of V-HeFT III. Circulation 82:1940–1953, 1990.
40. Inouye I, Massie B, Loge D, et al: Abnormal left ventricular filling: An early finding in mild to moderate systemic hypertension. Am J Cardiol 53:120–126, 1984.
41. Walsh RA, O'Rourke RA: Direct and indirect effects of calcium entry blocking agents on isovolumic left ventricular relaxation in conscious dogs. J Clin Invest 75:1426–1434, 1985.
42. Smith VE, White WB, Meeran MK, Karimeddini MK: Improved left ventricular filling accompanies left ventricular mass during therapy of essential hypertension. J Am Coll Cardiol 8:1449–1454, 1986.
43. Szlachic J, Tubau JF, Vollmer C, Massie BM: Effects of diltiazem on left ventricular mass and diastolic filling in mild to moderate hypertension. Am J Cardiol 63:198–201, 1989.
44. Piepho RW: Calcium antagonist use in congestive heart failure: Still a bridge too far? J Clin Pharmacol 35:443–453, 1995.
45. Hansen JF: ACE inhibitors and calcium antagonists in the treatment of congestive heart failure. Cardiovasc Drugs Ther 9(Suppl 3):503–507, 1995.
46. Aronow WS, Koenigsberg M, Schwartz KS: Usefulness of echocardiographic left ventricular hypertrophy in predicting new coronary events and atherothrombotic brain infarction in patients over 62 years of age. Am J Cardiol 61:1130–1132, 1988.
47. Casale BN, Devereux RB, Milner M, et al: Value of echocardiographic measurement of left ventricular mass in predicting cardiovascular morbid events in hypertensive men. Am J Hypertens 3:8–12, 1990.
48. Cooper RS, Simmons BE, Castaner A, et al: Left ventricular hypertrophy is associated with worse survival independent of ventricular function and the number of coronary arteries severely narrowed. Am J Cardiol 65:441–445, 1990.
49. Kannel WB: Prevalence and natural history of electrocardiographic left ventricular hypertrophy. Am J Med 75(Suppl 3A):4–11, 1983.
50. Kannel WB, Gordon T, Castelli WP, Margolis JR: Electrocardiographic left ventricular hypertrophy and risk of coronary heart disease. The Framingham Study. Ann Intern Med 72:813–822, 1970.
51. Koren MJ, Devereux RB, Casale PN, et al: Relation of left ventricular mass and geometry to morbidity and mortality in uncomplicated hypertension. Ann Intern Med 114:345–352, 1991.
52. Le Heuzey J-Y, Guize L: Cardiac prognosis in hypertensive patients. Incidence of sudden death and ventricular arrhythmias. Am J Med 84(Suppl 1B):65–68, 1988.
53. Levy D, Garrison RJ, Savage DD, et al: Left ventricular mass and incidence of coronary heart disease in an elderly cohort. The Framingham Heart Study. Ann Intern Med 110:101–107, 1989.
54. Levy D, Garrison RJ, Savage DD, et al: Prognostic implications of echocardiographically determined left ventricular mass in the Framingham Heart Study. N Engl J Med 322:1561–1566, 1990.
55. Messerli FH, Ventura HO, Elizardi DJ, et al: Hypertension and sudden death: Increased ventricular ectopy activity in left ventricular hypertrophy. Am J Med 77:18–22, 1984.
56. Cruickshank JM, Lewis J, Moore V, Dodd C: Reversibility of left ventricular hypertrophy by differing types of antihypertensive therapy. J Hum Hypertens 6:85–90, 1992.
57. Michalewicz L, Messerli FH: Cardiac effects of calcium antagonists in hypertension. Am J Cardiol 79:39–46, 1997.
58. Muiesan G, Agabiti-Rosei E, Romanelli G, et al: Adrenergic activity and left ventricular function during treatment in essential hypertension with calcium antagonists. Am J Cardiol 57: 44d–49d, 1986.
59. Schulman SP, Weiss JL, Becker LC, et al: The effects of antihypertensive therapy on left ventricular mass in elderly patients. N Engl J Med 322:1350–1356, 1990.
60. Granier P, Douste-Blazy MY, Tredez P, et al: Improvement in left ventricular hypertrophy and left ventricular diastolic function following verapamil therapy in mild to moderate hypertension. Eur J Clin Pharmacol 39(Suppl 1):S45–S46, 1990.
61. Weiss RJ, Bent B: Diltiazem-induced left ventricular mass regression in hypertensive patients. J Clin Hypertens 3:135–143, 1987.

62. Senda Y, Tohkai M, Shida Y, et al: ECG-gated cardiac scan and echocardiographic assessments of left ventricular hypertrophy: reversal by 6-month treatment with diltiazem. J Cardiovasc Pharmacol 16:298–304, 1990.
63. Gottdiener J, Reda D, Notargiacomo A, et al, for the VA Cooperative Study Group: Comparison of monotherapy on LV mass regression in mild-to-moderate hypertension: Echocardiographic results of a multicenter trial [abstract]. J Am Coll Cardiol 17(Suppl A): 178A, 1991.
64. van Leeuwen JTM, Smit AJ, May JF, et al: Comparative effects of diltiazem and lisinopril on left ventricular structure and filling in mild-to-moderate hypertension. J Cardiovasc Pharmacol 26:983–989, 1995.
65. Leenen FH, Fourney A: Comparison of the effects of amlodipine and diltiazem on 24-hour blood pressure, plasma catecholamines, and left ventricular mass. Am J Cardiol 78:203–207, 1996.
66. Ferrara LA, De Simone G, Mancini M, et al: Changes in left ventricular mass during a double-blind study with chlorthalidone and slow-release nifedipine. Eur J Clin Pharmacol 27:525–528, 1984.
67. Phillips RA, Ardeljan M, Shimabukuro S, et al: Effect of nifedipine GITS on left ventricular mass and diastolic function in severe hypertension. J Cardiovasc Pharmacol 17(Suppl 2):S172–S174, 1991.
68. Totteri A, Scopelliti G, Bertini M, et al: Evaluation of regression of left ventricular hypertrophy after antihypertensive therapy. Comparative echo-Doppler study of ACE inhibitors and calcium antagonists [Italian]. Minerva Cardioangiol 41:231–237, 1993.
69. Yamakado T, Teramura S, Oonishi T, et al: Regression of left ventricular hypertrophy with long-term treatment of nifedipine in systemic hypertension. Clin Cardiol 17:615–618, 1994.
70. Kirpizidis HG, Papazachariou GS: Comparative effects of fosinopril and nifedipine on regression of left ventricular hypertrophy in hypertensive patients: A double-blind study. Cardiovasc Drugs Ther 9:141–143, 1995.
71. Myers MG, Leenen FHH, Tanner J: Differential effects of felodipine and nifedipine on 24-h blood pressure and left ventricular mass. Am J Hypertens 8:712–718, 1995.
72. Ferrara LA, Fasano ML, de Simone G, et al: Antihypertensive and cardiovascular effects of nitrendipine: A controlled study vs. placebo. Clin Pharmacol Ther 38:434–438, 1985.
73. Drayer JI, Hall WD, Smith VE, et al: Effect of the calcium channel blocker nitrendipine on left ventricular mass in patients with hypertension. Clin Pharmacol Ther 40:679–685, 1986.
74. Giles TD, Sander GE, Roffidal LC, et al: Comparison of nitrendipine and hydrochlrothiazide for systemic hypertension. Am J Cardiol 60:103–106, 1987.
75. Machnig T, Henneke KH, Engels G, et al: Nitrendipine vs. captopril in essential hypertension: Effects on circadian blood pressure and left ventricular hypertrophy. Circulation 85:101–110, 1994.
76. Costantino G, Di Lorenzo L, Castaldo A, De Simone G: Changes in left ventricular structure and function during therapy with slow-release nicardipine in arterial hypertension. Curr Therapeut Res 44:547–553, 1988.
77. Gökçe Ç, Oram A, Kes S, et al: Effects of nicardipine on left ventricular dimensions and hemodynamics in systemic hypertension. Am J Cardiol 65:680–682, 1990.
78. Sumimoto T, Hiwada K, Ochi T, et al: Effects of long-term treatment with sustained-release nicardipine on left ventricular hypertrophy and function in patients with essential hypertension. J Clin Pharmacol 34:266–269, 1994.
79. Langan J, Rodriquez-Manas L, Sareli P, Heinig R: Nisoldipine CC: Clinical experience in hypertension. Cardiology 88(Suppl):56–62, 1997.
80. Kloner RA, Sowers JR, DiBona GF, et al: Effect of amlodipine on left ventricular mass in the Amlodipine Cardiovascular Community Trial. J Cardiovasc Pharmacol 26:471–476, 1995.
81. Bignotti M, Grandi AM, Gaudio G, et al: One-year antihypertensive treatment with amlodipine: Effects on 24-hour blood pressure and left ventricular anatomy and function. Acta Cardiol 50:135–142, 1995.
82. Skoularigis J, Strugo V, Weinberg J, et al: Effects of amlodipine on 24-hour ambulatory blood pressure profiles, electrocardiographic monitoring, and left ventricular mass and function in black patients with very severe hypertension. J Clin Pharmacol 35:1052–1059, 1995.
83. Carr AA, Prisant LM: The new calcium antagonist isradipine. Effect on blood pressure and the left ventricle in black hypertensive patients. Am J Hypertens 3:8–15, 1990.
84. Saragoca MA, Portela JE, Abreu P, et al: Regression of left ventricular hypertrophy in the short-term treatment of hypertension with isradipine. Am J Hypertens 4:188S–190S, 1991.
85. Vyssoulis GP, Karpanou EA, Pitsavos LE, et al: Regression of left ventricular hypertrophy with isradipine antihypertensive therapy. Am J Hypertens 6:82S–85S, 1993.
86. Manolis AJ, Kolovou G, Handanis S, et al: Regression of left ventricular hypertrophy with isradipine in previously untreated hypertensive patients. Am J Hypertens 6:86S–88S, 1993.

87. Modena MG, Masciocco G, Rossi R, et al: Evaluation of the effectiveness of isradipine SRO in the treatment of hypertensive patients with left ventricular hypertrophy. Cardiovasc Drugs Ther 8:153–160, 1994.
88. Galderisi M, Celentano A, Garofalo M, et al: Reduction of left ventricular mass by short-term antihypertensive treatment with isradipine: A double-blind comparison with enalapril. Int J Clin Pharmacol Therapeut 32:312–316, 1994.
89. Grandi AM, Bignotti M, Gaudio G, et al: Ambulatory blood pressure and left ventricular changes during antihypertensive treatment: Perindopril versus isradipine. J Cardiovasc Pharmacol 26:737–741, 1995.
90. Pringle SD, Barbour M, Simpson IA, et al: Effect of felodipine on left ventricular mass and Doppler-derived hemodynamics in patients with essential hypertension [abstract]. Proceedings of the 4th International Symposium on Calcium Antagonists: Pharmacology and Clinical Research, Florence, Italy, May 1989, p 174.
91. Cerasola G, Cottone S, Nardi E, et al: Reversal of cardiac hypertrophy with the calcium antagonist felodipine in hypertensive patients. J Hum Hypertens 4:703–708, 1990.
92. Wetzchewald D, Klaus D, Garanin G, et al: Regression of left ventricular hypertrophy during long-term antihypertensive treatment—A comparison between felodipine and the combination of felodipine and metoprolol. J Int Med 231:303–308, 1992.
93. Leenen FH, Holliwell DL: Antihypertensive effect of felodipine associated with persistent sympathetic activation and minimal regression of left ventricular hypertrophy. Am J Cardiol 69:639-645, 1992.
94. Nalbantgil I, Önder R, Killiçcioglu B, et al: The efficacy of felodipine ER on regression of left ventricular hypertrophy in patients with primary hypertension. Blood Pressure 5:285–291, 1996.
95. Bache RJ, Tockman BA: Effect of nitroglycerin and nifedipine on subendocardial perfusion in the presence of flow-limiting coronary stenosis in the awake dog. Circ Res 50:678–687, 1982.
96. Schmieder RE, Martus P, Klingbeil A: Reversal of left ventricular hypertrophy in essential hypertension. A meta-analysis of randomized double-blind studies. JAMA 275:1507–1513, 1996.
97. Campbell SE, Turek Z, Rakusan K, Kazda S: Cardiac structural remodelling after treatment of spontaneously hypertensive rats with nifedipine and nisoldipine. Cardiovasc Res 27:1350–1358, 1993.
98. Suzuki M, Yamanaka K, Nabata H, Tachibana M: Long term effects of amlodipine on organ damage, stroke and life span in stroke prone spontaneously hypertensive rats. Eur J Pharmacol 228:269–274, 1993.
99. Vogt M, Strauer BE: Response of hypertensive left ventricular hypertrophy and coronary microvascular disease to calcium antagonists. Am J Cardiol 76:24D–30D, 1995.
100. De Servi S, Ferrario M, Ghio S, et al: Effects of diltiazem on regional coronary hemodynamics during atrial pacing in patients with stable exertional angina: Implications for mechanisms of action. Circulation 73:1248–1253, 1986.
101. Emanuelsson H, Homberg S: Mechanisms of angina relief after nifedipine: A hemodynamic and myocardial metabolic study. Circulation 68:124–130, 1983.
102. Matsuzaki M, Gallagher KP, Patritti J, et al: Effects of a calcium-entry blocker (diltiazem) on regional myocardial flow and function during exercise in conscious dogs. Circulation 69:801–814, 1984.
103. Subramanian VB, Bowles MJ, Davies AB, et al: Calcium channel blockade as primary therapy for stable angina pectoris. A double-blind placebo-controlled comparison of verapamil and propranolol. Am J Cardiol 50:1158–1163, 1982.
104. Maseri A: Pathogenetic mechanisms of angina pectoris: Expanding views. Br Heart J 43:648–660, 1980.
105. Brown BG, Bolson EL, Dodge HT: Dynamic mechanisms in human coronary stenosis. Circulation 70:917–922, 1984.
106. Connon RO III, Matson RM, Rosing DR, Epstein SE: Angina caused by reduced vasodilator reserve of the small coronary arteries. J Am Coll Cardiol 1:1359–1373, 1983.
107. Epstein SL, Talbot TL: Dynamic coronary tone in precipitation, exacerbation, and relief of angina pectoris. Am J Cardiol 48:797–803, 1981.
108. Stone PH, Muller JE, Turi ZG, et al: Efficacy of nifedipine therapy in patients with refractory angina pectoris: Significance of the presence of coronary vasospasm. Am Heart J 106:644–652, 1983.
109. Boden WE, Korr KS, Bough EW: Nifedipine-induced hypotension and myocardial ischemia in refractory angina pectoris. JAMA 253:1131–1135, 1985.
110. Schanzenbächer P, Deeg P, Liebau G, Kochsiek K: Paradoxical angina after nifedipine: Angiographic documentation. Am J Cardiol 53:345–346, 1984.

111. Schanzenbächer P, Liebau G, Deeg P, Kochsiek K: Effect of intravenous and intracoronary nifedipine on coronary blood flow and myocardial oxygen consumption. Am J Cardiol 51:712–717, 1983.
112. Wilson DC, Schwarts GL,Textor SC, et al: Precipitous fall in blood pressure in the treatment of chronic hypertension. Presented at the 5th International Symposium on Calcium Antagonists: Pharmacology and Clinical Research, Houston, TX, September 1991.
113. Opie LH, Messerli FH: Nifedipine and mortality: Grave defects in the dossier. Circulation 92:1068–1073, 1995.
114. Psaty BM, Heckbert SR, Koepsell TD, et al: The risk of incipient myocardial infarction associated with antihypertensive drug therapies. JAMA 271:620–625, 1995.
115. National Heart, Lung and Blood Institute Statement: New analyses regarding the safety of calcium-channel blockers: A statement for health professionals from the National Heart, Lung and Blood Institute. Bethesda, MD, National Institutes of Health, 1995.
116. Laragh JH, Held C, Messerli FH, et al: Calcium antagonists and cardiovascular prognosis: A homogenous group? Am J Hypertens 9:99–109, 1996.
117. The DEFIANT-II Research Group: Doppler flow and echocardiography in functional cardiac insufficiency: Assessment of nisoldipine therapy. Results of the DEFIANT-II Study. Eur Heart J 18:31–40, 1997.
118. Rehnqvist N, Hjemdahl P, Billing E, et al: Effects of metoprolol vs verapamil in patients with stable angina pectoris. The Angina Prognosis Study in Stockholm (APSIS). Eur Heart J 17:76-81, 1996.
119. Gong L, Zhang W, Zhu Y, et al: Shanghai Trial of Nifedipine in the Elderly (STONE). J Hypertens 14:1237–1245, 1996.
120. Braun S, Boyko V, Behar S, et al: Calcium antagonists and mortality in patients with coronary artery disease: A cohort study of 11,575 patients. J Am Coll Cardiol 28:7–11, 1996.
121. Jick H, Derby LE, Gurewich V, Vasilakis C: The risk of myocardial infarction in persons with uncomplicated essential hypertension associated with antihypertensive drug treatment. Pharmacotherapy 16:321–326, 1996.
122. Aursnes I, Litleskare I, Froyland H, Abdelnoor M: Association between various drugs used for hypertension and risk of acute myocardial infarction. Blood Pressure 4:157–163, 1995.
123. Dargie HJ, Ford I, Fox KM, on behalf of the TIBET Study Group: Effects of ischaemia and treatment with atenolol, nifedipine SR and their combination on outcomes in patients with chronic stable angina. Eur Heart J 17:104–112, 1996.
124. Opie LH: Calcium channel antagonists in the management of anginal syndromes: Changing concepts in relation to the role of coronary vasospasm. Prog Cardiovasc Dis 38:291–314, 1996.
125. Chahine RA: Coronary Artery Spasm. Mt. Kisco, NY, Futura, 1983.
126. Chahine RA, Feldman RL, Giles TD, et al: Efficacy and safety of amlodipine in vasospastic angina: An interim report of a multicenter, placebo controlled trial. The Investigators of Study 160. Am Heart J 118:1128–1130, 1989.
127. Weiner DA, Klein MD, Cutler SS: Efficacy of sustained-release verapamil in chronic angina pectoris. Am J Cardiol 59:215–218, 1987.
128. Held PH, Yusuf S, Furberg CD: Calcium channel blockers in acute myocardial infarction and unstable angina: An overview. BMJ 299:1187–1192, 1989.
129. The Danish Verapamil Infarction Trial II–DAVIT II: Effect of verapamil on mortality and major events after acute myocardial infarction. Am J Cardiol 66:779–785, 1990.
130. Yusuf S, Held P, Furberg C: Update of effects of calcium antagonists in myocardial infarction or angina in light of the Second Danish Verapamil Infarction Trial (DAVIT-II) and other recent studies [editorial]. Am J Cardiol 67:1295–1297, 1991.
131. Messerli FH, Weiner DA: Are all calcium antagonists equally effective for reducing reinfarction rate? Am J Cardiol 72:818–820, 1993.
132. Rengo F, Carbonin P, Pahor M, et al: A controlled trial of verapamil in patients after an acute myocardial infarction. Results of the Calcium Antagonist Reinfarction Italian Study (CRIS). Am J Cardiol 77:365–369, 1996.
133. Gibson RS, Boden WE, Theroux P, et al: Diltiazem and reinfarction in patients with non–Q-wave myocardial infarction: Results of a double-blind, randomized multicenter trial. N Engl J Med 315:423–429, 1986.
134. The Multicenter Diltiazem Postinfarction Trial Research Group: The effect of diltiazem on mortality and reinfarction after myocardial infarction. N Engl J Med 319:385–392, 1988.
135. Fischer Hansen J, Hagerup L, Sigurd B, et al: The Danish Verapamil Infarction Trial (DAVIT) Study Group. Cardiac event rates after myocardial infarction in patients treated with verapamil and trandorapril versus trandorapril alone. Am J Cardiol 79:738–741, 1997.

136. Moss AJ, Oakes D, Rubison M, et al: Effects of diltiazem on long-term outcome after acute myocardial infarction in patients with and without a history of systemic hypertension. The Multicenter Diltiazem Postinfarction Trial Research Group. Am J Cardiol 68:429–433, 1991.
137. Jaspersen CM, Hansen JF, Mortensen LS: The prognostic significance of post-infarction angina pectoris and the effect of verapamil on the incidence of angina pectoris and prognosis. The Danish Study Group on Verapamil in Myocardial Infarction. Eur Heart J 15:270–276, 1994.
138. Jaspersen CM, Hagerup L, Holländer N, et al, and the Danish Study Group on Verapamil in Myocardial Infarction: Does exercise-induced ST-segment depression predict benefit of medical intervention in patients recovering from acute myocardial infarction? J Intern Med 233:33–37, 1993.
139. Persson S: Update on the use of angiotensin converting enzyme inhibitors and calcium antagonists in postinfarction patients. J Hypertens 13 (Suppl):57–63, 1995.
140. Blumlein SL, Sievers R, Kidd P, Parmley WW: Mechanism of protection from atherosclerosis by verapamil in the cholesterol-fed rabbit. Am J Cardiol 54:884–889, 1984.
141. Habib JB, Bossaller C, Wells S, et al: Preservation of endothelium-dependent vascular relaxation in cholesterol-fed rabbits by treatment with the calcium blocker PN 200 110. Circ Res 58:305–309, 1996.
142. Henry PD, Bentley KI: Suppression of atherogenesis in cholesterol-fed rabbits treated with nifedipine. J Clin Invest 68:1366–1369, 1981.
143. Ohata I, Sakomoto N, Nagano K, Maeno H: Low density lipoprotein-lowering and high density lipoprotein-elevating effects of nicardipine in rats. Biochem Pharmacol 33:2199–2205, 1984.
144. Rouleau J-L, Parmley WW, Stevens J, et al: Verapamil suppresses atherosclerosis in cholesterol-fed rabbits. J Am Coll Cardiol 1:1453–1460, 1983.
145. Sugano N, Nakashimo Y, Matsushima T, et al: Suppression of atherosclerosis in cholesterol-fed rabbits by diltiazem injection. Arteriosclerosis 6:237–241, 1986.
146. Walldius G: Effect of verapamil on serum lipoproteins in patients with angina pectoris. Acta Med Scand Suppl 681(Suppl):43–48, 1984.
147. Willis AL, Nagel B, Churchill V, et al: Antiatherosclerotic effects of nicardipine and nifedipine in cholesterol-fed rabbits. Arteriosclerosis 5:250–255, 1985.
148. Naito M, Kuzuya F, Asai K-I, et al: Ineffectiveness of Ca^{++}-antagonists nicardipine and diltiazem on experimental atherosclerosis in cholesterol-fed rabbits. Angiology 35:622–627, 1984.
149. Stender S, Stender I, Nordestgaard B, Kjeldsen K: No effect of nifedipine on atherogenesis in cholesterol-fed rabbits. Arteriosclerosis 4:389–394, 1984.
150. Van Niekerk JL, Hendriks T, De Boer HH, Van't Laar A: Does nifedipine suppress atherogenesis in WHHL rabbits? Atherosclerosis 53:91–98, 1984.
151. Zanchetti A: The antiatherogenetic effects of antihypertensive drugs: experimental and clinical evidence. Clin Exp Hypertens 14A(1-2): 307–331, 1992.
152. Waters D, Lesperance J, Francetich M, et al: A controlled clinical trial to assess the effect of a calcium channel blocker on the progression of coronary atherosclerosis. Circulation 82:1940–1953, 1990.
153. Lichtlen PR, Hugenholtz PG, Rafflenbeul W, et al: Retardation of angiographic progression of coronary artery disease by nifedipine. Results of the International Nifedipine Trial on Antiatherosclerotic Therapy (INTACT). INTACT Group Investigators. Lancet 335:1109–1113, 1990.
154. Hoberg E, Schwarz F, Schoemig A, et al: Prevention of restenosis by verapamil. The Verapamil Angioplasty Study (VAS) [abstract]. Circulation 82(Suppl III):428, 1990.
155. Schroeder JS, Gao SZ, Alderman EL, et al: A preliminary study of diltiazem in the prevention of coronary artery disease in heart-transplant recipients. N Engl J Med 328:164–170, 1993.
156. Borhani NO, Mercuri M, Borhani PA, et al: Final outcome results of the Multicenter Isradipine Diuretic Atherosclerosis Study (MIDAS). A randomized controlled trial. JAMA 276:785–791, 1996.
157. Schneider W, Kober G, Roebruck P, et al: Retardation of development and progression of coronary atherosclerosis: a new indication for calcium antagonists? Eur J Clin Pharmacol 39(Suppl 1):S17–S23, 1990.
158. Zanchetti A, Magnani B, Dai Palù C, on behalf of the Verapamil-Hypertension Atherosclerosis Study (VHAS) Investigators: Atherosclerosis and calcium antagonists: The VHAS. J Hum Hypertens 6(Suppl 2):S45–S48, 1992.
159. Bond G, Dal Palù C, Hansson L, et al: European Lacidipine Study on Atherosclerosis (ELSA) [abstract]. J Hypertens 11(Suppl 5):S405, 1993.
160. Zanchetti A: Antiatherosclerotic effects of antihypertensive drugs: recent evidence and ongoing trials. Clin Exp Hypertens 18:489–499, 1996.

161. Folkow B: The Fourth Volhard Lecture. Cardiovascular structural adaptation: Its role in the initiation and maintenance of primary hypertension. Clin Sci Mol Med 55(Suppl 4):3S–22S, 1978.
162. Strauer BE: Ventricular function and coronary hemodynamics in hypertensive heart disease. Am J Cardiol 44:999–1006, 1979.
163. Messerli FH, Ketelhut R: Left ventricular hypertrophy: A pressure-independent cardiovascular risk factor. J Cardiovasc Pharmacol 22(Suppl 1):S7–S13, 1993.
164. Tubau JF, Szlachcic J, Braun S, Massie BM: Impaired left ventricular functional reserve in hypertensive patients with left ventricular hypertrophy. Hypertension 14:1–8, 1989.
165. Schmieder RE, Messerli FH: Reversal of left ventricular hypertrophy: a desirable therapeutic goal? J Cardiovasc Pharmacol 16(Suppl 6): S16–S22, 1990.
166. Schmieder RE, Rüddel H, Grube E, Schulte W: Depressed myocardial contractility in early left ventricular hypertrophy (LVH) [abstract]. Circulation 78(Suppl II):II75, 1988.
167. de Simone G, Devereux RB, Roman MJ, et al: Assessment of left ventricular function by the midwall fractional fiber shortening/end systolic stress relation in human hypertension. J Am Coll Cardiol 23:1444–1451, 1994.
168. Marcus ML, Harrison DG, Chilian WM, et al: Alterations in the coronary circulation in hypertrophied ventricles. Circulation 75:I19–I25, 1987.
169. Opherk D, Mall G, Zebe H, et al: Reduction of coronary reserve: A mechanism for angina pectoris in patients with hypertension and normal coronary arteries. Circulation 69:1–7, 1984.
170. Schwartzkopff B, Frenzel H, Vogt M, et al: Myocardial structure in patients with reduced coronary reserve in hypertensive heart disease [abstract]. Circulation 80(Suppl II):II-539, 1989.
171. Levy D, Anderson KM, Savage DD, et al: Risk of ventricular arrhythmias in left ventricular hypertrophy: The Framingham Heart Study. Am J Cardiol 60:560–565, 1987.
172. Siegel D, Cheitlin MD, Black DM, et al: Risk of ventricular arrhythmias in hypertensive men with left ventricular hypertrophy. Am J Cardiol 65:742–747, 1990.
173. Widimsky J, Cifkova R: Hypertension and arrhythmias. A review. Cor Vasa 31:157–163, 1989.
174. Messerli FH, Nunez BD, Ventura HO, Snyder DW: Overweight and sudden death: Increased ventricular ectopy in cardiopathy of obesity. Arch Intern Med 147:1725–1728, 1987.
175. Nunez BD, Messerli FH, Garavaglia GE, Schmieder RE: Exaggerated atrial and ventricular excitability in hypertensive patients with isolated septal hypertrophy (ISH) [abstract]. J Am Coll Cardiol 9:225A, 1987.
176. Papademetriou V, Notagariacomo A, Heine D, et al: Ventricular arrhthmias [sic] in patients with essential hypertension [abstract]. J Am Coll Cardiol 13:105A, 1989.
177. Bikkina M, Larson MG, Levy D: Asymptomatic ventricular arrhythmias and mortality risk in subjects with left ventricular hypertrophy. J Am Coll Cardiol 22:1111–1116, 1993.
178. Almendral J, Villacastin JP, Arenal A, et al: Evidence favoring the hypothesis that ventricular arrhythmias have prognostic significance in left ventricular hypertrophy secondary to systemic hypertension. Am J Cardiol 76:60D–63D, 1995.
179. Messerli FH, Nunez BD, Nunez MM, et al: Hypertension and sudden death: Disparate effects of calcium entry blocker and diuretic therapy on cardiac dysrhythmias. Arch Intern Med 149: 1263–1267, 1989.
180. Davis BR, Cutler JA, Gordon DJ, et al, for the ALLHAT study: Rationale and design for the Antihypertensive and Lipid Lowering Treatment to Prevent Heart Attack Trial (ALLHAT). Am J Hypertens 9:342–360, 1996.
181. Hansson L, Zanchetti A, HOT Study Group: The Hypertension Optimal Treatment (HOT) Study: Patient characteristics: randomization, risk profiles, and early blood pressure results. Blood Pressure 3:322–327, 1994.
182. Dahlof B, Hansson L, Lindholm LH, et al: STOP-Hypertension 2: A prospective intervention trial of "newer" versus "older" treatment alternatives in old patients with hypertension. Blood Pressure 2:136–144, 1993.
183. Staessen J, Bert P, Bulpitt C, et al: Nitrendipine in older patients with isolated systolic hypertension: Second progress report on the SYST-EUR trial. J Hum Hypertens 7:265–271, 1993.

GEORGE W. VETROVEC M.D.

6

Calcium Antagonists and Myocardial Ischemia

Calcium antagonists represent a significant component of the medical management of angina pectoris. Nearly one-half of all patients with angina are treated with calcium antagonists as monotherapy or in combination with nitrates and/or beta blockers. The calcium antagonists have made a significant contribution to the medical management of angina pectoris because of their overall excellent efficacy and tolerability. This chapter focuses on considerations in the treatment of coronary artery disease, including the mechanisms specific to the effects of calcium antagonists in patients with ischemic heart disease. Emphasis is placed on direct application to clinical management and the potential secondary benefits of calcium antagonists. Included are the specific advantages and limitations of using calcium antagonists in the management of ischemic heart disease. Finally, attention is paid to safety issues specific to coronary events.

ISCHEMIA: TREATMENT CONSIDERATIONS

Table 1 summarizes the specific circumstances in which calcium antagonists are beneficial in management of coronary artery disease. Calcium antagonists not only reduce symptoms; they also reduce ischemia, as documented by changes in ST segment shifts with treadmill testing and ambulatory electrocardiographic recordings.[1]

Chronic stable angina is a broad category covering the majority of patients with symptomatic coronary disease. Calcium antagonists offer specific benefits to patients with vasoactive angina, including coronary spasm, Prinzmetal's angina, and "mixed" angina. In addition to reducing myocardial oxygen demand, calcium antagonists prevent coronary spasm and vasoreactive changes and maximize coronary blood flow, an important factor in optimizing myocardial blood supply. Coronary vasodilatation is unique to calcium antagonists compared with beta blockers and represents a major potential benefit in patients with so-called mixed angina. In fact, the first significant report of the efficacy of calcium antagonists in the management of coronary ischemia was made by Antman et al.,[2] who described the efficacy of nifedipine capsules in Prinzmetal's angina. Weekly angina rates were markedly reduced in 127 patients, with complete control of angina in over one-half and a significant reduction in angina in 87%. This study specifically documented the efficacy of nifedipine in the management of coronary vasospasm. Subsequently, other calcium antagonists have

TABLE 1. Uses of Calcium Antagonists in the Management of Coronary Disease

- Chronic stable angina pectoris
- Vasoreactive angina, including coronary spasm and Prinzmetal's angina
- Mixed angina
- Silent ischemia (?)

been shown to reduce the symptoms of Prinzmetal's angina, including diltiazem, verapamil, and amlodipine.[3,4] Thus, in terms of the the usual supply/demand imbalance described for myocardial ischemia, calcium antagonists affect both the demand and the supply side, particularly in patients who have a significant propensity to vasoreactive changes.

Silent or asymptomatic ischemia is still a controversial issue, but calcium antagonists may have a potential benefit in its management. Significant evidence indicates that patients with coronary artery disease often experience episodes of ischemia, as documented by objective measures of myocardial function or perfusion, in the absence of symptoms.[5] Although subgroups of patients, such as diabetics, may have a particular propensity to silent ischemia, most patients with angina have more episodes of asymptomatic than painful ischemia. In practical terms, as one treats clinical angina pectoris, the incidence of silent ischemia declines. Evidence indicates, however, that patients with increased silent ischemia have a greater risk of coronary events; thus, the potential benefit of suppressing ischemia rather than angina remains an important issue. The Asymptomatic Cardiac Ischemia Pilot (ACIP) study[6] suggests that coronary events tend to be decreased in patients with ischemic disease who are managed with the goal of eliminating ischemia rather than the goal of simply eliminating angina pectoris. Further studies are needed, but calcium antagonists have been shown to have 24-hour efficacy and to reduce ischemia as well as angina pectoris. The 24-hour efficacy is particularly applicable to the long-acting forms and may have particular relevance to the fact that most coronary events occur in the early morning hours when asymptomatic ischemia appears to be most common (see chapters 3 and 10).[7]

In summary, the calcium antagonists reduce myocardial ischemia and thus successfully treat angina pectoris. They have the ability both to maximize coronary blood flow and to reduce myocardial oxygen demand, predominantly through reduction of peripheral resistance.

MECHANISMS OF ACTION SPECIFIC TO ISCHEMIA

The calcium antagonists are classified according to their effects on the L-type calcium channels.[8] L-type channels are considered long-lasting, large-current, or slow, so-called voltage-dependent channels. In clinical practice, agents selective for the L-type channels are the most commonly used calcium antagonists. They inhibit the influx of extracellular calcium through the L-type channel, resulting in relaxation of vascular smooth muscle and reduction of vascular resistance in both the peripheral and the coronary circulation. The calcium antagonists, which are heterogeneous in chemical structure and function, are traditionally divided into three classes: the dihydropyridines, which account for the majority of currently available clinical agents; the phenylalkylamines (verapamil); and the benzothiazepines (diltiazem).

Mibefradil, a recently introduced calcium antagonist, affects not only the L-type receptors but also the T-type calcium channels.[9] The T-type calcium channels are low voltage-activated channels that are inhibited at drug concentrations lower than those needed to inhibit the L-type high voltage-activated calcium channels. Mibefradil is much more effective at blocking T-type than L-type channels. The physiologic effects of T-type channel blockade are yet to be fully elucidated. However, the L- and T-type calcium channels differ in terms of their various effects in cardiovascular tissue as well as their threshold of blocking activity. The L-type channels are abundant in cardiac muscle cells as well as in the conducting system, and their activation seems intimately involved in myocardial contraction and atrioventricular (AV) nodal conduction. Conversely, T-type channels are found in much lower density in cardiac muscle but are quite abundant in vascular smooth muscle. Thus, the clinical effect of a given calcium antagonist depends on its relative effects on vascular smooth, cardiac myocytes, and cardiac conducting cells (see chapter 1).

Original studies of the calcium antagonist prototypes compared short-acting nifedipine capsules with diltiazem and verapamil. It was clinically recognized that each of these agents has an innate potential in vitro to suppress myocardial contractility. In acute studies, however, nifedipine was associated with maintenance of acute myocardial contractility, presumably secondary to reflex effects from marked peripheral vasodilatation. As discussed later, however, this acute improvement was secondary to catecholamine stimulation, and the long-term effects were not favorable in patients with compromised ventricular function. Recently, dihydropyridines that are more vascular-specific have been developed, including felodipine, nisoldipine, and amlodipine. These agents promote vascular dilatation with less or no apparent clinical effect on myocardial contractility. Because of its combined L-type and predominantly T-type channel-blocking properties, mibefradil is thought to be clinically beneficial by reducing heart rate (as a consequence of sinoatrial [SA] nodal depression) and producing peripheral vasodilatation without significant myocardial depression. The clinical impact of these findings, however, has not been fully tested. Furthermore, the current calcium antagonists are long-acting because of longer half-lives or extended dose delivery systems. Such advances have improved the flexibility of their use.

From an antianginal standpoint, calcium antagonists have variable beneficial effects on reducing myocardial ischemia (Table 2). Peripheral vasodilatation, which is characteristic of all calcium antagonists, reduces myocardial work by reducing peripheral resistance. Furthermore, the degree to which afterload is reduced without increases in heart rate is a favorable factor in preventing myocardial ischemia. Current dihydropyridines, as well as verapamil and diltiazem, do not significantly increase resting heart rate.

In addition to peripheral vasodilatation, coronary vasodilatation is associated with the potential to increase myocardial oxygen delivery. To the extent that calcium antagonists prevent coronary vasospasm and, in particular, reduce vasomotor changes, they have a favorable effect.[10,11] The effect of small arteriolar vasodilatation is less clear. In the past concern has been raised about the possibility of creating a coronary steal syndrome with calcium antagonists.

TABLE 2. Mechanisms by Which Calcium Antagonists Reduce Myocardial Ischemia

Mechanism	Agents
Primary mechanisms	
Decreased contractility	Verapamil, diltiazem
Decreased heart rate	Verapamil, diltiazem, mibefradil
Afterload reduction (decreased peripheral resistance)	All, to variable degrees
Increased coronary blood flow (coronary vasodilatation)	All, to variable degrees
Secondary mechanisms	
Regression of left ventricular hypertrophy	All
Improved diastolic function	All; verapamil and diltiazem preferable in hyperdynamic left ventricular dysfunction
Retarded development of atherosclerosis (theoretical)	All studied (predominantly animal work)

Clinically, this potential is probably not important, although such changes may cause episodes of increased angina, which have been reported with calcium antagonists. In our study, wall motion changes actually improved in response to acute administration of short-acting nifedipine in patients taking high doses of a beta blocker. This finding suggests that the coronary steal phenomenon was not clinically operative.[12]

Lastly, direct depression of myocardial contractility is potentially an additional antianginal effect in patients treated with diltiazem and verapamil. Both agents have modest but definitive effects on myocardial contractility, although most clinical studies have not shown measurable differences in ejection fraction. However, both agents also appear to have the potential to minimize exercise-related increases in heart rate, which may contribute to antianginal efficacy. In contrast, mibefradil appears to have neutral effects on myocardial contractility (like the new dihydropyridines) but reduces heart rate, which may help to prevent angina. The impact of these effects on cardiac conduction and the propensity for higher degrees of AV block, particularly in elderly patients with sick sinus syndrome, is unknown.

In summary, calcium antagonists exert their antianginal effect primarily by reducing afterload. They also improve coronary blood flow and prevent vasomotion changes. Their effect on heart rate and myocardial contractility is variable and probably weak in terms of antianginal efficacy.

APPLICATIONS TO CLINICAL CONDITIONS

As noted earlier, calcium antagonists are the treatment of choice for patients with Prinzmetal's angina or angina with evidence of acute vasoreactivity. The frequency of true coronary spasm is limited; thus, the most common indication for calcium antagonists is management of chronic stable angina. Choice of therapy in chronic angina depends on multiple associated conditions. For instance, in patients with atrial fibrillation and/or recurrent episodes of supraventricular tachyarrhythmias, verapamil, diltiazem, and perhaps mibefradil are the best choices because of their antianginal efficacy as well as potential control of AV nodal conduction. Furthermore, in patients with severe hypertrophic myopathies and/or hyperdynamic concentric left ventricular hypertrophy, the same agents are ideal for treating ischemic disease because they also reduce the overall degree of contractility. Likewise, they appear to improve diastolic function, probably by improved diastolic relaxation.[13] All of these changes are beneficial in terms of reducing myocardial oxygen demand.

For patients who have strictly demand-type ischemia (i.e., at peak exercise performance), agents that reduce myocardial demand by reducing contractility, heart rate, and afterload are preferable. Conversely, patients with mixed angina (i.e., episodes with exercise and at rest) or variable threshold angina often respond well to calcium antagonists and/or nitrates, which reduce the vasoreactive component usually associated with both conditions. Under optimal conditions, combined reductions in myocardial oxygen demand maximize efficacy. Thus, combination therapy with a calcium antagonist and beta blocker may be ideal in such patients. For example, we evaluated a series of patients with angiographically defined atherosclerotic coronary artery disease and angina pectoris that was incompletely responsive to maximally tolerated therapy with nitrates and beta blockers.[14] When nifedipine capsules were added to the regimen and titrated as tolerated, the overall frequency of angina decreased significantly. However, when the patients were subdivided according to whether their angina was predominantly exercise-related or included a significant component at rest (i.e., > 50% of episodes at nonpeak activity), patients with a significant component at rest had the greater probability of becoming totally asymptomatic with the addition of a calcium antagonist. Patients with exercise-induced angina improved, but some degree of angina persisted in most. Thus, in deciding between a beta blocker or calcium antagonist as primary therapy, it is often helpful to profile patients according to the possibility of a vasoreactive component.

Decisions about therapy also should take into account associated conditions. Table 3 summarizes a medical management strategy that takes into consideration other medical disorders that are frequently associated with angina and directly influence the long-term outcome of cardiovascular disease. For example, of patients undergoing coronary angiography at our institution, approximately 50% have associated hypertension, 50% have left ventricular dysfunction with a reduced systolic ejection fraction, and 25% have diabetes mellitus.

TABLE 3. Use of Antianginal Agents in Chronic Stable Angina With and Without Associated Disease

		Angina Plus		
	Angina Only	Hypertension	Decreased Ejection Fraction	Diabetes Mellitus
Calcium antagonists	+++	+++	+++	+++
Beta blockers	+++	+++	++	+
Nitrates	++	+	+++	++
Angiotensin-converting enzyme inhibitors*			+++	

+++ = optimal, ++ = second line, + = as needed for unresolved symptoms not controlled by other agents.

* Not an antianginal agent but included because of survival benefits in patients with reduced left ventricular function.

As Table 3 indicates, calcium antagonists are often the ideal choice in all subcategories. Overall, I favor calcium antagonists and beta blockers as first-line therapy for angina because both are available in once-daily formulations, which improve convenience and compliance. Traditionally, nitrates have been effective in the management of angina but are potentially limited by tolerance and require some form of pulsed dosing; therefore, they may not provide true 24-hour efficacy. Either a calcium antagonist or beta blocker seems the optimal first-line treatment of angina, with nitrates used for patients not responsive to one or both of the other agents. Nitrates remain the treatment of choice for acute episodes of angina. For patients with associated hypertension, this scheme is even more relevant because the calcium antagonist or beta blocker should treat both angina and hypertension and may simplify the regimen with once-daily monotherapy. Nitrates do not treat hypertension in such patients and thus are beneficial only as adjunctive antianginal therapy.

Treatment of patients with compromised left ventricular function is more controversial. Although recent studies suggest that beta blockers improve symptoms and possibly survival, they may pose difficulties for the average practitioner because they require careful and slow titration in the setting of heart failure to prevent acute decompensation. Furthermore, beta blockers may actually cause a transient exacerbation before improvement occurs. Lastly, the beta blockers may blunt exercise performance by producing a fixed cardiac output with decreased overall contractility. In such circumstances, patients may have less angina but no improvement in exercise performance. Conversely, the calcium antagonists, particularly the newer, long-acting dihydropyridines, do not appear to impair ventricular function significantly and are thus ideal antianginal agents in the absence of catecholamine-induced worsening of heart failure. Likewise, in patients with impaired ventricular function, nitrates should be used earlier because many patients with a decreased ejection fraction also have elevated filling pressures. The venous capacitance reduction of the nitrates reduces filling pressure and wall tension, thus improving the threshold for angina. Although angiotensin-converting enzyme (ACE) inhibitors are not antianginal agents, they are included in Table 3 because improved survival makes them a necessary adjunct in any antianginal regimen for patients with significantly impaired ventricular function.

Lastly, patients with diabetes mellitus—particularly if they take insulin—may be adversely affected by beta blockers because of decreased warning signs of hypoglycemia.

Based on these considerations, it is clear that calcium antagonists provide an important contribution to the effective management of angina pectoris.

SECONDARY BENEFITS OF CALCIUM ANTAGONISTS IN THE MANAGEMENT OF ANGINA PECTORIS

As noted above, hypertension is commonly associated with atherosclerotic coronary artery disease. It is particularly significant in elderly patients, over 50% of whom have hypertension; as a result, the risk of subsequent coronary events is quite high. The risk of coronary events appears to be related to the development of secondary left ventricular hypertrophy.[15] Although a direct correlation between reduction in hypertension and risk of coronary events has been difficult to prove, improved control of blood pressure should reduce the risk. The potential to reduce left ventricular hypertrophy may be a secondary benefit of calcium antagonists,[16,17] particularly as opposed to nitrate therapy. Furthermore, even in the absence of a prognostic benefit, the regression of left ventricular hypertrophy appears to be associated with an improvement in left ventricular hemodynamics.[18] To that extent, improved diastolic function should enhance the management of refractory angina pectoris by increasing the angina threshold through reduced myocardial workload. Thus, better hypertensive treatment in patients with angina appears to have short- and long-term benefits.

In addition, a number of experimental studies have suggested that the calcium antagonists may help to reduce the development of new or progressive atherosclerotic lesions.[19,20] However, in the International Nifedipine Trial on Antiatherosclerotic Therapy (INTACT),[20] although new lesions were reduced in patients treated with calcium antagonists, coronary events were more common in patients treated with nifedipine capsules. The results may have been limited by the potential adverse effects of short-term nifedipine capsules. Whether calcium antagonists can retard atherosclerosis is at present an unanswered question. Of note, the calcium antagonists appear to be neutral in their effects on serum cholesterol, whereas beta blockers are often associated with a slight increase in cholesterol of unknown clinical significance. Overall, the long-term benefits of the secondary effects of calcium antagonists have not been proved but pose interesting theoretical possibilities.

POTENTIAL ADVERSE EFFECTS OF CALCIUM ANTAGONISTS IN THE MANAGEMENT OF ANGINA PECTORIS

Table 4 summarizes the usual side effects of calcium antagonists, including the subclass most likely to precipitate the side effects. Vascular side effects, including headaches and dizziness, may occur with any calcium antagonist but are worse with the dihydropyridines, specifically the short-acting, rapid-onset

TABLE 4. Potential Adverse Effects of Calcium Antagonists

Side Effect	Agent
Headache, dizziness	All; worse with dihydropyridines
Hypotension	All; potentially more with dihydropyridines, especially short-acting agents
Congestive heart failure	Short-acting agents
Edema	All
Heart block (second or third degree)	Diltiazem, verapamil
Gastrointestinal effects	
Vague nausea	Dihydropyridines
Constipation	Verapamil, diltiazem
Angina and/or myocardial infarction	? All; short-acting dihydropyridines

agents. The same is true of hypotension and peripheral edema. Of importance, the edema is not related to heart failure; it is a local phenomenon associated with relative differences in venous and arteriolar dilatation. Recent combination therapy with a low-dose ACE inhibitor and a dihydropyridine appears to reduce the edema to some degree. Furthermore, the incidence of edema is lower with the more recent long-acting calcium antagonists.

Gastrointestinal side effects include vague nausea, which may occur with any calcium antagonist, particularly the dihydropyridines. Constipation is most commonly associated with verapamil, although it may be seen with diltiazem.

Side effects of major concern include second-degree and, on rare occasions, third-degree heart block. Conduction system abnormalities are most common with diltiazem, particularly in older patients who appear to have a propensity to sick sinus syndrome. True complete heart block is quite rare. Recently increases in angina and acute myocardial infarction have been discussed at length because of reports of adverse outcomes in patients receiving nifedipine capsules. The risk is small but real and has been documented most clearly with short-acting agents. The greatest risk is probably in patients with unstable angina who receive monotherapy with a short-acting agent, which is likely to cause a sudden increase in heart rate and catecholamines (see chapters 28 and 29). To avoid this risk, calcium antagonists should not be first-line therapy in patients with unstable angina and/or recent acute myocardial infarction. If a calcium antagonist is required for control of angina, long-acting diltiazem and/or verapamil or perhaps mibefradil should be the agent of choice for patients not taking a beta blocker. Conversely, long-acting dihydropyridines may be used optimally in patients already taking a beta blocker, which should reduce the risk of reflux tachycardia and acceleration of angina.

SPECIAL ISSUES IN ISCHEMIC HEART DISEASE

Unstable Angina and Acute Myocardial Infarction

Concern about the risk of calcium antagonist therapy in unstable ischemic patients is most clearly related to the short-acting agent, capsule nifedipine. In the Holland Interuniversity Nifedipine/Metoprolol Trial (HINT) of unstable angina[21] and the Secondary Prevention Reinfarction Israeli Nifedipine Trial (SPRINT) I and II,[22,23] acute administration of nifedipine early in the course of unstable angina and acute myocardial infarction was associated with an increase in myocardial ischemic events and adverse outcomes. In contrast, the Danish Verapamil Infarction Trial (DAVIT) II[24] demonstrated an improvement in overall cardiovascular outcome for patients treated with verapamil after myocardial infarction compared with placebo. One may conclude from these data that the short-acting dihydropyridines are associated with an increased risk of adverse events, whereas the nondihydropyridines may be associated with lower or no risk.

Opie[25] defines the potential causes of adverse ischemic events due to short-acting nifedipine, including acute hypotension immediately after dosing. Nonhypotensive proischemic effects include coronary steal and repetitive neurohumoral stimulation, which in fact may become blunted over time by baroreflex desensitization. In support of these concepts, most complications appeared earlier in the course of treatment, particularly in patients with low blood pressure in the HINT[21] and SPRINT[22,23] trials.

What do these data mean clinically? One should avoid calcium antagonists in unstable ischemic syndromes unless they are needed for symptoms unresponsive to other classes of antianginal agents. In such circumstances, one must compare the potential risk of the calcium antagonist with the risk of alternative therapy, such as coronary intervention or bypass surgery. If calcium antagonists are believed to be indicated, long-acting agents, particularly the nondihydropyridines, appear to have the lowest overall risk. If a dihydropyridine is required, combining the calcium blocker with a beta blocker may reduce the risk of neurohormone-mediated catecholamine reactions. Furthermore, in such circumstances calcium antagonists should be avoided in patients with significant hypotension, who may be the subgroup at highest risk.

Left Ventricular Function

The role of calcium antagonists in the setting of impaired ventricular function has received much attention over the years. Negative inotropic effects were recognized for all of the original agents and also apply to verapamil and diltiazem. As noted earlier, more recent dihydropyridines, including nisoldipine, amlodipine, mibefradil, isradipine, and felodipine, seem to be associated with neutral effects on left ventricular function. Capsule nifedipine initially was shown to preserve left ventricular function when acutely administered even to patients taking high doses of a beta blocker. However, because of issues related to reflex sympathetic stimulation, particularly in patients who are not taking a beta blocker, short-acting calcium antagonists were associated with less ideal outcomes when used chronically in patients with congestive heart failure (CHF).

Recent data about the newer agents, including amlodipine (PRAISE trial)[26] and felodipine (V-HEFT III),[27] demonstrate no significant adverse events in patients with CHF who were treated with additional background CHF therapy. In the PRAISE trial, 1,153 patients with severe chronic heart failure (New York Heart Association Class III or IV, with ejection fraction $< 30\%$) were blindly randomized to either placebo or amlodipine; their usual therapy, including ACE inhibitors and digoxin, was maintained. The primary endpoints were death from any cause and hospitalization for major cardiovascular events. Overall, patients treated with amlodipine had a 9% lower risk of fatal and nonfatal events, although the difference was not statistically significant. The 16% reduction in mortality with amlodipine was of borderline statistical significance ($p \leq 0.07$).

Conversely, nondihydropyridines have been associated with adverse outcomes in the setting of poor ventricular function after myocardial infarction. The Multicenter Diltiazem Postinfarction Trial (MDPIT)[28] reported no overall improvement in cardiovascular outcome among patients randomized to placebo or diltiazem after myocardial infarction. However, a subgroup of patients with pulmonary congestion suggestive of left ventricular failure fared worse with the calcium antagonist.

This finding has raised concerns about the appropriateness of calcium antagonists in patients with compromised ventricular function, particularly in the setting of recent myocardial infarctions. Issues related to recent myocardial infarction have been discussed above, but chronic ventricular dysfunction is somewhat different. Nearly one-half of the patients undergoing cardiac catheterization in our institution have compromised left ventricular function—a significant proportion of patients with angina. Based on the neutral cardiac effects of the newer long-acting dihydropyridines and bolstered by the V-HEFT III[27] and PRAISE[26] studies, it appears that appropriately chosen calcium antagonists are not associated with increased adverse outcomes. Furthermore, the use of beta blockers may prevent acute decompensation in patients whose left ventricular function is already depressed. Thus, calcium antagonists, if chosen appropriately, represent a useful treatment for ischemia in patients with compromised ventricular function.

SAFETY CONSIDERATIONS

Although safety issues are discussed elsewhere in this volume, several points are worth mentioning in relation to reports of increased risk of ischemic events with the use of calcium antagonists. Among the major relevant studies, Furberg et al.[29] demonstrated that short-acting nifedipine in moderate-to-high doses was associated with an increased risk of coronary events. Significant adverse events were seen predominantly in patients taking doses greater than 80 mg/day. The authors suggested that this finding may apply to other agents, but their suggestion is not confirmed. The commonly cited report by Psaty et al.[30] describes a risk of adverse events in hypertensive patients treated predominantly with short-acting calcium antagonists, of which nifedipine was a major component. This case-control study suggested a relative risk ratio of 1.6–1.0 for calcium antagonists compared with beta blockers and diuretics. Although such reports are concerning, two other studies of large patient populations failed to show similar risks. In a study of over 11,000 patients, Braun et al.[31] failed to find a significant difference in outcome for patients treated with or without calcium antagonists. Pahor et al.[32] also failed to find a significant difference in hypertensive patients treated with or without calcium antagonists.

Thus, if a risk exists, it is related to the short-acting calcium antagonists and perhaps most specifically to capsule nifedipine. Although the terms *short-acting* and *long-acting* are popular, the most important issue is the rate of drug delivery. Because cardiovascular mortality risk is postulated to be secondary to increased catecholamines, the more rapid the increase in blood level and subsequent hypotension, the greater the risk. Kleinbloesem et al.[33] compared an exponential rate of intravenous administration of nifedipine, which produced an immediately high blood level, with a gradual development of drug level over 5–7 hours. Both infusion regimens achieved similar therapeutic steady-state serum concentrations. During the rapid infusion phase, however, subjects experienced an increase in heart rate with no change in diastolic blood pressure. Conversely, during slower development of blood level, a significant drop in diastolic blood level was accompanied by minimal change in heart rate. These data suggest that differences in the release of calcium antagonists may be important in determining the potential baroreceptor response and ultimate release of catecholamines. Thus, it appears that long-acting agents are preferable because of their low slow development of a peak blood level, which avoids a neurohormonal receptor response.

TABLE 5. Advantages of Calcium Antagonists in the Management of Angina Pectoris

- Effective treatment of stable angina symptoms
- Favorable safety profile
- Neutral effect on lipids
- Effective treatment for concomitant hypertension
- Potential regression of left ventricular hypertrophy
- Safety in patients with diabetes mellitus (particularly insulin-dependent diabetes mellitus)
- Favorable effects on diastolic dysfunction
- Neutral effect on left ventricular dysfunction (long-acting dihydropyridines only)

At present, no data indicate an adverse risk for long-acting calcium antagonists. Furthermore, data from high-risk populations with compromised left ventricular function, such as the PRAISE[26] and V-HEFT III[27] studies, support the safety of calcium antagonists. Other ongoing studies will help to determine the potential safety of other agents such as mibefradil (MACH trial) (see chapter 29).

WHY USE CALCIUM ANTAGONISTS IN ISCHEMIC HEART DISEASE?

Tables 5 and 6 summarize the favorable characteristics of the calcium antagonists. As noted earlier, although the beta blockers have theoretical antiischemic advantages related to blocking catecholamines, they are significantly less well tolerated by patients than calcium antagonists. A major component of patients' response to drug therapy is how well they tolerate the agent. The calcium antagonists have a low side-effect profile and provide 24-hour efficacy if the long-acting agents are used. Although safety issues have been raised with short-acting agents, use of long-acting agents should avoid adverse consequences.

Furthermore, calcium antagonists have many potential benefits in the management of ischemia, particularly in relation to their vasoactive blocking properties. The additional, more subtle benefits, such as neutral effects on lipids, safety in patients with diabetes mellitus (particularly insulin-dependent diabetes mellitus), and ease of use in patients with compromised ventricular function, clearly make calcium antagonists a major contributor to the well-being and effective management of patients with ischemic heart disease.

TABLE 6. Advantages of Long-acting Dihydropyridine Calcium Antagonists

- Once-daily therapy (convenience, compliance)
- Safety in stable angina and hypertensive patients with left ventricular dysfunction and/or congestive heart failure
- Infrequent side effects (severe side effects rare)
- High rate of acceptance by patients

References

1. Tzivoni D, Kadr H, Braat S, et al: Efficacy of mibefradil in comparison to amlodipine in suppessing exercise-induced and daily silent ischemia: Results of a multicenter, placebo-controlled trial. Circulation [in press].
2. Antman E, Muller J, Goldberg S, et al: Nifedipine therapy for coronary artery spasm. Experience in 127 patients. N Engl J Med 302:1269–1273, 1980.
3. Taylor SH: Usefulness of amlodipine for angina pectoris. Am J Cardiol 73:28A–33A, 1994.
4. Deanfield JE, Detry J-M RG, Lichtlen PR, et al, for the CAPE Study Group: Amlodipine reduces transient myocardial ischemia in patients with coronary artery disease: Double-blind Circadian Anti-Ischemia Program in Europe (CAPE Trial). J Am Coll Cardiol 24:1460–1467, 1994.
5. Muller JE: Morning increase of onset of myocardial infarction: Implications concerning triggering events. Cardiology 76:96–104, 1989.
6. Rogers W, Bourassa M, Andrews T, et al: Asymptomatic Cardiac Ichemia Pilot (ACIP) Study: One-year follow-up. Circulation 90:1–17, 1994.
7. Cohn PF, Vetrovec GW, Nesto R, Gerber FR, and Total Ischemia Awareness Program investigators: The Nifedipine-Total Ischemia Awareness Program: A national survey of painful and antiischemic therapy. Am J Cardiol 63:534–539, 1989.
8. Schwartz A: Calcium antagonists: Review and perspective on mechanism of action. Am J Cardiol 64:31–91, 1989.
9. Abernethy DR: Pharmacologic and pharmacokinetic profile of mibefradil, a T- and L-type calcium channel antagonist. Am J Cardiol 80(4B):4C–11C, 1997.
10. Egstrup K, Anderson PE: Transient myocardial ischemia during nifedipine therapy in stable angina pectoris, and its relation to coronary collateral flow and comparison with metoprolol. Am J Cardiol 71:177–183, 1993.
11. Rousseau MF, Vincent MF, Van Hoof F, et al: Effects of nicardipine and nisoldipine on myocardial metabolism, coronary blood flow, and oxygen and oxygen supply in angina pectoris. Am J Cardiol 54:1189–1194, 1984.
12. Vetrovec GW, Parker VE: Nifedipine–beta blocker interactions: Acute effects on electrophysiologic, hemodynamic, global and regional left ventricular function. Am J Cardiol 55:21E–26E, 1985.
13. Bonow RO, Leon MB, Rosing DR, et al: Effects of verapamil and propranolol on left ventricular systolic function and diastolic filling in patients with coronary artery disease: Radionuclide angiographic studies at rest and during exercise. Circulation 63:1337–1380, 1981.
14. Vetrovec GW: Angina pectoris and calcium channel blockers: Practical guidelines. Cardiol Illustr II(No. 3):20–25, 1987.
15. Levy D, Garrison RJ, Savage DD, et al: Prognostic complications of echocardiographically determined left ventricular mass in the Framingham Heart Study. N Engl J Med 322:1561–1566, 1990.
16. Schmieder RE, Martus P, Klingbeil A: Reversal of left ventricular hypertrophy in essential hypertension. A metaanalysis of randomized double-blind studies. JAMA 275:1507–1513, 1996.
17. Leenen FHH, Fourney A: Comparison of the effects of amlodipine and diltiazem on 24-hour blood pressure, plasma catecholamines, and left ventricular mass. Am J Cardiol 78:203–207, 1996.
18. Phillips RA, Ardeljan M, Shimabukuro S, et al: Effect of nifedipine GITS on left ventricular mass and diastolic function in severe hypertension. J Cardiovasc Pharmacol 17(Suppl 2): S172–S174, 1991.
19. Waters D, Lesperance J, Francetich M, et al: A controlled clinical trial to assess the effect of a calcium channel blocker on the progression of coronary atherosclerosis. Circulation 82:1940–1953, 1990.
20. Lichtlen PR, Hugenholtz PG, Rafflenbeul W, et al, on behalf of the INTACT Group: Retardation of angiographic progression of coronary artery disease by nifedipine: Results of International Nifedipine Trial on Antiatherosclerotic Therapy (INTACT). Lancet 335:1109–1113, 1990.
21. HINT Research Group (Holland Interuniversity Nifedipine/Metoprolol Trial): Early treatment of unstable angina in the coronary care unit: A randomized double-blind placebo-controlled comparison of recurrent ischaemia in patients treated with nifedipine or metoprolol or both. Br Heart J 56:400–413, 1986.
22. Israeli SPRINT Study Group: Secondary Prevention Reinfarction Israeli Nifedipine Trial (SPRINT), a randomized intervention trial of nifedipine in patients with acute myocardial infarction. Eur Heart J 9:354–364, 1988.
23. SPRINT Study Group: The Secondary Prevention Reinfarction Israeli Nifedipine Trial (SPRINT-II): Design and methods, results [abstract]. Eur Heart J 9(Suppl):350A, 1988.

24. Danish Study Group on Verapamil in Myocardial Infarction: Effect of verapamil on mortality and major events after acute myocardial infarction (The Danish Verapamil Infarction Trial II—DAVIT II). Am J Cardiol 66:779–785, 1990.
25. Opie LH, Messserli FH: Nifedipine and mortality: Grave defects in dossier. Circulation 92:1068–1073, 1995.
26. Packer M, O'Connor CM, Ghali JK, et al: Effect of amlodipine on morbidity and mortality in severe chronic heart failure. N Engl J Med 335:1107–1114, 1996.
27. Cohn JN, Ziesche SM, Loss LE, and the V-HEFT Study Group: Effect of felodipine on short-term exercise and neurohormones and long-term mortality in heart failure: Results of V-HEFT III. Circulation 82:1940–1953, 1990.
28. Multicenter Diltiazem Postinfarction Trial Research Group: The effect of diltiazem on mortality and reinfarction after myocardial infarction. N Engl J Med 319:385–392, 1988.
29. Furberg CD, Psaty BM, Meyer JV: Nifedipine dose-related increase in mortality in patients with coronary heart disease. Circulation 92:1326–1331, 1995.
30. Psaty BM, Heckbert SR, Koepsell TD, et al: The risk of myocardial infarction associated with antihypertensive drug therapies. JAMA 274:620–625, 1996.
31. Braun S, Boyko V, Behar S, et al: Calcium antagonists and mortality in patients with coronary artery disease: A cohort study of 11,575 patients. J Am Coll Cardiol 28:7–11, 1996.
32. Pahor M, Guralnik JM, Corti MC, et al: Long-term survival and use of antihypertensive medications in older persons. J Am Geriatr Soc 43:1191–1197, 1995.
33. Kleinbloesem CH, van Brummelen P, Danlöf M, et al: Rate of increase in the plasma concentration of nifedipine as a major determinant of its hemodynamic effects in humans. Clin Pharmacol Ther 41:26–30, 1987.

GEORGE E. BILLMAN Ph.D.

7

Potential Usefulness of Calcium Antagonists in Ventricular Arrhythmias: Can Their Actions Be Targeted in Myocardial Ischemia?

Calcium antagonists represent a large, structurally diverse group of chemicals that share the ability to inhibit calcium entry into muscle cells through actions on the L-type calcium channel. Since their discovery in the 1960s,[1–3] calcium antagonists have become increasingly more important in the treatment of various cardiovascular diseases, most notably hypertension and angina pectoris.[4] With respect to cardiac arrhythmias, their usefulness appears largely restricted to the treatment of supraventricular arrhythmias (see chapter 9); their therapeutic potential in the management of life-threatening ventricular arrhythmias, particularly during myocardial ischemia, is not widely appreciated. This chapter describes the effects of myocardial ischemia on the calcium homeostasis of cardiac myocytes, discusses the role of calcium in the genesis of life-threatening (malignant) arrhythmias (e.g., ventricular fibrillation), summarizes experimental and clinical studies that illustrate the antiarrhythmic potential of individual calcium channel antagonists, and concludes by describing the potential usefulness of ischemia-selective calcium channel antagonists.

ISCHEMIA AND THE REGULATION OF MYOCARDIAL CALCIUM

Myocardial ischemia provokes abnormalities in the biochemical homeostasis of individual cardiac cells. These intracellular changes culminate in the disruption of cellular electrophysiologic properties, and life-threatening alterations in cardiac rhythm, such as ventricular fibrillation, frequently occur. Various chemical substances have been proposed as possible causative factors in the genesis of ventricular fibrillation during myocardial ischemia, including catecholamines, amphiphilic products of lipid metabolism, various peptides, cytosolic calcium accumulation, and increases in extracellular potassium.[5–7] This section focuses on the role that changes in cellular calcium play in the induction of cardiac arrhythmias during myocardial ischemia.

Under normal conditions ventricular muscle cells maintain resting levels of calcium approximately 5000 times lower than the extracellular calcium concentration.[8] Several important regulatory mechanisms are responsible for maintaining the low intracellular calcium levels vital for normal cardiac function (Fig. 1). In brief, calcium influx is restricted by voltage-sensitive calcium channels that

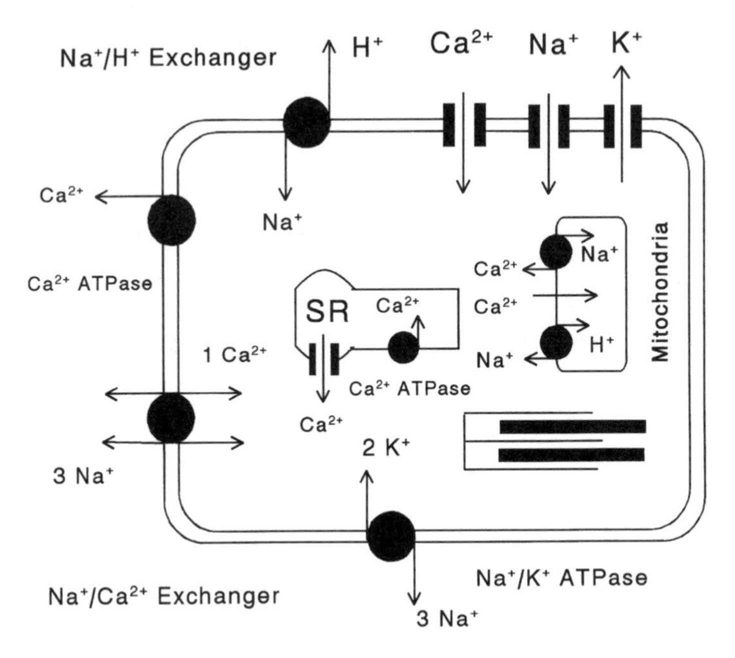

FIGURE 1. Mechanisms that regulate myocardial cell calcium levels (see text for details). The direction of the ion transport is indicated by the arrow. A double arrow indicates transport in both directions. SR = sarcoplasmic recticlum.

are activated by the cardiac action potential and regulated by intracellular messengers (e.g., phosphorylation).[8–10] Calcium is also extruded from the cell by an electrogenic Na^+/Ca^{2+} exchanger ($3Na^+$ in, $1Ca^{2+}$ out) and sarcolemmal Ca^{2+} adenosine triphosphatase (ATPase). Inside the cell, a second Ca^{2+} ATPase pumps calcium into the lumen of the sarcoplasmic reticulum. These systems rapidly decrease the elevations in cytosolic free calcium concentration brought about by excitation and induce relaxation during diastole. In addition, mitochondria can take up calcium, and a number of calcium-binding proteins also serve to buffer intracellular calcium levels.[8]

Intracellular calcium rises dramatically with the induction of ischemia, exceeding peak systolic calcium levels within 5–10 minutes.[11–14] Myocardial ischemia may provoke large increases in cellular calcium both directly (by alteration of the cellular calcium homeostatic mechanisms) and indirectly (by activation of the autonomic nervous system). Myocardial ischemia profoundly affects the autonomic regulation of the heart.[15] Coronary artery occlusion elicits reflex increases in cardiac sympathetic activity, accompanied by reductions in parasympathetic tone.[15,16] In fact, Collins and Billman[16] demonstrated that acute myocardial ischemia provoked larger increases in sympathetic activity, coupled with greater reductions in cardiac vagal tone, in animals subsequently shown to be susceptible to ventricular fibrillation (Fig. 2).

Alterations in autonomic regulation trigger a cascade of intracellular events that ultimately increase cytosolic calcium levels. Release of catecholamines from sympathetic nerve terminals activates both α- and β_1- and β_2-adrenergic receptors on cardiac myocytes. Stimulation of the β_1-adrenergic

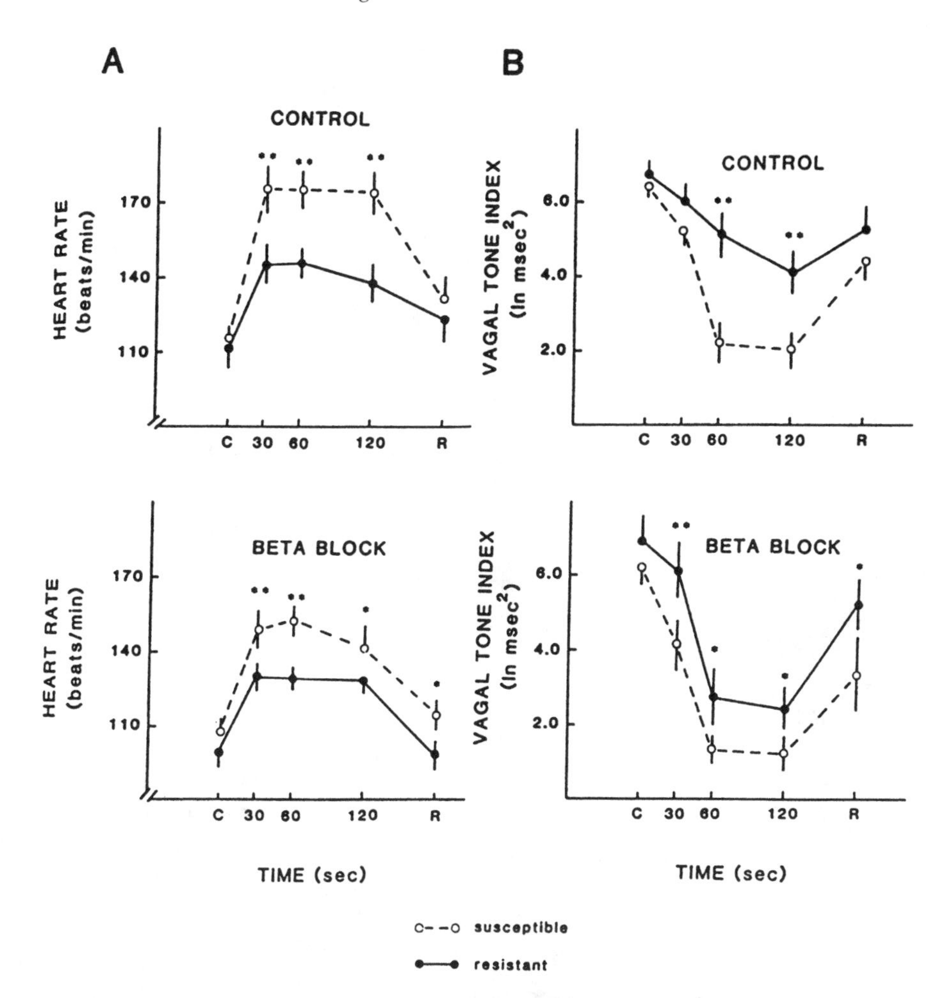

FIGURE 2. The heart rate (A) and cardiac vagal tone (B) index response to coronary occlusion before and after beta-adrenergic receptor blockade in animals susceptible and resistant to ventricular fibrillation. Note the greater increase in heart rate and greater reduction in cardiac vagal tone in the susceptible animals. * $p < 0.05$, ** $p < 0.01$ susceptible vs. resistant animals, C = control before occlusion, R = recovery 1 min after occlusion. (From Collins MN, Billman GE: Autonomic response to coronary occlusion in animals susceptible to ventricular fibrillation. Am J Physiol 257:H1886–H1894, 1987, with permission.)

receptor activates adenylyl cyclase, which, in turn, increases cellular levels of cyclic adenosine monophosphate (cAMP).[17] This cyclic nucleotide activates a cAMP- dependent protein kinase (PKA) that phosphorylates a variety of proteins, including the voltage-dependent calcium channel[18] and the calcium release channel of the sarcoplasmic reticulum.[19–22] It also phosphorylates the sarcoplasmic reticulum Ca-ATPase inhibitor, phospholamban, relieving its inhibition of calcium sequestration.[23] These reactions culminate in increased calcium entry into cardiac cells and increased uptake and release from intracellular stores. The mammalian myocardium also contains functional β_2-adrenergic receptors, which may become particularly important in cardiac disease.[24] Zinterol, a β_2-adrenergic receptor agonist, elicits significantly

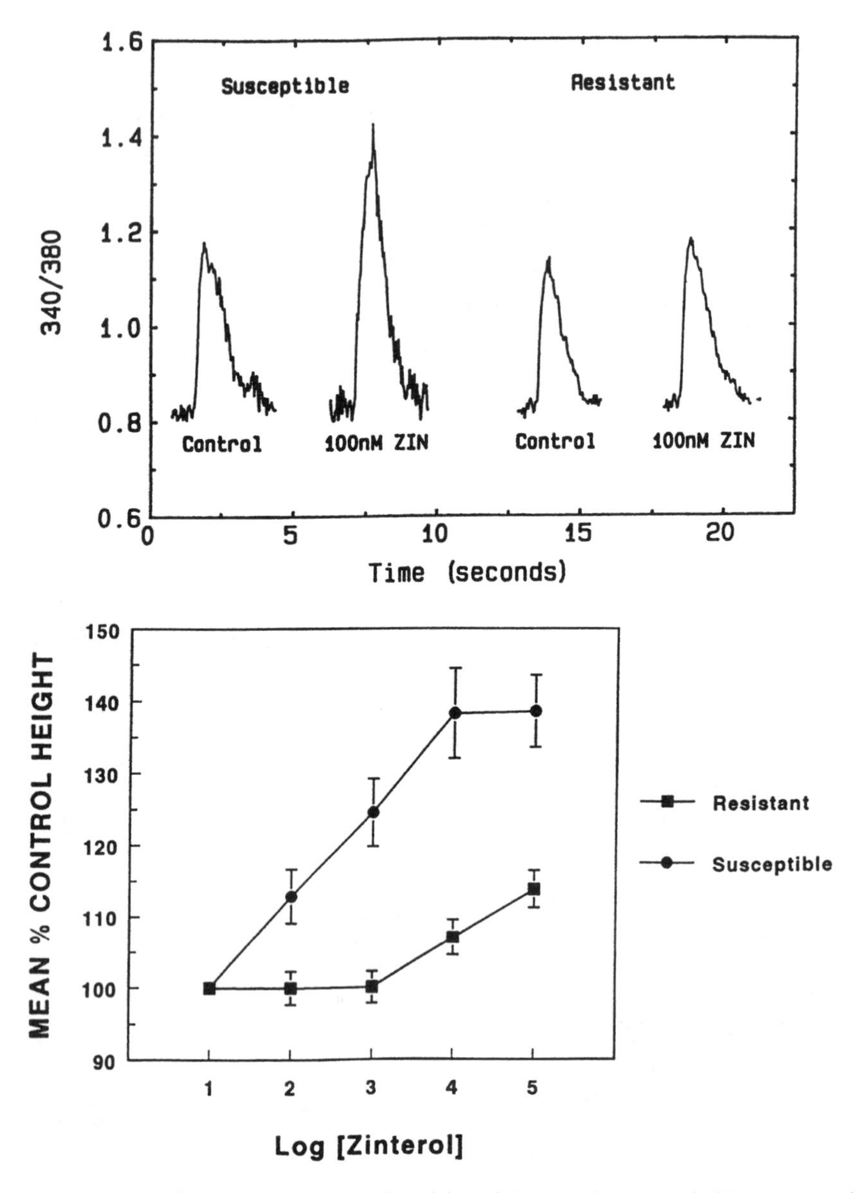

FIGURE 3. *Top panel,* Representative examples of the calcium transients recorded in myocytes obtained from one susceptible and one resistant animal before and after administration of the β_2-adrenergic receptor agonist zinterol. Note that the same dose of zinterol elicits a much larger increase in the transient amplitude in susceptible compared with resistant animals. *Bottom panel,* The effects of increasing doses of zinterol on the calcium transient amplitude recorded in ventricular myocytes obtained from hearts of susceptible (n = 5) and resistant animals (n = 5). $p < 0.01$ susceptible vs. resistant animals.

greater increases in calcium transient amplitude in myocytes isolated from animals susceptible to ventricular fibrillation than in myocytes obtained from animals resistant to malignant arrhythmias[25] (Fig. 3). It has recently been shown that β_2-adrenergic receptors elicit a cAMP-independent increase in cytosolic calcium,[24] which is perhaps mediated by direct G-protein coupling to the sarcolemmal L-type calcium channels.[26]

In a similar manner, α-adrenergic stimulation of the heart results in activation of a phospholipase that hydrolyzes phosphatidyl inositol into two second messengers, diacylglycerol and inositol trisphosphate.[27] Inositol trisphosphate facilitates calcium release from the sarcoplasmic reticulum, whereas diacylglycerol activates the important regulatory protein, protein kinase C (PKC). Thus, α- and β-adrenergic stimulation act synergistically to increase cytosolic calcium during ischemia.

Conversely, parasympathetic nerve activation opposes the action of sympathetic nerve stimulation, reduces cAMP levels, and increases levels of cyclic guanosine monophosphate (cGMP).[28] Cyclic GMP, in turn, decreases the open time of calcium channels independently of changes in cAMP levels[29] and activates a sarcolemmal calcium pump.[30] Parasympathetic stimulation, therefore, lowers intracellular calcium and halts the response to sympathetic stimulation; in general, however, the alterations in autonomic function elicited by myocardial ischemia favor the accumulation of cytosolic calcium.

Myocardial ischemia also directly alters several of the important calcium regulatory pathways noted above. As ischemia progresses, cellular ATP levels decline. As a consequence of ATP depletion, several energy-dependent functions are impaired. Sodium (Na^+)-potassium (K^+) ATPase (sodium pump) can no longer function properly, and cellular Na^+ levels increase.[8] Increased Na^+ reverses the normal direction of the Na^+/Ca^{2+} exchanger so that sodium is extruded and calcium is taken up by the cell.[8] In addition, the Ca^{2+} ATPases (calcium pumps) of the sarcolemma and sarcoplasmic reticulum are impaired so that less calcium is pumped out of the cell or into the sarcoplasmic reticulum during diastole (relaxation is delayed). The net result of this impairment of cellular calcium homeostatic mechanisms and enhanced sympathetic outflow to the heart is a significant rise in cytosolic calcium levels.[11–14] Indeed, Billman et al.[31] indirectly demonstrated that cytosolic calcium may be elevated in animals particularly susceptible to ventricular fibrillation. They found that calcium-dependent kinase activity was significantly greater in tissue obtained from animals that developed life-threatening arrhythmias during myocardial ischemia. Specifically, calcium-calmodulin–dependent phosphorylation was 2–3-fold higher in ventricular tissue obtained from animals that had ventricular fibrillation compared with animals that did not develop arrhythmias during ischemia (Fig. 4).

THE ROLE OF CALCIUM IN ARRHYTHMIA FORMATION

Disturbances in cardiac rhythm may result from perturbations in impulse generation, impulse conduction, or a combination of both.[32] Elevations in cellular calcium induced by myocardial ischemia may produce abnormalities in cardiac electrical properties and thereby trigger malignant arrhythmias.

Abnormal Impulse Generation

During coronary artery occlusion the resting potential of ischemic cardiac tissue becomes progressively less negative than the resting potential of surrounding

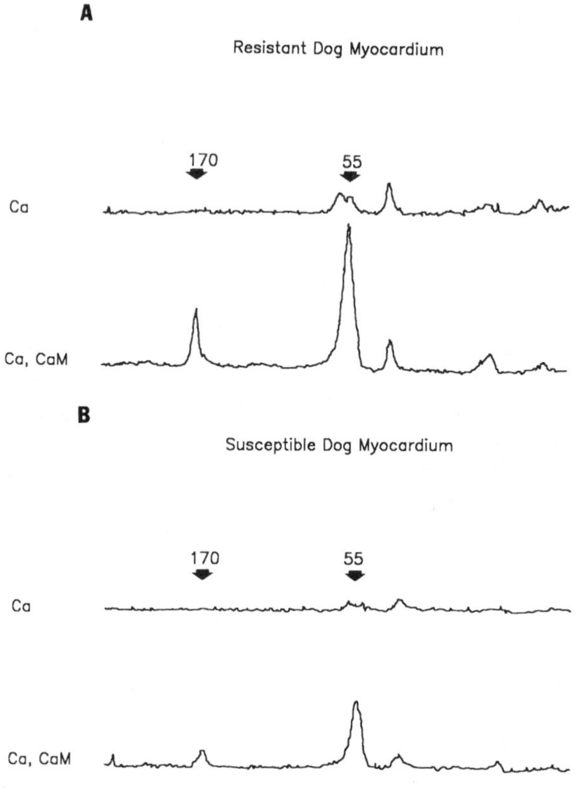

FIGURE 4. Calcium- and calmodulin (CaM)-dependent phosphorylation in resistant *(A)* and susceptible *(B)* dog posterior papillary muscle. Note the greater incorporation of P^{32} (i.e., phosphorylation) in 170- and 55-kDa proteins in the resistant myocardium. Pretreatment with the calcium ionophore A23187 increases calcium entry and thereby cytosolic calcium; thus it converts a resistant phosphorylation pattern to a susceptible pattern. Identical amounts of protein were used for both susceptible and resistant animals. (From Billman GE, McIlroy B, Johnson JD: Elevated myocardial calcium and its role in sudden cardiac death. FASEB J 5:2586–2592, 1991, with permission.)

nonischemic tissue.[32] The spread of this injury current tends to depolarize the surrounding tissue. Under normal conditions ventricular cells do not display a spontaneous rhythm; when such cells become partially depolarized, however, they may display an automatic rhythm.[32–34] This ischemia-induced ectopic rhythm is critically dependent on calcium entry; it is abolished by lowering extracellular calcium[34] or exposing the cardiac cells to a calcium channel antagonist.[33] Therefore, some forms of ventricular ectopic automaticity appear to depend on a slow inward calcium current.

Other mechanisms may lead to abnormal impulse formation. As noted above, myocardial ischemia results in elevations of cytosolic calcium, which, in turn, provoke oscillations in membrane potential.[35,36] These oscillations or fluctuations in membrane voltage occur either immediately after the plateau of the cardiac action potential (early afterdepolarizations [EADs]) or after repolarization of the membrane (delayed afterdepolarizations [DADs]). When the amplitude of the afterpotential is large enough to reach threshold, repetitively sustained action potentials are generated. This form of ectopic automaticity is known as triggered activity, because it does not occur unless preceded by at least one action potential. Afterdepolarizations have been recorded in isolated cardiac cells or tissue in response to interventions that favor calcium loading (hypoxia, cocaine, catecholamines, digitalis, calcium channel agonist BAY K 8644) and can be suppressed by calcium channel antagonists and the intracellular calcium chelator, BAPTA-AM.[31,35–40]

The initiation of ventricular fibrillation may depend on inward movement of calcium.[41] Ryanodine, a plant alkaloid that renders the sarcoplasmic reticulum leaky and unable to retain normal amounts of calcium,[42,43] suppresses cytosolic calcium oscillations but fails to prevent ventricular fibrillation in isolated rabbit hearts.[41] In contrast, verapamil and nifedipine, L-type calcium channel blockers, terminate ventricular fibrillation.[41] In related studies, Billman[44–46] demonstrated that several organic (verapamil, flunarizine, nifedipine, diltiazem, mibefradil) and inorganic (magnesium [Mg^{2+}]) calcium channel antagonists prevent malignant ventricular arrhythmias induced by ischemia (Fig. 5). Conversely, the L-type calcium channel agonist, BAY K 8644,[47] induced ventricular fibrillation in animals resistant to the development of arrhythmias.[44] Ryanodine failed to prevent malignant arrhythmias despite large reductions in peak cytosolic calcium, as indicated by corresponding reductions in contractile force development[48] (Fig. 6). These data indicate that calcium influx

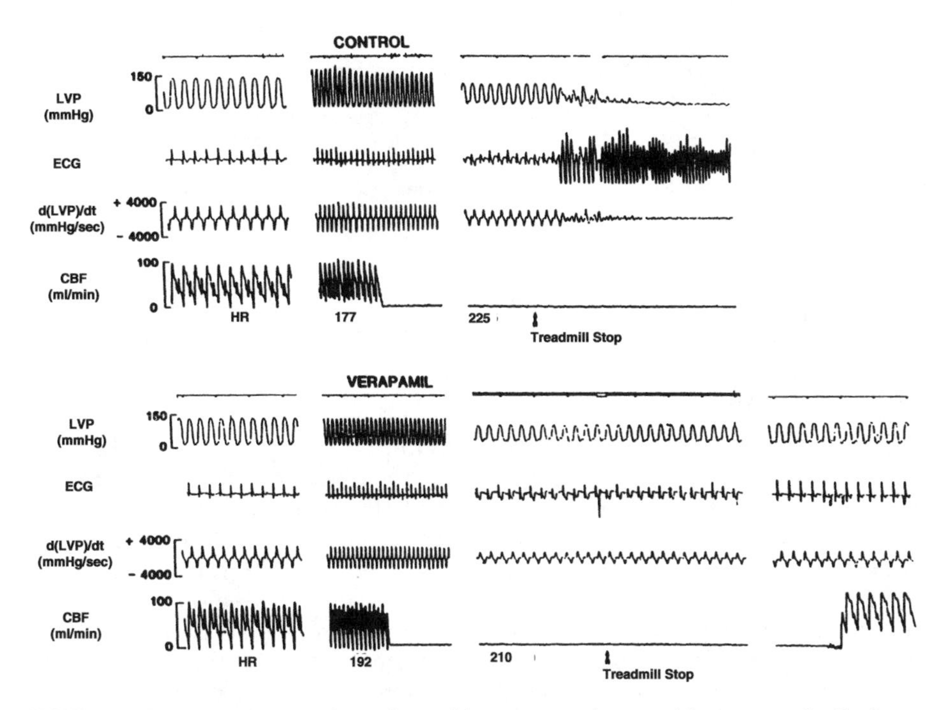

FIGURE 5. Representative recordings obtained from an animal susceptible to ventricular fibrillation induced by an exercise plus ischemia test before and after pretreatment with the calcium antagonist verapamil (250 μg/kg IV). LVP = left ventricular pressure, CBF = coronary blood flow, HR = heart rate. (From Billman GE: Effect of calcium channel antagonists on susceptibility to sudden death: Protection from ventricular fibrillation. J Pharmacol Exp Ther 248:1334–1342, 1989, with permission.)

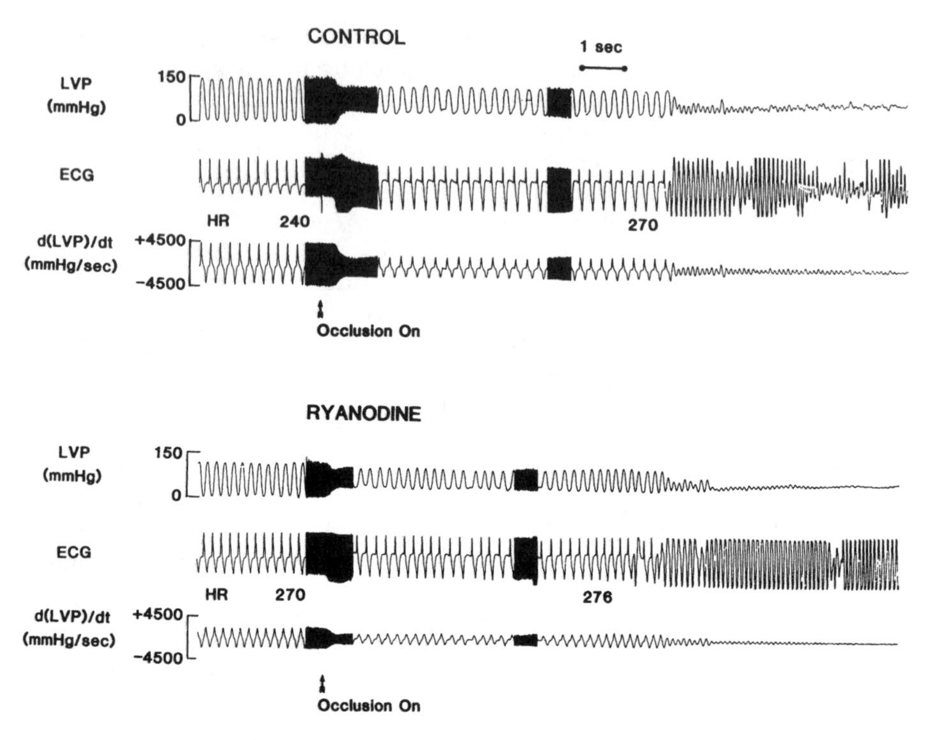

FIGURE 6. Representative recordings from an animal susceptible to ventricular fibrillation induced by an exercise plus ischemia before and after pretreatment with ryanodine (10 μg/kg IV). Note that ryanodine failed to prevent ventricular fibrillation despite large reductions in left ventricular dp/dt maximum (an indirect marker of reductions in cytosolic calcium levels). LVP = left ventricular pressure, CBF = coronary blood flow, HR = heart rate. (From Lappi MD, Billman GE: Effect of ryanodine on ventricular fibrillation induced by myocardial ischaemia. Cardiovasc Res 27:2152–2159, 1993, with permission.)

across the sarcolemma rather than calcium release from the sarcoplasmic reticulum may be critical for the induction of ventricular fibrillation.

Abnormal Impulse Conduction

Calcium also may contribute to changes in impulse conduction. Conduction abnormalities may result from simple conduction block or more complex forms of reentry.[32] In the normal heart, action potentials generated in the sinus node terminate after the sequential activation of the atria and the ventricles, because the surrounding tissue has become refractory or nonexcitable after depolarization. If, however, the impulse conduction is slowed in one region of the heart and the surrounding tissue has repolarized, it may be possible to reexcite the surrounding tissue before the next impulse is conducted from the sinus region. This phenomenon, known as reentrant excitation, is responsible for the generation of extrasystoles (reentrant arrhythmias). Calcium channel antagonists exert their most obvious effects on the conduction of action potentials through the atrioventricular (A-V) node. Because A-V nodal tissue generates slow-response (i.e., calcium-dependent) action potentials, calcium antagonists prolong

A-V conduction time and refractory period.[49,50] These actions attenuate the ventricular response to rapid atrial arrhythmias (atrial flutter or fibrillation) and terminate supraventricular tachycardias in which the A-V node forms part of the reentrant circuit.[49]

The effects of calcium on conduction abnormalities in the ventricles are, however, equivocal. Ordered or simple reentrant arrhythmias in which the impulse is conducted in a finite and well-circumscribed loop may occur in an ischemic heart.[32] Reentrant arrhythmias require decremental conduction and unidirectional block as preconditions for arrhythmia formation.[32] Conduction velocity in cardiac tissue depends on the rate of depolarization (dV/dt_{max} or V_{max}) and action potential amplitudes, factors mediated by the fast sodium channels.[32] As noted above, myocardial ischemia results in depolarization of the resting membrane potential, which may lead to inactivation of sodium channels.[32,51] Consequently, conduction velocity decreases, and unidirectional block may occur. In acute ischemia many conduction disturbances that produce reentrant arrhythmias are not mediated by slow-response action potentials but rather by reduced sodium entry through fast channels.[52] It is therefore not surprising that calcium channel antagonists are not effective against ordered reentrant arrhythmias.[50,53] However, conduction velocity also depends on a low electrical resistance between cells.[31,51] As ischemia progresses, intracellular Ca^{2+} and hydrogen (H^+) increase.[8,11–14,54] High concentrations of these ions reduce conductance across the gap junctions that form the low electrical resistance pathway that facilitates cell-to-cell coupling.[55] Thus, during later stages of ischemia or in chronically ischemic hearts, conduction disturbances may result from the uncoupling of cardiac cells due to the cellular accumulation of calcium. Calcium channel antagonists can diminish calcium accumulation and thereby improve conduction in ischemic hearts. Verapamil reduces, whereas BAY K 8644 exacerbates, the slowing of ventricular conduction induced by global ischemia in the isolated rabbit heart.[56]

In contrast to ordered reentrant circuits, calcium may contribute significantly to random or irregular reentrant circuits. Random reentry is characterized by multiple irregular pathways that change continuously, producing an unpredictable, chaotic conduction pattern. Ventricular fibrillation is the epitome of random reentry. A major factor contributing to ventricular fibrillation, particularly during myocardial ischemia, is a spatial dispersion or nonuniformity of the refractory period,[32] which allows impulse conduction to become fragmented during ensuing heartbeats and thus sets the stage for random reentry. Dispersion of refractory periods results, at least in part, from disturbances in action potential duration, which can be recorded as alterations in the S-T segment (electrical alternans).[32,57–60] Recently, Lee and coworkers[13] demonstrated that alterations in the amplitude of calcium transients accompanied corresponding changes in action potential duration. The pattern of alternans was stable at a given recording site but varied from site to site in a given preparation. They concluded that "the alternans behavior of the calcium transients in a particular region is independent of the behavior of other regions, which results in spatial heterogeneity of the calcium transients during ischemia."[13] Calcium channel antagonists have been shown to reduce calcium transient and electrical alternans[61] and the spatial dispersion of refractory period from the endocardium to epicardium during ischemia.[62] These data indicate that nonhomogeneity of refractory periods may result from a calcium-mediated oscillation of action potential duration and, in turn, form a substrate for irregular reentry.

In summary, abnormalities in cellular Ca^{2+} may contribute significantly to the development of malignant ventricular arrhythmias by inducing various forms of ectopic automaticity, by changing conduction, or by a combination of both automaticity and conduction disturbances. If, for example, an extrasystole occurs in a region of nonuniform refractory period, irregular reentrant pathways and ventricular fibrillation may result (Fig. 7).

CALCIUM ANTAGONISTS IN THE TREATMENT OF ARRHYTHMIAS INDUCED BY MYOCARDIAL ISCHEMIA

Interventions that alter calcium ion flux across the sarcolemma should also alter the potential for arrhythmias induced by myocardial ischemia. In particular, one might predict that calcium channel antagonists would protect against the formation of ischemia-associated malignant arrhythmias. The following section briefly reviews experimental and clinical experience with selected calcium ion channel antagonists. Numerous drugs affect other ions and are also useful in ischemia-related arrhythmias. Recent reviews provide relevant information about the classes of antiarrhythmic drugs.[6,63,64]

Verapamil

Verapamil, a benzeneacetonitrite, is structurally similar to papaverine and was first synthesized in 1962.[65] Verapamil blocks calcium entry equally in both cardiac and smooth muscle. Calcium-entry blockade in smooth muscle elicits vasodilation with corresponding reductions in arterial pressure and increases in coronary blood flow. In cardiac muscle, verapamil tends to depress both contractile force and impulse conduction. The second action is particularly obvious at the A-V node.

In 1968, Kaumann and Aramendia[66] were the first to demonstrate that verapamil can protect against ventricular fibrillation induced by ligation of a coronary artery. These findings have been confirmed by many other animal studies.[44,67–69] In intact animals, verapamil prevents ventricular arrhythmias induced by myocardial ischemia[44,69] and attenuates the reduction in ventricular fibrillation threshold that accompanies coronary artery occlusion.[67,70] Billman and coworkers further demonstrated that verapamil completely suppresses ventricular fibrillation induced by either cocaine[71] or the combination of exercise and acute myocardial ischemia.[44] In contrast, verapamil fails to prevent ordered reentrant arrhythmias induced by programmed electrical stimulation.[49,53,72,73]

Clinical experience with verapamil to date has been somewhat inconsistent. For example, verapamil is generally ineffective in the treatment of either stable reentrant arrhythmias or arrhythmias induced by programmed electrical stimulation,[49,53,72,73] which also may result from reentry. In contrast, some reports indicate that arrhythmias associated with acute myocardial ischemia or exercise-induced ventricular tachycardia respond favorably to verapamil.[72,74] Verapamil also significantly decreased the frequency and severity of ventricular arrhythmias in patients with left ventricular hypertrophy.[75] Finally, in a large Danish

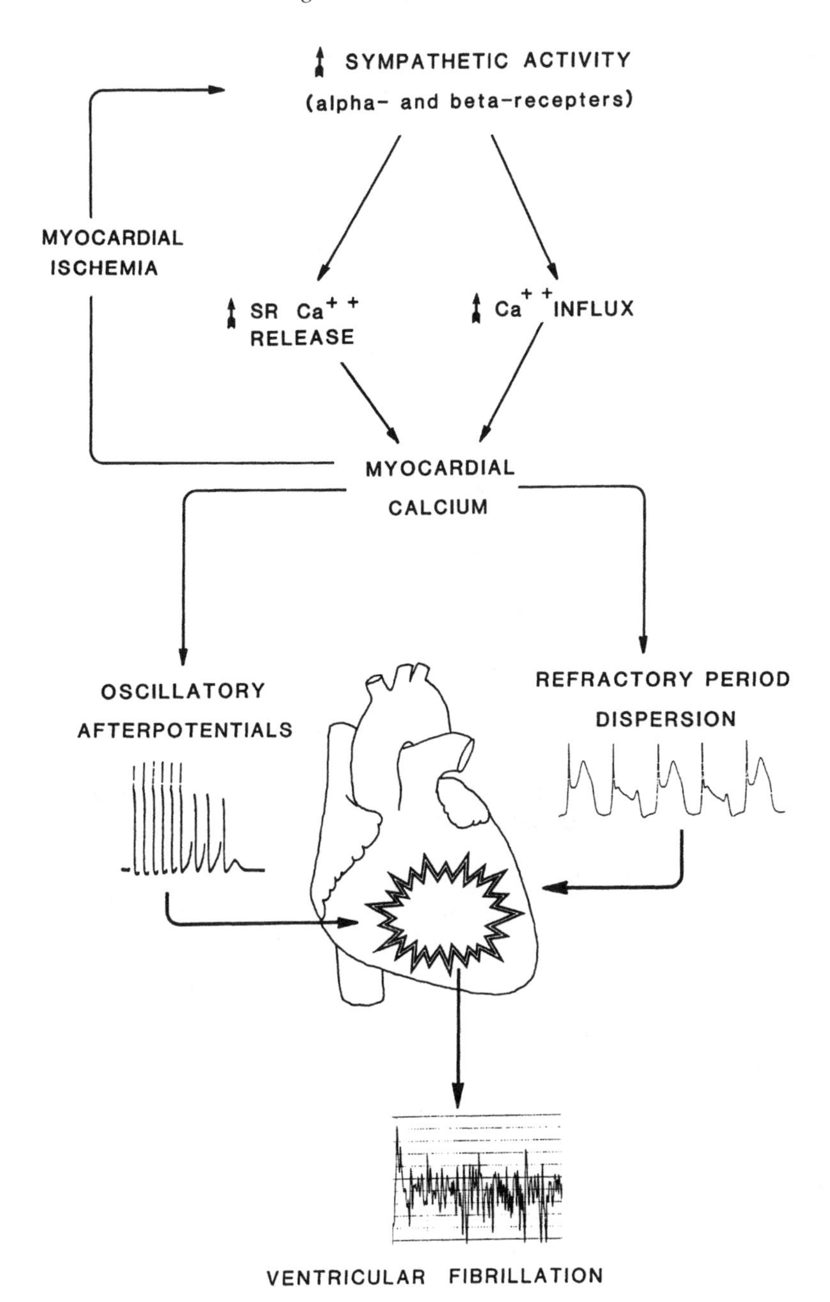

FIGURE 7. Mechanisms by which elevations in intracellular calcium may trigger malignant arrhythmias. Myocardial ischemia can increase calcium levels both directly and indirectly (sympathetic nerve activation). Increased myocardial calcium may trigger oscillatory afterpotential and extrasystoles. In addition, calcium transient alternans due to myocardial calcium overload provokes disturbances in action potential duration (electrical alternans) that may lead to dispersion of refractory period. The result is inhomogeneity of repolarization, which forms the substrate for random reentrant pathways. Therefore, when an extrasystole is triggered by oscillatory afterpotentials in a region of refractory period dispersion, lethal tachyarrhythmias (e.g., ventricular fibrillation) may result. (From Billman GE: The antiarrhythmic and antifibrillatory effects of calcium antagonists. J Cardiovasc Pharmacol 18(Suppl 10): S107–S117, 1991, with permission.)

multicenter study[76] of patients recovering from myocardial infarction, verapamil significantly reduced the frequency of major cardiac events (cardiac death or second myocardial infarction). Sudden death was reduced by 20–26% in patients treated with verapamil—results comparable to those obtained with β-adrenergic antagonists.[77] This protection, however, was noted only in patients without evidence of heart failure. Verapamil had no beneficial effects on major cardiac events in patients with heart failure and, in fact, may have increased cardiac mortality.[76] The reduction in sudden death may reflect inhibition or reduction of calcium loading in ventricular cells, but no direct data yet support this suggestion.

Diltiazem

Diltiazem, a benzodiazepine derivative, is structurally quite distinct from verapamil and was first identified as a calcium antagonist in 1971.[78] Diltiazem also has pronounced effects on both vascular smooth and cardiac muscle; it relaxes smooth muscle, increasing coronary blood flow and reducing arterial pressure. Diltiazem also depresses A-V conduction but has a lesser negative inotropic effect than verapamil. Less consistent antiarrhythmic actions have been noted for diltiazem than for verapamil, perhaps because of its less potent cardiac actions. Diltiazem reduces regional differences in impulse conduction in ischemic tissue,[33,79] prevents slow-response action potentials,[10] and abolishes calcium transient or electrical alternans.[13,61] Diltiazem also protects against ventricular fibrillation in anesthetized animals.[61,80] In unanesthetized canine preparations, diltiazem delays the time to onset of malignant arrhythmias, but it fails to prevent ventricular fibrillation induced by irreversible coronary artery occlusion.[81] Recently, Billman demonstrated that diltiazem prevents ventricular fibrillation induced by either cocaine[71] or exercise plus acute myocardial ischemia.[44] In contrast, diltiazem failed to prevent ischemic changes in the ventricular fibrillation threshold[82] as well as arrhythmias associated with reperfusion.[83]

Similar mixed results have been reported in clinical studies of the antiarrhythmic potential of diltiazem. Diltiazem was found to reduce the reinfarction rate by approximately 51% in patients with myocardial infarction but, in contrast to verapamil, did not affect overall mortality rates.[84] Furthermore, diltiazem failed to reduce stable ventricular arrhythmias during a 12-hour recording period.[85] However, in a study involving over 2,400 patients with myocardial infarction, diltiazem produced a significant decrease in mortality in a subgroup of patients without radiographic evidence of pulmonary congestion (i.e., heart failure).[86] It is unclear whether the reduction in mortality arises from blockade of cardiac calcium channels and the resultant prevention of cellular calcium overload. The subgroup analysis further demonstrated that diltiazem provoked a dramatic increase in mortality in patients with myocardial infarction complicated by pulmonary congestion.[86] Calcium antagonists, therefore, may have deleterious effects in patients with compromised cardiac function (see chapter 29).

Nifedipine

Nifedipine, a dihydropyridine, was first synthesized in 1971.[87] In contrast to either diltiazem or verapamil, therapeutic concentrations of nifedipine act primarily on vascular smooth muscle. Nifedipine also acts as a more potent vasodilator than either diltiazem or verapamil; cardiac actions are noted only at much higher concentrations. Therefore, it is not surprising that nifedipine exhibits few antiarrhythmic properties in intact animals or patients. Nifedipine, with a few exceptions (see below), is generally ineffective in the treatment of experimentally induced ischemic arrhythmias.[44,88,89] Nifedipine fails to alter the reduction in ventricular fibrillation threshold induced by coronary artery occlusion[89] and to prevent reperfusion arrhythmias.[83] However, it has been reported that nifedipine prevents ventricular fibrillation in the ischemic rat heart. This protection is attributed to improved coronary perfusion (i.e., less myocardial ischemia) rather than to direct cardiac actions of the drug.[90] Nifedipine also protects against malignant arrhythmias induced by exercise plus myocardial ischemia, but only at very high concentrations.[44] Nifedipine also fails to reduce mortality in patients recovering from myocardial infarction.[91] A reflex tachycardia induced by nifedipine's reduction of arterial blood pressure may worsen the condition of such patients by placing a higher metabolic demand on the damaged heart. Indeed, much of the recent controversy[92] surrounding its use may be related to this reflex action. The increased metabolic demand placed on the heart may counteract the beneficial actions of reduced arterial blood pressure (see chapter 29).

Flunarizine

Flunarizine is a difluoronated piperazine derivative that, under physiologic conditions, interacts weakly with both sodium and calcium channels.[93] However, during hypoxia flunarizine inhibits increases in cytosolic calcium and prevents tissue damage, particularly in neural tissue.[93] It seems reasonable that flunarizine may act as a calcium overload antagonist and thereby prevent malignant arrhythmias during myocardial ischemia. Flunarizine protects against isoproterenol-induced cardiac lesions and ischemic injury of the rat heart.[93] It also reduces ventricular fibrillation by 100% and ectopic beats during occlusion and reperfusion of the left anterior descending coronary artery;[93] however, a much higher dose was required for protection equivalent to that noted with verapamil.[93] Flunarizine also prevents ventricular tachycardia due to delayed afterdepolarizations induced by ouabain toxicity but not arrhythmias due to activation of reentrant circuits.[94] Based on these findings, Vos et al.[94] proposed that flunarizine may be used to differentiate between arrhythmias arising from calcium overload (triggered activity) and arrhythmias due to reentry. Recently, Billman[45,71] demonstrated that flunarizine completely suppresses ventricular fibrillation induced by either cocaine or exercise plus myocardial ischemia; however, large reductions in the inotropic state were noted. The author is unaware of any clinical reports of the use of flunarizine in the management of ventricular arrhythmias. However, given the potential for severe cardiac mechanical depression, it is likely that flunarizine may produce deleterious effects, particularly in patients with compromised cardiac function.

Mibefradil

Experimental and clinical evidence described above suggests that calcium antagonists have the potential to protect against malignant arrhythmias induced by myocardial ischemia. Calcium antagonists that exert demonstrable direct cardiac actions in therapeutic concentrations (verapamil, flunarizine) are, in general, more effective antiarrhythmic agents in myocardial ischemia than antagonists that are more selective for vascular smooth muscle (nifedipine). However, the more effective calcium channel antagonists also depress cardiac mechanical function and often produce deleterious effects in patients with compromised cardiac function[76,86]—the very patients at greatest risk for malignant arrhythmias.[95] The challenge, therefore, is to develop calcium antagonists that modulate myocardial calcium, particularly during ischemia, without compromising cardiac function.

Mibefradil (Ro 40-5967), a benzimidazolyl-substituted tetraline derivative, may represent a new class of calcium channel antagonists with these clinically important characteristics.[96,97] Mibefradil binds at or near the same membrane sites as verapamil but exerts few negative inotropic effects on the isolated heart.[98–101] For example, mibefradil has the same affinity for the [^{3}H]desmethoxyverapamil binding site as verapamil, but it elicits an approximately 10-fold less reduction in myocardial force.[96] In addition, mibefradil improves myocardial function during experimental myocardial ischemia, whereas in the same study verapamil not only failed to protect against ischemia but also produced severe cardiac depression.[98] Furthermore, in marked contrast to either diltiazem or verapamil, mibefradil appears not to alter contractile function in animal models of heart failure or myocardial infarction.[46,98–103] Recently, it has been reported that mibefradil significantly reduces ventricular fibrillation brought about by the combination of exercise and acute ischemia.[46] In contrast to either verapamil or diltiazem, this protection was afforded without adversely affecting A-V conduction or inotropic state.[46] Mibefradil is well tolerated by patients,[104] reducing arterial pressure without adverse effects on cardiac contractile function, even in patients with heart failure.[105]

It is still unclear how mibefradil, in contrast to verapamil or diltiazem, can inhibit myocardial calcium entry without adverse actions on contractile function or impulse conduction, even at high doses. However, a unique electropharmacologic property of mibefradil may be responsible for the cardioprotection. At physiologic cell membrane resting potentials, verapamil is 20 times more potent than mibefradil in the inhibition of calcium entry into cardiac myocytes.[106] However, in depolarized cells, verapamil and mibefradil are equipotent in the inhibition of calcium entry[106] (Fig. 8). More recent studies have confirmed these findings.[107–111] Bezprozvanny and Tsien[110] demonstrated that mibefradil is a more effective inhibitor of the L-type calcium current at reduced (depolarized) membrane potentials. They concluded that this voltage dependency resulted from a preferential binding to the open and inactivated state of the L-type calcium channel. As noted above, significant numbers of ventricular cells depolarize during myocardial ischemia.[32] Therefore, mibefradil may be particularly selective for ischemic tissue—the tissue most vulnerable to calcium overload and arrhythmia formation. As such, mibefradil would exert little negative effect on normal tissue but during ischemia would become an effective myocardial calcium channel antagonist. Perhaps because of its membrane potential sensitivity, mibefradil may act as an ischemia-selective calcium channel antagonist. Indeed, mibefradil completely suppresses ventricular tachycardia induced by programmed electrical stimulation during myocardial ischemia but fails to prevent electrically induced arrhythmias under nonischemic conditions.[112]

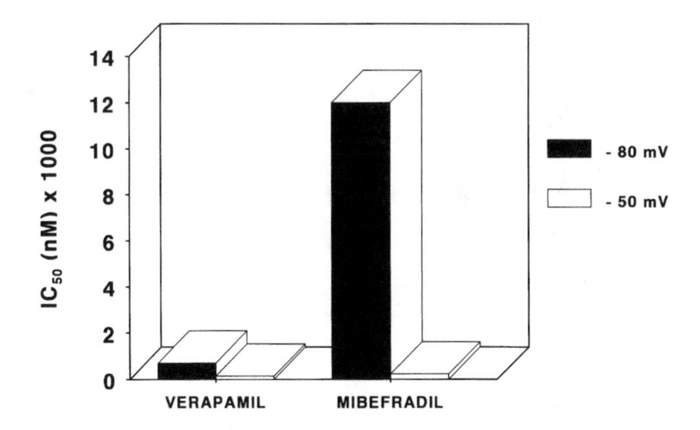

FIGURE 8. Comparison of the effects of mibefradil and verapamil on the inhibition of calcium entry into guinea pig ventricular myocytes at a holding potential of –50 mV (depolarized) and –80 mV (physiologic resting membrane potential). Note the large decrease in the concentration of mibefradil necessary to block calcium entry into the depolarized tissue. (Data from Fang L, Osterrieder W: Potential-dependent inhibition of cardiac Ca^{2+} inward currents by Ro 40-5967 and verapamil: Relation to negative inotropy. Eur J Pharmacol 196:205–207, 1991, with permission.)

In fact, mibefradil, as well as other calcium channel antagonists, exacerbated the nonischemic arrhythmias induced by programmed electrical stimulation[112] (Fig. 9). For example, mibefradil (as well as verapamil and diltiazem) increased both duration and rate of ventricular tachycardia in approximately 50% of the animals tested.[112] Finally, mibefradil has recently been shown to inhibit preferentially calcium entry through T-type calcium channels.[111,113] The number of such channels may increase in cardiac disease.[114,115] Activation of T-type calcium channels after myocardial infarction also may contribute significantly to the induction of ventricular fibrillation, particularly during myocardial ischemia.

CONCLUSION

Alterations in cellular calcium homeostasis induced by myocardial ischemia may be responsible for the perturbations in cardiac electrical stability that culminate in malignant ventricular arrhythmias. Calcium channel antagonists, by reducing calcium movement across the sarcolemma, may reduce the propensity for induction of ventricular fibrillation during myocardial ischemia. The attenuated calcium entry may directly abolish afterdepolarizations, reduce the dispersion of refractory period, and improve cell-to-cell coupling (conduction). In addition to these direct actions on the ischemic myocardium, smooth muscle relaxation may produce a more favorable balance between oxygen supply (i.e., increased coronary blood flow) and oxygen demand (i.e., reduced afterload).

The experimental and clinical evidence cited above demonstrate that individual calcium channel antagonists have variable antiarrhythmic properties (Table 1). In general, calcium channel antagonists with the most obvious direct

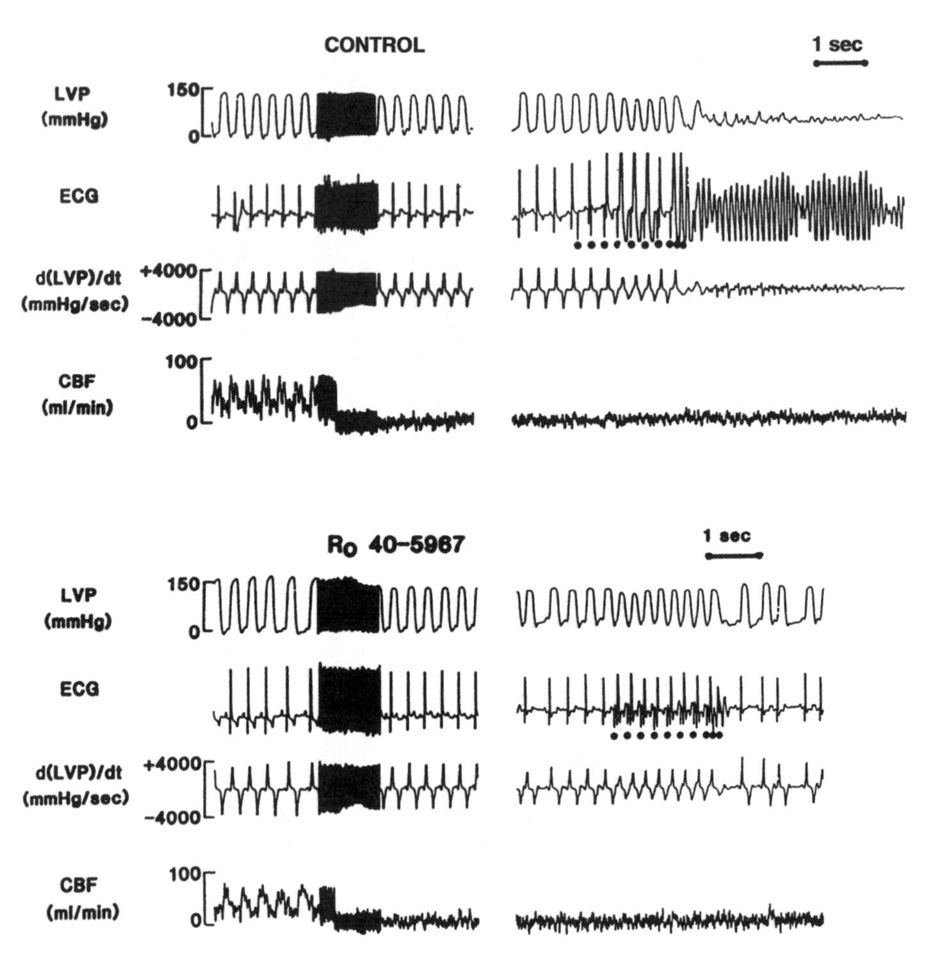

FIGURE 9. Representative recordings of the response to programmed electrical stimulation before and after pretreatment with mibefradil (Ro-40 5967, 1.0 mg/kg IV.). The dark circles indicate the electrical stimulation (8 paced beats, S1, followed by 2 progressively shorter extrastimuli, S2 and S3). Arrhythmias were not inducible in this animal without ischemia. Note that mibefradil completely suppressed the induction of the electrically induced arrhythmias. LVP = left ventricular pressure, CBF = coronary artery blood flow. (From Billman GE, Hamlin RL: The effects of a novel calcium channel antagonist, mibefradil, on refractory period and arrhythmias induced by programmed electrical stimulation and myocardial ischemia: A comparison with diltiazem and verapamil. J Pharmacol Exp Ther 277:1517–1526, 1996, with permission.)

cardiac actions (flunarizine, verapamil) are the most effective antiarrhythmic agents. In contrast, the drugs that exert a greater influence on vascular smooth muscle (nifedipine) are relatively ineffective in the prevention of ischemic arrhythmias. However, the most effective antiarrhythmic agents also depress cardiac function and may provoke deleterious responses in patients with compromised cardiac function—the very patients most at risk for lethal cardiac arrhythmias.[94] Effective antiarrhythmic therapies in such high-risk patients, therefore, must await the development of calcium channel antagonists that can modulate cardiac cellular calcium levels without adversely altering cardiac pump function. Mibefradil, which appears to act selectively on the ischemic

TABLE 1. Effect of the Modulation of Myocardial Calcium on Ventricular Fibrillation

Drug	Myocardial Calcium	Ventricular Arrhythmias
Calcium antagonists		
Verapamil	↓	↓
Diltiazem (high dose)	↓	↓
Flunarizine	↓	↓
Nifedipine (high dose)	↓	↓
Mibefradil	↓	↓
BAPTA-AM	↓	↓
Diltiazem (low dose)	—	—
Nifedipine (low dose)	—	— or ↑
Calcium agonists		
BAY K8644	↑	↑
Digitalis	↑	↑

↑ = increase; ↓ = decrease; — = unchanged.

myocardial tissue (i.e., depolarized tissue), prevents ventricular fibrillation without altering either A-V nodal conduction or inotropic state. As such, mibefradil may represent the first of the next generation of ischemia-selective calcium channel antagonists. Such drugs should prove to be particularly effective in the management of life-threatening arrhythmias in high-risk populations, such as patients with heart failure. Only time will tell whether this exciting prospect is confirmed in clinical settings, in which the unique pharmacokinetic properties of the various classes of calcium channel antagonists[116] may be modified by disease.

Acknowledgment

The author expresses his sincere gratitude to Arthur Bassett, Ph.D., for critical evaluation of the manuscript. His insightful suggestions led to a much better chapter.

References

1. Fleckenstein A: History of calcium antagonists. Circ Res 52(Suppl 1):I3–I16, 1983.
2. Fleckenstein A, Kammermeier H, Doring H, Freund HJ: Zum Wirkungs-Mechanismus Neuartiger Koronardilatoren Mit Gleichzeitig Sauerstoff-einsparenden Myodardeffekten, Prenylamin und Iproveratril. Z Kreislaufforsch 56:716–744, 835–853, 1967.
3. Kohlhardt M, Bauer P, Krause H, Fleckenstein A: Differentiation of transmembrane Na and Ca channel in mammalian cardiac fibres by the use of specific inhibitors. Pflügers Arch 335:309–322, 1972.
4. Struyker-Boudier HAJ, Smits JFM, DeMey JGR: The pharmacology of calcium antagonists: A review. J Cardiovasc Pharmacol 15(Suppl 4):S1–S10, 1990.
5. Opie LH, Nathan D, Lubbe WF: Biochemical aspects of arrhythmogenesis. Am J Cardiol 43:131–148, 1979.
6. Curtis MJ, Prigsley MK, Walker MJA: Endogenous chemical mediators of ventricular arrythmias in ischaemic heart disease. Cardiovasc Res 27:703–719, 1993.
7. Billman GE: The antiarrhythmic and antifibrillatory effects of calcium antagonists. J Cardiovasc Pharmacol 18(Suppl 10):S107–S117, 1991.
8. Barry WH: Calcium and ischemic injury. Trends Cardiovasc Med 1:162–166, 1991.
9. Reuter H: Calcium channel modulation by neurotransmitters, enzymes and drugs. Nature (Lond) 301:569–574, 1983.
10. Sperelakis N, Wahler GM, Bkaily G: Properties of myocardial calcium slow channels and mechanism of action of calcium antagonistic drugs. Curr Top Membrane Transport 25:43–76, 1985.

11. Kihara Y, Morgan JP: Intracellular calcium and ventricular fibrillation: studies in the aequorin-loaded isovolumic ferret heart. Circ Res 68:1378–1389, 1991.
12. Koretsune Y, Marban E: Cell calcium in the pathophysiology of ventricular fibrillation and in the pathogenesis of post-arrhythmic contractile dysfunction. Circulation 80:369–379, 1989.
13. Lee HC, Mohabir R, Smith N, et al: Effect of ischemia on calcium-dependent fluorescence transients in rabbit hearts containing Indo I: Correlation with monophasic action potentials and contraction. Circulation 78:1047–1059, 1988.
14. Steenbergen C, Levy L, Murphy E, London RE: Elevations in cytosolic free calcium early in myocardial ischemia in perfused rat heart. Circ Res 60:700–707, 1987.
15. Corr PB, Yamada KA, Witkowski FX: Mechanisms controlling cardiac autonomic function and their relationships to arrhythmogenesis. In Fozzard HA, Haber E, Jennings RB, et al (eds): The Heart and Cardiovascular System. New York, Raven Press, 1986, pp 1597–1612.
16. Collins MN, Billman GE: Autonomic response to coronary occlusion in animals susceptible to ventricular fibrillation. Am J Physiol 257:H1886–H1894, 1987.
17. Evans DB: Modulation of cAMP: Mechanism for positive inotropic action. J Cardiovasc Pharmacol 8(Suppl 9):S22–S29, 1986.
18. Sperelakis N, Wahler GM: Regulation of Ca^{2+} influx in myocardial cells by beta-adrenergic receptors, cyclic nucleotides and phosphorylation. Mol Cell Biochem 82:19–28, 1988.
19. Yoshida A, Takahashi M, Imagawa T, Shigekawa M, et al: Phosphorylation of ryanodine receptors in rat myocytes during beta-adrenergic stimulation. J Biochem (Tokyo) 111:186–190, 1992.
20. Strand MA, Louis CF, Mickelson JR: Phosphorylation of the porcine skeletal and cardiac muscle sarcoplasmic reticulum ryanodine receptor. Biochim Biophys Acta Mol Cell Res 1175:319–326, 1993.
21. Hawkins C, Xu A, Narayanan N: Comparison of the effects of the membrane-associated Ca^{2+}/calmodulin-dependent protein kinase on Ca^{2+}-ATPase function in cardiac and slow-twitch skeletal muscle sarcoplasmic reticulum. J Biol Chem 269:31198–31206, 1994.
22. Hain J, Onoue H, Maryleitner M, Fleischer S, Shindler H: Phosphorylation modulates the function of the calcium release channel of sarcoplasmic reticulum from cardiac muscle. J Biol Chem 270:2074–2081, 1995.
23. Kranias EG, Gupta RC, Jakab G, et al: The role of protein kinases and protein phosphatases in the regulation of the cardiac sarcoplasmic reticulum function. Mol Cell Biochem 82:37–44, 1988.
24. Altschuld RA, Starling RC, Hamlin RL, et al: Response of failing canine and human heart cells to β_2-adrenergic stimulation. Circulation 92:1612–1618, 1995.
25. Billman GE, Castillo LC, Hensley J, et al: Enhanced sensitivity to beta2-adrenergic receptor stimulation in myocytes from dogs susceptible to ventricular fibrillation. Circulation 92(Suppl I): I450, 1995.
26. Brown Am, Yatani A, Imoto Y, et al: Direct G-protein regulation of Ca^{2+} channels. Ann NY Acad Sci 560:373–386, 1989.
27. Berridge MJ: Inositol trisphosphate and diacylglycerol: Two interacting second messengers. Annu Rev Biochem 56:159–193, 1987.
28. Watanabe AM: Cellular mechanisms of muscarinic regulation of cardiac function. In Randall WC (ed): The Nervous Control of Cardiovascular Function. New York, Oxford University Press, 1984, pp 130–164.
29. Wahler GM, Rusch NJ, Sperelakis N: 8-Bromo-cyclic GMP inhibits the calcium current in embryonic chick ventricular myocytes. Can J Physiol Pharmacol 68:531–534, 1990.
30. Rashatwar SS, Cornwell TL, Lincoln TM. Effects of 8-bromo-cGMP on Ca^{2+}-ATPase by cGMP-dependent protein kinase. Proc Natl Acad Sci USA 84:5685–5689, 1987.
31. Billman GE, McIlroy B, Johnson JD: Elevated myocardial calcium and its role in sudden cardiac death. FASEB J 5:2586–2592, 1991.
32. Wit AL, Janse MJ: The Ventricular Arrhythmias of Ischemia and Infarction. Mount Kisco, NY, Futura, 1993.
33. Clusin WT, Buchbinder M, Ellis AK, et al: Reduction of ischemic depolarization by the calcium blocker diltiazem—Correlation with improvement of ventricular conduction and early arrhythmias in the dog. Circ Res 54:10–20, 1984.
34. Katzung BG: Effects of extracellular calcium and sodium on depolarization-induced automaticity in guinea pig papillary muscle. Circ Res 37:118–127, 1975.
35. Kass RS, Tsien RW: Fluctuations in membrane current driven by intracellular calcium in cardiac Purkinje fibers. Biophysics J 38:259–269, 1982.
36. Ferrier GR: Digitalis arrhythmias: Role of oscillatory afterpotentials. Prog Cardiovasc Dis 19:459–474, 1977.
37. Adamantidis MM, Caron JF, Dupuis BA: Triggered activity induced by combined mild hypoxia and acidosis in guinea-pig Purkinje fibers. J Mol Cell Cardiol 18:1287–1289, 1986.

38. Coetzee WA, Dennis SC, Opie LH, Muller CA: Calcium channel blockers and early ischemic ventricular arrhythmias: Electrophysiological versus anti-ischemic effects. J Mol Cell Cardiol 19(Suppl II):77–97, 1987.
39. Marban E, Robinson SW, Wier WG: Mechanisms of arrhythmogenic delayed and early afterdepolarizations in ferret ventricular muscle. J Clin Invest 78:1185–1192, 1986.
40. Kimura S, Cameron JS, Kozlovskis PL, et al: Delayed afterdepolarizations and triggered activity induced in feline Purkinje fibers by alpha-adrenergic stimulation in the presence of elevated calcium levels. Circulation 70:1074–1082, 1984.
41. Merillat JC, Lakatta EG, Hano O, Guarnieri T: Role of calcium and the calcium channel in the initiation and maintenance of ventricular fibrillation. Circ Res 67:1115–1123, 1990.
42. Hansford RG, Lakatta G: Ryanodine releases calcium from sarcoplasmic reticulum in calcium-tolerant rat cardiac myocytes. J Physiol (Lond) 390:453–467, 1987.
43. Lewartowski B, Hansford RG, Langer GA, Lakatta EG: Contraction and sarcoplasmic reticulum Ca^{2+} content in single myocytes of guinea pig heart: Effect of ryanodine. Am J Physiol 259:H1222–H1229, 1990.
44. Billman GE: Effect of calcium channel antagonists on susceptibility to sudden death: protection from ventricular fibrillation. J Pharmacol Exp Ther 248:1334–1342, 1989.
45. Billman GE: The calcium channel antagonist flunarizine protects against ventricular fibrillation. Eur J Pharmacol 212:231–235, 1992.
46. Billman GE: Ro 40-5967, a novel calcium channel antagonist, protects against ventricular fibrillation. Eur J Pharmacol 229:179–187, 1992.
47. Schramm M, Thomas G, Towart R, Franckowiak G: Novel dihydropyridines with positive inotropic action through activation of Ca^{2+} channels. Nature 303:535–537, 1983.
48. Lappi MD, Billman GE: Effect of ryanodine on ventricular fibrillation induced by myocardial ischaemia. Cardiovasc Res 27:2152–2159, 1993.
49. Akhtar M, Tchou P, Jazayeri M: Use of calcium entry blockers in the treatment of cardiac arrhythmias. Circulation 80(Suppl IV):IV31–IV39, 1989.
50. Singh BN, Nademanee K: Use of calcium antagonists for cardiac arrhythmias. Am J Cardiol 59:153B–162B, 1987.
51. Elharrar V, Zipes DP: Voltage modulation of automaticity in cardiac Purkinje fibers. In Zipes DP, Bailey JC, Elharrar V(eds): The Slow Inward Current and Cardiac Arrhythmias. The Hague, Martinus Nijhoff, 1980 pp 357–373.
52. Kléber AG: Conduction of the impulse in the ischemic myocardium—Implications for malignant ventricular arrhythmias. Experientia 43:1056–1061, 1987.
53. Sung RJ, Shapiro WA, Shen EN, et al: Effects of verapamil on ventricular tachycardias possibly caused by reentry, automaticity and triggered activity. J Clin Invest 72:350–360, 1983.
54. Gettes LS: Effect of ischemia on cardiac electrophysiology. In Fozzard HA, Haber E, Jennings RB, et al (eds): The Heart and Cardiovascular System. New York, Raven Press, 1986, pp 1317–1341.
55. Spray DC, Bennett MVL: Physiology and pharmacology of gap junctions. Annu Rev Physiol 47:281–303, 1985.
56. Kabell G: Modulation of conduction slowing in ischemic rabbit myocardium by calcium-channel activation and blockade. Circulation 77:1385–1394, 1988.
57. Dilly SC, Lab MJ: Electrophysiological alternans and restitution during acute regional ischaemia in myocardium of anesthetized pig. J Physiol (Lond) 402:315–333, 1988.
58. Hellerstein HK, Liebow IM: Electrical alternation in experimental coronary artery occlusion. Am J Physiol 160:366–374, 1950.
59. Russell DC, Smith HJ, Oliver MF: Transmembrane potential changes and ventricular fibrillation during repetitive myocardial ischaemia in the dog. Br Heart J 42:88–96, 1979.
60. Kléber AG, Janse MJ, van Capelle FJL, Durrer D: Mechanism and time course of S-T and T-Q segment changes during acute regional myocardial ischemia in the pig heart determined by extracellular and intracellular recordings. Circ Res 42:603–613, 1978.
61. Hashimoto H, Suzuki K, Miyake S, Nakashima M: Effects of calcium antagonists on the electrical alternans of the ST segment and on associated mechanical alternans during acute coronary occlusion in dogs. Circulation 68:667–672, 1983.
62. Kimura S, Bassett AL, Kohya T, et al: Regional effects of verapamil on recovery of excitability and conduction time in experimental ischemia. Circulation 75:1146–1154, 1987.
63. Billman GE: Role of ATP sensitive potassium channel in extracellular potassium accumulation and cardiac arrhythmias during myocardial ischaemia. Cardiovasc Res 28:762–769, 1994.
64. Messerli FH (ed): Cardiovascular Drug Therapy. Philadelphia, W.B. Saunders, 1990.
65. Haas H, Hartfelder G: Alpha-isopropyl-alpha-[(N-methyl-N-homoveratryl)gamma aminopropyl]-3,4-dimethoxy-phenylacetonitril, eine substanz mit coronargefasserweiterden Eigenschflen. Arzneimittelforschung 12:549–558, 1962.

66. Kaumann AJ, Aramendia P: Prevention of ventricular fibrillation induced by coronary ligation. J Pharmacol Exp Ther 164:326–332, 1968.
67. Fondacaro JD, Han J, Yoon MS: Effect of verapamil on ventricular rhythm during acute coronary occlusion. Am Heart J 96:81–86, 1978.
68. Mohabir R, Ferrier GR: Effects of verapamil and nifedipine on mechanisms of arrhythmia in an in vitro model of ischemia and reperfusion. J Cardiovasc Pharmacol 17:74–82, 1991.
69. Temsey-Armos PN, Legenza M, Southworth SR, Hoffman BF: Effects of verapamil and lidocaine in a canine model of sudden coronary death. J Am Coll Cardiol 6:674–681, 1985.
70. Brooks WW, Verrier RL, Lown B: Protective effect of verapamil on vulnerability to ventricular fibrillation during myocardial ischaemia and reperfusion. Cardiovasc Res 14:295–302, 1980.
71. Billman GE: The effect of calcium channel antagonists on cocaine-induced malignant arrhythmias: protection against ventricular fibrillation. J Pharmacol Exp Ther 266:407–416, 1993.
72. Buxton AE, Waxman HL, Marchlinski FE, Josephson ME: Electropharmacology of non-sustained ventricular tachycardia: effects of class I anti-arrhythmic agents, verapamil and propranolol. Am J Cardiol 53:738–744, 1984.
73. Wellens HJJ, Tan SL, Bar FWH, et al: Effect of verapamil studied by programmed electrical stimulation of the heart in patients with paroxysmal re-entrant supraventricular tachycardia. Br Heart J 39:1059–1066, 1977.
74. Schamroth L, Krikler DM, Garrett C: Immediate effects of intravenous verapamil in cardiac arrhythmias. BMJ 1:660–662, 1972.
75. Messerli FH, Nunez BD, Ventura HO, Snyder DW: Overweight and sudden death: Increased ventricular ectopy in cardiopathy of obesity. Arch Intern Med 147:1725–1728, 1987.
76. Hansen JF, the Danish Study Group on verapamil in myocardial infarction: Treatment with verapamil during and after an acute myocardial infarction: A review based on the Danish verapamil infarction trials I and II. J Cardiovasc Pharmacol 18(Suppl 6):S20–S25, 1991.
77. Yusef S, Wittes J, Griedman L: Overview of results of randomized clinical trials in heart disease. I: Treatments following myocardial infarction. JAMA 260:2088–2093, 1988.
78. Sato M, Nagao T, Yamaguchi I, et al: Pharmacological studies on a new 1,5-benzothiazepine derivative (CRD-401). Arzneimittelforschung 21:1338–1343, 1971.
79. Peter T, Fujimoto T, Hamamoto H, et al: Electrophysiologic effects of diltiazem in canine myocardium with special reference to conduction delay during ischemia. Am J Cardiol 49:602–605, 1982.
80. Anastasiou-Nana M, Anderson JL, Nanas J: Experimental antifibrillatory effects of calcium channel blockade with diltiazem. Comparison with β-blockade and nitroglycerin. J Cardiovasc Pharmacol 6:780–787, 1984.
81. Patterson E, Eller BT, Lucchesi BR: The effects of diltiazem upon experimental ventricular arrhythmias. J Pharmacol Exp Ther 225:224–233, 1983.
82. Lynch JJ, Montgomery DG, Lucchesi BR: The effects of calcium entry blockade on the vulnerability of infarcted canine myocardium toward ventricular fibrillation. J Pharmacol Exp Ther 239:340–345, 1986.
83. Sheenan FH, Epstein SE: Effects of calcium channel blocking agents on reperfusion arrhythmia. Am Heart J 103:973–977, 1982.
84. Mulitcenter Diltiazem Post-infarction Trial Research Group: The effect of diltiazem on mortality and reinfarction after myocardial infarction. N Engl J Med 319:385–392, 1988.
85. Bigger JT, Coromilas J, Rolnitzky LM, et al: Multicenter Diltiazem Post-infarction Trial: Effect of diltiazem on cardiac rate and rhythm after acute myocardial infarction. Am J Cardiol 65:539–543, 1990.
86. Boden WE, Krone RJ, Kleiger RE, et al: Electrocardiographic subset analysis of diltiazem on long-term outcome after acute myocardial infarction. Am J Cardiol 67:335–342, 1991.
87. Bossert E, Vater W: Dihydropyridine, eine neue Gruppe stark wirksamer Coronartherapeuteika. Naturwissenschaften 58:598, 1971.
88. Bergey JL, Wendt, RL, Nocella K, McCallum JD: Acute coronary artery occlusion-reperfusion arrhythmias in pigs. Antiarrhythmic and antifibrillatory evaluation of verapamil, nifedipine, prenylamine and propranolol. Eur J Pharmacol 97:95–103, 1984.
89. Naito M, Michelson EM, Kimetzo JJ, et al: Failure of antiarrhythmic drugs to prevent experimental reperfusion ventricular arrhythmias. Circulation 63:70–79, 1981.
90. Lubbe WF, McLean JA, Nguyen T: Antiarrhythmic actions of nifedipine in acute myocardial ischemia. Am Heart J 105:331–333, 1983.
91. Israeli Sprint Study Group Secondary Prevention Reinfarction: Israeli nifedipine trial (SPRINT): A randomized intervention trial of nifedipine in patients with acute myocardial infarction. Eur Heart J 9:354–364, 1988.

92. Furberg CD, Psaty BM, Meyer JV: Nifedipine: Dose-related increase in mortality in patients with coronary heart disease. Circulation 92:1326–1331, 1995.
93. Holmes B, Brogden RN, Heel RC, Speight TM, Avery GS: Flunarizine: a review of its pharmacodynamics and pharmacokinetic properties and theraperutic use. Drugs 27:6–44, 1984.
94. Vos MA, Gorgels PM, Leunissen JDM, Wellens HJJ: Flunarizine allows differentiation between mechanisms of arrhythmias in the intact heart. Circulation 81:343–349, 1990.
95. Bigger JT, Fleiss JL, Kleiger RE, et al, for the Multicenter Post-Infarction Research Group: The relationship among ventricular arrhythmias, left ventricular dysfunction and mortality in the two years after myocardial infarction. Am J Cardiol 69:250–258, 1984.
96. Osterrieder W, Holck M: The in vitro pharmacological properties of Ro 40-5967: A novel Ca^{2+} channel blocker with potent vasodilation but weak inotropic action. J Cardiovasc Pharmacol 13:754–759, 1989.
97. Clozel J-P, Osterrieder W, Kleinbloesem CH, et al: Ro 40-5967: a new nondihydropyridine calcium antagonist. Cardiovasc Drug Rev 9:4–17, 1991.
98. Clozel J-P, Banken L, Osterrieder W: Effects of Ro 40-5967, a novel calcium antagonist, on myocardial function during ischemia induced by lowering coronary perfusion pressure in dogs: comparison with verapamil. J Cardiovasc Pharmacol 14:713–721, 1989.
99. Ezzaher A, Bouanani NEH, Su JB, et al: Increased negative inotropic effect of calcium channel blockers in hypertrophied and failing rabbit heart. J Pharmacol Exp Ther 257:466–471, 1989.
100. Su J, Renaud N, Carayon A, et al: Effects of the calcium channel blockers, diltiazem and Ro 40-5967, on systemic hemodynamics and plasma noradrenaline levels in conscious dogs with pacing-induced heart failure. Br J Pharmacol 113:395–402, 1994.
101. Clozel J-P, Véniant M, Osterrieder W: The structurally novel Ca^{2+} channel blocker Ro 40-5967, which binds to the [3H] desmethoxyverapamil receptor, is devoid of the negative inotropic effects of verapamil in normal and failing rat hearts. Cardiovasc Drugs Ther 4:731–736, 1990.
102. Véniant M, Clozel J-P, Hess P, Wolfgang R: Ro 40-5967, in contrast to diltiazem, does not reduce left ventricular contractility in rats with chronic myocardial infarction. J Cardiovasc Pharmacol 17:277–284, 1991.
103. Véniant M, Clozel J-P, Hess P, Wolfgang R: Hemodynamic profile of Ro 40-5967 in conscious rats: comparison with diltiazem, verapamil, and amlodipine. J Cardiovasc Pharmacol 18 (Suppl):S55–S58, 1991.
104. Schmitt R, Kleinbloesem CH, Belz GG, et al: Hemodynamic and humoral effect of the novel calcium antagonist Ro 40-5967 in patients with hypertension. Clin Pharmacol Ther 52:314–323, 1992.
105. Rousseau MF, Hayashida W, van Eyll C, et al: Hemodynamic and cardiac effects of the selective T-type and L-type calcium channel blocker, mibefradil, in patients with varying degree of left ventricular systolic dysfunction. J Am Coll Cardiol 28:972–979, 1996.
106. Fang L, Osterrieder W: Potential-dependent inhibition of cardiac Ca^{2+} inward currents by Ro 40-5967 and verapamil: Relation to negative inotropy. Eur J Pharmacol 196:205–207, 1991.
107. Hensley J, Billman GE, Johnson JD, et al: Effects of calcium channel antagonists on Ca^{2+} transients in rat and canine cardiomyocytes. J Mol Cell Cardiol 29:1037–1043, 1997.
108. Rutledge A, Triggle DJ: The binding interactions of Ro 40-5967 at the L-type Ca^{2+} channel in cardiac tissue. Eur J Pharmacol 280:155–158, 1995
109. Bian K, Hermsmeyer K: Ca^{2+} channel activation of the nondihydropyridine Ca^{2+} channel antagonist Ro 40-5967 in vascular muscle cells cultured from dog coronary and saphenous arteries. Arch Pharmacol 348:191–196, 1993.
110. Bezprozvanny I, Tsien RW: Voltage-dependent blockade of diverse types of voltage-gated Ca^{2+} channels expressed in Xenopus oocytes by the Ca^{2+} channel antagonist mibefradil (Ro 40-5967). Mol Pharmacol 48:540–549, 1995.
111. Mehrke G, Zong XG, Flockerzi V, Hofmann F: The Ca^{++}-channel blocker Ro 40-5967 blocks differently T-type and L-type Ca^{++} channels. J Pharmacol Exp Ther 271:1483–1488, 1994.
112. Billman GE, Hamlin RL: The effects of a novel calcium channel antagonist, mibefradil, on refractory period and arrhythmias induced by programmed electrical stimulation and myocardial ischemia: A comparison with diltiazem and verapamil. J Pharmacol Exp Ther 277:1517–1526, 1996.
113. Mishra SK, Hermsmeyer K: Inhibition of signal Ca^{2+} in dog coronary arterial vascular muscle cells by Ro 40-5967. J Cardiovasc Pharmacol 24:1–7, 1994.
114. Nuss HB, Houser SR: T-type Ca^{2+} current is expressed in hypertrophied adult feline left ventricular myocytes. Circ Res 73:777–782, 1993.
115. Sen L, Smith TW: T-type Ca^{2+} channels are abnormal in genetically determined cardiomyopathic hamster hearts. Circ Res 75:149–155, 1994.
116. Bean BP: Classes of calcium channels in vertebrate cells. Annu Rev Physiol 51:367–384, 1989.

SIMON CHAKKO M.D. / ARTHUR L. BASSETT Ph.D.

8

The Clinical Antiarrhythmic Effectiveness of Calcium Antagonists

CLINICAL USE OF CALCIUM ANTAGONISTS IN THE TREATMENT OF ARRHYTHMIAS

Calcium antagonists compose class 4 in the Vaughan Williams classification of antiarrhythmic drugs.[1] Nine calcium antagonists are approved for clinical use in the United States. Verapamil, the first calcium antagonist, is a phenylalkylamine; diltiazem is a benzothiazepine; amlodipine, felodipine, isradipine, nifedipine, nicardipine, and nimodipine are dihydropyridines; and bepridil is a diarylaminopropylamine ester. All of the agents may block cardiac calcium channels when given in sufficiently high quantities. Although all of the calcium antagonists are coronary vasodilators, their electrophysiologic effects and antiarrhythmic actions in humans vary greatly (Table 1) because of structural differences that influence their affinity for cellular binding sites and specific tissues and different pharmacokinetics, including that of their potentially active metabolites. In therapeutic doses, phenylalkylamines (e.g., verapamil, gallopamil) and benzothiazepines (e.g., diltiazem) may reduce the automaticity of the sinus node and decrease atrioventricular (A-V) nodal conduction. The dihydropyridine compounds (e.g., amlodipine, nifedipine, nicardipine, isradapine, felodipine) preferentially block calcium channels in vascular smooth muscle. In vitro, their cardiac effects are similar to those of verapamil, but these effects are nullified by the reflex sympathetic stimulation caused by their induction of peripheral vasodilation. Thus the dihydropyridine compounds have little significant antiarrhythmic effects and are not used clinically for that purpose.

Verapamil and bepridil are also mild blockers of the fast sodium channel in the heart. Bepridil has a moderate potassium channel-blocking effect and may cause prolongation of the QRS duration and QT interval on the surface electrocardiogram.[2] Antiarrhythmic effects of phenylalkylamine and benzothiazipine types of calcium antagonists originate from the ability to block calcium-dependent conduction in the A-V node in a dose-dependent fashion. In doses that slow A-V conduction, they have no significant effect on the atrial, ventricular, or His-Purkinje refractory periods or on conduction velocity in these tissues.[4] Thus, their use has been restricted mainly to the treatment of supraventricular arrhythmias. An exception is bepridil, which prolongs ventricular and His-Purkinje refractory periods. However, bepridil is not used in the treatment of ventricular arrhythmias; unlike the conventional calcium antagonists, it may induce torsades de pointes, especially in the presence of hypokalemia.[5] Some

TABLE 1. Electrophysiologic Effects of Calcium Antagonists

Effects	Verapamil	Nifedipine	Diltiazem	Bepridil
PR interval	↑↑	0	↑	↑
AH interval	↑↑↑	0	↑↑	↑
HV interval	0	0	0	↑
A-V node functional refractory period	↑↑↑	0	↑↑	↑↑
QRS duration	0	0	0	↑
QT interval	0	0	0	↑

0 = no effect; ↑ = mild increase; ↑↑ = moderate increase; ↑↑↑ = marked increase.

of the antiarrhythmic drugs belonging to other classes also have calcium channel-blocking effects in addition to their primary action. An example is amiodarone, a class 3 drug with significant calcium-blocking activity.[3]

SUPRAVENTRICULAR TACHYARRHYTHMIAS

The term supraventricular tachyarrhythmia (SVT) includes all tachyarrhythmias originating above the bifurcation of the bundle of His with an atrial rate above 100/min. Examples are atrial fibrillation, atrial flutter, A-V nodal reentrant tachycardias (AVNRTs), A-V reciprocating tachycardias (AVRT), ectopic atrial tachycardia, sinus node reentry tachycardia, and multifocal atrial tachycardia. From a therapeutic point of view, SVTs may be classified into two types:

1. SVTs that require the A-V node for initiation and maintenance. Examples include AVNRT and AVRT. AVNRT is characterized by a reentrant circuit within the A-V node, usually antegrade via a slow pathway and retrograde via a fast pathway. In AVRT, the reentrant circuit includes the A-V node, ventricle, an accessory pathway, and the atrium. These SVTs may be terminated by blocking conduction over any portion of the reentrant circuit. Calcium antagonists terminate this type of SVT by reducing calcium entry during the action potential in A-V nodal cells. Thus, conduction over the A-V node is slowed.

2. SVTs generated in the atrium and conducted to the ventricle through the A-V node or an accessory pathway. Examples are atrial fibrillation, atrial flutter, ectopic atrial tachycardia, and multifocal atrial tachycardia. Ventricular rate is increased during these disorders. Calcium antagonists do not terminate this type of SVT, but they are useful in slowing the rapid ventricular rate. By slowing A-V nodal conduction, they effectively reduce the number of impulses traversing that structure.

Verapamil and diltiazem are the calcium antagonists commonly used for termination of SVT. A-V nodal conduction delay occurs within 1–2 minutes after intravenous administration of verapamil and may last up to 6 hours. The recommended dose for verapamil is 5–10 mg given as an intravenous bolus over at least 2 minutes. If the response is not adequate, a repeat dose of 10 mg may be given. Because verapamil is a potent peripheral vasodilator, hypotension is a potential adverse effect; thus, verapamil should be avoided in hypotensive patients. Afterload reduction may compensate for the negative inotropic effects of verapamil, but it should be avoided or used with care in patients with moderate-to-severe left ventricular dysfunction.

Sung et al.[7] evaluated the effectiveness of intravenous verapamil in terminating SVTs in a double-blind, placebo-controlled study. Verapamil was effective in terminating 80% of AVNRTs. Although it had no effect on the A-V bypass tract, it also was effective in terminating AVRT by reduction of A-V conduction. Intravenous diltiazem has similar effectiveness in terminating reentrant SVT. In a double-blind, placebo-controlled study of 54 patients with AVNRT or AVRT, Huycke et al. reported that diltiazem was effective in terminating 86% of the tachycardias.[8] The recommended dose is 0.25 mg/kg as an intravenous bolus injection; if response is inadequate, a second bolus of 0.35 mg/kg may be administered. Plasma elimination half-life is approximately 3.4 hours. Hypotension is the most common adverse effect but occurs in less than 5% of the patients. Congestive heart failure may worsen in patients with impaired left ventricular function. Adenosine is presently the drug of choice for termination of SVTs, because it is equally effective and has an extremely short elimination half-life of 10 seconds.[6,9]

Oral therapy with verapamil and diltiazem has been used in the prevention of SVTs. The bulk of the clinical experience has been with verapamil.[4] The response to oral prophylactic therapy with verapamil cannot be predicted from the effectiveness of the intravenous form in terminating the SVT. In a double-blind, placebo-controlled trial, orally administered verapamil was tested in 11 patients with frequent paroxysmal SVTs.[10] Verapamil reduced the frequency and duration of episodes of SVT. During the 4 months of blinded therapy, 35 episodes of SVT required additional therapy; of these, 33 occurred during the placebo period. Propranolol and digoxin have been reported to have efficacy similar to verapamil in the chronic prophylactic therapy of SVT.[11]

Verapamil and diltiazem are of little value in the prevention or conversion of atrial flutter or fibrillation to sinus rhythm. However, as noted above, they are useful in the control of ventricular rate in the presence of atrial flutter or atrial fibrillation. Intravenous injection of verapamil or diltiazem provides more rapid control of ventricular rate than digoxin, which has a slow onset of action. In a placebo-controlled trial, an intravenous bolus injection of diltiazem, followed by infusion of 10–15 mg/hr, provided rapid and adequate rate control in 83% of 47 patients with atrial fibrillation or atrial flutter.[12] Diltiazem has negative inotropic effects. Although in general it should be avoided in the presence of congestive heart failure, its negative inotropic effects may be compensated by control of ventricular rate control and vasodilation. Diltiazem was well tolerated in a study of 9 patients with atrial fibrillation accompanied by congestive heart failure.[13] In patients with accessory pathways between atrial and ventricles and preexcitation syndromes, atrial flutter or fibrillation should not be treated with calcium antagonists. A-V nodal conduction is reduced by calcium antagonists, but conduction over the accessory pathway is not affected. Thus, the ventricular rate in fact may increase, and ventricular fibrillation may be precipitated.[4]

Oral forms of the calcium antagonists are also useful in chronic control of ventricular rate in atrial flutter or fibrillation. Although digoxin is effective in controlling the ventricular rate at rest in such atrial tachycardias, it is often not effective during exercise because its beneficial electrophysiologic effects are mediated through increased vagal tone.[6] The vagal effects are easily overridden by the increase in sympathetic tone during exercise. Recent reports indicate that calcium channel blockers alone or in combination with digoxin may provide better control of ventricular rate during exercise of patients with atrial fibrillation.[14] Verapamil, diltiazem, and beta-adrenergic blockers alone or in combination with digoxin appear to be preferable to digoxin alone for control of ventricular rate in atrial fibrillation in the absence of left ventricular systolic dysfunction. In a placebo-controlled study of 20 patients with chronic atrial fibrillation receiving digoxin, addition of verapamil improved both rate control and exercise capacity.[15] In another study of patients with chronic atrial fibrillation, control of ventricular rate with diltiazem, 240 mg/day, was comparable to that obtained with therapeutic doses of digoxin at rest but was superior during peak exercise. However, higher doses were associated with more adverse effects. Combined therapy with digoxin and diltiazem, 240 mg/day, enhanced the effects of digoxin and increased rate control at rest and during peak exercise.[16]

Multifocal atrial tachycardia is often caused by respiratory failure in patients with chronic obstructive pulmonary disease. Correction of hypoxia, hypercarbia, and electrolyte abnormalities is the mainstay of treatment.[6] Occasionally, multifocal atrial tachycardia is difficult to terminate, and some patients may not tolerate the accompanying rapid ventricular rate. Verapamil is moderately effective in controlling ventricular rate during multifocal atrial tachycardia.[17,18] The beta-1 receptor antagonist, metoprolol, is more effective than verapamil, but beta receptor blockade is contraindicated in obstructive pulmonary disease.[6,17]

VENTRICULAR TACHYCARDIA AND FIBRILLATION

Calcium antagonists are not useful in the treatment of common forms of ventricular tachycardias (VTs). Intravenous verapamil has been reported to be ineffective in the treatment of sustained[19] and nonsustained VT.[20] Although experimental data suggest that calcium antagonists may be effective in the treatment of ventricular arrhythmias in the setting of acute myocardial infarction and reperfusion, clinical studies have been disappointing (see chapter 9).[4]

Most VTs occur in the presence of structural heart disease.[6] However, certain VTs may occur in the absence of structural heart disease, and some of these unusual forms appear to respond well to calcium antagonists.[21] One such idiopathic VT originates in the right ventricular outflow tract, displays a left bundle branch morphology on the surface electrocardiogram, and is usually initiated by exercise. The electrophysiologic mechanism is believed to be cyclic adenosine monophosphate-mediated triggered activity[22] (see chapter 7 for description of triggered activity). In a study of 42 patients with such idiopathic VT, verapamil was reported to be effective in two-thirds of the patients.[23] Adenosine also was effective. Another unusual type of VT in young patients with structurally normal hearts has a right bundle branch morphology.[21] Initiation by exercise is less common. Reentry involving the slow channel tissue and triggered activity resulting from delayed afterdepolarizations are possible mechanisms. This type of VT responds to intravenous verapamil, whereas beta-adrenergic blockers and class 1 antiarrhythmic agents are ineffective.[24] However, oral verapamil may not be effective in preventing recurrences.

Silent myocardial ischemia is associated with increased risk for adverse events in patients with coronary artery disease.[25] Among patients who suffered cardiac arrest and were found to have no structural heart disease, a small minority had life-threatening ventricular arrhythmias that were induced by coronary artery spasm.[26] Reperfusion rather than spasm may be responsible. In any case, a calcium antagonist (verapamil, diltiazem, or nifedipine) was effective in preventing spasm and arrhythmias.

Cocaine causes alpha-adrenergic stimulation, which may result in coronary vasoconstriction, myocardial ischemia, and infarction.[27] Myocardial calcium entry may play a critical role in cocaine-induced ventricular fibrillation.[28] In studies of cocaine intoxication in animals, calcium antagonists prevented malignant arrhythmias and protected against myocardial infarction.[28,29]

A double-blind, placebo-controlled study was conducted to establish the efficacy of nimodipine, a dihydropyridine-calcium antagonist, in the prevention of anoxic brain injury among patients resuscitated from ventricular fibrillation. During the 24-hour treatment, ventricular fibrillation recurred in 15% of placebo-treated patients but was significantly less frequent in nimodipine-treated patients (incidence = 1.3%).[30] Whether other calcium channel antagonists have similar effects is unknown.

CONCLUSION

Calcium antagonists that reduce A-V nodal conduction (e.g., verapamil, diltiazem) are effective in the treatment of many supraventricular tachycardias. Calcium antagonists are not used clinically for the treatment of ventricular arrhythmias. Rare forms of ventricular arrhythmias, such as idiopathic ventricular tachycardia, may be terminated by verapamil. Experimental studies suggest that coronary vasodilation caused by calcium antagonists may prevent arrhythmias caused by coronary spasm and reperfusion.

References

1. Vaughan William EM: A classification of antiarrhythmic actions reassessed after a decade of new drugs. J Clin Pharmacol 24:129–147, 1984.
2. Singh BN, Nademanee K: Use of calcium antagonists for cardiac arrhythmias. Am J Cardiol 59:153B–162B, 1987.
3. January CT, Cunningham PM, Zhou Z: Pharmacology of L- and T-type calcium channels in the heart. In Zipes DP, Jalife J (eds): Cardiac Electrophysiology. Philadelphia, W.B. Saunders, 1995, pp 269–277.
4. Singh BN: Beta-blockade and calcium channel blockers as antiarrhythmic drugs. In Zipes DP, Jalife J (eds): Cardiac Electrophysiology. Philadelphia, W.B. Saunders, 1995, pp 1317–1330.
5. Singh BN, Hecht HS, Nademanee K, Chew CYC: Electrophysiological and hemodynamic actions of slow-channel blocking compounds. Prog Cardiovasc Dis 25:103–132, 1982.
6. Chakko S, Kessler KM: Recognition and management of cardiac arrhythmias. Curr Probl Cardiol 20:53–120, 1995.
7. Sung RJ, Elser B, McAllister RG: Intravenous verapamil for termination of re-entrant supraventricular tachycardias. Ann Intern Med 93:682–689, 1980.
8. Huycke EC, Sung RJ, Dias VC, et al, for the Multicenter Diltiazem PSVT Study Group: Intravenous diltiazem for termination of reentrant supraventricular tachycardia: A placebo-controlled, randomized, double-blind, multicenter study. J am Coll Cardiol 13:538–544, 1989.
9. Emergency Cardiac Care Committee and Subcommittees, American Heart Association: Guidelines for cardiopulmonary resuscitation and emergency cardiac care. III: Adult advanced cardiac life support. JAMA 268:2172–2241, 1992.

10. Mauritson DR, Winniford MD, Walker WS, et al: Oral verapamil for paroxysmal supraventricular tachycardia. A long-term, double-blind randomized trial. Ann Intern Med 96:409–412, 1982.
11. Winniford MD, Fulton KL, Hillis LD: Long-term therapy of supraventricular tachycardia: A randomized, double-blind comparison of digoxin, propranalol and verapamil. Am J Cardiol 54:1138–1139, 1984.
12. Ellenbogen KA, Dias VC, Plumb VJ, et al: A placebo-controlled trial of continuous intravenous diltiazem infusion for 24 hour heart rate control during atrial fibrillation and atrial flutter: A multicenter study. J Am Coll Cardiol 18:891–897, 1991.
13. Heywood JT, Graham B, Marais GE, Jutzy KR: Effects of intravenous diltiazem on rapid atrial fibrillation accompanied by congestive heart failure. Am J Cardiol 67:1150–1152, 1991.
14. Zarowitz BJ, Gheorghiade M: Optimal heart rate control for patients with chronic atrial fibrillation: Are pharmacologic choices truly changing? Am Heart J 123:1401–1403, 1992.
15. Lang R, Klein HO, Di Segni E, et al: Verapamil improves exercise capacity in chronic atrial fibrillation: Double-blind crossover study. Am Heart J 105:820–825, 1983.
16. Roth A, Harrison E, Mitani G, et al: Efficacy and safety of medium- and high-dose diltiazem alone and in combination with digoxin for control of heart rate at rest and during exercise in patients with chronic atrial fibrillation. Circulation 73:316–324, 1986.
17. Arsura EL, Lefkin AS, Scher DL, et al: A randomized, double-blind, placebo- controlled study of verapamil and metoprolol in treatment of multifocal atrial tachycardia. Am J Med 85:519–524, 1988.
18. Levine JH, Michael JR, Guarnieri T: Treatment of multifocal atrial tachycardia with verapamil. N Engl J Med 312:21–25, 1985.
19. Sung RJ, Shapiro WA, Shen EW, et al: Effects of verapamil on ventricular tachycardia possibly caused by re-entry, automaticity, and triggered activity. J Clin Invest 72:350–360, 1983.
20. Buxton AE, Waxman HL, Marchlinski FE, Josephson ME: Electropharmacology of nonsustained ventricular tachycardia: Effects of class I antiarrhythmic agents, verapamil and propranolol. Am J Cardiol 53:738–744, 1984.
21. Wellens HJJ, Rodriguez LM, Smeets JL: Ventricular tachycardia in structurally normal hearts. In Zipes DP, Jalife J (eds): Cardiac Electrophysiology. Philadelphia, W.B.Saunders, 1995, pp 780–788.
22. Lerman BB, Belardinelli L, West GA, et al: Adenosine-sensitive ventricular tachycardia: Evidence suggesting cyclic AMP-mediated triggered activity. Circulation 74:270–280, 1986.
23. Gill JS, Blaszyk K, Ward DE, Camm AJ: Verapamil for the suppression of idiopathic ventricular tachycardia of left bundle branch block-like morphology. Am Heart J 126:1126–1133, 1993.
24. Bellhassen B, Shapira I, Pelleg A, et al: Idiopathic recurrent sustained ventricular tachycardia responsive to verapamil: An ECG-electrophysiologic entity. Am Heart J 108:1034–1037, 1984.
25. de Marchena E, Asch J, Martinez J, et al: Usefulness of persistent silent myocardial ischemia in predicting a high cardiac event rate in men with medically controlled stable angina pectoris. Am J Cardiol 73:390–392, 1994.
26. Myerburg RJ, Kessler KM, Mallon SM, et al: Life-threatening ventricular arrhythmias in patients with silent myocardial ischemia due to coronary artery spasm. N Engl J Med 326:1451–1455, 1992.
27. Chakko S, Myerburg RJ: Cardiac complications of cocaine abuse. Clin Cardiol 18:67–72, 1995.
28. Billman GE: Effect of calcium channel antagonists on cocaine-induced malignant arrhythmias: Protection against ventricular fibrillation. J Pharmacol Exp Ther. 266:407–416, 1993.
29. Hollander JE: The management of cocaine-associated myocardial ischemia. N Engl J Med 333:1267–1272, 1995.
30. Roine RO, Kaste M, Kinnunen A, et al: Nimodipine after resuscitation from out-of-hospital ventricular fibrillation. A placebo-controlled, double-blind, randomized trial. JAMA 264:3171–3177, 1990.

ARTHUR L. BASSETT Ph.D. / SIMON CHAKKO M.D.
MURRAY EPSTEIN M.D., F.A.C.P.

9

Are Calcium Antagonists Proarrhythmic?

Clinical and experimental studies demonstrate that calcium (Ca^{2+}) overload in myocardial cells is an important factor in the genesis of various serious arrhythmias. Calcium antagonists block voltage-dependent channels and thus reduce entry of Ca^{2+} into heart cells. Because of their specificity for atrioventricular (A-V) nodal cells, verapamil and diltiazem are used clinically to treat supraventricular arrhythmias involving transmission in the A-V node. Both drugs, as well as the dihydropyridine (DHP) calcium antagonists, have been shown to prevent ventricular ischemic and reperfusion arrhythmias in the laboratory. Despite data indicating that calcium antagonists are antiarrhythmic, a recent controversy has raised the possibility that certain calcium antagonists are unsafe, especially in patients with coronary heart disease. Proarrhythmia has been proposed as a mechanism contributing to potentially adverse outcomes. Excessive concentrations of verapamil and diltiazem may cause sinoatrial (S-A) nodal asystole and varying degrees of A-V block, but little direct evidence indicates that they contribute to significant proarrhythmia—for example, ventricular tachyarrhythmias.

Although it appears paradoxical that agents which block Ca^{2+} entry into heart cells may be considered arrhythmogenic, in specific circumstances dosage with certain calcium channel antagonists potentially leads to myocardial Ca^{2+} overload. For example, bouts of neurohormonal activation due to abrupt reductions in blood pressure induced by calcium antagonists may be accompanied by significant β-adrenergic-enhanced influx of Ca^{2+} through the L-type cardiac calcium channels. This influx elevates Ca^{2+}_i and disturbs Ca^{2+} regulation, especially in diseased hearts in which intracellular Ca^{2+} regulation is already compromised, and may induce alterations in cardiac electrical activity. This chapter discusses interaction among cardiac calcium channels, classes of calcium antagonists, and specific formulations of certain antagonists in relation to directly induced ventricular arrhythmogenesis. Indirect potentially proarrhythmic actions of the calcium antagonists are also discussed, along with some of the many questions about the actions of DHP on the heart, such as whether β-adrenergic stimulation modifies the degree of cardiac Ca^{2+} channel inhibition by DHP calcium antagonists.

BACKGROUND

A cardiac arrhythmia that is drug-induced or drug-aggravated is described as proarrhythmia.[1] The definition of proarrhythmia has been expanded to include aggravation of an existing arrhythmia, appearance of a new asymptomatic arrhythmia, or a change in inducibility.[2] Proarrhythmia has been suggested as a

TABLE 1. Calcium-channel Modulations

1. Membrane potential (voltage): depolarization allows Ca^{2+} entry.
2. Hormones and neurotransmitters (e.g., antiotensin II, epinephrine) can regulate the channel:
   - by activating protein kinases (A or C) that phosphorylate the channel
   - through guanine nucleotide-binding (G) proteins that directly link β-receptors to calcium channels
3. Inorganic ions: cobalt (Co^{2+}), nickel (Ni^{2+}), and cadmium (Cd^{2+}) are inhibitors.
4. Biologic ions: magnesium (Mg^{2+}) and $Ca^{2+}{}_i$ may inhibit.
5. Dihydropyridine calcium agonists increase calcium influx.

mechanism contributing to potentially life-threatening effects of calcium channel blockers (CCBs). Arrhythmias may reflect direct actions of a drug on cardiac cellular electrophysiology, indirect actions that ultimately affect myocardial electrophysiology, or a combination of both. In any case, intracellular Ca^{2+} plays a significant role in the regulation of normal cardiac and smooth muscle physiology and biochemistry and in disease-induced alterations in cellular processes. Calcium current flowing into the cardiac cell through voltage-dependent Ca^{2+}-specific channels is a major source of intracellular Ca^{2+}. The regulation of these cardiac calcium channels is complex (Table 1). However, clinical and experimental studies demonstrate that Ca^{2+} overload within ventricular cells is an important factor in the genesis of various serious arrhythmias.[3] CCBs are sometimes used to prevent such overload.

In addition to the direct deleterious actions of elevated myocardial $Ca^{2+}{}_i$, the effects of Ca^{2+} on coronary tone may indirectly affect arrhythmia formation. For example, catecholamine-induced coronary spasm involves increased $Ca^{2+}{}_i$ within the coronary artery cells and may result in myocardial ischemia and reperfusion; several biochemical changes associated with these conditions favor the accumulation and possible overload of myocardial cytosolic Ca^{2+}, especially during reperfusion.

Coronary steal due to dilation of coronary vessels by CCBs highly selective for the coronary system and/or negative inotropy sometimes induced by CCBs also may lead to regions of myocardial ischemia. In turn, $Ca^{2+}{}_i$ overload may result in rhythm disturbances.

In terms of electrophysiologic mechanisms, disturbances in cardiac rhythm may result from abnormalities in the generation or conduction of impulses or from a combination of both. Calcium ions directly affect impulse conduction, especially in the S-A and A-V nodes. Many supraventricular arrhythmias are caused by reentry circuits within the A-V node or circuits that include Ca^{2+} influx during Ca^{2+}-dependent action potentials within the A-V node. Certain CCBs (specifically verapamil and diltiazem) are effective in the treatment of A-V nodal-associated arrhythmias because they have an affinity for Ca^{2+} channels in A-V nodal cells (Table 2). By reducing Ca^{2+} entry, verapamil and diltiazem also reduce the amplitude and conduction of such action potentials, interrupting the reentry circuits to terminate the rhythm disturbances. The DHP CCBs have less affinity for Ca^{2+} channels in cardiac cells (including those in the A-V node) than verapamil and diltiazem and in therapeutic concentrations have little effect on such arrhythmias. Consequently, DHPs are rarely used clinically for A-V node-related rhythm disorders.

TABLE 2. Reasons for Tissue Selectivity of Calcium Antagonists

- Influence of external Ca^{2+} on tissue responses
- Importance of Ca^{2+}-channel subtypes on responsiveness to disparate agents
- Voltage dependence of binding
- Frequency dependence of drug effects

Calcium also is involved in normal pacemaker activity originating in the S-A node. In both clinical and experimental studies, excessive quantities of CCBs may cause S-A nodal asystole, a potentially serious arrhythmia. This effect, however, is rare.

In contrast, all classes of CCBs have been used to suppress abnormal impulse generation in experimental laboratory studies. Abnormal impulse generation may arise from excessive elevations in intracellular myocardial Ca^{2+}, including within working atrial and ventricular muscle cells and Purkinje fibers. In turn, Ca^{2+}_i overload-induced disturbance of the myocardial cell Ca^{2+} cycle may result in oscillations of membrane potential, known as afterdepolarizations.[3] Experimental studies show that when the amplitude of afterdepolarizations reaches a threshold voltage, single or repetitive firing of a cell or group of cells may occur and cause triggered activity and arrhythmias.[4] In vitro, all CCBs block voltage-dependent Ca^{2+}-specific channels and thus reduce entry of Ca^{2+} into myocardial muscle cells. They have been shown to prevent experimentally induced ischemic and reperfusion arrhythmias. However, neither verapamil nor diltiazem is highly effective for clinical treatment of acute ischemic-reperfusion arrhythmias, and the DHP CCBs have been used largely without therapeutic effectiveness in acute ischemic arrhythmias. Verapamil and diltiazem are used mainly against supraventricular arrhythmias. Their clinical indications are summarized in Table 3.

TABLE 3. Antiarrhythmic Effectiveness of Verapamil and Diltiazem

A-V nodal reentrant tachycardia	Highly effective in acute termination. Moderately effective in chronic prophylaxis.
A-V reciprocating tachycardia (bypass tract tachycardia)	Highly effective in acute termination. Use with caution for chronic prophylaxis because ventricular rate will be very rapid if atrial flutter or fibrillation develops.
Atrial flutter and fibrillation	Not effective in acute termination or chronic prophylaxis. Highly effective in controlling ventricular rate.
Multifocal atrial tachycardia	Not effective in termination. Limited use in controlling ventricular rate
Ventricular arrhythmia	Not used clinically. Experimental data suggest effectiveness in preventing reperfusion-induced and cocaine-induced ventricular arrhythmias. Not effective in common forms of VT. Uncommon idiopathic VTs may terminate with verapamil.

A-V = atrioventricular, VT = ventricular tachycardia.
Adapted from Chakko S, Basset AL: The clinical antiarrhythmic effectiveness of calcium antagonists. In Epstein M (ed): Calcium Antagonists in Clinical Medicine, 2nd ed. Philadelphia, Hanley & Belfus, 1997, pp 135–140.

TABLE 4. Classification of Calcium Antagonists by Structure

Phenylalkylamines	Dihydropyridines	Benzothiazepines	Others
Verapamil	Nifedipine	Diltiazem	Flunarizine
Gallapamil	Nitrendipine		Cinnarizine
Tiapamil	Nisoldipine		Lidoflazine
Anipamil	Nimodipine		Perhexiline
	Felodipine		Bepridil
	Nicardipine		
	Amlodipine		
	Isradipine		
	Nilvaldipine		
	Lacidipine		

Adapted from Morris AD, Meredith PA, Reid JL: Pharmacokinetics of calcium antagonists: Implications for therapy. In Epstein M (ed): Calcium Antagonists in Clinical Medicine. Philadelphia, Hanley & Belfus, 1992, pp 49–68.

The CCBs are a heterogeneous group in terms of their target channels in various cells (e.g., A-V node vs. working myocardium vs. vascular smooth muscle). This heterogeneity reflects their tissue affinities (see Table 2), their structural (Table 4) and physicochemical properties (Table 5), and various other factors, including pharmacokinetics. For example, the CCBs may differ with respect to absorption after oral administration and rates of metabolism and elimination. Thus, the onset and duration of clinical activity may vary among the phenylalkylamine, verapamil; the benzothiazepine, diltiazem; and the DHP CCBs, including the prototype, nifedipine. Furberg and associates[10] focused on nifedipine in developing their formulation that all CCBs have potentially life-threatening actions in patients with coronary heart disease. However, the CCBs are highly diverse and cannot be considered as a homogeneous group in terms of their actions on heart and vasculature.

TABLE 5. Pharmacokinetics of Calcium Antagonists

	Nifedipine	Verapamil	Diltiazem
Oral absorption	> 90%	> 90%	> 90%
Bioavailability	30–60%	10–30%	30–60%
Protein binding	> 90%	> 90%	> 90%
Elimination half-life	3.4 ± 10.4 hr (capsule) 5.4 ± 2.8 hr (tablet)	3–7 hr*	3–6 hr
Hepatic metabolism	+++	+++	+++
Active metabolites	? Yes	Yes	Yes

* May be prolonged to 10–12 hr on chronic dosing.
+++ = extensive hepatic metabolism.
Adapted from Morris AD, Meredith PA, Reid JL: Pharmacokinetics of calcium antagonists: Implications for therapy. In Epstein M (ed): Calcium Antagonists in Clinical Medicine. Philadelphia, Hanley & Belfus, 1992, pp 49–68.

TABLE 6. Potential Mechanisms for Increased Relative Risk

1. Negative inotropic effects	4. Prohemorrhagic effects
2. Proarrhythmic effects	5. Transient asystole
3. Proischemic effects (from coronary steal)	6. Enhanced sympathetic activity

Adapted from Furberg C: Calcium antagonists: Not appropriate as first line antihypertensive agents. Am J Hypertens 9:122–125, 1996.

ARE CALCIUM ANTAGONISTS PROARRHYTHMIC?

A series of retrospective analyses led Furberg et al.[5–9] to their controversial suggestion that CCBs may promote adverse cardiovascular events.[5,10] Despite the data indicating that certain CCBs are antiarrhythmic, Furberg et al. raised the possibility that CCBs may be proarrhythmic. They included proarrhythmia among several other mechanisms to account for the increased incidence of adverse events putatively attributed to CCBs (Table 6). Although much attention has centered on the proischemic and prohemorrhagic effects, relatively little critical consideration has been devoted to the postulate that CCBs are proarrhythmic (see Table 6).

CCBs as a class and DHPs in particular are much less likely to cause proarrhythmia than class IA, IB, IC, and III antiarrhythmic drugs.[11] Furthermore, in the relatively rare instance of such drug-induced rhythm disturbances, the arrhythmia is usually supraventricular.[12] It should be noted, however, that bepridil, a diarylaminopropylamine ether, possesses significant Na^+ and K^+ channel-blocking actions. Bepridil causes significant and life-threatening ventricular proarrhythmias (i.e., torsades de pointes); thus, its clinical use is discouraged. Furthermore, within the broad definition of proarrhythmia, verapamil at times also may be proarrhythmic. Significant worsening of ventricular tachycardia may accompany the use of intravenous verapamil prescribed for erroneously diagnosed supraventricular tachycardia.[12a]

This review briefly considers both the electrophysiologic and molecular basis of calcium-related arrhythmia formation in the heart and how CCBs may modulate cardiac cellular events to promote arrhythmogenic effects. Another section considers the relevant clinical data. The final section describes a number of questions bearing on the cardiac electrophysiologic effects of certain CCBs; the answers may further elucidate their clinical properties.

MECHANISMS WHEREBY CALCIUM ANTAGONISTS MAY PROMOTE ARRHYTHMIAS

In specific circumstances, dosage with certain calcium channel antagonists may lead to myocardial Ca^{2+} overload. For example, activation of cardiac β_1 receptors favors the entry of Ca^{2+} ions through L-type calcium channels located in cardiac cell membranes[3] (see Table 1). Such receptor activation may accompany significant sympathetic nervous system arousal and repetitive, abrupt, marked falls in arterial blood pressure, both of which are associated with the release of epinephrine and norepinephrine at myocardial sites. The β-adrenergic receptor-enhanced influx of Ca^{2+} through the L-type calcium channels, in turn, may elevate $Ca^{2+}{}_i$. This disturbance in myocardial Ca^{2+} regulation, especially in diseased hearts in which intracellular Ca^{2+} regulation is already compromised, may lead ultimately to development of afterdepolarizations, triggered activity, and arrhythmias. In addition, β-adrenergic activation favors enhanced pacemaking activity, both at the normally dominant sinus node pacemaker but also at latent, potentially ectopic pacemakers.

Dosage with a short elimination half-life calcium antagonist such as nifedipine (or a short half-lived formulation; see below) may cause significant vasodilation several times per day. Each bout of reduced blood pressure may elicit sympathetic activity that disturbs cardiac electrical activity, perhaps resulting in proarrhythmia, on each occasion that blood pressure falls. This effect, however, may vary with the clinical circumstance and patient population. For example, although increased levels of circulating catecholamines accompany congestive heart failure,[13] cardiac stores of norepinephrine may be lower in the failing heart, and, of equal importance, reflex sympathetic regulation (i.e., sympathetic activation after an intervention such as nitroprusside-decreased blood pressure) may be blunted.[14] The cardiac channel responses induced by α-adrenergic agonists, however, are not completely defined.

In addition, Ca^{2+} is essential for transmission of impulses (action potentials) across the A-V node. Thus, blockade of A-V calcium channels induced by diltiazem and verapamil may slow A-V conduction or cause varying degrees of block. Conduction block favors the emergence of ectopic ventricular rhythms, as does calcium antagonist-induced asystole in the S-A node (see Table 6). Although such actions may be considered proarrhythmic, they most often occur with excessive dosage or intravenous administration of diltiazem or verapamil or when either agent is given in conjunction with another drug that has similar effects on the S-A and A-V nodes (i.e., beta blockers and digitalis).

As noted above, in pharmacologic and therapeutic doses, benzothiazepines (e.g., diltiazem) and phenylalkylamines (verapamil) diminish automaticity of the sinus node and slow conduction in the A-V node. However, these electrophysiologic properties vary greatly from one drug to another and also seem to depend on the form of administration.[3,15] In laboratory animals, the DHPs generally have less effect on cardiac automaticity and A-V conduction in situ than non-DHP CCBs.[3]

In addition to direct actions on calcium channels in specialized cardiac tissues, it has been proposed that the CCBs with major actions on vasculature (vascular selective CCBs) may indirectly provoke cardiac ischemia or reperfusion via coronary steal. This phenomenon may induce regional ischemia-reperfusion and lead to myocardial $Ca^{2+}{}_i$ overload and its adverse sequelae. The negative inotropic effect of verapamil and diltiazem on working ventricular muscle also may reduce coronary artery perfusion (see Table 6) . Thus, in either drug-induced coronary steal or negative inotropy, ischemia and reperfusion, if sufficient to provoke sympathetic activation, favor generation of ectopic impulses.[3,16] Such transient ischemia and ectopic impulse formation may occur more readily in a diseased heart (e.g., significant coronary atherosclerosis, diminished contractility).

Role of Catecholamines

Compelling evidence indicates that sympathetic activation persists during long-term treatment when rapid-onset, short-acting DHPs are administered.[17]

A recent analysis of 63 clinical studies showed that acutely, after single dosing, short-acting CCBs increased heart rate and plasma norepinephrine levels.[18] The change in the norepinephrine level had a positive correlation with the change in heart rate and an inverse correlation with the change in arterial pressure in patients receiving DHPs. The change in norepinephrine levels was less pronounced in patients receiving non-DHPs. With sustained therapy, both DHP and non-DHP preparations of short-acting calcium antagonists increased norepinephrine levels.

Different Formulations as a Determinant of Proarrhythmic Risk

The fact that CCBs are heterogeneous and consist of chemically dissimilar agents is well established.[19–21] Less appreciated is the fact that different formulations of the same chemical moiety produce markedly differing hemodynamic and neurohormonal effects.[21,22] The earliest CCBs were short-acting. Subsequently, the drug delivery systems for the short-acting agents were modified to provide more fully and consistently maintained calcium antagonist activity.

As detailed in a recent review,[9] the rate of drug delivery into the systemic circulation has profound effects on the hemodynamic and neurohumoral responses to DHPs. Studies by Kleinbloesem et al.[22] emphasize the importance of rate of attainment of plasma levels in determining the consequent adrenergic and cardioaccelatory response. Different formulations may influence proarrhythmic risk in two ways:

1. The newer slow-release and slow-acting formulations are less likely to provoke intermittent activation of the sympathetic nervous system and thus are less likely to be proarrhythmic.

2. Short-acting formulations cause wide fluctuations in drug concentrations and vasodilator effects during apparent steady-state treatment. Consequently, medications that produce intermittent fluctuations in blood pressure may exacerbate ischemia in patients with basal myocardial ischemia and are more apt to be proarrhythmic. Indeed, a recent editorial emphasized that slow- and long-acting CCBs effectively lower blood pressure without causing reflex sympathetic activation and tachycardia.[23]

In addition to formulation and mode of administration, other factors that influence the potential therapeutic usefulness and adverse effects of CCBs include changes in bioavailability and potential increases in half-life as hepatic metabolism saturates during repetitive oral administration. As with many drugs metabolized by the liver, the half-life of CCBs may increase in the elderly.

Calcium Antagonists and Left Ventricular Mass

The potential proarrhythmic effects of CCBs are exceedingly complex and often confounded by their other significant actions. For example, another mechanism, albeit indirect, whereby DHPs may affect arrhythmias is their ability to regress left ventricular mass.[24] The decrease in left ventricular mass documented with DHPs is associated with improved ventricular filling, diminished ventricular ectopy, and preserved contractility.[15,24–26] In addition to myocyte growth, left ventricular hypertrophy (LVH) is characterized by activation of nonmyocyte cells. Proliferation of fibroblasts that generate collagen, augmentation in local fibroblast activity, and reduced local collagenolysis contribute to interstitial and perivascular fibrosis and coronary microangiopathy and macroangiopathy. Evidence from experimental studies indicates that nifedipine, nisoldipine, and amlodipine lead to regression of interstitial and perivascular myocardial fibrosis, which may contribute to the improvement in diastolic function and coronary reserve seen with administration of CCBs.[25,27,28]

Although CCBs promote regression of LVH, not all antihypertensive agents are equally effective. Schmieder et al.[26] recently conducted a metaanalysis to determine the ability of various antihypertensive agents to reduce LVH, which is a strong, blood pressure-independent cardiovascular risk factor, in persons with essential hypertension. Of importance, only published reports of double-blind, randomized, controlled clinical studies with parallel-group design were included in the pooled analysis. Decline in blood pressure, duration of drug treatment, and drug class determined the reductions in left ventricular mass index. The angiotensin-converting enzyme (ACE) inhibitors and, to a lesser extent, CCBs emerged as first-line candidates to reduce the risk associated with LVH.

An additional mechanism whereby CCBs may theoretically mediate antiarrhythmic properties derives from the fact that they are antioxidants. Such a formulation is attractive, but additional data are required for corroboration.[29]

In summary, it is possible that the potential proarrhythmic actions of CCBs reflect some combination of the factors listed above: inappropriate vasodilation, excessive decrease in A-V conduction, transient ischemia, diminished contractility, and transient asystole. Such untoward effects ultimately arise from numerous factors, such as individual pharmacokinetic characteristics of the various classes of CCBs, their formulations, intrinsic affinities for channels (see Tables 2, 4, and 5), and hepatic function. However, the indirect mechanisms described above have not been demonstrated in clinical studies.

RECENT PROSPECTIVE STUDIES

Large prospective studies such as ALLHAT, NORDIL, and HOT are now in progress and may provide clearer answers about the risk of treatment of hypertension with various CCBs, especially cardiovascular outcomes.[9,30] Although the results of these studies will not be available for some time, some randomized prospective studies provide useful information about potential arrhythmogenic actions of calcium channel antagonists.

A double-blind, placebo-controlled study attempted to establish the efficacy of nimodipine (DHP derivative) in the prevention of anoxic brain injury in 155 patients resuscitated after ventricular fibrillation.[31] Nimodipine or placebo was infused intravenously for 24 hours, and most patients also received intravenous lidocaine. Recurrent ventricular fibrillation during the treatment period occurred in 1 patient in the nimodipine group compared with 12 patients in the placebo group ($p < 0.01$). There was no significant difference in recurrent cardiac arrest or survival at 3 and 12 months.

The Multicenter Diltiazem Postinfarction Trial evaluated the effectiveness of diltiazem in patients with recent myocardial infarction. Patients received diltiazem (60 mg 4 times/day) or a corresponding placebo.[32] After 3 months of follow-up, 1,546 of the 2,466 patients had a 24-hour continuous electrocardiographic

(EKG) recording.[33] There were no significant differences between the diltiazem and placebo groups in the prevalence of A-V block, frequency of atrial arrhythmias, or frequency or repetitiveness of ventricular arrhythmias; heart rate was significantly lower with diltiazem. Comparison with placebo revealed no evidence either for anti- or proarrhythmic effects of diltiazem. Prevalence of sudden and arrhythmic deaths was similar in the diltiazem and placebo groups. The percentage of total arrhythmic deaths (by Hinkle classification) was 41% in the placebo group and 42% in the diltiazem group.[33]

Patients with congestive heart failure are at increased risk for proarrhythmias.[34] The PRAISE trial[35] evaluated the efficacy of amlodipine compared with placebo added to conventional therapy in 1153 patients with New York Heart Association class IIIB or IV heart failure. The mean left ventricular ejection fraction was 0.21. Mortality rates were similar in the two treatment arms. Among patients with heart failure secondary to ischemic heart disease, the incidence of sudden deaths was not altered by amlodipine. In nonischemic heart disease, the incidence of sudden deaths and pump failure deaths was significantly reduced in patients receiving amlodipine. These studies[31,33,36] do not support the hypothesis that CCBs cause proarrhythmias.

The adverse effects reported with short-acting nifedipine may not be applicable to long-acting nifedipine or other CCBs. Different CCBs or different formulations of the same antagonist produce markedly differing pharmacodynamic effects.[9] During treatment with short-acting DHPs, major fluctuations in blood pressure may result from the rapid onset and offset of antihypertensive effects. Periodic, sudden, or rapid decreases in blood pressure are accompanied by increases in sympathetic activity, which may attenuate the regression of LVH and the prevention of progression of coronary artery disease. In contrast, slow-release formulations of otherwise rapidly absorbed DHPs achieved a more gradual and sustained antihypertensive effect.[9] Preliminary data from many recent studies indicate that long-acting CCBs do not have significant adverse cardiac effects associated with lowering of blood pressure.

Preliminary analysis of two of the largest hypertension trials of long-acting nifedipine and amlodipine were recently reported.[37] Clinical databases for amlodipine and long-acting nifedipine from Pfizer pharmaceuticals were reviewed. No deaths were reported, and the rate of myocardial infarction was 5.2 per 1000 patient years for long-acting nifedipine and 3.9 per 1000 patient years for amlodipine. These rates are similar to those reported for diuretic and beta-adrenergic blocking drugs in other studies.[1] In addition, Messerli[38] recently summarized the preliminary results from studies that used the newer long-acting CCBs for various cardiovascular disorders: amlodipine for heart failure (PRAISE), nisoldipine in patients with myocardial infarction (DEFIANT II), felodipine for heart failure (VHeFT III), and long-acting nifedipine for hypertension (STONE). These studies suggest that the long-acting DHPs do not increase mortality. Data from the trials of non-DHPs have not demonstrated an increase in mortality or myocardial infarction in patients with coronary artery disease and good left ventricular function.[30] The DAVIT II study[39] indicated that verapamil reduced major cardiac events in patients with recent myocardial infarction during the 18-month follow-up. Diltiazem has been reported to reduce the incidence of reinfarction in patients with non-Q infarction.[40] When the data from DAVIT[39] and MDPIT[32] studies were pooled, the incidence of cardiac events was lower in hypertensive patients receiving verapamil or diltiazem compared with placebo (21.4% vs. 27.4%, $p = 0.004$).[41]

CONCLUSION

In considering the adverse effects of CCBs, it should be remembered that they are a heterogeneous group of drugs.[20] They vary in structure, elimination half-lives, blocking activity of metabolites, relative affinities for myocardium and smooth muscles, degree of blockade of calcium channels in specialized cardiac conducting tissue, cardiac electrophysiologic effects, and formulations. Their indirect actions may vary quantitatively with mode of administration. Although excessive concentrations of various CCBs may cause varying degrees of A-V block, little direct evidence indicates that they are proarrhythmic. An exception is beperidil, which has potassium channel-blocking effects and prolongs QT interval.[3] CCBs are contraindicated in (1) patients with both accessory pathways between the atria and ventricles and preexcitation syndrome (see chapter 7) and (2) patients with diseased hearts also treated with β-adrenergic blockers. Despite these contraindications and the possibility of inducing A-V block with overdosage, CCBs appear to be safe. However, despite many limitations in clinical studies and analyses, reports of increased incidence of myocardial infarction and mortality associated with the use of short-acting nifedipine are a cause for both concern and reevaluation of choice of calcium antagonist. Short-acting nifedipine is not approved by the Food and Drug Administration for the treatment of hypertension; long-acting formulations of nifedipine or other long-acting DHPs are more appropriate choices. Although, in general, use of CCBs has been avoided in patients with heart failure, recent studies indicate that amlodipine and felodipine are well tolerated. Continued reevaluation and more informed therapeutic choice of the various CCBs are, in any case, good medical practice.

Future Prospectives

There are many glaring voids in our knowledge of the electrophysiologic and molecular basis of arrhythmia formation. Careful consideration of disparate and intriguing observations from several laboratories have led us to formulate a number of questions that we believe should be investigated to elucidate understanding of the precise role of DHPs in modulating the electrical activity of the heart.

Does Beta-adrenergic Stimulation Modify the Degree of Inhibition by Dihydropyridines?

The interaction of β-adrenergic agonists and the voltage-gated L-type Ca^{2+} channel are not completely understood.[42,43] Beta agonists increase L-type Ca^{2+} current in vitro, and this effect may be reduced or blocked by sufficient concentration of Ca^{2+} antagonists. Numerous laboratory studies of heart tissue clearly demonstrate that various CCBs block Ca^{2+} current that is potentiated by prior exposure to β-agonists.[44] On the other hand, the specificity of DHPs for vascular smooth muscle suggests that at the therapeutic concentrations required for blood pressure control, the newer DHPs probably have little action on Ca^{2+} channels in atrial and ventricular muscle in situ. However, this postulate may need reassessment. Few clinical data are relevant to this issue.

In one recent study, action potential duration was shortened and Ca^{2+} current was depressed in isolated human atrial muscles and myocytes after cessation of chronic treatment with nifedipine, nicardipine, or diltiazem.[45] Whether similar doses of these CCBs also shorten action potential duration and alter patterns of refractoriness in the human ventricle is unknown. From an electrophysiologic point of view, such drug-induced alteration of action potential duration and refractoriness may be especially detrimental in diseased ventricles with preexisting temporal and spatial dispersion of refractoriness and conduction. In diseased hearts, CCB-induced changes in cardiac electrophysiologic properties may provide a favorable substrate for the formation of reentrant circuits and arrhythmias, particularly, during periods of enhanced sympathetic activity, which also provokes dispersion of refractoriness in the ventricle. In other words, is a diseased heart exposed to CCBs, which may shorten action potential duration,[45] more likely to demonstrate reentrant circuits during sympathetic nervous system activation? As noted previously, however, CCBs with specific affinity for Ca^{2+} channels rarely induce ventricular arrhythmias. Nevertheless, reassessment of the significance of cardiac action potential changes and refractoriness during CCB dosage is warranted in regard to potential proarrhythmic actions in patients with cardiomyopathy and patients exposed to increased quantities of catecholamines.

Le Grand et al.[45] also demonstrated that the Ca^{2+} current response to application of the DHP Bay K 8644 was reduced in myocytes by chronic exposure of CCBs. In contrast, receptor-mediated regulation of Ca^{2+} current was not altered by chronic treatment with CCBs. Thus Ca^{2+} current increased similarly in both control myocytes and myocytes isolated from humans receiving CCBs with either isoproterenol or angiotensin II.

Laboratory studies indicate DHP-type antagonists (mixtures of stereoisomers) may transiently activate L-type Ca^{2+} channels,[41,42] and it has been shown that exposure to a combination of Bay K 8644 and isoproterenol results in less inactivation of the Ca^{2+} current and thus greater Ca^{2+} influx in heart muscle. Tsien et al.[46] have suggested investigations to determine whether the degree of inhibition of cardiac calcium channels by DHPs changes with β-adrenergic stimulation. This possibility appears particularly reasonable in view of the reflex sympathetic activity induced by short-acting preparations of the DHPs. It also remains to be determined whether the L-type Ca^{2+} channel is altered in disease, possibly further modifying the relations between DHPs and β-adrenergic stimulation.

Do Changes in Voltage-dependent Binding Characteristics Affect Cardiac Arrhythmogenicity?

In general, the mechanisms by which DHPs act appear more complex than originally described.[42,47,48] It has been suggested that their voltage-dependent binding characteristics in the heart should be further explored.[48] We also recommend studies in diseased myocardium. Whether changes in calcium antagonist binding characteristics significantly alter the responses to DHP-type antagonists remains to be determined.

Do Changes in T-Type Channel Density and Current Modulate the Effect of Dihydropyridines on the Heart?

There are two identifiable voltage-gated channels in the cardiac plasma membrane:

1. The L-type channel is abundant in adult hearts. Ca^{2+} current flowing through such channels is large and relatively slowly inactivating.[47] L-type Ca^{2+} current is important for myocardial excitation contraction coupling. So far, this chapter has considered Ca^{2+} currents flowing only through L-type channels.

2. The T-type calcium channel is more sparsely distributed in ventricular muscle cells. T-type channels open briefly at negative potentials, and their activation produces a transient small current. T-type current may be involved in pacemaking activity. Information about adrenergic regulation of cardiac T-type calcium channels is sparse,[42,48] although there are reports of β-agonist-induced potentiation of T-type Ca^{2+} current in specialized mammalian cardiac tissue.[49] Perhaps of greater importance, the number of T-type Ca^{2+} channels has been shown to increase in experimental cardiac disease (e.g., LVH in cats[41]). Although verapamil and diltiazem appear to have little effect on cardiac T-type Ca^{2+} channels, some DHPs appear to block them.[42,48] Because T-type channels appear to be more numerous in experimental cardiac hypertrophy,[50] we suggest a reappraisal of DHP interactions with the T-type channels in diseased myocardium. Whether the increased T-type Ca^{2+} current is more or less sensitive to β-agonist and/or DHP Ca^{2+} antagonists is unknown. One may speculate about a number of clinical scenarios wherein changes in T-type Ca^{2+} current or its modulation might alter the arrhythmogenic potential of DHPs (e.g., potentially greater risk of asystole in S-A node).

What Are the Determinants of Dihydropyridine-induced Blood Pressure Lowering?

It is unclear why DHPs appear to have greater effects on arterial blood pressure when the pressure is more elevated. Are the peripheral arterioles more Ca^{2+}-loaded and contracted from the actions of circulating mediators such as endothelin, angiotensin II, and norepinephrine? Does the degree of vasodilation by DHPs reflect an altered interaction with binding sites at smooth muscle Ca^{2+} channels and/or effects on smooth muscle Ca^{2+} channel kinetic properties because of background neurohormonal and circulating peptides? Exploration of this question may lead to prediction of inordinate blood pressure lowering and possibly more rational dosing

References

1. Chakko S, Kessler KM: Recognition and management of cardiac arrhythmias. Curr Probl Cardiol 20:53–120, 1995.
2. Myerburg RJ, Kessler KM, Chakko S, et al: Future evaluation of antiarrhythmic therapy. Am Heart J 127:1111–1118, 1994.

3. Billman GE: The antiarrhythmic effects of calcium antagonists. In Epstein M (ed): Calcium Antagonists in Clinical Medicine, 2nd ed. Philadelphia, Hanley & Belfus, 1997.
4. Levy MN: Role of calcium in arrhythmogenesis. Circulation 80:IV23–IV30, 1983.
5. Psaty BM, Heckbert SR, Koepsell TD, et al: The risk of myocardial infarction associated with antihypertensive drug therapies. JAMA 274:620–625, 1995.
6. Yusuf S: Calcium antagonists in coronary artery disease and hypertension. Time for reevaluation? Circulation 92:1079-1082, 1995.
7. Buring JE, Glynn RJ, Hennekens CH: Calcium channel blockers and myocardial infarction. A hypothesis formulated but not tested. JAMA 274:654–655, 1995.
8. Epstein M: Calcium antagonists should continue to be used for first line treatment of hypertension. Arch Intern Med 155:2150–2156, 1995.
9. Epstein M: Calcium antagonists: Still appropriate as first line antihypertensive agents. Am J Hypertens 9:110–121, 1996.
10. Furberg C: Calcium antagonists: Not appropriate as first line antihypertensive agents. Am J Hypertens 9:122–125, 1996.
11. Chinushi M, Aizawa Y, Miyajima S, et al: Proarrhythmic effects of antiarrhythmic drugs assessed by electrophysiologic study in recurrent sustained ventricular tachycardia. Jpn Circ J 55:133–141, 1991.
12. Saikawa T, Inoue K, Ohmura I, et al: The paradoxical adverse effect of verapamil for treating clinical paroxysmal supraventricular tachycardia. Jpn Heart J 28:107–113, 1987.
12a. Millar RNS: It looks like SVT—The misuse of intravenous verapamil in broad complex tachycardia. S Afr Med J 76:296–297, 1989.
13. Eisenhofer G, Friberg P, Rundqvist B, et al: Cardiac sympathetic nerve function in congestive heart failure. Circulation 93:1667–1676, 1996.
14. Grassi G, Seravalle G, Cattaneo BM, et al: Sympathetic activation and loss of reflex sympathetic control in mild congestive heart failure. Circulation 92:3206–3211, 1995.
15. Michalewicz L, Messerli FH: Cardiac effects of calcium antagonists in hypertension. Am J Cardiol 79(Suppl 10A):39–46, 1997.
16. Chakko S, Bassett AL: The clinical antiarrhythmic effectiveness of calcium channel antagonists. In Epstein M (ed): Calcium Antagonists in Clinical Medicine, 2nd ed. Philadelphia, Hanley & Belfus, 1997, pp 135–140.
17. Leenen FHH: Dihydropyridine calcium antagonists and sympathetic activity: Relevance to cardiovascular morbidity and mortality. In Epstein M (ed): Calcium Antagonists in Clinical Medicine, 2nd ed. Philadelphia, Hanley & Belfus, 1997, pp 527–552.
18. Grossman E, Messerli FH: Calcium antagonists and sympathetic activity. Am J Cardiol 1997 [in press].
19. Morris AD, Meredith PA, Reid JL: Pharmacokinetics of calcium antagonists: Implications for therapy. In Epstein M (ed): Calcium Antagonists in Clinical Medicine. Philadelphia, Hanley & Belfus, 1992, pp 49–68.
20. Triggle DJ: Mechanisms of action of calcium channel antagonists. In Epstein M (ed): Calcium Antagonists in Clinical Medicine, 2nd ed. Philadelphia, Hanley & Belfus, 1997, pp 1–26.
21. Ruzicka M, Leenen FHH: Relevance of intermittent increases in sympathetic activity for adverse outcome on short-acting antagonists. In Laragh JH (ed): Diagnosis, and Management, 2nd ed. New York, Raven, 1995, pp 2815–2825.
22. Kleinbloesem CH, van Brummelen P, Danhof M, et al: Rate of increase in the plasma concentration of nifedipine as a major determinant of its hemodynamic effects in humans. Clin Pharmacol Ther 41:26–30, 1987.
23. van Zwieten PA, Hansson L, Epstein M: Slowly acting calcium antagonists and their merits. Blood Press 6:78–80, 1997.
24. Messerli FH, Nuñez BD, Nuñez MM, et al: Calcium entry blocker and diuretic therapy on cardiac dysrhythmias. Arch Intern Med 149:1263–1267, 1989.
25. Vogt M, Strauer BE: Response of hypertensive left ventricular hypertrophy and coronary microvascular disease to calcium antagonists. Am J Cardiol 76:24D–30D, 1995.
26. Schmieder RE, Martus P, Klingbeil A: Reversal of left ventricular hypertrophy in essential hypertension. JAMA 275:1507–1513, 1996.
27. Campbell SE, Turek Z, Rakusan K, Kazda S: Cardiac structural remodelling after treatment of spontaneously hypertensive rats with nifedipine and nisoldipine. Cardiovasc Res 27:1350–1358, 1993.
28. Suzuki M, Yamanaka K, Nabata H, Tachibana M: Long term effects of amlodipine on organ damage, stroke and life span in stroke prone spontaneously hypertensive rats. Eur J Pharmacol 228:269–274, 1993.
29. Nayler WG: The role of oxygen radicals during reperfusion. J Cardiovasc Pharmacol 20:S14–S17, 1992.

30. Yusuf S: Calcium antagonists in coronary artery disease and hypertension. Time for reevaluation? Circulation 92:1079–1082, 1995.
31. Roine RO, Kaste M, Kinnunen A, et al: Nimodipine after resuscitation from out-of-hospital ventricular fibrillation. A placebo-controlled, double-blind, randomized trial. JAMA 264:3171–3177, 1990.
32. The Multicenter Diltiazem Post-Infarction Trial research Group: The effect of diltiazem on mortality and reinfarction after myocardial infarction. N Engl J Med 319:385–392, 1988.
33. Bigger JT, Coromilas J, Rolnitzky LM, et al, and the Multicenter Diltiazem Post-Infarction Trial investigators: Effect of diltiazem on cardiac rate and rhythm after myocardial infarction. Am J Cardiol 65:539–546, 1990.
34. Chakko S, de Marchena E, Kessler KM, Myerburg RJ: Ventricular arrhythmias in congestive heart failure. Clin Cardiol 12:525–530, 1989.
35. Packer M, O'Conner CM, Ghali JK, et al, for the PRAISE Study Group: Effect of amlodipine on morbidity and mortality in chronic heart failure. N Engl J Med 335:1107–1114, 1996.
36. Braun S, Boyko V, Behar S, et al, on behalf of the Bezafibrate Infarction Prevention Study participants: Calcium antagonists and mortality in patients with coronary artery disease: A cohort study of 11,575 patients. J Am Coll Cardiol 28:7–11, 1996.
37. Kloner RA, Vetrovec G: Long acting calcium channel blockers nifedipine GITS and amlodipine in hypertension: Rate of myocardial infarction [abstract].J Am Coll Cardiol 27:177A, 1996.
38. Messerli FH: Case-control study, meta-analysis, and bouillabaisse: Putting the calcium antagonist scare into context. Ann Intern Med 123:888–889, 1995.
39. The Danish Study Group on Verapamil in Myocardial Infarction: Effect of verapamil on mortality and major events after acute myocardial infarction (The Danish Verapmil Infarction Trial II—DAVIT II). Am J Cardiol 66:779–785, 1990.
40. Gibson RS, Boden WE, Theroux P, et al, and the Diltiazem Reinfarction Study Group: Diltiazem and reinfarction in patients with non-Q myocardial infarction: results of double-blind, randomized, multicenter trial. N Engl J Med 315:423–429, 1986.
41. Messerli FH, Boden WE, Fischer-Hansen J, Schechtman KB: Heart rate lowering calcium antgonists in hypertensive post-MI patients [abstract]. J Am Coll Cardiol 27:178A, 1996.
42. McDonald TF, Pelzer S, Trautwein W, Pelzer DJ: Regulation and modulation of calcium channels in cardiac, skeletal and smooth muscle cells. Physiol Rev 74:365–507, 1994.
43. Tiaho F, Richard S, Lory P, et al: Cyclic AMP-dependent phosphorylation modulates the stereospecific activation of cardiac calcium channels by Bay 8649. Pflügers Arch 417:58–66, 1990.
44. Trautwein W, Hescheler J: Regulation of cardiac L-type calcium current by phosphorylation and G proteins. Annu Rev Physiol 52:257–274, 1990.
45. Le Grand B, Hatem S, Deroubaix E, et al: Calcium current depression in isolated human atrial myocytes after cessation of chronic treatment with calcium antagonists. Circ Res 69:292–300, 1991.
46. Tsien RW, Bean BP, Hess P, et al: Mechanism of calcium channel modulation by β-adrenergic agents and dihydropyridine calcium antagonists. J Mol Cell Cardiol 18:691–710, 1986.
47. Marban E, O'Rourke B: Calcium channels: Structure, function, and regulation. In Zipes DP, Jalife J (eds): Cardiac Electrophysiology: From Cell to Bedside, 2nd ed. Philadelphia, W.B. Saunders, 1995, pp 11–21.
48. January CT, Cunningham PM, Zhou Z: Pharmacology of L- and T-type calcium channels in the heart. In Zipes DP, Jalife J (eds): Cardiac Electrophysiology: From Cell to Bedside, 2nd ed. Philadelphia, W.B. Saunders, 1995, pp 269–277.
49. Tseng GM, Boyden PA: Multiple types of Ca^{2+} currents in single canine Purkinje cells. Circ Res 65:1735–1750, 1989.
50. Nuss HB, Houser SR: T-type Ca^{2+} current is expressed in hypertrophied adult feline left ventricular myocytes. Circ Res 73:777–782, 1993.

MURRAY EPSTEIN M.D., F.A.C.P.

10

Calcium Antagonists in the Management of Hypertension

Calcium antagonists have assumed an important role in the treatment of patients with ischemic heart disease and in a variety of cardiovascular and noncardiovascular disorders. Similarly, calcium antagonists are used extensively in the United States and elsewhere as antihypertensive agents, and their availability has been an important advance in the management of hypertension. The major hemodynamic abnormality in most patients with essential hypertension is an increase in peripheral vascular resistance. Considerable evidence suggests that the elevation of peripheral vascular resistance is mediated in part by abnormal transmembrane influx of calcium. To the extent that abnormal calcium flux constitutes a determinant of elevated peripheral resistance, the major mechanism whereby calcium antagonists lower blood pressure—blockade of calcium-mediated electromechanical coupling in contractile tissue produces arteriolar vasodilation—is particularly apropos (see chapters 1 and 2). As a result of reducing total peripheral vascular resistance (PVR), systemic blood pressure decreases. Calcium antagonists cause widespread arterial and arteriolar vasodilation but vary in their ability to dilate different vascular beds. In addition to their effects on peripheral blood vessels, the hormonal and renal actions of calcium antagonists have been postulated to contribute to lowering of blood pressure (see chapter 24).

DIFFERENCES AMONG CALCIUM ANTAGONISTS

Calcium antagonists are a heterogeneous group of compounds with diverse chemical structures and pharmacologic actions. Presently, ten calcium antagonists are available in the United States; more will be marketed soon (Table 1). Eight have indications as antihypertensive agents. The order in which the three prototypic agents were released into the market was verapamil (Calan), nifedipine (Procardia), and diltiazem (Cardizem). Discussion in this chapter necessarily focuses on these three medications. Subsequently, nicardipine (Cardene), isradipine (DynaCirc), and felodipine (Plendil) were released. Seven of the currently marketed agents—amlodipine, nifedipine, nicardipine, isradipine, felodipine, nisoldipine, and nimodipine—are members of the dihydropyridine

TABLE 1. Calcium Antagonists Indicated for the Treatment of Hypertension

			Dosage for Hypertension	
Generic	**Brand**	**Strengths (mg)**	**Initial (mg)**	**Maximum (mg)**
Amlodipine	Norvasc	2.5, 5, 10	5 once daily	10 once daily
Bepridil	Vascor	200, 300, 400	Not for hypertension	
Diltiazem	Cardizem	30, 60, 90, 120	30 four times/day	360 in divided doses, 3 or 4 times/day
Diltiazem SR	Dilacor XR	120, 180, 240, 300	180–240 once daily	540 once daily
	Cardizem SR	60, 90, 120	60–120 twice daily	360/day
	Cardizem CD	120, 180, 240, 300	180–240 once daily	480 once daily
	Tiazac	120, 180, 240, 300, 360	120–240 once daily	540/day
Felodipine	Plendil	2.5, 5, 10	5 once daily	10 once daily
Isradipine	DynaCirc	2.5, 5	2.5 twice daily	20/day
Mibefradil	Posicor	50, 100	50–100 once daily	100/day
Nicardipine	Cardene	20, 30	20 three times/day	40 three times/day
Nicardipine SR	Cardene SR	30, 45, 60	30 twice daily	60 twice daily
Nifedipine	Adalat	10, 20	10 three times/day	30 three times/day
	Procardia	10, 20	10 three times/day	30 three times/day
Nifedipine SR	Adalat CC	30, 60, 90	30 once daily	90 once daily
	Procardia XL	30, 60, 90	30 or 60 once daily	120 once daily
Nimodipine	Nimotop	30	Not for hypertension	
Verapamil	Isoptin	40, 80, 120	80 three times/day	Up to 480/day; 360 recommended
	Calan	40, 80, 120		
Verapamil SR	Isoptin SR*	120, 180, 240	180*–240/day	Up to 480/day
	Calan SR*			
	Verelan			
Nisoldipine	Sular	10, 20, 30, 40	20/day	60/day

Data based on several sources, including Drug Facts and Comparison, and Micromedex.
* Isoptin SR and Calan SR are different trade names for an identical product.

group of calcium antagonists. All calcium antagonists retard the entry of calcium into the cell, be it cardiac muscle, peripheral vascular smooth muscle, or endocrine. The calcium antagonists are not alpha blockers, but some of them prevent the increase in calcium influx resulting from stimulation of alpha receptors.

Mibefradil (Posicor), the newest of the calcium antagonists to become available, has been proposed to be the first representative of a new class of calcium antagonists. Its status as such is justified by its novel molecular structure (a benzimidazolyl-substituted tetraline derivative) and unique mechanism of action, selective T-type calcium channel blockade. Preclinical investigation demonstrated potent vasodilatory activity with greater selectivity for coronary over peripheral vasculature. In various in vitro and animal models, mibefradil showed less negative inotropism than verapamil or diltiazem. In contrast to the dihydropyridines and in common with verapamil and diltiazem, the vasodilatory effects of mibefradil were accompanied by a slight decrease in heart rate.

The clinical pharmacokinetic features of mibefradil include high bioavailability and long half-life, which together make it suitable for once-daily dosing. The pharmacokinetic characteristics of mibefradil are not altered by such demographic factors as gender, age, or race. Clinical studies have demonstrated that mibefradil is an effective antihypertensive, antianginal, and antiischemic drug.

PHARMACOKINETICS

In general, conventional preparations of calcium antagonists have relatively short half-lives and extensive hepatic first-pass metabolism. Nevertheless, the available agents not only differ in their pharmacodynamic effects, as discussed below, but also manifest several differences in the way that they are handled by the body (see chapter 3). Some of these pharmacokinetic variables, which are summarized in Table 2, may significantly influence clinical decisions, including choice of agent, optimal dose, and mode and frequency of administration. Furthermore, as discussed in chapter 28, the short time to peak blood levels produced by the earlier rapid-, short-acting agents may account for their tendency to promote sympathetic neurohormonal activation; this tendency may be detrimental to patients with underlying myocardial disease.

The onset of action of the available calcium antagonists differs. Orally administered nifedipine has a more rapid onset of action and attainment of maximal blood pressure reduction than verapamil. Initial response times differ by about 15 minutes. The conventional preparation of verapamil also achieves a maximal lowering of the blood pressure about 1 hour later than nifedipine. On the other hand, the maximal drop in blood pressure and the duration of the hypotensive response after a single dose are similar with each agent.

The duration of action of the conventional forms of all three prototypic drugs is between 6 and 8 hours after a single dose. Therefore, they are usually administered either three or four times daily.

The relatively low bioavailability of verapamil relates mainly to the high hepatic first-pass effect. Verapamil is metabolized in the liver to norverapamil. The half-life of verapamil is prolonged in patients with liver disease, and smaller doses should be used when initiating therapy in such patients. Careful individual titration of dosage (not only for verapamil but also for the other calcium antagonists) is thus required. Verapamil (and to a lesser degree nifedipine and diltiazem) also undergoes unexpected accumulation, despite the general rule that a steady-state blood level of a medication usually occurs within 3 to 5 half-lives. With chronic treatment, the half-life of verapamil may double, perhaps as a result of stereoselective metabolism or from verapamil's inducing an effect on its own metabolism.

Another important index of the pharmacokinetic properties of calcium antagonists is the trough-to-peak ratio. Currently, high trough-to-peak blood pressure-lowering effects are regarded as a desirable property of antihypertensive agents. Drugs such as short-acting (instant-release) nifedipine have a very low trough-to-peak ratio. In contrast, some of the newer extended-release formulations have better trough-to-peak ratios. According to Zannad et al., ratios > 70% were found with only one angiotensin-converting enzyme (ACE) inhibitor and only three calcium antagonists (nifedipine, gastrointestinal therapeutic system [GITS] with

TABLE 2. Some Important Pharmacokinetic Properties of the Nine Currently Available

	Nifedipine	Verapamil	Diltiazem	Nicardipine
Time to peak blood level (min)	30–60 (360)†	90–120	120–180‡	30–120
Systemic bioavailability (%)	60	20	40–50	35
Protein binding (%)	90	95	80	95
Volume of distribution (L/kg)	3	7	4	0.6 to 63 L?
Plasma elimination half-life (hr)	3–5	4–5§	4–5	0.7–1.8
Elimination (major organ)	Liver	Liver	Liver	Liver
Unexpected accumulation	No	Yes	Yes	No

* Extended-release formulation.
† For nifedipine GITS (Procardia XL).
‡ For extended-release diltiazem (Cardizem CD).
§ May be prolonged in patients with liver disease.
// Half-life of the terminal elimination phase.

77% and sustained-release [SR] verapamil with 82%. Nisoldipine CC has a trough-to-peak ratio of 70–100%, according to the Physician's Desk Reference).

Slow-release Formulations

In an attempt to obviate some of the problems posed by the pharmacokinetics of short-acting calcium antagonists, manufacturers have developed innovative slow-release delivery systems as well as longer-acting formulations to allow once-daily dosing. Examples include Procardia XL (nifedipine, Pfizer), Covera HS (verapamil, G.D. Searle), Cardizem CD (diltiazem, Hoechst Marion Roussel), Verelan (verapamil, Wyeth-Ayerst and Lederle Laboratories), Plendil (felodipine, Astra Merck), and Sular (nisoldipine, Zeneca Pharmaceuticals) (Table 3).

Controlled-release drug delivery systems allow once- and twice-daily dosing of shorter-acting antihypertensive agents, which may enhance patient compliance and reduce side effects encountered with the basic formulations. The goal, although not always realized, is to attain zero-order drug kinetics with a constant rate of drug release over time. This goal is inevitably affected by several variables, including the pattern of gastrointestinal motility in the fed and fasted states, gastrointestinal transit time from the stomach to the ileocecal valve, splanchnic blood flow, increasing pH from the stomach to the colon, and differing areas of gastrointestinal absorption throughout the gastrointestinal tract. The most common oral controlled-release formulations include a polymeric-coated reservoir or homogenous drug-polymer matrix formulation dissolutional system via bioerosion or degradation and osmotic pump systems (GITS).

Procardia XL (nifedipine) uses the gastrointestinal therapeutic system (GITS) developed by the Alza Corporation. The basis of this system is a push-pull

Calcium Antagonists *(Continued from opposite page)*

Isradipine	Felodipine	Nisoldipine*	Amlodipine	Mibefradil
90	150–300	360–720	360–720	60–120
15–24	20	5	60–65	> 90
95	99	99	99	≥ 99
3	10	3	21	2.5
8//	11–16	7–12	35–50	17–27
Liver	Liver	Liver	Liver	Liver
No	No	No	No	No

osmotic pump that releases medication over 24 hours and maintains relatively constant blood levels over this period when taken once daily. One of the two layers of the tablet contains an osmotic driving substance, whereas the other layer contains a suspension of nifedipine. As water is absorbed through a semipermeable membrane into the tablet, expansion of the osmotic driving agent pushes the medication into the gastrointestinal tract through a laser-drilled opening. This GITS formulation of nifedipine is available as Procardia XL (Pfizer) in the United States and as Adalat Oros (Bayer AG) elsewhere in the world.

Covera HS (verapamil) also uses Alza's novel delivery system (GITS), an osmotic pump within a bilayer tablet, to delay onset of drug release and to provide sustained release of verapamil over a predetermined period at a virtually constant rate. The constant rate of delivery minimizes side effects from extreme drug plasma fluctuations and is substantially independent of gastrointestinal pH and motility. The tablet is taken at bedtime. It is designed to delay release of drug and to provide peak plasma concentrations of verapamil during the early morning hours when catecholamines are released and blood pressure elevations are typically observed. After a 5-hour delay, Covera HS delivers verapamil at a uniform rate for approximately 10 hours. Based on circadian variation in exacerbation of cardiovascular disease, this formulation is an attempt to curtail the early morning surge of blood pressure and to lessen the more frequent occurrence of myocardial events in the early morning hours.

The most recent calcium antagonist to use the GITS delivery system is DynaCirc CR (controlled-release isradipine). Cardizem CD uses two types of sustained-release beads to achieve controlled delivery of diltiazem. The two types of beads differ in the thickness of their copolymer membrane coatings, resulting in a two-phased, water-dependent process for controlled timing of drug release. The bead blend provides release of approximately 40% of the total diltiazem content in the first 12 hours and 60% in the second 12 hours. Verelan, on the other hand, uses sustained-release pellets that ensure wide distribution of verapamil. The formulation consists of a newly patented spheroidal oral

TABLE 3. Delayed-release Delivery Systems of Long-acting Calcium Antagonists

Agent	System
Nifedipine GITS	Osmolar pump system: Fluid from the gastrointestinal tract enters the capsule's semipermeable coating, putting the drug into suspension, expanding the inert "push" compartment, and thereby pushing the drug out a laser-drilled hole at a fairly constant rate.
Verapamil	
Calan SR, Isoptin SR	Diffusional system: Verapamil is incorporated into a matrix of the natural polysaccharide, sodium alginate, which swells in contact with gastrointestinal fluid. Verapamil diffuses out over 7 hours.
Verelan	Dissolutional system: The spheroidal oral drug absorption system (SODAS) consists of multiple 1-mm spheres surrounded by rate-controlling-polymers stored in a hard gelatin capsule. A fraction of the beads are released immediately, and the rest are distributed within the gastrointestinal tract. Dissolution is independent of pH and food.
Covera HS	A new controlled-onset, extended-release verapamil delivery system that uses the Alza GITS osmotic pump system within a bilayer tablet to delay onset of drug release and to provide sustained release of verapamil over a predetermined period at a virtually constant rate.
Felodipine ER	Diffusional system: Upon contact with gastrointestinal fluid, felodipine diffuses over 12 hours, first from an outer hydrophilic gel layer and then from an inner reservoir, permitting once-daily dosing.
Diltiazem	
Cardizem SR	Dissolutional system: Diltiazem beads contain a variably thick pharmaceutical coating that dissolves over 3–12 hours in the gastrointestinal tract. Twice-daily dosing required.
Cardizem CD	Diffusional system: In a water-dependent process, two populations of SR beads, thin- and thick-coated, release diltiazem over the first 12 hours and then the second 12 hours, respectively, for once-daily dosing.
Dilacor XR	Dissolutional system: Tablet swelling causes release of 3–4 60-mg diltiazem tablets beyond an outer hard gelatin coating over 24 hours for once-daily dosing. The outer slow-hydrating coat ensures drug release at a constant rate.
Nisoldipine (Sular)	Nisoldipine is contained in a coat-core formulation that allows once-daily dosing. The tablet contains 80% of the nisoldipine in a slow-release formulation in the outer coat and 20% in an immediate-release form within the core. The coat-core construction takes advantage of the differential absorption of nisoldipine as the tablet traverses the gastrointestinal tract. In the upper portions of the tract, where nisoldipine is rapidly absorbed, drug is released slowly, whereas in the lower portions of the tract, where absorption is slow, release is rapid. The result is continuous 24-hour drug delivery.
Nicardipine SR	Dissolutional system: A two-component capsule contains an immediate-release powder and a slow-release spherical granule, the former containing 25% and the latter 75% of the total nicardipine dose. The granule's polymer of methacrylic acid ensures insolubility at $pH < 5.0$, and the limited spheroidal surface area available for dissolution creates a sustained-release effect once the capsule is past the duodenum.

drug absorption system (SODAS), wherein several hundred 1-mm microspheroidal beads are encased in such a way as to control medication release precisely. The coating consists of a water-insoluble polymer that controls water influx and medication efflux.

The drug delivery formulation used for Plendil is based on the hydrophilic gel principle, whereby solubilized felodipine is embedded in a gel-forming matrix. The surface of the tablet gradually forms a swelling gel layer when it comes in contact with gastrointestinal fluids. Felodipine is released at a predetermined rate by diffusion through the hydrated gel layer and by slow attrition of the gel matrix.

A new entry into the dihydropyridine calcium channel blocker market is Sular, which contains nisoldipine in a coat-core formulation to allow once-daily dosing. The tablet contains 80% of the nisoldipine in a slow-release formulation in the outer coat and the remaining 20% in an immediate-release form within the core. The coat-core construction takes advantage of the differential absorption of nisoldipine as the tablet traverses the gastrointestinal tract. In the upper portions of the tract, where nisoldipine is rapidly absorbed, drug is released slowly, whereas in the lower portions of the tract, where absorption is slow, release is rapid. The result is continuous 24-hour drug delivery.

In contrast to the agents listed above, which require a delivery system to maintain a 24-hour duration of action, two new slow-onset, long-acting agents have become available—amlodipine and lacidipine. Norvasc (amlodipine) is a true once-daily calcium antagonist owing to its half-life of 30–50 hours. Lacidipine, another dihydropyridine, has an intrinsically long duration of action. Currently lacidipine is available in only a few countries (Aponil, Italy; Caldine, France; Lacipil, Italy, Spain, and France; Lacirex, Italy; Motens, United Kingdom, Netherlands, Switzerland, and Belgium; Viapres, Italy; Zascal, South Africa); it is not available in the United States.

Whether a drug is a true long-acting formulation or is delivered using a sustained-release system, studies have shown that once-daily preparations reduce the peak/trough fluctuations in plasma drug concentrations seen with short-acting preparations and therefore deliver a more controlled antihypertensive effect with the potential for fewer side effects. A caveat is in order, however. As pointed out by Leenen (chapter 28), not all once-daily, slow-release formulations consistently reduce the peak/trough fluctuations in plasma drug concentrations. Leenen suggests that once-daily dosing with several of the extended-release (ER) tablets exhibits a clear peak-to-trough variation in plasma drug concentration, which may result in fluctuations in the magnitude of antihypertensive effect over 24 hours.

Circadian Variation and Controlled-release Formulations

Circadian rhythms are among the most prominent bioperiodicities in humans, and many diseases and their symptoms display distinct day-night patterns. It is known, for example, that the risk of ischemic heart disease (angina, acute myocardial infarction, and sudden cardiac death) and hemorrhagic and thrombotic stroke is greatest at the start of daily activity. The morning increase in cardiac events corresponds to activation of the sympathetic nervous system and is a result of triggering of onset. Such triggers may be due to external stimuli such as heavy physical exertion, sexual activity, and anger. Surges in blood pressure, heart rate, and catecholamines are believed to contribute to the hemodynamic and vasoconstrictive forces that promote thrombogenic activity. Increased understanding of the chronobiology of cardiac disease has impelled several pharmaceutical companies to develop new drug delivery systems that align drug level and pharmacologic actions to biologic rhythm and cyclic triggers of disease and may be beneficial in the management of hypertension and angina.

The Controlled Onset Verapamil Investigation for Cardiovascular Endpoints (CONVINCE) is an ongoing double-blind, parallel-group control study with a goal of enrolling 1,500 patients to investigate whether the Covera HS formulation of verapamil is at least as effective as the standard-of-care agents (diuretics and beta blockers). The combined endpoints of fatal cardiovascular disease, nonfatal stroke, and nonfatal myocardial infarction will be compared.

PHARMACODYNAMICS (Table 4)

In vitro, all of the calcium antagonists have negative inotropic and chronotropic actions. Nevertheless, in patients and intact animals, these direct effects are offset variably by reflex sympathetic responses (see chapter 28).

Of the three prototypic agents, nifedipine is the most potent vasodilator and has the least effect on sinoatrial (SA) and atrioventricular (AV) nodal function. It is more likely to provoke reflex stimulation of the heart, resulting initially in increases in heart rate, myocardial contractility, and cardiac output. In patients with impaired left ventricular function, however, nifedipine may reduce myocardial contractility. Verapamil has more potent direct negative chronotropic and inotropic actions; therefore, the end result is usually a lack of change in heart rate and cardiac output in normal and hypertensive subjects. In patients with depressed left ventricular (systolic) function, however, verapamil is more likely to produce undesirable further deterioration. Diltiazem seems to have a somewhat intermediate effect. It produces less peripheral vasodilation and less

TABLE 4. Cardiac and Hemodynamic Effects of Calcium Antagonists in Patients with

Effect	Nifedipine	Nicardipine	Isradipine	Diltiazem
Arteriolar dilation	++++	++++	++++	++
Coronary vasodilation	++++	++++	++++	++
Cardiac preload	0/↓[†]	?	0	?
Cardiac afterload	↓↓	↓↓	↓↓	↓
Cardiac contractility*	0/↓[‡]	0	0	↓
Heart rate	↑	↑ (acute), then →	→ or ↑ (acute), then →	↓ or →
AV conduction	0	0	0	↓↓
SA automaticity	0	0	↓ (slight)	↓↓

* Net effect (direct cardiac effect plus influence of activation of sympathetic nervous system, etc.)

[†] In patients with an increased preload, afterload reduction may result in lowering of preload.

of a reflex sympathetic response than nifedipine. Diltiazem has mild negative chronotropic activity, which is usually more apparent during exercise than at rest, and relatively little negative inotropic action. The latter is clinically negligible in patients with normal left ventricular function. Although verapamil and diltiazem influence the activity of both the SA and AV nodes, in general, verapamil has a greater effect on AV nodal function and diltiazem on SA function.

Vascular and Ischemia Selectivity

An interesting concept that may be relevant for future therapy with calcium antagonists is vascular and ischemia selectivity. For example, nisoldipine, the most recently introduced dihydropyridine, is reported to have a ratio of 1000:1 for coronary-to-myocardial effects; such an agent may be the most vascular-selective. According to the concept of ischemia selectivity, which was reviewed recently by Opie, agents that are highly vascular-selective are also likely to be ischemia-selective; the common basis is their capacity to interact with depolarized tissue. The concept implies that not only vascular but also myocardial benefit may be obtained from drugs that are highly vascular-selective when they are given to patients with ischemic heart disease. This interesting concept awaits confirmation by randomized, controlled studies.

DIFFERENCES BETWEEN CALCIUM ANTAGONISTS AND OTHER VASODILATORS

Drugs that directly reduce peripheral vascular resistance, such as hydralazine, have been used for antihypertensive therapy for many years. Nevertheless, the effectiveness of these agents is often limited by the reactive stimulation of renal and hormonal responses that counteract their antihypertensive actions (Fig. 1). These responses tend to produce tolerance to hydralazine's vasodilating action

Normal Cardiac Function *(Continued from opposite page)*

Verapamil	Felodipine	Nisoldipine	Amlodipine	Mibefradil
++	++++	++++	++++	++++
++	++++	++++	++++	++++
↓[‡]	0/↓[‡]	0/↓[‡]	0/↓[‡]	?
↓	↓↓	↓↓	↓↓	↓↓
↓↓	0	0	0	0
↓ or →	↓ or → (acute), then →	→ or ↑ (acute), then →	→	↓
↓↓↓	0	0	0	↓↓
↓	?	0	0	↓↓

[‡] Depends on type of patient and extent of sympathetic stimulation; nifedipine may decrease contractility in susceptible patients.

[§] With the extended-release preparation, the heart rate increase is modest and transient (returning to normal within 1 week due to baroreceptor resetting.)

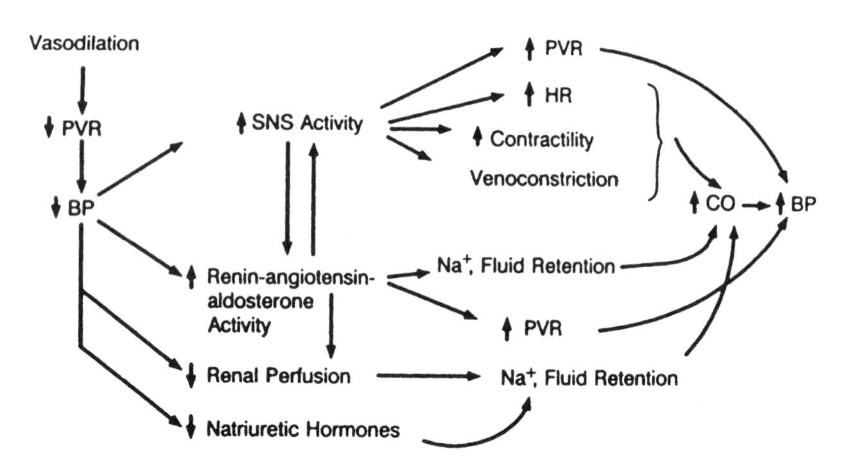

FIGURE 1. Known and postulated compensatory mechanisms whereby vasodilation induced by nonspecific vasodilators such as hydralazine is offset by the reactive stimulation of renal and hormonal responses. Examples include an increase in the activity of the sympathetic nervous system and renin-angiotensin-aldosterone system and a decrease in renal perfusion. Although the precise mechanisms remain uncertain, calcium antagonists attenuate such adaptive changes. (From Epstein M, Loutzenhiser R (eds): Calcium Antagonists and the Kidney. Philadelphia, Hanley & Belfus, 1990, with permission.)

and cause volume-expansion-induced pseudotolerance to its antihypertensive effects.

The precise mechanism(s) whereby calcium antagonists interfere with angiotensin II or alpha-adrenergic-mediated vasoconstriction remains uncertain. Nevertheless, in the presence of calcium antagonists (as indicated in Figure 1 by the interruptions of the arrows), the expected adaptive changes in peripheral vascular resistance, heart rate, cardiac output, and extracellular volume that eventually lead to a reduction in the blood pressure-lowering response to vasodilators are mitigated. An intriguing speculation is the possibility that calcium antagonists may countervail the sodium-retaining renal effects of decreased perfusion and decreased levels of natriuretic hormones.

In contrast to other vasodilators such as hydralazine and minoxidil, the calcium antagonists blunt the reflex increase in sympathetic activity that typically occurs in association with blood pressure reduction. It is not known whether this phenomenon occurs by virtue of indirect alpha-adrenergic inhibitory action. Calcium antagonists also may interfere with the action of angiotensin II on vascular smooth muscle. Despite the early increase in plasma renin activity (PRA) during calcium antagonist therapy, aldosterone levels fail to rise proportionally (and occasionally decrease), perhaps because of the role of calcium in hormonal release. Recent evidence indicates that acute administration of calcium antagonists, unlike most other nondiuretic antihypertensive medications, produces diuresis and natriuresis. In addition, when calcium antagonists are given over the long term, they are not sodium retaining (see chapter 24).

The calcium antagonists differ from previous vasodilators because of their favorable accompanying effects on the heart and kidney. As antihypertensive agents, the calcium antagonists thus appear considerably more versatile than previous vasodilators.

USE OF CALCIUM ANTAGONISTS AS ANTIHYPERTENSIVE AGENTS

To date, all calcium antagonists have been shown to be similarly effective and safe antihypertensive drugs. For this reason, calcium antagonists have been recommended as first-line drugs by many national and international authorities (World Health Organization, International Society of Hypertension, Canadian Hypertension Society Consensus Conference, the British Hypertension Society, and the Australian Consensus Panel). In the aftermath of the recent calcium antagonist controversy, a few investigators have suggested that short-acting formulations of verapamil and diltiazem may be less apt to produce adverse cardiovascular events than short-acting dihydropyridines. Such proposals have not been validated in long-term randomized controlled trials. On the other hand, no increase in coronary risks or serious adverse effects for patients on long-acting formulations have been reported. Of interest, calcium antagonists generally have a lesser effect on the blood pressure of normotensive patients. In hypertensive patients, however, some evidence suggests that their effect on blood pressure increases with the degree of hypertension. These observations, taken together, lend support to the notion that calcium antagonists act by specifically reversing pathophysiologic perturbations.

There are few contraindications to the use of calcium antagonists (as opposed to beta blockers) as antihypertensive agents: severely depressed left ventricular systolic function (frank congestive heart failure or history thereof), sick sinus syndrome, and AV conduction disturbances.

All of the calcium antagonists currently available in the United States are efficacious as initial antihypertensive monotherapy. In general, a lower dose (given less frequently) of a calcium antagonist is required to control blood pressure than to manage angina. The antihypertensive response demonstrates a dose-response curve, but small doses have been used with success.

Studies that compared the antihypertensive effect of calcium antagonists and other agents reveal that the various calcium antagonists are usually equally efficacious in unselected patients with mild to moderate hypertension. In addition, calcium antagonists appear to be at least as good as beta blockers in lowering blood pressure in patients with both hypertension and angina. For example, comparison of conventional nifedipine with hydralazine has shown that the medications are equivalent in both efficacy and side effects; diltiazem, at dosages of 60–120 mg three times daily, has been demonstrated to produce an antihypertensive response similar to that of hydrochlorothiazide.

In general, the blood pressure-lowering effects of calcium antagonists are sustained. To our knowledge, little or no resistance has been reported for any of the agents during studies lasting for several months. Although there have been rare reports of tolerance with calcium antagonists, an increase in dosage reestablished blood pressure control, and noncompliance could not be excluded.

As mentioned, several studies have shown that generally all calcium antagonists have similar antihypertensive properties, although some reduce cardiac output and total peripheral resistance (TPR), whereas others reduce only TPR. Adverse or salutary effects thus dictate the choice of a specific agent. For example, dihydropyridines may be preferred in a patient with mild ventricular systolic dysfunction, conduction system disease, or slow heart rate. In contrast, verapamil or diltiazem may be more suitable in a patient with tachycardia.

DEMOGRAPHIC CONSIDERATIONS (AGE, RACE, PLASMA RENIN ACTIVITY)

It has been estimated that with calcium antagonist monotherapy, approximately 50% of the general population with essential hypertension can achieve blood pressure control. Calcium antagonists appear to work equally well for both men and women. The degree of efficacy of calcium antagonists may relate, at least in part, to differing demographic features of hypertensive patients: age, pretreatment blood pressure, PRA, and race. Several lines of evidence, however, suggest that calcium antagonists are equally effective in all age groups, whereas beta-adrenoceptor blockers show decreased efficacy in patients older than 60 years. The blood pressure-lowering effect of calcium antagonists may be inversely proportional to the baseline PRA: in general, the lower the PRA, the greater the response to calcium antagonists. In addition, preliminary reports suggest that in patients with low PRA, a high dietary sodium intake does not impair the antihypertensive effect of calcium antagonists. Perhaps this phenomenon is related to the observation that the combined use of calcium antagonists and diuretics may not provide an additive antihypertensive response. Furthermore, whereas converting enzyme inhibitors (depending on the dose) and beta blockers tend to work less well as antihypertensive agents in black patients, calcium antagonists are equally efficacious in black and white patients (see chapter 16). Finally, preliminary evidence also suggests that calcium antagonists preferentially lower blood pressure in salt-sensitive patients and patients with lower levels of serum ionized calcium.

ANTIHYPERTENSIVE THERAPY IN PATIENTS WITH ACCOMPANYING DISEASES (Table 5)

Originally, the major therapeutic use of calcium antagonists was for treatment of coronary artery vasospasm, angina pectoris, and supraventricular arrhythmias. Not surprisingly, much of the initial experience with calcium antagonists as antihypertensive agents was accumulated in patients with these coexisting problems. Certainly, the combination of hypertension and ischemic heart disease is not uncommon. The calcium antagonists can be used to treat both problems and often provide a welcome alternative for patients who are not good candidates for beta blockers.

TABLE 5. Complicating Problems That Commend the Selection of Calcium Antagonists as Antihypertensive Drugs

1. Ischemic heart disease	4. Peripheral vascular insufficiency
2. History of recent non-Q-wave myocardial infarction (diltiazem only)	5. Asthma, chronic pulmonary disease
	6. Chronic renal failure (?)
3. Left ventricular hypertrophy	7. Diabetes: insulin-dependent or non-dependent

Calcium antagonists have an important therapeutic role in the management of patients with concomitant ischemic heart disease (IHD) and hypertension. IHD and systemic hypertension coexist in a large number of patients. More than one-half of all patients with angina pectoris have a history of hypertension, which has long been recognized as as major risk factor for the development of atherosclerotic coronary artery disease (CAD). Long-standing hypertension is also the primary cause of left ventricular hypertrophy (LVH) and hypertrophic cardio- myopathy, both of which aggravate IHD by increasing myocardial oxygen demand and decreasing myocardial blood supply. Thus, hypertension both promotes and exacerbates IHD by multiple mechanisms. Because the two conditions coexist in such a large number of patients, monotherapy that would treat both effectively is desirable. Calcium antagonists and beta-adrenergic blockers are the primary therapeutic agents that have been assessed in this context.

Although the findings certainly require further confirmation, calcium antagonists, like beta blockers, appear to have a cardioprotective effect in experimental ischemia and infarction. A few intriguing studies also indicate that they may decrease experimental atherogenesis.

A limited number of studies have addressed the question whether calcium antagonists provide secondary protection in patients with myocardial infarction. Trials with verapamil and nifedipine indicate that, at least in the short term, neither agent reduces postinfarction mortality or reinfarction rates. In contrast, studies carried out by the Multicenter Diltiazem Postinfarction Trial Research Group have shown that in postinfarction patients without pulmonary congestion, diltiazem reduces the number of cardiac events (death from cardiac causes or nonfatal myocardial infarction). When pulmonary congestion is present, however, no benefit is conferred. Furthermore, in patients with previous non-Q-wave infarction, diltiazem decreases the rate of early reinfarction, refractory angina, and angina with EKG changes.

Several recent metaanalyses indicate that calcium antagonists decrease left ventricular mass. The benefits of short-acting dihydropyridines may be counteracted by the stimulatory effects on the sympathetic and renin-angiotensin systems. A limited number of studies with felodipine, isradipine, and amlodipine are available, but each has been shown to decrease left ventricular mass. In a recent study performed by Sareli and reported by Langan et al., a new vascular-selective dihydropyridine calcium antagonist, nisoldipine coat-core (CC), reduced left ventricular mass index with no deleterious effects on left ventricular systolic performance. The decrease in left ventricular mass documented with calcium antagonists is associated with improved ventricular filling, diminished ventricular ectopy, and preserved contractility (see chapter 5 for detailed discussion).

All of the above-mentioned salutary features, if substantiated, commend the use of calcium antagonists in many patients. In addition, an important observation is that calcium antagonists do not induce the undesirable changes in serum concentration of lipoproteins, uric acid, or potassium that are characteristically noted with other antihypertensive agents.

Calcium antagonists may be particularly desirable antihypertensive agents in other situations. They are unlikely to worsen symptomatic peripheral vascular disease. Whether they are beneficial in patients with bronchospastic pulmonary disease is debatable, but certainly they are not detrimental. In addition,

TABLE 6. Potential Considerations in Choosing Initial Antihypertensive Therapy

Accompanying Illness	Disadvantages	Advantages
Coronary disease	Hydralazine, $alpha_1$ blocker, diuretic	Beta blocker, calcium antagonist
Arrhythmias	Diuretic, hydralazine	Beta blocker, calcium antagonist
Congestive heart failure	Beta blocker, calcium antagonist	Diuretic ACE inhibitor, $alpha_1$ blocker, calcium antagonist*
Peripheral vascular disease	Beta blocker, diuretic	Diuretic, ACE inhibitor, calcium antagonist, $alpha_1$ blocker
Bronchospastic disease	Beta blocker	Calcium antagonist, ACE inhibitor, $alpha_1$ blocker, central sympatholytic
Diabetes	Beta blocker, diuretic	ACE inhibitor, calcium antagonist, $alpha_1$ blocker
Demographics		
Elderly patients	Beta blocker	Diuretic, calcium antagonist, central sympatholytic
Black patients	Beta blocker, ACE inhibitor (depending on dose)	Diuretic, calcium antagonist

* Newer, more vascular-selective agents are currently under evaluation in patients with heart failure.

whereas some data indicate that calcium antagonists inhibit pancreatic insulin release, they do not cause clinically important metabolic problems in insulin-dependent diabetic patients. Finally, some calcium antagonists (e.g, verapamil) tend to improve ventricular compliance and may be used with caution advantageously in certain patients with left ventricular diastolic dysfunction. Table 6 lists some of the diseases that frequently accompany hypertension and how their presence may affect the selection of an antihypertensive drug. Clearly, the calcium antagonists appear to be useful in many such patients.

An intriguing salutary effect, not fully established, is the influence of calcium antagonists on renal function. Several lines of evidence indicate that calcium antagonists are capable of inducing a dramatic reversal of (or protection from) acute renal ischemia under a number of experimental conditions. Calcium antagonists preferentially attenuate afferent arteriolar vasoconstriction in response to diverse agonists, including norepinephrine, angiotensin II, and thromboxane. This salutary effect of calcium antagonists on intrarenal hemodynamics suggests that they are particularly well suited for the management of hypertension; moreover, they may have a future role in managing certain types of acute renal insufficiency. To the extent that renal ischemia underlies many disease states, including hypertension, the potential beneficial effects of such agents in this setting are of great interest. For example, calcium antagonists have recently been shown to confer a protective effect against radiocontrast-induced nephrotoxicity, transplant-associated acute renal insufficiency, and acute cyclosporine nephrotoxicity (for detailed discussion, see chapter 24).

One final aspect of calcium antagonists is their atypical action on renal sodium handling. Unlike many nondiuretic antihypertensive drugs, calcium antagonists have the direct acute renal effect of enhancing sodium excretion. Although the natriuretic effects of calcium antagonists are not sustained during long-term administration, it is nevertheless clear that with long-term administration calcium antagonists are not sodium-retentive. It should be emphasized, however, that the net effect of a medication on sodium homeostasis is determined by the sum of the separate effects on blood pressure, cardiac performance, systemic vascular resistance, and renal tubular transport.

COMBINATION THERAPY

Calcium antagonists have been used successfully in combination with other antihypertensive drugs as second-line agents in the treatment of hypertension or as third-step agents in patients with refractory hypertension. Calcium antagonists are best combined with beta blockers and ACE inhibitors for increased efficacy, whereas the addition of a thiazide diuretic may provide less than an additive blood pressure-lowering effect. Nevertheless, patients who are receiving a beta blocker and a thiazide diuretic may have an additional hypotensive action with a calcium antagonist. In patients with severe hypertension or patients with mild left ventricular (diastolic) dysfunction, the combination of a calcium antagonist and ACE inhibitor appears to be efficacious and safe. On the other hand, several investigators have cautioned against the combined use of an alpha blocker and calcium antagonist because of an unacceptable risk of hypotension (see chapter 26 for detailed discussion).

Combined Therapy with Beta Blockers

The rationale and safety of combined therapy with calcium antagonists and beta blockers (for the treatment of angina) have been reviewed by Bala Subramanian. Historically, the concomitant use of these two classes of drugs (especially verapamil) was contraindicated. The reservations were based on both theoretical factors and data about the effect of intravenous rather than oral use. The contraindication was related to the putative risk of inducing acute left ventricular failure, advanced heart block (or asystole), severe bradycardia, or hypotension. Calcium antagonists of the dihydropyridine class are, however, vascular-selective to varying degrees. All currently available dihydropyridines, therefore, may be used in combination with beta blockers. In particular, the combination of beta blockers and highly vascular-selective calcium antagonists—nisoldipine, isradipine, and felodipine—may be preferred because they do not add to the cardiodepressant effect of the beta blocker.

The compensatory reflex increase in sympathetic tone induced by the decrement in blood pressure is one reason to combine a dihydropyridine-type calcium antagonist with a beta blocker. The increase in sympathetic tone may produce a symptomatic increase in heart rate and an increment in PRA. These two effects may blunt the antihypertensive effect, but both are preventable with concomitant use of a beta blocker. In addition, calcium antagonists may counteract some of the adverse effects induced by beta blockers (particularly the initial vasoconstriction induced by many). The recent report that beta blockade may decrease the clearance of verapamil requires confirmation and delineation of clinical relevance.

Combination of Diltiazem and Dihydropyridines

In 1995 Materson suggested that the combination of diltiazem with a dihydropyridine may efficaciously reduce blood pressure and be useful in treating severe hypertension. Most of the experience has been with diltiazem and nifedipine, but it is reasonable to anticipate that any dihydropyridine may be used. Although patients generally tolerate the combination well, pedal edema occurs in about 15–20%. It is not clear that the combination of verapamil with dihydropyridine is clinically incorrect. Indeed, data from Kaesemeyer and colleagues (1994) suggest that, in high doses, it works quite well. They did, however, see a high (30%) incidence of pedal edema. According to Kiowski, the combination of amlodipine with verapamil seems to be enhancing. Verapamil and diltiazem should not be combined because of the possibility of cardiac depression with little gain in therapeutic efficacy.

SIDE EFFECTS OF CALCIUM ANTAGONISTS

The major adverse reactions to calcium antagonists are exaggerated pharmacologic responses. In striking contrast, the incidence of orthostatic hypotension, central nervous side effects, sexual dysfunction, and bronchospasm and the worsening or masking of hypoglycemia are negligible during treatment with calcium antagonists.

As with other potent antihypertensive agents, an excessive decrement in blood pressure is possible when calcium antagonists are used. Frank hypotension is rare when calcium antagonists are used as monotherapy, but it may occur during initial titration or with subsequent upward adjustment of dosage and may be more likely in patients receiving beta blockers concomitantly. It appears to be in part because of hypotension (and the resultant decrease in coronary artery perfusion) that some patients with underlying coronary artery disease paradoxically develop increased frequency, duration, or severity of angina when they start or increase the dose of nifedipine (or, rarely, of other calcium antagonists).

An important potential side effect of calcium antagonists is periorbital and, more commonly, peripheral edema. The edema is attributable to arteriolar vasodilation rather than to decreased urinary sodium excretion.

As mentioned, the major side effects of calcium antagonists often can be predicted from their pharmacologic actions. Because dihydropyridine calcium antagonists are the most potent vasodilators, they more commonly cause side effects attributable to vasodilation, such as headache, flushing, and edema, than verapamil or diltiazem. Although vasodilator side effects are also induced by diltiazem and verapamil, their frequency appears to be less than with nifedipine. Of note, it appears that when nifedipine or nicardipine is ingested along with food, their absorption is slowed, resulting in a less abrupt fall in blood pressure (but to a similar nadir level) and fewer vasodilator-related side effects. As expected, some of the side effects of calcium antagonists, particularly hypotension and peripheral edema, may be dose-related.

Similarly, a smoothing out of the peak-vs.-trough blood levels is the presumed mechanism whereby the extended-release formulations of dihydropyridines (e.g., nifedipine) may produce fewer vasodilator-related side effects.

Adverse events during felodipine treatment are similar to those seen with nifedipine. Vasodilation-related side effects (headache, ankle edema, flushing) are dose-dependent, generally well tolerated, and often transient.

Verapamil may produce important effects on cardiac conduction. In most cases, prolongation of AV conduction time causes no symptoms, but more serious conduction disturbances have been reported. Examples include asystole and higher forms of AV block, which usually are induced by intravenous administration. Diltiazem affects cardiac conduction similarly. The usual doses of nifedipine, nicardipine, and isradipine have little effect on SA or AV conduction. The effects of calcium antagonists on myocardial contractility are discussed above (see also chapters 5 and 6).

Verapamil has the greatest effect on the gastrointestinal tract; constipation and nausea are caused more commonly by verapamil than by diltiazem, nifedipine, or isradipine.

Reversible renal deterioration in patients with chronic renal failure has been described in rare cases with both nifedipine and diltiazem. Although investigators postulated that this effect may be attributable to perturbation of compensatory renal autoregulation, this concept is difficult to reconcile with the known ability of calcium antagonists to vasodilate preferentially the preglomerular resistance bed (see chapter 22).

In general, the side-effect profiles of the new slow-release formulations are similar to those of the conventional parent counterparts. In the case of dihydropyridines, several of the side effects related to acute, potent peripheral vasodilation, including headache, tachycardia, and flushing, are reduced in frequency and/or severity by the long-acting preparation. The tendency to produce peripheral edema does not seem to be diminished.

Metabolic Effects

Hypertension is often accompanied by impaired glucose tolerance, insulin resistance, clinically evident type II diabetes mellitus, and dyslipidemia (see chapters 14 and 20). The dyslipidemia accompanying hypertension is often characterized by low concentrations of high-density lipoprotein (HDL) cholesterol, elevated triglycerides, and abnormal, dense, small, more atherogenic low-density lipoprotein (LDL) cholesterol particles. Because hypertension is often accompanied by metabolic abnormalities, the metabolic effects of antihypertensive agents and their impact on cardiovascular risk reduction as a result of treatment of hypertension are important concerns. In contrast to the metabolic abnormalities associated with diuretic and beta blocker antihypertensive therapy, such as hypokalemia, hypercalcemia, and hyperuricemia, calcium antagonists are metabolically neutral.

DOSAGE

The enormous interpatient variation in the rates of absorption, protein binding, and clearance of calcium antagonists, coupled with the variation in plasma concentration-response relationships, precludes hard and fast dosing schedules. In practice, the desired therapeutic effect is achieved by dose titration in the

TABLE 7. Dosages of Calcium Antagonists for the Treatment of Hypertension

Formulation	Form Supplied	Initial Dose (mg/day)	Maximal Dose (mg/day)	Frequency
Conventional				
Verapamil	40, 80, 120 mg tablets	240	480*	bid or tid
Nifedipine	10, 20 mg capsules	30	120–180	tid
Diltiazem	30, 60, 90, 120 mg tablets	90	240–360 (?)	tid
Nicardipine	20, 30 mg capsules	60	120	tid
Isradipine	2.5, 5 mg capsules	5	20†	bid
Slow-release				
Plendil	2.5, 5, 10 mg extended-release tablets	5	10	qd
Verelan	120, 240 mg sustained-release capsules	120 or 240	480	qd
Calan SR‡	120, 180, 240 mg sustained-release caplets	120 or 180	480	qd or bid
Isoptin SR‡	120, 180, 240 mg sustained-release tablets	120 or 180	480	qd or bid
Procardia XL	30, 60, 90 mg extended-release tablets	30 or 60	90–120	qd
Cardizem CD	180, 240, 300 mg sustained-release capsules	180–240	360 (?)	qd
Sular	10, 20, 30, 40 mg tablets	10 or 20	60	qd
Covera	180, 240 mg tablets	180 or 240	480	qd
DynaCirc CR	5, 10 mg tablets	5	20	qd
Intrinsically long-acting				
Norvasc	2.5, 5, 10 mg tablets	5§	10	qd
Mibefradil	50, 100 mg tablets	50	100	qd

bid = 2 times/day, tid = 3 times/day, qd = each day.

* Most patients are controlled with no more than 120 mg tid; rarely, dosages as high as 720 mg/day may be used.

† Most patients are controlled at total daily dosages between 5 and 10 mg.

‡ Calan SR and Isoptin SR are different tradenames for an identical product.

§ Small, fragile, or elderly patients or patients with hepatic insufficiency may be started on 2.5 mg/day.

individual patient. The dosages and schedules that are generally recommended for both conventional (immediate-release) and sustained-release formulations of calcium antagonists as monotherapeutic antihypertensive agents are shown in Table 7. Of course, as additional data become available about dose/duration and dose/side effect interactions as well as 24-hour blood pressure monitoring, these recommendations may change. As is the case with other blood pressure-lowering medications, the dosages of calcium antagonists needed for combination therapy are often lower than the doses commonly used for monotherapy (see chapter 26).

DRUG INTERACTIONS

Drug interactions between calcium antagonists have been described. In particular, the concomitant use of verapamil and diltiazem is to be avoided because of the considerable risk of marked bradycardia. Drug interactions between calcium antagonists and quinidine, digoxin, phenytoin, theophylline, prazosin, propranolol, and rifampin have been observed. When verapamil is used together with quinidine, patients may experience severe bradycardia and/or hypotension.

Nifedipine and verapamil increase the plasma concentration of coadministered digoxin, and verapamil appears to increase the plasma level of prazosin. Isradipine and felodipine, on the other hand, do not affect digoxin clearance or serum digoxin levels. However, isradipine may increase the "area under the curve" and the maximal concentration of propranolol when the two are coadministered. Verapamil and diltiazem also reduce the clearance of theophylline. Phenytoin and rifampin tend to decrease the blood level of verapamil, and propranolol tends to increase it. Verapamil and diltiazem may enhance the effects of digoxin on atrioventricular conduction. Nicardipine may increase the blood levels of digoxin and cyclosporine. Although there are no absolute contraindications to the use of these combinations, patients who receive them should be monitored closely (see chapter 3).

CONCLUSION

Over the past decade multiple lines of research have implicated transmembrane calcium ion fluxes as an important mediator of the increased systemic vascular resistance that characterizes the majority of hypertensive states. Calcium antagonists exhibit a potent vasodilating effect by directly relaxing vascular smooth muscle. In contrast with previous vasodilators, reflex stimulation of heart rate appears to be minimal.

Numerous studies indicate that calcium antagonists effectively lower blood pressure in hypertensive subjects. Calcium antagonists are as effective as the other medications currently used to initiate antihypertensive therapy. They seem to work in all populations. Because of their relatively short half-lives, frequent administration of the conventional preparations is required. Consequently, slow-release formulations should be used to circumvent this problem.

Because of their unique spectrum of pharmacologic effects as well as actions on the cardiovascular system, calcium antagonists have potentially important advantages in several groups of patients. Included are patients with coexisting coronary disease, variant angina, and supraventricular arrhythmias and probably (for some but not other calcium antagonists) some patients with left ventricular diastolic dysfunction ("stiff heart syndrome"). Some of the newly developed and forthcoming calcium antagonists may prove to have fewer cardiodepressant properties than the currently available agents. For example, isradipine, amlodipine, nisoldipine, and felodipine in particular induce little decrease in myocardial contractile force. The role of certain calcium antagonists in patients with previous myocardial infarction requires further study. Another unsettled question is whether the actions of calcium antagonists on the coronary circulation and their putative ability to reverse the progression of left ventricular hypertrophy and possibly prevent or delay atherogenesis will prove to be of clinical importance.

Finally, there is mounting concern that the choice of an antihypertensive agent should be based in part on its effect on electrolytes, glucose, and lipid profile. The demonstration that calcium antagonists do not exert adverse effects in these areas has thus assumed considerable importance. The paucity of adverse effects increasingly commends calcium antagonists as initial monotherapy in the management of hypertension.

Bibliography

National and International Guidelines

Guidelines Sub-Committee: 1993 Guidelines for the management of mild hypertension: Memorandum from a World Health Organization/International Society of Hypertension meeting. J Hypertens 11:905–918, 1993.

Joint National Committee on Detection, Evaluation and Treatment of High Blood Pressure: The fifth report of the Joint National Committee on Detection, Evaluation and Treatment of High Blood Pressure (JNC-V). Arch Intern Med 153:154–183, 1993.

Ogilvie R, Burgess E, Cusson J, et al: Report of the Canadian Hypertension Society Consensus Conference. 3: Pharmacological treatment of essential hypertension. Can Med Assoc J 149:575–584, 1993.

Sever P, Beevers G, Bulpitt C, et al: Management guidelines in essential hypertension: Report of the Second Working Party of the British Hypertension Society. BMJ 306: 983–987, 1993.

Australian Consensus Panel: The management of hypertension: A consensus statement. Med J Aust 160(Suppl):S1–S16, 1994.

Specific Drugs

McTavish D, Sorkin EM: Verapamil. An updated review of its pharmacodynamic and pharmacokinetic properties and therapeutic uses in hypertension. Drugs 38:19–76, 1989.

Buckley MT, Grant SM, Goa KL, et al: Diltiazem: A reappraisal of its pharmacologic properties and therapeutic use. Drugs 39:757–806, 1990.

Murdoch D, Brogden RN: Sustained-release nifedipine formulations. An appraisal of their current uses and prospective roles in the treatment of hypertension, ischaemic heart disease and peripheral vascular disorders. Drugs 41:737–779, 1991.

Sorkin EM, Clissold DP: Nicardipine. A review of its pharmacodynamic and pharmacokinetic properties and therapeutic efficacy in the treatment of angina pectoris, hypertension and related cardiovascular disorders. Drugs 33:296–345, 1987.

Fitton A, Benfield P: Isradipine. A review of its pharmacodynamic and pharmacokinetic properties and therapeutic use in cardiovascular disease. Drugs 40:31–74, 1990.

Haria M, Wagstaff AJ: Amlodipine: A reappraisal of its pharmacological properties and therapeutic uses in cardiovascular disease. Drugs 50:560–586, 1995.

Meredith PA, Elliot HL: Clinical pharmacokinetics of amlodipine. Clin Pharmacokinet 22:22–31, 1992.

Plosker GL, Faulds D: Nisoldipine coat-core. A review of its pharmacology and therapeutic efficacy in hypertension. Drugs 52:232–253, 1996.

Langan J, Rodriguez-Mañas L, Sareli P, Heinig R: Nisoldipine CC: Clinical experience in hypertension. Cardiology 88(Suppl):56–62, 1997.

Lee CR, Bryson HM: Lacidipine. A review of its pharmacodynamic and pharmacokinetic properties and therapeutic potential in the treatment of hypertension. Drugs 48:274–296, 1994.

Todd PA, Faulds D: Felodipine. A review of the pharmacology and therapeutic use of the extended-release formulation in cardiovascular disorders. Drugs 44:251–277, 1992.

Chrysant SG, Cohen M: Sustained blood pressure control with controlled-release isradipine. Am J Hypertens 8:87–89, 1995.

Oparil S, Kobrin I, Abernethy DR, et al: Dose-response characteristics of mibefradil, a novel calcium antagonist, in the treatment of essential hypertension. Am J Hypertens 10:735–742, 1997.

Frishman WH, Sonnenblick EH: Calcium channel blockers. In Schlant RC, Alexander RW (eds): Hurst's The Heart, 8th ed. New York, McGraw-Hill, 1994, pp 1271–1290.

Opie LH, Frishman WH, Thadani U: Calcium channel antagonists. In Opie LH (ed): Drugs for the Heart, 4th ed. Philadelphia, W.B. Saunders, 1995, pp 50–83.

Boden WE: Anti-ischaemic therapy during the follow-up phase of acute coronary syndromes. Is there a role for calcium channel blockers? Drugs 52(Suppl 4):20–30, 1996.

Sleight P: Calcium antagonists during and after myocardial infarction. Drugs 51:216–225, 1996.

Miller K: Pharmacological management of hypertension in paediatric patients. A comprehensive review of the efficacy, safety and dosage guidelines of the available agents. Drugs 48:868–887, 1994.

Frishman WH, Sonnenblick EH: Cardiovascular uses of calcium antagonists. In Messerli FH (ed): Cardiovascular Drug Therapy, 2nd ed. Philadelphia, W.B. Saunders, 1996, pp 891–901.

Brown MJ, Castaigne A, Ruilope LM, et al: INSIGHT: International Nifedipine GITS Study Intervention as a Goal in Hypertension Treatment. J Hum Hypertens 10(Suppl 3):157–160, 1996.

Mancia G, Seravalle G, Vailati S, Grassi G: Benefit versus risk of calcium antagonists in hypertensive patients with concomitant risk factors. J Hypertens 14(Suppl 4):S33–S38, 1996.

Borhani NO, Mercuri M, Borhani PA, et al: Final outcome results of the Multicenter Isradipine Diuretic Atherosclerosis Study (MIDAS). JAMA 276:785–791, 1996.

Opie LH: Calcium channel antagonists in the treatment of coronary artery disease: Fundamental pharmacological properties relevant to clinical use. Prog Cardiovasc Dis 38:273–290, 1996.
Wilson TW, Quest DW: Comparative pharmacology of calcium antagonists. Can J Cardiol 11:243–249, 1995.

Trough-to-Peak Ratio

Elliot HL, Meredith PA: Methodological considerations in calculation of the trough : peak ratio. J Hypertens 12(Suppl 8):S3–S7, 1994.
Zannad F, Matzinger A, Larche J: Trough/peak ratios of once daily angiotensin converting enzyme inhibitors and calcium antagonists. Am J Hypertens 9:633–643, 1996.

Delivery Systems for Extended Release and/or Controlled Onset

Prisant ML, Bottini B, DiPiro JT, Carr AA: Novel drug-delivery systems for hypertension. Am J Med 93(Suppl 2A):45S–55S, 1992.
van Zwieten PA, Hansson L, Epstein M: Slowly acting calcium antagonists and their merits. Blood Pressure 6:78–80, 1997.
White WB: A chronotherapeutic approach to the management of hypertension. Am J Hypertens 9:29S–33S, 1996.
White WB, Anders R, MacIntyre J, et al: Nocturnal dosing of a novel delivery system of verapamil for systemic hypertension. Am J Cardiol 76:375–380, 1995.
Neutel JM, Alderman M, Anders RJ, Weber MA: Novel delivery system for verapamil designed to achieve maximal blood pressure control during the early morning. Am Heart J 132:1202–1206, 1996.

Safety Issues

Marwick C: FDA gives calcium channel blockers clean bill of health but warns of short-acting nifedipine hazards. JAMA 275:423–424, 1996.
Laragh JH, Held C, Messerli F, et al: Calcium antagonists and cardiovascular prognosis: A homogeneous group? Am J Hypertens 9:99–109, 1996.
Braun S, Boyko V, Behar S, et al: Calcium antagonists and mortality in patients with coronary artery disease: A cohort study of 11,575 patients. J Am Coll Cardiol 28:7–11, 1996.
Jick H, Derby LE, Gurewich V, Vasilakis C: The risk of myocardial infarction associated with antihypertensive drug treatment in persons with uncomplicated essential hypertension. Pharmacotherapy 16:321–326, 1996.
Ruzicka M, Leenen FHH: Relevance of intermittent increases in sympathetic activity for adverse outcome on short-acting calcium antagonists. In Laragh JH, Brenner BM (eds): Hypertension: Pathophysiology, Diagnosis, and Management, 2nd ed. New York, Raven Press, 1995, pp 2815–2825.
Ruzicka M, Leenen FH: Relevance of 24 h blood pressure profile and sympathetic activity for outcome on short- versus long-acting 1,4-dihydropyridines. Am J Hypertens 9:86–94, 1996.
Kleinbloesem CH, van Brummelen P, Danlof M, et al: Rate of increase in the plasma concentration of nifedipine as a major determinant of its hemodynamic effects in humans. Clin Pharmacol Ther 41:26–30, 1987.
Mancia G, van Zwieten PA: How safe are calcium antagonists in hypertension and coronary heart disease? J Hypertens 14:13–17, 1996.
Calcium Channel Blockers: Lessons Learned From MIDAS and Other Clinical Trials [editorial]. JAMA 276:829–830, 1996.
Effects of calcium antagonists on the risk of coronary heart disease, cancer and bleeding: Ad Hoc Subcommittee of the Liaison Committee of the World Health Organization and the International Society of Hypertension. J Hypertens 15:105–115, 1997.
Alderman MH, Cohen H, Roque R, Madhaven S: Effect of long-acting and short-acting calcium antagonists on cardiovascular outcomes in hypertensive patients. Lancet 349:594–598, 1997.
Husten L: Calcium antagonists: "Not guilty." Lancet 349:1818, 1997.

Salutary Effects in Coronary Artery Disease, Left Ventricular Hypertrophy, and Congestive Heart Failure

Frishman WH, Michaelson D: Use of calcium antagonists in patients with ischemic heart disease and systemic hypertension. Am J Cardiol 79(10A):33–38, 1997.
Michalewicz L, Messerli FH: Cardiac effects of calcium antagonists in systemic hypertension. Am J Cardiol 79(10A):39–46, 1997.
Messerli FH, Aepfelbacher FC: Cardiac effects of calcium antagonists in hypertension. In Messerli FH (ed): Cardiovascular Drug Therapy, 2nd ed. Philadelphia, W.B. Saunders, 1996, pp 908–915.
Ferrari R: Calcium antagonists and left ventricular dysfunction. Am J Cardiol 75:71E–76E, 1995.
Knorr AM: Why is nisoldipine a specific agent in ischemic left ventricular dysfunction? Am J Cardiol 75:36E–40E, 1995.
Packer M, O'Connor CM, Ghali JK, et al, for the Prospective Randomized Amlodipine Survival Evaluation Study Group (PRAISE): Effect of amlodipine on morbidity and mortality in severe chronic heart failure. N Engl J Med 335:1107–1114, 1996.

The DEFIANT-II Research Group: Doppler flow and echocardiography in functional cardiac insufficiency: Assessment of nisoldipine therapy. Results of the DEFIANT-II study. Eur Heart J 18:31–40, 1997.

Renal Effects

Zanchi A, Brunner HR, Waeber B, Burnier M: Renal haemodynamic and protective effects of calcium antagonists in hypertension. J Hypertens 13:1363–1375, 1995.

Epstein M: The benefits of ACE inhibitors and calcium antagonists in slowing progressive renal failure: Focus on fixed-dose combination antihypertensive therapy. Renal Failure 18:813–832, 1996.

ter Wee P, De Micheli AG, Epstein M: Effects of calcium antagonists on renal hemodynamics and progression of non-diabetic chronic renal disease. Arch Intern Med 154:1185–1202, 1994.

Epstein M: Calcium antagonists and renal protection: Current status and future perspectives. Arch Intern Med 152:1573–1584, 1992.

Rabelink TJ, Koomans HA: Endothelial function and the kidney. An emerging target for cardiovascular therapy. Drugs 53(Suppl 1):11–19, 1997.

Prevention of Atherosclerosis

Parmley WW: Calcium antagonists in the prevention of atherosclerosis. In Messerli FH (ed): Cardiovascular Drug Therapy, 2nd ed. Philadelphia, W.B. Saunders, 1996, pp 901–907.

Herrman JP, Hermans WR, Vos J, Serruys PW: Pharmacological approaches to the prevention of restenosis following angioplasty. The search for the Holy Grail? (Part I). Drugs 46:18–52, 1993.

Rafflenbeul W: Hypertension treatment and prevention of new atherosclerotic plaque formation. Drugs 48(Suppl 1): 11–15, 1994.

Metabolic Effects

Giordano M, Matsuda M, Sanders L, et al: Effects of angiotensin-converting enzyme inhibitors, Ca^{2+} channel antagonists, and alpha-adrenergic blockers on glucose and lipid metabolism in NIDDM patients with hypertension. Diabetes 44:665–671, 1995.

Cabezas-Cerrato J, Garcia-Estevez DA, Araujo D, Iglesias M: Insulin sensitivity, glucose effectiveness, and beta-cell function in obese males with essential hypertension: Investigation of the effects of treatment with a calcium channel blocker (diltiazem) or an angiotensin-converting enzyme inhibitor (quinapril). Metabolism 46:173–178, 1997.

Giugliano D, DeRosa N, Marfella R: Comparison of nifedipine and cilazapril in patients with hypertension and non-insulin-dependent diabetes mellitus. Am J Hypertens 6:927–932, 1993.

Ferrari P, Giachino D, Weidmann P: Unaltered insulin sensitivity during calcium channel blockage with amlodipine. J Clin Pharmacol 41:109–113, 1991.

Zanetti-Elshater F, Pingitore R, Beretta-Piccoli C, et al: Calcium antagonists for treatment of diabetes-associated hypertension. Metabolic and renal effects of amlodipine. Am J Hypertens 7:36–45, 1994.

Sowers JR: Effects of ACE inhibitors and calcium channel blockers on insulin sensitivity and other components of the syndrome. Nephrol Dial Transplant 10(Suppl 9):52–55, 1995.

Side Effects

Stern R, Khalsa JH: Cutaneous adverse reactions associated with calcium channel blockers. Arch Intern Med 149:829–832, 1989.

Krebs R: Adverse reactions with calcium antagonists. Hypertension 5(4 Pt 2):II125–II129, 1983.

Kubota K, Pearce GL, Inman WH: Vasodilation-related adverse events in diltiazem and dihydropyridine calcium antagonists studied by prescription-event monitoring. Eur J Clin Pharmacol 48: 1–7, 1995.

Steele RM, Schuna AA, Schreiber RT: Calcium antagonist-induced gingival hyperplasia. Ann Intern Med 120:663–664, 1994.

Fattore L, Stablein M, Bredfeldt G, et al: Gingival hyperplasia: A side effect of nifedipine and diltiazem. Spec Care Dentistry 11:107–109, 1991.

Treatment of Noncardivascular Disorders

Belch JJ, Ho M: Pharmacotherapy of Raynaud's phenomenon. Drugs 52: 682–695, 1996.

Combination of Chemically Dissimilar Calcium Antagonists

Busse JC, de Velasco RE, Pellegrini EL: Combined use of nifedipine and diltiazem for the treatment of severe hypertension. Southern Med J 84: 502–504, 1991.

Kaesemeyer WH, Carr AA, Bottini PB, Prisant LM: Verapamil and nifedipine in combination for the treatment of hypertension. J Clin Pharmacol 34:48–51, 1994.

Kiowski W, Erne P, Linder L, Behler FR: Arterial vasodilator effects of the dihydropyridine calcium antagonist amlodipine alone and in combination with verapamil in systemic hypertension. Am J Cardiol 66:1469–1472, 1990.

Materson BJ: Calcium channel blockers: Is it time to split the lump? Am J Hypertens 8:325–329, 1995.

PER LUND-JOHANSEN M.D.

11

Hemodynamic Effects of Calcium Antagonists in Hypertension

During the past two years the calcium antagonists have caused more hot debates and editorials in the medical literature and lay press than any other class of antihypertensive drugs. Some authors have claimed that the calcium antagonists, particularly short-acting formulations of the the dihydropyridines, may be harmful for patients with hypertension, and acute hemodynamic responses have been suggested as one of the possible mechanisms.[1,2] This debate has renewed interest in the hemodynamic effects of the different classes and formulations of the calcium antagonists for treatment of hypertension. The present chapter updates present knowledge about the acute and chronic effects of the different classes and formulations of the calcium antagonists.

So-called essential or primary hypertension may be due to several pathophysiologic mechanisms: (1) increased sympathetic activity, (2) abnormalities in sodium handling, and (3) disturbances in calcium transport, leading to increased intracellular calcium concentration and augmented contraction of smooth muscles in the walls of resistance vessels.[3,4] Whether this mechanism is important in most patients with essential hypertension is still unclear.[5] However, it is well documented that in nearly all patients who need drug therapy for essential hypertension, total peripheral resistance is elevated and arteriolar resistance is increased in most tissues.[4,6] Cardiac output, the other main determinant of blood pressure, is usually normal or slightly reduced at rest. During exercise, however, cardiac output is usually subnormal because of insufficient increase in stroke volume.[6]

Logical treatment of hypertension should aim at reducing total peripheral resistance without lowering cardiac output either at rest or during exercise. Calcium antagonists have attracted great interest as antihypertensive agents because they decrease vascular resistance without reducing cardiac output, at least in patients with good left ventricular function.[7–9] However, in patients with heart failure they should be used with caution and only in association with other conventional heart failure drugs such as diuretics, angiotensin-converting enzyme (ACE) inhibitors, and digitalis.[10,11]

The first part of this chapter reviews the central hemodynamic effects of seven calcium antagonists—nifedipine, nisoldipine, felodipine, amlodipine, verapamil, diltiazem, and tiapamil—partly on the basis of studies in our own laboratory of 110 patients with mild to moderately severe essential hypertension. The second part discusses newer calcium antagonists not studied in our laboratory: nitrendipine, isradipine, lacidipine, nilvadipine, and nicardipine.

CENTRAL HEMODYNAMICS AT REST AND DURING EXERCISE (INVASIVE STUDIES)

Details about invasive studies of central hemodynamics at rest and during exercise have been published elsewhere.[12–19] In recent years noninvasive methods such as echocardiography and impedance cardiography have been used by other authors, sometimes with rather dubious results.[20,21] In our studies blood pressure was recorded intraarterially, and cardiac output was measured by the old-fashioned but accurate dye-dilution method, using Cardiogreen and a linearily responding densitometer.[19,22] This method probably still gives the most accurate measurement of cardiac output during exercise.[19,22]

Hemodynamics were studied at rest, during the supine and seated position, and during steady-state bicycle exercise at 50, 100, and 150 watts. In the nisoldipine and tiapamil series, the acute effect of the first dose was studied, and only one work load of 100 watts was used. After the first hemodynamic study, the patients were started on monotherapy with one of the seven calcium antagonists. After 9–12 months the hemodynamic study was repeated about 3 hours after the last dose.

Table 1 shows the patient's age, casual blood pressure, and body surface area (BSA) as well as the mean daily dose during the chronic study. Most patients were relatively young (mean age of the two groups: 43 and 52 years). Generally, the starting dose was approximately half of the later daily maintenance dose to minimize problems related to initial reflex tachycardia. The aim of treatment in the chronic studies was a casual blood pressure ≤ 140/90 mmHg without side effects.

Hemodynamic Results

In all series, the initial predrug studies (before any treatment or after at least 8 weeks of wash-out) showed the typical hemodynamic pattern of established essential hypertension: elevated total peripheral resistance index (TPRI)and subnormal cardiac index (CI), particularly during exercise. None of the patients had symptoms suggestive of angina pectoris.

TABLE 1. Patients' Age, Casual Blood Pressure, Body Surface Area, and Daily Dose

		Age		Casual BP (mmHg)		BSA (m^2)	Daily Dose (mg)	
Calcium Antagonist	**n**	**Range**	**(Mean)**	**1**	**2**		**Range**	**(Mean)**
Verapamil	9	35–55	(45)	166/106	143/92	2.01	120–240	(220)
Diltiazem	16	38–64	(52)	174/106	144/86	1.99	180–360	(278)
Tiapamil	19	19–64	(45)	166/106	148/92	1.97	600–1200	(980)
Nifedipine	15	20–64	(44)	160/104	140/94	1.99	40–80	(52)
Nisoldipine	17	29–61	(43)	167/108	141/88	2.01	10–40	(25)
Felodipine	16	37–64	(48)	166/105	139/86	1.98	10–20	(15)
Amlodipine	18	22–61	(45)	172/108	143/89	2.04	5–10	(9)

BSA = body surface area; 1 = before drug; 2 = on chronic therapy.

Acute Effects

The acute effects of the calcium antagonists during the first hours after the first dose are of particular relevance to the ongoing discussion of possible harmful effects. The typical acute effects of a conventional dihydropyridine tablet are seen in our study of nisoldipine.[14]

Dihydropyridine Derivatives One hour after a conventional 10-mg tablet of nisoldipine was given orally, TPRI declined by 19%. The fall in blood pressure was partly counteracted by reflex tachycardia (heart rate increase of 9%), which was the main mechanism behind a 12% increase in CI. Despite these counterregulatory reactions, blood pressure was reduced by 9% within 1 hour because of the marked reduction in TPRI. After 2–3 hours the effects leveled off. Figure 1 shows the acute effects of nisoldipine compared with placebo. Although the mean heart rate (HR) increase after 1 hour was only 9% (or 6 beats/min), 10 subjects had an increase of more than 10 beats/min, and in 2

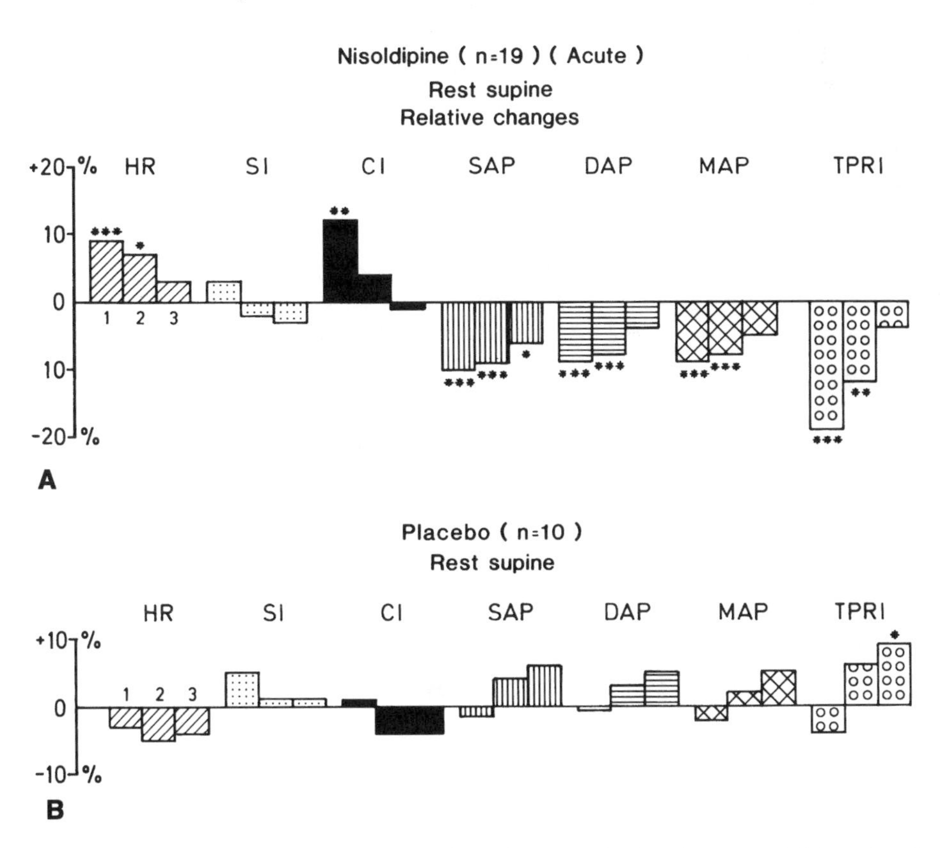

FIGURE 1. Hemodynamic changes during 3 hours following the first dose of nisoldipine (*A*) and placebo (*B*). Bars show changes in mean values. HR = heart rate, CI = cardiac index, SI=stroke index, SAP = systolic arterial pressure, DAP = diastolic arterial pressure, MAP = mean arterial pressure, TPRI = total peripheral resistance index. Mean values 1, 2, 3 indicate hours after dose. *$p < 0.05$, **$p < 0.01$, ***$p < 0.001$. Recordings in the supine position. (From Omvik P, Lund-Johansen P, Haugland H: Nisoldipine. Central hemodynamics at rest and during exercise in essential hypertension. Acute and chronic studies. J Hypertens 6:95–103, 1988, with permission.)

subjects HR increased more than 20 beats/min—an increase that may be unpleasant or possibly harmful in certain patients (e.g., patients with unknown coronary artery stenotic lesions). No harmful reactions were seen in our small group. In principle, similar acute responses have been reported for other dihydropyridines (e.g., nifedipine, felodipine) and also for amlodipine when doses were extremely high (about 80 mg or 8 times normal dose).[23–25]

The response depends on the speed of increase in plasma concentration. Thus, different hemodynamic patterns in terms of heart rate response can be demonstrated for rapid bolus injection vs. slow infusion of low doses of nifedipine. With rapid bolus injections, heart rate increased by about 20 beats/min within few minutes and partly counteracted the fall in blood pressure. With slow infusion, blood pressure fell slowly with no increase in heart rate.[26] It also has been shown that nifedipine in capsules (10–40 mg 3 times daily) caused fluctuation in blood pressure and heart rate, whereas the retard tablets induced stable blood pressure reduction without heart rate changes.[27] In recent years it has been stressed that the dihydropyridines should be used in slow-release preparations such as the GITS (gastrointestinal therapeutic system) formulation of nifedipine. Another alternative is to use drugs with slow onset and long duration of effect such as amlodipine, which has a rather unique pharmacodynamic profile.[28]

There is general agreement that in all forms of mild or moderately severe hypertension, blood pressure therapy should be started with low doses and all vasodilators should be given in formulations and doses that do not induce a rapid increase in serum concentration. Thus, reflex tachycardia and rapid falls in blood pressure are avoided. In special situations, such as acute elevations of blood pressure after surgery (e.g., coronary bypass), dihydropyridines may be used intravenously with constant monitoring of blood pressure.[29]

Nondihydropyridines Nondihydropyridine calcium antagonists, which have electrophysiologic effects on the sinus and atrioventricular node, induce some reduction in heart rate during chronic use, but after the first dose they may cause reflex tachycardia. Thus, in our study on tiapamil (600 mg in conventional tablet given orally as the first dose), HR increased by 7% and CI by 11%, but because of the marked reduction in TPRI, systolic arterial (SAP), diastolic arterial (SAP), and mean arterial (MAP) pressures were reduced by 13–14%. After 3 hours HR and CI returned to baseline, whereas MAP was still reduced by 9%.[14]

Studies by others have shown that the acute effects of verapamil[30,31] and diltiazem[32–34] include the same type of hemodynamic changes as tiapamil—primarily reduction in TPRI. These agents usually have no marked effect on stroke volume and CI.

Chronic Effects

Dihydropyridines The long-term hemodynamic responses to all dihydropyridines are similar. In our series the hemodynamic study was repeated after 8–12 months of treatment. Patients received their normal daily dose in the morning, and the hemodynamic study was performed approximately 3 hours later.

The response to nisoldipine (administered as extended-release tablets) is shown in Figure 2. MAP was reduced at rest as well as during exercise by 17%,

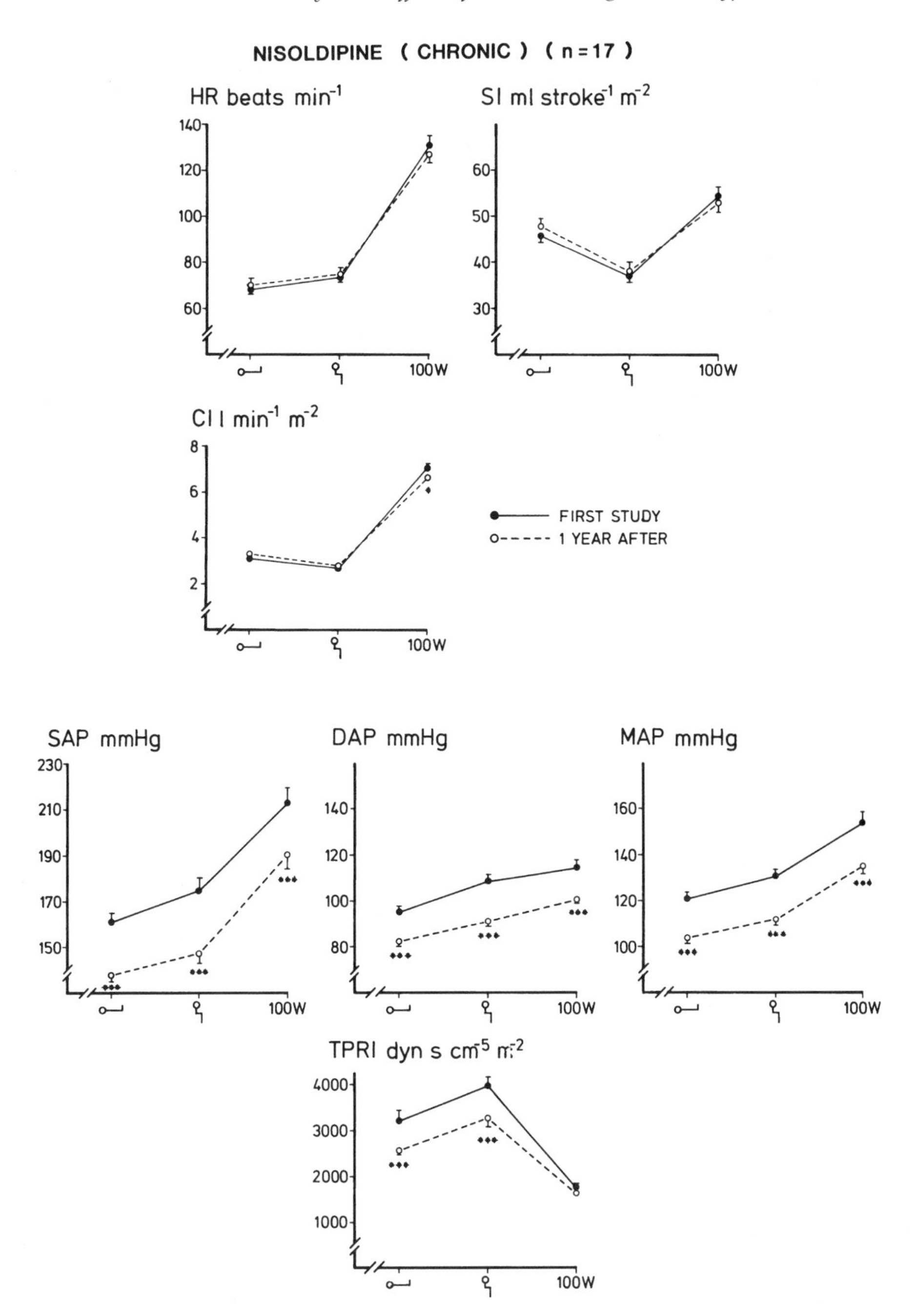

FIGURE 2. Hemodynamic changes during chronic treatment with nisoldipine 3 hours after ER-tablet (extended release). Mean values and SEM. HR = heart rate, SI = stroke index, CI = cardiac index, SAP = systolic arterial pressure, DAP = diastolic arterial pressure, MAP = mean arterial pressure, TPRI = total peripheral resistance index. o—┘ = supine position, ⅂ = sitting position, 100 W = 100 watt exercise. *$p < 0.05$, **$p < 0.01$, ***$p < 0.001$. (From Omvik P, Lund-Johansen P, Haugland H: Nisoldipine. Central hemodynamics at rest and during exercise in essential hypertension. Acute and chronic studies. J Hypertens 6:95–103, 1988, with permission.)

whereas HR values before and during treatment overlapped completely. Thus, blood pressure control was similar during exercise and at rest (see Fig. 2). No significant changes in stroke index (SI) or CI were noted; in other words, the heart pump function was unchanged. Because HR is not changed, the effect of the HR-pressure product is relatively modest.

Our study of nifedipine in 1983[9] disclosed similar results. Because nifedipine has attracted so much attention recently, one should note that 3 hours after administration of an extended-release formulation during chronic therapy, blood pressure control was achieved without HR changes either at rest or during exercise (Fig. 3).

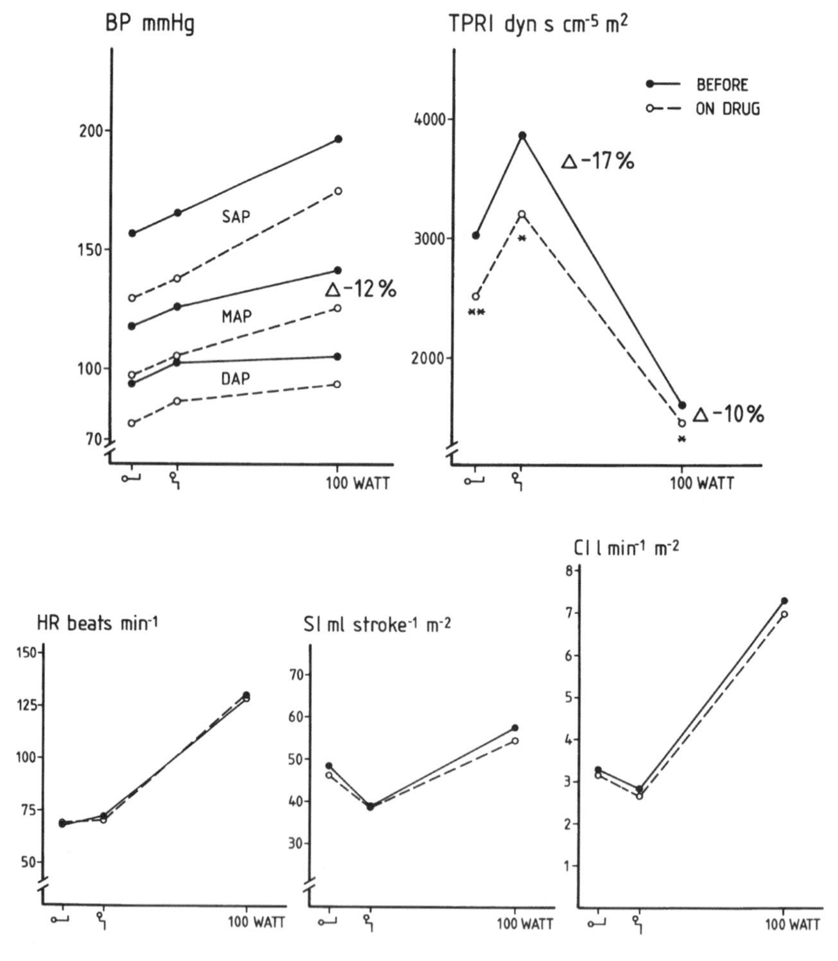

FIGURE 3. Hemodynamic changes during chronic treatment with nifedipine 3 hours after ER-tablet (extended release). Mean values. HR = heart rate, SI = stroke index, CI = cardiac index, SAP = systolic arterial pressure, DAP = diastolic arterial pressure, MAP = mean arterial pressure, TPRI = total peripheral resistance index. ○— = supine position, ⅂ = sitting position, 100 W = 100 watt exercise.*p < 0.05, *p < 0.01, ***p < 0.001. (From Lund-Johansen P, Omvik P: Haemodynamic effects of nifedipine in essential hypertension at rest and during exercise. J Hypertens 1:159–163, 1983, with permission.)

Amlodipine is unique because of its long-acting mode of action. It lowers blood pressure gradually, and with ordinary doses (5–10 mg tablet daily) reflex tachycardia is not seen.[35] After 11 months of treatment with 5–10 mg once daily, a hemodynamic study was performed approximately 3 hours after the morning dose. Amlodipine induced a significant reduction in BP, which was completely due to reduction in TPRI, with no significant changes in CI, SI, or HR (Fig. 4). The heart rate-pressure product was reduced by approximately 18% (similar to the reduction in pressure because HR was unchanged).

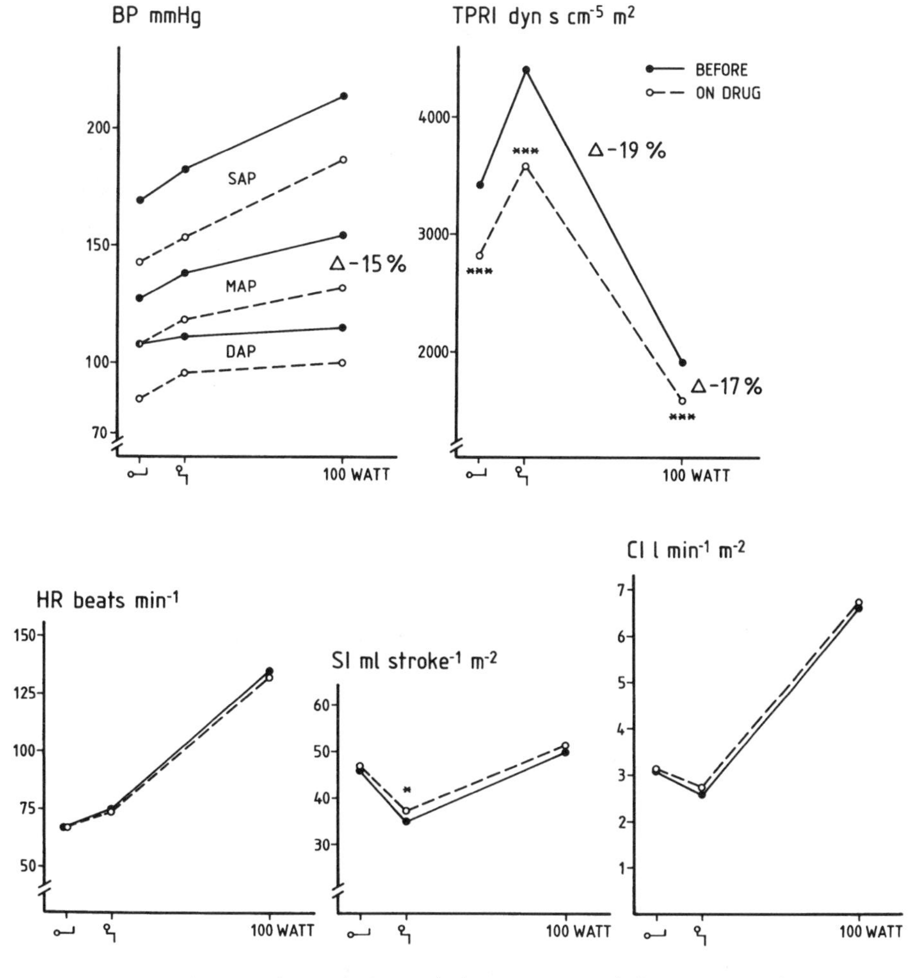

FIGURE 4. Hemodynamic changes before and after 1 year on amlodipine. Mean values. Note the similarity in the long-term changes of nifedipine and amlodipine. HR = heart rate, SI = stroke index, CI = cardiac index, SAP = systolic arterial pressure, DAP = diastolic arterial pressure, MAP = mean arterial pressure, TPRI = total peripheral resistance index. ⟜ = supine position, ⟟ = sitting position, 100 W = 100 watt exercise.*p < 0.05, *p < 0.01, ***p < 0.001. (From Lund-Johansen P, Omvik P: Haemodynamic effects of nifedipine in essential hypertension at rest and during exercise. J Hypertens 1:159–163, 1983, with permission.)

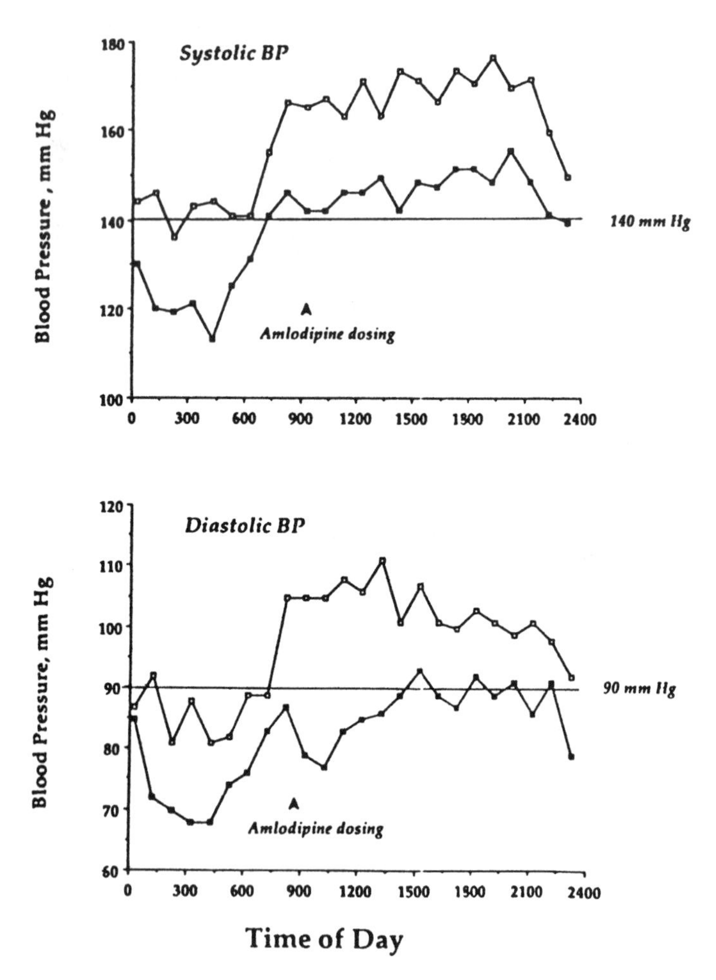

FIGURE 5. Ambulatory 24 hour blood pressure recordings before (□ on placebo) and on chronic amlodipine (■) treatment. (From Lund-Johansen P, Omvik P, White W, et al: Long-term haemodynamic effects of amlodipine at rest and during exercise in essential hypertension. J Hypertens 8:1129–1136, 1990, with permission.)

To investigate the efficacy of amlodipine for 24-hour blood pressure control, patients were studied with the Accutracker ambulatory 24-hour blood pressure recorder.[36] Amlodipine once daily induced good blood pressure control during 24 hours—daytime as well as nighttime (Fig. 5). HR did not change significantly. It is reasonable to assume that amlodipine induces blood pressure control through chronic reduction in TPRI.

In our chronic series felodipine[16] was compared with tiapamil.[18] Felodipine clearly and significantly reduced TPRI at rest by 15% (supine) and 16% (sitting) and by 13% during exercise. There were no significant changes in CI, HR ,or SI. No reflex tachycardia was seen during chronic treatment. Three hours after the dose the mean HR values before and during drug administration were almost identical: 62.9 vs. 63.9 (supine), 68.2 vs. 68.9 (sitting), and 126.6 vs 126.9 (100 watts) beats/min.

Reduction in blood pressure was more impressive with felodipine than with tiapamil in the same group of patients. The fall in blood pressure with tiapamil was caused partly by a fall in TPRI and partly by reduction in CI. These results support the concept that felodipine is a more vascular-selective calcium antagonist than tiapamil.[24]

Nondihydropyridines Like the dihydropyridine derivatives, all of the nondihydropyridines reduced blood pressure through chronic reduction in total peripheral resistance. However, the drugs differ from the dihydropyridines in their effect on HR. All of the nondihydropyridines produced a significant reduction in HR—at rest as well as during exercise—on the order of 10%. As discussed later, this finding may be important.

At the first hemodynamic study, the following mean values were found at rest in the sitting position: SAP/DAP, 167/102 mmHg; CI, 2.70 L/min/m^2; and TPRI, 3848 dyn sec cm^{-5} m^2. With doses from 120–240 mg daily (mean dose: 220 mg), MAP was reduced about 10% at rest and slightly less during exercise. Reduction in blood pressure was associated with a significant reduction in TPRI (14%) at rest, but during exercise the effect was more modest (5–7%). At rest HR was almost unchanged, but during exercise it decreased by 8–11%.[8] Of interest, the reduction in HR was partly compensated by an increase in stroke volume; consequently, CI was practically unchanged compared with pretreatment values (Fig. 6). The doses used in this study were relatively small. Other studies using a daily dose of 320 mg or more have generally reported a greater reduction in blood pressure.[19,20]

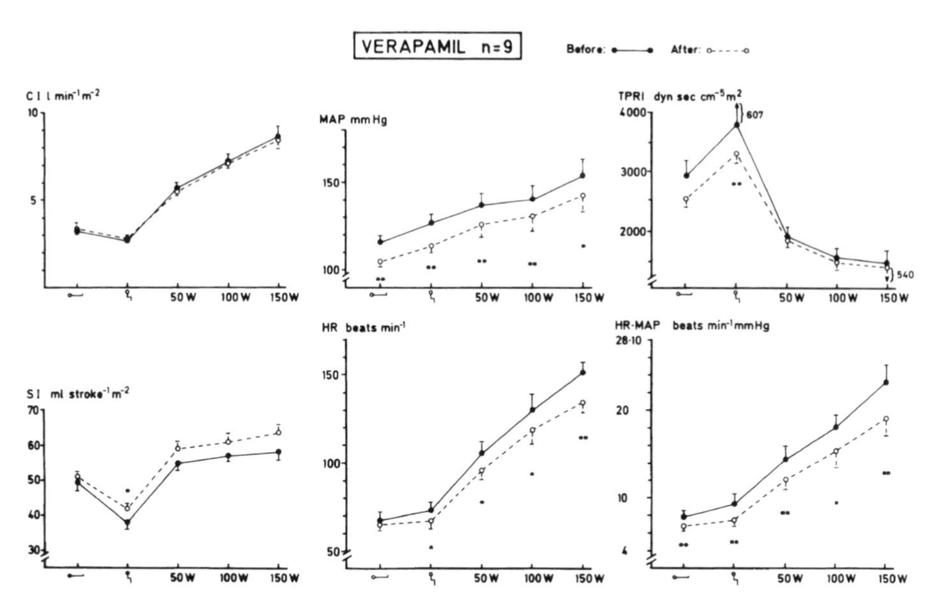

FIGURE 6. Hemodynamic changes during chronic treatment with verapamil. Mean values and SEM. Note reduction in HR, but increase in SI. HR = heart rate, SI = stroke index, CI = cardiac index, SAP = systolic arterial pressure, DAP = diastolic arterial pressure, MAP = mean arterial pressure, TPRI = total peripheral resistance index. o—┘ = supine position, ⅄ = sitting position, 100 W = 100 watt exercise. *$p < 0.05$, **$p < 0.01$, ***$p < 0.001$. (From Lund-Johansen P: Haemodynamic long-term effects of verapamil in essential hypertension at rest and during exercise. Acta Med Scand 681(suppl): 109–115, 1984, with permission.)

The 16 patients participating in the diltiazem study[15] had slightly higher blood pressure and TPRI values than patients participating in the other series. During the first hemodynamic study, the intraarterially recorded blood pressure in the sitting position was 183/108 mmHg; CI as 2.43 L/min/m^2; and TPRI was 4630 dyn sec cm^{-5} m^2.

The hemodynamic profile of diltiazem (Fig. 7) was similar to the profile of verapamil, but the reduction in blood pressure was greater: 12–14% during exercise. Diltiazem also induced a fall in HR during exercise on the order of 10% ($p < 0.001$) at all work levels (50, 100, and 150 watts). No patient developed bradycardia, however. Of interest, SI increased; during exercise, the increase was 9% ($p < 0.05$). Because of the increase in SI, CI was unchanged. The reduction in blood pressure was due entirely to the reduction in TPRI, which was most pronounced at rest in the sitting position (19 %) ($p < 0.01$).

Based on our acute observations tiapamil appeared to be a highly effective antihypertensive agent, because TPRI was reduced by 21% and MAP by 14%. Unfortunately, during the chronic study,[14] the effect of tiapamil leveled off. MAP was reduced by only 8–12%. The fall in blood pressure was associated with a reduction in TPRI, which was less than the reduction with the other calcium antagonists. There was a nonsignificant reduction in HR during rest, but during exercise HR was reduced by 8%. These findings clearly emphasize that long-term studies are needed to evaluate the efficacy of antihypertensive agents. Despite promising acute effects, the long-term effects may differ.

Side Effects

In general, patients with mild to moderately severe hypertension tolerated the calcium antagonists well. With verapamil and diltiazem no complaints were recorded. In particular, no patients had problems with constipation. For the other drugs, the results were as follows:

Tiapamil	Palpitations	1
	Flushing	3
Nifedipine	Palpitations	1
	Flushing	3
Nisoldipine	Palpitations	2
	Flushing	4
	Edema	4
Felodipine	Edema	3
Amlodipine	Edema	2

Palpitations and flushing were the most common side effects, but the complaints disappeared within 3 weeks. However, edema necessitated withdrawal of 2 patients in the nisoldipine group and 1 patient in the amlodipine group.

FIGURE 7 (*Facing page*). Hemodynamic changes during chronic treatment with diltiazem. Mean values and SEM. HR = heart rate, SI = stroke index, CI = cardiac index, SAP = systolic arterial pressure, DAP = diastolic arterial pressure, MAP = mean arterial pressure, TPRI = total peripheral resistance index. ○—┘ = supine position, ┗┐ = sitting position, 100 W = 100 watt exercise. *$p < 0.05$, **$p < 0.01$, ***$p < 0.001$. (●—● = before; ○......○ = on treatment.) (From Lund-Johansen P, Omvik P: Effect of long-term diltiazem treatment on central haemodynamics and exercise endurance in essential hypertension. Eur Heart J 11:543–551, 1990, with permission.)

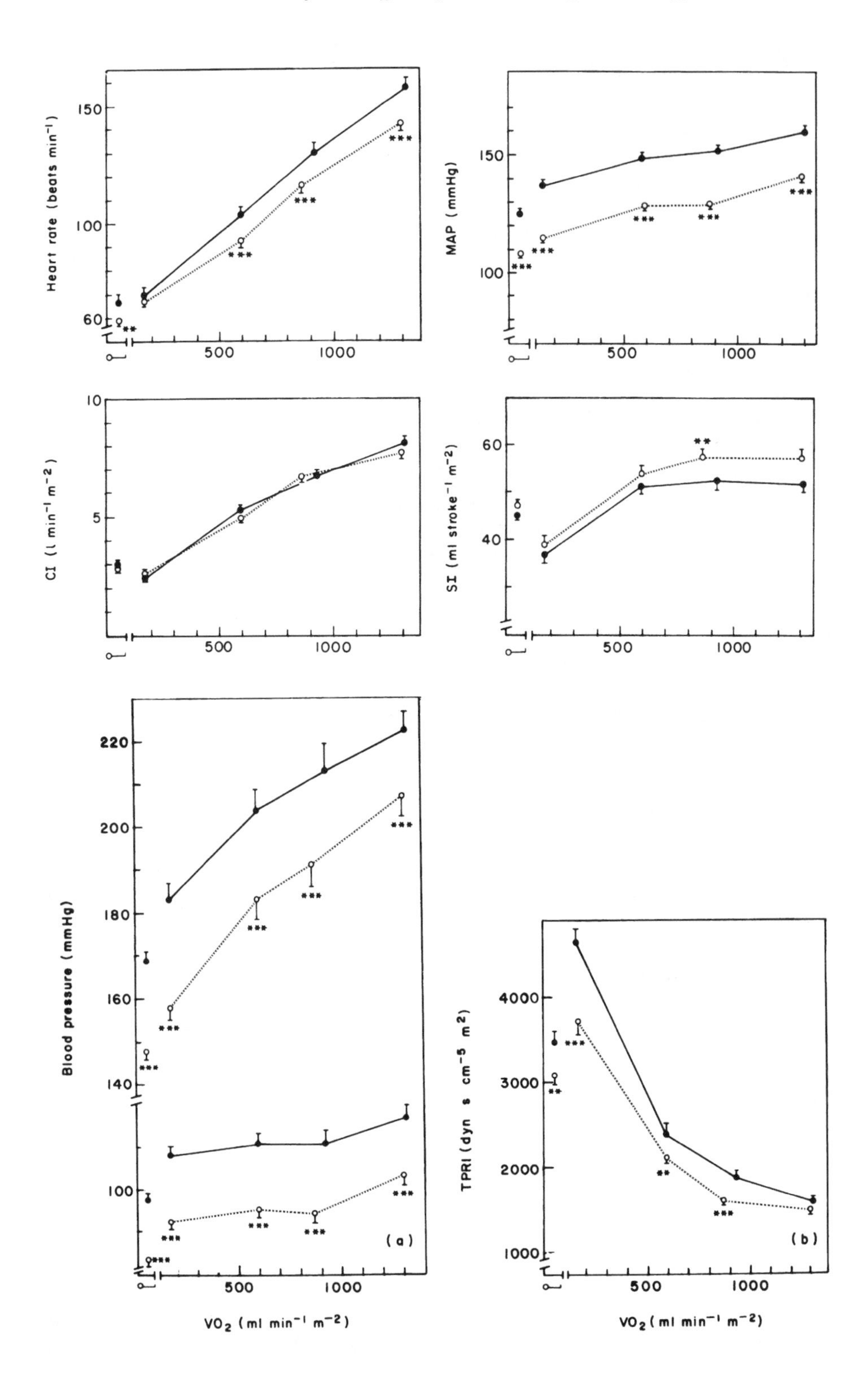
Heart rate (beats min⁻¹)
MAP (mmHg)
CI (l min⁻¹ m⁻²)
SI (ml stroke⁻¹ m⁻²)
Blood pressure (mmHg)
TPRI (dyn s cm⁻⁵ m²)
VO₂ (ml min⁻¹ m⁻²)
(a)
(b)

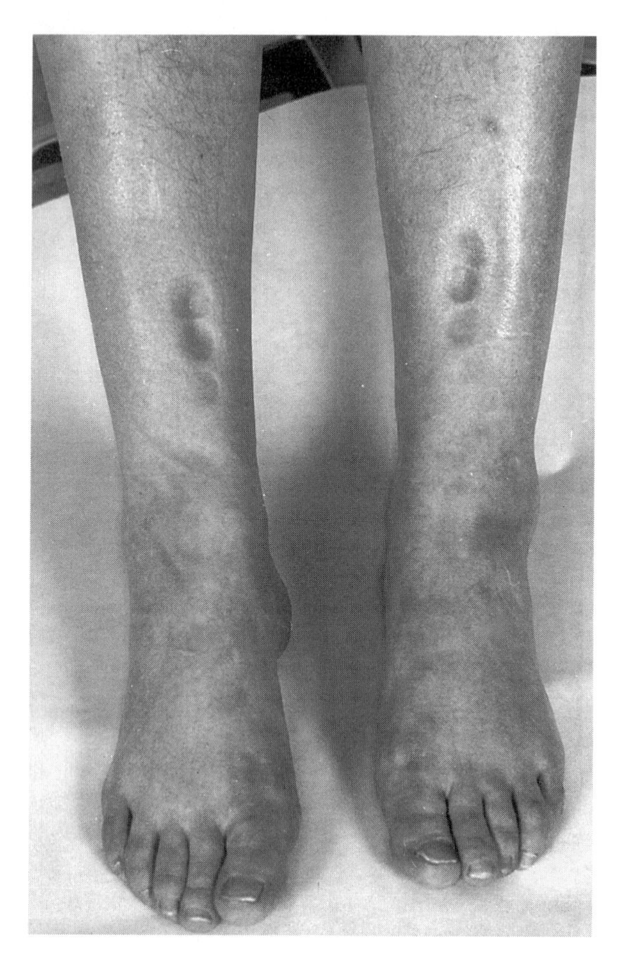

FIGURE 8. Ankle and leg edema in patient after 3 weeks on nisoldipine. The edemas disappeared with 2 weeks after withdrawal. (From Omvik P, Lund-Johansen P, Haugland H: Nisoldipine. Central hemodynamics at rest and during exercise in essential hypertension. Acute and chronic studies. J Hypertens 6:95–103, 1988, with permission.)

One patient in the nisoldipine group reported leg edema (Fig. 8). Edema resolved after withdrawal.

Plasma Volume and Extracellular Fluid Volume

In the tiapamil, diltiazem, nisoldipine, and amlodipine series, the effects on plasma volume and extracellular fluid volume were measured after 9–12 months. In none of the series did the calcium antagonist change body weight, plasma volume, or extracellular fluid volume. The edemas are related to the vascular effect of the calcium antagonist and not to general fluid retention. In some patients (5–10% in most series), ankle edemas may be substantial and necessitate reduction in the dose or withdrawal of the drug.

OTHER CALCIUM ANTAGONISTS

Several new dihydropyridines have been used in treatment of hypertension during the past few years. Basically they reduce blood pressure by the same mechanisms as nifedipine and nisoldipine. The first-dose effect and reflex tachycardia depend largely on the size of the first dose and how it is given—intravenously, as an oral solution, capsule, or slow-release tablet). The larger the dose and the quicker the increase in plasma concentration, the more pronounced the reflex tachycardia and (in most cases) the drop in blood pressure.

Nitrendipine given acutely lowers blood pressure, but side effects include baroreflex-mediated activation of the sympathetic nervous and renin-angiotensin systems and increased HR.[37,38] However, after treatment for 1 month, sympathetic activity, plasma renin levels, and HR were generally not increased because of resetting of the baroreflex.[38,39] Nitrendipine has proved effective in patients with both mild and severe hypertension.[38]

Isradipine also reduces blood pressure via reduction in total peripheral resistance[40–43] and, like felodipine, has a high affinity for the smooth muscles in the peripheral vessels compared with heart muscle.[40,41] Isradipine selectively inhibits the sinus node but has no effect on atrioventricular conduction. Hence, reflex tachycardia should be inhibited.[41] In studies of atherosclerotic rabbits, isradipine caused redistribution of cardiac output in favor of the heart and brain.[42] In hypertensive patients, twice-daily doses of 7.5-mg tablets during 9 weeks caused a fall in blood pressure from 184/96 mmHg to 162/83 mmHg. Total peripheral resistance was markedly reduced (37%), and brachial artery compliance increased. Cardiac output was increased by 11%.[41] Isradipine has been used as an intravenous solution for blood pressure control after surgery, particularly aortocoronary bypass.[29] In animal studies isradipine appears to have a pronounced antiatherogenic effect,[42] and a large study of its efficacy in counteracting atherosclerosis in the carotid arteries of humans showed less progression in wall changes with isradipine than with diuretics.[44]

Lacidipine is a lipophilic dihydropyridine and, like some of the previous dihydropyridines, has a strong selectivity for vascular smooth muscle over cardiac tissue. With short-term administration this selectivity results in reflex tachycardia, which usually disappears with long-term use.[45] In patients with essential hypertension, a single 1–5-mg tablet induced a fall of about 20% in systolic and diastolic blood pressure associated with a 20% increase (14 beats/min) in HR, with peak effect after 1–3 hours. In chronic studies the results have been largely similar to results with isradipine, nifedipine, and other dihydropyridines. The side-effect profile is similar (headache, flushing, ankle edema, dizziness, palpitations). The starting dose should not exceed 2 mg/day. Studies of possible effects on human atherosclerosis are in progress.

Nilvadipine is a dihydropyridine with a high degree of vascular selectivity.[46] In normotensive controls 16 mg in oral solution caused a rapid fall in blood pressure of 33%, and heart rate increased by 46%.[47] In hypertensive patients, sustained-release tablets of 4 mg decreased blood pressure by 16% after 2 hours, whereas HR increased by 10%. In other words, nilvadipine produces a response typical of other dihydropyridine calcium antagonists given in a similar formulation. During long-term use blood pressure control is largely the same as with other dihydropyridines, and the side-effect profile is similar. It has been suggested that nilvadipine may be particularly useful in patients surviving cerebral infarction, but its putative advantages are not well documented, nor is it clear whether they are related to the effect on cerebral blood flow.[46]

Nicardipine is another second-generation dihydropyridine that selectively inhibits vascular smooth muscle contraction. It has no effect on the atrioventricular node. As with similar dihydropyridines, acute administration (intravenously or in ordinary tablets) reduces blood pressure and produces reflex tachycardia.[48] In low dose (8.75 mg intravenously) HR was increased by only 4% in one study, but in other series an increase of 28% has been reported. As in previous studies of dihydropyridines such as nifedipine and felodipine, reflex tachycardia has been shown to disappear during long-term use (60–90 mg for 8 weeks or more). Reduction in blood pressure is due to fall in TPR and reduction in vascular resistance in the coronary, renal, and cerebral circulations. In elderly as well as younger patients, the side effects are similar to those of other dihydropyridines: headache, flushing, palpitation, and ankle edema. The side effects related to reflex tachycardia tend to disappear during 1–3 weeks.[48]

DISCUSSION

Despite some disagreement about the cardinal hemodynamic disorders during the initial phase of essential hypertension, it has been demonstrated by numerous methods that total peripheral resistance as well as regional vascular resistance in most areas is increased when hypertension becomes established.[3–6,49] In young people, discussion centers on whether high cardiac output during rest is a typical phenomenon or merely related to the experimental procedure. Even people who demonstrate increased CI at rest (by accurate invasive methods) do not demonstrate excessive cardiac output during exercise. On the contrary, cardiac output related to oxygen consumption is slightly subnormal.[6] The reason for subnormal cardiac output during exercise is an insufficient increase in stroke volume. With increasing age and duration of hypertension, the insufficient increase in stroke volume during exercise becomes more pronounced and total peripheral resistance increases.[50]

Our observations about the acute effects of the dihydropyridine derivatives agree with what many other laboratories have reported about nifedipine,[34] felodipine,[24] and the newer second-generation dihydropyridines. Reductions in total as well as regional vascular resistance have been demonstrated in humans and spontaneously hypertensive rats.[51–54] Chronic studies from other laboratories have also shown that dihydropyridines, diltiazem, and verapamil reduce blood pressure via reduction in TPRI.

One argument against the use of calcium antagonists for treatment of hypertension has been reflex tachycardia.[1,2,8] The recent discussion of whether calcium antagonists may be potentially dangerous in hypertensive patients with symptomless, unknown coronary heart disease has focused to large degree on the reflex increase in heart rate.[1,2] As this overview indicates, any dihydropyridine that is given quickly by intravenous infusion, oral solution, or capsule—and in doses sufficiently high—increases heart rate. On the other hand, if the compound is given in low doses as slow intravenous infusion or in a slow-release formulation (extended release or GITS), the fall in blood pressure is gradual and the increase in heart rate is minimal or absent. In our series of 110 patients, no patient reported chest pain during the study.

During chronic use of conventional doses, blood pressure may be permanently reduced within 24 hours, and heart rate is not increased compared with pretreatment values either at rest or during exercise, as shown in our long-term studies of nifedipine, nisoldipine, felodipine, and amlodipine. Possibly a resetting of the baroreceptors takes place.[52–54] A more extensive discussion of hemodynamic issues and their relation to possible harmful cardiac effects has been published recently.[55]

In contrast to the dihydropyridines, long-term use of diltiazem, verapamil, and tiapamil induced a 10% reduction in heart rate, mainly during exercise. This reduction is far less than that with ordinary beta blockers, which often reduce heart rate by 25–30%.[56] However, a practical therapeutic consequence is that in patients with hypertension and high heart rate or atrial fibrillation, verapamil and diltiazem are preferred. On the other hand, in patients with a tendency to bradycardia or A-V block, the dihydropyridines should be used. In contrast to the dihydropyridines, verapamil was found to reduce mortality in survivors of myocardial infarction without congestive heart failure.[57–59] Whether reduction in heart rate affects the reduction in mortality is not known.

Control of Blood Pressure during Exercise and Physical Performance

In isolated muscle strips from the heart, all calcium antagonists in high concentrations reduce contractility—the so-called negative inotrophic effect. During severe muscular exercise the heart pump must meet a manifold increase in load due to increase in cardiac output, heart rate, and blood pressure.[4,6] Does the negative inotrophic effect possibly reduce heart pump function during exercise in ordinary therapeutic dosages? In our series the reduction in heart rate by verapamil[17] or diltiazem[15] was completely compensated by an increase in stroke volume during exercise; as a consequence, cardiac output was maintained. Thus, there were no changes in the arteriovenous oxygen difference, and oxygen reserve in venous blood was maintained. This finding is in striking contrast to the effect of most beta blockers, wherein stroke volume does not increase and cardiac output during exercise is reduced on the same order as heart rate—often by 25–30% of pretreatment values[56] (Fig. 9). Such a drastic reduction in cardiac output leads to an increase in the arteriovenous oxygen difference and a reduction in physical performance.[60,61] Several studies of the effects of beta blockers have demonstrated a reduction in endurance, sometimes by as much as 50% (exercise time reduced from 2 hours to 1 hour). This drastic reduction is due not only to cardiovascular effects but also to the metabolic effects of beta blockers.[61]

In contrast, several investigators have shown that calcium antagonists do not reduce physical endurance.[34,62–64] In our study of diltiazem, we estimated the patients' endurance. During the last exercise load of 150 watts, when hemodynamic parameters were recorded after 7 minutes, the patients continued to exercise until exhaustion or until they reached a total of 20 minutes. The results indicated that chronic treatment with diltiazem slightly increased compared with the control period. We also studied central hemodynamics to a point just before exhaustion (Fig. 10). The stroke volume did not decline during the endurance test either before or after treatment. Thus, in patients with mild to

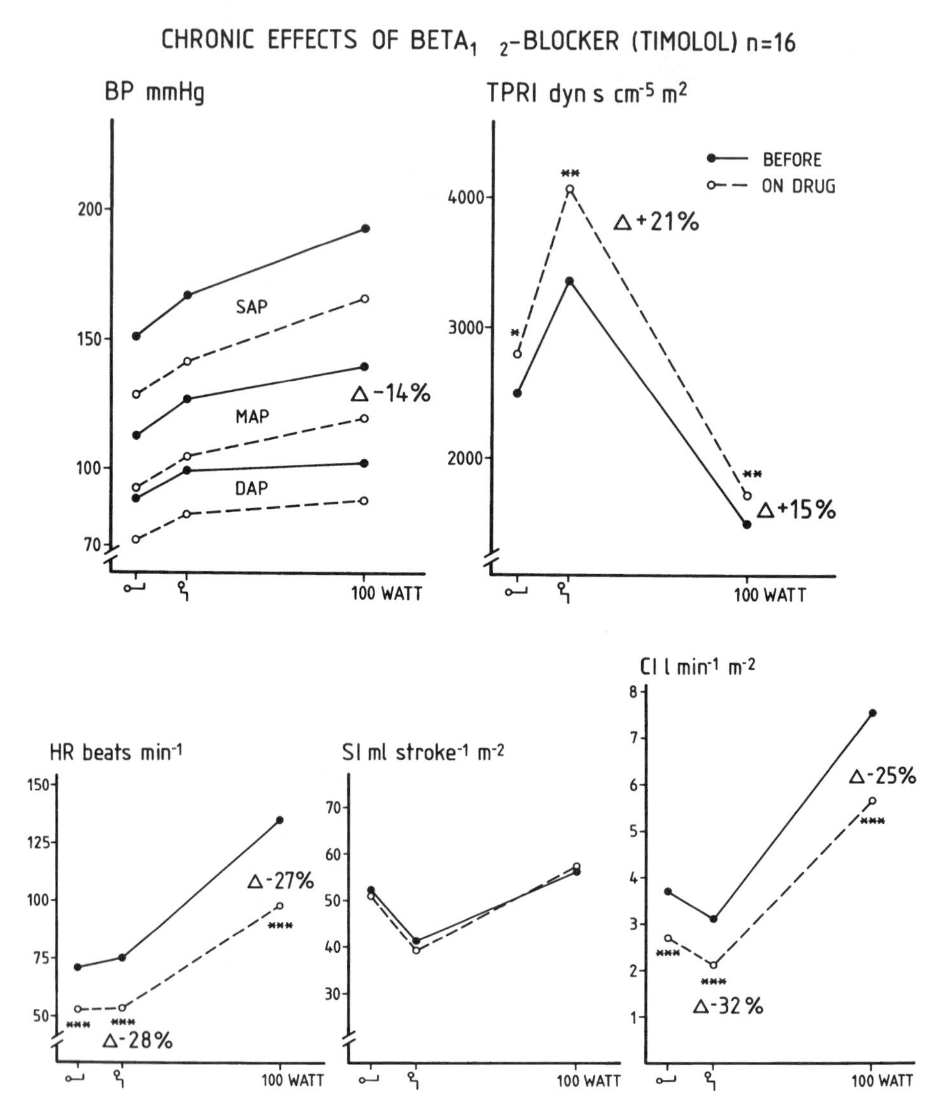

FIGURE 9. Hemodynamic changes during chronic (1 year) therapy on the beta-blocker timolol. Mean values and SEM. HR = heart rate, SI = stroke index, CI = cardiac index, SAP = systolic arterial pressure, DAP = diastolic arterial pressure, MAP = mean arterial pressure, TPRI = total peripheral resistance index. o—┘ = supine position, ╘ = sitting position, 100 W = 100 watt exercise. *$p < 0.05$, **$p < 0.01$, ***$p < 0.001$. (From Lund-Johansen P: Central haemodynamic effects of beta-blockers in hypertension. A comparison between atenolol, metoprolol, timolol, penbutolol, alprenolol, pindolol and bunitrolol. Eur Heart J 4(suppl D):1–12. 1984, with permission.)

moderately severe hypertension and good left ventricular function, calcium antagonists do not reduce stroke volume even during hard physical exercise for 20 minutes. Thus no negative inotrophic effect was found. Because no intracardiac pressure measurements were performed, however, we cannot exclude a possible increase in filling pressure. Blood pressure control was maintained, and reduction in total peripheral resistance remained the same.

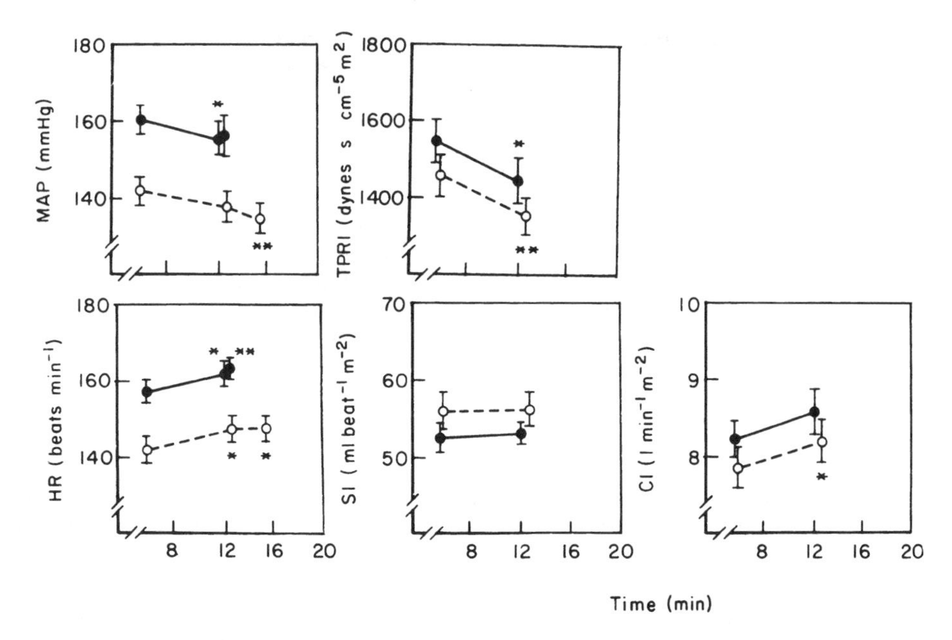

FIGURE 10. Hemodynamic changes during 150 W exercise load after 6 minutes and before exhaustion at approximately 13 minutes. Note that SI shows no tendency to decline, either before (●—●) or during (o—o) chronic treatment with diltiazem. Mean values. HR = heart rate, SI = stroke index, CI = cardiac index, SAP = systolic arterial pressure, DAP = diastolic arterial pressure, MAP = mean arterial pressure, TPRI = total peripheral resistance index. o—˩ = supine position, ᒪ = sitting position, 100 W = 100 watt exercise. *$p < 0.05$, **$p < 0.01$, ***$p < 0.001$. (From Lund-Johansen P, Omvik P: Effect of long-term diltiazem treatment on central haemodynamics and exercise endurance in essential hypertension. Eur Heart J 11:543–551, 1990, with permission.)

Other investigators have compared exercise performance in the same patients while taking beta blockers or calcium antagonists. In general, exercise performance was improved by calcium antagonists and reduced by beta blockers.[64] A placebo-controlled study found that diltiazem, 360 mg/day, reduced blood pressure and heart rate and increased maximal duration of exercise. Studies of the regional hemodynamics of diltiazem have shown dilatation of both large and small arteries.[32]

Hemodynamic Effects of Calcium Antagonists Compared with other Antihypertensive Agents

We studied most of the commonly used antihypertensive agents with the same methods.[65] The alpha-receptor blockers prazosin[66] and doxazosin[67] reduced blood pressure and resistance to much the same degree as calcium antagonists, and the reduction was due entirely to reduction in total peripheral resistance. The alpha-blockers showed a clear tendency to increase stroke volume during exercise, and postreatment cardiac output during exercise tended to exceed pretreatment levels.

In addition, ACE inhibitors, such as enalapril,[68] captopril,[69] and lisinopril,[70] reduced blood pressure through vasodilatation without reducing heart pump function. In our series using enalapril alone or in combination with hydrochlorothiazide, the pressure reduction was induced entirely through a fall in total peripheral resistance. Posttreatment stroke volume did not increase, however.[68]

In contrast, in all our series using beta blockers, with or without intrinsic sympathicomimetic activity, no significant reduction in total peripheral resistance was seen after chronic treatment for 1 or 5 years. In general, total peripheral resistance values were higher than pretreatment levels. Cardiac output during exercise was reduced on the order of 25–30% in most series.[56] Beta blockers with vasodilating activity, such as labetalol[71] or carvedilol,[72] produced a smaller reduction in cardiac output; blood pressure reduction was due partly to alpha blockade.

Calcium Antagonists in Patients with Reduced Left Ventricular Function

Patients with reduced left ventricular function were not included in our series, but discussion of the use of calcium antagonists in such patients (generally patients with heart failure due to coronary heart disease) is ongoing.[7] Deterioration of heart failure has been observed in several studies using calcium antagonists in contrast to treatment with ACE inhibitors[73] or isosorbide dinitrate.[74] Thus, although calcium antagonists appear safe in patients with mild to moderately severe hypertension at rest as well as during exercise—and appear even to improve exercise performance—they should be used with caution in patients with more severe hypertension and increased risk of reduced left ventricular function. If calcium antagonists are used in such patients, they should be treated also with ACE inhibitors, diuretics, and possibly digitalis. Calcium antagonists should be added as additional therapy when needed. A recent study of amlodipine (the PRAISE study)[11] showed that amlodipine was tolerated but did not improve mortality rates (participants, however, did not have hypertension). In patients with left ventricular hypertrophy, calcium antagonists have been shown to induce regression of left ventricular mass, and presumably to decrease the risk of ventricular arrhythmias.[75]

CONCLUSION

The different classes of antihypertensive agents have advantages and disadvantages. In treating patients with calcium antagonists, the starting dose should be low, and a slow-release form or an agent with a slow mode of action should be used. With such an approach, problems with reflex tachycardia are usually minimized or avoided. During chronic use with a slow-release preparation or use of a long-acting calcium antagonist, reflex tachycardia presents no problem,[76] and smooth control of blood pressure is possible.[77] One argument in favor of calcium antagonists as the first choice for treatment of hypertension, is the preliminary observation that they seem to have an antiatherosclerotic effect in animal models.[42,43,78] Preliminary observations from coronary artery angiography indicate that the same effect may be found in humans, but confirmation is needed.[79] On the other hand, a large percentage of hypertensive patients die suddenly from heart arrest. Calcium antagonists, unfortunately, are less protective in this respect than beta blockers.[58,59] However, it is well documented that calcium antagonists control blood pressure well in a large percentage of patients of all ages with primary hypertension. There are few contraindications to their use, and they tend to normalize the cardinal hemodynamic disorder—increased total peripheral resistance.[80] So far, no controlled studies are available to determine whether the long-term prognosis in hypertensive patients is improved or worsened by treatment with calcium antagonists compared with other antihypertensive agents. At present several large controlled studies are underway, such as the NORDIL study,[81] the HOT study,[82] and the CAPP study.[83] However, we must wait until year 2000 before the results of these trials are available. Meanwhile, the calcium antagonists should continue to be used as first-line antihypertensive drugs.[55] One should respect contraindications to the different classes in special patient groups and start with a low dose of a slow-acting compound.

References

1. Psaty BM, Heckbert SR, Koepsell TD, et al: The risk of myocardial infarction associated with antihypertensive drug therapies. JAMA 274:620–625, 1995.
2. Furberg CD, Psaty BM, Meyer JV: Nifedipine: Dose-related increase in mortality in patients with coronary heart disease. Circulation 92:1326–1331 1995.
3. Folkow B: Physiological aspects of primary hypertension. Physiol Rev 62: 347-504, 1982.
4. Conway J: Haemodynamic aspects of essential hypertension in humans. Physiol Rev 64: 617–660, 1984.
5. Kazda S: Future prospects for calcium antagonists. Drugs 48(Suppl 1):32–39, 1994.
6. Lund-Johansen P, Omvik P: Hemodynamic patterns of untreated hypertensive disease. In Laragh JH, Brenner BM (eds): Hypertension: Pathophysiology, Diagnosis and Management. New York, Raven Press, 1995, pp 323–342.
7. Halperin AK, Cubeddu LX: The role of calcium channel blockers in treatment of hypertension. Am Heart J 111:363–382, 1986.
8. Doyle AE: An overview of indication and choice for the use of calcium antagonists for the treatment of essential hypertension. In Aoki K, Frohlich ED (eds): Calcium in Essential Hypertension. New York, Academic Press, 1989, pp 601–619.
9. Nayler WG: The effect of amlodipine on hypertension-induced cardiac hypertrophy and reperfusion-induced calcium overload. J Cardiovasc Pharmacol 12(Suppl 7):41–44, 1988.
10. Editorial: Calcium antagonist caution. Lancet 337:885–886, 1991.
11. Packer M, Nicod P, Khandheria BR, et al: Randomized, multicenter, double-blind, placebo-controlled evaluation of amlodipine in patients with mild-to-moderate heart failure (abstract). J Am Coll Cardiol 17:274A, 1991.
12. Lund-Johansen P: Haemodynamic long-term effects of verapamil in essential hypertension at rest and during exercise. Acta Med Scand 681(Suppl): 109–115, 1984.
13. Lund-Johansen P, Omvik P: Haemodynamic effects of nifedipine in essential hypertension at rest and during exercise. J Hypertens 1:159–163, 1983.
14. Omvik P, Lund-Johansen P, Haugland H: Nisoldipine. Central hemodynamics at rest and during exercise in essential hypertension. Acute and chronic studies. J Hypertens 6:95–103, 1988.
15. Lund-Johansen P, Omvik P: Effect of long-term diltiazem treatment on central haemodynamics and exercise endurance in essential hypertension. Eur Heart J 11:543–551, 1990.
16. Lund-Johansen P: Hemodynamic effects of felodipine in hypertension: A review. J Cardiovasc Pharmacol 15(Suppl 4):34–39, 1990.
17. Lund-Johansen P, Omvik P, White W, et al: Long-term haemodynamic effects of amlodipine at rest and during exercise in essential hypertension. J Hypertens 8:1129–1136, 1990.
18. Omvik P, Lund-Johansen P: Acute and long-term hemodynamic effects of tiapamil at rest and during exercise in essential hypertension. Cardiovasc Drugs Ther 3:517–523, 1989.
19. Lund-Johansen P: Hemodynamics in early essential hypertension. Acta Med Scand 482(Suppl), 1968.
20. van Hooft IMS, Grobbe DE, Waal-Manning HJ, Hofman A: Hemodynamic characteristics of the early phase of primary hypertension. The Dutch hypertension and offspring study. Circulation 87:1100–1106, 1993.

21. Lund-Johansen P: Newer thinking on the hemodynamics of hypertension. Curr Opin Cardiol 9:505–511, 1994.
22. Lund-Johansen P: The dye dilution method for measurement of cardiac output. Eur Heart J 11(Suppl 1):6–12, 1990.
23. Bühler FR, Bolli P, Erne P, et al: Position of calcium antagonists in antihypertensive therapy. J Cardiovasc Pharmacol 7(Suppl 4):21–27, 1985.
24. Lorimer AR, McAlpine HM, Rae AP, et al: Effects of felodipine on rest and exercise heart rate and blood pressure in hypertensive patients. Drugs 29 (Suppl 2):154–164, 1985.
25. Burges RA, Dodd MG: Amlodipine. Cardiovasc Drug Rev 8:25–44, 1990.
26. Kleinbloesem CH, van Brummelen P, Danhof M, et al: Rate of increase in the plasma concentration of nifedipine as a major determinant of its hemodynamic effects in humans. Clin Pharmacol Ther 41:26–30, 1987.
27. Frohlich ED, McLoughlin MJ, Losem CJ, et al: Hemodynamic comparison of two nifedipine formulations in patients with essential hypertension. Am J Cardiol 68:1346–1350, 1991.
28. Burges R, Moisey D: Unique pharmacologic properties of amlodipine. Am J Cardiol 73:2A–9A, 1994.
29. Ruegg PC, David D, Loria Y: Isradipine for the treatment of hypertension following coronary artery bypass graft surgery: A randomized trial versus nitroprusside. Eur J Anaesthesiol 9:293–305, 1992.
30. Herpin D, Amiel A, Boutaud P, et al: Effect of a calcium antagonist verapamil on resting blood pressure and pressor response to dynamic exercise. Acta Cardiol 40:277–290, 1985.
31. Laragh JH: Calcium antagonists in hypertension—focus on verapamil. A symposium. Am J Cardiol 57:1–107, 1986.
32. Safar ME, Simon ACH, Levenson JA, Cazor JL: Haemodynamic effects of diltiazem in hypertension. Circ Res 52(Suppl 1):169–173, 1983.
33. Aoki K, Sato K, Kondo S, Yamamoto M: Hypotensive effects of diltiazem to normals and essential hypertension. Eur J Clin Pharmacol 25:475–480, 1983.
34. Yamakado T, Oonishi N, Nakano T, Takezawa H: Effects of nifedipine and diltiazem on haemodynamic responses at rest and during exercise in hypertensive patients. Jpn Circ J 4:415–421, 1985.
35. Stopher DA, Beresford AP, Macrae PV, Humphrey MJ: The metabolism and pharmacokinetics of amlodipine in humans and animals. J Cardiovasc Pharmacol 12(Suppl 7):55–59, 1988.
36. White WB, Lund-Johansen P, McCabe EJ, Omvik P: Clinical evaluation of the Accutracker II ambulatory blood pressure monitor: Assessment of performance in two countries and comparison with sphygmomanometry and intra-arterial blood pressure at rest and during exercise. J Hypertens 7:967–975, 1989.
37. Eichelbaum M, Mikus G, Mast V, et al: Pharmacokinetics and pharmacodynamics of nitrendipine in healthy subjects and patients with kidney and liver disease. J Cardiovasc Pharmacol 12(Suppl 4):6–10, 1988.
38. Hulthén UL, Katzman PL: Review of long-term trials with nitrendipine. J Cardiovasc Pharmacol 12(Suppl 4):11–15, 1988.
39. Nannan ME, Melin JA, Vanbutsele RJ, et al: Acute and long-term effects of nitrendipine on resting and exercise hemodynamics in essential hypertension. J Cardiovasc Pharmacol 6(Suppl 7): 1943–1048, 1984.
40. Hof RP, Salzmann RU, Siegl H: Selective effects of isradipine on the peripheral circulation and the heart. Am J Cardiol 59:30–36, 1987.
41. Andersson OK, Persson B, Widgren BR, Wysocki M: Central hemodynamics and brachial artery compliance during therapy with isradipine, a new calcium antagonist. J Cardiovasc Pharmacol 15(Suppl 1):87–89, 1990.
42. Hof RP, Hof A, Takiguchi Y: Comparative hemodynamic studies of isradipine and dihydralazine in atherosclerotic and normal rabbits. J Cardiovasc Pharmacol 15(Suppl 1):13–22, 1990.
43. Lund-Johansen P: Cardiac effects of isradipine in patients with hypertension. Am J Hypertens 6:S294–S299, 1993.
44. Borhani NO, Miller ST, Brugger SB, et al: MIDAS: Hypertension and atherosclerosis. A trial of the effects of antihypertensive drug treatment on atherosclerosis. J Cardiovasc Pharmacol 19(Suppl 3):S16–S20, 1992.
45. Lee CR, Bryson HM: Lacidipine—A review of its pharmacodynamic and pharmacokinetic properties and therapeutic potential in the treatment of hypertension. Drugs 48(2):274–296, 1994.
46. Brogden RN, McTavish D: Nilvadipine—A review of its pharmacodynamic and pharmacokinetic properties, therapeutic use in hypertension and potential in cerebrovascular disease and angina. Drugs Aging 6(2):150–171, 1995.
47. Cheung WK, Sia LL, Woodward DL, et al: Importance of oral dosing rate on the hemodynamic and pharmacokinetic profile of nilvadipine. J Clin Pharmacol 28:1000–1007, 1988.

48. Frampton JE, Faulds D: Nicardipine. A review of its pharmacology and therapeutic efficacy in older patients. Drugs Aging 3(2):165–187, 1993.
49. Mulvany MJ: The structure of the resistance vasculature in essential hypertension. J Hypertens 5:129–136, 1987.
50. Lund-Johansen P: Twenty year follow-up of hemodynamics in essential hypertension during rest and exercise. Hypertens 18(Suppl 3):54–61, 1991.
51. Nordlander M: Haemodynamic effects of short and long term administration of felodipine in spontaneously hypertensive rats. Drugs 29(Suppl 2):90–101, 1985.
52. Eichelbaum M, Echizen H: Clinical pharmacology of calcium antagonists 4: A critical review. J Cardiovasc Pharmacol 6:963–967, 1984.
53. McAllister RG Jr, Schloemer GL, Hamann SR: Kinetics and dynamics of calcium entry antagonists in systemic hypertension. Am J Cardiol 57:16–21, 1986.
54. Reid JL, Elliott HL: Calcium antagonists in essential hypertension: Clinical pharmacological aspects. In Aoki K, Frohlich ED (eds): Calcium in Essential Hypertension. New York, Academic Press, 1989, pp 575–600.
55. Epstein M: Calcium antagonists should continue to be used for first-line treatment of hypertension. Arch Intern Med 155:2150–2156, 1995.
56. Lund-Johansen P: Central haemodynamic effects of beta-blockers in hypertension. A comparison between atenolol, metoprolol, timolol, penbutolol, alprenolol, pindolol and bunitrolol. Eur Heart J 4(Suppl D):1–12. 1984.
57. Danish Study Group on Verapamil in myocardial infarction. Effect of verapamil on mortality and major events after acute myocardial infarction (The Danish Verapamil Infarction Trial II-DAVIT II). Am J Cardiol 66:779–785, 1990.
58. Yusuf S: Verapamil following uncomplicated myocardial infarction: Promising, but not proven. Am J Cardiol 77:421–422, 1996.
59. Held PH, Yusuf S: Effects of beta-blockers and calcium channel blockers in acute myocardial infarction. Eur Heart J 14(Suppl F):18–25, 1993.
60. Kaiser P: Physical performance and muscle metabolism during beta-adrenergic blockade in man. Acta Physiol Scand 536(Suppl):1–44, 1984.
61. Lundborg P, Åström H, Bengtsson C, et al: Effect of beta-adrenoceptor blockade on exercise performance and metabolism. Clin Sci 61:299–305, 1981.
62. Cody RJ, Kubo SH, Covit AB, et al: Exercise haemodynamics and oxygen delivery in human hypertension. Response to verapamil. Hypertension 8:3–10, 1986.
63. Pool PE, Seagren SC, Salel AF, Skalland ML: Effects of diltiazem on serum lipids, exercise performance and blood pressure: Randomized, double-blind, placebo-controlled evaluation for systemic hypertension. Am J Cardiol 56:86–91, 1985.
64. Franz I-W, Wiewel D: Antihypertensive effects on blood pressure at rest and during exercise of calcium antagonists, beta-receptor blockers, and their combination in hypertensive patients. J Cardiovasc Pharmacol 6:1037–1042, 1984.
65. Lund-Johansen P: Hemodynamic effects of antihypertensive agents. In Doyle AE: Handbook of Hypertension, vol 11. Clinical Pharmacology of Antihypertensive Drugs. Amsterdam, Elsevier Science, 1988, pp 41–72.
66. Lund-Johansen P: Haemodynamic changes at rest and during exercise in long-term prazosin therapy for essential hypertension. Postgrad Med J 58 (Suppl 1):45–52, 1975.
67. Lund-Johansen P, Omvik P, Haugland H: Acute and chronic haemodynamic effects of doxazosin in hypertension at rest and during exercise. Br J Clin Pharmacol 21:45–54, 1986.
68. Lund-Johansen P, Omvik P: Long-term haemodynamic effect of enalapril (alone and in combination with hydrochlorothiazide) at rest and during exercise in essential hypertension. J Hypertens 2(Suppl 2):49–56, 1984.
69. Omvik P, Lund-Johansen P: Combined captopril and hydrochlorothiazide therapy in severe hypertension: Long-term haemodynamic changes at rest and during exercise. J Hypertens 2:73–80, 1984.
70. Omvik P, Lund-Johansen P: Lisinopril plus sodium restriction versus hydrochlorothiazide: Comparison of long-term hemodynamic effects at rest and during exercise in patients with essential hypertension. Am J Cardiol 65:331–338, 1990.
71. Lund-Johansen P, Omvik P: The role of multiple action agents in hypertension. Eur J Clin Pharmacol 38:1–7, 1990.
72. Lund-Johansen P, Omvik P: Chronic haemodynamic effects of carvedilol in essential hypertension at rest and during exercise. Eur Heart J 13:281–286, 1992.
73. Dunselman PHJM, van der Mark TW, Kuntze CEE, et al: Different results in cardiopulmonary exercise tests after long-term treatment with felodipine and enalapril in patients with congestive heart failure due to ischaemic heart disease. Eur Heart J 11:200–206, 1990.

74. Elkayam U, Amin J, Mehra A, et al: A prospective, randomized, double-blind, crossover study to compare the efficacy and safety of chronic nifedipine therapy with that of isosorbide dinitrate and their combination in the treatment of chronic congestive heart failure. Circulation 82:1954–1961, 1990.
75. Messerli FH, Kaesser UR, Losem CJ: Effects of antihypertensive therapy on hypertensive heart disease. Circulation 80(Suppl 4):145–150, 1989.
76. Zannad F: Clinical pharmacology of nisoldipine coat core. Am J Cardiol 75:41E–45E, 1995.
77. Zanchetti A, on behalf of the Italian Nifedipine GITS Study Group: The 24-hour efficacy of a new once-daily formulation of nifedipine. Drugs 48(Suppl 1):23–31, 1994.
78. Fleckenstein A, Frey M, Fleckenstein-Grün G: Antihypertensive and arterial anticalcinotic effects of calcium antagonists. Am J Cardiol 57:1–10, 1986.
79. Bond MG, Purvis C, Mercuri M: Antiatherogenic properties of calcium antagonists. J Cardiovasc Pharmacol 17(Suppl 4):87–93, 1991.
80. Zanchetti A: Introduction: Why a new calcium antagonist? J Cardiovasc Pharmacol 17(Suppl 4): 5–7, 1991.
81. The Nordil Group. The Nordic Diltiazem Study (NORDIL). A prospective intervention trial of calcium antagonist therapy in hypertension. Blood Pressure 2:312–321, 1993.
82. Hansson L, Zanchetti A, for the HOT Study Group. The Hypertension Optimal Treatment Study (The HOT Study)—patient characteristics, randomization, risk profiles, and early blood pressure results. Blood Pressure 3:322–327, 1994.
83. The CAPP group. The Captopril Prevention Project: A prospective intervention trial of angiotensin converting enzyme inhibition in the treatment of hypertension. J Hypertens 8:985–990, 1990.

GERARD M. LONDON M.D. / MICHEL E. SAFAR M.D.

12

Arterial Compliance and Effect of Calcium Antagonists

Epidemiologic studies have emphasized the close relationship between the height of blood pressure and the incidence of cardiovascular diseases.[1,2] Usually, clinical hypertension is classified on the basis of diastolic blood pressure, and the hemodynamic characteristics of hypertension are attributed to a reduction in the caliber of small arteries with a resulting increase in vascular resistance and secondary adaptive changes in the structure and function of the heart. This definition does not account for the fact that the arterial system plays an important role in determining the shape and amplitude of the blood pressure wave, influencing directly the level of systolic, diastolic, and pulse pressures.[3]

The above facts have been constantly underestimated in the definition, diagnosis, and therapeutic evaluation of clinical hypertension. Nevertheless, recent prospective epidemiologic studies have directed attention to systolic pressure as a better guide than diastolic pressure to cardiovascular and all mortality.[4,5] It also has been shown that increased pulse pressure is an independent cardiovascular risk factor in hypertension.[6–10] These studies have focused attention on arterial compliance and how it and other factors determine the level of systolic and pulse pressures.[11,12] The implication of large arteries in the morbid events of patients treated for hypertension has resulted in an extensive investigation of the effects of antihypertensive drugs on arterial compliance.[13–23]

Calcium antagonists have been used for a long time in the treatment of coronary insufficiency. Their therapeutic action is related to a dual mechanism: (1) they reduce the contractile activity of the heart, and (2) they promote vasodilation of the coronary vessels and reduction of ventricular afterload, which historically has been attributed to peripheral arteriolar vasodilatation. In fact, ventricular afterload, which describes the relationship between pressure and flow in the arterial tree downstream, depends on the properties of the vascular bed characterized as the **ascending aortic impedance.**[3,11] Ascending aortic impedance relates not only to peripheral resistance but also to capacitance properties of the arterial tree. Therefore, studies of vasodilators should not be limited to their effect on peripheral resistance but also should include their effect on vascular capacitance.

Calcium antagonists are expected to have important acute and long-term effects on large arteries. Indeed, calcium ions play an important role in controlling arteriolar tone. Increased intracellular calcium activates the contraction of smooth muscle fibers and increases vascular reactivity to vasoconstrictor substances. By blocking the transport of calcium across cell membranes, calcium antagonists reduce the tone of vascular smooth muscle and produce vasorelaxation.

Moreover, calcium antagonists protect and slow down experimental arterial calcinosis in rats.[24] Reversal of fibrous plaques, along with reduction of calcium and collagen, has been demonstrated in rabbits.[25] Furthermore, calcium entry blockers reduce diet-induced atherosclerosis in rabbits and cynomolgus monkeys without reducing elevated plasma lipids and with effect on arterial wall connective tissue.[26]

This chapter summarizes experimental and clinical studies of the effect of calcium antagonists on the capacitive properties of large arteries to the exclusion of their antiatherogenic action.

BASIC FUNCTIONS OF ARTERIES

The arterial system has two distinct, interrelated functions: (1) to deliver an adequate supply of blood to body tissues (**conduit function**) and (2) to smooth out the pulsations occurring with intermittent ventricular ejection (**cushioning function**).[3,11]

Cushioning Function

The principal role of arteries as cushions is to dampen the pressure oscillations resulting from intermittent ventricular ejection. Hemodynamically the cushioning function of arteries is characterized by pulsatile flow and pulsatile pressure.[3,11] Indeed, large arteries can instantaneously accommodate the volume of blood ejected from the heart, storing part of the stroke volume during systolic ejection and draining the stored volume during diastole, thereby ensuring continuous perfusion of organs and tissues. This Windkessel effect is due to the viscoelastic properties of arterial walls and the geometric characteristic of arteries, including their diameter and length. The principal alteration in cushioning function is due to the stiffening of arterial walls, with increased systolic and pulse pressure as the principal consequence. Two mechanisms are involved:[3,11] (1) the direct mechanism, which involves generation of a higher pressure wave by the left ventricle as it ejects blood into a stiff arterial system, and (2) the indirect mechanism, which is due to the influence of increased arterial stiffness on pulse wave velocity (PWV) and the timing of incident and reflected pressure waves.

Indeed, ejection of blood into the aorta generates a pressure wave that is propagated to other arteries throughout the body. The forward travelling (incident) pressure wave is reflected at any points of structural and functional discontinuity of the arterial tree, generating a reflected echo wave that travels backward toward the ascending aorta.[3,11] Incident and reflected pressure waves are in constant interaction and are summed in a measured pressure wave (Fig. 1). The final amplitude and shape of the measured pulse pressure wave are determined by the phase relationship (timing) between the component waves. The timing of incident and reflected pressure waves depends on pulse wave velocity (Fig. 2), traveling distance of pressure waves, and duration of ventricular ejection. The shape and amplitude of measured pressure waves depend on the site of pressure recording in the arterial tree. Peripheral arteries are close to reflecting sites, and the incident and reflected waves in peripheral arteries are in phase, producing an additive effect. The ascending aorta and

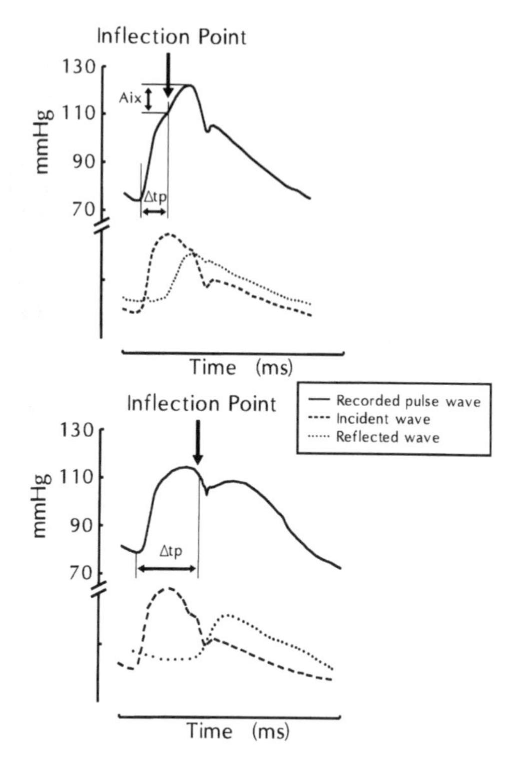

FIGURE 1. Recorded (—), incident (---), and reflected (···) pressure waves in the aorta of a subject with stiff arteries (*upper panel*) and subject with distensible arteries (*lower panel*). The algebraic sum of incident and reflected waves yields the recorded pressure wave, on which the reflected wave appears at the inflection point. ΔTp is the travel time of pressure wave to reflecting sites and back.

central arteries are distant from reflecting sites and, depending on PWV and arterial length, the return of the reflected wave is variably delayed; therefore, the incident and reflected waves are not in phase.[3,11,27]

In young people with distensible arteries and low PWV, the reflected waves affect central arteries during diastole after ventricular ejection has ceased (see Fig. 1). Such timing is desirable, because the reflected wave causes an increase in ascending aortic pressure in early diastole but not during systole, resulting in

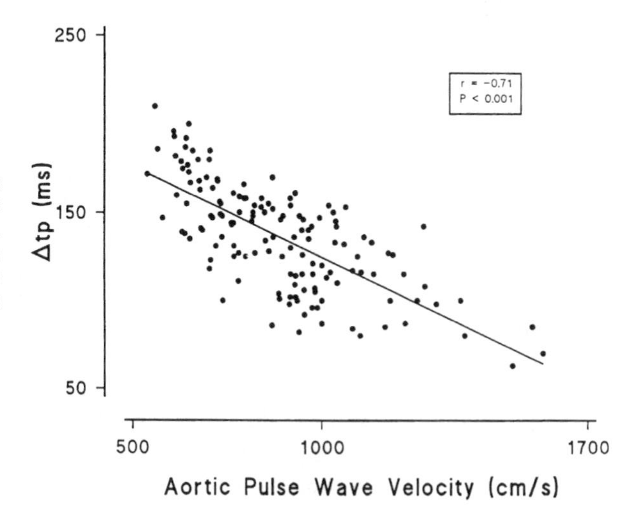

FIGURE 2. Scatterplot showing the correlation between aortic pulse wave velocity and ΔTp. The travel time of pressure wave to reflecting sites and back is inversely proportional to pulse wave velocity (personal data).

aortic systolic and pulse pressures that are lower than in peripheral arteries (only mean blood pressure is almost constant throughout the arterial system). This arrangement is physiologically advantageous, because the increase in early diastolic pressure has a boosting effect on coronary perfusion without increasing left ventricular afterload.[3,11,27] The desirable timing is disrupted by increased PWV due to arterial stiffening. With increased PWV, the reflecting sites appear closer to the ascending aorta, and the reflected waves occur earlier, more closely in phase with incident waves in the same region. The earlier return means that the reflected wave affects the central arteries during systole rather than diastole, amplifying aortic and ventricular pressures during systole and reducing aortic pressure during diastole (see Fig. 1).[3,11,27] Decreased end-diastolic pressure is also due to decreased reservoir effect, with an increase in stroke volume run-off during systole and less blood volume to be drained during diastole. The combination of decreased reservoir effect and diminished elastic recoil increases diastolic pressure decay, with lower pressure at the end of diastole. Hence the viscoelastic properties of the arterial system influence the level of systolic as well as diastolic pressure. Through promoting early wave reflections and increased incident pressure wave amplitude, increased arterial stiffness is disadvantageous to left ventricular function. Increases in mean, peak, and end systolic blood pressures in the ascending aorta lead to increasing myocardial oxygen consumption, and a decrease in mean diastolic blood pressure tends to impair coronary blood supply.[3,27] Furthermore, increased systolic blood pressure induces myocardial hypertrophy and impairs diastolic myocardial function and ventricular ejection.[28,29] In addition, increases in systolic blood pressure and pulse pressure accelerate arterial damage by increasing fatigue, degenerative changes, and arterial stiffening and thus feed a vicious cycle.[3,11]

As a consequence of the dual function of arteries, arterial pressure has two components that reflect the influence of different factors. The steady component (mean blood pressure) is determined exclusively by cardiac output and peripheral resistance, which in turn are determined by the caliber and number of small arteries and arterioles. The pulsatile component (pulse pressure) represents the oscillation around the mean pressure, with systolic and diastolic pressures as the highest and lowest points, respectively.[3,11] The magnitude of pulse pressure is determined by the pattern of left ventricular ejection and the viscoelastic and propagative properties of large arteries. The two components of blood pressure are tightly interrelated, because the changes in mean blood pressure influence directly the viscoelastic and propagative properties of the arterial system (Fig. 3).[3,11]

DEFINITIONS AND CONCEPTUAL FRAMEWORK

The capability of arteries to accommodate instantaneously the volume ejected by the left ventricle may be described in terms of compliance, distensibility, or stiffness of the aorta or an individual artery. These terms relate the contained volume of the vasculature (total or segmental) to a given transmural pressure over the physiologic range of pressure.

Compliance describes the absolute amount of change in strain following change in stress. In physiology compliance (C) is defined as the change in volume (dV) due to a change in pressure (dP), that is $C = dv/dP$. The reciprocal

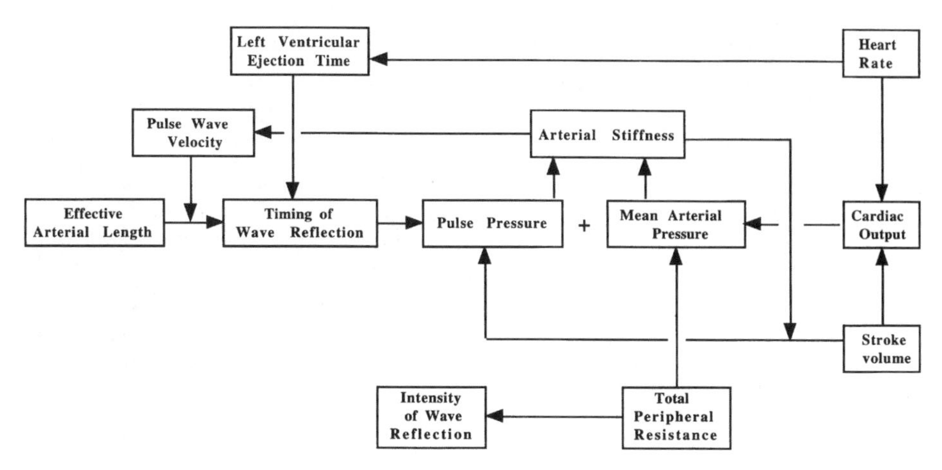

FIGURE 3. Schematic representation of major determinants of arterial pressure considered in terms of steady (mean blood pressure) and oscillatory (pulse pressure) components.

value of compliance is the **elastance** (E = dP/dV). Compliance represents the slope of the pressure-volume relationship (Figs. 4 and 5) at a specified point on the pressure-volume curve. The media of the arteries is responsible for its physical properties.[3,11] Because the arterial media is composed of a mixture of smooth muscle cells and connective tissue containing elastin and collagen fibers, the pressure-volume relationship is nonlinear.[3,30,31] At low distending pressure the tension is borne by elastin fibers, whereas at a high distending pressure the tension is predominantly borne by less extensible collagen fibers and the arterial wall becomes stiffer (less compliant). Thus the compliance can be defined only in terms of a given pressure. The compliance depends on blood pressure levels and mostly on the intrinsic elastic properties and amount of material that composes the arterial wall. To facilitate comparisons of viscoelastic properties of structures with different initial dimensions, the compliance may be expressed relative to the initial volume as a **coefficient of distensibility**: dV/dPV, where dV/dP is compliance and V is the initial volume.[3,11] The proper determination of distensibility in theory requires the determination of volume in condition of zero transmural pressure, i.e., unstressed volume. The reciprocal values of distensibility defines the coefficient of **stiffness**: dP·V/dV.

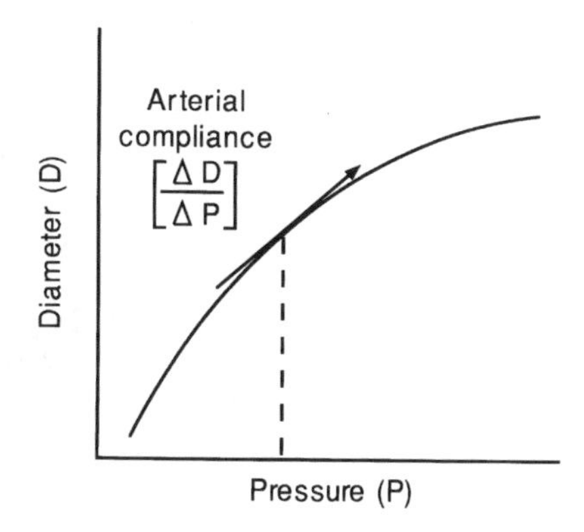

FIGURE 4. Pressure-volume relationship in a large artery. The slope of the curve at a given pressure defines arterial compliance.

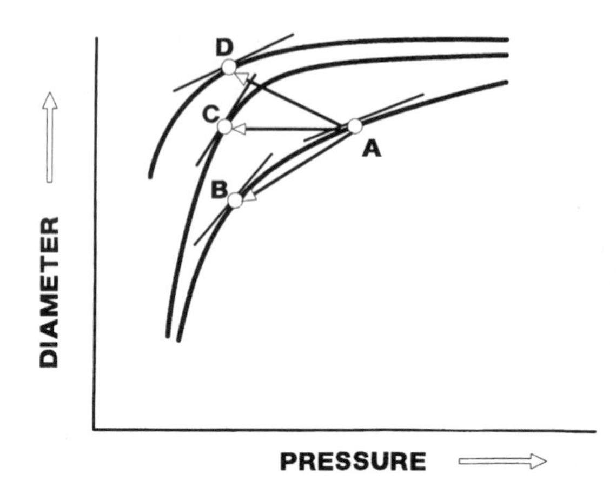

FIGURE 5. Possible changes in static pressure-diameter relationship in a large artery induced by antihypertensive drugs. Increase in compliance is a passive consequence of blood pressure decrease. The pressure-diameter curve is unchanged, and compliance moves from A to B. Drug has induced active changes in diameter and/or viscoelastic properties of the arterial wall. The pressure-diameter curve changed, and compliance moves from A to C or D.

Arterial volume per unit length is equal to the arterial cross-sectional area, depending on arterial diameter (D). Arterial compliance is determined from the pressure-diameter relationship as C = dD/dP, and arterial distensibility is expressed as dD/dPD. The reciprocal value of distensibility may be expressed as elastic modulus Ep (wall tension for 100% increase of diameter from resting diameter) = dPD/dD or as Young's modulus E (wall tension per cm wall thickness for 100% diameter increase) = dPD/dDh, where dP and dD are the changes in pressure and diameter about the resting diameter D in an artery with wall thickness h.[3,11,30,31]

The mutual interrelationships among blood pressure, geometric characteristics of the arteries (diameter at a fixed vessel length), and physical properties of the arterial walls (compliance) are complex. Complexity is attributable to nonlinear elastic behavior of the arterial wall,[3,11] differing effects of smooth muscle contraction on arterial stiffness and diameter,[32] and the direct effect of blood pressure changes on diameter of the arteries.[3,11] The complexity of the diameter and compliance relationships complicates the analysis of capacitance changes under various physiologic conditions or after administration of vasoactive drugs (see Fig. 5). Changes in diameter and compliance may be the passive consequence of blood pressure changes or the result of a direct action of drugs and physiologic stimuli on the arterial walls, principally on smooth muscle cells.[3]

Under normal circumstances and with constant blood pressure, an artery becomes less compliant as it is dilated.[3,30–32] Decreased compliance is related to an increase in wall tension as diameter increases and a transfer of tension from elastin to the less extensible collagen fibers.[3] Arterial capacitance may be increased by increase in diameter or by relaxation of smooth muscle cells with an eventual increase in compliance. Nevertheless, vasoactive agents, including calcium antagonists, nitrates, and angiotensin-converting enzyme inhibitors, can dilate medium-sized arteries in parallel with an increase in their compliance.[18,20] This paradoxical increase in arterial compliance with increase in diameter is explained on the basis of the alignment of smooth muscle in the arterial wall in series with collageneous elements and in parallel with elastic fibers; relaxation of smooth muscle transfers wall stress from collagen to elastin fibers.[33] One of the major points of debate about the action of drugs is whether the arterial smooth muscle relaxation is due to an an in situ effect or also to arteriolar vasodilation. Indeed, drug-induced arteriolar dilatation may change arterial diameter and compliance through the mechanism of high-flow dilation, which is endothelium-dependent.[34]

METHODOLOGIC CONSIDERATIONS

Measurement of Arterial Diameter and Intima–Media Thickness

The study of pressure-diameter relationships in the large arteries has necessitated the development of noninvasive techniques for arterial diameter measurements. Several techniques based on pulsed Doppler velocimetry and echography were developed in the past several years.[35] With the use of pulsed Doppler velocimetry, the inner diameter of straight superficial arteries—for example, brachial or carotid artery—may be measured in situ. The pulsed Doppler technique was the first used noninvasively in humans, but it was relatively insensitive, the measurement of arterial diameters could be performed only sequentially, and the measured values were mean diameters. The recently developed multigate pulsed Doppler systems and computer-assisted echo-tracking systems have a higher resolution and are characterized by sufficient linearity, dynamic range, and tracking speed, even when the signal-to-noise ratio of the original signal is not high. The phase-locked echo-tracking Walltrack system[36] allows reliable measurements of wall motions of large arteries, such as the brachial and common carotid arteries, and of systolic-diastolic changes in arterial diameters (dD). The Diarad system was developed to obtain diameter-pressure curves of the radial artery by coupling the measurement of systolic-diastolic variations of arterial diameter to those of digital pulse pressure.[37] These techniques have a high intra- and inter-observer variability below 10%. To quantify Young's modulus, the arterial intima–media thickness of superficial blood vessels is usually measured by high-resolution B-mode ultrasound imaging.[38] Computer-assisted reading with specially designed software is used to improve the reproducibility of measurements.

Measurement of Distending Pressure

The stress applied on an arterial segment is usually expressed in terms of pulse pressure. In peripheral arteries (e.g., brachial or radial arteries) the pulse pressure is measured by mercury sphygmomanometry or automated methods. Because of the nonhomogeneous viscoelastic properties of successive arterial segments and the effect of wave reflections, pulse and systolic pressures are amplified from the ascending aorta toward peripheral arteries, and in young and middle-aged people peripheral pulse and systolic pressures overestimate corresponding pressures in the aorta and central arteries.[3,11,27] To improve the accuracy of noninvasive pulse pressure measurements in central arteries, the techniques of applanation tonometry have recently been improved by the use of probes that incorporate high-fidelity strain-gauge transducers[39] and computer-assisted sphygmocardiography, which permit assessment of aortic or carotid pressure from peripheral pulse analysis.[40] Another advantage of applanation tonometry is the possibility of an accurate assessment of arterial pulse pressure waveforms and determination of the timing and effect of arterial wave reflections (see Fig. 1).[39]

Measurement of Arterial Compliance and Distensibility

Compliance and distensibility of arteries are usually determined from systolic-diastolic changes in diameter of superficial arteries coupled with measurement of local pulse pressure. Several alternative possibilities are frequently used, based on propagative and nonpropagative models of arterial pulse transmission. In nonpropagative models the arteries are viewed as a system of interconnected tubes with fluid storage capacity. This model assumes that all pressure changes within the arterial system occur simultaneously (i.e., pulse wave velocity is infinite). However, pulse pressure is propagated at a given velocity (PWV), which varies greatly according to the site of the measurements.[3,11,41] Therefore, propagative models are more suitable for assessment of arterial stiffness in humans, and the most common methods to evaluate arterial distensibility are based on the study of the PWV along a given large artery.[3,11] The most widely accepted relationship of PWV and elastic modulus (E) is the Moens-Korteweg equation:[3,11] $PWV^2 = Eh/2r\rho$, where E is elastic modulus, r is the radius, h is wall thickness, and ρ is fluid density. Bramwell and Hill[42] derived the equation, $PWV = VdP/dV$, where dP and dV are the changes in pressure and volume and V is the initial volume. Thus PWV is equal to the reciprocal value of volume distensibility.[3,11]

CHANGES IN ARTERIAL FUNCTION WITH AGE AND HYPERTENSION

Aging and Arterial Function

With aging the arterial wall thickens, and the arteries become less compliant.[3,11,24,41,43,44] The principal changes occur in the media and intima and are related to the elastic fibers and laminae, which are principally responsible for vessel distensibility. The orderly arrangement of elastic laminae is replaced by thinning, splitting, and fragmentation. The degeneration of elastic fibers is associated with an increase in collagen fibers and ground substance and depositions of calcium.[3,45] Deposition of calcium salts in the walls of human arteries is a physiologic process that starts in childhood and progresses with age. Blumenthal et al.[46] were among the first to present evidence that calcium concentration in the human aorta increased with age even before the development of fibrous or atherosclerotic plaques. Calcinosis appears to be an inevitable consequence of aging, even in the absence of atherosclerotic plaques and occlusive lesions.[24,47] Arterial calcinosis per se was considered a phenomenon of secondary importance because it does not compromise downstream perfusion.

Nevertheless, cross-sectional studies in normal populations have shown that calcinosis decreases the distensibility of the aorta even after adjustment for age and blood pressure.[48] Of interest, several studies have demonstrated that calcium antagonists prevent the development of experimental arterial calcinosis and retard the calcium accumulation in aging, hypertensive, and alloxan-diabetic rats.[24,47] The loss of distensibility is partly compensated by dilatation of arteries. These age-related changes are most marked in the aorta and are attributed to the fatiguing effect of cyclic stress over many decades.[3,11,39,41] A major manifestation of such changes is the rise in systolic and pulse pressures, an increase in aortic PWV, and the disappearance of pressure amplification between aorta and peripheral arteries due to the earlier return wave reflections.[3,11,39,42] Arterial stiffening and early wave reflections are important factors in left ventricular hypertrophy in aging humans.[3,11,41]

Large Arteries and Hypertension

Experimental studies have shown that compliance is reduced in hypertensive animals, independently of blood pressure levels.[49,50] It has also been demonstrated that both functional and structural factors are responsible for increased arterial stiffness. Abolition of vascular smooth muscle tone with potassium cyanide solution results in a significant increase in carotid artery compliance both in normotensive and hypertensive animals. Nevertheless, in hypertensive animals the compliance remained lower even after abolition of smooth muscle activity, pointing to the role of structural modifications, principally an increased proportion of collagen in the connective tissue of arterial walls.[49,50]

Several studies in humans have shown that the cushioning function of the arteries is altered in hypertension.[3,43,44,51–57] Arterial changes concern both diameter and compliance. Investigations of the effect of blood pressure on arterial diameters in the general population have produced inconsistent results, but the most recent studies found that common carotid artery diameter was increased in patients with untreated essential hypertension,[22,51,52,55–57] principally in relation to increased pulsatile pressure,[52–57] perhaps as a result of passive distention of the arterial wall under the influence of high blood pressure. Similarly, brachial and aortic diameters were also increased in hypertensive patients proportionally to blood pressure level.[53,54] Differences in the extent to which the different arteries are involved are probably due to variations in regional histologic composition of the arterial wall and in magnitude of pulse pressure in central and peripheral arteries.[3,11,58,59]

Arterial compliance is reduced in isolated systolic hypertension in elderly people, in sustained systolic-diastolic hypertension in middle-aged people, and even in borderline hypertension in young people.[3,60,61] The decrease in arterial compliance observed in hypertensive patients has often been attributed both to the nonlinearity of the pressure-diameter relationship (see Figs. 4 and 6) and to structural changes in arteries, mainly increased thickness of arterial walls due to smooth muscle hypertrophy and increased collagen content. Recent studies, however, found that distensibility and compliance of the medium-sized muscular arteries in patients with essential hypertension was not significantly different from those in normotensive controls when the two groups were studied at their respective mean blood pressure; when calculated for a similar pressure (isobaric compliance), however, compliance and distensibility were even higher in hypertensive patients.[62] Similar results have been observed in large elastic arteries.[63] Therefore, decreased arterial compliance and distensibility in middle-aged and young patients with essential hypertension may be explained by the increase in distending pressure. In patients with isolated systolic hypertension, the reduction of arterial compliance results from intrinsic alterations of arterial walls and is not the passive consequence of increased blood pressure.[60]

Increased arterial stiffness has been demonstrated in normotensive and hypertensive patients with end-stage renal disease compared with nonuremic patients matched for age and mean blood pressure.[48,64] In patients with end-stage renal disease the content of calcium in the arterial walls and the frequence and importance of arterial calcinosis are increased[48,64] and associated with decreased arterial distensibility.[48,64,65] The role played by calcium in the alterations of arterial distensibility was demonstrated in patients with end-stage renal disease. Hypercalcemia induced by dialysis with high-calcium dialysate significantly decreased arterial distensibility in the presence of decreased blood pressure.[66]

RESPONSE OF THE LARGE ARTERIES TO CALCIUM ANTAGONISTS

Experimental Data

Because increased pulse pressure has a specific impact on cardiovascular risk, it seems reasonable to develop drugs capable of increasing the compliance of large arteries and selectively or predominantly decreasing the pulse pressure. Because arterial compliance depends on blood pressure, it is obvious that any antihypertensive drug that reduces arteriolar tone may increase arterial compliance passively by displacement to a lower blood pressure level on the same curvilinear pressure-diameter curve (see Fig. 5).[3,11] However, an active increase in arterial compliance may be attained independently of the decrease in blood pressure with drugs that alter the mechanical properties of the arteries by modification of the structure of large arteries or by changes in smooth muscle activity (see Fig. 5). Because of the interaction of passive and active changes of arterial diameter and compliance, the analysis of the effects of antihypertensive drugs, including calcium antagonists, is difficult and complex.[3,11,67–70] Moreover, the calcium antagonists are heterogeneous, and their effects vary with their class. Furthermore, the response to calcium antagonists is difficult to predict because of the diversity of physiologic and pharmacologic responses of different large arteries; as they become smaller, their intrinsic tone increasingly depends on calcium in the extracellular space.[69] Finally, the heterogeneity of response of large arteries may be related to alterations in the baroreflex control of circulation. Because baroreflex mechanisms are partly initiated by changes in tension of the arterial wall, they are important factors to consider with drugs, such as calcium blockers, that are capable of modifying the diameter of the arteries. Studies in animals have demonstrated that calcium antagonists, in addition to direct action on blood vessels and the heart, exert an important influence on autonomic nervous control.[70,71] Studies in normotensive and hypertensive men showed that acute and chronic administration of dihydropyridines resets the sinoaortic baroreflex control of heart rate and vascular resistances and increases its sensitivity.[72–75] In contrast, such an increase in baroreceptor activity was not observed with verapamil or diltiazem.[13,71]

Few reports have evaluated responses of large arteries to calcium antagonists. The usual response evaluated is the aortic pressure-diameter or volume relationship during acute administration of drugs. Yano et al.[76] studied the effects of diltiazem on the aortic pressure-diameter relationship in anesthetized dogs. They found that the aortic diameter was reduced by diltiazem and that the aortic pressure-diameter curve was shifted toward higher diameters for any given pressure. Furthermore, the dP/dD-diameter curve was shifted downward, with reduced values for dP/dD in the physiologic range of diameter, suggesting that diltiazem increases aortic wall distensibility through a decrease in aortic smooth muscle tone.

Using an experimental model for analysis of the volume-pressure relationship of the in situ isolated carotid artery,[77] Levy et al.[78] studied the acute effects of calcium blockade by the diltiazem-like substance TA-3090 in spontaneously hypertensive rats (SHRs) and Wistar-Kyoto (WKY) rats. With intact or removed endothelium, calcium blockade induced in both strains a significant shift of the volume-pressure curve; at each given value of transmural pressure, volume was significantly higher after calcium blockade than under control conditions. Levy et al.[79] also studied the effects of chronic calcium antagonist therapy on the arterial wall in SHRs. Using the volume-pressure analysis of an in situ isolated carotid artery, they showed that isradipine, a dihydropyridine calcium antagonist, given daily for 12 weeks in 15-week-old SHRs significantly increased both systemic and carotid arterial compliance compared with SHR and WKY rats that were given placebo. This hemodynamic effect was paralleled by a significant decrease in medial thickness and increase in the elastin content of the arterial wall. Similar structural alterations were observed by Lacolley et al.[80] using in vivo echo-tracking technics in SHRs. Isradipine had a short-term effect on blood pressure, suggesting that observed structural changes may be in part independent of pressure reduction.

Clinical Data

Calcium antagonists can modify vascular capacitance by increasing the diameter and/or compliance of the vessels.

Effect of Calcium Antagonists on Arterial Diameter

The fall in mean arterial pressure produced by antihypertensive drugs may be associated with active changes in the geometry of large arteries, as demonstrated in healthy volunteers as well as hypertensive humans.[13,81,82]

In a double-blind, placebo-controlled study of healthy volunteers maintained on an ad-libitum sodium intake, acute administration in increasing doses of the calcium antagonist nicardipine augmented the diameter of the brachial and common carotid arteries in a dose-dependent manner.[81] No significant modification was observed with placebo. Systemic blood pressure did not change, suggesting that the calcium antagonist acted on the arterial wall independently of pressure-induced mechanical factors. When healthy volunteers were given verapamil, arterial diameter did not increase, but blood pressure significantly decreased also suggesting a direct action on the arterial wall.[82]

Similar observations about the brachial artery were made in patients with essential hypertension taking diltiazem.[13] For an equipotent antihypertensive effect, dihydralazine constricted the brachial artery diameter, whereas diltiazem caused an enhancement. No alteration in carotid artery diameter, despite a decrease in arterial pressure, was observed in hypertensive patients receiving verapamil vs. placebo during a 4-week period.[22]

In patients with sustained essential hypertension, acute oral administration of dihydropyridine derivatives such as nifedipine, nicardipine, and nitrendipine caused a significant increase in arterial diameter in the brachial artery[14–16,83] but not in carotid arteries.[14] When nicardipine was given orally, the increase in brachial artery diameter persisted for several weeks.[83] Because such findings were accompanied by a significant reduction of blood pressure, it appeared that the mechanical effects of blood pressure were offset by the vasodilating effect of the calcium antagonists. In a 24-week, double-blind, placebo-controlled study of nitrendipine in patients with end-stage renal disease, the active treatment was not associated with changes in aortic diameters despite a significant decrease in blood pressure.[23]

Several studies have shown that dihydropyridine derivatives significantly increase forearm and brachial blood flow; the increase was more pronounced in hypertensive than in normotensive subjects.[84,85] Because blood flow is directly proportional to blood flow velocity and the cross-sectional area of the artery, changes in forearm blood flow after calcium inhibition are related to changes of either or both factors. Because calcium antagonists induce an increase in blood flow, the increase in the diameter of large arteries may be related to the mechanism of high-flow dilation. At the site of the brachial artery, this possibility seems likely with dihydropyridines[15,16,22] but less plausible with diltiazem, which does not significantly increase flow velocity.[13] At the site of the common carotid artery, the calcium antagonists nitrendipine and verapamil did not modify arterial diameter, although blood flow velocity markedly increased.[16] Such findings illustrate the heterogeneity of response according to type of calcium antagonist and characteristics of different large arteries. The possible role of high-flow dilation in arterial diameter alterations after calcium blockade is less important than converting enzyme inhibition.[86] Endothelial factors play a major role in the high-flow dilatation. Experimental studies in SHRs have shown that acute calcium blockade increases the arterial compliance independently of endothelium integrity,[78] whereas an opposite result was observed with converting enzyme inhibition.[86]

Effect of Antagonists on Arterial Compliance

Short- and long-term administration of dihydropyridine derivatives to patients with essential and secondary hypertension caused a significant increase in arterial compliance, in both systemic and brachial circulations.[14–16,82] However, increased arterial compliance may result from several mechanisms and interactions.[3,67,68] First, blood pressure reduction may favor compliance enhancement through a lower stretch on the arterial wall (see Figs. 4 and 5). Second, the drug effects on arterial smooth muscle may favor relaxation of the arterial wall. Finally, a chronic decrease in blood pressure may induce a remodeling of the arterial wall, decreasing arterial hypertrophy, modifying elastin vs. collagen content, and contributing to increased compliance of the vessels, as observed in experimental studies.[79]

Therefore, to demonstrate that calcium antagonists affect arterial stiffness through their own actions, it is important to evaluate simultaneously the alterations of arterial pressure, arterial diameter, and arterial compliance. Such a study was performed in patients with essential hypertension and normotensive controls of the same age using three pharmacologic agents: cadralazine, a dihydralazine-like compound; nicorandil, a nicotinamide derivative; and the calcium antagonist nitrendipine.[87] In comparison with the two other antihypertensive agents, nitrendipine induced changes in arterial distensibility that could not be due entirely to pressure changes and were mediated either by geometric modifications, a relaxing effect of the drug on arterial smooth muscle, or a combination of both factors.

Because nitrendipine improved the viscoelastic properties of large arteries independently of geometric and blood pressure modifications, it was important to evaluate the degree of compliance enhancement compared with that produced by other antihypertensive agents. Concomitantly, three groups of patients with sustained essential hypertension of the same age, sex, and mean arterial pressure were studied before and after acute vasodilatation caused by nitrendipine; medroxalol, an alpha- and beta-blocking agent; or isosorbide dinitrate.[88] Each drug was given in a dosage that would cause a similar significant decrease in mean blood pressure. The three drugs also caused a significant similar increase in brachial artery diameter. Because the three drugs produced comparable mechanical and geometric changes, the observed modifications in arterial elastic modulus reflected changes in the tone of smooth muscle cells of the arterial wall. The calcium blockade caused by nitrendipine improved the arterial elasticity to a more significant extent than autonomic blockade or administration of nitrates.[88] Similar results were observed with the calcium antagonist isradipine in comparison with the beta blocker, metoprolol: isradipine increased operating as well as isobaric compliance, whereas metoprolol produced the opposite effect.[89]

In agreement with observations made during acute and short-term administration of calcium antagonists, long-term treatment also increases the distensibility of large arteries, including the aorta. In a double-blind, placebo-controlled study of hypertensive patients with end-stage renal disease, nitrendipine, given for a period of 24 weeks, effectively lowered blood pressure, and induced a significant decrease in aortic and femoral pulse wave velocity.[23] The decrease in aortic pulse wave velocity was not related to alteration in aortic diameter, which remained unchanged. During the first 8–16 weeks of treatment, the decrease in aortic pulse wave velocity was parallel to the decrease in blood pressure. The decrease of aortic pulse wave velocity continued until the end of the therapeutic trial at the 24th week, but late changes in aortic distensibility were no longer related to antihypertensive effect, which remained constant. These results suggest that the long-term improvement of aortic distensibility should be related to changes in elastic modulus, wall thickness, or both. The antihypertensive effect of nitrendipine and the improvement of aortic distensibility were more marked in patients with pronounced arterial calcinosis[90] and patients with more pronounced salt and water retention (Fig. 6).[23] In another study of patients undergoing chronic hemodialysis, long-term administration of nifedipine decreased the time-related increase in PWV, suggesting a decrease in progression of arteriosclerosis.[91] Similar findings were observed in patients with essential hypertension.[92]

Besides the increase in peripheral resistance and arterial stiffness, the pressure overload in hypertensive patients is characterized by an increase in and altered timing of arterial wave reflections. Antihypertensive drugs in general have desirable effects on these hemodynamic abnormalities through reduction of blood pressure, but not all antihypertensive treatments exert the same effect on hemodynamic factors of pressure overload. Antihypertensive agents such

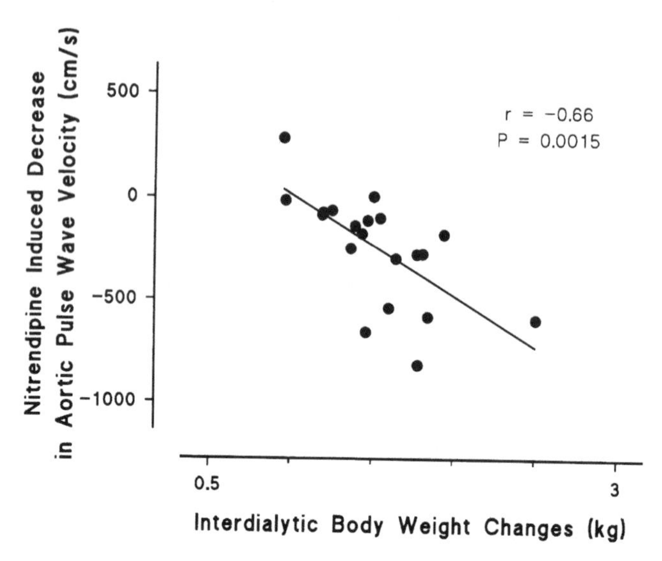

FIGURE 6. Correlation between the effect of nitrendipine on aortic pulse wave velocity and body fluid volume changes in end-stage renal disease (personal data from reference 23).

as angiotensin-converting enzyme inhibitors and calcium antagonists decrease the stiffness of conduit arteries in parallel with dilation of peripheral arteries,[23,93,94] thereby reducing wave reflections.[94–96] The effect on wave reflections is not observed with all antihypertensive drugs.[95–98] Beta-blocking drugs increase the intensity of wave reflections and, by increasing duration of left ventricular ejection (i.e., prolonging the interval available for reflected waves to merge with incident waves) amplify the effect of reflected waves in the aorta and central arteries. For these reasons beta blockers have less effect on aortic systolic blood pressure than on brachial systolic blood pressure.[95,97] The best therapeutic effect is observed with drugs that dilate small conduit arteries or both small arteries and arterioles (such as calcium antagonists and angiotensin-converting enzyme inhibitors). These drugs not only decrease mean blood pressure and reduce PWV but also reduce the reflection coefficient and amount of reflected pressure.[94–97] This effect leads to a larger reduction of aortic than brachial SBP and may account for their superiority in regression of left ventricular hypertrophy.[94]

CONCLUSION

On the whole, the results reported above show that calcium antagonists have a pronounced effect on both conduit arteries and arterioles. Such dual action is responsible for improvement of the conduit as well as the cushioning function of arteries. The changes in conduit function result from a decrease in peripheral resistance and mean blood pressure and an increase in blood flow to the tissues. The improvement in cushioning function with a decrease in pulsatile components of blood pressure results from an increase in arterial wave reflections. The latter effect explains the reduction of systolic and blood pressure in peripheral arteries and, more importantly, in the aorta and central arteries. The consequence is a decrease in left ventricular systolic stress and in the cyclic stress imposed on the arterial walls. These effects on large arteries explain the beneficial effect of calcium antagonists on regression of left ventricular hypertrophy and improvement of cardiac function in elderly patients.[99–101] Large peripheral arteries may be a preferential site of action for calcium blockers, causing an increase in arterial compliance and/or diameter along with blood pressure reduction in patients treated for sustained essential or secondary hypertension. Because increased arterial stiffness and altered wave reflections are the primary mechanism of systolic hypertension, we suggest that the arterial changes brought about by calcium entry blockers could be the major factor contributing to the decrease in cardiovascular morbidity and mortality, as observed in the elderly.[102]

References

1. Kannel WB, Stokes J III: Hypertension as a cardiovascular risk factor. In Robertson JIS (ed): Handbook of Hypertension Epidemiology of Hypertension, vol. 6. New York, Elsevier Science, 1985, pp 15–34.
2. Tverdal A: Systolic and diastolic blood pressure as predictor of coronary heart disease in middle-aged Norwegian men. BMJ 294:671–673, 1987.
3. O'Rourke MF: Vascular impedance: The relationship between pressure and flow. In Arterial Function in Health and Disease. Edinburgh, Churchill Livingston, 1982.
4. Kannel WB, Gordon T, Schwartz MJ: Systolic versus diastolic blood pressure and risk of coronary heart disease: The Framingham Study. Am J Cardiol 27:335–346, 1971.
5. Curb JD, Borhani NO, Entwisle G, et al: Isolated systolic hypertension in 14 communities. Am J Epidemiol 121:362–370, 1985.
6. Dyer AR, Stamler J, Shekelle RB, et al: Pulse Pressure. III: Prognostic significance in four Chicago epidemiologic studies. J Chron Dis 35:283–294, 1985.
7. Rutan GH, Kuller LH, Neaton JD, et al: Mortality associated with diastolic hypertension and isolated systolic hypertension among men screened for the Multiple Risk Factor Intervention Trial. Circulation 77:504–514, 1988.
8. Darne B, Girerd X, Safar M, et al: Pulsatile versus steady component of blood pressure: A cross-sectional and a prospective analysis on cardiovascular mortality. Hypertension 13:392–400, 1989.
9. Madhavan S, Ooi WL, Cohen H, Alderman MH: Relation of pulse pressure and blood pressure reduction to the incidence of myocardial infarction. Hypertension 23:395–401, 1994.
10. Scuteri A, Cacciafesta M, Di Bernardo MG, et al: Pulsatile versus steady-state component of blood pressure in elderly females: An independent risk factor for cardiovascular disease? J Hypertens 13:185–191, 1995.
11. Nichols WW, O'Rourke MF: Vascular impedance. In McDonald's Blood Flow in Arteries: Theoretic, Experimental and Clinical Principles, 3rd ed. London, Edward Arnold, 1991.
12. O'Rourke MF, Avolio AP: Pulsatile flow and pressure in human systemic arteries: Studies in man and in multibranched model of the human systemic arterial tree. Circ Res 46:363–372, 1980.
13. Safar ME, Simon AC, Levenson JA, Cazor JL: Hemodynamic effect of diltiazem in hypertension. Circ Res 52(Suppl 1):169–173, 1983.
14. Levenson JA, Safar ME, Simon AC, et al: Systemic and arterial hemodynamic effect of nifedipine (20 mg) in mild-to-moderate hypertension. Hypertension 5(Suppl V):57–60, 1983.
15. Levenson JA, Simon AC, Safar ME, et al: Large arteries in hypertension: Acute effect of a new calcium entry blocker, nitrendipine. J Cardiovasc Pharmacol 6(Suppl 7):1006–1010, 1984.
16. Bouthier JD, Safar ME, Benetos A, et al: Haemodynamic effects of vasodilating drugs on the common carotid and brachial circulations of patients with essential hypertension. Br J Clin Pharmacol 21:137–142, 1986.
17. Yaginuma T, Avolio A, O'rourke MF, et al: Effects of glyceryl trinitrate on peripheral arteries alters left ventricular load in man. Cardiovasc Res 20:153–160, 1986.
18. Safar ME, Laurent SL, Bouthier JD, et al: Effect of converting enzyme inhibitors on hypertensive large arteries in human. J Hypertens 4(Suppl 5):285–289, 1986.
19. Smulyan H, Mookerherjee S, Warner RA: The effect of nitroglycerin on forearm arterial distensibility. Circulation 73:1264–1269, 1986.

20. Simon AC, Levenson JA, Levy BI, et al: Effect of nitroglycerin on peripheral large arteries in hypertension. Br J Clin Pharmacol 14:241–246, 1982.
21. Asmar RG, Pannier B, Santoni JP, et al: Reversion of cardiac hypertrophy and reduced arterial compliance after converting enzyme inhibition in essential hypertension. Circulation 78:941–950, 1988.
22. Van Merode T, Van Bortel L, Smeets FA, et al: The effect of verapamil on carotid artery distensibility and cross-sectional compliance in hypertensive patients. J Cardiovasc Pharmacol 15:109–103, 1990.
23. London GM, Marchais SJ, Guerin AP, et al: Salt and water retention and calcium blockade in uremia. Circulation 82:105–113, 1990.
24. Fleckenstein A, Frey M, Fleckenstein-Grn G: Protection by calcium antagonists against experimental arterial calcinosis. In Pyrl K, Rapaport E, Knig K, et al: Secondary Prevention of Coronary Heart Disease. Stuttgart, Georg Thieme, 1983.
25. Dramsch DM: Calcium antagonists and atherosclerosis. In Kritchevsky D, Holmes WL, Paoletti R (eds): Drugs Affecting Lipid Metabolism, vol. 8. London, Plenum, 1985.
26. Kramsch DM, Aspen AJ, Rozler LJ: Atherosclerosis: Prevention by agents not affecting abnormal level of blood lipids. Science 213:1511–1512, 1981.
27. O'Rourke MF: Arterial stiffness, systolic blood pressure and logical treatment of hypertension. Hypertension 15:339–347, 1990.
28. Marchais SJ, Guerin AP, Pannier BM, et al: Wave reflections and cardiac hypertrophy in chronic uremia: Influence of body size. Hypertension 22:876–883, 1993.
29. Bouthier JD, De Luca N, Safar ME, Simon AC: Cardiac hypertrophy and arterial distensibility in essential hypertension. Am Heart J 109:1345–1352, 1985.
30. Gow BS: Circulatory correlates: Vascular impedance, resistance and capacity. In Bohr DF, Somlyo AP, Sparks HV (eds): The Cardiovascular System, vol. 2. Bethesda, MD, American Physiological Society, 1980, pp 353–408.
31. Dobrin PB: Vascular mechanics. In Shepherd JT, Abboud FM (eds): Handbook of Physiology. II. The Cardiovascular System, vol. 3. Bethesda, American Physiological Society, 1983, pp 65–102.
32. Dobrin PB, Rovick AA: Influence of vascular smooth muscle on contractile mechanics and elasticity of arteries. Am J Physiol 217:1644–1652, 1969.
33. O'Rourke MF, Avolio AP: Structural basis for increased distensibility of systemic muscular arteries with arterial vasodilator agents. J Molec Cell Cardiol 18(Suppl 2): 374, 1986.
34. Pohl U, Holtz J, Busse R, Bassenge E: Crucial role of endothelium in the vasodilator response to increased flow in vivo. Hypertension 8:37–44, 1986.
35. Safar M, Peronneau J, Levenson J, Simon A: Pulsed Doppler: Diameter, velocity and flow of brachial artery in sustained essential hypertension. Circulation 63:393–400, 1981.
36. Hoeks APG, Brands PJ, Smeets FAM, Reneman RS: Assessment of distensibility of superficial arteries. Ultrasound Med Biol 16:121–128, 1990.
37. Tardy Y, Meister JJ, Perret F, et al: Noninvasive estimate of mechanical properties of peripheral arteries from ultrasonic and photoplethysmographic measurements. Clin Phys Physiol Measure 12:39–54, 1991.
38. Bonithon-Kopp C, Ducimetière P, Touboul JP, et al: Plasma converting-enzyme activity and carotid wall thickening. Circulation 89:952–954, 1994.
39. Kelly R, Hayward C, Avolio A, O'Rourke M: Noninvasive determination of age-related changes in the human arterial pulse. Circulation 80:1652–1659, 1989.
40. O'Rourke MF: Appendix. In O'Rourke M, Safar M, Dzau V (eds): Arterial Vasodilation: Mechanisms and Therapy. London, Edward Arnold, 1993.
41. Avolio AO, Chen SG, Wang RP, et al: Effects of aging on changing arterial compliance and left ventricular load in a northern Chinese urban community. Circulation 68:50–58, 1983.
42. Bramwell JV, Hill AV: Velocity of transmission of the pulse wave and elasticity of arteries. Lancet 1:892–892, 1922.
43. Mitchell JRA, Schwartz CJ: Arterial Disease. Oxford, Bramwell, 1965.
44. Wolinsky H: Long term effects of hypertension on rat aortic wall and their relation to concurrent aging changes: Morpho logical and chemical studies. Circ Res 30:301–309, 1972.
45. O'Rourke MF, Avolio AP, Lauren PD, Yong J: Age related changes of elastic lamellae in the human thoracic aorta. J Am Coll Cardiol 9:53A, 1987.
46. Blumenthal HT, Lansing AI, Wheeler PA: Calcification of human aorta and its relation to intimal atherosclerosis, aging and disease. Am J Pathol 20:665–687, 1944.
47. Fleckenstein A: Calcium antagonism: History and prospect for a multifaceted pharmacodynamic principle. In Opie LH (ed): Calcium Antagonists and Cardiovascular Disease. New York, Raven Press, 1984, pp 9–28.

48. London GM, Marchais SJ, Safar ME, et al: Aortic and large artery compliance in end-stage renal failure. Kidney Int 37:137–142, 1990.
49. Levy BI, Michel J-B, Salzmann J-L, et al: Effects of chronic inhibition of converting enzyme on mechanical and structural properties of arteries in rat renovascular hypertension. Circ Res 63:227–239, 1988.
50. Levy BI, Michel J-B, Salzmann J-L, et al: Remodeling of heart and arteries by chronic converting enzyme inhibition in spontaneously hypertensive rats. Am J Hypertens 4:240S–245S, 1991.
51. Laurent S, Lacolley P, London G, Safar M: Hemodynamics of the carotid artery after vasodilation in essential hypertension. Hypertension 11:134–140, 1988.
52. London GM, Guerin AP, Marchais SJ, et al: Cardiac and arterial interactions in end-stage renal disease. Kidney Int 50:600–608, 1996.
53. Isnard RN, Pannier BM, Laurent S, et al: Pulsatile diameter and elastic modulus of the aortic arch in essential hypertension: A noninvasive study. J Am Coll Cardiol 13:339–406, 1989.
54. Hugue CJ, Safar ME, Aleferakis MC, et al: The ratio between ankle and brachial systolic pressure in patients with sustained uncomplicated essential hypertension. Clin Sci 74:179–182, 1988.
55. Roman MJ, Saba PS, Pini R, et al: Parallel cardiac and vascular adaptation in hypertension. Circulation 86:1909–1918, 1992.
56. Boutouyrie P, Laurent S, Girerd X, et al: Common carotid artery stiffness and patterns of left ventricular hypertrophy in hypertensive patients. Hypertension 25(Part 1):651–569, 1995.
57. Benetos A, Laurent S, Hoeks AP, et al: Arterial alterations with aging and high blood pressure: A noninvasive study of carotid and femoral arteries. Arterioscler Thromb 13:90–97, 1993.
58. Murgo JP, Westerhof N, Giolma JP, Altobelli SA: Aortic input impedance in normal man: Relationship to pressure wave forms. Circulation 62:105–116, 1980.
59. Latham RD, Westerhof N, Sipkema P, et al: Regional wave travel and reflection along the human aorta: A study with six simultaneous micromanometric pressures. Circulation 72:1257–1269, 1985.
60. Safar ME, Simon AC: Hemodynamics in systolic hypertension. In Zanchetti A, Tarazi RC (eds): Pathophysiology of Hypertension, Cardiovascular Aspects, vol. 7. Amsterdam, Elsevier, 1986, pp 225–241, 1986.
61. Safar ME, Laurent S, Pannier BM, London GM: Structural and functional modifications of peripheral large arteries in hypertensive patients. J Clin Hypertens 3:360–367, 1987.
62. Hayoz D, Rutschmann B, Perret F, et al: Conduit artery compliance and distensibility are not necessarily reduced in hypertension. Hypertension 20:1–6, 1992.
63. Laurent S, Caviezel B, Beck I, et al: Carotid artery distensibility and distending pressure in hypertensive humans. Hypertension 23(Part 2):878–883, 1994.
64. London GM, Guerin AP, Pannier B, et al: Increased systolic pressure in chronic uremia: Role of arterial wave reflections. Hypertension 20:10–19, 1992.
65. Ibels LS, Alfrey AL, Huffer WE, et al: Arterial calcification and pathology in uremic patients undergoing dialysis. Am J Med 66:790–796, 1979.
66. Marchais S, Guerin A, Safar M, London G: Arterial compliance in uremia. J Hypertens 7(Suppl 6):84–85, 1989.
67. Milnor WR: Hemodynamics. Baltimore, Williams & Wilkins, 1982, pp 56–96.
68. Safar ME, London GM: Arterial and venous compliance in sustained essential hypertension. Hypertension 10:133–139, 1987.
69. Bevan JA, Bevan RD, Hwa JJ, et al: Calcium regulation in vascular smooth muscle: Is there a pattern to its variability within the arterial tree? J Cardiovasc Pharmacol 8 (Suppl 8):S71–S75, 1986.
70. Millard RW, Lathrop DA, Grupp G, et al: Differential cardiovascular effects of calcium channel blocking agents: Potential mechanisms. Am J Cardiol 49:499–507, 1982.
71. Nakaya H, Schwartz A, Millard RW: Reflex chronotropic and inotropic effects of calcium channel-blocking agents in conscious dogs: Diltiazem, verapamil, and nifedipine compared. Circ Res 52:302–308, 1983.
72. McLeay RAB, Stallard TJ, Watson RDS, Littler WA: The effect of nifedipine on arterial and reflex cardiac control. Circulation 67:1084–1091, 1983.
73. Young MA, Watson RDS, Littler WA: Baroreflex setting and sensitivity after acute and chronic nicardipine therapy. Clin Sci 66:233–240, 1984.
74. Ferguson DW, Dorsey JK: Effects of nifedipine on baroreflex modulation of vascular resistance in man. Am Heart J 109:55–62, 1985.
75. Ferguson DW: Influence of nifedipine on arterial baroreflex modulation of heart rate control during dynamic increase in arterial pressure: Studies in man. Am Heart J 114:773–781, 1987.
76. Yano M, Kumada T, Matsuzaki M, et al: Effect of diltiazem on aortic pressure-diameter relationship in dogs. Am J Physiol 256:H1580–H1587, 1989.

77. Levy BI, Michel JB, Salzmann JL, Poitevin P, et al: Effects of chronic inhibition of converting enzyme on mechanical and structural properties of arteries in rat renovascular hypertension. Circ Res 63:227–239, 1988.
78. Levy BI, El Fertak L, Pieddeloup C, et al: Role of endothelium in the mechanical response of the carotid arterial wall to calcium blockade in SHR and WKY rat. J Hypertens 11:57–63, 1993.
79. Levy BI, Duriez M, Phillipe M, et al: Effect of chronic dihydropyridine (isradipine) on the large arterial walls of spontaneously hypertensive rats. Circulation 90:3024–3033, 1994.
80. Lacolley P, Ghodsi N, Glazer E, et al: Influence of graded changes in vasomotor tone on the carotid arterial mechanics in live spontaneously hypertensive rats. Br J Pharmacol 115:1235–1244, 1995.
81. Thuillez C, Gueret M, Duhaze P, et al: Nicardipine: Pharmacokinetics and effects on carotid and brachial blood flows in normal volunteers. Br J Clin Pharmacol 18:837–847, 1984.
82. Thuillez C, Duhaze P, Fournier C, et al: Arterial and venous effects of verapamil in normal volunteers. Fund Clin Pharmacol 1:35–44, 1987.
83. Levenson J, Simon A, Bouthier J, et al: The effect of acute and chronic nicardipine therapy on forearm arterial hemodynamics in essential hypertension. Br J Clin Pharmacol 20:107–113, 1985.
84. Robinson BF, Collier JG, Dobbs RJ: Comparative dilator effect of verapamil and sodium nitroprussiate in forearm vascular bed and dorsal hand veins in man: Functional differences between vascular smooth muscles in arterioles and veins. Cardiovasc Res 13:16–21, 1979.
85. Hulthen UL, Bolli P, Amann FW, et al: Enhanced vasodilatation in essential hypertension by calcium channel blockade with verapamil. Hypertension 4 (Suppl II):26–31, 1982.
86. Levy BI, Benessiano J, Poitevin P, Safar ME: Endothelium-dependent mechanical properties of the carotid artery in WKY and SHR: Role of angiotensin-converting enzyme inhibition. Circ Res 66:321–328, 1990.
87. Safar ME, London GM, Bouthier JA, et al: Brachial artery cross-sectional area and distensibility before and after vasodilatation in men with sustained essential hypertension. J Cardiovasc Pharmacol 9:734–742, 1987.
88. Safar ME, London GM, Asmar RG, et al: An indirect approach for the study of the elastic modulus of the brachial artery in patients with essential hypertension. Cardiovasc Res 20:563–567, 1986.
89. Merli I, Simon A, Del Pino M, et al: Intrinsic effect of antihypertensive treatment with isradipine and metoprolol on large artery geometric and elastic properties. Clin Pharmacol Ther 54:76–83, 1993.
90. Marchais SJ, Boussac I, Guerin AP, et al: Arteriosclerosis and antihypertensive response to calcium antagonists in end-stage renal failure. J Cardiovasc Pharmacol 18(Suppl 5):14–18, 1991.
91. Saito Y, Shirai K, Uchino J, et al: Effect of nifedipine administration on pulse wave velocity (PWV) of chronic hemodialysis patients—2-years trial. Cardiovasc Drug Ther 4:987–990, 1990.
92. Asmar R, Benetos A, Brahimi M, et al: Arterial and antihypertensive effects of nitrendipine: A double-blind comparison versus placebo. J Cardiovasc Pharmacol 20:858–863, 1992.
93. Asmar RG, Pannier B, Santoni JP, et al: Reversion of cardiac hypertrophy and reduced arterial compliance after converting enzyme inhibition in essential hypertension. Circulation 78:941–950, 1988.
94. London GM, Pannier B, Guerin AP, et al: Cardiac hypertrophy, aortic compliance, peripheral resistance, and wave reflections in end-stage renal disease: Comparative effects of ACE inhibition and calcium channel blockade. Circulation 90:2786–2796, 1994.
95. Guerin AP, Pannier BM, Marchais SJ, et al: Effects of antihypertensive agents on carotid pulse contour in humans. J Hum Hypertens 6(Suppl 2):S37–S40, 1992.
96. Pannier BM, Lafleche AB, Girerd XJM, et al: Arterial stiffness and wave reflections following acute calcium blockade in essential hypertension. Am J Hypertens 7:168–176, 1994.
97. Ting CT, Chen JW, Chang MS, Yin FCP: Arterial hemodynamics in human hypertension: Effect of the calcium channel antagonist nifedipine. Hypertension 25:1326–1332, 1995.
98. Kelly R, Daley J, Avolio A, O'Rourke M: Arterial dilation and reduced wave reflection: Benefit of dilevalol in hypertension. Hypertension 14:14–21, 1989.
99. Schulman SP, Weiss JL, Becker LC, et al: The effects of antihypertensive therapy on left ventricular mass in elderly patients. N Engl J Med 322:1350–1356, 1990.
100. SHEP cooperative research group: Prevention of stroke by antihypertensive drug treatment in older persons with isolated systolic hypertension: Final results of the Systolic Hypertension in the Elderly Program (SHEP). JAMA 265:3255–3264, 1991.
101. Zusman RM, Christensen DM, Federman EB, et al: Nifedipine but not propranolol improves left ventricular systolic and diastolic function in patients with hypertension. Am J Cardiol 64:51–61, 1989.
102. Staessen J: The systolic hypertension in Europe trial (SYST-EUR): Principal results. Presented at the meeting of European Society of Hypertension, Milan, June 13–16, 1997. J Hypertens 15(Suppl 4), 1997.

DALILA B. CORRY M.D. / MICHAEL L. TUCK M.D.

13

Insulin and Glucoregulatory Hormones: Implications for Antihypertensive Therapy

INSULIN AND GLUCOREGULATORY HORMONES IN ESSENTIAL HYPERTENSION

Essential hypertension (EH) has a high association with metabolic abnormalities such as insulin resistance (IR), elevated insulin levels, impaired glucose tolerance, elevated triglycerides, and decreased high-density lipoprotein (HDL). Multiple clinical and cross-sectional epidemiologic studies have reported an independent positive correlation between plasma insulin and blood pressure.[1–20] Some studies either have not found this relationship or have noted its dependency on other factors such as obesity.[16] These studies are confounded by many variables, such as gender, age, ethnic background, and adjustment for other risk factors.[16] In addition, insulin and IR have been proposed as possible factors in the pathogenesis of EH. Several cardiovascular, renal, and nervous system actions of insulin may contribute to blood pressure regulation.

The list of risk factors that may be components of the insulin resistance syndrome continues to grow. The relation of insulin and IR to coronary heart disease (CHD) has been recently reviewed by Wingard et al.,[21] who noted that some studies show a positive relationship between insulin and CHD, especially in men, although the data are inconsistent. The Insulin Resistance Atherosclerosis Study,[22] one of the first studies to measure IR directly in a large population by using the minimal model technique, reports in a cohort analysis that IR correlates with carotid intimal medial wall thickness. These data suggest that accelerated atherosclerosis may be part of the insulin resistance syndrome. Left ventricular hypertrophy, an independent risk factor in EH, is associated with IR.[23] Elevated levels of plasminogen-activating inhibitor-1 (PAI-1) are found in people with IR.[24] Insulin stimulates PAI-1, which regulates fibrinolytic activity and promotes thrombus formation. However, the high oxidative stress found in EH does not relate to insulin.[25] Insulin resistance is also related to microalbuminuria, which may be predictive of CHD.[26]

One of the more interesting associations of the insulin resistance syndrome and EH is the role of intestinal hormones.[27] Both normotensive people with a family history of EH and people with established hypertension have increased insulin and gastrointestinal inhibitory polypeptide (GIP) responses to a mixed meal.[27] GIP is one of the strongest insulinotropic gut factors. In addition, a decline in growth hormone and insulin-like growth factor is associated with the age-related decline in insulin sensitivity and is often accompanied by a decline in muscle mass, an increase in central fat mass, and an increased risk for cardiovascular events.[28]

IR, defined as reduced insulin-mediated glucose uptake, is found in 20–40% of people with EH. It has been noted in both obese and nonobese people who do not have diabetes mellitus. Thus, most definitions of IR are confined to its effects on glucose disposal and do not address insulin action on cardiovascular, renal, and other tissues. In a recent report, IR was directly measured in cardiac muscle in patients with CHD. Positron emission tomography was used to measure cardiac blood flow during an insulin clamp.[29] In patients with CHD, the degree of IR in cardiac muscle was similar to the degree in skeletal muscle as measured by the insulin clamp. This study implies that IR may occur in different tissues, including skeletal muscle and heart. In fact, further evidence indicates that resistance to the vasodilator effect of insulin may be found in obese people and in patients with non–insulin-dependent diabetes mellitus (NIDDM).[30] This evidence supports the presence of insulin resistance in vascular tissue.

Several questions remain about the relationship of IR to EH, including the role of insulin as a predictor of EH and the causal role of insulin in elevation of blood pressure. Not all people with EH have metabolic abnormalities, and the exact incidence of IR in EH is not well established. In a study of 420 patients with EH, Lind et al.[31] noted that about 25% had IR. The San Antonio Heart Study also provided numerical data about the incidence of metabolic abnormalities in EH.[14,15] Among 2,930 Hispanic and non-Hispanic white subjects, hypertension was present in 287 (9.8%). Five metabolic disorders, including obesity, NIDDM, impaired glucose tolerance, and high serum cholesterol and triglycerides, were associated with EH. Almost three-fourths of subjects with EH were obese, most had higher insulin levels and over 20% had NIDDM or impaired glucose tolerance. Of interest, seven patients had all five metabolic conditions, whereas 44 subjects with EH (15%) were free of metabolic dysfunction.

FACTORS DETERMINING INSULIN RESISTANCE IN ESSENTIAL HYPERTENSION

The factors that determine insulin resistance in people with EH include body weight, heredity, ethnicity, gender, age, and sodium intake.

Body Weight

Body weight is probably the most important variable determining the association of IR with EH. A substantial number of people with EH are overweight. In addition, obesity with or without hypertension is almost always associated with insulin resistance and hyperinsulinemia. One explanation for inconsistent findings in studies of the relationship between insulin levels and blood pressure is that obesity acts as the major confounder. Although early reports found IR to be independent of body weight in people with EH,[3,4] others showed the opposite.[32,33] Insulin responses to oral glucose are more accentuated in obese than in lean people with EH,[35] and the accentuation progresses in a stepwise fashion from lean to obese to obese diabetic.[34]

In the Paris Prospective Study,[36] mean glucose concentrations were higher in hypertensive men at all levels of body mass index, whereas hyperinsulinemia was noted only in the more obese men. Conversely, some studies find higher insulin levels in lean rather than obese people with EH. In the San Antonio Heart Study, fasting insulin levels were a better predictor of the 8-year incidence of hypertension in lean than in obese subjects.[14,15] Mykkanen et al.[37] found that high fasting insulin concentration and high diastolic blood pressure were more pronounced in lean than in obese subjects.

A possible explanation of this discrepancy is that different mechanisms may control blood pressure in lean vs. obese people. Long-term insulin exposure in lean people with EH may activate sympathetic nervous system effects or structural changes that supersede the role of insulin in blood pressure control. Fat distribution is also an important determinant of the relationship between insulin and blood pressure. People with central or abdominal obesity have a much higher incidence of hypertension and hyperinsulinemia than people with peripheral obesity.[38] Another form of IR—resistance to the capacity of insulin to suppress lipolysis—is seen predominantly in people with abdominal obesity and hypertension. The major signs are increased nonesterified free fatty acid flux[38] and elevations of the renin-angiotensin system. These mechanisms may be active in sustaining high blood pressure in people with central obesity.

Ikeda et al.[39] have recently shown that improvement in insulin sensitivity contributes to the reduction in blood pressure after weight loss in obese hypertensive subjects. In their study, insulin sensitivity improved by 46.2% and showed a distinct relation to the decrease in blood pressure. Reduction in body weight also reduces catecholamine levels[40] and plasma renin activity.[41] In another study, rapid weight loss due to a very low calorie diet produced an immediate depressor response and a decline in insulin that was highly correlated with blood pressure change.[42] Thus, weight loss studies indicate that the high insulin levels in obese hypertensive people contribute to blood pressure maintenance.

Heredity

The genetic influences on EH and IR are complex and do not follow classic Mendelian patterns. Most studies examining the genetic role of IR in EH have used normotensive offspring from families with a high incidence of hypertension (FH+). Ferrari et al.[43] noted IR and high insulin responses to glucose in young normotensive FH+ males. In a study of two generations of normotensive FH+ subjects, Widgren et al.[44] reported that waist-hip ratio was a key determinant of blood pressure and insulin levels. Neutel et al.[45] found that plasma insulin levels are higher in normotensive FH+ subjects and also predict other risk factors, such as higher levels of norepinephrine, renin, and cholesterol. Facchini et al.[46] noted greater insulin responses, IR, and higher triglyceride levels in 38 normotensive FH+ subjects. Ishibashi found higher insulin levels in 152 FH+ school-aged girls than in 131 girls with no family history of hypertension;[47] the higher insulin levels correlated most strongly with body weight. Beatty et al.[48] noted higher blood pressure and IR in FH+ offspring.

Ionic factors may be important. FH+ offspring with IR also have higher sodium-lithium countertransport activity in erythrocytes[49] and higher levels of calcium in platelets, a finding also common in EH.[50]

Wing et al.[51] reported that women with a family history of either diabetes or EH had higher insulin levels and that the highest values were found in women with a family history of both diseases. The Heureka Study of 11,001 participants reported that plasma insulin is correlated with blood pressure only in FH+ subjects.[52] Masuo et al.[53] noted increases in both IR and sympathetic nervous system activity in young normotensive Japanese who were FH+, suggesting a possible genetic link between IR and SNS activity in people with EH. In a study of 175 normotensive and hypertensive subjects grouped by positive or negative family history of EH, both normotensive and and hypertensive FH+ subjects had higher levels of insulin and C-peptide and a lower glucose/insulin ratio.[54] The study concluded that IR has a familial basis independent of hypertension and that both IR and EH are components of a common familial pattern. A study of insulin resistance in Mexico City concluded that the insulin resistance in FH+ offspring was more closely related to obesity than family history.[55] In general, family studies offer strong evidence that in young normotensive subjects with an increased familial risk of EH, insulin resistance precedes the rise in blood pressure.

Ethnicity

Ethnic background is another strong determinant of how IR relates to blood pressure. Studies recording the strongest correlation were carried out mainly in Caucasians. In general, studies relating IR to blood pressure in African-Americans,[56–58] Mexican-Americans,[59] and Asian Indians[60,61] are less consistent. In Pima Indians, the incidence of obesity and metabolic dysfunction is extremely high, but there is no correlation between IR and blood pressure.[56] Thus, certain ethnic groups (Pima Indians, African-Americans, Hispanics) may have a much higher incidence of metabolic disorders but a much lower correlation of IR with blood pressure than Caucasians.

In smaller clinical studies using more precise methods (insulin clamp) to define IR, the relationship between IR and EH has been stronger. Falkner et al.[62,63] found significant IR in African-Americans with established and borderline EH.[51] Haffner et al.[64] and Buchanan et al.[65] found a significant correlation between IR and blood pressure in Mexican-American families. The relationship between IR and blood pressure in Asian populations appears to be similar to that in Caucasians. For example, in Japanese subjects with EH, IR is strongly correlated with blood pressure.[66–68] In a study of 491 normotensive Japanese men, Miura et al.[69] found that blood pressure was significantly and independently related to plasma insulin, and in an 11-year follow-up of normotensive Japanese subjects, Tsuruta et al.[70] found that hyperinsulinemia is a good predictor of hypertension. As most Japanese subjects were lean compared with other populations, obesity may not be as important a contributor to the insulin-blood pressure relationship in Japan. The correlation of insulin with blood pressure also has been well documented in Taiwan, but body weight is a more closely related factor. Ongoing studies indicate that the correlation between insulin and blood pressure seems to hold up for most other Southeastern Asian countries, although data are limited. In select Pacific Island settings, the relationship between insulin and blood pressure does not appear to be strong.[71]

In certain ethnic populations, migration may be an important determinant of a more positive relationship between metabolic and vascular variables. A study of Asian communities in the United Kingdom found that Asian men had a greater tendency to central obesity, IR, and lipid abnormalities than the more settled non-Asian groups.[61] Ethiopian immigrants to Israel have a three-fold increase in hypertension associated with an increase in the insulin index; the increased incidence is related to duration of residence in Israel.[72] Thus, environmental factors may bring out the expression of metabolic risk factors in ethnically and genetically distinct populations.

Gender

There may be gender differences in the relationship between IR and blood pressure. According to data from the San Antonio Heart Study, plasma insulin levels are higher in Hispanic women than in Hispanic men.[73] In contrast, insulin levels may be lower in Caucasian women than in Caucasian men.[74] African-American women at all blood pressure levels have higher insulin levels and a greater degree of IR than African-American men.[63] Gender differences may be due to less skeletal muscle mass, more adipose mass, and shorter height in women or to other variables such as androgen/estrogen balance.[75] The relationship between insulin and blood pressure may vary geographically; the European Fat Distribution Study showed differences among women from different countries in Europe.[76]

Age

Among healthy children, insulin levels may be correlated positively with both systolic and diastolic blood pressure. In one study, insulin levels in children predicted systolic blood pressure at 3- and 6-year follow-ups, even when blood pressure was within the normal range.[77] The Bogalusa study of 3313 children found positive correlations between fasting insulin and systolic and diastolic blood pressure, but the relationship weakened when adjusted for age and weight.[57] Because aging is associated with increasing IR, it has been difficult to separate the contributions of age, hypertension, and drug therapy to the etiology of IR in older populations. In the Rotterdam Study, IR was higher in patients with EH, but the increase in IR with age was much more apparent in normotensive than in hypertensive patients, resulting in similar levels of IR at older ages.[78] Subjects using antihypertensive drugs had higher insulin levels at all ages.

Sodium Intake

An interaction between IR and sodium intake has been noted. Several studies of patients whose sodium intake was restricted to 10–80 mmol/day have shown increased insulin levels, increased lipid levels, and reduced insulin sensitivity.[79–83] The mechanism for this effect is not clear, but activation of the sympathetic nervous system or increases in the renin-angiotensin system may be involved. Milder salt restriction (80 mmol/day) does not affect IR in young white males.[84] The salt sensitivity of blood pressure is also associated with changes in insulin sensitivity. Rocchini et al.[85] have shown that in obese adolescents, weight reduction, accompanied by decreases in insulin and norepinephrine, corrects abnormal salt-sensitive blood pressure. Mexican-Americans with salt sensitivity are more obese and have more pronounced IR.[65] Salt-loading may result in exaggerated insulin responses in people with salt-sensitive EH. Salt- sensitive subjects also have greater insulin responses during salt restriction.[82,86,87] In general, there appears to be a clustering of cardiovascular risk factors in salt-sensitive patients with EH.[88] It remains to be determined whether salt-sensitive blood pressure is a component of the insulin resistance syndrome.

MECHANISMS OF INSULIN RESISTANCE IN ESSENTIAL HYPERTENSION

Insulin resistance, which refers to reduced insulin-mediated glucose disposal, may be demonstrated by a rightward shift in the dose-response relation of insulin concentration to glucose disposal. To maintain a normal fasting glucose level, nondiabetic subjects with IR show a compensatory increase in insulin secretion and higher insulin levels. Rare causes of IR include insulin antibodies, insulin receptor mutations, and receptor auto-antibodies.[89,90] Because the receptors for insulin and insulin-like growth factor-1 (IGF-1) are more than 80% homologous, some of the findings in IR relate to IGF-1. The most common forms of insulin resistance, as seen in NIDDM, obesity, and EH, usually involve unknown defects in postreceptor signaling pathways, such as the tyrosine kinase cascade and the insulin receptor substrate (IRS)-1 and IRS-2 pathways.[91] The IRS proteins act as docking sites where kinases and phosphatases bind, leading to phosphorylation of other substrates that activate the glucose transport proteins.[63–65] Important cell pathways by which insulin stimulates glucose transport proteins include the PI3 kinase system and the MAP kinase cascade. Others believe that the postreceptor defect in IR may reside in the glucose transporter proteins.[92]

Laboratory diagnosis of IR in patients with EH is usually not attempted because no precise and simple methods of measurement are yet available. In clinical research laboratories, IR is measured directly by such procedures as the insulin suppression test, euglycemic hyperinsulinemic clamp,[93] and minimal model program using the intravenous glucose tolerance test.[94] The diagnostic tests that indirectly suggest IR in clinical settings include elevated fasting or postprandial insulin levels or exaggerated glucose and insulin responses to glucose tolerance testing. These tests are often used in population studies of IR in EH.

Although EH, like NIDDM and obesity, is associated with a decrease in whole body glucose uptake, the pattern of IR in EH is different. In EH, the reduced glucose disposal is almost exclusively through the nonoxidative or glycogen storage pathway.[3] The effects of insulin on hepatic glucose output appear not to be altered in EH, and some of the insulin elevation in EH may be due to reduced hepatic clearance.[95] Thus, IR in EH seems to be largely confined to skeletal muscle and may be due to reduced skeletal muscle blood flow or abnormal muscle fiber type. Slow- and fast-twitch fiber distribution is altered in hypertension with a relative reduction in slow-twitch fibers that have high oxidative, low glycolytic activity and are more insulin-sensitive.[95] In turn, insulin itself may alter skeletal muscle fiber composition.[97] Skeletal muscle blood flow may be an important determinant of insulin-mediated glucose uptake,[98,99] and reduced blood flow may contribute to IR in EH. The microcirculation around insulin-sensitive sites is diminished in EH and may limit insulin delivery in skeletal muscle.[98] One of the early pathologic changes in EH is capillary rarefaction in skeletal muscle. Increased blood flow to these areas may recruit new vessels, thereby exposing more insulin-sensitive sites. However, short- or long-term hypertension per se does not modify insulin action as studied in induced models of hypertension.[100]

MECHANISMS FOR REGULATION OF BLOOD PRESSURE BY INSULIN

Insulin has effects on the kidney, sympathetic nervous system, ion transport, and blood vessels that may contribute to the mechanisms whereby blood pressure is elevated in EH. In addition, the growth effects of insulin and IGF-1 also may alter blood vessel structure and initiate hypertension.

Renal Effects

Infusion of insulin in humans[101–104] produces an acute reduction in sodium excretion. Muscelli et al.[104] have shown that even in patients with EH and documented insulin resistance, the effects of insulin infusion on renal handling of sodium, potassium, and uric acid are normal. They conclude that there is no renal resistance to insulin action in patients with EH. In normotensive people with a family history of EH and IR, the infusion of insulin normally reduces fractional excretion of sodium.[50] In vitro studies show that insulin can act directly on sodium transport in the proximal convoluted tubule,[105] the distal tubule,[102] and the loop of Henle. Takahashi et al.[106] recently demonstrated that insulin directly increases resorption of sodium chloride in the medullary thick ascending limbs of Henle's loop. Octreotide, a somatostatin analog, reduces insulin and increases sodium excretion in normal subjects.[107] Elevated insulin may cause sodium retention in EH, but the long-term effects of insulin on sodium balance are less known. In dogs, chronic insulin infusion has no long-term effects on sodium excretion.[108] Insulin also may alter renal hemodynamics[109] and increase mesangial cell growth and volume, thus affecting protein excretion.[110] Several conditions associated with IR and hypertension, such as diabetes, obesity, EH, and aging, also share abnormal ion transport (erythrocyte sodium-lithium countertransport) and renal abnormalities (mesangial cell growth, microalbuminuria), suggesting a link between IR, blood pressure, and renal function.[111]

Sympathetic Nervous System Activity

Acute insulin infusion in normal subjects increases plasma norepinephrine (NE)[112,113] and activates regional sympathetic nervous system responses.[114,115]

Insulin infusion in patients with EH evokes exaggerated sympathetic responses and raises vasoconstrictor nerve activity in skeletal muscle.[116] Evidence also has been presented against the hypothesis that hyperinsulinemia alters SNS activity.[117] However, increased sympathetic activity theoretically may worsen insulin resistance in EH; in turn, the higher insulin levels would set up a cycle that further increases sympathetic activity and blood pressure.[113] In obesity-associated hypertension, high insulin levels correlate with both NE and blood pressure; all three are reduced by weight loss.[113] The Normative Aging Study of 752 men showed that insulin levels and sympathetic activity were correlated with blood pressure among middle-aged and older men.[118] Insulin levels and urinary NE excretion were independent predictors of hypertension. A population survey of 383 young normotensive Danes found that sympathetic activity was an important determinant of systolic blood pressure, whereas insulin levels and insulin sensitivity were minor determinants.[119] In animal models, the effect of insulin on the sympathetic nervous system is less clear. Rats show a dose-dependent increase in sympathetic activity[120], but in dogs long-term insulin does not alter catecholamine levels.[121]

Ion Transport

Insulin increases the membrane sodium-potassium (Na^+,K^+) ATPase pump activity when added in vitro to tissue systems.[122] It also increases mRNA encoding the alpha 2-subunit of the Na^+,K^+ ATPase pump in vascular cells.[123] Studies in vascular tissue show that part of the vasodilatory effect of insulin is mediated through the pump.[124] Insulin also increases the sodium-hydrogen (Na^+,H^+) antiporter,[125] which may alter blood pressure by increasing intracellular sodium and calcium ions. Aviv[126] suggests that abnormal Na^+,H^+ antiporter activity, protein kinase C, and cytosolic free Ca^{2+} interactions play a major role in linking IR to EH. In impaired glucose tolerance, erythrocyte sodium-lithium countertransport correlates with IR.[127] Alterations in ion transport have been described in diabetes and other states of insulin resistance.[128,129] Thus, in blood vessels insulin may raise blood pressure by causing accumulation of intracellular sodium and calcium ions. Insulin raises levels of intracellular calcium ions in isolated rat adipocytes, and high levels of intracellular calcium ions have been proposed as a mechanism for IR.[130]

Vascular Effects

During the past several years, it has been well documented that vascular smooth muscle and probably vascular endothelium are insulin-sensitive tissues similar to skeletal muscle and fat cells. Acute administration of insulin produces a vasodilator effect.[131–133] Infusion of insulin in normal subjects reduces peripheral arterial resistance with little change in blood pressure.[131] Of interest, infusion of insulin in subjects with IR produces the opposite effect—vasoconstriction.[134–136] Baron et al.[137–140] have demonstrated incremental leg blood flow and a fall in vascular resistance during graded insulin infusion. In obesity and NIDDM, the insulin dose-response curve is shifted to the right, suggesting resistance to insulin-mediated blood flow. Because L-NMMA, an inhibitor of nitric oxide (NO) production, completely prevents insulin-induced increase in leg blood flow, insulin-mediated vasodilatation appears to be affected by a significant NO-dependent mechanism.[141] Scherrer et al.[142] also demonstrated that NO release accounts for the action of insulin in humans. Because NO is derived from the endothelium, endothelial dysfunction may be a major factor in the insulin resistance syndrome. Studies have shown that skeletal muscle blood flow itself can modulate insulin-mediated glucose uptake.[116]

Studies of the in vitro effects of insulin have documented its vasodilator mechanisms in vascular tissue. Insulin attenuates vascular responses to angiotensin II and other pressor agents in isolated vascular segments in a dose-response manner.[143] In human studies, intraarterial infusion of insulin attenuates pressor-induced vasoconstriction and decreases venous tone in vessels preconstricted with phenylephrine.[134,144] The intracellular mechanisms for insulin suggest several direct actions on calcium ions in the vascular system. In cultured vascular smooth muscle cells (VSMCs), Saito et al.[145] reported a dose-dependent attenuation by insulin of angiotensin II-stimulated cytosolic Ca^{2+} by blockade of cytosolic Ca^{2+} mobilization from inositol triphosphate (IP3)-sensitive calcium stores.

In a series of studies, Kahn et al.[146] showed that insulin can simultaneously inhibit serotonin-induced contractions and cytosolic Ca^{2+} responses in vascular tissue. Verapamil and ouabain block these effects, indicating that insulin can modulate voltage-operated calcium channels and the Na^+,K^+ ATPase pump. Standly et al.[147] demonstrated that vasopressin-induced contractility and cytosolic Ca^{2+} responses are blocked by insulin secondary to a shift in voltage current in VSMCs. Touyz et al.[148] found that insulin can inhibit cytosolic Ca^{2+} responses to several pressor agents in VSMCs including angiotensin II, NE, and vasopressin. Thus, insulin has several different and direct mechanisms to control vascular smooth muscle tone, mostly through control of cytosolic Ca^{2+} responses. In addition, insulin appears to have major effects on vascular tone through the vascular endothelium. There is specific insulin binding to endothelial cells,[149] insulin can increase NO production,[150] and blockade of NO attenuates the vascular actions of insulin.[141,142]

Sowers[151] recently reviewed the evidence comparing the effects of insulin and IGF-1 on vascular smooth muscle glucose and cation metabolism. IGF-1 simulates many of the same actions as insulin, including glucose uptake, attenuation of cytosolic calcium, and vasoconstrictive responses. Both insulin and IGF-1 stimulate NO production from the endothelium and vascular smooth muscle cells. The fact that only IGF-1 is synthesized by vascular smooth muscle cells makes it of growing interest as a physiologic regulator. Insulin and IGF-1 also have many direct effects on blood vessels that promote tissue growth. Such effects may contribute to hypertension through structural changes as well as promote atherosclerosis. Insulin at low doses causes proliferation of vascular smooth muscle cells in culture[152] and augments the action of other growth factors, such as platelet-derived growth factor.[153] In animal models of diabetes, the rate of atherosclerosis is higher in animals treated with insulin than in untreated animals.[154] Insulin stimulates cholesterol synthesis in vascular tissue and alters LDL receptor activity to enhance cholesterol transport in vascular cells.

IMPLICATIONS FOR ANTIHYPERTENSIVE THERAPY

Based on substantial information about the metabolic risks associated with EH, a careful reexamination of the effects of different classes of antihypertensive agents has been carried out. Many diuretics and beta blockers increase circulating levels of glucose and insulin, worsen IR, and alter lipid profiles, whereas angiotensin-converting enzyme (ACE) inhibitors, calcium channel blockers, and $alpha_1$ blockers have neutral to beneficial effects. In addition, within classes of drugs, the metabolic effects are not always uniform and may depend on the properties of the specific agent. Because treatment of EH should strive for reduction not just of blood pressure but of total risk to the patient, one must be familiar with the metabolic properties of all antihypertensive agents. Patients with EH are at increased risk of developing NIDDM and lipid abnormalities, and treated patients have a higher rate of glucose intolerance and diabetes than untreated subjects.[155–158] Harano et al.[159] noted the vicious cycle of drug-induced generation of cumulative risk factors and emphasized that efforts to alleviate insulin resistance are crucial to primary and secondary prevention of cardiovascular disease.

The goals of treating EH in patients with metabolic risk include lifestyle modifications and, if they are not successful, selection of blood pressure-lowering agents that do not alter metabolic function.[160–162] Lifestyle modification should be the primary treatment of mild-to-moderate hypertension for 3–6 months. Weight management is perhaps the most important factor because it can correct IR and glucose intolerance and its blood pressure-lowering effect can be dramatic, even at modest weight reductions (10–15% initial body weight).[163] Results from trials of lifestyle changes (diet, exercise) show that they can result in a prolonged and sustained reduction in insulin resistance in EH.[144] However, Fagerberg et al.[164] noted that diet modification was inferior to drug treatment for controlling hypertension but superior for lowering plasma insulin and improving lipid profiles. Thus, both pharmacologic and nonpharmacologic therapies are often needed for control of blood pressure and other risk factors in high-risk patients with EH. At present, however, many of the above concerns remain theoretical because no prospective randomized trials have been done to control for the other effects of antihypertensive drugs.[158,165]

Diuretic Agents

Thiazide diuretics have recently undergone considerable scrutiny for their effects on carbohydrate and lipid metabolism in patients with EH. Swislocki et al.[166] reported significantly higher plasma glucose and insulin responses to oral glucose in thiazide-treated patients with EH. Pollare et al.[167] found an 18% decrease from baseline in insulin sensitivity after 4 months of thiazide diuretic administration. Thiazides impair insulin secretion by potassium depletion, but lower doses and control of potassium balance may correct this effect.[168] Long-term diuretic therapy in patients with EH increases the risk ratio for diabetes mellitus.[155] Yet in a large retrospective analysis of prescription patterns of hypoglycemic agents for treatment of EH, diuretics resulted in no greater use of antidiabetic therapy than any other class of antihypertensive drug.[169] Thus, the long-term risk of drug-induced metabolic changes with diuretics has not been resolved, and some of the consideration may be hypothetical. In fact, many patients with EH and diabetes have salt-sensitive hypertension and eventually need some form of diuretic therapy[170] for adequate control of blood pressure.

Neutel recently reviewed the renewed interest in the safety and efficacy of low-dose thiazide diuretics.[171] Several studies show that hydrochlorothiazide used in doses of 6.25–12.5 mg/day has no significant metabolic side effects.[172–174] Neither plasma insulin levels nor insulin resistance change during monotherapy with 12.5 mg of hydrochlorothiazide.[174]

Thiazide diuretics increase levels of total and low-density lipoprotein (LDL) cholesterol and triglycerides.[175,176] In a recent multilinear regression analysis of the effects of antihypertensive therapy on serum lipids, diuretics caused a relative dose-related increase in cholesterol that was worse in black than nonblack hypertensive patients.[176] Thus, as with glucose, low-dose diuretics have less effect on lipid metabolism; several studies assessing the effect of hydrochlorothiazide, 6.25–12.5 mg, found no significant changes.[172–174] Specific concerns have been raised about diuretics and sudden cardiac death, especially in hypertensive patients with underlying cardiovascular problems,[177] in addition to general concern about myocardial infarction and sudden death with other classes of agents.[178] These issues are highly controversial and unresolved. As noted by Freis,[179] lower-dose diuretics are associated with a reduced number of sudden deaths, and small doses seem prudent and often necessary in treating hypertension.

Beta-adrenergic Blocking Agents

The potential for beta-adrenergic blockade to cause metabolic disarray is quite impressive because insulin release from the pancreas is modulated by $beta_2$-adrenergic receptors and intermediary metabolism is under adrenergic control. Blockade of beta-adrenergic receptors inhibits insulin secretion by as much as 50% in animal studies and may reduce insulin sensitivity. In human studies, beta blockers have been shown to worsen insulin action and glucose tolerance, to increase insulin and glucose levels, and to alter lipid profiles. In a series of studies in which beta blockers were systematically evaluated for effects on insulin sensitivity in hypertension, the nonselective agent propranolol reduced insulin sensitivity by 32%, whereas the cardioselective agent atenolol reduced sensitivity by 13% and metoprolol by 23%.[158,180] Beta blockers with additional properties have less pronounced effects on metabolism. Pindolol, which has intrinsic sympathomimetic effects, has less effect on insulin sensitivity, carbohydrate metabolism, and lipids.[181] Dilevolol, a nonselective beta blocker with $beta_2$ agonist properties, improves insulin sensitivity by 10% and reduces serum triglycerides by 25% in patients with EH.[158,182]

In one study of patients with EH, carvedilol, a nonselective beta blocker with $alpha_1$-blocking properties, had more favorable effects on insulin sensitivity and glucose metabolism than metoprolol.[183] The combined properties of celiprolol are similar to those of dilevolol, but celiprolol is less toxic and also has favorable effects on metabolism.[184] The large variation in effects of beta-blocker subclasses on metabolic parameters has important implications for use in at-risk patients in whom secondary cardioprotection is indicated. The varied effects of beta blocker subclasses have been ascribed to differential actions on peripheral and nutritional blood flow.[158]

Combination therapy with diuretics and beta blockers has been a popular choice for years. However, studies clearly show additive effects on deterioration in insulin sensitivity and glucose tolerance.[166,185] Schneider et al.[186] reported that the combination of atenolol and chlorthalidone aggravated insulin resistance in hypertensive patients with NIDDM, whereas the combination of a calcium channel blocker (verapamil) and an ACE inhibitor (trandolapril) had much less adverse effect.

Through combined effects of impairing insulin action and decreasing synthesis of lipoprotein lipase, lecithin cholesterol acyltransferase, and cholesterol, beta-adrenergic blockade can have substantial effects on lipid profiles in patients with EH. In general, the major effect of beta blockers is to increase VLDL lipoprotein triglyceride levels and to reduce HDL cholesterol. Beta blockers result in less pronounced increases in total and LDL cholesterol. In a recent metaanalysis of studies using beta blockers in EH, the major significant effect was to increase triglyceride levels.[176] Agents with intrinsic sympathomimetic activity combined with cardioselectivity led to less adverse lipid profiles and less suppression of HDL cholesterol in diabetic patients.

Alpha$_1$-adrenergic Blocking Agents

Alpha$_1$-adrenergic antagonists, such as prazosin, terazosin, and doxazosin, have favorable effects on glucose and lipid profiles in patients with EH. Short-term (3-month) treatment with prazosin markedly improved insulin sensitivity by as much as 25%, and glucose concentrations were reduced by 30%.[187] These changes were so large as to restore almost normal insulin action. Suzuki et al.[188] and Harano et al.[159] found significant improvement in insulin sensitivity in EH with the alpha$_1$-blocking agent bunazosin, which also increases the metabolic clearance of glucose. Doxazocin lowers plasma insulin response to oral glucose, and improves insulin sensitivity in hypertensive patients and also may increase the clearance of glucose.[189] Interesting data related to doxazocin suggest that the effects on glucose and insulin may be dose- and time-dependent; better results are seen at 9–12 months of therapy and are most pronounced in patients with more profound metabolic dysfunction.[158] Larger studies, such as the Finnish Multicenter Study Group, report that an alpha$_1$ blocker consistently reduces insulin and glucose levels in patients with EH.[190]

The effects of alpha blockers on serum lipids may be among the most beneficial of all classes of antihypertensive agents. In a metaanalysis of 158 studies using alpha blockers, all lipid fractions were favorably affected, including reductions in total and LDL cholesterol and serum triglycerides as well as increases in HDL cholesterol.[176] A study comparing doxazocin to the ACE inhibitor enalapril in hypertriglyceridemic, hypertensive men found that doxazosin had superior metabolic effects, increasing insulin sensitivity by 21% and lowering serum triglycerides by 23%.[191] Similar beneficial effects of alpha$_1$ blockers on lipids have been reported by larger population studies. In the Treatment of Mild Hypertension Study (TOMHS),[192] alpha blocker monotherapy had the only significant beneficial effect on the total cholesterol/HDL cholesterol ratio compared with a diuretic, beta blocker, calcium channel blocker, or ACE inhibitor.

There are several interesting mechanisms whereby $alpha_1$-blocking agents may favorably alter metabolic function. As vasodilators, alpha blockers may open up more capillaries in skeletal muscle, thus providing more sites for insulin action. In addition, they may activate heparin-releasable lipoprotein lipase, which improves the removal capacity for triglyceride-rich lipoproteins[158] and may increase the metabolic clearance of glucose.[191] On the other hand, alpha blockers do not alter insulin receptor binding or tyrosine kinase activity.[193]

Calcium Channel Blockers

Theoretically, calcium channel blockade can create metabolic dysfunction because calcium is crucial in the temporal release of insulin. Its blockade may lead to glucose intolerance. In addition, tissue levels of calcium and magnesium are determinants of insulin action. Draznin[130] offered evidence that abnormal intracellular calcium may contribute to insulin resistance. In the ionic hypothesis,[194] the intracellular balance of calcium, magnesium, sodium, and pH may be major determinants of cardiovascular risk in several tissues. However, these theoretical considerations do not appear to be a problem; therapeutic doses of most calcium channel blockers have not been associated with adverse effects on glucose and insulin in patients with EH.[195,196] Early studies implicated the short-acting forms of diltiazem and nifedipine as causes of insulin resistance and glucose intolerance.[197,198] Euglycemic clamp studies show that diltiazem does not alter insulin sensitivity. Its effect may be dose-dependent, with more favorable effects on insulin sensitivity at higher doses.[158] Because calcium channel blockers act as direct vasodilators, they should also have the potential to improve insulin action, but this effect appears to vary among the different compounds. According to one study, short-acting nifedipine reduces insulin sensitivity,[198] whereas the longer-acting forms of nifedipine are either neutral or improve insulin action.[200] This difference in effect may relate to the greater reflex activation of the sympathetic nervous system with short-acting compared with long-acting calcium channel blockers. The Modern Approach to Treatment of Hypertension (MATH) study[201] showed that the long-acting nifedipine GITS had no negative effects on glucose and lipids in 926 obese and 157 diabetic subjects. Another study of nifedipine GITS reported significant improvement in insulin sensitivity.[202] Most formulations of verapamil have neutral-to-mildly favorable effects on metabolic parameters.[203]

Most of the new longer-acting dihydropyridines have been analyzed for their effects on insulin and glucose. Amlodipine has no effect on insulin resistance in obese patients[204] but may improve glucose tolerance and insulin sensitivity in patients with EH.[205] Another study reports that it also improves insulin sensitivity.[159] Nitrendipine improves glucose tolerance and lowers insulin levels in humans[206] and improves glucose tolerance and deoxyglucose uptake in rats.[207] A different action may be seen with nicardipine, which reduced insulin secretion but did not alter insulin sensitivity.[208] Isradipine has neutral effects on insulin sensitivity and serum lipids in patients with EH.[209] It also may have an effect on cytosolic Ca^{2+}. Touyz and Schiffrin[210] have shown that isradipine decreases fasting serum insulin and increases platelet calcium sensitivity to insulin in diabetic patients.

Almost every study has shown that calcium channel blockers have neutral effects on lipid profiles. A metaanalysis of 100 studies found that calcium antagonists have less effect on serum lipids than any other class of antihypertensive drugs.[176]

Angiotensin-converting Enzyme Inhibitors

ACE inhibitors have neutral-to-beneficial effects on glucose, insulin, and lipid metabolism. They may improve insulin sensitivity, insulin secretion, potassium balance, and intermediary metabolism, as shown in patients with EH and hypertensive diabetics. An increase in insulin-mediated glucose disposal is noted with acute ACE inhibitor administration,[211,212] and more prolonged therapy with captopril in patients with EH increased insulin-mediated glucose disposal by 15–20%.[167] Paolisso et al.[213] reported that five ACE inhibitors—captopril, enalapril, quinapril, ramipril, and lisinopril—lowered blood pressure and improved insulin sensitivity in 86 elderly patients with EH. These data may suggest that improvement in insulin sensitivity is a class effect. Yet other studies show diverse effects on metabolic function of ACE inhibitors. Shionoiri et al.[214] found neutral effects on glucose tolerance with cilazapril, whereas Santoro et al.[215] reported that oral glucose tolerance but not insulin sensitivity was improved. Ramipril and perindopril appear to have a neutral influence on glucose tolerance and insulin sensitivity in patients with EH,[216,217] whereas fosinopril improves glucose tolerance and insulin sensitivity.[218] In obese patients with hypertension, Reaven et al.[219] found that lisinopril had no untoward effect on multiple aspects of glucose, insulin, and lipoprotein metabolism. In a study primarily of African-American patients with mild hypertension, Falkner et al.[220] found that lisinopril improves blood pressure, insulin sensitivity, insulin levels, and total cholesterol counts. In addition, these changes were associated with a reduction in several indicators of abnormal erythrocyte sodium transport. After captopril, enalapril is the most studied ACE inhibitor in terms of metabolic effects; several reports characterize its effects as ranging from neutral to beneficial.[221] In high-risk patients with EH, IR may improve by as much as 60% after switching from a beta blocker to an ACE inhibitor.[222] The greatest improvement was noted in hypertensive patients with the most marked IR. No studies have shown deterioration in insulin and glucose metabolism during ACE inhibitor therapy. ACE inhibitors have generally neutral effects on lipid profiles, although some report a modest but significant reduction in total cholesterol, and on triglycerides in diabetics and younger patients with hypertension.[176]

ACE inhibitors are widely used in both normotensive and hypertensive patients with diabetes mellitus because of their proven renal protective effects and other beneficial properties. As in patients with EH, they have either neutral or beneficial effects on metabolic parameters such as fasting glucose, postprandial glucose, glucose tolerance, and glycosylated hemoglobin.[223] Shamiss et al.[224] examined 10 hypertensive patients with NIDDM using sequential clamp procedures during treatment with 20 mg of enalapril. Insulin sensitivity improved, and levels of insulin and glucose declined by 16 weeks of therapy. Coadministration of a low dose of hydrochlorothiazide did not adversely affect these parameters. In a study of 130 patients with NIDDM and hypertension, captopril lowered fasting plasma glucose levels in a significant stepwise fashion during each month for 4 months.[224] In 16 patients with NIDDM and hypertension, Torlone et al.[225] noted that captopril improved postprandial insulin action in the liver, adipose tissue, and muscle. Prince et al.[226] found that enalapril improved glycosylated hemoglobin and cholesterol levels as well as glucose and insulin sensitivity in diabetic patients with hypertension. Many other studies using ACE inhibitors in diabetic patients with hypertension have found generally favorable effects on glucose metabolism and insulin action.[196] Important studies have also shown that addition of ACE inhibitors to low-dose hydrochlorothiazide treatment offsets the adverse effects of diuretics on glucose and lipoprotein metabolism.[224,228–230]

Several mechanisms may account for the improvement of insulin action and glucose metabolism. Forearm perfusion studies show that the primary site at which ACE inhibitors act to increase insulin-mediated glucose uptake is in skeletal muscle.[231,232] Because infusion of bradykinin enhances insulin-mediated glucose uptake, ACE inhibitor-induced accumulation of tissue kinins is one proposed mechanism. Glucose clamp testing in rats shows that enalapril augments the kinin inhibitor HOE 140.[233] Ferannini et al.[234] have shown that the ACE inhibitor cilazapril improves glucose-induced insulin release and oral glucose tolerance by blunting the hypokalemic response to insulin. Because ACE inhibition lowers angiotensin II levels, it might be predicted that it would also improve blood flow and IR. In fact, quite the opposite seems to be true. Numerous studies have shown that angiotensin II infusion at both subpressor and pressor doses increases insulin-mediated glucose transport and oxidation in skeletal muscle.[235–237] In a review by Morris and Donnely,[238] it is proposed that angiotensin II may be an insulin-sensitizing vasoactive hormone. The authors suggest that angiotensin II causes redistribution of blood flow with increased perfusion of skeletal muscle flow. However, it is also proposed that angiotensin II may have unknown direct effects on insulin-mediated glucose transport.

Angiotensin II Receptor Antagonists

Several studies in patients with EH indicate that losartan, the first orally active, selective angiotensin II receptor antagonist, lowers blood pressure and has neutral-to-beneficial effects on insulin sensitivity, glucose metabolism, and lipids.[239–244] Moan et al.[239] found that in mild hypertension losartan has a neutral effect on insulin sensitivity, glucose metabolism, and serum lipids. The same group has also shown that losartan has no effect on fibrinolytic variables and catecholamines.[240] Tikkanen et al.[241] studied a large group of patients (n = 407) with mild-to-moderate EH to compare losartan with enalapril. Both agents had similar effects on blood pressure and neutral effects on most metabolic parameters. Laakso et al.[242] compared the effect on insulin sensitivity and glucose oxidation of losartan vs. metoprolol in 20 patients with EH and hyperinsulinemia. In this already identified high risk group, both agents demonstrated metabolic neutrality. One study in 13 Japanese patients with EH demonstrated that both the ACE inhibitor delapril and an angiotensin II receptor antagonist improved insulin resistance. The authors concluded that this effect may depend on suppression of angiotensin II action.[243] In fructose-fed rats, losartan improves glucose tolerance.[244]

Insulin-sensitizing Agents

The biguanide agent metformin improves insulin sensitivity in both the liver and skeletal muscle, and in diabetic patients with other risk factors it has beneficial effects on body weight, lipid profiles, and blood pressure. The beneficial effects on blood pressure are the least certain. Some studies have shown a significant reduction,[245] whereas others have noted less effect.[246]

The thiazolidinedione agents are a novel group of oral hypoglycemic compounds that directly improve IR.[247] Because IR may contribute to hypertension, its specific reversal by these agents may offer evidence for a specific effect of IR on blood pressure control. Thiazolidinediones attenuate the development of hypertension in Dahl salt-sensitive rats, one-kidney–one clip hypertensive rats, obese Zucker rats, and fructose-fed rats.[247] They appear to have little effect on blood pressure in the spontaneously hypertensive rat.[247] They also may lower blood pressure, serum triglycerides, and IR in humans. Ogihara et al.[248] have shown that the compound troglitazone enhances insulin sensitivity and lowers blood pressure in diabetic hypertensive patients. The compound pioglitazone increases insulin sensitivity and reduces blood pressure, insulin, and lipid levels in obese, insulin-resistant rhesus monkeys.[249] However, in vitro studies show that it also may have direct vascular effects. Pioglitazone reduces directly the norepinephrine- and arginine-induced contractions in aortic ring preparations.[250] These agents also may alter calcium channels in vascular smooth muscle. Vanadyl sulfate improves insulin sensitivity and prevents fructose-induced hyperinsulinemia and hypertension in Sprague-Dawley rats.[251]

References

1. Modan M, Halkin H, Almog S, et al: Hyperinsulinemia: A link between hypertension obesity and glucose intolerance. J Clin Invest 75:809–817, 1985.
2. Shen DC, Shieh SM, Fuh MT, et al: Resistance to insulin-stimulated-glucose uptake in patients with hypertension. J Clin Endocrinol Metab 66:580–583, 1988.
3. Ferrannini E, Buzzigoli G, Bonadonna R, et al: Insulin resistance in essential hypertension. N Engl J Med 17:350–357, 1987.
4. Pollare T, Lithell H, Berne C: Insulin resistance is a characteristic feature of primary hypertension independent of obesity. Metabolism 39:167–174, 1990.
5. Natali A, Santoro D, Palombo C, et al: Impaired insulin action on skeletal muscle metabolism in essential hypertension. Hypertension 17:170–178, 1991.
6. Christlieb AR, Krolewski AS, Warram JH, Soeldener JS: Is insulin the link between hypertension and obesity? Hypertension 7(Suppl 2):54–57, 1985.
7. Landsberg L: Diet, obesity and hypertension: An hypothesis involving insulin, the sympathetic nervous system, and adaptive thermogenesis. Q J Med 236:1081–1090, 1986.
8. Reaven GM, Hoffman BB: A role for insulin in the aetiology and course of hypertension. Lancet 2:435–436, 1987.
9. Singer P, Godicke W, Voight S, et al: Postprandial hyperinsulinemia in patients with mild essential hypertension. Hypertension 7:182–186, 1985.
10. Fuh MM-T, Shieh S-M, Wu D-A, et al: Abnormalities of carbohydrate and lipid metabolism in patients with hypertension. Arch Intern Med 147:1035–1038, 1987.
11. Marigliano A, Tedde R, Sechi LA, et al: Insulinemia and blood pressure: relationships in patients with primary and secondary hypertension, and with or without glucose metabolism impairment. Am J Hypertens 3:521–526, 1990.
12. Salvatorre T, Cozzolino D, Giunta R, et al: Decreased insulin clearance as a feature of essential hypertension. J Clin Endocrinol Metab 74:144–149, 1992.
13. Denker PS, Pollock VE: Fasting serum insulin levels in essential hypertension. A meta-analysis. Arch Intern Med 152(8):1649-1651, 1992.

14. Mitchell BD, Stern MP, Haffner SM, et al: Risk factors for cardiovascular mortality in Mexican Americans and non-Hispanic whites: The San Antonio Heart Study. Am J Epidemiol 131:423–433, 1990.
15. Haffner SM, Ferrannini E, Hazuda HP, Stern MP: Clustering of cardiovascular risk factors in confirmed prehypertensive individuals. Hypertension 20:38–45, 1992.
16. Meehan WP, Darwin CH, Maalouf TA, et al: Insulin and hypertension: Are they related? Steroids 58:621–630, 1993.
17. Lithell HO: Hyperinsulinemia, insulin resistance, and the treatment of hypertension. Am J Hypertens 9:150S–154S, 1996.
18. Mbanya JC, Thomas TH, Wilkinson R, et al: Hypertension and hyperinsulinemia: A relation in diabetes but not essential hypertension. Lancet 1:733–734, 1988.
19. Laakso M, Sarlund H, Mykkanen L: Essential hypertension and insulin resistance in non-insulin dependent diabetes. Eur J Clin Invest 19:519–526, 1989.
20. Reaven G, Landsberg L, Lithell H: The insulin resistance syndrome. N Engl J Med 334:2345, 1966.
21. Wingard DL, Barrett-Conner EL, Ferara A: Is insulin really a heart disease factor? Diabetes Care 18:1299–1304, 1995.
22. Howard G, O'Leary DH, Zaccaro D, et al, for the IRAS Investigators: Insulin sensitivity and atherosclerosis. Circulation 93:1809–1817, 1996.
23. Lind L, Andersson P-E, Andren B, et al: Left ventricular hypertrophy in hypertension is associated with the insulin resistance metabolic syndrome. J Hypertens 13:433–438, 1995.
24. Jeng JR, Wayne H, Sheu H, et al: Impaired fibrinolysis and insulin resistance in patients with hypertension. Am J Hypertens 9:484–490, 1996.
25. Parik T, Allikmets K, Teesalu R, Zilmer M: Oxidative stress and hyperinsulinemia in essential hypertension: Different facets of increased risk. J Hypertens 14:407–410, 1996.
26. Kuusisto J, Mykkanen L, Pyorala K, Laakso M: Hyperinsulinemia microalbuminuria: A new risk indicator for coronary heart disease. Circulation 91:831–837, 1995.
27. Tedde R, Pala A, Melis A, et al: Hyperinsulinemia and hypertension. Do intestinal hormones play a role? Am J Hypertens 8:99–103, 1995.
28. Hintz RL: Current and potential therapeutic uses of growth hormone and insulin-like growth factor 1. Endocrinol Metabol Clin North Am 25:759–773, 1996.
29. Paternostro G, Camici PG, Lammerstma AA, et al: Cardiac and skeletal muscle insulin resistance in patients with coronary heart disease. J Clin Invest 98:2094–2099, 1996.
30. Baron AD, Laakso M, Brechtel G, et al: Reduced postprandial skeletal muscle blood flow contributes to glucose intolerance in human obesity. J Clin Endocrinol Metab 70:1525–1533, 1990.
31. Lind L, Berne C, Lithell H: Prevalence of insulin resistance in essential hypertension. J Hypertens 13:1457–1462, 1995.
32. Istfan NW, Plaisted CS, Bistrian BR, Blackburn GL: Insulin resistance versus insulin secretion in the hypertension of obesity. Hypertension 19:385–392, 1992.
33. Landin K, Lindgarde F, Saltin B, Smith U: The skeletal muscle Na:K ratio is not increased in hypertension: Evidence for the importance of obesity and glucose intolerance. J Hypertens 9:65–69, 1991.
34. Donatelli M, Scarpinato A, Bucalo ML, et al: Stepwise increase in plasma insulin and C-peptide concentrations in obese, obese hypertensive and in obese hypertensive diabetic subjects. Diabetes Res 17:125–129, 1991.
35. Sechi LA, Melis A, Pala A, et al: Serum insulin, insulin sensitivity and erythrocyte sodium metabolism in normotensive and essential hypertensive subjects with or without overweight. Clin Exp Hypertens 13A:261–276, 1991.
36. Filipovsky J, Ducimetiere P, Eschwege E, et al: The relationship of blood pressure with glucose, insulin, heart rate,free fatty acids and plasma cortisol levels according to degree of obesity in middle-aged men. J Hypertens 14:229–235, 1996.
37. Mykkanen L, Haffner SM, Ronnema T, et al: Relationship of plasma insulin concentration and insulin sensitivity to blood pressure. Is it modified by obesity? J Hypertens 14:399–405, 1996.
38. Hennes MM, O'Shaughnessy IM, Kelly TM, et al: Insulin-resistant lipolysis in abdominally obese hypertensive individuals. Role of the renin-angiotensin system. Hypertension 28:120–126, 1996.
39. Ikeda T, Gomi T, Hirawa N, et al: Improvement of insulin sensitivity contributes to blood pressure reduction after weight loss in hypertensive subjects with obesity. Hypertension 27:1180–1186, 1996.
40. Sowers JR,Whitfield LA, Catania RA, et al: Role of the sympathetic nervous system in blood pressure maintenance in obesity. J Clin Endocrinol Metab 54:1181–1186, 1982.

41. Tuck ML, Sowers JR, Dornfeld L, et al: The effect of weight reduction on blood pressure, plasma renin activity, and plasma aldosterone levels in obese patients. N Engl J Med 304:930–933, 1981.
42. Maxwell M, Heber D, Waks U, et al: Role of insulin and catecholamines in the hypotensive response to very-rapid weight reduction. Am J Med 7:23–29, 1994.
43. Ferrari P, Weidman P, Shaw S, et al: Altered insulin sensitivity, hyperinsulinemia, and dyslipidemia in individuals with a hypertensive parent. Am J Med 91:589–596, 1991.
44. Widgren BR, Herlitz H, Wikstrand J, et al: Increased waist/hip ratio, metabolic disturbances, and family history of hypertension. Hypertension 20:563–568, 1992.
45. Neutel JM, Smith DH, Graettinger WF, et al: Metabolic characteristics of hypertension: importance of positive family history. Am Heart J 126:924–929, 1993.
46. Facchini F, Chen YD, Clinkingbeard C, et al: Insulin resistance, hyperinsulinemia, and dyslipidemia in nonobese individuals with a family history of hypertension. Am J Hypertens 5:94–99, 1992.
47. Ishibashi F: Higher serum insulin level due to greater total body fat mass in offspring of patients with essential hypertension. Diabetes Res Clin Pract 20:63–68, 1993.
48. Beatty DL, Harper R, Sheridan B, et al: Insulin resistance in offspring of hypertensive parents. BMJ 307:92–96, 1993.
49. Ohno Y, Suzuki H, Yamakawa H, et al: Impaired insulin sensitivity in young, lean normotensive offspring of essential hypertensives: possible role of disturbed calcium metabolism. J Hypertens 11:421–426, 1993.
50. Grunfeld B, Balzareti M, Romo M, et al: Hyperinsulinemia in normotensive offspring of hypertensive subjects. Hypertension 23(Suppl I):112–115, 1994.
51. Wing RR, Mathews KA, Kuller LH, et al: Environmental and familial contributions to insulin levels and change in insulin levels in middle-aged women. JAMA 268:1890–1895, 1992.
52. Weisser B, Grune S, Spuhler T, et al: Plasma insulin is correlated with blood pressure only in subjects with a family history of hypertension or diabetes. J Hypertens 11(Suppl 5):S308–S309, 1993.
53. Masuo K, Mikami H, Ogihara T, Tuck ML: Differences in insulin and sympathetic responses to glucose ingestion due to family history of hypertension. Am J Hypertens 9:739–745, 1996.
54. Grandi AM, Gaudo G, Fachineti A, et al: Hyperinsulinemia, family history of hypertension, and essential hypertension. J Hypertens 9:732–738, 1996.
55. Mino D, Wacher N, Amato D, et al: Insulin resistance in offspring of hypertensive subjects. J Hypertens 14:1189–1193, 1996.
56. Saad MF, Lillioja S, Nyomba BL, et al: Racial differences in the relation between blood pressure and insulin resistance. N Engl J Med 324:733–739, 1991.
57. Burke GL, Webber LS, Srinvasan ST, et al: Fasting plasma glucose and insulin levels and their relationship to cardiovascular risk factors in children: Bogalusa Heart Study. Metabolism 35:441–446, 1986.
58. Cruikshank JK, Cooper J, Burnett M, et al: Ethnic differences in fasting plasma C-peptide and insulin in relation to glucose tolerance and blood pressure. Lancet 338:842–847, 1991.
59. Haffner SM, Mitchell BD, Stern MP, et al: Decreased prevalence of hypertension in Mexican Americans. Hypertension 16:225–232, 1990.
60. Dowse GK, Collins VR, Alberti KG, et al: Insulin and blood pressure levels are not independently related in Mauritians of Asian Indian, Creole or Chinese origin. The Mauritius Noncommunicable Disease Study Group. J Hypertens 11:297–307, 1993.
61. Knight TM, Smith Z, Whittles A, et al: Insulin resistance, diabetes, and risk markers for ischemic heart disease in Asian men and non-Asian in Bradford. Br Heart J 67:343–350, 1992.
62. Falkner B, Hulman S, Tannenbaum J, Kushner H: Insulin resistance and blood pressure in young black men. Hypertension 16:706–711, 1990.
63. Falkner B, Hulman S, Kushner H: Insulin-stimulated glucose utilization and borderline hypertension in young adult blacks. Hypertension 22:18–25, 1993.
64. Haffner SM, Stern MP, Dunn J, et al: Diminished insulin sensitivity and insulin response in non-obese, non-diabetic Mexican Americans. Metabolism 39:842–847, 1990.
65. Buchanan T, Raffel L. Cantor R, et al: Factors associated with salt sensitivity in Latino families. Hypertension 28:454, 1995.
66. Baba T, Kodama T, Tomiyama T, et al: Hyperinsulinemia and blood pressure in non-obese middle-aged subjects with normal glucose tolerance. Tokohu J Exp Med 165:229–235, 1991.
67. Iimura O: Insulin resistance and hypertension in Japan. Hypertens Res 19(Suppl I):S1–S8, 1996.
68. Shimamoto K, Hirata A, Fukuoka M et al: Insulin sensitivity and the effects of insulin on renal sodium handling and pressor systems in essential hypertensive patients. Hypertension 23 (Suppl I):I29–I33, 1994.

69. Miura K, Nakagawa H, Nishijo M, et al: Plasma insulin and blood pressure in normotensive Japanese men with normal glucose tolerance. J Hypertens 13:427–432, 1995.
70. Tsuruta M, Hashimoto R, Adachi H, et al: Hyperinsulinemia as a predictor of hypertension: an 11-year follow-up study in Japan. J Hypertens 14:483–488, 1966.
71. Collins VR, Dowse GK, Finch CF, Zimmet PZ: An inconsistent relationship between insulin and blood pressure in three Pacific Island populations. J Clin Epidemiol 43:1369–1378, 1990.
72. Bursztyn M, Raz I: Prediction of hypertension by the insulinogenic index in young Ethiopian immigrants. J Hypertens 13:57–61, 1995.
73. Haffner SM, Valdez R, Morales PA, et al: Greater effect of glycemia on incidence of hypertension in women than in men. Diabetes Care 15:1277–1284, 1992.
74. McKeigue PM, Laws A, Chen YD, et al: Relation of plasma triglyceride and ApoB levels to insulin-mediated suppression of nonesterified fatty acids. Arterioscler Thromb 13:1187–1192, 1993.
75. Os I, Kjeldsen SE, Nordby G, et al: Sex differences in essential hypertension. J Intern Med 233:13–19, 1993.
76. Cigolini M, Seidell JC, Charzewska J, et al: Fasting serum insulin in relation to fat distribution, serum lipid profile, and blood pressure in European woman: The European Fat Distribution Study. Metabolism 40:781–787, 1991.
77. Taittonen L, Ulhari M, Nuutinen M, et al: Insulin and blood pressure among children: Cardiovascular risk in young Finns. Am J Hypertens 9:193–199, 1996.
78. Stolk RP, Hoes AW, Pols HAP, et al: Insulin, hypertension and antihypertensive drugs in elderly patients: The Rotterdam Study J Hypertens 14:237–242, 1996.
79. Donovan DS, Solomon CG, Seely EW, et al: Effect of sodium intake on insulin sensitivity. Am J Physiol 264:E730–E734, 1993.
80. Lind L, Lithell H, Gustafsson IB, et al: Metabolic cardiovascular risk factors and sodium sensitivity in hypertensive subjects. Am J Hypertens 5:502–505, 1992.
81. Egan BM, Weder AB, Petrin J, et al. Neurohumoral and metabolic effects of short-term dietary NaCl restriction in men. Relationship to salt-sensitivity status. Am J Hypertens 4:416–421, 1991.
82. Ruppert M, Diehl J Kolloch R, et al: Short-term dietary sodium restriction increases serum lipids and insulin in salt-sensitive and salt-resistant normotensive adults. Klin Wochenschr 4 (Suppl 25):51–57, 1991.
83. Del Rio A, Rodriquezvillamil JL: Metabolic effects of strict salt restriction in essential hypertensive patients. J Intern Med 233:409–414, 1993.
84. Grey A, Braatveldt G, Holdaway I: Moderate dietary salt restriction does not alter insulin resistance or serum lipids in normal man. Am J Hypertens 9:317–322, 1996.
85. Rocchini AP, Key J, Bondie D, et al: The effect of weight loss on the sensitivity of blood pressure to sodium in obese adolescents. N Engl J Med 321:580–585, 1989.
86. Sharma AM, Schorr U, Distler A: Insulin resistance in young salt-sensitive normotensive subjects. Hypertension 21:273–279, 1993.
87. Egan BM, Stepniakowski K, Nazzaro P: Insulin levels are similar in obese salt-sensitive and salt-resistant hypertensive subjects. Hypertension 23(Suppl I):I1–7, 1994.
88. Bigazzi R, Bianchi S, Baldari G, Campese VM: Clustering of cardiovascular risk factors in salt-sensitive patients with essential hypertension: Role of insulin. Am J Hypertens 9:24–32, 1996.
89. Moller DE, Flier JS: Insulin-resistance—Mechanisms, syndromes and implications. N Engl J Med 32S:938–948, 1991.
90. Moller DE, Flier JS: Detection of an alteration in the insulin receptor gene in a patient with insulin resistance, acanthosis nigricans and polycystic ovary syndrome (type A insulin resistance). N Engl J Med 379:1526–1529, 1988.
91. White MF, Khan RF: The insulin signaling system J Biol Chem 269:1–4, 1994.
92. Bell GI, Kayano T, Buse JB, et al: Molecular biology of mammalian glucose transporters. Diabetes Care 13:198–208, 1990.
93. Bonora E, Bonadonna RC, Del Prato S, et al: In vivo glucose metabolism in obese and type II diabetic subjects with or without hypertension. Diabetes 42:764–772, 1993.
94. Bergman RN, Beard J, Chen M: The minimal modeling method. Assessment of insulin sensitivity and B-cell function in vivo. In Larner J, Pohl S (eds): Methods in Clinical Diabetes Research. New York, Wiley International, 1986, pp 13–20.
95. Giugliano D, Quatraro A, Minei A, et al: Hyperinsulinemia in hypertension: Increased secretion, reduced clearance or both? J Endocrinol Invest 16:315–321, 1993.
96. Lillioja S, Young AA, Culter CL, et al: Skeletal muscle capillary density and fiber type are possible determinants of in vivo insulin resistance in man. J Clin Invest 80:415–424, 1987.

97. Holmag A, Brezinska Z, Bjorntorp P: Effects of hyperinsulinemia on muscle fiber composition and capitalization in rats. Diabetes 42:1073–1081, 1993.
98. Julius S, Gudbrandsson T: Early association of sympathetic overactivity, hypertension, insulin resistance, and coronary risk. J Cardiovasc Pharmacol 20 (Suppl 8):S40–S48, 1992.
99. Julius S, Gudbrandsson T, Jamerson K, Anderson O: The interconnection between sympathetics, microcirculation, and insulin resistance in hypertension. Blood Pressure 1:9–19, 1992.
100. Reaven GM, Ho H. Renal vascular hypertension does not lead to hyperinsulinemia in Sprague-Dawley rats. Am J Hypertens 5:314–317, 1992.
101. DeFronzo RA, Cooke CR, Andres R, et al: The effects of insulin on renal handling of sodium, potassium, calcium and phosphate in man. J Clin Invest 55:845–855, 1975.
102. Skott P, Hother-Nielsen O, Bruun NE, et al: Effect of insulin on kidney function and sodium excretion in healthy subjects. Diabetologia 32:694–699, 1989.
103. Gans ROB, Toorn L, Bilo HJG, et al: Renal and cardiovascular effects of exogenous insulin in healthy volunteers. Clin Sci 80:219–225, 1991.
104. Muscelli E, Natali A, Bianchi S, et al: Effect of insulin on renal sodium and uric acid handling in essential hypertension. Am J Hypertens 9:746–752, 1996.
105. Baum M: Insulin stimulates volume absorption in the proximal convoluted tubule. J Clin Invest 79:1104–1109, 1987.
106. Takahashi N, Ito O, Abe K: Tubular effects of insulin. Hypertens Res 19(Suppl I):S41–S45, 1996.
107. Ferri C, DeMattia G, Belllini C, et al: Octreotide, a somatostatin analog, reduces insulin secretion and increases renal Na^+ excretion in lean essential hypertensive patients. Am J Hypertens 6(Suppl 4):276–281, 1993.
108. Hall JE, Coleman TG, Mizelle HL, Smith MJ: Chronic hyperinsulinemia and blood pressure regulation. Am J Physiol 258:F722–F731, 1990.
109. Mogensen CE, Christensen NJ, Gundersen HJG: The acute effect of insulin on renal hemodynamics and protein excretion in diabetics. Diabetologia 15:153–157, 1978.
110. Viberti GC: Introduction to a structural basis for renal and vascular complications in diabetes and hypertension. J Hypertens 10:S1–S4, 1992.
111. Corry DB, Tuck ML: Hypertension and diabetes. Semin Nephrol 11:561–570, 1991.
112. O'Hare JA, Minaker KL, Meneilly GS, et al: Effect of insulin on plasma norepinephrine and 3,4-dihydroxyphenylalanine in obese men. Metabolism 38:322–329, 1989.
113. Tuck ML: Obesity, the sympathetic nervous system and essential hypertension. Hypertension 19(Suppl I):I67–I77, 1992.
114. Anderson E, Sinkey CA, Lawton WJ, Mark AL: Elevated sympathetic nerve activity in borderline hypertensive humans: Evidence from direct intraneuronal recordings. Hypertension 14:177–183, 1989.
115. Berne C, Fagius J, Pollare T, Hjemdahl P: The sympathetic response to euglycemic hyperinsulinemia. Evidence from microelectrode nerve recordings in healthy subject. Diabetologia 35:873–879, 1992.
116. Lembo G, Napoli R, Capaldo B, et al: Abnormal sympathetic overactivity evoked by insulin in the skeletal muscle of patients with essential hypertension. J Clin Invest 90:24–29, 1992.
117. Mitrakou A, Mokan M, Bolli G, et al: Evidence against the hypothesis that hyperinsulinemia increases sympathetic nervous system activity in man. Metabolism 41:198–200, 1992.
118. Ward KD, Sparrow D, Landsberg L, et al: Influence of insulin, sympathetic nervous system, and obesity on blood pressure: The Normative Aging Study. J Hypertens 14:301–308, 1996.
119. Clausen JO, Isben H, Dige-Petersen H, et al: The importance of adrenaline, insulin, and insulin sensitivity as determinants for blood pressure in young Danes. J Hypertens 13:499–505, 1995.
120. Moreau P, Lamarche L, Laflamme AK, et al: Chronic hyperinsulinemia and hypertension: the role of the sympathetic nervous system. J Hypertens 13:333–340, 1995.
121. Hall JE, Brands MW, Kivlighn SD, et al: Chronic hyperinsulinemia and blood pressure. Interaction with catecholamines. Hypertension 15:519–527, 1990.
122. Rosic NK, Srandaert ML, Pollet RJ: The mechanism of insulin stimulation of Na^+/K^+ ATPase activity in muscle. J Biol Chem 260:6202–6212, 1985.
123. Tirupattur PR, Ram JL, Standley PR, Sowers JR: Regulation of Na^+,K^+-ATPase gene expression by insulin in vascular smooth muscle cells. Am J Hypertens 6:626–629, 1993.
124. Kahn A, Song T: Effects of insulin on vascular smooth muscle contraction. In Sowers JR (ed): Endocrinology of the Vasculature. Totowa, NJ, Humana Press, pp 215–223.
125. Pontremoli R, Zerbini G, Rivera A, Canessa M: Insulin activation of red blood cell Na^+/H^+ exchange decreases the affinity of sodium sites. Kidney International 46:365–375, 1994.
126. Aviv A: The roles of cell Ca^{2+}, protein kinase C and the Na^+-H^+antiport in the development of hypertension and insulin resistance [editorial]. J Am Soc Nephrol 3:1049–1063, 1992.

127. Mattiasson I, Berntorp K, Lindgarde F: Sodium-lithium countertransport and platelet cytosolic free calcium concentration in relation to peripheral insulin sensitivity in postmenopausal woman. Clin Sci 83:319–324, 1992.
128. Herman WH, Prior DE, Yassine MD, Weder AB: Nephropathy in NIDDM is associated with cellular markers for hypertension. Diabetes Care 16:815–818, 1993.
129. Canessa M, Falkner B, Hulman S: Red blood cell sodium-proton exchange in hypertensive blacks with insulin-resistant glucose disposal. Hypertension 22:204–213, 1993.
130. Draznin B, Kao M, Sussman KE: Insulin and glyburide increase cytosolic free Ca^{2+} concentrations in isolated rat adipocytes. Diabetes 36:174–178, 1987.
131. Anderson EA, Mark AL: The vasodilator action of insulin. Implications for the insulin hypothesis of hypertension. Hypertension 21:136–141, 1993.
132. Sakai K, Imaizmi T, Masaki H, Takeshita A: Intra-arterial infusion of insulin attenuates vasoreactivity in human forearm. Hypertension 22:67–73, 1993.
133. Randin D, Vollenweider P, Tappy L, Jequier E, et al: Effects of adrenergic and cholinergic blockade on insulin-induced stimulation of calf blood flow in humans. Am J Physiol 266:R809–R816, 1994.
134. Feldman RD, Bierbrier GS: Insulin-mediated vasodilation: impairment with increased blood pressure and body mass. Lancet 342:707–709, 1993.
135. Gudbjornsdottir S, Elam M, Sellgren J, Anderson E: Insulin increases forearm resistance in obese, insulin resistant hypertensives. J Hypertens 14:91–97, 1966.
136. Vollenweider P, Randin D, Tappy L, et al: Impaired insulin-induced sympathetic neural activation and vasodilation in skeletal muscle in obese humans. J Clin Invest 93:2365–2371, 1994.
137. Laakso M, Edelman SV, Brechtel G, Baron AD: Decreased effect of insulin to stimulate skeletal muscle blood flow in obese man. A novel mechanism for insulin resistance. J Clin Invest 85:1844–1852, 1990.
138. Baron AD, Steinberg H, Brechtel G, Johnson A: Skeletal muscle blood flow independently modulates insulin-mediated glucose uptake. Am J Physiol 226:E284–E253, 1994.
139. Baron AD, Brechtel G: Insulin differentially regulated systemic and skeletal muscle vascular resistance. Am J Physiol 265:E61–E67, 1993.
140. Baron AD, Brechtel-Hook G, Johnson A, Hardin D: Skeletal muscle blood flow. A possible link between insulin reisitance and blood pressure. Hypertension 21:129–135, 1993.
141. Baron AD: Insulin and the vasculature—old actors, new roles. J Invest Med 8:406–412, 1996.
142. Scherrer U, Randin D, Vollenweinder P, Nicod P: Nitric oxide release accounts for insulin's vascular effects in humans. J Clin Invest 94:2511–2515, 1994.
143. Yagi S, Takata S, Hiyokawa H, et al: Effects of insulin on vasoconstrictive response to norepinephrine and angiotensin II in rabbit femoral artery and vein. Diabetes 37:1064–1067, 1988.
144. Sakai E, Imaizmi T, Masaki H, Takashita A: Intra-arterial infusion of insulin attenuates vasoreactivity in human forearm. Hypertension 22:67–73, 1993.
145. Siato F, Hori MT, Fittingoff M, et al: Insulin attenuates agonist-mediated calcium mobilization in cultured rat vascular smooth muscle cells. J Clin Invest 92:1161–1167, 1993.
146. Kahn AM, Seidel CL, Allen JC, et al: Insulin reduces contraction and intracellular calcium concentration in vascular smooth muscle. Hypertension 22(5):735–742, 1993.
147. Standley PR, Zhang F, Ram JL, et al: Insulin attenuates vasopressin-induced calcium transients and a voltage-dependent calcium response in rat vascular smooth muscle cells. J Clin Invest 88:1230–1236, 1991.
148. Touyz RM, Tolloczko B, Schiffrin EL: Insulin attenuates agonist-evoked calcium transients in vascular smooth muscle cells. Hypertension 23(Suppl I):I25–I28, 1994.
149. Bar RS, Boes M, Dake BL, et al: Insulin, insulin-like growth factors and vascular endothelium. Am J Med 85(Suppl 5A):59–70, 1988.
150. Wu HY, Jeng YY, Yue CJ, et al: Endothelial dependent vascular effects of insulin and insulin-like growth factor-1 in the perfused rat mesenteric artery and aortic ring. Diabetes 43:1027–1031, 1994.
151. Sowers JR: Effects of insulin and IGF-1 on vascular smooth muscle glucose and cation metabolism. Diabetes 45 (Suppl 3):S47–S51, 1996.
152. Pfeifle B, Dutschuneit H: Effect of insulin on the growth of cultured smooth muscle cells. Diabetologia 120:155–158, 1980.
153. Banskota NK, Taub R, Zellner K, King GL: Insulin, insulin-like growth factor 1 and platelet derived growth factor interact additively in the induction of protooncogenes c-myc and cellular proliferation in cultured bovine aortic smooth muscle cells. Mol Endocrinol 89:1182–1190, 1989.
154. Stout RW: Insulin and atheroma: A 20-year perspective. Diabetes Care 13:631–654, 1990.

155. Bengtsson C, Blohme G, Lapidus L, Lundgren L: Diabetes in hypertensive women: An effect of antihypertensive drugs or the hypertensive state per se? Diabetes Med 5:261–264, 1988.
156. Skarfors ET, Selinus KI, Lithell HO: Risk factors for development of non-insulin dependent diabetes in middle-aged men. Results from a 10-year follow-up of participants in an Uppsala health survey. BMJ 303:755–760, 1991.
157. Samuelsson O, Hedner T, Persson B, et al: The role of diabetes mellitus and hypertriglyceridemia as coronary risk factors in treated hypertension: 15 years of follow-up of antihypertensive treatment in middle-aged men in the Primary Prevention Trial in Goteborg, Sweden. J Intern Med 235:217–227, 1994.
158. Lithell HO: Hyperinsulinemia, insulin resistance, and the treatment of hypertension. Am J Hypertens 9:150S–154S, 1996.
159. Harano Y, Suzuki M, Shinozaki K, et al: Clinical impact of insulin resistance syndrome in cardiovascular disease and its therapeutic approach. Hypertens Res 19(Suppl I): S81–S85, 1996.
160. The Fifth Report of the Joint National Committee on Detection, Evaluation, and Treatment of High Blood Pressure (JNCV): Arch Intern Med 153:154–183, 1993.
161. National High Blood Pressure Education Program Working Group Report on Hypertension in Diabetes. Hypertension 23:145–158, 1993.
162. Pyorala K, DeBacker G, Grahm I, et al, on behalf of the Task Force: Prevention of coronary heart disease in clinical practice. Recommendations of the Task force of the European Society of Cardiology, European Atherosclerosis Society, and European Society of Hypertension. Eur Heart J 15:1300–1331, 1994.
163. Nilsson PM, Lindholm LH, Schersten BF: Life style changes improve insulin resistance in hyperinsulinaemic subjects: a one-year intervention study of hypertensives and normotensive in Dalby. J Hypertens 10:1071–1078, 1992.
164. Fagerberg B, Berglund A, Andersson OK, Berglund G: Weight reduction versus antihypertensive drug therapy in obese men with high blood pressure: Effects upon plasma insulin levels and association with changes in blood pressure and serum lipids. J Hypertens 10(9):1053–1061, 1992.
165. Moser M, Hebert P, Hennekens CH: An overview of the meta-analyses of the hypertension treatment trials. Arch Intern Med 151:1277–1279, 1991.
166. Swislocki A: Insulin resistance and hypertension. Am J Med Sci 300:104–115, 1990.
167. Pollare T, Lithell H, Berne C: A comparison of the effects of hydrochlorothiazide and captopril on glucose and lipid metabolism in patients with hypertension. N Engl J Med 321:868–873, 1989.
168. Helderman JH, Elahi D, Andersen DK: Prevention of the glucose intolerance of thiazide diuretics by maintenance of body potassium. Diabetes 32:106–111, 1983.
169. Gurwitz JH, Bohn RL, Glynn RJ, et al: Antihypertensive drug therapy and the initiation of treatment for diabetes mellitus. Ann Intern Med 118:273–278, 1993.
170. Tuck ML, Corry DB, Trujillo A: Salt sensitive blood pressure and enhanced vascular reactivity in non-insulin dependent diabetes mellitus. Am J Med 88:210–216, 1990.
171. Neutel JM: Metabolic Manifestations of low-dose diuretics. Am J Med 101 (Suppl 3A):71S–82S, 1966.
172. Fernandez M, Madero R, Gonzalez D, et al: Combined versus single effect of fosinopril and hydrochlorothiazide in hypertensive patients. J Hypertens 231:207–210, 1994.
173. Mersey J, D'Hemecourt P, Blaze K: Once-daily fixed combination of captopril and hydrochlorothiazide as first line therapy for mild to moderate hypertension. Curr The Res 53:502–512, 1993.
174. Reaven GM, Clinkingbeard C, Jeppesen J, et al: Comparison of the hemodynamic and metabolic effects of low-dose hydrochlorothiazide and lisinopril treatment in obese patients with high blood pressure. Am J Hypertens 8:461–466, 1995.
175. Ames RP: The effects of antihypertensive drugs on serum lipids and lipoproteins. I: Diuretics. Drugs 32:260–271, 1986.
176. Kasiske BL, Ma JZ, Kalil RSN, Louis TA: Effects of antihypertensive therapy on serum lipids. Ann Intern Med 122:133–141, 1995.
177. Hoes AW, Grobbee DE, Lubsen J, et al: Diuretis, beta-blockers, and the risk for sudden cardiac death in hypertensive patients. Ann Intern Med 123:481–487, 1995.
178. Psaty BM, Heckbert SR, Koepsell TD, et al: The risk of myocardial infarction associated with antihypertensive drug therapies. JAMA 274:620–625, 1995.
179. Freis ED: The efficacy and safety of diuretics in treating hypertension. Ann Intern Med 122:223–226, 1995.
180. Pollare T, Lithell H, Selinus I, Berne C: Sensitivity to insulin during treatment with atenolol and metoprolol: A randomized double-blind study of the effects on carbohydrate and lipoprotein metabolism in hypertensive patients. BMJ 298:1152–1157, 1989.

181. Lithell HO, Pollare T, Vessby: Metabolic effects of pindolol and propranolol in a double-blind cross-over study in hypertensive patients. Blood Pressure 1:92–101, 1992.
182. Haenni A, Lithell H: Treatment with a beta-blocker with beta 2-agonism improves glucose and lipid metabolism in essential hypertension. Metabolism 43:455–461, 1994.
183. Jacob S, Rett K, Wicklmayr M, et al: Differential effect of chronic treatment with two beta-blocking agents on insulin sensitivity: the carvedilol-metoprolol study. J Hypertens 14:489–494, 1996.
184. Malminiemi K, Lahtela JT, Huupponen R: Effect of celiprolol on insulin sensitivity and glucose tolerance in dyslipidemic hypertension. Int J Clin Pharmacol 32:156–163, 1995.
185. Dornhorst A, Powell SH, Pensky J: Aggravation by propranolol of hyperglycemic effect of hydrochlorothiazide in type II diabetics without alteration of insulin secretion. Lancet 1:123–126, 1985.
186. Schneider M, Lerch M, Papiri M, et al: Metabolic neutrality of combined verapamil-trandolapril treatment in contrast to beta-blocker-low-dose chlorthalidone treatment in hypertensive type 2 diabetes. J Hypertens 14:669–677, 1996.
187. Pollare T, Lithell H, Selinus I, Berne C: Application of prazosin is associated with an increase in insulin sensitivity in obese patients with hypertension. Diabetologia 31:415–420, 1988.
188. Suzuki M, Hirose J, Asakura Y, et al: Insulin insensitivity in nonobese, nondiabetic essential hypertension and its improvement by an alpha 1-blocker (bunazosin). Am J Hypertens 5:869–874, 1992.
189. Shieh SM, Sheu WH, Shen DC, et al: Glucose, insulin, and lipid metabolism in doxazosin-treated patients with hypertension. Am J Hypertens 5:827–831, 1992.
190. Lehtonen A: Doxazosin effects on insulin and glucose in hypertensive patients. The Finnish Multicenter Study Group. Am Heart J 121(4 Pt 2):1307–1311, 1991.
191. Anderson P-E, Lithell HO: Metabolic effects of doxazocin and enalapril in hypertriglyceridemic, hypertensive men. Relationship to changes in skeletal muscle blood flow. Am J Hypertens 9:323–333, 1996.
192. The Treatment of Mild Hypertension Research Group: The Treatment of Mild Hypertension Study: A randomized, placebo-controlled trial of a nutritional-hygienic regimen along with various drug monotherapies. Arch Intern Med 151:1413–1423, 1991.
193. Dominguez LJ, Weinberger MH, Cefalu WT, et al: Doxazocin lowers blood pressure and improves insulin responses to a glucose load with no changes in tyrosine kinase activity or insulin binding. Am J Hypertens 8:528–532, 1995.
194. Baragallo M, Resnick LM: Calcium and magnesium in the regulation of smooth muscle function and blood pressure: The ionic hypothesis of cardiovascular and metabolic diseases and vascular aging. In Sowers JR (ed): Endocrinology of the Vasculature. Totowa, NJ, Humana Press, 1995, pp 283–300.
195. Trost BN, Weidmann P: Effects of calcium antagonists on glucose homeostasis and serum lipids in non-diabetic and diabetic subjects: A review. J Hypertens 5(Suppl 4):S81–S104, 1987.
196. Stern N, Tuck ML: Drug therapy of hypertension in diabetic patients. J Hum Hypertens 5:295–305, 1991.
197. Iversen E, Jeppesen D, Steensgaard-Hansen F: Direct diabetogenic effect of diltiazem. J Intern Med 227:285–286, 1990.
198. Bhatnagar SK, Amin MHA, Al-Yusuf AR. Diabetogenic effects of nifedipine. BMJ 289:19, 1984.
199. Pollare T, Lithell H, Morlin C: Metabolic effects of diltiazem and atenolol: Results from a randomized, double-blind study with parallel groups. J Hypertens 7:551–559, 1989.
200. Lind L, Berne C, Pollare T, Lithell H: Metabolic effects of antihypertensive treatment with nifedipine or furosemide: A double-blind, cross-over study. J Hum Hypertens 9:137–141, 1995.
201. Tuck ML, Bravo EL, Krakoff LR, Friedman CP, and the Modern Approach to the Treatment of Hypertension Study Group: Endocrine and renal effects of nifedipine gastrointestinal therapeutic system in patients with essential hypertension: Results of a multicenter trial. Am J Hypertens 3:333S–341S, 1990.
202. Sheu WH, Swislocki AL, Hoffman B, et al: Comparison of the effects of atenolol and nifedipine on glucose, insulin, and lipid metabolism in patients with hypertension. Am J Hypertens 4:199–205, 1991.
203. Andersson D, Rojdmark S: Improvement of glucose tolerance by verapamil in patients with non-insulin-dependent diabetes mellitus. Acta Med Scand 210:27–33, 1981.
204. deCourten M, Ferrari P, Bohlen L, et al: Lack of effect of long-term amlodipine on insulin sensitivity and plasma insulin in obese patients with essential hypertension. Eur J Clin Pharmacol 44:457–462, 1993.
205. Beer NA, Jakubowicz DJ, Beer RM, Nestler JE: The calcium channel blocker amlodipine raises serum dehydroepiandrosterone sulfate and androstenedione, but lowers serum cortisol, in insulin-resistant obese and hypertensive men. J Clin Endocrinol Metab 76:1464–1469, 1993.

206. Beer NA, Jakubowicz DJ, Beer RM, et al: Effects of nitrendipine on glucose tolerance and serum insulin and dehydroepiandrosterone sulfate levels in insulin resistant obese men. J Clin Endocrinol Metab 76:178–183, 1993.
207. Bursztyn M, Raz I, Meckler J, Ben-Ishay D: Nitrendipine improves glucose tolerance and deoxyglucose uptake in hypertensive rats. Hypertension 23(Pt 2):1051–1053, 1994.
208. Gugliano D, Saccomanno F, Paolisso G, et al: Nicardipine does not cause deterioration of glucose homeostasis in man. Eur J Clin Pharmacol 43:39–45, 1992.
209. Lind L, Berne C, Pollare T, Lithell H: Metabolic effects of isradipine as monotherapy or in combination with pindolol during long-term antihypertensive therapy. J Intern Med 236:37–42, 1994.
210. Touyz RM, Schiffrin EL: Treatment of non-insulin dependent diabetic hypertensive patients with Ca^{2+} channel blockers is associated with increased platelet sensitivity to insulin. Am J Hypertens 8:1214–1221, 1995.
211. Rett K, Jauch KW, Wicklmayr M, et al: Angiotensin converting enzyme inhibitors in diabetes: Experimental and human experience. Postgrad Med J 62(Suppl 1):59–64, 1986.
212. Jauch KW, Hartl W, Guenther HN, et al: Captopril enhances insulin responsiveness of forearm muscle tissue in non-insulin-dependent diabetes mellitus. Eur J Clin Invest 17:448–454, 1987.
213. Paolisso G, Gambardella A, Verza M, et al: ACE inhibition improves insulin-sensitivity in aged insulin-resistant hypertensive patients. J Hum Hypertens 6(3):175–179, 1992.
214. Shionoiri H, Sugimoto K, Minamisawa K, et al: Glucose and lipid metabolism during long-term treatment with cilazapril in hypertensive patients with or without impaired glucose metabolism. J Cardiovasc Pharmacol 15:933–938, 1990.
215. Santoro D, Galvan AQ, Natali A, Ferranini E: Some metabolic aspects of essential hypertension and its treatment. Am J Med 94(4A):32S–39S, 1993.
216. Ludvik B, Kueenburg E, Brunnbauer M, et al: The effects of ramipril on glucose tolerance, insulin secretion, and insulin sensitivity in patients with hypertension. J Cardiovasc Pharmacol 18(Suppl 2):S157–S159, 1991.
217. Bak JF, Gerdes LU, Sorenson NS, Pederson D: Effects of perindopril on insulin sensitivity and plasma lipid profile in hypertensive non-insulin dependent diabetic patients. Am J Med 92(4B):69S–72S, 1992.
218. Allemann Y, Baumann S, Jost M, et al: Insulin sensitivity in normotensive subjects during angiotensin converting enzyme inhibition with fosinopril. Eur J Clin Pharmacol 42:275–280, 1992.
219. Reaven GM, Clinkingbeard C, Jeppesen J, et al: Comparison of the hemodynamic and metabolic effects of low-dose hydrochlorothiazide and lisinopril treatment in obese patients with high blood pressure. Am J Hypertens 8:461–466, 1995.
220. Falkner B, Canessa M, Anzalone D: Effect of angiotensin converting enzyme inhibitor (lisinopril) on sensitivity and sodium transport in mild hypertension. Am J Hypertens 8:454–460, 1995.
221. Donnelly R: Angiotensin-converting enzyme inhibitors and insulin sensitivity: Metabolic effects in hypertension, diabetes, and heart failure. J Cardiovasc Pharmacol 20 (Suppl 11):S38–S44, 1992.
222. Bjorntorp K, Lindgarde F, Mattiasson I: Long-term effects on insulin sensitivity and sodium transport in glucose-intolerant hypertensive subjects when beta-blockade is replaced by captopril treatment. J Hum Hypertens 6(4):291–298, 1992.
223. Lithell HO: Effect of antihypertensive drugs on insulin, glucose, and lipid metabolism. Diabetes Care 14:203–209, 1991.
224. Shamiss A, Carroll J, Peleg E, et al: The effect of enalapril with and without hydrochlorothiazide on insulin sensitivity and other metabolic abnormalities of hypertensive patients with NIDDM. Am J Hypertens 8:276–281, 1995.
225. Alkharouf J, Nalinikumari K, Corry DB, Tuck ML: Long-term effects of the angiotensin converting enzyme inhibitor captopril on metabolic control in noninsulin dependent diabetes mellitus. Am J Hypertens 6:337–343, 1993.
226. Torlone E, Britta M, Rambotti AM, et al: Improved insulin action and glycemic control after long-term angiotensin-converting enzyme inhibition in subjects with arterial hypertension and type II diabetes. Diabetes Care 16:1347–1355, 1993.
227. Prince MJ, Stuart CA, Padia RN, et al: Metabolic effects of hydrochlorothiazide and enalapril during treatment of the hypertensive diabetic patient: Enalapril for hypertensive diabetics. Arch Intern Med 148:2363–2368, 1988.
228. Perani G, Martignono A, Muggia C, et al: Metabolic effects of the combination of captopril and hydrochlorothiazide in hypertensive subjects. J Clin Pharmacol 30:1031–1035, 1990.
229. Weinberger MH: Angiotensin converting enzyme inhibitors enhance the antihypertensive efficacy of diuretics and blunt or prevent adverse metabolic effects. J Cadiovasc Pharmacol 13:51–54, 1989.

230. Graham RD: Treating mild-to-moderate hypertension: A comparison of lisinopril-hydrochlorothiazide fixed combination with captopril and hydrochlorothiazide free combination. J Hum Hypertens 5(Suppl 2):59–61, 1991.
231. Gans ROB, Bilo HJG, Nauta JJP, et al: The effect of angiotensin-1 converting enzyme inhibition on insulin action in healthy volunteers. Eur J Clin Invest 21:527–533, 1991.
232. Jauch KW, Hartl W, Guenther B, et al: Captopril enhances insulin responsiveness of forearm muscle tissue in non-insulin dependent diabetes mellitus. Eur J Clin Invest 17:448–454, 1987.
233. Tomiyama H, Kushiro T, Abeta H, et al: Kinins contribute to the improvement of insulin sensitivity during treatment with angiotensin converting enzyme inhibitor. Hypertension 23:450–455, 1994.
234. Ferrannini E, Galvan AQ, Santoro D, Natali A: Potassium as a link between insulin and the renin-angiotensin-aldosterone system. J Hypertens 10:S5–S10, 1992.
235. Townsend RR, DiPette DJ: Pressor doses of angiotensin II increase insulin-mediated glucose uptake in normotensive men. Am J Physiol 265:E362–E366, 1993.
236. Widgren BR, Urbanavicius V, Wikstrand J, et al: Low-dose angiotensin II increases glucose disposal rate during euglycemic hyperinsulinemia. Am J Hypertens 6:892–895, 1993.
237. Buchanan TA, Thawani H, Kade W, et al: Angiotensin II increases glucose utilization during acute hyperinsulinemia via a hemodynamic mechanism. J Clin Invest 92:720–726, 1993.
238. Morris AD, Donnelly R: Angiotensin II: An insulin-sensitizing vasoactive hormone? 81:1303–1306, 1996.
239. Moan A. Hoieggen A, Seljeflot I, et al: The effects of angiotensin II receptor antagonism with losartan on glcose metabolism and insulin sensitivity. J Hypertens 14:1093–1097, 1966.
240. Seljeflot I, Moan A, Kjeldsen S, et al: Effect of angiotensin II receptor blockade on fibrinolysis durine acute hyperinsulinemia in patients with essential hypertension. Hypertension 27:1299–1304, 1966.
241. Tikkanen I, Omvik P, Jensen HAE, for the Scandinavian Study Group: Comparison of the angiotensin II antagonist losartan with the angiotensin converting enzyme inhibitor enalapril in patients with esssential hypertension. J Hypertens 13:133–1351, 1995.
242. Laakso M, Karjalainen L, Lempiainen-Kuosa P: Effect of losartan on insulin sensitivity in hypertensive subjects. Hypertension 28:392–396, 1966.
243. Iimura O, Shimamoto K, Matsuda K, et al: Effects of angiotensin receptor antagonist and angiotensin converting enzyme inhibitor on insulin sensitivity in fructose-fed hypertensive rats and essential hypertensives. Am J Hypertens 8:353–357, 1995.
244. Iyer SN,Katovich J: Effect of acute and chronic losartan treatment on glucose tolerance and insulin sensitivity in fructose-fed rats. Am J Hypertens 9:662–669, 1996.
245. Gugliano D, DeRosa N, DiMaro G, et al: Metformin improves glucose, lipid metabolism, and reduces blood pressure in hypertensive, obese women. Diabetes Care 16:1387–1390, 1993.
246. Ngai DK, Yudkin JS: Effects of metformin on insulin resistance, risk factors for cardiovascular disease and plasminogen activator inhibitor in NIDDM subjects. Diabetes Care 16:621–629, 1993.
247. Kothen TA: Attenuation of hypertension by insulin-sensitizing agents. Hypertension 28:219–223, 1996.
248. Ogihara T, Rakugi H, Ikegami H, et al: Enhancement of insulin sensitivity by troglitazone lowers blood pressure in hypertensive diabetics. Am J Hypertens 8:316–320, 1995.
249. Nolan JJ, Ludvik B, Beerdsen P, et al: Improvement in glucose tolerance and insulin resistance in obese subjects treated with torglitazone. N Engl J Med 331:1188–1193, 1994.
250. Kemnitz JW, Elson DF, Roecker EB, et al: Pioglitazone increases insulin sensitivity, reduces blood glucose, insulin, and lipid levels, and lowers blood pressure in obese, insulin-resistant rhesus monkeys. Diabetes 43:204–211, 1994.
251. Bhanot S, McNeill JH, Bryer-Ash M: Vanadyl sulfate prevents fructose-induced hyperinsulinemia and hypertension in rats. Hypertension 23:308–312, 1994.

MURRAY EPSTEIN M.D., F.A.C.P. / MARK E. COOPER M.D., F.R.A.C.P.

14

Diabetic Nephropathy: Focus on ACE Inhibition and Calcium Channel Blockade

Diabetes is one of the major causes of endstage renal failure. Recent data from the U.S. Renal Data System Registry indicate that despite good blood pressure control the incidence of endstage renal disease from diabetic nephropathy has not decreased.[1] Indeed, diabetic nephropathy has become the leading cause of endstage renal disease in the United States. Hypertension is acknowledged to be a major risk factor in the progression of diabetic renal disease. Consequently, consideration of the interplay of diabetes, hypertension, and nephropathy is crucial to developing a rational therapeutic regimen for diabetic renal disease.

Diabetic nephropathy, defined as the appearance of proteinuria, elevated arterial blood pressure, and diminished glomerular filtration rate (GFR), develops in as many as 40% of patients with insulin-dependent diabetes mellitus (IDDM).[2–4] In patients with the onset of diabetes at an early age, renal disease is an important contributor to mortality, accounting for up to 31% of all deaths.[5] Renal disease also complicates non–insulin-dependent diabetes mellitus (NIDDM) in adults and contributes significantly to morbidity and mortality.[6] A smaller percentage of patients with NIDDM develop renal disease; the incidence of endstage renal disease is strongly related to the duration of diabetes. NIDDM is usually insidious in onset and thus can remain asymptomatic for many years. In contrast to IDDM, in which the onset is often abrupt, it is not unusual for NIDDM to be recognized only after many years when secondary complications have developed. As a consequence, information about the prevalence and incidence of renal disease in NIDDM is difficult to interpret. Prospective studies of the natural history of NIDDM are necessary to investigate the evolution of nephropathy in NIDDM. Such studies have been conducted in the Pima Indian population in Arizona, a group with an extraordinarily high incidence of NIDDM and endstage renal disease.[7]

In both types of diabetes mellitus, the appearance of clinically detectable proteinuria (>200 μg/min of urinary albumin excretion) signals the onset of the relentless progression of diabetic nephropathy, which is typically followed by deterioration to endstage renal disease.[8] Currently, it is believed that diabetic nephropathy develops as a result of the interplay of metabolic abnormalities inherent to diabetes (e.g. hyperglycemia) and hemodynamic abnormalities of the renal microcirculation that result in progressive structural and functional glomerular abnormalities.[9–11]

PATHOGENESIS OF DIABETIC NEPHROPATHY

Several studies have suggested that diabetic nephropathy may be genetically determined. Indeed, diabetic siblings of probands with diabetic nephropathy have a high incidence of renal disease.[12] Epidemiologic studies have revealed that the risk of diabetic nephropathy is three times higher in diabetic patients in whom parental hypertension was present.[13] Furthermore, the rate of sodium lithium countertransport in red blood cells, a marker of risk for essential hypertension, has been demonstrated to be higher in patients with diabetic nephropathy than in patients with other forms of renal disease.[14,15] The mechanisms involved in this genetically determined end-organ susceptibility have not been fully defined.

Investigators in Copenhagen proposed the Steno hypothesis to account for the generalized macroangiopathy observed in diabetes.[16] They postulate that albuminuria reflects a more generalized vascular process (endothelial dysfunction), which affects the glomerulus, retina, and intima of large vessels. This hypothesis explains why patients with NIDDM and microvascular complications such as nephropathy have a high incidence of atherosclerotic vascular disease.[16,17]

Familial and genetic factors may predispose to diabetic nephropathy.[18] For example, one study, as outlined previously, examined the incidence of diabetic nephropathy (defined as abnormal rates of albumin excretion) in sibling pairs concordant for insulin-dependent diabetes. In diabetic siblings who had an elevated rate of albuminuria, 83% had excessive albuminuria. On the other hand, only 17 percent of diabetic siblings in probands with a normal rate of albumin excretion had excessive albuminuria. Thus, this study suggests the occurrence of familial clustering of diabetic nephropathy.[12]

Hollenberg and Williams have demonstrated that in diabetic patients prone to develop nephropathy, a family history of hypertension and increased red blood cell sodium and lithium countertransport are associated with a clinical constellation consisting of disturbed renal sodium handling and sodium-sensitive hypertension, abnormalities in the renal vascular response to changes in sodium intake and angiotensin II (AII), blunted decrements of renin release in response to saline or AII, and an accentuated renal vasodilator response to inhibition of angiotensin-converting enzyme (ACE).[19]

Increasing investigative efforts have focused on the pivotal role of renal microcirculatory derangements that characterize diabetic nephropathy both as a basis for development of nephropathy and as a rational framework for therapy. In experimental models of diabetes, glomerular hyperfiltration associated with increased intraglomerular pressures and flow appears to be responsible for the development and progression of diabetic nephropathy.[20] Experimental maneuvers known to increase intraglomerular pressure, such as contralateral uninephrectomy[21] and high-protein feeding,[20,22] have been shown to accelerate renal injury in experimental diabetic nephropathy. In fact, in experimental diabetes, glomerular flows and pressures are increased even when systemic blood pressure is normal.[22] This finding may be due to an imbalance in the interaction between hemodynamic determinants that control the glomerular microcirculation—namely, afferent and efferent vascular resistances—and the glomerular capillary ultrafiltration coefficient. Moreover, in diabetes a relative decrease in preglomerular resistance may permit disproportionate rises in glomerular pressures, even when the elevation in systemic hypertension is small,[23–25] thereby explaining the exquisite sensitivity of the diabetic kidney to mild hypertension.

Several studies have suggested a role for endothelium-derived nitric oxide (NO) in the pathogenesis of the renal hemodynamic changes of experimental diabetes.[26,27] In one study renal hemodynamic changes were correlated with changes in urinary excretion of NO_2/NO_3 in response to the NO inhibitor nitro-L-arginine methyl ester (L-NAME) and the NO-donating agent, glyceryl trinitrate (GTN). The results of these studies suggest that increased renal production and/or sensitivity to endothelium-derived relaxing factor/NO may play a role in the genesis of diabetic hyperfiltration.[28] A better understanding of renal hemodynamic and glomerular microcirculatory changes has not only pathophysiologic but also important therapeutic implications.

Glucose-dependent Biochemical Processes

Altered Polyol Metabolism

Although chronic hyperglycemia has been linked to the development of the long-term complications of diabetes, the precise mechanisms are incompletely defined. A large body of clinical and experimental evidence supports a role for the hyperglycemia-induced acceleration of polyol pathway metabolism in mediating the development of diabetes-induced neuropathy, retinopathy, and nephropathy. In tissues in which glucose uptake is independent of insulin—i.e., the lens, retina, peripheral nerves, and kidney—diabetic hyperglycemia results in increased levels of tissue glucose.[29–31] The excessive glucose is subsequently reduced to sorbitol by aldose reductase, the first enzyme in the polyol pathway.[29–31] Aldose reductase is dependent on reduced nicotinamide adenine dinucleotide phosphate (NADPH). The increased formation and accumulation of sorbitol in these tissues is accompanied by a depletion of free myoinositol, loss of sodium-potassium-adenosine triphosphatase (Na^+,K^+-ATPase) activity, and increased consumption of the enzyme cofactors NADPH and oxidized nicotinamide adenine dinucleotide (NAD^+), leading to changes in cellular redox potential.[32,33] These metabolic derangements have been postulated to result in cellular dysfunction and, ultimately, the morphologic lesions that characterize diabetic neuropathy, nephropathy, retinopathy, and cataract formation.[30,34–36] Indeed, prevention of cataracts, preservation of retinal and neural ultrastructure, and prevention and reversal of nerve conduction abnormalities in animal models of diabetes by aldose reductase inhibitors have been firmly established.[37–41]

Accumulating experimental data have implicated increased polyol pathway metabolism in mediating the loss of renal vascular tone in animals with chronic hyperglycemia. Several investigators have demonstrated that glomerular hyperfiltration in diabetic rats could be prevented by treatment with aldose reductase inhibitors.[42–44] The role of altered polyol pathway metabolism in mediating the impairment of afferent arteriolar tone that supervenes in diabetes has been investigated in galactose-fed rats.[45] The galactose-fed rat constitutes an ideal model in which to study the relative participation of increased polyol metabolism in the chronic complications of diabetes without the confounding effects of hyperglycemia and insulin deficiency observed in streptozocin (STZ)-induced diabetic rats.[46] In contrast to control rats with hydronephrotic kidneys, rats fed a diet of 50% galactose lose the intrinsic ability for the afferent arteriole to vasoconstrict in response to graded increases in renal perfusion pressure.[45] This striking defect in myogenic contractility in galactose-fed rats was identical to the myogenic defect that has been documented previously in STZ-diabetic rats.[47]

Further studies have indicated that interventions with aldose reductase inhibitors may prevent the loss in renal microvascular resistance in galactose-fed rats.[48] In a similar group of rats fed a 50% galactose diet supplemented with the aldose reductase inhibitor, tolrestat (25 mg/kg), pressure-induced afferent arteriolar vasoconstriction did not differ from that in control rats. Thus, the impairment of myogenic responsiveness by galactose feeding and its prevention by tolrestat implicate elevated polyol metabolism in the pathogenesis of deranged autoregulation in experimental diabetes and suggest future therapeutic strategies using aldose reductase inhibitors to retard the development of diabetic nephropathy.

Several clinical trials have demonstrated improvement in a number of variables relating to motor, sensory, and autonomic neuropathy as well as glomerular hyperfiltration in diabetic patients treated with aldose reductase inhibitors, thereby complementing and confirming the sorbitol hypothesis.[49–54] In a double-blind placebo-controlled study[53] the administration of tolrestat for 6 months to 20 hyperfiltering patients with IDDM and early nephropathy produced a normalization of GFR. Concomitantly, there was a fourfold decrease in urinary albumin excretion (UAE). No significant change in renal plasma flow (RPF) was seen during tolrestat treatment. Filtration fraction decreased significantly in the tolrestat-treated group. Pedersen et al.[54] reported similar results in normoalbuminuric, hyperfiltering patients with IDDM. They demonstrated that 6 months of treatment with the aldose reductase inhibitor ponalrestat reduced the hyperfiltration in patients with IDDM without concomitant changes in RPF, renal vascular resistance, or AER.

Advanced Glycation

A major glucose-dependent pathway that is now an area of active interest is the process of advanced glycation. This pathway involves a spontaneous reaction between glucose and proteins and lipids, particularly long-lived structural proteins, leading to the formation of advanced glycation (glycosylation) end products (AGEs).[55] This process involves a series of biochemical reactions, the first of which leads to formation of Amadori products known as early glycation products. The best-known example of such a protein is HbA1c. This is followed by a series of poorly defined reactions that generate a range of intermediates, including 3-deoxyglucosone,[56] and ultimately result in various products, some of which are fluorescent. N-fluorescent AGEs that involve not only glycation but also oxidation steps, such as carboxymethylysine, also have been detected.[57] In vivo evidence that this glucoxidation pathway is operative in diabetes has been clearly documented by several groups. A radioimmunoassay that detects an epitope on the carboxymethylysine molecule has demonstrated increased levels of these products in blood vessels and kidneys from diabetic animals.[58,59] Studies involving inhibition of AGE formation with the hydrazine derivative, aminoguanidine, have allowed investigators to explore the relationship between AGE formation and diabetic vascular injury. Aminoguanidine does not inhibit the generation of early glycated products but acts at a subsequent site, not clearly defined, to reduce AGE formation. Aminoguanidine treatment has been shown to decrease AGE formation in rat tissues, including the aorta[60] and kidney.[61,62] These results have been an important stimulus for clinical studies in diabetic nephropathy, which are now in progress. Aminoguanidine has been administered to humans and shown to inhibit hemoglobin-AGE levels.[63]

New, more potent inhibitors of advanced glycation have been developed recently. Two of these, ALT462 and ALT486, are approximately 5 and 20 times, respectively, more potent than aminoguanidine in their ability to inhibit fluorescence generated from the reaction of lysozyme with ribose.[64] Of particular interest is the thiazolium compound, phenacylthiazolium bromide (PTB), a new compound that reacts with and cleaves covalent, AGE-derived protein cross-links.[65] Daily intraperitoneal injections of PTB in streptozotocin diabetic rats have been shown to ameliorate AGE cross-links in rat tail collagen over 32 weeks and to halve IgG binding to red blood cell surfaces over 4 weeks.[65] If PTB can be shown to have similar effects in the kidney, researchers would have a conceptual basis for the reversal of AGE-mediated tissue damage, which till now has been regarded as irreversible. Such an agent may be particularly useful in diabetic patients with established renal disease.

Protein Kinase C

The adverse effects of hyperglycemia have been attributed to activation of protein kinase C (PKC),[66] a family of serine-threonine kinases that regulates diverse vascular functions, including contractility, blood flow, cellular proliferation,[67] and vascular permeability.[68] PKC activity, especially of the membrane- bound form, is increased in the glomeruli of diabetic animals, probably because of an increase in de novo synthesis of diacylglycerol, a major endogenous activator of PKC.[69] The observation of preferential activation of the βII isoform of PKC in diabetes[66] led to the synthesis of an orally effective PKCβ-selective inhibitor, LY333531.[70] LY333531 is a competitive, reversible inhibitor of $PKC\beta_I$ and $PKC\beta_{II}$, with a 50-fold lesser effect on other PKC isoenzymes. In studies over 2–8 weeks in streptozotocin diabetic rats, LY333531 ameliorated glomerular hyperfiltration and albuminuria.[70]

Transforming Growth Factor-β

Recent investigative attention has focused on the possible role of the prosclerotic cytokine, transforming growth factor-β (TGF-β), in mediating the mesangial matrix expansion that characterizes diabetic nephropathy. At least three actions of TGF-β are central to normal tissue repair: (1) stimulation of matrix synthesis, (2) inhibition of matrix degradation, and (3) modulation of matrix receptors. These actions have been demonstrated to underlie matrix accumulation in experimental glomerulonephritis.[71] Although fundamentally different in pathogenesis from glomerulonephritis, diabetic nephropathy shares the common feature of mesangial matrix expansion. Based on this association, it was determined whether TGF-β expression might be involved in diabetic kidney disease.[72] Onset of diabetes was followed by a progressive increase in the levels of TGF-β mRNA and TGF-β protein in the glomeruli from STZ-diabetic rats. Insulin treatment tended to reduce the expression of TGF-β, although the levels were still abnormal compared with those of age-matched normal rats with or without identical treatment with insulin. The investigators demonstrated elevated levels of proteoglycans and other matrix components known to be induced by TGF-β, indicating increased TGF-β biologic activity in the diabetic glomeruli. The relevance of these findings in the experimental model of human disease was confirmed by the demonstration of similarly elevated levels of TGF-β protein and matrix in the glomeruli of patients with diabetic nephropathy.[72] In contrast, glomeruli from normal kidneys or from patients with nonprogressive glomerular diseases (minimal change disease and thin basement membrane disease) were negative for TGF-β. Thus, TGF-β may play a key role in the development of glomerular matrix accumulation in both glomerulonephritis and diabetic nephropathy.[73,74]

The possible role of TGF-β in the pathogenesis of diabetic nephropathy is currently under active investigation in several laboratories. Further evidence of a role for TGF-β in diabetic renal disease is provided by the results of a study showing that administration of antibodies to TGF-β to diabetic mice was associated with attenuation of renal hypertrophy.[75] Of particular interest is the possibility that agents with a renoprotective role in diabetic vascular complications may mediate their effects via TGF-β–dependent pathways. Angiotensin II (AII) has been shown in vitro[76] to be a potent stimulator of TGF-β production in mesangial cells. This, in turn, leads to collagen formation. Recent in vivo studies in the model of subtotal nephrectomy, which has many renal hemodynamic similarities to experimental diabetes,[20] have indicated that ACE inhibitor and AII receptor antagonist therapies were associated with a reduction in TGF-β1 gene expression in the kidney, both in glomeruli and in the tubulointerstitium.[77] This decrease in TGF-β1 gene expression closely correlated with a decrease in the mRNA levels for the extracellular matrix protein, type IV collagen. Recent studies indicate that the increase in TGF-β1 gene expression is also observed in diabetic vessels and that these changes can be ameliorated not only by ACE inhibition but also by the inhibitor of advanced glycation, aminoguanidine.[59] These studies provide further evidence for a pivotal role of TGF-β in the pathogenesis of diabetic microvascular complications.

TREATMENT OF HYPERTENSION IN PATIENTS WITH DIABETES MELLITUS

General Goals of Therapy

No large, population-based, randomized trials of hypertension treatment in diabetic patients have been conducted. Nevertheless, as proposed by a recent consensus statement,[78] the goal of treating hypertension in diabetic patients should be to prevent death and disability associated with high blood pressure. In addition, other reversible risk factors for cardiovascular disease need to be addressed. For example, target-organ involvement should be taken into consideration when formulating a treatment plan. Clinical and investigative efforts have focused on retarding progression of diabetic nephropathy (see below) and, to a lesser extent, reducing cardiovascular morbidity and mortality.

The diagnosis of hypertension should be based on multiple blood pressure measurements obtained in a standardized fashion on at least three occasions.

Automated ambulatory blood pressure (ABP) monitoring can be valuable in resolving certain clinical situations (e.g., "white coat" hypertension, or high blood pressure in the medical office with normal self-measurement at home). In addition, ABP monitoring may be especially helpful in diabetic patients for documenting blood pressure control over an entire 24-hour period to confirm the absence of the usual nocturnal fall in blood pressure in diabetes (especially with autonomic dysfunction or nephropathy) and for documenting episodic hypertension, orthostatic hypotension, or resistant hypertension.[79]

It is important to emphasize, however, that many patients with diabetic glomerulopathy exhibit orthostatic hypotension because of attendant autonomic neuropathy. Consequently, blood pressure should be measured with the patient in both recumbent and upright positions,[79] and the latter measurement should be used for defining the target of control.

What Is the Optimal Blood Pressure Level During Treatment?

The optimal blood pressure level during antihypertensive treatment in patients with diabetic nephropathy has not been defined. According to the Fifth Report of the Joint National Committee on Detection, Evaluation, and Treatment of High Blood Pressure (JNC-V), hypertension is defined as an average blood pressure of 140 mmHg systolic or 90 mmHg diastolic.[78] Diabetes is associated with a high incidence of hypertension, according to the JNC-V criteria. For example, the proportion of type I diabetic patients with hypertension attending a diabetes clinic in Denmark was 42%, 52%, and 79% in normo-, micro-, and macroalbuminuric patients, respectively, confirming the strong association of elevated blood pressure with evolving nephropathy in type I diabetic patients.[80] Type II diabetes is also associated with a high incidence of hypertension, but in this setting it occurs even in the absence of renal disease.[81] The prevalence of hypertension increases with increasing albuminuria in normo-, micro-, and macroalbuminuric type II diabetic patients. In the same Danish outpatient clinic, the frequency of hypertension was 71%, 90%, and 93%, respectively.[80]

The JNC-V further states that "in nondiabetic patients, nonfatal and fatal cardiovascular disease including coronary heart disease and stroke, as well as renal disease and all-cause mortality, increase progressively with higher levels of both systolic and diastolic blood pressure."[78] These relationships are strong, continuous, graded, consistent, independent, predictive, and etiologically important. In the general population, risks are lowest for adults with an average systolic blood pressure < 120 mmHg and an average diastolic blood pressure < 80 mmHg. Higher levels of either systolic or diastolic blood pressure are associated with increased risks of morbidity, disability, and mortality.[78]

Although the optimal blood pressure level during antihypertensive treatment in patients with diabetic nephropathy has not been defined, a review of the relationship between the rate of fall in GFR and blood pressure level during antihypertensive treatments suggests that we should strive for lower goal blood pressure than recommended by some of the current guidelines.[82–85] The benefit of reducing blood pressure has been demonstrated most clearly when treatment is instituted before the GFR is markedly reduced. This emphasizes the concept that efforts to reduce the blood pressure should begin before the serum creatinine is elevated. The best results appear to have been achieved by reducing blood pressure below conventionally accepted levels. The study by Parving et al.[82] suggests that the smallest decline in renal function is found in patients with blood pressure levels around 130/85, a level that is readily attainable during treatment in incipient diabetic nephropathy before the decline in GFR has started.

Large, prospective, randomized clinical trials, such as the Appropriate Blood Pressure Control in Type II Diabetes (ABCD) Trial[86] and the Effect of Intensity of Blood Pressure Management on the Progression of Type I Diabetic Nephropathy Trial,[87] may provide more substantive data defining the optimal blood pressure levels. The ABCD Trial, currently in progress, in addition to comparing an ACE inhibitor with a calcium channel blocker, aims to answer two further questions.[86] Firstly, the study plans to assess whether reducing diastolic blood pressure to 75 mmHg rather than 89 mmHg will confer added benefit in regard to complications in hypertensive diabetic patients. Secondly, the study will examine the effect of a 10-mmHg reduction in blood pressure in normotensive people. The primary outcome to be examined is renal function as assessed by creatinine clearance. The secondary outcome measures include albumin excretion rate, left ventricular hypertrophy, retinopathy, and neuropathy. The results of this 5-year prospective trial are keenly awaited.

Isolated systolic hypertension is defined as a systolic blood pressure > 140 mmHg with a diastolic blood pressure < 90 mmHg and is more common in diabetic patients of all ages. The initial goal of therapy is to reduce systolic blood pressure to < 160 mmHg for patients with systolic blood pressure of 180 mmHg and to lower blood pressure by 20 mmHg for patients with a systolic blood pressure of 160–179 mmHg. If such therapy is well tolerated, it may be appropriate to reduce the blood pressure even further. At isolated systolic blood pressure levels of 140–160 mmHg, lifestyle modifications may be considered as adjunctive or definitive therapy.

In young diabetic patients, age-related normal blood pressures should be used as a goal for treatment, and a persistent elevation of blood pressure greater than the 95th percentile for age should be considered hypertensive.

An Ad Hoc Committee of the Council on Diabetes Mellitus of the National Kidney Foundation has issued a consensus statement recommending that in diabetic patients with microalbuminuria, blood pressure should be reduced to 130/85 mmHg or lower.[88] The recommendations of six different organizations with respect to blood pressure and microalbuminuria have been recently summarized.[89] The authors made further modifications to the strategy outlined above.[88]

Metabolic Control and Lifestyle Factors: Concerns in Diabetic Nephropathy

The diagnosis of nephropathy is a crucial time to assist diabetic patients in refocusing their efforts to make lifestyle changes that are necessary to prevent additional morbidity and early mortality. The most important lifestyle change that can influence the progression of nephropathy is to optimize metabolic control. Initiatives include glycemic control, protein restriction, and attempts to influence lipid abnormalities favorably.

Lifestyle modifications may serve as definitive therapy for mild hypertension in diabetic patients or as an adjunct to pharmacologic therapy to lower the number and dose of antihypertensive drugs.[90] The diet recommended by the American Diabetes Association, which is low in calories and fat, high in carbohydrate and soluble fiber, and moderately low in protein, has been reported to lower blood pressure in diabetic patients.[90] Moderate salt restriction reduces systolic blood pressure, which is often inordinately elevated in diabetic patients.[90,91] The importance of salt restriction in hypertensive, diabetic patients may be of particular relevance in patients at risk or with evidence of renal disease. Although not formally tested in humans, it has been shown experimentally that salt restriction is associated with attenuation of the development of renal hypertrophy and albuminuria.[92] Weight reduction is important, particularly in type II diabetic patients, and improves glucose tolerance as well as reduces blood pressure.[90] For example, for each 10-lb reduction in weight systolic and diastolic pressures can be expected to decrease by 10 and 5 mmHg, respectively.[90,93] Moderate but regular aerobic exercise improves glycemic and lipid control and helps with weight reduction. It also helps to lower blood pressure by decreasing peripheral vascular resistance.[90]

Although for some time it has been believed that tighter control of blood glucose concentrations results in fewer micro- and macrovascular complications, definitive evidence was obtained from the results of the Diabetes Control and Complications Trial (DCCT).[94] This 7-year study of more than 1,440 patients examined the effects of aggressive control of blood sugar to "physiologic levels" on the development of nephropathy and retinopathy among patients with type I diabetes.[94] It demonstrated that daily, multiple insulin injections to keep fasting blood sugar values less than or equal to 120 mg/dl provided protection against the progression of renal disease. Specifically, intensive therapy reduced the occurrence of microalbuminuria by 39%, and the occurrence of albuminuria by 54%. In addition, compared with patients receiving standard therapy, patients receiving intensified therapy had lower rates of serious retinopathy requiring photocoagulation (27% vs. 52%), lower rates of decreased visual acuity (14% vs. 35%), and fewer cases of overt nephropathy (1 patient vs. 9 patients). These findings suggest that complications that may lead to long-term disability, blindness, and renal dialysis in patients with diabetes may be prevented or at least delayed by an intensified insulin treatment that includes educating the patient in diabetes control. However, the results of the DCCT study were not as clearcut in the group that had microalbuminuria at the commencement of the trial.[94] In addition, a large British trial also failed to show a clear benefit of intensified glycemic control in type I diabetic patients with persistent microalbuminuria.[95]

The role of intensified glycemic control in patients with overt renal disease remains unresolved. Initially, it was thought that aggressive blood sugar control had no significant effect on the progression of established diabetic nephropathy.[96] However, more recent studies suggest that even in the setting of macroproteinuria, tight glycemic control has a renoprotective role.[97]

The results of an innovative study designed to test the hypothesis that optimized glycemic control in type I diabetic recipients of renal allografts will prevent or delay diabetic renal lesions in the allograft were recently reported.[98] The study was a prospective, controlled, and randomized trial of glycemic control in an inception cohort (i.e., all patients were at stage 0 for diabetic renal lesions in the graft when randomized to the trial) of type I diabetic renal allograft recipients. The experimental group had maximized glycemic control, and the control group received standard clinical diabetic care. Allograft biopsies were performed at baseline (before transplant) and 5 years after transplant. The primary endpoint of the trial was the difference between the groups in glomerular mesangial expansion as determined by electron microscopy. The investigators noted a more than twofold increase in the volume fraction of mesangial matrix per glomerulus in the standard therapy group compared with the maximized therapy group and a threefold increase in arteriolar hyalinosis and greater widening of the glomerular basement in patients receiving standard treatment. These results approached statistical significance. The trial indicates a causal relationship between hyperglycemia and an important lesion of diabetic nephropathy, mesangial matrix expansion, in renal allografts transplanted into diabetic recipients.

Patients with type II diabetes (NIDDM) were not studied in the DCCT trial but are currently under study in a trial in Great Britain.[99] Prospective studies in patients with NIDDM suggest that, as in patients with IDDM, the rate of increase in albuminuria correlates with long-term glycemic control.[100] Therefore, it is probable that, in general, the effects of better control of blood glucose also apply to patients with NIDDM. The eye, kidney, and nerve abnormalities appear quite similar in IDDM and NIDDM, and it is likely that the same or similar underlying mechanisms of disease apply. A recent prospective study performed in Japan over 6 years has suggested a role for intensified glycemic control in preventing the development of microvascular complications in NIDDM subjects.[101]

It is anticipated that the results of the ongoing trial in Great Britain will help to determine whether rigid glucose control affects atherosclerotic vascular disease. However, additional problems may be experienced in the NIDDM cohort. The weight gain characteristic of intensive therapy[102] may be greater in patients with NIDDM because many are already overweight or frankly obese. Also, because of the increased prevalence of macrovascular disease, older patients with NIDDM may be more vulnerable to serious side effects of hypoglycemia, including falls, stroke, seizures, silent ischemia, myocardial infarctions, and even sudden death.

In summary, glycemic control appears to affect the rate of progression of diabetic nephropathy, particularly in the early stages. Achieving optimal glycemic control in patients with diabetic nephropathy requires the care of a team of experts, including nephrologists, endocrinologists, and hypertension specialists.[103]

Blood Pressure Control and Progression of Diabetic Nephropathy

Hypertension invariably complicates the course of patients with diabetic nephropathy. Of the 35–40% of patients with either IDDM or NIDDM who ultimately develop nephropathy, all at some time in their natural history will be hypertensive.[81,104] It is well established that hypertension constitutes a major risk factor in the progression of diabetic renal disease.[105] Consequently, much investigative effort has focused on attempts to attenuate or retard diabetic nephropathy by lowering blood pressure.

Although evidence indicates that diabetic nephropathy cannot be cured, persuasive data suggest that its clinical course can be substantially modified not only by metabolic control and lifestyle modifications but also by normalization of blood pressure, often involving pharmacologic therapies.

Numerous clinical trials of hypertensive patients with either IDDM or NIDDM and nephropathy have assessed diverse forms of blood pressure-lowering therapy on progression of renal disease.[83–85] These trials have demonstrated that aggressive reduction of elevated blood pressure to levels below 140/90 mmHg retards progression of diabetic renal disease. Furthermore, data suggest that some antihypertensive drugs may confer unique beneficial effects in attenuating the progression of this disease independently of their blood pressure-lowering effects.[106]

Pharmacologic Treatment

General Concepts

Pharmacologic therapy should be initiated when lifestyle modifications are unsuccessful in controlling hypertension in diabetic patients. All classes of antihypertensive drugs are effective in controlling blood pressure in diabetic patients. The ADA Consensus Panel[107] recommended five classes of drugs as effective first-line, single-agent therapy. Nevertheless, only a limited number of diabetic patients have been included in the large-scale hypertension treatment trials that evaluate the role of blood pressure control in reducing cardiovascular mortality. Only diuretics and beta blockers have been shown to reduce cardiovascular morbidity and mortality in hypertensive patients. Approximately 10% of the participants in both the Hypertension Detection and Follow-up Program (HDFP)[108] and Systolic Hypertension in the Elderly Program (SHEP)[109] had diabetes. Although the diabetic subgroup was not specifically analysed, the relative reductions in mortality rates (HDFP) and stroke (SHEP) appeared to be similar in the diabetic cohort compared with other subjects. Most other large-scale trials have not included significant numbers of patients with diabetes, not screened for its presence, or not performed an appropriate subgroup analysis. As yet, no studies have specifically examined the efficacy of antihypertensive therapy in reducing cardiovascular morbidity and mortality in diabetic patients. However, many groups have demonstrated the ability of antihypertensive therapy to reduce albuminuria in diabetic patients. Thus, given the link between micro- and macroalbuminuria and both cardiovascular disease and nephropathy in diabetes, antihypertensive efficacy in patients with diabetes should be assessed in terms of both reduction in blood pressure and the surrogate endpoint of reduction in albuminuria. Such studies have led to the concept that certain antihypertensive drugs may be "organ-protective" in addition to their hypotensive action.

The ADA Consensus Panel reached no consensus that any single class of antihypertensive drugs was preferred as initial drug therapy for hypertension in diabetes in the absence of nephropathy.[107] Each class of drugs has potential advantages and disadvantages. Recent data from several large-scale hypertension treatment trials suggest that some classes may be preferred in diabetic patients with nephropathy. These considerations are discussed in more depth below.

The major focus is on ACE inhibitors, which are viewed by many investigators and consensus panels as the drugs of first choice in diabetic patients, particularly in the setting of renal disease, and calcium channel blockers, a widely used group of antihypertensive drugs that, although not as extensively studied in diabetes, may have important renoprotective properties, at least from a theoretical point of view.[110]

Angiotensin-converting Enzyme Inhibitors

ACE inhibitors have no adverse effects on lipid levels or glycemic control.[111] In addition to the compelling importance of blood pressure reduction per se, increasing evidence indicates that some classes of antihypertensive medication may confer a greater effect than others in slowing progression of diabetic renal disease despite similar levels of blood pressure reduction. Such studies have provided a theoretic framework for anticipating that ACE inhibition may preferentially retard the progression of diabetic renal disease. Studies over the past decade have demonstrated that the sustained increase in glomerular capillary pressure evoked in response to loss of renal mass produces a destructive sclerosing reaction.[20,112] Administration of ACE inhibitors decreases glomerular capillary pressure with a resultant reduction of glomerular sclerosis, suggesting that ACE inhibitor therapy may protect the injured kidney from hemodynamically mediated glomerular damage.[106,112]

ACE inhibitors may confer their beneficial effects by modulating events independent of their ability to attenuate glomerular hypertension. Studies conducted in rodent models of diabetes have established that, in addition to a persistent increase in intraglomerular capillary pressure compared with control animals, diabetic animals show an increase in glomerular volume,[92,113–115] expansion of the mesangial matrix, and ultimately development of focal glomerulosclerosis.[25,113,116] Indeed, a recent study clearly demonstrated the importance of AII-dependent mechanisms in the genesis of diabetes-associated glomerular hypertrophy.[117] Both the ACE inhibitor, ramipril, and the AII receptor anatagonist, valsartan, prevented the development of glomerular hypertrophy after 6 months of diabetes in rats. This reduction in glomerular volume was associated with retardation in glomerular basement membrane thickening and development of albuminuria, consistent with the inportance of trophic processes in the genesis of progressive renal injury.[118] Other actions of ACE inhibitors relevant to diabetes include effects on heparan sulphate synthesis[119] and growth factor expression.[77]

ACE inhibitors also may confer a beneficial role by attenuating vascular hypertrophy. Angiotensin II has been shown to promote growth in vascular smooth muscle cells and also to be mitogenic in mesangial cells.[120] ACE inhibitors have an antitrophic action in experimental diabetes.[121] The ability of angiotensin II blockade to inhibit vascular growth may be mediated via modulation of protein kinase C-dependent mechanisms. For example, Williams et al. have shown that AII stimulation of vascular endothelial growth factor, a cytokine expressed in the glomerulus that promotes vascular permeability and is angiogenic, is mediated by protein kinase C-dependent pathways.[122]

The earliest clinical studies were open and compared baseline renal hemodynamics before initiation of ACE inhibition with subsequent events during treatment. Unfortunately, the results of studies using less than rigorous protocols were further complicated by the fact that most were of short duration. Only in the past 5–10 years have appropriate studies compared ACE- inhibition with alternative therapies. Bjørck et al. studied 40 patients with IDDM and diabetic nephropathy randomized to treatment with either enalapril or metoprolol, generally combined with furosemide.[123] Their major outcome measure was the rate of decline in GFR as assessed with a widely accepted isotopic method. Treatment with enalapril, in comparison with metoprolol, resulted in a highly statistically significant reduction in the rate of decline of GFR and in the level of proteinuria. Overall, there was no statistical difference in the degree of blood pressure reduction or level of blood pressure achieved with the two treatments.

Until recently, however, a well-designed, prospective, double-blind, randomized trial with a sufficient number of patients to assess the issue had not been performed. In 1993, Lewis and the Collaborative Study Group reported the results of a trial designed to determine whether the angiotensin-converting enzyme inhibitor captopril is more effective in slowing the progression of diabetic nephropathy than conventional blood pressure-lowering agents.[124] The investigators performed a randomized, controlled trial comparing captopril with placebo in patients with IDDM in whom urinary protein excretion was > 500 mg/day. Blood pressure goals were defined to achieve control during a median follow-up of 3 years. The primary endpoint was a doubling of the baseline serum creatinine concentration. The investigators reported that serum creatinine concentrations doubled in 25 patients in the captopril group compared with 43 patients receiving conventional therapy (termed "placebo" by the investigators). The reduction in risk of a doubling of the serum creatinine concentration was 48% in the captopril group as a whole. Captopril therapy was associated with a 50% reduction in the risk of the combined endpoints of death, dialysis, and transplantation.

Additional insights from this study included confirmation of previous reports that a reduction in urinary protein excretion predicts remission of diabetic renal disease. The large number of nephrotic-range proteinuric patients enrolled in the Collaborative Study Group[126] provided an opportunity to test the widely believed notion[125] that the onset of nephrotic-range proteinuria in diabetic patients heralds the onset of inexorable progression to endstage kidney failure. In a subgroup analysis of the data, the investigators reported that remission of nephrotic-range proteinuria occurred in 7 of 42 patients assigned to captopril (16.7%; mean follow-up: 3.4 ± 0.8 years) but in only 1 of 66 patients assigned to placebo.[124] The findings were interpreted as suggesting that both blood pressure control and reduced proteinuria contribute to the reduced rate of GFR loss in the remission group. The possibility that reducing proteinuria per se may be an important determinant in slowing progression of renal disease is suggested by studies in experimental models of proteinuria and in humans.[127] These studies indicate that filtered proteins or substances accompanying the filtered proteins may cause injury to the glomerular mesangium or epithelium or to the renal tubules.[128]

Of interest, the Collaborative Study Group found that a greater effect of the converting enzyme inhibitor to preserve renal function was seen among patients with more advanced renal disease (baseline serum creatinine concentration >1.5 mg/dl or 133 μmol/L). This subgroup accounted for 25% of the study population. This finding may have resulted from a faster rate of decline in GFR, which allowed more stopping points during the 3-year study, or from a greater impact of the drug in this population. The study was sufficiently compelling that the Food and Drug Administration granted an indication for captopril as a renoprotective agent in diabetes. It should be emphasized that the study was carried out in patients with IDDM, who account for an increasingly smaller number of patients who start dialysis in the United States. Whether these findings can be generalized to patients with NIDDM remains to be established.

A number of caveats are in order when reviewing the results from the Collaborative Study. Although patients were randomized to the control or captopril group, there was a trend for proteinuria to be higher in the placebo group. Furthermore, mean systolic pressure was 4 mmHg lower in the captopril group than in the placebo group. Although these values were not significantly different, it is conceivable that such a difference may affect the target organs at risk. The possibility that the effect of captopril involves blood pressure reduction is further suggested by the lack of a significant effect of ACE inhibition in the minority of patients who were normotensive. A significant number of type I diabetic patients from the study by Lewis et al. have been recruited into a new protocol that will evaluate the role of intensive blood pressure reduction (mean arterial pressure < 92 mmHg) with low and high doses of ramipril on the progression of renal disease.[85] This protocol should provide further insight into the relative contributions of inhibition of the renin-angiotensin system and blood pressure reduction in mediating the renoprotective effects of ACE inhibition. Finally, the captopril dosing schedule (3 times/day) conceivably may have resulted in even better control of nocturnal blood pressures, thereby favorably affecting the progression of diabetic nephropathy. No results of ambulatory blood pressure monitoring, which could address this issue, have been reported.

More recently, in two multicenter studies involving both European and American investigators, a similar effect of the ACE inhibitor, captopril, on microalbuminuria was reported.[129,130] The study groups examined the effects of 2-year treatment with captopril on the progression to persistent albuminuria in normotensive patients with IDDM and microalbuminuria, defined as an albumin excretion rate (AER) of 20–200 μg/min. Albuminuria was defined as an AER > 200 μg/min with at least a 30% increase over the previous baseline on two consecutive occasions. The major findings of the European study are outlined below.[129] AER was reduced by approximately 31% in the captopril group but increased by approximately 25% in the placebo group. Thus, 21.9% of placebo-treated patients, in comparison with only 7.2% of captopril-treated patients, progressed to persistent albuminuria. Life-table analysis showed the risk of progression to be significantly lowered by captopril with a risk reduction of 69.2% over 2 years. This decrease in risk remained of a similar magnitude even after adjustment for time, varying mean arterial pressure (MAP), or other covariates, such as baseline AER and glycosylated hemoglobin. There were no significant time and between-group changes in glycosylated hemoglobin or urine urea nitrogen excretion. Changes in MAP, serum cholesterol, and serum creatinine correlated significantly with progression. The investigators concluded that captopril reduces the risk of progression to overt nephropathy by approximately 60%. Similar findings have been reported by the North American investigators. In view of the similarity in the study protocols from both Europe and North America, a recent report has analyzed data combining the results of both studies.[131] This analysis also found a trend toward improvement in the rate of decline in renal function compared with the placebo groups.

Elving et al.[132] conducted a 2-year prospective, randomized study to compare the effect of an ACE-inhibitor (captopril) and a beta blocker (atenolol) on retarding the progression of established diabetic nephropathy in 29 patients with IDDM. They observed that both drugs lowered MAP equally. Of interest, there was no difference in the decline of GFR over time, and albumin excretion and total protein at 2-year follow-up were similarly and significantly lowered. In view of the much smaller number of patients compared with the Lewis study, it is difficult to determine whether the study had adequate power to detect differences between the two drug treatments.

Whereas the majority of available clinical trials have assessed the effects of ACE inhibition in patients with IDDM, a recent long-term clinical trial in patients with NIDDM has been reported. Ravid et al.[133] conducted a 5-year study evaluating the effects of ACE inhibition on proteinuria and rate of decline in renal function in patients with NIDDM and microalbuminuria. Ninety-four patients were studied for 5 years in a randomized, double-blind, placebo-controlled protocol. The investigators reported that ACE inhibition during the early stage of diabetic nephropathy results in long-term stabilization of plasma creatinine levels and degree of urinary loss of albumin. A follow-up report after 7 years of treatment confirmed a renoprotective effect of the ACE inhibitor, enalapril.[134]

Lebovitz et al. reported the results of a 3-year prospective, double-blind, placebo-controlled trial in patients with NIDDM.[135] The investigators demonstrated that an antihypertensive regimen including the ACE inhibitor enalapril preserves renal function to a greater extent than antihypertensive regimens excluding ACE inhibitors. The rate of loss of GFR was significantly greater in patients with overt proteinuria at baseline (UAE > 300 mg/24 hr) compared with patients with baseline subclinical proteinuria (UAE < 300 mg/24 hr). Antihypertensive treatment with enalapril was superior in terms of preservation of GFR in patients with subclinical proteinuria at baseline than antihypertensive treatments that excluded the ACE inhibitor. Furthermore, only 7% of the enalapril-treated group progressed to clinical albuminuria compared with 21% of control patients. Based on these findings, the investigators suggested that ACE inhibitors should be used as initial treatment for hypertensive patients with NIDDM, with or without microalbuminuria, and not held in reserve until clinical albuminuria or proteinuria develops.

Despite such promising results, several caveats are in order. ACE inhibitors are not free of side effects. An infrequent but important risk of ACE inhibitors is acceleration of renal insufficiency, particularly in patients with bilateral renal artery stenosis.[136] Under conditions in which filtration pressure is angiotensin II-dependent, the converting enzyme inhibitors may cause a precipitous fall in GFR. This complication is most likely in the presence of bilateral renal artery stenosis due to atheromatous plaques or severe congestive cardiac failure. Renal function and serum potassium should be monitored closely in the first week after initiation of therapy or if bilateral renal artery disease is suspected. Some investigators have suggested that ACE inhibitors should be avoided in diabetic patients.[137] However, in view of their renoprotective action and the low incidence of clinically significant renal artery stenosis, careful introduction of these agents is suggested in diabetic patients at high risk of renal artery stenosis.[138] Risk factors include the known presence of widespread atheromatous vascular disease and, in particular, peripheral vascular disease and a history of refractory hypertension. Of importance, ACE inhibitors may provoke hyperkalemia, particularly in patients with decrements in GFR or hyporeninemic hypoaldosteronism. Finally, care must be exercised in initiating ACE inhibitor therapy in patients receiving diuretics, who may have a profound drop in blood pressure and a decline in renal function. However, diuretics should not be totally avoided in diabetic patients receiving ACE inhibitors because they may potentiate not only hypotensive but also antialbuminuric effects. Indeed, diuretics decrease albuminuria in diabetic patients[139] and in several studies have been shown to promote the effect of ACE inhibitors in postponing the development of overt diabetic nephropathy.[140]

Angiotensin II Receptor Antagonists

Although most studies involving agents that interrupt the renin-angiotensin system have involved ACE inhibitors, the advent of AII receptor antagonists such as losartan provides an alternative therapeutic approach to lowering blood pressure as well as inhibiting AII-dependent mechanisms.[141] As yet, it is not clear whether this new class of antihypertensive agents will reproduce the renoprotective effects of ACE inhibitors.[142] ACE inhibitors not only decrease formation of AII but also inhibit degradation of the vasodilatory kinins.[143] Whether the potentiation of the vasodilator bradykinin (BK) by ACE inhibitors is an important mechanism for their antihypertensive and renoprotective effects remains controversial. It is likely that the side effect of cough induced by ACE inhibitors is mediated by non–AII-dependent mechanisms because AII receptor antagonists do not appear to have this adverse effect.[144] Several experimental studies have suggested that ACE inhibitors and AII antagonists have different effects on renal hemodynamic parameters, including GFR and renal blood flow.[145,146] However, no convincing data indicate that BK-dependent mechanisms play an important role in mediating the long-term renoprotection afforded by ACE inhibitors. Several studies have indicated that AII antagonists attenuate the development of albuminuria in experimental diabetes.[117,147,148]

Recently, several small studies have been performed in hypertensive and diabetic patients.[149,150] In a study of elderly Chinese hypertensive patients, some of whom had diabetes, losartan reduced albuminuria whereas the calcium channel blocker, felodipine, was ineffective.[149] A similar finding has been suggested by a preliminary analysis of a multicenter study in European patients with NIDDM comparing amlodipine with losartan.[150] The AII antagonist but not amlodipine reduced albuminuria. Several large clinical trials are in progress to assess renoprotective effects of AII receptor antagonists in patients with NIDDM and macroproteinuria.

Calcium Antagonists

Calcium channel blockers (CCBs) are generally well tolerated by patients with diabetes and have been shown to be effective in decreasing their blood pressure.[151,152] Most studies have demonstrated no significant effect on glucose or lipoprotein levels.[111] Adverse reactions to CCBs include flushing, headache, peripheral edema, and constipation.[111]

As detailed previously, much investigative attention has focused on the attributes of ACE inhibitors in diabetes mellitus because of their apparent ability to decrease glomerular capillary pressure, thereby protecting the injured kidney from hemodynamically mediated glomerular damage.[10] Conversely, it has been proposed that because calcium antagonists preferentially dilate the afferent arteriole,[110,153,154] theoretically they should favor an increase in glomerular capillary pressure. Consequently, there has been concern that calcium antagonists may be detrimental to renal function in the long term. As previously reviewed, whether this actually occurs in patients with diabetes mellitus[154] or chronic renal disease[155] has not been established. At least one factor, systemic blood pressure, must be taken into account. Even in the face of afferent arteriolar vasodilatation, the resultant glomerular capillary pressure depends on the net effect of systemic and renal microvascular influences. Thus, a concomitant reduction in MAP may eventuate in a negligible increase in glomerular pressure despite afferent arteriolar vasodilatation.[110,153,156] Furthermore, in addition to their renal microcirculatory effects, calcium antagonists have properties that may contribute to their ability to afford renal protection under diverse experimental conditions and perhaps in clinical disease.[110,153] Such properties include the ability to lessen injury by retarding renal growth, to attenuate mesangial entrapment of macromolecules, and to countervail the mitogenic effect of diverse mediators, including platelet-derived growth factor and platelet-activating factor.[153,157] An additional mechanism by which calcium antagonists may exert protective effects is amelioration of mitochondrial calcium overload, which results in mitochondrial malfunction and eventual cell death.[153] Indeed, a number of studies in animal models of diabetes have suggested that calcium antagonists may be renoprotective in retarding progression of diabetic nephropathy,[158–160] although this finding has not been universal.[113,161]

The published studies of calcium antagonists and diabetic renal disease have been widely divergent in design as well as results. The results are summarized in Tables 1 and 2.[151,162–183] Although calcium antagonist therapy has been found to diminish proteinuria significantly in some studies,[151,180,181] others have shown either no effect,[162,168,171,172] or an actual worsening of proteinuria.[169,177,179,182] A number of design flaws in several of the studies have confounded their interpretation.

Mimran et al.[179] reported that captopril and nifedipine induced diametrically opposite effects on the urinary excretion of albumin. They studied 22 normotensive patients with IDDM and incipient diabetic nephropathy and observed the AER response to two treatment regimens. AER decreased by 40% in the captopril-treated group and increased by 40% in patients treated with nifedipine.

The Melbourne Diabetic Nephropathy Study Group reported the 12-month results of their initial prospective, randomized study comparing the effects of the ACE inhibitor perindopril with those of the calcium antagonist nifedipine on blood pressure and microalbuminuria in diabetic patients with persistent microalbuminuria.[174] After 12 months of therapy, the investigators observed that both regimens were equally efficacious in reducing blood pressure and albumin excretion in hypertensive patients. Further analysis indicated that the

TABLE 1. Effects of Calcium Channel Blockers and Other Agents on Renal Function and Albuminuria in Type I and Type II Diabetic Patients with Microalbuminuria

Agent	Type DM	HT/NT	Duration	n	AER%	GFR	Reference
Nifedipine	I	NT	6 wk	7	+42	→	179
Captopril				8	–41	→	
Nifedipine	I + II	NT	12 mo	13	→	→	174
Perindopril				17	→	→	
Cilazapril	I	HT	3 mo	16	–47	ND	170
Nifedipine				16	→		
Verapamil				16	–22		
Cilazapril + verapamil				16	–80		
Nifedipine	I + II	HT	12 mo	10	–30	→	174
Perindopril				3	–28	→	
Amlodipine	I + II	HT	8 mo	15	→	ND	183
Enalapril	II	HT	12 mo	16	–70	→	168
Nifedipine	II			15	→	→	
Nicardipine	II	HT	4 wk	6	–29	→	197
Lisinopril	II	HT	12 mo	156	–37	→	162
Nifedipine				158	–8	→	
Enalapril	II	HT	1 yr	8	–28	↑	178
Nitrendipine				8	–17	↑	
Cilazapril	II		3 yr	9	–27	↓	181
Amlodipine		HT		9	–31	↓	
Enalapril + nifedipine	II	HT + NT	48 mo	11	–42	↓	182
Nifedipine				13	+29	↓	
Enalapril				12	–47	↓	
Untreated				12	→	→	

DM = diabetes mellitus, HT = hypertensive, NT = normotensive, AER = albumin excretion rate, GFR = glomerular filtration rate, ND = not determined, → = unchanged, ↑ = increased, ↓ = decreased.

antialbuminuric effect of nifedipine was dependent on sodium intake.[184] In contrast to ACE inhibition, sodium intake did not influence the hypotensive response to calcium channel blockade. A similar dependency of sodium intake on albuminuria in response to the calcium channel blocker, diltiazem, has been recently reported in a prospective study.[185] The effects of both antihypertensive regimens were clearly observed in the hypertensive rather than the normotensive subgroup.

As noted in Table 1, the role of CCBs in normotensive patients with microalbuminuria has been less well examined. To address this issue, the Melbourne Diabetic Nephropathy Study Group has instituted a placebo-controlled study comparing the ACE inhibitor, perindopril, with the dihydropyridine calcium antagonist, nifedipine, in a group of type I and type II normotensive, microalbuminuric diabetic patients. Preliminary analysis indicates that at both 12 and 24 months perindopril reduced albuminuria.[186,187] A reduction in albuminuria was not observed in the placebo or nifedipine-treated groups. A more recent

TABLE 2. Effects of Calcium Channel Blockers and Other Agents on Renal Function and Albuminuria in Type I and Type II Diabetic Patients with Macroproteinuria

Agent	Type DM	HT/NT	Duration	n	AER%	GFR	Reference
Nifedipine	I	HT	3 wk	12	→	→	173
Lisinopril				12	–35	→	
Lisinopril	I + II	HT	19 wk	12	→ –13	↓	175
Nifedipine				14	→ –15	↓	
Captopril	II	HT	4 wk	12	–48	→	180
Nicardipine				12	–61	→	
Captopril + nicardipine				12	–75	→	
Diltiazem	II	HT	18 wk	8	–38	→	164
Lisinopril				8	–43	→	
Nifedipine	II	HT	6 wk	14	+89	→	169
Diltiazem					–52		
Lisinopril	II	HT	18 mo	10	–42	→	151
Diltiazem				10	–45	→	
Furosemide + atenolol				10	↓	↓	
Ramipril	II	HT	6 mo	19	–28	→	172
Nitrendipine				19	→	→	
Enalapril	II	HT	12 mo	19	–87	→	171
Nifedipine				12	→	→	
Enalapril	II	HT	12 mo	7	–71	→	168
Nifedipine				10	→	→	
Lisinopril	II	HT	12 mo	8	–59	↓	165
Verapamil				8	–50	↓	
Lisinopril + verapamil				8	–78	↓	
Captopril	II	HT	6 mo	10	–50	↓	176
Nifedipine				10	+18	→	
Lisinopril	I	HT	1 yr	24	–47	↓	177
Nisoldipine				25	+11	→	
Verapamil (slow release)	II	HT	> 4 yr	18	–60	↓	166
Atenolol				16	–20	↓↓	
Lisinopril	II	HT	5 yr	18	–25	↓	167
Atenolol				16	→	↓↓	
Verapamil or diltiazem				16	–18	↓	

DM = diabetes mellitus, HT = hypertensive, NT = normotensive, AER = albumin excretion rate, GFR = glomerular filtration rate, → = unchanged, ↓ = decreased, ↓↓ = markedly decreased.

analysis of this study observed that after 3 years no patients in the IDDM subgroup that received perindopril therapy had developed macroproteinuria, whereas in the placebo and nifedipine-treated groups 3 and 4 patients developed overt nephropathy, respectively.[188] Whether this effect on urinary albumin excretion ultimately will be reflected by prevention or delay in the development of overt nephropathy and, in particular, decline in renal function remains to be determined.

In a large multicenter study,[162] over 300 hypertensive type II diabetic patients were randomized to lisinopril or nifedipine. Although the ACE inhibitor was clearly superior to the CCB in terms of albuminuria, no difference in effects on renal function were detected. In contrast, the recently published report by Velussi et al. detected no difference between the CCB, amlodipine, and the ACE inhibitor, cilazapril.[181] Over at least 3 years, 18 of the subjects had microalbuminuria in association with hypertension. Both treatment groups showed a similar decrease in albuminuria and decline in GFR. It remains to be determined whether all CCBs have similar effects on AER and whether there may be differences even within the dihydropyridine CCBs (see below).

Chan et al. conducted a 1-year, double-blind study of hypertensive patients with NIDDM assigned to either enalapril or nifedipine with matching placebos for the alternative drug.[168] Of 102 Chinese patients, 52 were randomized to nifedipine and 50 to enalapril. At baseline 44 patients had normoalbuminuria, 36 microalbuminuria, and 22 macroalbuminuria. Endogenous creatinine clearance rather than radionuclide clearance was used to approximate changes in GFR. At 1 year mean arterial blood pressures were similar in both groups. Albuminuria fell by 54% in the enalapril group and by 11% in the nifedipine group ($p = 0.006$). Creatinine clearance fell similarly in both groups, but plasma creatinine concentration was increased by 20% in the enalapril group vs. 8% in the nifedipine group ($p = 0.001$). Although enalapril reduced proteinuria significantly more than nifedipine in the microalbuminuric and macroalbuminuric patients, it was more likely to increase plasma creatinine concentrations.

Recently, Rossing et al. initiated an important long-term study comparing the effects of a dihydropyridine calcium antagonist (nisoldipine) and lisinopril on albuminuria and kidney function in hypertensive patients with IDDM and nephropathy. Its rigorous design suggests that the study may permit definitive comparisons between the two classes of agents. This study included ambulatory blood pressure monitoring to document accurately the level of blood pressure control achieved with both drug regimens. The primary endpoint is the rate of decline in GFR after 4 years. Secondary endpoints include sequential determinations of changes in left ventricular mass index, urinary excretion of albumin and IgG, and fractional clearance of albumin and IgG. A recently published analysis after 12 months of treatment has shown that the ACE inhibitor was more effective in reducing albuminuria. However, there was a disparity in terms of effects on renal function: an acute drop in GFR was observed with the ACE inhibitor but not with the CCB. Whether this finding ultimately translates to differences between the effects of the two drugs on long-term renal function cannot yet be ascertained. Although, in general, increasing proteinuria and declining renal function have been viewed as inextricably linked, it is possible that in the setting of pharmacologic treatment and, in particular, CCBs, this link is uncoupled. If so, the measurement of urinary albumin excretion would be an inappropriate surrogate of declining renal function in clinical studies. This issue has not yet been resolved.

Ferder et al.[171] compared the effects of enalapril and nifedipine on renal function, proteinuria, and blood pressure in 30 type II diabetic patients with proteinuria. In a double-blind trial, patients received either enalapril, 40 mg/day, or nifedipine, 40 mg/day, for 12 months. They also received a low protein diet with 0.8 g/kg wt/day of protein. Both drugs induced a similar hypotensive effect. Enalapril produced an antiproteinuric effect, whereas proteinuria persisted in the nifedipine group.

An additional mechanism that may account for these discrepancies relates to the effects of calcium antagonists on proximal tubular function. Holdaas et al.[173] conducted a randomized, crossover study to compare the effects of lisinopril and nifedipine on renal hemodynamics, albumin excretion, and segmental tubular reabsorption in hypertensive patients with IDDM. No differences between the drugs were observed with regard to their effect on renal hemodynamic parameters. The investigators demonstrated an inhibitory effect of nifedipine but not lisinopril on several proximal tubular function markers such as beta-2-microglobulin and N-acetyl-D-glucosaminidase (NAG) and postulated that nifedipine also may inhibit proximal tubular albumin transport mechanisms, thereby accounting for the lack of an antiproteinuric effect.

Although this review considered all CCBs as essentially similar in terms of renal effects, Bakris and colleagues have suggested, based on a range of studies, that the various CCBs may act differently.[151,165,169,189] In particular, the authors suggested that the dihydropyridines may be less efficacious than other CCBs, such as verapamil and diltiazem, in affording renal protection. This issue has not been fully resolved. Furthermore, the possibility that nifedipine—in particular, the shorter-acting form—may be less effective in reducing AER has been suggested by findings from a metaanalysis of studies of diabetic patients with microalbuminuria or overt nephropathy.[190] Whether these findings can be extrapolated to other dihydropyridines or to the newer slow-release formulation of nifedipine remains unknown. However, no other groups have published findings comparing dihydropyridines with other CCB classes in terms of renoprotection in diabetes.

Because of the conflicting results of many clinical trials of the effects of antihypertensive agents on renal function, several metaregression analyses of the relative effects of different antihypertensive agents on proteinuria and renal function in patients with diabetes have been reported.[190–192] The investigators assessed at least 100 controlled and uncontrolled studies that provided data about renal function, proteinuria, or both, before and after treatment with an antihypertensive agent. Multiple linear regression analysis indicated that ACE inhibitors decrease proteinuria independently of changes in blood pressure, treatment duration, and type of diabetes or stage of nephropathy as well as study design. The investigators concluded that, among all antihypertensive agents, ACE inhibitors had a unique ability to decrease proteinuria independently of the reduction in proteinuria caused by changes in systemic blood pressure. The reductions in protein excretion by other agents were less impressive and could be attributed entirely to decreases in systemic blood pressure. However, a recent update of one of these metaanalyses suggested that at maximal hypotensive doses no significant difference was observed in the antiproteinuric effects of ACE inhibitors and other antihypertensive drugs.[193]

Combination of Calcium Antagonist and ACE Inhibitor

It has been proposed that the combination of a calcium antagonist with an ACE inhibitor should result in a greater reduction in urinary protein excretion and slowed morphologic progression of nephropathy.[114] Bakris et al. compared the renal hemodynamic and antiproteinuric effects of a calcium antagonist, verapamil, and an ACE inhibitor, lisinopril, alone and in combination, in three groups of patients with NIDDM and documented nephrotic-range proteinuria, hypertension, and renal insufficiency.[165] Patients treated with the combination of a calcium antagonist and an ACE inhibitor manifested the greatest reduction in albuminuria. In addition, the decline in GFR was the lowest in this group. Similar findings are suggested by another study performed in microalbuminuric patients using the combination of verapamil and cilazapril.[170] Although such an approach is extremely attractive, additional studies are required to confirm and extend initial observations. Experimental studies using a combination of trandolopril and verapamil have suggested that the combinations are superior as renoprotective agents to either agent alone in stroke-prone spontaneously hypertensive rats[194] and more recently in a diabetic animal model.[195] A preliminary report by Bakris et al. suggests that the combination of verapamil and trandolapril, administered in a fixed-dose combination, is more effective at reducing proteinuria than either drug alone, despite similar effects on blood pressure.[196]

Although captopril was renoprotective in the study by Lewis et al., patients continued to progress toward renal failure, albeit at a slower rate.[124] Therefore, additional therapy is needed to assist in retarding the progression of diabetes-associated renal injury. The above findings with combination therapy provide an exciting and novel approach for optimizing antihypertensive therapy in diabetic patients with renal disease.

References

1. U.S. Renal Data System: 1994 Annual Data Report. National Institutes of Health, National Institute of Diabetes and Digestive and Kidney Diseases, Bethesda, MD, 1994.
2. Andersen AR, Christiansen JS, Anderson JK, et al: Diabetic nephropathy in type I (insulin-dependent) diabetes: An epidemiological study. Diabetologia 25:496–501, 1983.
3. Mogensen CE: Natural history of renal functional abnormalities in human diabetes mellitus: From normoalbuminuria to incipient and overt nephropathy. Contemp Issues Nephrol 20:19–49, 1989.
4. Myers BD, Bennett PH: Clinical evolution of renal disease in insulin-dependent and non-insulin dependent diabetes. In Jacobson HR, Striker GE, Klahr S (eds): The Principles and Practice of Nephrology. Philadelphia, B.C. Decker, 1991, pp 463–470.
5. Deckert T, Pouken JE, Larsen M: Prognosis of diabetics with diabetes onset before age 31. Diabetologia 14:363–370, 1978.
6. Knowles HCJ: Magnitude of the renal failure problem in diabetic patients. Kidney Int 6(Suppl 1) 1:S2–S7, 1974.
7. Nelson RG, Kunzelman CL, Pettit DJ, et al: Albuminuria in type 2 (non-insulin-dependent) diabetes mellitus and impaired glucose tolerance in Pima Indians. Diabetologia 32:870–876, 1989.
8. Mogensen CE: Microalbuminuria as a predictor of clinical diabetic nephropathy. Kidney Int 31:673–689, 1987.
9. Tolins JP, Raij L: Concerns about diabetic nephropathy in the treatment of diabetic hypertensive patients. Am J Med 87(Suppl 6A):29S–33S, 1989.
10. Zatz R, Dunn BR, Meyer TW, Brenner B: Prevention of diabetic glomerulopathy by pharmacological amelioration of glomerular capillary hypertension. J Clin Invest 77:1925–1930, 1986.
11. Cooper ME, Jerums G, Gilbert RE: Diabetic vascular complications. Clin Exp Pharmacol Physiol 24:770–775, 1997.
12. Seaquist E, Goetz F, Rich S, Barbosa J: Familial clustering of diabetic kidney disease. N Engl J Med 320:1161–1165, 1989.
13. Viberti GC, Keen H, Wiseman MJ: Raised arterial pressure in parents of proteinuric insulin-dependent patients. BMJ 295:575–577, 1987.
14. Krolewski AS, Canessa M, Warram JH, et al: Predisposition to hypertension and susceptibility to renal disease in insulin-dependent diabetes mellitus. N Engl J Med 318:140–145, 1988.

15. Mangili R, Bending J, Scott G, et al: Increased sodium–lithium countertransport activity in red cells of patients with insulin-dependent diabetes and nephropathy. N Engl J Med 318:146–150, 1988.
16. Deckert T, Feldt-Rasmussen B, Borch-Johnsen T, Kofoed Enevoldsen A: Albuminuria reflects widespread vascular damage: The Steno hypothesis. Diabetologia 32:219–226, 1989.
17. Parving H-H, Nielsen FS, Bang LE, et al: Endothelial dysfunction in non-insulin dependent diabetic patients with and without diabetic nephropathy. Diabetologia 39:1590–1597, 1996.
18. Tolins J, Raij L: Genetic factors and susceptibility to diabetic nephropathy. N Engl J Med 319:180–181, 1989.
19. Hollenberg NK, Williams GH: Abnormal renal function, sodium-volume homeostasis, and renin system behavior in normal-renin essential hypertension. In Laragh JH, Brenner BM (eds): Hypertension: Pathophysiology, Diagnosis and Management. New York, Raven Press, 1990, pp 1349–1370.
20. Hostetter T, Rennke H, Brenner B: The case for intrarenal hypertension in the initiation and progression of diabetic and other glomerulopathies. Am J Med 72:375–380, 1982.
21. Mauer SM, Steffes MW, Azar S, et al: The effects of Goldblatt hypertension on development of the glomerular lesions of diabetes mellitus in the rat. Diabetes 27:738–744, 1978.
22. Brenner BM, Anderson S: Glomerular function in diabetes mellitus. Adv Nephrol 19:135–144, 1990.
23. Raij L, Azar S, Keane W: Role of hypertension in progressive glomerular injury. Hypertension 7:398–404, 1985.
24. Bidani AK, Schwartz MM, Lewis EJ: Renal autoregulation and vulnerability to hypertensive injury in the remnant kidney. Am J Physiol 252:F1103–F1110, 1987.
25. Griffin KA, Picken M, Bidani AK: Radiotelemetric BP monitoring, antihypertensives and glomeruloprotection in remnant kidney model. Kidney Int 46:1010–1018, 1994.
26. Tolins JP, Shultz PJ, Raij L, et al: Abnormal renal hemodynamic response to reduced renal perfusion pressure in diabetic rats: Role of NO. Am J Physiol 265(6 Pt 2):F886–F895, 1993.
27. Bank N, Aynedjian HS: Role of EDRF (nitric oxide) in diabetic renal hyperfiltration. Kidney Int 43:1306–1312, 1993.
28. Komers R, Allen TJ, Cooper ME: Role of endothelium-derived nitric oxide in the pathogenesis of the renal hemodynamic changes of experimental diabetes. Diabetes 43:1190–1197, 1994.
29. Cogan DG, Kinoshita JH, Kador PF, et al: Aldose reductase and complications of diabetes. Ann Intern Med 101:82–91, 1984.
30. Kador PF, Kinoshita JH: Role of aldose reductase in the development of diabetes-associated complications. Am J Med 79(Suppl 5A):8–12, 1985.
31. Greene DA, Lattimer SA, Sima AAF: Sorbitol, phosphoinositides, and sodium-potassium-ATPase in the pathogenesis of diabetic complications. N Engl J Med 316:599–606, 1987.
32. Williamson JR, Chang K, Frangos M, et al: Hyperglycemic pseudohypoxia and diabetic complications. Diabetes 42:801–813, 1993.
33. Tilton RG, Baier LD, Harlow JE, et al: Diabetes-induced glomerular dysfunction: Links to a more reduced cytosolic ratio of NADH/NAD^+. Kidney Int 41:778–788, 1992.
34. Cohen MP: Aldose reductase: Glomerular metabolism and diabetic nephropathy. Metabolism 35:55–59, 1986.
35. Finegold D, Lattimer S, Nolle S, et al: Polyol pathway activities and myo-inositol metabolism: A suggested relationship in the pathogenesis of diabetic nephropathy. Diabetes 32:988–992, 1983.
36. McCaleb ML, McKean ML, Hohman TC, et al: Intervention with the aldose reductase inhibitor, tolrestat, in renal and retinal lesions of streptozotocin-diabetic rats. Diabetologia 34:695–701, 1991.
37. Greene DA, Lattimer-Greene S, Sima AAF: Pathogenesis and prevention of diabetic neuropathy. Diabetes Metab Rev 4:201–221, 1988.
38. Tomlinson DR: The pharmacology of diabetic neuropathy. Diabetes Metab Rev 8:67–84, 1992.
39. Kinoshita JH: Mechanism initiating cataract formation. Invest Ophthalmol 13:713–724, 1974.
40. Robison WG, Nagata M, Laver N, et al: Diabetic-like retinopathy in rats prevented with an aldose reductase inhibitor. Invest Ophthalmol Vis Sci 30:2285–2292, 1989.
41. Sima A, Prashar A, Zhang W-X, et al: Preventative effect of long-term aldose reductase inhibition (ponalrestat) on nerve conduction and sural-nerve structure in the spontaneously diabetic Bio-Breeding rat. J Clin Invest 85:1410–1420, 1990.
42. Goldfarb S, Ziyadeh FN, Kern EF, Simmons DA: Effects of polyol-pathway inhibition and dietary myo-inositol on glomerular hemodynamic function in experimental diabetes mellitus in rats. Diabetes 40:465–471, 1991.
43. Tilton RG, Chang K, Pugliese G, et al: Prevention of hemodynamic and vascular albumin filtration changes in diabetic rats by aldose reductase inhibitors. Diabetes 37:1258–1270, 1989.

44. Bank N, Mower P, Aynedjian H, et al: Sorbinil prevents glomerular hyperperfusion in diabetic rats. Am J Physiol 256:F1000–F1006, 1989.
45. Forster HG, ter Wee PM, Takenaka T, et al: Impairment of afferent arteriolar myogenic responsiveness in the galactose-fed rat. Proc Soc Exp Biol Med 206:365–374, 1994.
46. Dvornik D (ed): Aldose Reductase Inhibition: An Approach to the Prevention of Diabetic Complications. New York, McGraw-Hill, 1987, p 368.
47. Hayashi K, Epstein M, Loutzenhiser R, Forster H: Impaired myogenic responsiveness of the afferent arteriole in streptozotocin-induced diabetic rats: Role of eicosanoid derangements. J Am Soc Nephrol 2:1578–1586, 1992.
48. Forster HG, ter Wee PM, Hohman TC, Epstein M: Impairment of afferent arteriolar myogenic responsiveness in the galactose-fed rat is prevented by tolrestat. Diabetologia 39:907–914, 1996.
49. Judzewitsch RG, Jaspan JB, Polonsky KS, et al: Aldose reductase inhibition improves nerve conduction velocity in diabetic patients. N Engl J Med 308:119–125, 1983.
50. Santiago JV, Snksen PH, Boulton AJ, et al: Withdrawal of the aldose reductase inhibitor tolrestat in patients with diabetic neuropathy: Effect on nerve function. The Tolrestat Study Group. J Diabetes Compl 7:170–178, 1993.
51. Sima AAF, Bril V, Nathaniel V, et al: Regeneration and repair of myelinated fibers in sural nerve biopsy specimens from patients with diabetic neuropathy treated with sorbinil. N Engl J Med 319:548–555, 1988.
52. Giugliano D, Acampora R, Marfella R, et al: Tolrestat in the primary prevention of diabetic neuropathy. Diabetes Care 18:536–541, 1995.
53. Passariello N, Sepe J, Marrazzo G, et al: Effect of aldose reductase inhibitor (tolrestat) on urinary albumin excretion rate and glomerular filtration rate in IDDM subjects with nephropathy. Diabetes Care 16:789–795, 1993.
54. Pedersen MM, Christiansen JS, Mogensen CE: Reduction of glomerular hyperfiltration in normoalbuminuric IDDM patients by 6 mo of aldose reductase inhibition. Diabetes 40:527–531, 1991.
55. Brownlee M: Glycation and diabetic complications [Lilly Lecture 1993]. Diabetes 43:836–841, 1994.
56. Yamada H, Miyata S, Igaki N, et al: Increase in 3-deoxyglucosone levels in diabetic rat plasma. Specific in vivo determination of intermediate in advanced Maillard reaction. J Biol Chem 269:20275–20280, 1994.
57. Fu MX, Knecht KJ, Thorpe SR, Baynes JW: Role of oxygen in cross-linking and chemical modification of collagen by glucose. Diabetes 41(Suppl 2):42–48, 1992.
58. Soulis T, Cooper M, Vranes D, et al: The effects of aminoguanidine in preventing experimental diabetic nephropathy are related to duration of treatment. Kidney Int 50:627–634, 1996.
59. Rumble JR, Cooper ME, Soulis T, et al: Vascular hypertrophy in experimental diabetes: Role of advanced glycation end products. J Clin Invest 99:1016–1027, 1997.
60. Brownlee M, Vlassara H, Kooney A, et al: Aminoguanidine prevents diabetes-induced arterial wall protein cross-linking. Science 232:1629–1632, 1986.
61. Nicholls K, Mandel T: Advanced glycosylation end products in experimental murine diabetic nephropathy: Effect of islet grafting and of aminoguanidine. Lab Invest 60:486–489, 1989.
62. Soulis-Liparota T, Cooper M, Papazoglou D, et al: Retardation by aminoguanidine of development of albuminuria, mesangial expansion, and tissue fluorescence in streptozocin-induced diabetic rat. Diabetes 40:1328–1334, 1991.
63. Makita Z, Vlassara H, Rayfield E, et al: Hemoglobin-AGE: A circulating marker of advanced glycosylation. Science 258:651–653, 1992.
64. Kochakian M, Manjula B, Egan J: Chronic dosing with aminoguanidine and novel advanced glycosylation end product formation inhibitors ameliorates cross-linking of tail tendon collagen in streptozotocin-induced diabetic rats. Diabetes 45:1694–1700, 1996.
65. Vasan S, Zhang X, Zhang X, et al: An agent cleaving glucose-derived protein cross-links in vitro and in vivo. Nature 382:275–278, 1996.
66. Inoguchi T, Battan R, Handler E, et al: Preferential elevation of protein kinase C isoform beta II and diacylglycerol levels in the aorta and heart of diabetic rats: Differential reversibility to glycemic control by islet cell transplantation. Proc Natl Acad Sci USA 89:11059–11063, 1992.
67. Nishizuka Y: Intracellular signaling by hydrolysis of phospholipids and activation of protein kinase C. Science 258:607–614, 1992.
68. Lynch JJ, Ferro TJ, Blumenstock FA, et al: Increased endothelial albumin permeability mediated by protein kinase C activation. J Clin Invest 85:1991–1998, 1990.
69. Craven PA, Studer RK, Negrete H, DeRubertis FR: Protein kinase C in diabetic nephropathy. J Diabetes Compl 9:241–245, 1995.
70. Ishii H, Jirousek MR, Koya D, et al: Amelioration of vascular dysfunctions in diabetic rats by an oral PKC beta inhibitor. Science 272:728–731, 1996.

71. Border WA, Noble NA: Cytokines in kidney disease: The role of transforming growth factor-beta. Am J Kidney Dis 22:105–113, 1993.
72. Yamamoto T, Nakamura T, Noble NA, et al: Diabetic kidney disease is linked to elevated expression of transforming growth factor-β. Proc Natl Acad Sci USA 90:1814–1818, 1993.
73. Wolf G, Sharma K, Chen Y, et al: High glucose-induced proliferation in mesangial cells is reversed by autocrine TGF-beta. Kidney Int 42:647–656, 1992.
74. Ziyadeh FN, Goldfarb S: The renal tubulointerstitium in diabetes mellitus. Kidney Int 39:464–475, 1991.
75. Sharma K, Jin Y, Guo J, Ziyadeh FN: Neutralization of TGF-beta by anti-TGF-beta antibody attenuates kidney hypertrophy and the enhanced extracellular matrix gene expression in STZ-induced diabetic mice. Diabetes 45:522–530, 1996.
76. Kagami S, Border WA, Miller DE, Noble NA: Angiotensin II stimulates extracellular matrix protein synthesis through induction of transforming growth factor-beta expression in rat glomerular mesangial cells. J Clin Invest 93:2431–2437, 1994.
77. Wu L, Roe C, Cox A, et al: Transforming growth factor β1 and renal injury following subtotal nephrectomy in the rat: Role of the renin-angiotensin system. Kidney Int 51:1553–1567, 1997.
78. National High Blood Pressure Education Program Working Group: Report on hypertension and chronic renal failure. Arch Intern Med 151:1280–1287, 1991.
79. Gilbert RE, Jasik M, DeLuise M, et al: Diabetes and hypertension. Australian Diabetes Society position statement. Med J Aust 163:372–375, 1995.
80. Tarnow L, Rossing P, Gall M-A, et al: Prevalence of arterial hypertension in diabetic patients before and after JNC-V. Diabetes Care 17:1247–1251, 1994.
81. Williams B: Insulin resistance: The shape of things to come. Lancet 344:521–524, 1994.
82. Parving H-H, Andersen AR, Smidt VM, et al: Effect of antihypertensive treatment on kidney function in diabetic nephropathy. BMJ 294:1443–1447, 1987.
83. Mogensen CE: Long-term antihypertensive treatment inhibiting progression of diabetic nephropathy. BMJ 285:685–688, 1982.
84. Parving H-H, Hommel E, Smidt UM: Protection of kidney function and decrease in albuminuria by captopril in insulin-dependent diabetes with nephropathy. BMJ 297:1086–1091, 1988.
85. Björck S, Nyberg G, Mulec H: Beneficial effects of angiotensin converting enzyme inhibition on renal function in patients with diabetic nephropathy. BMJ 293:471–474, 1986.
86. Schrier RW, Savage S: Appropriate Blood Pressure Control in type II diabetes (ABCD Trial): Implications for complications. Am J Kidney Dis 20:653–657, 1992.
87. Rodby RA, Rohde R, Evans J, et al: The study of the effect of intensity of blood pressure management on the progression of type I diabetic nephropathy: Study design and baseline patient characteristics. J Am Soc Nephrol 5:1775–1781, 1995.
88. Bennett PH, Haffner S, Kasiske BL, et al: Screening and management of microalbuminuria in patients with diabetes mellitus: Recommendations to the Scientific Advisory Board of the National Kidney Foundation from an ad hoc committee of the Council on Diabetes Mellitus of the National Kidney Foundation. Am J Kidney Dis 25:107–112, 1995.
89. Mogensen CE, Keane WF, Bennett PH, et al: Prevention of diabetic renal disease with special reference to microalbuminuria. Lancet 346:1080–1084, 1995.
90. Kochar MS, Kalluru VB: Hypertension in the diabetic patient: Controlling its harmful effects. Postgrad Med 96:101–102, 1994.
91. Dodson PM, Beevers M, Hallworth R, et al: Sodium restriction and blood pressure in hypertensive type II diabetes: Randomized blind controlled and crossover studies of moderate sodium restriction and sodium supplementation. BMJ 298:227–230, 1989.
92. Allen TJ, Waldron MJ, Casley D, et al: Salt restriction reduces hyperfiltration, renal enlargement and albuminuria in experimental diabetes. Diabetes 46:119–124, 1997.
93. Jacobs DB, Sowers JR, Hmeidan A, et al: Effects of weight reduction on cellular cation metabolism and vascular resistance. Hypertension 21:308–314, 1993.
94. Diabetes Control and Complications Trial Research Group: The effect of intensive treatment on the development and progression of long-term complications in insulin-dependent diabetes mellitus. N Engl J Med 329:977–986, 1993.
95. Microalbuminuria Collaborative Study Group: United Kingdom. Intensive therapy and progression to clinical albuminuria in patients with insulin dependent diabetes mellitus and microalbuminuria. BMJ 311:973–977, 1995.
96. Bending JJ, Viberti GC, Watkins PJ, Keen H: Intermittent clinical proteinuria and renal function in diabetes: Evolution and the effect of glycaemic control. BMJ 292:83–86, 1993.
97. Björck S: Clinical trials in overt diabetic nephropathy. In Mogensen CE (ed): The Kidney and Hypertension in Diabetes Mellitus, 3rd ed. London, Kluwer Academic, 1996, pp 375–384.

98. Barbosa J, Steffes MW, Sutherland DE, et al: Effect of glycemic control on early diabetic renal lesions. A 5-year randomized controlled clinical trial of insulin-dependent diabetic kidney transplant recipients. JAMA 272:600–606, 1994.
99. UK Prospective Diabetes Study (UKPDS): VIII. Study design, progress and performance. Diabetologia 34:877–890, 1991.
100. Gilbert RE, Tsalamandris C, Bach L, et al: Glycemic control and the rate of progression of early diabetic kidney disease: A nine year longitudinal study. Kidney Int 44:855–859, 1993.
101. Ohkubo Y, Kishikawa H, Araki E, et al: Intensive insulin therapy prevents the progression of diabetic microvascualr complications in Japanese patients with non-insulin-dependent diabetes mellitus: A randomized prospective 6-year study. Diabetes Res Clin Pract 28:103–117, 1995.
102. Bloomgarden ZT: American Diabetes Association annual meeting 1996: The etiology of type II diabetes. Diabetes Care 19:1311–1315, 1996.
103. Couper JJ, Jones TW, Donaghue KC, et al: The Diabetes Control and Complications Trial. Implications for children and adolescents. Australasian Paediatric Endocrine Group. Med J Aust 162:369–372, 1995.
104. Parving H-H, Andersen AR, Smidt VM, et al: Diabetic nephropathy and arterial hypertension. Diabetologia 24:10–12, 1983.
105. Epstein M, Sowers JR: Diabetes mellitus and hypertension. Hypertension 19:403–418, 1992.
106. Meyer TW, Anderson S, Rennke HG, Brenner BM: Reversing glomerular hypertension stabilizes established glomerular injury. Kidney Int 31:752–759, 1987.
107. American Diabetes Association: Consensus development conference on the diagnosis and management of nephropathy in patients with diabetes mellitus. American Diabetes Association and the National Kidney Foundation. Diabetes Care 17:1357–1361, 1994.
108. Hypertension Detection and Follow-up Program Cooperative Group: Five-year findings of the Hypertension Detection and Follow-up Program. I: Reduction in mortality of persons with high blood pressure, including mild hypertension. JAMA 242:2562–2571, 1979.
109. Systolic Hypertension in the Elderly Program Cooperative Research Group: Implications of the Systolic Hypertension in the Elderly Program. Hypertension 21:335–343, 1993.
110. Epstein M: Calcium antagonists and renal protection. Current status and future perspectives. Arch Intern Med 152:1573–1584, 1992.
111. Stein P, Black H: Drug treatment of hypertension in patients with diabetes mellitus. Diabetes Care 14:425–448, 1991.
112. Anderson S, Rennke HG, Brenner BM: Therapeutic advantage of converting enzyme inhibitors in arresting progressive renal disease associated with systemic hypertension in the rat. J Clin Invest 77:1993–2000, 1986.
113. Anderson S, Rennke HG, Brenner BM: Nifedipine versus fosinopril in uninephrectomized diabetic rats. Kidney Int 41:891–897, 1992.
114. Brown SA, Walton CL, Crawford P, Bakris GL: Long-term effects of antihypertensive regimens on renal hemodynamics and proteinuria. Kidney Int 43:1210–1218, 1993.
115. Jyothirmayi GN, Reddi AS: Effect of diltiazem on glomerular heparan sulfate and albuminuria in diabetic rats. Hypertension 21:795–802, 1993.
116. Anderson S, Rennke HG, Garcia DL, Brenner BM: Short and long term effects of antihypertensive therapy in the diabetic rat. Kidney Int 36:526–536, 1989.
117. Allen TJ, Cao Z, Youssef S, et al: The role of angiotensin II and bradykinin in experimental diabetic nephropathy: Functional and structural studies. Diabetes 46:1612–1618, 1997.
118. Yoshida Y, Fogo A, Ichikawa I: Glomerular hemodynamic changes vs. hypertrophy in experimental glomerular sclerosis. Kidney Int 36:654–660, 1989.
119. Reddi AS, Ramamurthi R, Miller M, et al: Enalapril improves albuminuria by preventing glomerular loss of heparan sulfate in diabetic rats. Biochem Med Metab Biol 45:119–131, 1991.
120. Anderson PW, Do YS, Hsueh WA: Angiotensin II causes mesangial cell hypertrophy. Hypertension 21:29–35, 1993.
121. Cooper ME, Rumble J, Komers R, et al: Diabetes-associated mesenteric vascular hypertrophy is attenuated by angiotensin-converting enzyme inhibition. Diabetes 43:1221–1228, 1994.
122. Williams B, Baker AQ, Gallacher B, Lodwick D: Angiotensin II increases vascular permeability factor gene expression by human vascular smooth muscle cells. Hypertension 25:913–917, 1995.
123. Bjorck S, Mulec H, Johnsen SA, et al: Renal protective effect of enalapril in diabetic nephropathy. BMJ 304:339–343, 1992.
124. Lewis EJ, Hunsicker LG, Bain RP, Rohde RD: The effect of angiotensin converting enzyme inhibition on diabetic nephropathy. N Engl J Med 329:1456–1462, 1993.
125. Breyer JA: Diabetic nephropathy in insulin-dependent patients. Am J Kidney Dis 20:533–547, 1992.

126. Hebert LA, Bain RP, Verme D, et al: Remission of nephrotic range proteinuria in type I diabetes. Collaborative Study Group. Kidney Int 46:1688–1693, 1994.
127. Remuzzi G, Zoja C, Bertani T: Glomerulonephritis. Curr Opin Nephrol Hypertens 2:465–474, 1993.
128. Remuzzi G, Bertani T: Is glomerulosclerosis a consequence of altered glomerular permeability to macromolecules? Kidney Int 38:384–394, 1990.
129. Viberti GC, Mogensen CE, Groop LC, Pauls JF: Effect of captopril on progression to clinical proteinuria in patients with insulin-dependent diabetes mellitus and microalbuminuria. European Microalbuminuria Captopril Study Group. JAMA 271:275–279, 1994.
130. Laffel LM, McGill JB, Gans DJ: The beneficial effect of angiotensin-converting enzyme inhibition with captopril on diabetic nephropathy in normotensive IDDM patients with microalbuminuria. North American Microalbuminuria Study Group. Am J Med 99:497–504, 1995.
131. The Microalbuminuria Captopril Study Group: Captopril reduces the risk of nephropathy in IDDM patients with microalbuminuria. Diabetologia 39:587–593, 1996.
132. Elving LD, Wetzels JF, van Lier HJ, et al: Captopril and atenolol are equally effective in retarding progression of diabetic nephropathy. Results of a 2-year prospective, randomized study. Diabetologia 37:604–609, 1994.
133. Ravid M, Savin H, Jutrin I, et al: Long-term stabilizing effect of angiotensin-converting enzyme inhibition on plasma creatinine and on proteinuria in normotensive type II diabetic patients. Ann Intern Med 118:577–581, 1993.
134. Ravid M, Lang R, Rachmani R, Lishner M: Long-term renoprotective effect of angiotensin-converting enzyme inhibition in non-insulin-dependent diabetes mellitus. A 7-year follow-up study. Arch Intern Med 156:286–289, 1996.
135. Lebovitz HE, Wiegmann TB, Cnaan A, et al: Renal protective effects of enalapril in hypertensive NIDDM: Role of baseline albuminuria. Kidney Int 45(Suppl):S150–S155, 1994.
136. Hricik DE, Browning PJ, Kopelman R, et al: Captopril induced functional renal insufficiency in patients with bilateral renal-artery stenoses or renal-artery stenosis in a solitary kidney. N Engl J Med 308:373–376, 1983.
137. Kerr D, Tattersall R: Renal artery stenosis [letter]. BMJ 302:115–116, 1991.
138. Cooper ME, Williams B: Renal artery stenosis in diabetes. Diabetes Rev Int 1997 (in press).
139. Gambardella S, Frontoni S, Lala A, et al: Regression of microalbuminuria in type II diabetic, hypertensive patients after long-term indapamide treatment. Am Heart J 122:1232–1238, 1991.
140. Mathiesen ER, Hommel E, Giese J, Parving H-H: Efficacy of captopril in postponing nephropathy in normotensive insulin dependent diabetic patients with microalbuminuria. BMJ 303:81–87, 1991.
141. Johnston CI: Angiotensin receptor antagonists: focus on losartan. Lancet 346:1403–1407, 1995.
142. Ichikawa I, Madias NE, Harrington JT, et al: Will angiotensin II receptor antagonists be renoprotective in humans?[discussion]. Kidney Int 50:684–692, 1996.
143. Erdos EG: Angiotenin I-converting enzyme and the changes in our concepts through the years. Hypertension 16:363–370, 1990.
144. Lacourciere Y, Brunner H, Irwin R, et al: Effects of modulators of the renin-angiotensin-aldosterone system on cough. J Hypertens 12:1387–1393, 1994.
145. Kon V, Fogo A, Ichikawa I: Bradykinin causes selective efferent arteriolar dilatation during angiotensin I converting enzyme inhibition. Kidney Int 44:545–550, 1993.
146. Komers R, Cooper ME: Acute renal haemodynamic effects of angiotensin converting enzyme inhibition in diabetic hyperfiltration: The role of kinins. Am J Physiol 268:F588–F594, 1995.
147. Kohzuki M, Yasujima M, Kanazawa M, et al: Antihypertensive and renal-protective effects of losartan in streptozotocin diabetic rats. J Hypertens 13:97–103, 1995.
148. Remuzzi A, Perico N, Amuchastegui CS, et al: Short- and long-term effect of angiotensin II receptor blockade in rats with experimental diabetes. J Am Soc Nephrol 4:40–49, 1993.
149. Chan JCN, Critchley J, Tomlinson B, et al: Antihypertensive and anti-albuminuric effects of losartan potassium and felodipine in Chinese elderly hypertensive patients with or without non-insulin-dependent diabetes mellitus. Am J Nephrol 17:72–80, 1997.
150. Os I: Losartan vs. amlodipine in hypertensive patients with NIDDM. Proceedings of the 15th International Congress of Nephrology Symposium, 1997.
151. Slataper R, Vicknair N, Sadler R, Bakris GL: Comparative effects of different antihypertensive treatments on progression of diabetic renal disease. Arch Intern Med 153:973–980, 1993.
152. Parving H-H, Rossing P: Calcium antagonists and the diabetic hypertensive patient. Am J Kidney Dis 21(Suppl 3):47–52, 1993.
153. Epstein M: Calcium antagonists and the kidney: Implications for renal protection. In Epstein M (ed): Calcium Antagonists in Clinical Medicine. Philadelphia, Hanley & Belfus, 1992, pp 309–348.

154. Epstein M: Calcium antagonists and diabetic nephropathy. Arch Intern Med 151:2361–2364, 1991.
155. ter Wee PM, De Micheli AG, Epstein M: Effects of calcium antagonists on renal hemodynamics and progression of nondiabetic chronic renal disease. Arch Intern Med 154:1185–1202, 1994.
156. Anderson S: Renal hemodynamic effects of calcium antagonists in rats with reduced renal mass. Hypertension 17:288–295, 1991.
157. Sweeney C, Raij L: Interactions of the endothelium and mesangium in glomerular injury: Potential role of calcium antagonists. In Epstein M (ed): Calcium Antagonists in Clinical Medicine. Philadelphia, Hanley & Belfus, 1992, pp 413–426.
158. Bakris GL: Abnormalities of calcium and the diabetic hypertensive patient: Implications for renal preservation. In Epstein M (ed): Calcium Antagonists in Clinical Medicine. Philadelphia, Hanley & Belfus, 1992, pp 367–389.
159. Bakris GL: Renal effects of calcium antagonists in diabetes mellitus: An overview of studies in animal models and in humans. Am J Hypertens 4:487s–493s, 1991.
160. Rosenthal T, Rosenmann E, Cohen AM: Effects of nisoldipine on hypertension and glomerulosclerosis in Cohen diabetic rat with Goldblatt hypertension. Clin Exp Hypertens 15:395–408, 1993.
161. Rumble JR, Doyle AE, Cooper ME: Comparison of effects of ACE inhibition with calcium channel blockade on renal disease in a model combining genetic hypertension with diabetes. Am J Hypertens 8:53–57, 1995.
162. Agardh C-D, Garcia-Puig J, Charbonnel B, et al: Greater reduction of urinary albumin excretion in hypertensive type II diabetic patients with incipient nephropathy by lisinopril than by nifedipine. J Hum Hypertens 10:185–192, 1996.
163. Baba T, Murabayashi S, Takebe K: Comparison of the renal effects of angiotensin converting enzyme inhibitor and calcium antagonist in hypertensive type II (non–insulin-dependent) diabetic patients with micro albuminuria: A randomised controlled trial. Diabetologia 32:40–44, 1989.
164. Bakris GL: Effects of diltiazem or lisinopril on massive proteinuria associated with diabetes mellitus. Ann Intern Med 112:701–702, 1990.
165. Bakris GL, Barnhill BW, Sadler R: Treatment of arterial hypertension in diabetic humans: Importance of therapeutic selection. Kidney Int 41:912–919, 1992.
166. Bakris GL, Mangrum A, Copley JB, et al: Effect of calcium channel or beta-blockade on the progression of diabetic nephropathy in African Americans. Hypertension 29:744–750, 1997.
167. Bakris GL, Copley JB, Vicknair N, et al: Calcium channel blockers versus other antihypertensive therapies on progression of NIDDM associated nephropathy. Kidney Int 50:1641–1650, 1996.
168. Chan JC, Cockram CS, Nicholls MG, et al: Comparison of enalapril and nifedipine in treating non-insulin dependent diabetes associated with hypertension: One year analysis. BMJ 305:981–985, 1992.
169. Demarie BK, Bakris GL: Effects of different calcium antagonists on proteinuria associated with diabetes mellitus. Ann Intern Med 113:987–988, 1990.
170. Fioretto P, Frigato F, Velussi M, et al: Effects of angiotensin converting enzyme inhibitors and calcium antagonists on atrial natriuretic peptide release and action and on albumin excretion rate in hypertensive insulin-dependent diabetic patients. Am J Hypertens 5:837–846, 1992.
171. Ferder L, Daccordi H, Martello M, et al: Angiotensin converting enzyme inhibitors versus calcium antagonists in the treatment of diabetic hypertensive patients. Hypertension 19:II237–II242, 1992.
172. Fogari R, Zoppi A, Pasotti C, et al: Comparative effects of ramipril and nitrendipine on albuminuria in hypertensive patients with non–insulin-dependent diabetes mellitus and impaired renal function. J Hum Hypertens 9:13–15, 1995.
173. Holdaas H, Hartmann A, Lien MG, et al: Contrasting effects of lisinopril and nifedipine on albuminuria and tubular transport functions in insulin-dependent diabetics with nephropathy. J Intern Med 229:163–170, 1991.
174. Melbourne Diabetic Nephropathy Study Group: Comparison between perindopril and nifedipine in hypertensive and normotensive diabetic patients with microalbuminuria. BMJ 302:210–216, 1991.
175. O'Donnell MJ, Rowe BR, Lawson N, et al: Comparison of the effects of an angiotensin converting enzyme inhibitor and a calcium antagonist in hypertensive, macroproteinuric diabetic patients: A randomised double-blind study. J Hum Hypertens 7:333–339, 1993.
176. Romero R, Salinas I, Lucas A, et al: Comparative effects of captopril versus nifedipine on proteinuria and renal function of type 2 diabetic patients. Diabetes Res Clin Pract 17:191–198, 1992.

177. Rossing P, Tarnow L, Boelskifte S, et al: Differences between nisoldipine and lisinopril on glomerular filtration rates and albuminuria in hypertensive IDDM patients with diabetic nephropathy during the first year of treatment. Diabetes 46:481–487, 1997.
178. Ruggenenti P, Mosconi L, Bianchi L, et al: Long-term treatment with either enalapril or nitrendipine stabilizes albuminuria and increases glomerular filtration rate in non-insulin-dependent diabetic patients. Am J Kidney Dis 24:753–761, 1994.
179. Mimran A, Insua A, Ribstein J, et al: Contrasting effects of captopril and nifedipine in normotensive patients with incipient diabetic nephropathy. J Hypertens 6:919–923, 1988.
180. Stornello M, Valvo EV, Scapellato L: Hemodynamic, renal, and humoral effects of the calcium entry blocker nicardipine and converting enzyme inhibitor captopril in hypertensive type II diabetic patients with nephropathy. J Cardiovasc Pharmacol 14:851–855, 1989.
181. Velussi M, Brocco E, Frogato F, et al: Effects of cilazapril and amlodipine on kidney function in hypertensive NIDDM patients. Diabetes 45:216–222, 1996.
182. Sano T, Kawamura T, Matsumae H, et al: Effects of long-term enalapril treatment on persistent micro-albuminuria in well-controlled hypertensive and normotensive NIDDM patients. Diabetes Care 17:420–424, 1994.
183. Zanetti-Elshater F, Pingitore R, Beretta-Piccoli G: Calcium antagonists for treatment of diabetes-associated hypertension. Metabolic and renal effects of amlodipine. Am J Hypertens 7:36–45, 1994.
184. Jerums G, Allen TJ, Tsalamandris C, Cooper ME, Melbourne Diabetic Nephropathy Study G: Angiotensin converting enzyme inhibition and calcium channel blockade in incipient diabetic nephropathy. Kidney Int 41:904–911, 1992.
185. Bakris GL, Smith A: Effects of sodium intake on albumin excretion in patients with diabetic nephropathy treated with long-acting calcium antagonists. Ann Intern Med 125:201–204, 1996.
186. Gilbert RE, Jerums G, Allen T, Hammond J, Cooper ME, on behalf of the Melbourne Diabetic Nephropathy Study Group: Effect of different antihypertensive agents on normotensive microalbuminuric patients with IDDM and NIDDM. J Am Soc Nephrol 5:377, 1994.
187. Melbourne Diabetic Nephropathy Study Group, Cooper ME, Allen T, et al: Effects of different antihypertensive agents on normotensive microalbuminuric type I and type II diabetic patients. Proceedings of the XIIth International Congress of Nephrology, Jerusalem, 1993, p 424.
188. Jerums G: Angiotensin converting enzyme inhibition and calcium channel blockade in diabetic patients with microalbuminuria. Nephrology 3 (Suppl 1):S41, 1997.
189. Bakris GL: Diabetic renal disease: An overview of concepts and intervention. Focus Opin Intern Med 1:7–11, 1995.
190. Bohlen L, de Courten M, Weidmann P: Comparative study of the effect of ACE-inhibitors and other antihypertensive agents on proteinuria in diabetic patients. Am J Hypertens 7:84s–92s, 1994.
191. Kasiske BL, Kalil RS, Ma JZ, et al: Effect of antihypertensive therapy on the kidney in patients with diabetes: A meta-regression analysis. Ann Intern Med 118:129–138, 1993.
192. Remuzzi G, Ruggenenti P, Benigni A: Understanding the nature of renal disease progression. Kidney Intl 51:2–15, 1997.
193. Weidmann P, Schneider M, Bohlen L: Therapeutic efficacy of different antihypertensive drugs in human diabetic nephropathy: An updated meta-analysis. Nephrol Dialysis Transplant 10(Suppl):39–45, 1995.
194. Munter K, Hergenroder S, Jochims K, Kirchengast M: Individual and combined effects of verapamil or trandolapril on attenuating hypertensive glomerulopathic changes in the stroke-prone rat. J Am Soc Nephrol 7:681–686, 1996.
195. Kirchengast M, Hergenröder S, Munter K, Schult S: Development of renal impairment in experimental diabetes: effects of chronic calcium antagonism and/or ACE inhibition [abstract]. Diabetologia 39(Suppl 1):A305, 1996.
196. Bakris G, Weir M, De Quattro V, et al: Renal effects of a long acting ACE inhibitor, trandolapril (T) or nondihydropyridine calcium blocker, verapamil (V) or in a fixed dose combination in diabetic nephropathy: A randomized double blind placebo controlled multicenter study. Nephrology 3 (Suppl 1):S271, 1997.
197. Baba T, Tomiyama T, Murabayashi S, Takebe K: Renal effects of nicardipine, a calcium antagonist, in hypertensive type 2 (non-insulin-dependent) diabetic patients with and without nephropathy. Eur J Clin Pharmacol 38:425–429, 1990.

KONSTANTINOS MAKRILAKIS M.D., Ph.D. / GEORGE L. BAKRIS M.D., F.A.C.P.

15

Calcium Antagonists: Are They All Created Equal in Regard to Slowing Progression of Diabetic Nephropathy?

Numerous clinical studies of varying duration in patients with established renal disease from either diabetes or hypertension reinforce the concept that blood pressure must be reduced to certain levels to slow the progression of renal disease.[1–28] Moreover, the evidence from both animal models of insulin-dependent diabetes mellitus (IDDM) and patients with IDDM-associated nephropathy demonstrate an advantage of angiotensin-converting enzyme (ACE) inhibitors over other antihypertensive agents in slowing the progression of diabetic nephropathy, given comparable blood pressure control.[2,4,6,9–20]

Conversely, the renal benefit of ACE inhibitors is less consistent among patients with nephropathy due to non–insulin-dependent diabetes mellitus (NIDDM).[3,7,22–28] Possible reasons for these inconsistent results include (1) differences in the level of blood pressure reduction among randomized patients or (2) differences in the effects of individual drug classes on specific pathophysiologic processes within the kidney.

In contrast to the data that demonstrate improvement in renal survival with ACE inhibitors, preservation of renal function and effects on albuminuria clearly diverge within the different subclasses of calcium channel blockers (CCBs). The potential reasons for the differences among CCBs are discussed in the context of surrogate markers, such as changes in albuminuria, in both clinical and animal studies.

SURROGATE MARKERS OF RENAL DISEASE

The surrogate markers used to assess the effect of blood pressure reduction with CCBs on progression of diabetic renal disease include an increase in glomerular membrane permeability as measured by albuminuria and increases in mesangial matrix expansion as measured by renal biopsy. Surrogate markers serve as a guide to the direction in which renal function is headed rather than a true endpoint of dialysis or death. Moreover, changes in albuminuria depend on various factors, including (1) the agents used to lower blood pressure, (2) sodium intake, and (3) the level to which blood pressure is reduced. Of interest, changes in albuminuria also appear to be a far better predictor of outcome in diabetic vs. nondiabetic renal disease.[29–31] This difference relates largely to the fact that albumin is glycated in diabetes because of high glucose levels.

Glycation transforms the albumin molecule into a new protein that, when it encounters glomerular cells, increases mesangial matrix expansion, the earliest morphologic change associated with diabetic nephropathy. These cellular and related effects of glycated albumin are discussed below. Because matrix expansion has not been described for nonglycated albumin, a decrease in membrane permeability—and hence albuminuria—reduces cellular exposure to glycated albumin, thus minimizing renal injury in diabetes. Long-term animal and clinical studies that support this hypothesis are also discussed below.

Changes in membrane permeability occur early in the natural history of diabetes and are related to both the level of glucose and arterial pressure. These changes in membrane permeability are assessed by infusion of different-sized dextran particles along with insulin to measure glomerular filtration rate (GFR). The clearance of these dextrans is calculated on the basis of GFR. It is clear that ACE inhibitors and nondihydropyridine calcium channel blockers (non-DHPCCBs), under appropriate conditions such as a low salt diet, reduce membrane permeability. The majority of antihypertensive agents, however, have no direct effects on membrane permeability independent of blood pressure reduction. The effects of various antihypertensive agents on glomerular permeability are summarized in Table 1.

Both increases in albuminuria and failure to decrease preexisting albuminuria correlate with poor renal outcomes. Albuminuria in diabetes is related primarily to increases in glomerular capillary membrane permeability.

TABLE 1. Intrarenal Hemodynamic and Morphologic Effects of Antihypertensive Drug Classes in Animal Models of Diabetes

		Calcium Blockers				
	ACEIs	DHPs	NDHPs	α Blockers	Diuretics	β blockers
Hemodynamics						
P_{GC}	↓	→	→	→	→	→
R_A (tone)	↑	↑↑	↑	↑↑	→	→↓
E_A (tone)	↓	→	→↓	→	↑	→
ΔP	→↓	→	→↓	→	→	→
Autoregulation	→	↓↓	→	→	↓*	→
Glomerular permeability	↓	→↑	↓	→	?	?
Morphology						
V_V	↓	→	↓	→	?	?
Glomerulosclerosis	↓↓	→	↓	→	→	↓
Matrix proteins						
Fibronectin	↓↓	→	→	?	?	?
Laminin	↓↓	→	→	?	?	?
Collagen IV	↓↓	→	↓	?	?	?
Heparan sulfate	↑	→	↑	?	?	?
TGFβ expression	↓	→	?	?	?	?

ACEIs = angiotensin-converting enzyme inhibitors, DHPs = dihdyropyridines, NDHPs = nondihydrophyridines, V_V = mesangial matrix expansion, K_F = ultrafiltration coefficient, P_{GC} = intralomerular pressure, → = no effect, ↑ = increase, ↓ = decrease, ? = unknown, * = loop diuretics only, RA = afferent arteriolar tone, EA = efferent arteriolar tone.

The mechanisms that portend such changes in permeability are complex and beyond the scope of this chapter. The interested reader is referred to relevant reviews.[32,33]

Albuminuria reduction by antihypertensive agents has been investigated in a number of animal models of diabetes as well as in clinical trials. Several meta-analyses indicate that ACE inhibitors have the beneficial effects of reducing albuminuria and thus preserving renal function in patients with diabetic nephropathy.[34–37] Preservation of renal function is perhaps best demonstrated by the largest and most comprehensive clinical trial to date, performed by the Collaborative Study Group.[2] The investigators found that ACE inhibition slowed the progression of renal disease to a greater extent than placebo treatment. In this study of 409 patients with type I diabetes, the group randomized to captopril as part of the antihypertensive "cocktail" showed a significant reduction in the combined endpoints of progression to end-stage renal disease, transplantation, or death. In addition, serum creatinine was doubled at a significantly slower rate in the captopril-treated group compared with placebo. Moreover, the greatest benefit in preservation of renal function was demonstrated among patients with the most advanced renal disease (baseline serum creatinine >1.5 mg/dl), possibly because this group had already lost a large amount of renal function and hence had a greater propensity to progress to end-stage renal disease.

A decrease in albuminuria also correlated to a significant degree with slowed progression of renal disease in the captopril trial, as evidenced by the results of a subgroup analysis of patients with baseline proteinuria values > 3.5 gm/day (i.e., nephrotic range). This post hoc analysis demonstrated that patients randomized to captopril had a significantly higher percentage of remission of nephrotic range proteinuria compared with the placebo group (7/42 vs. 1/66, $p = 0.005$).[30] There are several reasons, however, for these results:

1. The group with remission of proteinuria also had a significantly lower mean blood pressure level (by 4 mmHg) compared with the placebo group at study end.[30]
2. Diabetic patients at this stage of renal disease generally do not exhibit the normal nocturnal drop in blood pressure while they sleep;[38] that is, they are "nondippers." Did captopril, given at a dose of 25 mg thrice daily, convert a proportion of this group to "dippers"? Because 24-hour blood pressure measurements were not performed in this group, we do not know the answer.
3. The group in the study was heterogeneous. One hundred of the 409 patients had microalbuminuria, whereas the remainder had frank proteinuria of levels exceeding 3.5 gm/day.

In the context of the natural history of diabetic nephropathy, the above factors indicate that, although ACE inhibitors are helpful in slowing progression of nephropathy, the level of blood pressure reduction may have driven the statistical analysis to significance.

4. The number of patients with more advanced levels of nephropathy may have increased the number of primary end-point events (e.g., doubling of serum creatinine), thus increasing the chance for statistical significance.

Despite these factors, a blunted rise or marked fall in proteinuria is considered to be a surrogate marker for slowed progression of diabetic renal disease.

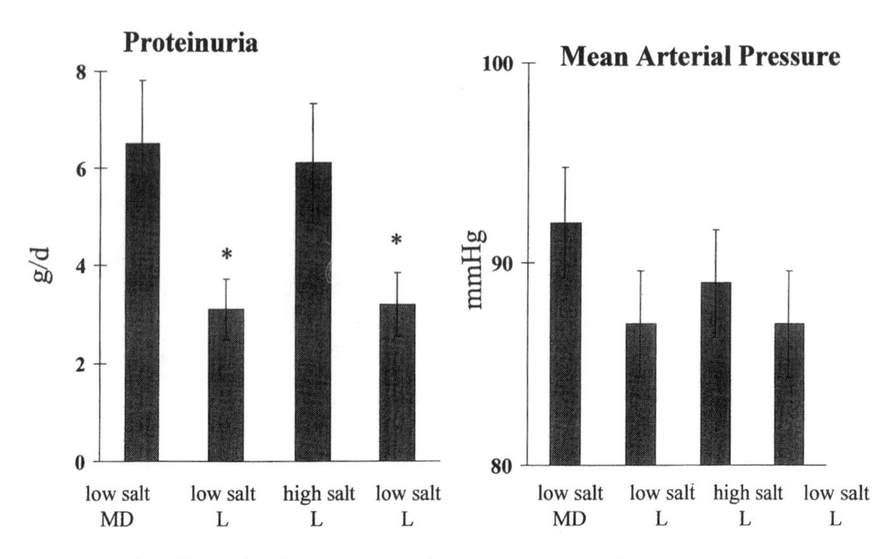

FIGURE 1. The effects of sodium intake on the antiproteinuric effects of angiotensin-converting enzyme (ACE) inhibition. MD = methyldopa, L = lisinopril. * $p < 0.05$ compared with MD or high-salt L. (Adapted from Heeg JE, deJong PE, van der Hem GK, de Zeeuw D: Efficacy and variability of the antiproteinuric effect of ACE inhibition by lisinopril. Kidney Int 36:272–279, 1989.)

INFLUENCE OF SODIUM ON ANTIPROTEINURIC EFFECTS OF CALCIUM CHANNEL BLOCKERS

Salt intake has been shown to be a major determinant of the antiproteinuric capabilities of ACE inhibitors and nondihydropyridine CCBs[39–41] (Figs. 1 and 2). Animal studies suggest that sodium intake ≥ 200 mEq/day (the average daily amount in the typical American diet) may be associated with an increase in glomerular membrane permeability to albumin through changes in the renin-angiotensin system.[42] Whereas this factor may be of relative importance in contributing to progression of nondiabetic renal disease, it is of critical importance in determining renal outcome in genetically susceptible diabetic patients.

The difference in predicting renal outcomes relates to the process of glycation of albumin in diabetic patients.[29,43] The glycation process transforms albumin into a relatively toxic antigenic particle that stimulates matrix proteins and alters glomerular permeability.[44–46] These changes in albumin chemistry contribute to an increase in mesangial matrix expansion and development of proteinuria.[45,46] Moreover, the effects of glycated albumin do not appear to be related to increases in intraglomerular pressure.[43,47,48]

The effects of sodium intake on CCB-associated changes in albuminuria have been examined. Dihydropyridine CCBs do not reduce albuminuria, regardless of the patients' sodium intake.[37,38,49–54] Moreover, a low sodium diet fails to express an antiproteinuric effect by dihydropyridine CCBs.[37,38,52,53] A recent randomized prospective study compared the effects of a dihydropyridine and nondihydropyridine CCB on albuminuria in patients with NIDDM nephropathy.[50] This study demonstrated that the dihydropyridine CCB, nifedipine XL, worsened glomerular membrane permeability despite reductions in arterial

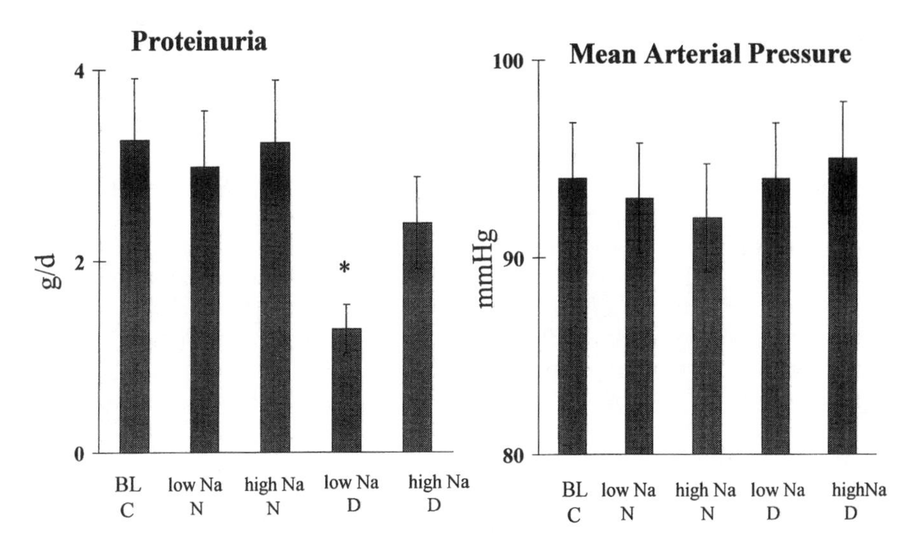

FIGURE 2. The effects of sodium intake on the antiproteinuric effects of calcium channel blockade. N = nifedipine XL, D = diltiazem, C = clonidine, BL = baseline. * $p < 0.05$ compared with high-sodium N and D and low-sodium N. (Adapted from Bakris GL, Smith AC: Effects of sodium intake on albumin excretion in patients with diabetic nephropathy treated with long-acting calcium antagonists. Ann Intern Med 125:201–203, 1996.)

pressure and low sodium diet. This effect was not observed for the nondihydropyridine diltiazem CD.

Two prospective, randomized long-term studies that compared the effects of dihydropyridine CCBs (nifedipine and nisoldipine) with the effects of an ACE inhibitor on both progression of renal disease and proteinuria demonstrated no reduction in proteinuria with either an ACE inhibitor or dihydropyridine CCB.[31,54] However, neither of these studies controlled for sodium intake, and only one measured sodium output, which averaged 184 mEq/day.[54] At this level one can predict no significant reduction in proteinuria. Thus, the studies are consistent with the hypothesis that high salt intakes blunt the antiproteinuric effects of ACE inhibitors. Of greater importance, the study of nondiabetic patients with renal insufficiency found no difference in overall renal outcome between the groups randomized to an ACE inhibitor or nifedipine.[31] However, in the last year of this study, over five times more patients started dialysis in the nifedipine group than in the ACE inhibitor group.[31] This difference could not be explained by changes in proteinuria, blood pressure control, or baseline serum creatinine level.

Conversely, nondihydropyridine CCBs improve membrane permeability, especially to large proteins. This is important because increased leakiness to such proteins has been shown recently to be an early manifestation of a glomerular membrane problem in diabetics.[55] Additional evidence to support this observation comes from other studies[41,53] as well as a recent review.[40,56] Of interest, high sodium intake mitigates against reductions in albuminuria associated with nondihydropyridine CCBs. The reasons are unclear.

Thus, in all patients with proteinuria and hypertension, it is critically important not only to lower blood pressure to levels recommended to slow progression of renal disease (< 130/85 mmHg) but also to use agents with known effects on membrane permeability (ACE inhibitors and non-DHPCCBs) in combination with a low salt diet (> 50 mEq/day, < 150 mEq/day).

NEPHROPATHY OF INSULIN-DEPENDENT DIABETES MELLITUS

Numerous long-term studies, ranging from 1–9 years, have evaluated the effects of different classes of antihypertensive medications on progression of nephropathy secondary to IDDM.[2,4–7,9–18,54,56–60] These studies clearly demonstrate the following points: (1) blood pressure reduction, regardless of the agent used, slows progression of nephropathy by as much as 90% compared with untreated patients; (2) whereas beta blockers and diuretics reduce the rate of decline in renal function from an average of 12 ml/min/yr to 3.5 ml/min/yr, ACE inhibitors reduce decline to an average of 1.5 –2 ml/min/year; and (3) the degree of antiproteinuric effect correlates with preservation of renal function. In this context, ACE inhibitors also have been shown to lengthen the time required to initiation of dialysis and to reduce mortality.

Animal studies provide insight into why these renal benefits are observed not only with ACE inhibitors but also with some other classes of antihypertensive agents (see Table 1). The early pathophysiologic changes observed in the diabetic kidney include an increase in intraglomerular pressure and mesangial matrix expansion. These studies uniformly demonstrate that a reduction in arterial pressure with ACE inhibitors confers relatively greater benefit at a given level of blood pressure reduction. The greater benefit is due to the fact that ACE inhibitors reduce intraglomerular capillary pressure, mesangial matrix expansion, and progression of glomerulosclerosis as well as preserve autoregulation compared with conventional antihypertensive therapies.[60] Such protection is exemplified by rat studies, using 24-hour blood pressure monitoring, that demonstrate prevention of glomerulosclerosis with ACE inhibitors at a systolic pressure of 140 mmHg, whereas the dihydropyridine CCB, nifedipine, required pressure reduction to 100 mmHg for similar effects.[61]

The effects of various antihypertensive medications as well as different CCBs on glomerular morphology are also presented in Table 1. ACE inhibitors prevent development of mesangial matrix expansion in diabetic nephropathy independently of blood glucose control.[19] This results from their effects on mesangial matrix proteins. ACE inhibitors have been shown specifically to inhibit increases in mesangial matrix proteins, such as laminin, collagen IV, and fibronectin, that contribute to mesangial matrix expansion.[62]

The nondihydropyridine CCBs (verapamil and diltiazem), like the ACE inhibitors, have been shown to reduce mesangial matrix expansion.[19,56,63] Recent animal studies in diabetic dogs demonstrate similar effects on slowing matrix expansion with use of either lisinopril or diltiazem.[19] Other studies that have evaluated development of glomerulosclerosis confirm this observation and show similar results for verapamil.[56] This effect has not been shown, however, for any dihydropyridine (nifedipine-like) CCB.[43,56] Four recent meta-analyses demonstrate a relatively greater impact on slowing progression of renal disease with nondihydropyridine CCBs compared with dihydropyridine CCBs.[34–37] However, these observations are predominantly based on short-term studies.

Unfortunately, there are no long-term (> 3 years) studies of nondihydropyridine CCBs in patients with IDDM-associated nephropathy to observe the effects on time to dialysis or death.

Studies with dihydropyridine CCBs in animal models of diabetic nephropathy have also examined both intrarenal hemodynamic and morphologic changes. A detailed review of these studies is presented elsewhere.[56,60] However, they uniformly demonstrate failure to reduce intraglomerular pressure and to prevent development of either glomerulosclerosis or albuminuria. Moreover, in clinical studies of patients with IDDM nephropathy, this subclass of agents failed to demonstrate a reduction in albuminuria.[54,56–59] All of these studies, however, are of less than two years' duration and used predominantly short-acting preparations of nifedipine or related CCBs, agents shown to be of no benefit in reducing cardiovascular mortality.[64]

Changes in the ability of the kidney to dampen high pressure gradients (i.e., autoregulation) also may affect the development of renal disease. The effects of various antihypertensive agents on renal autoregulation have been examined in many animal models (see Table 1). These studies indicate that dihydropyridine CCBs, regardless of their duration of action, totally abolish autoregulation.[65–68] These data confirm earlier reports that nifedipine markedly attenuated renal autoregulation.[61] Complete abolishment of autoregulation is observed only with dihydropyridine CCBs, whereas other antihypertensive medications, including nondihydropyridine CCBs, only partially or negligibly affect autoregulation[65] (Table 2). The consequence is that the kidney loses its natural ability to protect itself in contrast to the use of other antihypertensive agents.

Given the effects on autoregulation, blood pressure must be lowered to levels < 110/70 mmHg to provide the same level of renal protection with nifedipine as is seen at a blood pressure level of 130/80 mmHg with an ACE inhibitor.[61] In addition, a recent comparison of different CCBs demonstrated that nondihydropyridine CCBs provided protection against development of proteinuria and glomerulosclerosis at relatively higher pressures than dihydropyridine agents[67] (Fig. 3). These studies further demonstrated that at a given level of blood pressure, long-acting dihydropyridine CCBs offered relatively no protection compared with nondihydropyridine CCBs. Consequently, none of the dihydropyridine CCBs prevent increases in proteinuria or delay development of glomerulosclerosis.

TABLE 2. Antihypertensive Drugs and Renal Autoregulation*

Totally Abolish	Partially Abolish	Negligible Effects
Dihydropyridines	Angiotensin-converting enzyme inhibitors	Alpha blockers
	Loop diuretics	Beta blockers
	Nondihydropyridines	

* Agents not formally studied in either remnant kidney or diabetic models: potassium channel openers (e.g., minoxidil, hydralazine) and central alpha agonists (e.g., clonidine).

Adapted from Kwam FI, Iverson BM, Ofstad J: Autoregulation of glomerular capillary pressure in SHR during successive reduction of systemic blood pressure with antihypertensive agents [abstract]. J Am Soc Nephrol 7:1552, 1996.

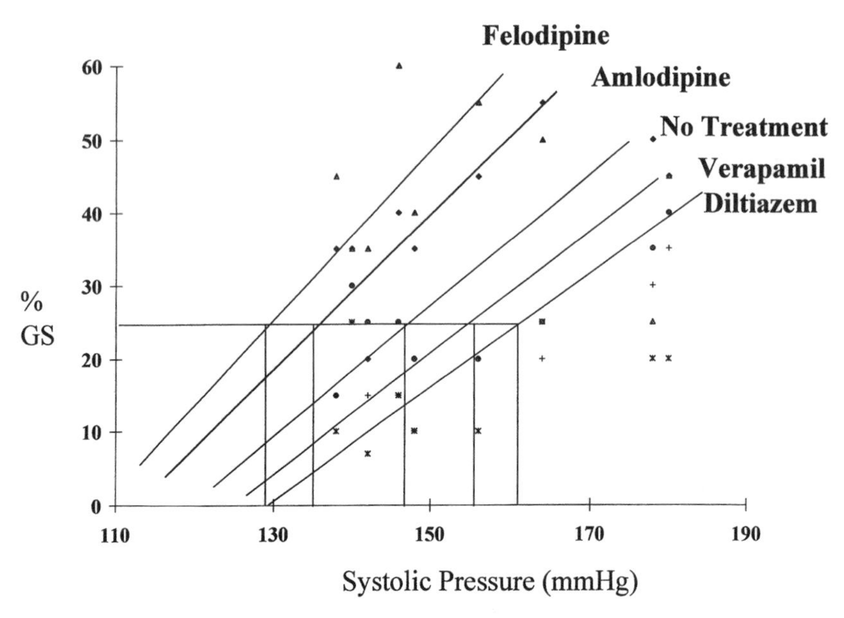

FIGURE 3. The slopes of 24-hour arterial pressure control vs. degree of glomerular scarring in the rat remnant kidney model. Note that 25% sclerosis occurs with dihydropyridine calcium channel blockers at systolic pressures between 130–138 mmHg. Similar levels of scarring occur with nondihydropyridine calcium channel blockers at systolic pressures between 155–165 mmHg. In the pressure range of 130–140 mmHg nondihydropyridine calcium channel blockers are associated with only 5–10% sclerosis. ▲ = felodipine, ● = amlodipine, + = no treatment, ✱ = verapamil, ◆ = diltiazem. (Adapted from Picken M, Griffin K, Bakris GL, Bidani A: Comparative effects of four different calcium antagonists on progression of renal disease in a remnant kidney model [abstract]. J Am Soc Nephrol 7:1586, 1996.)

NEPHROPATHY OF NON–INSULIN-DEPENDENT DIABETES MELLITUS

Fewer long-term studies examine the effects of different antihypertensive agents on progression of renal disease in patients with NIDDM-associated nephropathy. Hence, the data do not appear as consistent as those seen in patients with IDDM-associated nephropathy. Only five studies extending for 3 years or longer have compared ACE inhibitors with other antihypertensive therapies or placebo in terms of progression of renal disease in patients with NIDDM nephropathy.[22–24,27,28] Although all of these studies tend to show a greater benefit with ACE inhibitors, only two achieve statistically significant results in terms of renal disease progression.[22,27] Conversely, only two completed studies of NIDDM-associated nephropathy[27,28] and three ongoing studies[58,69,70] evaluate the effects of CCBs on progression of various nephropathies, including diabetes. The completed studies, however, demonstrate two clear benefits when nondihydropyridine CCBs are used rather than conventional therapy: prevention of progression of nephropathy and reduction in proteinuria.[27,28] One study of 52 patients with NIDDM-associated nephropathy and hypertension, randomized to receive an ACE inhibitor (lisinopril), nondihydropyridine CCB (verapamil or diltiazem SR), or beta blocker (atenolol) and followed for 6 years, demonstrated that the greatest mean rate of decline in creatinine clearance occurred in the atenolol group (–3.4 ml/min/yr/1.73 m^2) with no differences between the lisinopril (–1.0 ml/min/yr/1.73 m^2) and nondihydropyridine CCB (–1.4 ml/min/year/1.73 m^2) groups[27] (Fig. 4). The same result was noted in the serum creatinine change among the groups (Fig. 5). Moreover, although the degree of blood pressure reduction was similar in all groups, proteinuria was reduced to a similar extent only in the lisinopril and nondihydropyridine CCB groups (Fig. 6). These observations were supported in a 5-year randomized study among African-American patients with NIDDM nephropathy. Verapamil was associated with a 62% slower progression of renal disease compared with a beta blocker, given similar reductions in arterial pressure[28] (Fig. 7). Conversely, with one exception,[24] studies of the progression of NIDDM-associated renal disease in patients treated with dihydropyridine agents have thus far failed to show significant reduction in albuminuria or slowed progression of nephropathy. Given the relatively short duration of follow-up (3 years) in patients with minimal renal dysfunction (i.e., microalbuminuria with slight reductions in GFRs [70–110 ml/min]), one would not have predicted a significant difference in progression of renal disease.

The results of long-term clinical trials and three separate metaanalyses demonstrate that ACE inhibitors and nondihydropyridine CCBs have similar effects on both albuminuria and, indirectly, glomerular permeability in patients with NIDDM nephropathy.[27,34–37] Of interest, in all animal studies of diabetic nephropathy or models of elevated intraglomerular pressure in which ΔP (gradient across the capillary wall) is increased, both types of antihypertensive drugs demonstrate protection against glomerulosclerosis. This effect has not been described in animal models of dihydropyridine CCBs, which do not

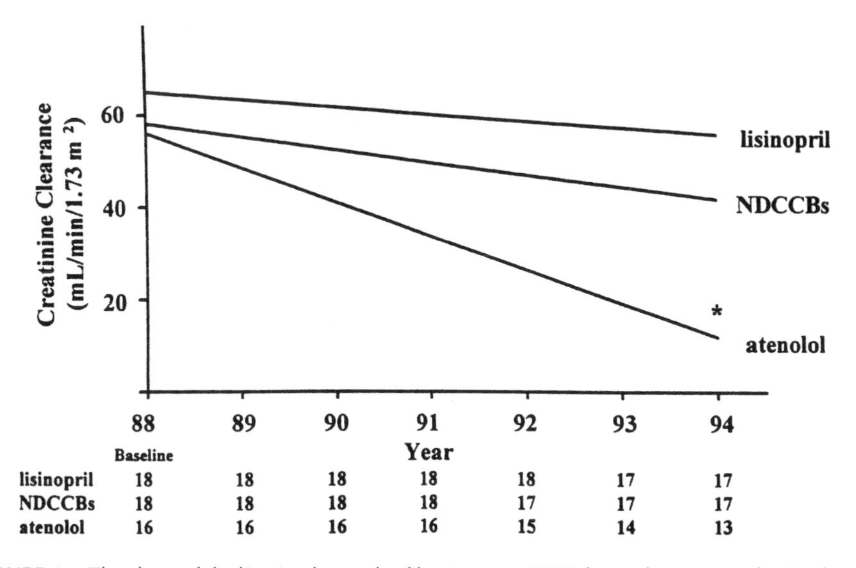

	88 (Baseline)	89	90	91	92	93	94
lisinopril	18	18	18	18	18	17	17
NDCCBs	18	18	18	18	17	17	17
atenolol	16	16	16	16	15	14	13

FIGURE 4. The slope of decline in glomerular filtration rate (GFR) for each group randomized to a different anithypertensive "cocktail" made up of an average of 4.3 medications per day to reduce blood pressure to levels of < 135/85 mmHg. * $p < 0.01$ compared with other two groups. (From Bakris GL, Copley JB, Vicknair N, et al: Calcium channel blockers versus other antihypertensive therapies on progression of NIDDM associated nephropathy: Kidney Int 50:1641–1650, 1996, with permission.)

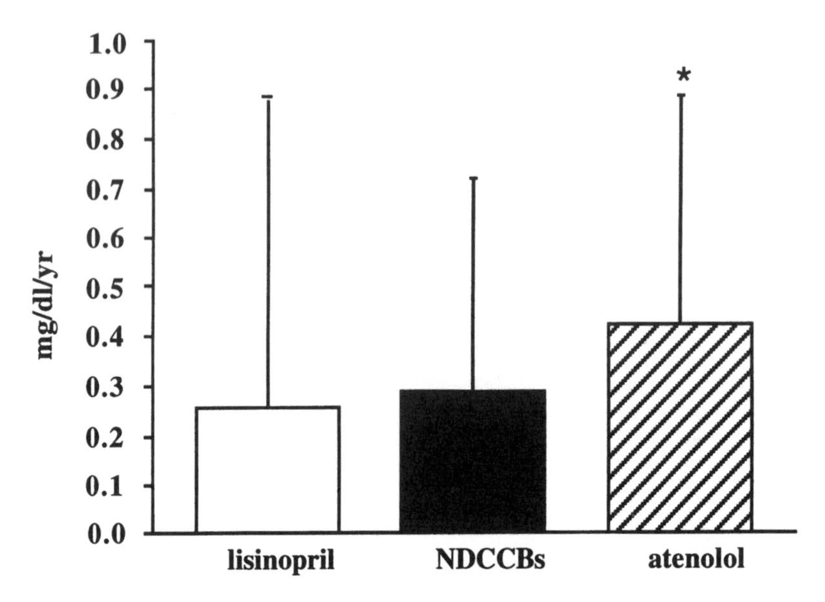

FIGURE 5. The mean change per year in serum creatinine at 6-year follow-up. * $p < 0.05$ compared with either of the other two groups. (Adapted from Bakris GL, Copley JB, Vicknair N, et al: Calcium channel blockers versus other antihypertensive therapies on progression of NIDDM associated nephropathy. Kidney Int 50:1641–1650, 1996, with permission.)

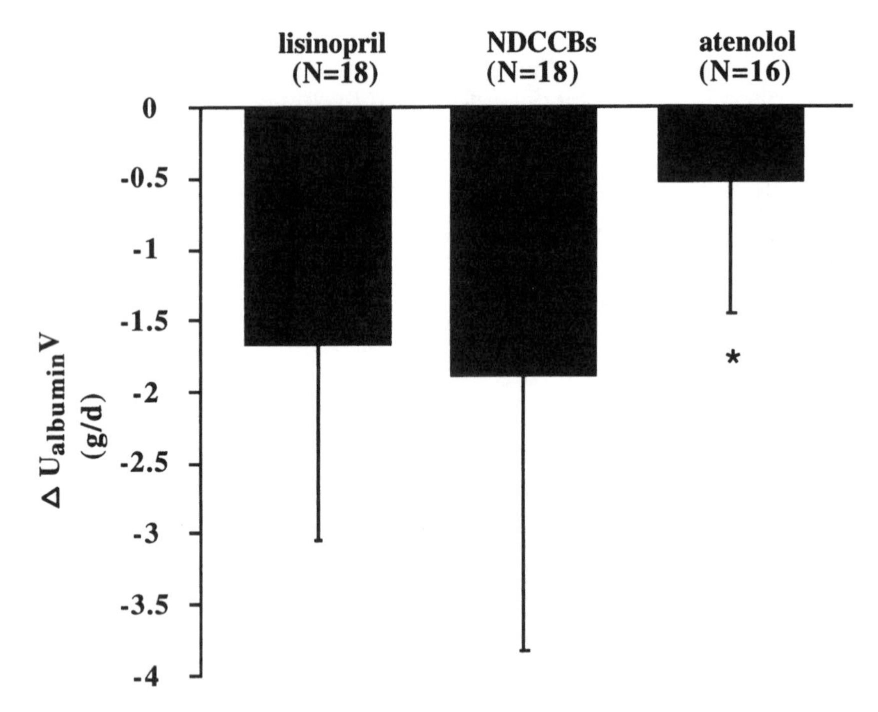

FIGURE 6. The change in proteinuria from baseline levels in each randomized group. * $p < 0.01$ compared with the other two groups. (Adapted from Bakris GL, Copley JB, Vicknair N, et al: Calcium channel blockers versus other antihypertensive therapies on progression of NIDDM associated nephropathy. Kidney Int 50:1641–1650, 1996, with permission.)

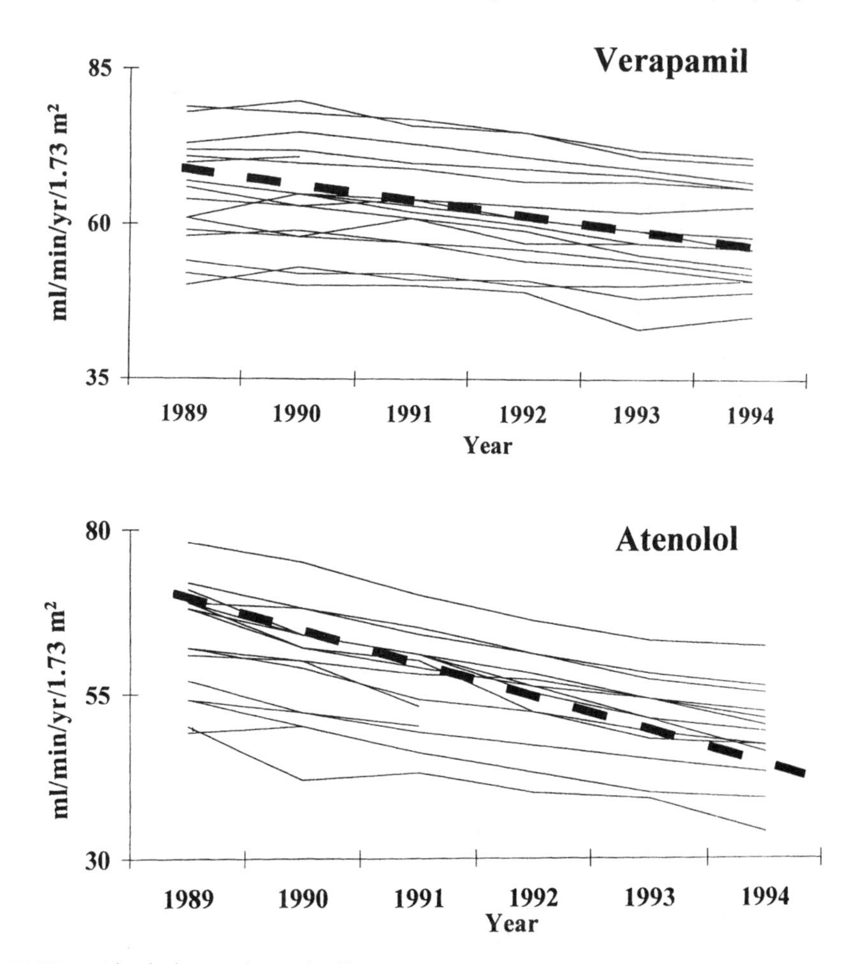

FIGURE 7. The decline in glomerular filtration rate (GFR) over 5 years in each African-American patient with established diabetic nephropathy. The slope of each group randomized to either verapamil or atenolol is shown with dashed lines. The slopes are significantly different ($p < 0.01$). (From Bakris GL, Mangrum A, Copley JB, et al: Calcium channel or beta blockade on progression of diabetic renal disease in African-Americans. Hypertension 29:743–750, 1997, with permission.)

reduce proteinuria.[60,66] Thus, reductions in albuminuria in patients with NIDDM nephropathy are associated with slowed declines in renal function. The mechanism of albuminuria reduction is related in part to blood pressure control but, more importantly, also to effects on glomerular permeability that are independent of blood pressure.

Conversely, dihydropyridine CCBs do not affect glomerular permeability, as noted in both diabetic and nondiabetic patients. Furthermore, it has been suggested that their use may be associated with poor renal outcomes in long-term studies of patients with heart failure.[71] However, a recent analysis of the time course of progression of renal insufficiency in patients with class IIIb-IV heart failure enrolled in the PRAISE trial disclosed that treatment with amlodipine did not worsen the progression of renal dysfunction.[94] These observations are consistent with the suggestion that dihydropyridines do not exacerbate progressive renal insufficiency. Further studies are needed to characterize the effects of dihydropyridine antagonists on renal function and their ability to retard progression of chronic renal disease.

Two separate studies with dihydropyridine CCBs in patients with NIDDM nephropathy demonstrate some reduction in albuminuria and renal function.[18,24] Unfortunately, the long-term follow-up of one study did not find sustained reductions in albuminuria.[59] As discussed earlier, the second study required a much longer follow-up period, given the stage of the disease. Hence, one may have predicted no difference from an ACE inhibitor in regard to changes in albuminuria, because blood pressures were reduced to levels < 130/80 mmHg.[24] Perhaps more disturbing is the fact that Velussi et al. demonstrated a significant reduction in GFR with amlodipine to a degree similar to that seen with the ACE inhibitor.[24] This finding is disturbing because it is contrary to all other published studies of CCBs that examine both renal blood flow changes and GFR.[17,60,72] This hemodynamic change alone could have predicted the observed change in albuminuria. Together with previous studies, these data support the concept that long-term use of dihydropyridine CCBs neither reduces albuminuria nor slows progression of nephropathy by mechanisms other than blood pressure reduction.

POTENTIAL REASONS FOR DIFFERENT RENAL OUTCOMES WITH CALCIUM CHANNEL BLOCKERS

Divergent Distribution and Type of Calcium Channels

CCBs as a class have quite divergent morphologic and, to a lesser extent, hemodynamic effects on the kidney (see Table 1), primarily because they do not all inhibit the same calcium channel to achieve their effect. Preliminary classification schemes divide calcium channels into low voltage-activated (T-type) and high voltage-activated types. The high voltage-activated types include the L-type, which is dihydropyridine-sensitive, and the N-type, which is ω-conotoxin GVIA-sensitive. Experimental use of various toxins has led to the further subclassification of high voltage-activated channels into P-, Q-, and R-types.[73] The L-type calcium channel is abundant in the cardiovascular system and within the past few years has been isolated and cloned. Its subunits have been identified and its amino-acid composition determined.[74] The L-type calcium channels consist of five subunits termed α_1, α_2, β, γ, and δ. The α_1 subunit appears to be responsible for channel opening and voltage dependency and contains receptors for CCBs. These geographically distinct receptors correspond to each of the three different chemical classes of antagonists, as exemplified by diltiazem, nifedipine, and verapamil. Diltiazem appears to have an inhibitory effect on mitochondrial sodium–calcium exchange that is unique among CCBs.[75] In addition, calcium channels have an unequal distribution as well as differential binding throughout the neural and cardiovascular systems.[76–78] Hence, a particular CCB has its own channel to block and thus yields different biologic effects.

LEVEL OF PRESSURE REDUCTION AND PRESERVATION OF RENAL FUNCTION

The primary cardiovascular events seen in patients with IDDM- or NIDDM-associated nephropathy are ischemic heart disease and death from cardiovascular causes. Both events are far more predominant in NIDDM- than IDDM-associated nephropathy.[79,80] Therefore, it is relatively more difficult to assess hard renal endpoints such as time to dialysis in patients with NIDDM- than IDDM-associated nephropathy, which has a relatively longer natural history. Recent evidence, however, suggests that the level of blood pressure reduction may be equally critical, if not the most critical factor, in slowing progression of renal disease and reducing overall cardiovascular deaths in patients with NIDDM.

Recent studies demonstrate that lowering blood pressure to levels < 135/85 mmHg yields a clear benefit in all groups in terms of progression of renal disease.[81–85] Moreover, these and other studies demonstrate that changes in mean arterial pressure as small as 4 mmHg can yield significantly different results in terms of progression of renal disease, even when the same classes of antihypertensive agents were used. In the Collaborative Research Study a difference in mean arterial pressure of 4 mmHg resulted in dramatic changes in urinary protein excretion within the group receiving captopril.[30] Likewise, in a recently published 6-year follow-up study in patients with NIDDM-associated nephropathy, a 4 mmHg higher systolic pressure in the group receiving beta blockers yielded a worse outcome compared with either ACE inhibitors or nondihydropyridine CCBs.[27] These benefits are reflected in cardiovascular endpoints as well as in other longer-term, larger studies such as the MRC study.[82] Therefore, the level of blood pressure reduction may be as important as the agents used to treat blood pressure.

The level of blood pressure reduction is also important for renal protection in choosing the type of CCB to lower arterial pressure. In a recent randomized, prospective study, using 24-hour blood pressure monitoring in a rat model of renal insufficiency, we compared four different CCBs in terms of development of glomerular sclerosis and urinary protein excretion.[67] This study demonstrated that the long-acting dihydropyridines, felodipine and amlodipine, failed to protect against development of glomerular scarring or to reduce albuminuria, despite the fact that systolic pressures were reduced to levels of 135–140 mmHg. Of interest, higher systolic pressure with a nondihydropyridine CCB yielded greater protection against glomerular scarring than a lower systolic pressure with a dihydropyridine (see Fig. 3). The mechanism for these differences in renal protection appears to be related to the relative loss of renal autoregulation and differential effects on albuminuria.

Lastly, recent recommendations by the National High Blood Pressure Education Group suggest that in patients with diabetic renal disease arterial pressure should be reduced to levels < 130/85 mmHg.[86] As clinicians know, it is extremely difficult to achieve such levels in clinical practice. Hence, the combination of a lower dose of an ACE inhibitor and a CCB, especially of the nondihydropyridine type, may play a major role not only in lowering pressure but also in providing additive protection against progression of renal disease. This approach is exemplified by three recent reports, two in an animal model of nephropathy,[87,88] the third in patients with NIDDM-associated nephropathy.[89] Two of the studies with a nondihydropyridine CCB/ACE inhibitor combination demonstrated almost additive effects on albuminuria reduction and prevention of glomerulosclerosis, independent of blood pressure level.[87,89] The third demonstrated that an ACE inhibitor in combination with amlodipine protected against the increase in albuminuria and glomerulosclerosis observed with amlodipine alone.[88] A recent review of combination agents clearly demonstrated that they can produce additive effects on reductions in both albuminuria and blood pressure.[90,91] In addition, combination therapy with such agents offers complementary renal hemodynamic and glomerular permeability effects as well as improvement in compliance by the patient.

CONCLUSION

This chapter may be summarized by the following points:

- Diabetes is a complicated metabolic disease with multiple effects on the kidney and vasculature, leading to renal failure in a percentage of patients.
- Albuminuria strongly correlates with progression of renal disease in diabetic patients. Large amounts of salt intake blunt the antiproteinuric effects of both ACE inhibitors and nondihydropyridine CCBs.
- ACE inhibitors and nondihydropyridine CCBs may be ideal antihypertensive agents in diabetic hypertensive patients, because they not only lower blood pressure, but also protect the kidney through other pathways, independent of blood pressure reduction.
- The different subclasses of CCBs have divergent effects on albuminuria and perhaps preservation of renal function. These differences may result from a differential distribution and binding of calcium channels, the blockade of which may not reduce glomerular permeability or prevent mechanisms that naturally protect the kidney (i.e., autoregulation).
- The level to which blood pressure is reduced may have a major impact on progression of diabetic nephropathy, especially if agents such as dihydropyridine CCBs are used. The current recommendation by the Fifth Joint National Committee, the American Diabetes Association, and the National Kidney Foundation is to reduce blood pressure to <130/85 mmHg in diabetic patients to preserve renal function.[86,92,93]

References

1. Peterson JC, Adler S, Burkart JM, et al: Blood pressure control, proteinuria and the progression of renal disease. The Modification of Diet in Renal Disease Study. Ann Intern Med 123:754–762, 1995.
2. Lewis EJ, Hunsicker LG, Bain RP, Rohde RD, for the Collaborative Study Group. The effect of angiotensin converting enzyme inhibition on diabetic nephropathy. N Engl J Med 329:1456–1462, 1993.
3. Walker WG, Hermann JA, Anderson JE: Randomized doubly blinded trial of enalapril vs. hydrochlorothiazide on glomerular filtration rate in diabetic nephropathy [abstract]. Hypertension 22:410, 1993.
4. Bjorck S, Mulec H, Johnsen SA, et al: Renal protective effect of enalapril in diabetic nephropathy. BMJ 304:339–343, 1992.
5. Bakris GL: Pathogenesis of hypertension in diabetes. Diabetes Rev 3:460–476, 1996.
6. Nielsen FS, Rossing P, Gall MA, et al: Impact of lisinopril and atenolol on kidney function in hypertensive NIDDM subjects with diabetic nephropathy. Diabetes. 43:1108–1113, 1994.

7. Lacourciere Y, Nadeau A, Poireier L, Tancrede G: Captopril or conventional therapy in hypertensive type II diabetics: Three-year analysis. Hypertension 21:786–794, 1993.
8. Walker WG, Hermann J, Yin D, et al: Diuretics accelerate diabetic nephropathy in hypertensive insulin dependent and non-insulin dependent subjects. Trans Assoc Am Physicians C305–C315, 1987.
9. Parving HH, Smidt UM, Hommel E: Effective antihypertensive treatment postpones renal insufficiency in diabetic nephropathy. Am J Kidney Dis 22:188–195, 1993.
10. Mulec H, Johnsen SA, Bjorck S: Long-term enalapril treatment in diabetic nephropathy. Kidney Int 45(Suppl):S141–S144, 1994.
11. Mimran A, Insua A, Ribstein J, et al: Comparative effect of captopril and nifedipine in normotensive patients with incipient diabetic nephropathy. Diabetes Care 11:850–853, 1988.
12. Parving H, Hommel E, Nielsen M, Giese J: Effect of captopril on blood pressure and kidney function in normotensive insulin dependent diabetics with nephropathy. BMJ 299:533–536, 1989.
13. Bakris GL, Slataper R, Vicknair N, Sadler R: ACE inhibitor mediated reductions in renal size and microalbuminuria in normotensive, diabetic subjects. J Diabetic Compl 8:2–6, 1994.
14. Parving H, Hommel E, Smidt U: Protection of kidney function and decrease in albumuniuria by captopril in insulin dependent diabetes with nephropathy. BMJ 297:1086–1091, 1988.
15. Laffel LMB, McGill JB, Gans DJ, on behalf of the North America Microalbuminuria Study Group: The beneficial effect of angiotensin converting enzyme inhibition with captopril on diabetic nephropathy in normotensive IDDM patients with microalbuminuria. Am J Med 99:497–504, 1995.
16. Ferder L, Daccordi H, Martello M, et al: Angiotensin converting enzyme inhibitors versus calcium antagonists in the treatment of diabetic hypertensive patients. Hypertension 9(Suppl 2): II237–II242, 1992.
17. Bakris GL, Stein JH: Diabetic nephropathy. Dis Month 39:573–611, 1993.
18. Melbourne Diabetic Nephropathy Study Group: Comparison between perindopril and nifedipine in hypertensive and normotensive diabetic patients with micoalbuminuria. BMJ 302:210–216, 1991.
19. Gaber L, Walton C, Brown S, Bakris GL: Effects of different antihypertensive treatments on morphologic progression of diabetic nephropathy in uninephrectomized dogs. Kidney Int 46:161–169, 1994.
20. Brown SA, Walton CL, Crawford P, Bakris GL. Long-term effects of different antihypertensive regimens on renal hemodynamics and proteinuria. Kidney Int 43:1210–1218, 1993.
21. Maschio G, Alberti D, Janin G, et al, for the Angiotensin-Converting-Enzyme Inibition in Progressive Renal Insufficiency Study Group: Effect of the angiotensin converting enzyme inhibitor benazepril on the progression of chronic renal insufficiency. N Engl J Med 334:939–945, 1996.
22. Ravid M, Lang R, Rachmani R, Lishner M: Long-term renoprotective effect of angiotensin-converting enzyme inhibition in non-insulin-dependent diabetes mellitus. Arch Intern Med 156:286–289, 1996.
23. Lebovitz HE, Wiegmann TB, Cnaan A: Renal protective effects of enalapril in hypertensive NIDDM: role of baseline albuminuria. Kidney Int Suppl 45:S150–S155, 1994.
24. Velussi M, Brocco E, Frigato F, et al: Effects of cilazapril and amlodipine on kidney function in hypertensive NIDDM patients. Diabetes 45:216–222, 1996.
25. Bakris G: Effects of diltiazem or lisinopril on massive proteinuria associated with diabetes mellitus. Ann Intern Med 112:707–708, 1990.
26. Baba T, Murabayashi S, Takebe K: Comparison of the renal effects of angiotensin converting enzyme inhibitor and calcium antagonist in hypertensive type 2 (non–insulin-dependent) diabetics with microalbuminuria: a randomized controlled trial. Diabetologia 32:40–44, 1989.
27. Bakris GL, Copley JB, Vicknair N, et al: Calcium channel blockers versus other antihypertensive therapies on progression of NIDDM associated nephropathy: Kidney Int 50:1641–1650, 1996.
28. Bakris GL, Mangrum A, Copley JB, et al: Calcium channel or beta blockade on progression of diabetic renal disease in African-Americans. Hypertension 29:743–750, 1997.
29. Bakris GL: Microalbuminuria: Prognostic implications. Curr Opin Nephrol and Hypertens 7:219–223, 1996.
30. Hebert LA, Bain RP, Verme D, et al for the Collaborative Study Group. Remission of nephrotic range proteinuria in type I diabetes. Kidney Int 46:1688–1693, 1994.
31. Zucchelli P, Auzccala A, Borghi M, Fusaroli M, Sasdelli M, Stallone C, Sanna G, Gaggi R: Long term comparison between captopril and nifedipine in the progression of renal insufficiency. Kidney Int 42:452–458, 1992.
32. Myers BD , Guasch A: Selectivity of the glomerular filtration barrier in healthy and nephrotic humans. Am J Nephrol 13:311–317, 1993.

33. Drumond MC, Kristal B, Myers BD, Deen WM: Structural basis for reduced glomerular filtration capacity in nephrotic humans. J Clin Invest 94:1187–1195, 1994.
34. Weidmann P, Schneider M, Bohlen L. et al: Therapeutic efficacy of different antihypertensive drugs in human diabetic nephropathy: An updated meta-analysis. Nephrol Dial Transplant 10:1963–1974, 1995.
35. Gansevoort RT, Sluiter WJ, Hemmelder MH, et al: Antiproteinuric effect of blood pressure lowering agents: a meta-analysis of comparative trials. Nephrol Dial Transplant 10:1963–1974, 1995.
36. Maki DD, Ma JZ, Louis TA, Kasiske BL: Effects of antihypertensive agents on the kidney. Arch Intern Med 155:1073–1082, 1995.
37. Remuzzi G, Ruggenenti P, Benigni A: Understanding the nature of renal disease progression. Kidney Int 51:2–15, 1997.
38. Nakano S, Ishii T, Kitazawa M, Kigoshi T, et al: Altered circadian blood pressure rhythm and progression of diabetic nephropathy in non-insulin dependent diabetes mellitus subjects: An average three year follow-up study. J Invest Med 44:247–253, 1996.
39. Heeg JE, deJong PE, van der Hem GK, de Zeeuw D: Efficacy and variability of the antiproteinuric effect of ACE inhibition by lisinopril. Kidney Int 36:272–279, 1989.
40. Bakris GL, Weir MR: Salt intake and reductions in arterial pressure and proteinuria: Is there a direct link? Am J Hypertens 9:200S–206S, 1996.
41. Bakris GL, Smith AC: Effects of sodium intake on albumin excretion in patients with diabetic nephropathy treated with long-acting calcium antagonists. Ann Intern Med 125:201–203, 1996.
42. Oyama TT, Thompson MM, Anderson S: Sodium restriction and the renin-angiotensin system in diabetic rats [abstract]. J Am Soc Nephrol 7:1876, 1996.
43. Kilaru P and Bakris GL: Microalbuminuria and progressive renal disease. J Hum Hypertens 8:809–817, 1994.
44. Cohen MP, Clements RS, Cohen JA, Shearman CW: Prevention of decline in renal function in the diabetic db/db mouse. Diabetologia 39:270–274, 1996.
45. Cohen MP, Hud E, Wu VY, Ziyadeh FN: Albumin modified by Amadori glucose adducts activates mesangial cell type IV collagen gene transcription. Mol Cell Biochem 151:61–67, 1995.
46. Cohen MP, Sharma K, Jin Y, et al: Prevention of diabetic nephropathy in db/db mice with glycated albumin antagonists: A novel treatment strategy. J Clin Invest 95:2338–2345, 1995.
47. Dworkin LD, Benstein JA, Tolbert E, Feiner HD: Salt restriction inhibits renal growth and stabilizes injury in rats with established renal disease. J Am Soc Nephrol 7:437–442, 1996.
48. Bank N, Lahorra MA, Aynedjian HS, Wilkes BM: Sodium restriction corrects hyperfiltration of diabetes. Am J Physiol 254:F668–F676, 1988.
49. Demarie BK, Bakris GL: Effects of different calcium antagonists on proteinuria associated with diabetes mellitus. Ann Intern Med 113: 987–988, 1990.
50. Capewell S, Collier A, Matthews D, et al: A trial of the calcium antagonist felodipine in hypertensive type II diabetic patients. Diabetic Med 6:809–812, 1989.
51. Hartmann A, Lund K, Hagel L, et al: Contrasting short term effects of nifedipine on glomerular and tubular functions in glomerulonephritic patients. J Am Soc Nephrol 5:1385–1390, 1994.
52. Abbott K, Smith AC, Bakris GL: Effects of dihydropyridine calcium antagonists on albuminuria in diabetic subjects. J Clin Pharmacol 36:274–279, 1996.
53. Smith AC, Bakris GL: Differential effects of calcium channel blockers on albuminuria and glomerular permeability in NIDDM subjects with nephropathy: Pilot results of a two year study [abstract]. J Am Soc Nephrol 7:1364, 1996.
54. Rossing P, Tarnow L, Boelskifte S, et al: Differences between nisoldipine and lisinopril in GFRs and albuminuria in hypertensive IDDM patients with diabetic nephropathy during the first year of treatment. Diabetes 46:481–487, 1997.
55. Moscorie L, Remuzzi A, Ruggenenti P, et al: Glomerular size-selective dysfunction is not ameliorated by ACE inhibition and Ca channel blockade [abstract]. Am Soc Nephrol 7:1362, 1996.
56. Bakris GL, Mehler P, Schrier R: Hypertension and diabetes. In Schrier RW, Gottschalk CW (eds): Diseases of the Kidney, 6th ed. Boston, Little, Brown, 1996, pp 1455–1464.
57. Hatov C, Pavlova M, Demetrakov D, Veleukov H, Bakalov B. The influence of nifedipine on renal function in patients with chronic renal disease at the initial stage of drug use. Ther Arch 297:88-89, 1990.
58. Parving HH, Tarnow L, Rossing P: Renal protection in diabetes: An emerging role for calcium antagonists. J Hypertens 14(Suppl 4):S21–S25, 1996.
59. Gilbert RE, Jerums G, Allen T, et al, on behalf of the Diabetic Nephropathy Study Group: Effect of different antihypertensive agents on normotensive microalbuminuric patients with IDDM and NIDDM [abstract]. J Am Soc Nephrol 5:377, 1994.

60. Bakris GL: Hypertension in diabetic patients: An overview of interventional studies to preserve renal function. Am J Hypertens 6:140S–147S, 1993.
61. Griffin KA, Picken MM, Bidani AK: Deleterious effects of calcium channel blockade on pressure transmission and glomerular injury in rat remnant kidneys. J Clin Invest 96:793–800, 1995.
62. Nakamura T, Takahashi T, Fukui M, et al: Enalapril attenuates increased gene expression of extracellular matrix components in diabetic rats. J Am Soc Nephrol 5:1492–1497, 1995.
63. Jyothirmayi GN, Reddi AS: Effect of diltiazem on glomerular heparan sulfate and albuminuria in diabetic rats. Hypertension 21 (6 Pt 1): 795–802, 1993.
64. Furberg CD, Psaty BM, Meyer JV: Nifedipine. Dose-related increase in mortality in patients with coronary heart disease. Circulation 92:1326–1331, 1995.
65. Kvam FI, Iverson BM, Ofstad J: Autoregulation of glomerular capillary pressure in SHR during successive reduction of systemic blood pressure with antihypertensive agents [abstract]. J Am Soc Nephrol 7:1552, 1996.
66. Anderson S, Rennke HG, Brenner BM: Nifedipine versus fosinopril in uninephrectomized diabetic rats. Kidney Int 41:891–897, 1992.
67. Picken M, Griffin K, Bakris GL, Bidani A: Comparative effects of four different calcium antagonists on progression of renal disease in a remnant kidney model [abstract]. J Am Soc Nephrol 7:1586, 1996.
68. Bakris GL, Picken M, Griffin K, Bidani A: Effects of a fixed dose combination of either amlodipine, benazapril or their combination on glomerulosclerosis and proteinuria. J Hypertens (in press).
69. Estacio RO, Savage S, Nagel NJ, Schrier RW: Baseline characteristics of participants in the Appropriate Blood Pressure Control in Diabetes Trials. Control Clin Trials 17:242–257, 1996.
70. Wright JR, Kusek J, Toto R, et al, for the AASK Pilot Study Investigators: Design and baseline characteristics of participants in the African American Study of Kidney Diseases and hypertension (AASK) Pilot Study. Control Clin Trials 17:3S–16S, 1996.
71. Packer M, Oconnor CM, Ghali JK, et al, for the Prospective Randomized Amlodipine Survival Evaluation study group: Effect on amlodipine on morbidity and mortality in severe chronic heart failure. N Engl J Med 335:1107–1114, 1996.
72. Abbott K, Bakris GL: Renal effects of antihypertensive medications: An overview. J. Clin Pharmacol 33:392–399, 1993.
73. Perez-Reyes E, Schneider T: Molecular biology of calcium channels. Kidney Int 48:1111–1124, 1995.
74. Nayler WG: Calcium channels and their involvement in cardiovascular disease. Biochem Pharm 43:39–46, 1992.
75. Schwartz A: Molecular and cellular aspects of calcium channel antagonism. Am J Cardiol 70:6F–8F, 1992.
76. Walsh KB, Bryant SH, Schwartz A: Effect of calcium antagonist drugs on calcium currents in mammalian skeletal muscle fibers. J Pharmacol Exp Ther 236:403–407, 1986.
77. Finkel MS, Patterson RE, Roberts WC, et al: Calcium channel binding characteristics in the human heart. Am J Cardiol 62:1281–1284, 1988.
78. Walker BR: Evidence for uneven distribution of L-type calcium channels in rat pulmonary circulation. Am J Physiol 269(6 Pt 2):H2051–2056, 1995.
79. Mogensen CE: Microalbuminuria predicts clinical proteinuria and early mortality in maturity-onset diabetes. N Engl J Med 310:356–360, 1984.
80. Selby JV, Fitzsimmon SC, Newman JM, et al: The natural history and epidemiology of diabetic nephropathy. JAMA 263:1954–1960, 1989.
81. Lazarus JM, Bourgoignie JJ, Buckalew VM, et al, for the MDRD Group. Achievement and safety of a low blood pressure goal in chronic renal disease: The modification of diet in renal disease study group. Hypertension 29:641–650, 1997.
82. MRC Working Party: Medical Research Council trial of treatment of hypertension in older adults: Principal results. BMJ 304:405–412, 1992.
83. Fletcher A, Donoghue M: What are optimum blood pressures? Contemp Intern Med 8:31–44, 1996.
84. Bakris GL: Is the level of arterial pressure reduction important for preservation of renal function? Nephrol Dial Transplant 11:2383–2384, 1996.
85. Toto R, Mitchell HC, Smith RD, et al: Strict blood pressure control and progression of renal disease in hypertensive nephrosclerosis. Kidney Int 48:851–858, 1995.
86. National High Blood Pressure Education Program Working Group: 1995 update of the Working Group reports on chronic renal failure and renovascular hypertension. Arch Intern Med 156:1938–1947, 1996.

87. Munter K, Hergenroder S, Jochims K, Kirchengast M: Individual and combined effects of verapamil or trandolapril on glomerular morphology and function in the stroke prone rat. J Am Soc Nephrol 7:681–686, 1996.
88. Bakris GL, Griffin K, Picken M, Bidani A: Similar levels of blood pressure reduction with a combination of amlodipine and benazapril protects the kidney against the effects of amlodipine alone [abstract]. Nephrology 3(Suppl 1):5506, 1997.
89. Bakris GL, Weir MR, DeQuattro V, et al: Renal hemodynamic and antiproteinuric response to an ACE inhibitor, trandolapril or calcium antagonist, verapamil alone or in a fixed dose combination in patients with diabetic nephropathy: A randomized multicentered study [abstract]. J Am Soc Nephrol 7:1546, 1996.
90. Epstein M, Bakris GL: Newer approaches to antihypertensive therapy: Use of fixed dose combination therapy. Arch Intern Med 156:1969–1978, 1996.
91. Bakris GL: Combination therapy for hypertension and renal disease in diabetes. In Mogensen CE (ed): The Kidney and Hypertension in Diabetes Mellitus, 3rd ed. Boston, Kluwer Academic, 1997, pp 561–568.
92. National High Blood Pressure Education Program Working Group: National High Blood Pressure Education Program working group report on hypertension and diabetes. Hypertension 23:145–158, 1994.
93. Bennett PH, Haffner S, Kasiske BL, et al: Screening and management of microalbuminuria in patients with diabetes mellitus: Recommendations to the Scientific Advisory Board of the National Kidney Foundation. Am J Kid Dis 25:107–112, 1995.
94. O'Connor GM: Personal communication, 1998.

JOHN M. FLACK M.D., M.P.H. / MATTHEW R. WEIR M.D.

16

Calcium Antagonists as Antihypertensive Agents in African-American Patients

Calcium antagonists are commonly used to treat hypertension and, in fact, are the most commonly prescribed class of cardiovascular drugs in the United States. They have been preferentially prescribed to hypertensive African-Americans because of their proven efficacy in lowering blood pressure when used as monotherapy. In addition, calcium antagonists have been used successfully to lower blood pressure in combination with other classes of antihypertensive drugs, including diuretics,[1–7] beta blockers,[8–10] angiotensin-converting enzyme (ACE) inhibitors,[11,12] and alpha antagonists.[13,14] This chapter focuses on the epidemiology of hypertension and pressure-related target-organ damage in African-Americans; the pathophysiologic profile of hypertensive African-Americans; and the unique pharmacologic profile of calcium antagonists, which reverse many of the hallmark physiologic aberrations of this high-risk demographic group.

EPIDEMIOLOGY OF HYPERTENSION AND PRESSURE-RELATED TARGET-ORGAN DAMAGE IN AFRICAN-AMERICANS

Hypertension has an earlier onset in African-Americans than in white or Mexican-American populations.[15] Middle-aged and older hypertensive African-Americans typically present with a longer duration of elevated blood pressure compared with their white counterparts. Moreover, the blood pressure distribution is shifted to the right, leading to an inordinately high prevalence of severe hypertension.[15] It has been well established that the presence of blood pressure-related target-organ damage significantly amplifies the risk for cardiovascular events at a given level of blood pressure.[16–19] African-Americans, particularly women, have a high prevalence of obesity, which undoubtedly contributes to the high prevalence of hypertension. In the southeast United States, 71.2% of African-American women are obese compared with 41% of African-American women residing elsewhere.[20] The higher levels of blood pressure in the African-American population reduce the likelihood of achieving normalization (minimally < 140/90 mmHg) during antihypertensive monotherapy. Indeed, only 45% of drug-treated hypertensive African-Americans have attained levels < 140/90 mmHg.[15] Moreover, the likelihood of blood pressure control is much lower in drug-treated hypertensive African-American men compared with women. Accordingly, blood pressure-related target-organ damage, such as stroke, renal insufficiency, and left ventricular hypertrophy, is excessively prevalent among African-Americans compared with other demographic groups.[21–25]

The risk for all stages of clinically measurable renal injury in African-Americans (from hypercreatinemia to end-stage renal disease and renal mortality) is much higher in African-Americans than in whites.[25–28] In fact, the relative risk for hypertensive end-stage renal disease ranges from 5–17 times greater in African-Americans compared with whites.[29,30] Diabetes mellitus is also found disproportionately among African-Americans[31] and unquestionably is a major contributor to the excessive burden of cardiovascular-renal disease.

PHYSIOLOGIC PROFILE OF HYPERTENSIVE AFRICAN-AMERICANS

Hypertensive African-Americans have several physiologic tendencies that may influence the success of antihypertensive drug therapy (Table 1). First, both normotensive and hypertensive African-Americans tend to be more salt-sensitive than their white counterparts with similar blood pressure levels;[32,33] that is, the renal pressure-natriuresis curve is shifted rightward so that a higher blood pressure level is necessary to achieve steady-state sodium balance. Secondly, hypertensive African-Americans often present with reduced renal natriuretic capacity with or without clinically detectable renal dysfunction (i.e., hypercreatinemia, albuminuria). They also have higher renal vascular resistance and lower renal blood flow than white hypertensives.[34] The smaller increase in renal perfusion in African-Americans during periods of dietary salt loading may be related to deficiencies of intrarenal vasodilatory hormones (i.e., angiotensin,[1–7] dopamine, kallikrein, prostaglandins, and/or nitric oxide).[35–38]

Salt sensitivity, a common intermediate blood pressure phenotype in African-Americans, has been linked to a constellation of intrarenal hemodynamic abnormalities—raised glomerular intracapillary pressure,[39,40] inadequate rise in renal blood flow,[41,42] and glomerular hyperfiltration[39,40]—as well as to abnormally

TABLE 1. Physiologic Profile of Hypertension in African-Americans*

- Salt sensitivity
- Obesity
- High prevalence of coexisting risk factors for cardiovascular disease (e.g., diabetes mellitus)
- Decreased renal vasodilator hormones
- Suppressed renin levels and (?) increased intrarenal angiotensin II
- Decreased renal natriuretic capacity
- Increased peripheral vascular resistance
- High renal vascular resistance
- Reduced renal flow
- Excessive blood pressure-related target-organ damage (e.g., left ventricular hypertrophy)
- Excessive risk of pressure-related complications (e.g., stroke, congestive heart failure, renal insufficiency)

* None of these anthropometric, hemodynamic, physiologic, and target-organ correlates are unique to hypertension in African-Americans; however, they are more common in hypertensive African-Americans than in hypertensive whites.

high renal protein excretion.[40,41,43] Obesity, which is highly prevalent among African-Americans (particularly women), appears to amplify the impact of elevated blood pressure on the kidneys because of the greater level of albuminuria in overweight compared with lean people, whether they are hypertensive or normotensive.[44] Moreover, salt-sensitivity has been linked to cardiovascular risk factor clustering.[45] Hypertensive African-Americans also tend to have lower circulating renin levels and higher peripheral vascular resistance (PVR) than their white counterparts.[46,47] Plasma volume expansion, however, has not been consistently demonstrated in hypertensive African-Americans compared with their white counterparts.[48,49] Thus, the pathophysiologic aberrations commonly encountered in hypertensive African-Americans underscore the need for modest dietary salt restriction and pharmacologic therapy that induce vasodilation as well as promote both salt and water excretion. Furthermore, it is likely that the kidney has a primary role both in the etiology and maintenance of elevated blood pressure and in the commonly encountered intermediate phenotype, salt-sensitivity, in hypertensive African-Americans.

PHARMACOLOGIC ATTRIBUTES THAT MAKE CALCIUM ANTAGONISTS PARTICULARLY SUITABLE FOR AFRICAN-AMERICANS

Reduced renal natriuretic capacity, a central pathophysiologic feature of hypertension in African-Americans, probably explains, at least in part, the rightward shift of the dose-response curve for certain drug classes, including ACE inhibitors, alpha receptor antagonists, angiotensin receptor antagonists, and beta blockers. Clinical studies in humans have demonstrated that inadequate augmentation of renal plasma flow and elevated glomerular filtration rate (GFR) in response to high levels of dietary sodium consumption are plausible explanations, at least in part, of the rightward shift. These changes in glomerular capillary hemodynamics amplify salt and water retention through reduction of peritubular capillary hydrostatic pressure and an increase in peritubular capillary colloid osmotic pressure.[50] Antihypertensive therapies that reverse these intrarenal pathophysiologic abnormalities should not only facilitate the attainment of optimal blood pressure control but also, in theory, augment renal target-organ protection.

In experimental studies of renal hemodynamics, calcium antagonists have been shown to reduce both afferent and efferent glomerular arteriolar tone by antagonizing the vasoconstrictor actions of norepinephrine (which primarily constricts the afferent arteriole) and angiotensin II (which primarily constricts the efferent arteriole).[51–54] In human studies, calcium antagonists consistently improve renal plasma flow and GFR.[51–54] The renal effects of calcium antagonists probably help to explain their ability to lower blood pressure in African-Americans with salt-sensitive hypertension and ad libitum sodium intake. Evidence also suggests that calcium antagonists possess intrinsic natriuretic properties because of their ability to augment renal blood flow.[54–57] Independent of their affect on intrarenal hemodynamics, calcium antagonists appear to inhibit tubular reabsorption of sodium.[55,57–59] The pharmacologic effects of calcium antagonists that augment renal blood flow and enhance salt and water excretion make them an attractive therapeutic option for lowering blood pressure in hypertensive African-Americans. Clinical studies in humans support the thesis that calcium antagonists have equivalent blood pressure-lowering efficacy in various racial and ethnic groups, regardless of whether the patients are salt-sensitive. Moreover, the blood pressure-lowering effect of calcium antagonists, even at low doses, is less attenuated (compared with other nondiuretic antihypertensive agents) in the setting of high levels of dietary salt consumption.[60–63] The fact that the antihypertensive efficacy of calcium antagonists is largely independent of the level of dietary salt consumption is perhaps their most important pharmacologic attribute. Long-term lowering of dietary salt consumption is not always feasible because it requires intensive and sustained patient education and support. Moreover, in many clinical settings, neither the requisite economic resources nor the personnel skilled in dietary education and behavior modification are available. The blood pressure-lowering efficacy of the calcium antagonists, unlike most other antihypertensive drug classes, is not attenuated by simultaneous use of nonsteroidal antiinflammatory drugs (NSAIDs).[64,65]

Control of blood pressure has been shown to be an important factor in obtaining regression of left ventricular hypertrophy (LVH) in hypertensive patients.[66,67] Although LVH is a powerful predictor of subsequent cardiovascular morbidity and mortality as well as all-cause mortality,[68,72] the intuitive appeal of the biologic plausibility that regression of LVH confers incremental reduction in the risk for cardiovascular disease over and above the reduction attributable to lowering of blood pressure remains an unproven (and untested) speculation in humans. Calcium antagonists improve ventricular relaxation and thereby reverse the LV filling abnormalities in hypertensive patients with diastolic dysfunction, an early manifestation of hypertensive heart disease.[73] Calcium antagonists also possess antianginal properties, in part because of their ability to both dilate the coronary circulation and to reduce afterload. Improved oxygen delivery to the myocardium improves both diastolic and systolic myocardial function. Calcium antagonists are also metabolically neutral; they do not adversely affect blood lipids, glucose, serum electrolytes, or uric acid.[74–78]

CLINICAL TRIALS OF CALCIUM ANTAGONISTS IN HYPERTENSIVE AFRICAN-AMERICANS

Racial differences in response to antihypertensive drug therapy have long been noted in clinical trials. A large body of data has led to the widely held opinion that hypertensive African-Americans respond better to monotherapy with diuretics and calcium antagonists than to beta blockers and ACE inhibitors.[60,79–86] Unfortunately, only a few clinical trials have focused solely on an African-American population, and only recently have such studies examined antihypertensive drug responses in different racial groups using salt-sensitivity profiling and/or variations in dietary salt consumption.[87,88]

Zing et al.[89] published a review of the blood pressure-lowering effect of calcium antagonists in African-Americans compared with other agents and in African-Americans compared with whites. This review confirmed that, like diuretics, calcium antagonist monotherapy lowered blood pressure more effectively than either ACE inhibitors or beta blockers. Moreover, the authors found no consistent difference in efficacy among African-Americans compared with whites. Other studies have shown that calcium antagonists effectively lower both systolic and diastolic blood pressure in hypertensive African-Americans. Calcium antagonists typically demonstrate similar efficacy compared with thiazide diuretics.[60,90–95] However, monotherapy with calcium antagonists has demonstrated relatively greater efficacy in African-Americans than either beta blockers or ACE inhibitors.[60,79,82] By unclear mechanisms salt-induced drug resistance appears to play a major role in the lesser response to ACE inhibitors compared with calcium antagonists.

One of the first large-scale clinical trials to assess response to antihypertensive drugs in a solely African-American population was conducted in 1989. Saunders and coworkers[79] compared antihypertensive responses to atenolol, captopril, and verapamil SR in 394 hypertensive African-Americans with a mean diastolic blood pressure of ~100 mmHg. All three monotherapies demonstrated clinically relevant therapeutic success (Table 2). Nevertheless, the calcium antagonist, verapamil SR (starting dose of 240 mg/day with titration to a maximal dose of 360 mg/day), lowered blood pressure significantly better than the other two drugs. The overall proportion of hypertensive African-Americans responding to drug monotherapy (diastolic blood pressure < 90 mmHg and/or a fall of at least 10 mmHg) was also significantly greater in the verapamil group—65.2% of patients in the group receiving 240 mg/day and 73% in the group receiving 360 mg/day compared with 55.1% in the atenolol group and 43.8% in the captopril group (Table 3). Neither salt sensitivity nor dietary salt

TABLE 2. Mean Changes from Baseline in Supine Diastolic and Systolic Blood Pressure*

	Atenolol	Captopril	Verapamil SR
Patients completing period 2			
Number	118	112	115
Period 1			
Supine diastolic† (mmHg)	110.4 ± 4.2	100.7 ± 4.3	100.4 ± 4.5
Supine systolic† (mmHg)	152.5 ± 13.9	151.8 ± 16.6	151.7 ± 15.5
Period 2 change			
Supine diastolic† (mmHg)	–9.7 ± 8.0	–8.5 ± 8.1	–11.0 ± 8.5
Supine systolic† (mmHg)	–8.3 ± 12.6	–8.2 ± 12.5	–10.3 ± 12.7
Patients completing period 3			
Number	109	96	100
Period 1			
Supine diastolic† (mmHg)	100.9 ± 5.9	100.4 ± 5.5	100.8 ± 4.6
Supine systolic† (mmHg)	152.0 ± 13.6	150.2 ± 15.4	151.0 ± 14.8
Period 3 change			
Supine diastolic† (mmHg)	–10.2 ± 8.6	–9.6 ± 9.1	–12.9 ± 7.7
Supine systolic† (mmHg)	–9.6 ± 13.8	–8.2 ± 14.2	–13.3 ± 12.7

* Data for assessable patients who completed period 2 (4 weeks of initial therapy) and period 3 (an additional 4 weeks of forced titration and maintenance therapy). Blood pressure change in period 3 represents the primary study endpoint.

† Mean ± SD

From Saunders E, Weir MR, Kong BW, et al: A comparison of the efficacy and safety of a beta-blocker, a calcium blocker and a converting enzyme inhibitor in hypertensive blacks. Arch Intern Med 150:1707–1713, 1990, with permission.

TABLE 3. Response to Treatment

	Atenolol	Captopril	Verapamil SR
Period 2 (n = 345)*			
Number in each treatment group	118	112	115
Success (%)	55.1	43.8	65.2
Failure (%)	44.9	56.3	34.8
Period 3 (low and high dose) (n = 307)*			
Number in each treatment group	109	98	100
Success (%)	59.6	57.1	73.0
Failure (%)	40.4	42.9	27.0
Period 3 (high dose only) (n = 157)*			
Number in each treatment group	56	47	54
Success (%)	58.9	61.7	83.3
Failure (%)	41.1	38.3	16.7

* Number of patients for whom data could be evaluated.
Responders are defined as patients with diastolic blood pressure < 90 mmHg and/or a drop of 10 mmHg or more.
From Saunders E, Weir MR, Kong BW, et al: A comparison of the efficacy and safety of a beta-blocker, a calcium blocker and a converting enzyme inhibitor in hypertensive blacks. Arch Intern Med 150:1707–1713, 1990, with permission.

consumption was assessed, and no attempt was made to modify dietary salt intake. These results support the thesis that middle-aged African-Americans with mostly stage 1–2 diastolic hypertension and ad libitum salt intake show a greater response to starting doses of a calcium antagonist compared with relatively lower starting doses of either an ACE inhibitor or beta blocker. However, both the calcium antagonist and the ACE inhibitor (but not the beta blocker) demonstrated increasing response rates with increasing doses. These data are consistent with other studies that have documented a rightward shift of the dose-response curve to calcium antagonists and ACE inhibitors in hypertensive African-Americans.[96-98] Consequently, clinical trials that compare calcium antagonists with comparably dosed newer ACE inhibitors need to be conducted.

Veterans Affairs Cooperative Study of Single-drug Therapy for Hypertension in Men

Materson and coworkers[82] conducted a randomized, double-blind, parallel group clinical trial in 1292 male veterans with diastolic blood pressure of 95–109 mmHg (stage 1–2 hypertension). The mean age of eligible participants was 59 years. Forty-eight percent were African-Americans. Baseline blood pressure averaged 152/99 mmHg; however, older African-Americans (≥ 60 years) were more likely to have stage 2 diastolic hypertension (100–109 mmHg) than older whites (40% vs. 25%, respectively). A similar but less striking trend was seen among younger participants: 43% of African-Americans compared with 36% of whites had stage 2 hypertension. Eligible participants received placebo or one of the six following drugs:

1. Hydrochlorothiazide, 12.5, 25, or 50 mg/day
2. Atenolol, 25, 50, or 100 mg/day
3. Captopril, 25, 50, or 100 mg/day

4. Sustained-release diltiazem, 120, 240, or 360 mg/day
5. Prazosin, 4, 10, or 20 mg/day
6. Clonidine, 0.2, 0.4, or 0.6 mg/day

Hydrochlorothiazide and atenolol were given once daily; all other drugs were given twice daily. After randomization participants entered a titration phase lasting 4–8 weeks. Medications were initiated at the lowest dose; doses were increased every 2 weeks until diastolic blood pressure was less than 90 mmHg on two consecutive visits. Participants achieving diastolic blood pressure < 90 mmHg during the titration phase entered a maintenance phase lasting at least 12 months, during which doses were adjusted to keep diastolic blood pressure at 99 mmHg or less and to minimize side effects. Blood pressure control during this phase was defined as diastolic blood pressure ≤ 95 mmHg. Among African-Americans sustained-release diltiazem was more successful in achieving the target blood pressure at the end of the titration phase and keeping the diastolic blood pressure at 95 mmHg or less at 1 year. Figure 1 shows the percentage of both younger and older African-Americans who responded to therapy. Sustained-release diltiazem was more successful than the other drugs in achieving antihypertensive responses among both younger and older African-Americans. Figure 2 shows corresponding data for younger and older white men.

The Treatment of Mild Hypertension Study

The Treatment of Mild Hypertension Study (TOMHS) was conducted in 902 hypertensive men and women aged 45–69 years with stage 1 diastolic hypertension (diastolic blood pressure < 100 mmHg).[60,99] African-Americans composed 19% (n = 177) of the TOMHS sample. The majority of African-Americans were women, whereas the majority of whites were men. All participants received sustained dietary counseling focused on weight loss, dietary salt and alcohol restriction, and increased physical activity. TOMHS participants were randomized to one of five active drug groups or placebo. Drugs were given once daily at the following doses:

1. Chorthalidone, 15 mg/day
2. Acebutolol, 400 mg/day
3. Amlodipine, 5 mg/day
4. Enalapril, 5 mg/day
5. Doxazosin, 2 mg/day

Follow-up averaged 4.4 years. The data reveal several important points. First, the hypotensive effect of most drugs (except diuretics, with which the opposite effect was seen) was attenuated by higher levels of urinary sodium, although the calcium antagonist, amlodipine, was less affected than other drugs.[60] Secondly, the blood pressure-lowering effect of the calcium antagonist, in both African-Americans and whites but more so in the former, was superior to that of the alpha antagonist, doxazosin, and the ACE inhibitor, enalapril. However, both of these agents were prescribed at relatively lower daily doses than other drugs. In the overall TOMHS cohort, amlodipine, a long-acting dihydropyridine calcium antagonist, was the best tolerated drug therapy over the 4.4 year average follow-up: 83% of participants remained on amlodipine compared with 78% on acebutolol, 67% on enalapril, 66% on chlorthalidone, and 65% on doxazosin.

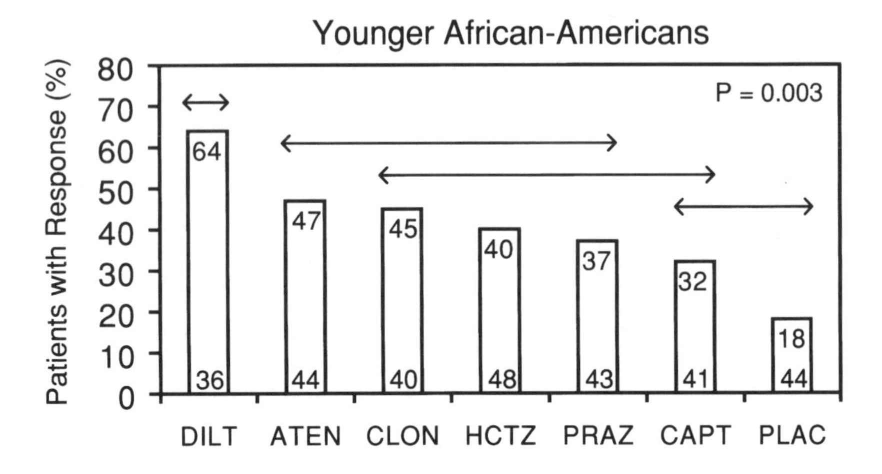

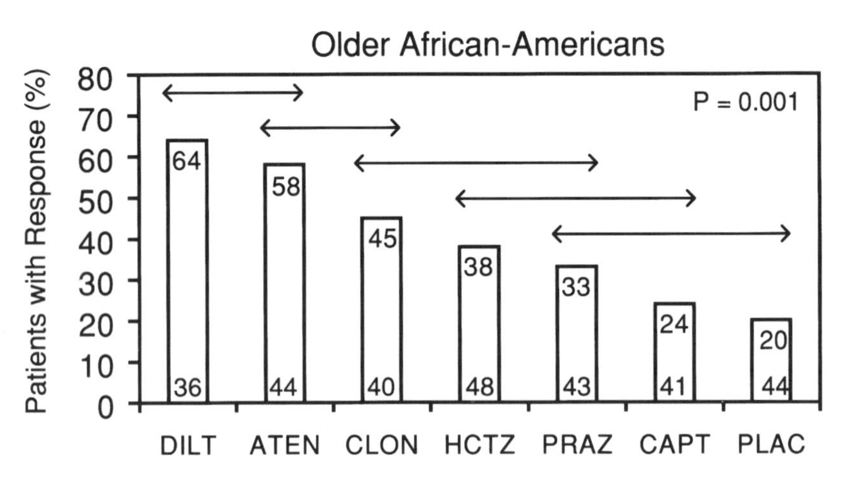

FIGURE 1. Single-drug therapy for hypertension in African-American men. Blood pressure response is defined as diastolic blood pressure < 90 mmHg at the end of titration and diastolic blood pressure < 95 mmHg at the end of 1 year of treatment. Chi-square p-values were derived from pairwise contrasts of the proportion of responders in each treatment group. Horizontal arrows indicate drug groups that did not differ statistically from each other. ATEN = atenolol, CAPT = captopril, CLON = clonidine, DILT = diltiazem, PLAC = placebo, PRAZ = prazosin. (From Materson BJ, Reda DJ, Cushman WC, et al: Single drug therapy for hypertension in men: A comparison of six antihypertensive agents with placebo. N Engl J Med 328:914–921, 1993, with permission.)

A large multicenter clinical trial[88] evaluating the influence of race and dietary salt consumption on the antihypertensive efficacy of a calcium antagonist and an ACE inhibitor was conducted in a hypertensive population selected for salt sensitivity (> 5 mmHg rise in diastolic blood pressure after going from a low- to a high-salt diet). To elucidate racial patterns of response as well as to assess the importance of salt restriction on pharmacologic lowering of blood pressure, 232 white and 96 African-American hypertensive patients completed a sextuple crossover clinical trial that determined blood pressure responses on high-salt (~250 mmol Na/day) and low-salt (~70 mmol Na/day) diets during

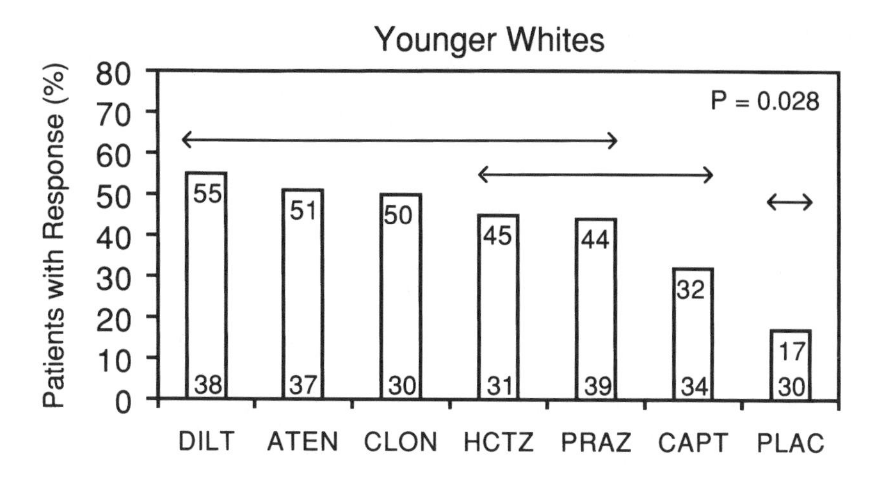

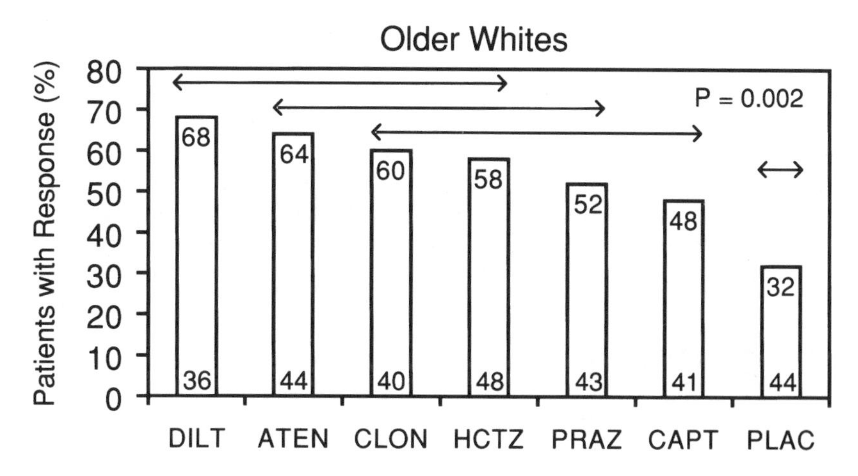

FIGURE 2. Single-drug therapy for hypertension in white men. See legend to Figure 1.

treatment with placebo, ACE inhibitor (enalapril, 5 or 20 mg 2 times/day), or calcium antagonist (isradapine, 5 or 10 mg 2 times/day). During the high salt diet, African-Americans demonstrated a greater reduction in blood pressure with isradipine (–15.9/–12.1 mmHg) than with enalapril (–10.3/–8.6 mmHg). Lower absolute blood pressure levels were attained with isradipine (139.3/90.4 mmHg) than with enalapril (146.2/96.8 mmHg). However, patients on the low-salt diet showed a comparable reduction in blood pressure with either isradipine (–7.1/–5.9 mmHg) or enalapril (–7.7/–5.5 mmHg). Similar absolute levels of blood pressure were achieved with both agents (135.8/88.8 mmHg with isradipine vs. 137.2/91.4 mmHg with enalapril). If the ACE inhibitor and calcium antagonist were titrated to a higher dose in salt-sensitive African-Americans, the absolute level of blood pressure achieved with the ACE inhibitor improved significantly. On the contrary, a low-salt diet lowered only slightly the level of blood pressure achieved with the calcium antagonist. With both therapies, however, the absolute level of blood pressure was lowest during the period of low sodium intake. In addition, the greatest differential in favor of the calcium antagonist vs. the ACE inhibitor occurred at the highest level of dietary sodium intake. This important study demonstrates that the dose-response curve of the calcium antagonist is shifted only slightly to the right by a high sodium diet compared with the relatively greater rightward shift of the ACE inhibitor dose-response curve.

Utility of Calcium Antagonists in Combination Therapy

Calcium antagonists provide a highly effective, well-tolerated therapeutic option for lowering blood pressure in hypertensive African-Americans; they correct many of the cardiorenal and systemic hemodynamic and structural abnormalities. Nevertheless, because of their higher baseline blood pressure and reduced renal natriuretic capacity, hypertensive African-Americans more often need combination therapy, even after titration of antihypertensive agents into the upper part of their usually effective dosing range, if blood pressure is to be normalized (minimally < 140/90 mmHg). The combination of a calcium antagonist and a thiazide diuretic has been shunned by some authorities. Nevertheless, compelling data indicate that this combination is highly effective,[1–4,6,7] and it has a strong theoretical rationale. One of the seldom mentioned pharmacologic properties of calcium antagonists is that they prevent or at least attenuate diuretic-induced activation of the renin-angiotensin-aldosterone cascade (Fig. 3). Substantial data also demonstrate high efficacy and excellent tolerability with the combined use of calcium antagonists and ACE inhibitors.100 This combination reduces the peripheral edema induced by calcium antagonist

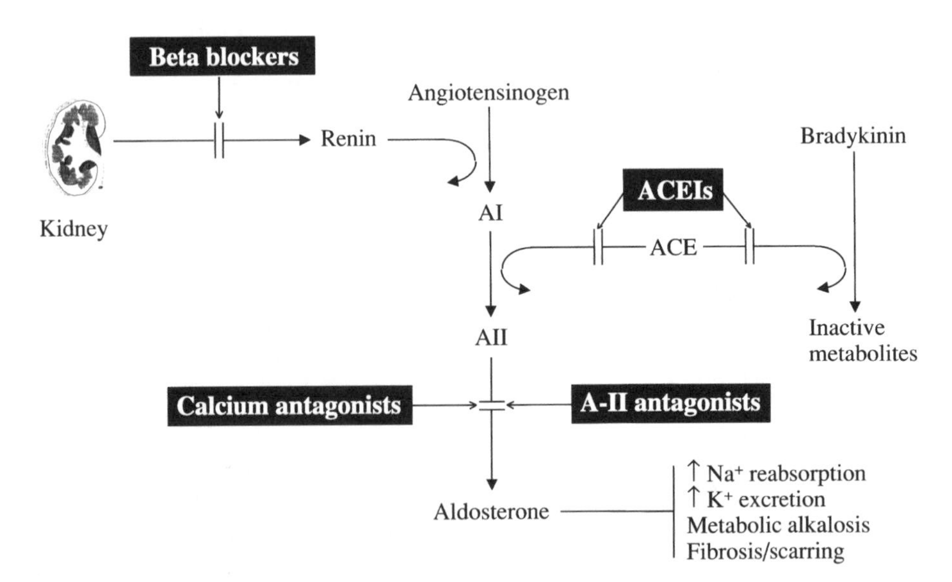

FIGURE 3. Sites of blockade of diuretic-induced activation of the renin–angiotensin-aldosterone system (RAAS) by various pharmacologic agents, including calcium antagonists. AI = angiotensin I, AII = angiotensin II; ACE = angiotensin-converting enzyme, ACEI = ACE inhibitors.

monotherapy. Calcium antagonists also have been safely used in combination with alphal antagonists[13,14,101] and beta blockers.[8–10]

CONCLUSION

Calcium antagonists are ideally suited for the treatment of hypertension in African-Americans because of their outstanding efficacy when used as monotherapy or in combination with other antihypertensive drug classes. Moreover, they reverse many of the pathophysiologic features of African-American hypertension, are metabolically neutral and well tolerated over the long term, and reverse target-organ remodeling. Lowering of blood pressure is minimally affected by high levels of sodium intake, a highly important pharmacologic attribute in salt-sensitive hypertensive African-Americans.

Acknowledgments

The authors thank Sergei Novikov, M.D., and Ms. Sharon Ireland for their diligent assistance in the preparation of this manuscript.

References

1. Schoenberger JA, Glasser SP, Ram CBS, et al: Comparison of nitrendipine combined with low-dose hydrochlorothiazide to hydrochlorothiazide alone in mild to moderate essential hypertension. J Cardiovasc Pharmacol 6:S1105–S1108, 1984.
2. Sever PS, Poulter NR: Calcium antagonists and diuretics as combined therapy. J Hypertens 5(Suppl 4):S123–S126, 1987.
3. Weinberger MH: Additive effects of diuretics or sodium restriction with calcium channel blockers in the treatment of hypertension. J Cardiovasc Pharmacol 10(Suppl 4):S72–S75, 1988.
4. Massie B, MacCarthy P, Ramanathan KB, et al: Diltiazem and propranolol in mild to moderate essential hypertension as monotherapy or with hydrochlorothiazide. Ann Intern Med 107:150–157, 1987.
5. Stergiou GS, Malakos JS, Achimastos AD, Mountokalakis TD: Additive hypotensive effect of a dihydropyridine calcium antagonist to that produced by a thiazide diuretic: A double-blind, placebo-controlled crossover trial with ambulatory blood pressure monitoring. J Cardiovasc Pharmacol 29:412–416, 1997.
6. Chrysant SG, Chysant C, Trus J, Hitchcock A: Antihypertensive effectiveness of amlodipine in combination with hydrochlorothiazide. Am J Hypertens 2:537–541, 1989.
7. Glasser SP, Chrysant SG, Graves J, et al: Safety and efficacy of amlodipine added to hydrochlorothiazide therapy in essential hypertension. Am J Hypertens 2:154–157, 1989.
8. Julius S: Amlodipine in hypertension: An overview of the clinical dossier. J Cardiovasc Pharmacol 12(Suppl 7):S27–S33, 1988.
9. Walton T, Symes LR: Felodipine and isradipine: New calcium channel-blocking agents for the treatment of hypertension [review]. Clin Pharmacol 12:261–275, 1993.
10. Menard J: Critical assessment of combination therapy development [review]. Blood Pressure 1(Suppl):5–9, 1993.
11. Cappucio FP, Markandu ND, Singer DR, MacGregor GA: Amlodipine and lisinopril in combination for the treatment of essential hypertension: Efficacy and predictors of response. J Hypertens 11:839–847, 1993.
12. Lüscher TF, Wenzel RR, Moreau P, Takase H: Vascular protective effects of ACE inhibitors and calcium antagonists: Theoretical basis for a combination therapy in hypertension and other cardiovascular diseases. Cardiovasc Drugs Ther 9(Suppl 3):509–523, 1995.
13. Lenz LM, Pool JL, Laddu AR, et al: Combined terazosin and verapamil therapy in essential hypertension. Hemodynamic and pharmacokinetic interactions. Am J Hypertens 8:133–145, 1995.
14. Elliott HL, Meredith PA, Cambell L, Reid JL: The combination of parzosin and verapamil in the treatment of essential hypertension. Clin Pharmacol Ther 43:554–560, 1988.

15. Burt VL, Cutler JA, Higgins M, et al: Trends in the prevalence, awareness, treatment and control of hypertension in the adult US population. Hypertension 26:60–69, 1995.
16. Sutton-Tyrrell K, Alcorn HG, Herzog H, et al: Morbidity, mortality, and antihypertensive treatment effects by extent of atherosclerosis in older adults with isolated systolic hypertension. Stroke 26:1319–1324, 1995.
17. The effect of treatment on mortality in "mild" hypertension: results of the hypertension detected and follow-up program. N Engl J Med 14:976–980, 1982.
18. Mo R, Omvik P, Lund-Johansen P: The Bergen Blood Pressure Study: Blood pressure changes, target organ damage and mortality in subjects with high and low blood pressure over 27 years. Blood Pressure 2:113–123, 1993.
19. Waal-Manning HJ, Paulin JM, Wallis AT, Simpson FO: Risk predictors in treated hypertension. J Cardov Pharmacol 7(16 Suppl):S87–S88, 1990.
20. Kumanyika S: Obesity in black women. Epidemiol Rev 9:31–50, 1987.
21. Gillum RF: Stroke in blacks. Stroke 19:1–9, 1988.
22. Rocella EJ, Lenfant C: Regional and racial differences among stroke victims in the United States. Clin Cardiol 12(Suppl 4):IV-18–IV-22, 1989.
23. Koren MJ, Mensah GA, Blake J, et al: Comparison of left ventricular mass and geometry in black and white patients with essential hypertension. Am J Hypertens 6:815–823, 1993.
24. Blythe WB, Maddux FW: Hypertension as a causative diagnosis of patients entering end-stage renal disease programs in the United States from 1980 to 1986. Am J Kidney Dis 18:33–37, 1991.
25. Klag MJ, Whelton PK, Neaton JD, et al: Higher incidence of ESRD in African-American men in the MRFIT screened cohort [abstract]. J Am Soc Nephrol 5:334, 1994.
26. Lopes AA, Horbuckle K, James SA, Port FK: The joint effects of race and age on the risk of end-stage renal disease attributed to hypertension. Am J Kidney Dis 24:554–560, 1994.
27. Brazy PC: Kidney disease and hypertension in blacks [review]. Curr Opin Nephrol Hypertens 3:554–557, 1994.
28. Flack JM, Neaton JD, Daniels B, Esunge P. Ethnicity and renal disease: Lessons from the Multiple Risk Factor Intervention Trial and the Treatment of Mild Hypertension Study. Am J Kidney Dis 21(Suppl I):31–40, 1993.
29. McClellan W, Tuttle E, Issa A: Racial differences in the incidence of hypertensive end-stage renal disease (ESRD) are not entirely explained by differences in the prevalence of hypertension. Am J Kidney Dis 4:285–290, 1988.
30. Rostand SG, Kirk KA, Rutsky EA, Pate BA: Racial differences in the incidence of end stage renal disease. N Engl J Med 306:1276–1279, 1982.
31. Division of Diabetes Translation: Diabetes Surveillance, 1991. Atlanta, U.S. Department of Health and Human Services, Public Health Service, Centers for Disease Control, 1991.
32. Flack JM, Ensrud KE, Mascioli S, et al: Racial and ethnic modifiers of the salt-blood pressure response. Hypertension 17(Suppl I):I115–I121, 1991.
33. Faulkner B, Kushner H: Effect of chronic sodium loading on cardiovascular responses in young blacks and whites. Hypertension 15:36–43, 1990.
34. Schwartz GL, Strong CG: Renal parenchymal involvement in essential hypertension [review]. Med Clin North Am 71:843–858, 1987.
35. Sowers JR, Zemel MB, Zemel P, et al: Salt sensitivity in blacks. Salt intake and natriuretic substances. Hypertension 12:485–490, 1988.
36. Gillum RF: Pathophysiology of hypertension in blacks and whites. Hypertension 1:468–475, 1979.
37. Zinner SH, Margoulious HS, Rosner B, et al: Familial aggregation of urinary kallikrein concentration in childhood: Relation to blood pressure, race and urinary electrolytes. Am J Epidemiol 104:124–132, 1976.
38. Holland OB, Chud JM, Braunstein H: Urinary kallikrein excretion in essential and mineralocorticoid hypertension. J Clin Invest 65:347–356, 1980.
39. Campese VM, Parise M, Karubian F, Bigazzi R: Abnormal renal hemodynamics in black salt-sensitive patients with hypertension. Hypertension 18:805–812, 1992.
40. Weir MR, Dengel DR, Behrens MT, Goldberg AP: Salt-induced increases in systolic blood pressure affect renal hemodynamics. Hypertension 25:1339–1344, 1995.
41. Redgrave JE, Rabinowe SL, Hollenberg NK, Williams GH: Correction of abnormal renal blood flow response to angiotensin II by converting-enzyme inhibition in essential hypertensives. J Clin Invest 75:1285–1290, 1985.
42. van Paassen P, de Zeeuw D, Navis G, de Jong P: Does the renin-angiotensin system determine the renal and systemic hemodynamic response to sodium in patients with essential hypertension? Hypertension 27:202–208, 1996.

43. Bigazzi R, Bianchi S, Baldari D, et al: Microalbuminuria in salt-sensitive patients. A marker for renal and cardiovascular risk factors. Hypertension 23:195–199, 1994.
44. Ribstein J, du Cailar G, Mimran A: Combined renal effects of overweight and hypertension. Hypertension 26:610–615, 1995.
45. Bigazzi R, Bianchi S, Baldari G, Campese V: Clustering of cardiovascular risk factors in salt-sensitive patients with essential hypertension: Role of insulin. Am J Hypertens 9:24–32, 1996.
46. Falkner B: Differences in blacks and whites with essential hypertension: Biochemistry and endocrine. State of the art lecture. Hypertension 15(6 Pt 2):681–686, 1990.
47. Weir MR, Saunders E: Pharmacologic management of systemic hypertension in blacks [review]. Am J Cardiol 61(16):46H–52H, 1988.
48. Schachter J, Kuller LJ: Blood volume expansion among blacks: A hypothesis. Med Hypothesis 14:1–19, 1984.
49. Chrysant SG, Danisa K, Kiem DC, et al: Racial differences in pressure, volume, and renin alter relationship in essential hypertension. Hypertension 1:136–141, 1979.
50. Hall JE, Guyton AC: Control of sodium excretion and arterial pressure by intrarenal mechanisms and the renin-angiotensin system. In Laragh JH, Brenner BM (eds): Hypertension: Pathophysiology, Diagnosis and Management. New York, Raven, 1990, pp 1105–1130.
51. Bauer JH, Sunderrajan S, Reams G: Effects of calcium entry blockers on renin-angiotensin-aldosterone system, renal function and hemodynamics, salt and water excretion and body fluid composition. Am J Cardiol 56:62H–67H, 1985.
52. Loutzenhiser R, Epstein M: Effects of calcium antagonists on renal hemodynamics. Am J Physiol 249:F619–F629, 1985.
53. Anderson S: Renal hemodynamics effect of calcium antagonists in rats with reduced renal mass. Hypertension 17:388–395, 1991.
54. Romero JC, Raij L, Granger JP, et al: Multiple effects of calcium entry blockers on renal function in hypertension. Hypertension 10:140–151, 1987.
55. Epstein M, De Micheli AG: Natriuretic effects of calcium antagonists. In Epstein M (ed): Calcium Antagonists In Clinical Medicine. Philadelphia, Hanley & Belfus, 1992, pp 349-366.
56. Nagao T, Yamaguchi I, Narita H, Nakajima H: Calcium entry blockers: Antihypertensive and natriuretic effects in experimental animals. Am J Cardiol 56:56H–61H, 1985.
57. Kubo SH, Cody RJ, Covit AB, et al: The effects of verapamil on renal blood flow, renal function, and neurohormonal profiles in patients with moderate to severe hypertension. J Clin Hypertens (3 Suppl):38S–46S, 1986.
58. Luft FC, Weinberger MH: Calcium antagonists and renal sodium homeostasis. In Epstein M, Loutzenhiser R (eds): Calcium Antagonists and the Kidney. Philadelphia, Hanley & Belfus, 1990, pp 203–212.
59. Hughes GS Jr, Cowart TD, Oexmann MJ, Conradi EC: Verapamil induced natriuretic and diuretic effects: Dependency on sodium intake. Clin Pharmacol Ther 44:400–407, 1988.
60. Flack JM, Grimm RH Jr, Yunis C, et al: Ethnic blood pressure response patterns in the Treatment of Mild Hypertension Study (TOMHS). [manuscript in preparation]
61. Nicholson JP, Resnick LM, Laragh JH: The antihypertensive effect of verapamil at extremes of dietary sodium intake. Ann Intern Med 107:329–334, 1987.
62. MacGregor GA, Markandu ND, Smith SJ, Sagnella GA: Does nifedipine reveal a function abnormality of arteriolar smooth muscle cell in essential hypertension? The effect of altering sodium balance. J Cardiovasc Pharmacol 7(Suppl 6):S178–S181, 1985.
63. MacGregor GA, Cappuccio FP, Markandu ND: Sodium intake, high blood pressure, and calcium channel blockers. Am J Med 82(Suppl 3B):16–22, 1987.
64. Polonia J, Boaventura I, Gama G, et al: Influence of nonsteroidal anti-inflammatory drugs on renal function and 24-hr ambulatory blood pressure: Reducing effects of enalapril and nifedipine gastrointestinal therapeutic system in hypertensive patients. J Hypertens 13:925–931, 1995.
65. Houston MC, Weir M, Gray J, et al: The effects of nonsteriodal anti-inflammatory drugs on blood pressure of patients with hypertension controlled with verapamil. Arch Intern Med 155:1049–1054, 1995.
66. Tarazi RC, Sen S, Fouad FM, et al: Regression of myocardial hypertrophy conditions and sequelae of reversal in hypertensive heart disease. In Alpert NR (ed): Myocardial Hypertrophy and Failure. New York, Raven Press, 1983.
67. Liebson PR, Savage DD: Echocardiography in hypertension: A review. II: Echocardiographic studies of the effects of antihypertensive agents on left ventricular wall mass and function. Echocardiography 4:215–249, 1987.
68. Kahn S, Frishman WH, Weissman S, et al: Left ventricular hypertrophy on electrocardiogram: Prognostic implications from a 10-year cohort study of older subjects: A report from the Bronx Longitudinal Aging Center, Bronx, New York, USA. J Am Geriatr Soc 44:524–529, 1996.

69. Tanser PH: Hypertrophy to failure [review]. J Hum Hypertens 8(Suppl 1):S17–S20, 1994.
70. Levy D, Garrison RJ, Savage DD, et al: Prognostic implications of echocardiographically determined left ventricular mass in the Framingham Heart Study. N Engl J Med 322:1561–1566, 1990.
71. Messerli FH, Ventura HO, Elizardi DJ, et al: Hypertension and sudden deaths: Increased ventricular ectopic activity in left ventricular hypertrophy Am J Med 77:18–22, 1984.
72. McLenachan JM, Henderson E, Morris KI, Dagie HJ: Ventricular arrhythmias in patients with hypertensive left ventricular hypertrophy. N Engl J Med 317:787–792, 1987.
73. Bonow RO, Dilsizian V, Rosing DR, et al: Verapamil-induced improvement in left ventricular diastolic filling and increased exercise tolerance in patients with hypertrophic cardiomyopathy: Short- and long-term effects. Circulation 72:853–864, 1985.
74. Houston MC: New insights and approaches to reduce end-organ damage in the treatment of hypertension: Subsets of hypertension approach. Am Heart J 123:1337–1367, 1992.
75. Kaiske BL, Ma JZ, Kalil RS, Louis TA: Effects of antihypertensive therapy on serum lipids. Ann Intern Med 122:133–141, 1995.
76. Ramsey LE, Yeo WW, Jackson PR: Influence of diuretics, calcium antagonists, and alpha-blockers on insulin sensitivity and glucose tolerance in hypertensive patients. J Cardiovasc Pharmacol 20(Suppl 11):S49–S53, 1992.
77. Leonetti G: Modification of cardiovascular risk factors during antihypertensive treatment; current and new approach trends. The urapidil experience [review]. Blood Pressure Suppl 4:49–52, 1994.
78. Forette F, Bert P, Rigaud AS: Are calcium antagonists the best option in elderly hypertensives? [review]. J Hypertens Suppl 12(6):S19–S23, 1994.
79. Saunders E, Weir MR, Kong BW, et al: A comparison of the efficacy and safety of a beta-blocker, a calcium blocker and a converting enzyme inhibitor in hypertensive blacks. Arch Intern Med 150:1707–1713, 1990.
80. Drayer JIM, Weber MA: Monotherapy of essential hypertension with a converting enzyme inhibitor. Hypertension 5(Suppl III):III-108–III-113, 1983.
81. Veterans Administration Cooperative Study Group on Antihypertensive Agents: Comparison of propranolol and hydrochlorothiazide for the initial treatment of hypertension. Results of short-term titration with emphasis on racial difference in response. JAMA 248:1996–2003, 1982.
82. Materson BJ, Reda DJ, Cushman WC, et al: Single drug therapy for hypertension in men: A comparison of six antihypertensive agents with placebo. N Engl J Med 328:914–921, 1993.
83. Buyamba-Kabangu JR, Lepira B, Fogard R, et al: Relative potency of a beta-blocking and calcium entry blocking agent as antihypertensive drugs in black patients. Eur J Clin Pharmacol 29:523–527, 1986.
84. Racial differences in response to low-dose captopril are abolished by the addition of hydrocholorothiazide. Veterans Administration Cooperative Study Group on Antihypertensive Agents. Br J Clin Pharmacol 14:97s–101s, 1982.
85. Seedat YK: Trial of atenolol and chlorthalidone for hypertension in black South Africans. BMJ 281:1241–1243, 1980.
86. Cubeddu LX, Aranda J, Singh B, et al: A comparison of verapamil and propranolol for the inital treatment of hypertension: Racial differences in response. JAMA 256:2214–2221, 1986.
87. Weder AB, Weinberger MH, McCarron DA: Salt-sensitivity of blood pressure and the antihypertensive effect of isradipine and enalapril. Am J Hypertens 9:178A, 1995.
88. Weir MR, McCarron DA, Canossa-Terris M, et al: Influence of race and dietary salt on the antihypertensive efficacy of an angiotensin converting enzyme inhibitor or a calcium channel antagonist in salt sensitive hypertensives. Am J Hypertens 1997 [in press].
89. Zing W, Fergusson RK, Vlasses PH: Calcium antagonists in elderly and black hypertensive patients. Therapeutic controversies. Arch Intern Med 151:2154–2162, 1991.
90. Hughes GS Jr, Cowart TD Jr, Conradi EC: Efficacy of verapamil-hydrochlorothiazide therapy in hypertensive black patients. Clin Pharmacy 6:322–326, 1987.
91. Frishman W, Zawada EJ, Smith L, et al: Comparison of hydrochlorothiazide and sustained-release diltiazem for mild-to-moderate systemic hypertension. Am J Cardiol 59:615–623, 1987.
92. Storm TL, Badskjaer J, Hammer R, Kogler P: Tiapamil and hydrochlorothiazide: A double-blind comparison of two antihypertensive agents. J Clin Pharmacol 27:18–21, 1987.
93. Moser M, Lunn J, Materson B: Comparative effects of diltiazem and hydrochlorothiazide in blacks with systemic hypertension. Am J Cardiol 56:101H–104H, 1985.
94. Leary WP, Maharaj B: Comparison of felodipine and hydrochlorothiazide for the treatment of mild to moderate hypertension in black Africans. J Cardiovasc Pharmacol 15(Suppl 4):S91–S93, 1990.
95. Leehey D, Hartman E: Comparison of diltiazem and hydrochlorothiazide for treatment of patients 60 years of age or older with systemic hypertension. Am J Cardiol 62:1218–1223, 1988.

96. Weir MW, Gray JM, Paster R, Saunders E: Differing mechanisms of actions of angiotensin-converting enzyme inhibition in black and white hypertensive patients. The Trandolapril Multicenter Study Group. Hypertension 26:124–130, 1995.
97. Matthews HW: Racial, ethnic and gender differences in response to medicines. Drug Metab Drug Interact 12(2):77–91, 1995.
98. Jamerson KA: Prevalence of complications and response to different treatments of hypertension in African-Americans and white Americans in the U.S. Clin Exp Hypertens 15:979–995, 1993.
99. Neaton JD, Grimm RH Jr, Prineas RJ, et al: Treatment of Mild Hypertension Study. Final results. JAMA 270:713–724, 1993.
100. Epstein M, Bakris G: Newer approaches to antihypertensive therapy. Arch Intern Med 156:1969–1978, 1996.
101. Reid JL: First-line and combination treatment of hypertension. Am J Med 86(Suppl 4A):2–5, 1989.

BRYAN BURNS M.D. / WILLIAM H. FRISHMAN M.D.

17

Calcium Antagonists in Elderly Patients with Systemic Hypertension

Strong evidence from large, well-controlled clinical trials indicates that elderly patients with either combined systolic and diastolic hypertension or isolated systolic hypertension benefit from pharmacologic blood pressure reduction.[1] These studies reveal the benefit of treatment in reducing the risk of both cardiovascular and cerebrovascular morbidity.

The hemodynamic profile of calcium antagonists suggests their efficacy in treating elderly hypertensive patients with and without angina pectoris. However, no data showing the benefit of calcium antagonists in reducing overall morbidity and mortality in this population are yet available. This chapter first reviews the pathophysiology of hypertension in the elderly and the rationale for drug treatment. The experiences with calcium antagonists in treating elderly hypertensive patients are then discussed.

HYPERTENSION IN THE ELDERLY

Elderly hypertensive patients differ from young hypertensive patients in several respects. It is important to be aware of these differences in planning therapeutic interventions. The aging process is associated with multiple anatomic and physiologic alterations in the cardiovascular system that can influence blood pressure regulation. Young hypertensive patients tend to be characterized by a hyperkinetic circulatory state, evidently resulting from increased sensitivity to circulating catecholamines.[2] This state is expressed as an increase in heart rate, contractility, and cardiac output with no significant change in systemic vascular resistance. In contrast, hypertension in elderly patients appears to be a consequence of structural cardiovascular changes. Decreased vascular compliance and increased resistance predominate and are associated with narrowing of the vascular radius and increases in wall-to-lumen ratios.[3] Histologically, the changes are apparent in the vascular subendothelial and media layers, which thicken with age and demonstrate increased connective tissue infiltration as well as calcification and lipid deposition.[4] With age, this process results in a progressive increase in wall stiffness and tortuosity of the aorta and large arteries, often reflected by a rise in systolic pressure.[5] The decreased vascular compliance seen in elderly patients also may result in reduced sensitivity of volume and baroreceptor function.

Aging also affects the vascular endothelium, the cells of which become smaller and less uniformly aligned. This change may result in decreased production of endogenous vasodilating substances (e.g., nitric oxide) and a decline in local control of vascular tone.[6,7]

The myocardium of elderly hypertensive patients is also altered by the aging process. Although total heart size generally remains unchanged, mild-to-moderate left ventricular hypertrophy often develops in response to increased afterload. This hypertrophy tends to offset the decline in myocardial contractile speed and strength seen with aging.[8] The elderly heart demonstrates microscopic foci of calcification and fibrosis as well as macroscopic calcification of the mitral and aortic valves and conduction system.[9,10] Thus the stiffened and hypertrophied left ventricle shows decreased diastolic filling and greater dependence on atrial contraction.

Renal function also shows an age-dependent decline, which is further accelerated by chronically elevated blood pressure. The decline in renal function often manifests as a decline in glomerular filtration rate.[11] Postglomerular increases in vascular resistance have been described as a possible compensatory mechanism and may further increase vascular pressure in the glomerulus.[12]

Despite the fact that elderly patients (including those with hypertension) tend to have lower plasma volumes than younger patients, their plasma renin levels are usually normal or low.[13] Plasma levels of angiotensin II and aldosterone are also decreased, and the response to antidiuretic hormone is blunted. In theory, these hormonal changes should help to lower or maintain arterial pressure in the elderly. Nonetheless, arterial blood pressure (usually systolic) appears to rise progressively with increasing age.[14–16]

It has been estimated that 80% of elderly hypertensive patients present with concomitant medical conditions, which must be taken into consideration in planning therapeutic intervention.[17] Examples include coronary artery disease, diastolic and systolic dysfunction, diabetes mellitus, hyperlipidemia, renal impairment, and chronic obstructive pulmonary disease. Elderly patients also frequently suffer from left ventricular hypertrophy, insulin resistance, obesity, depression, and cognitive dysfunction. Therapy for hypertensive elderly patients needs to account for these and other concomitant diseases. As discussed later, calcium antagonists are a logical and appropriate therapy in the absence of contraindications.

RATIONALE FOR TREATMENT

It has been estimated that 50–70% of elderly Americans have hypertension, which is defined by the National Institutes of Health as blood pressure > 140/90 mmHg.[1,18] Higher levels of either systolic or diastolic blood pressure are associated with an increased risk of cardiovascular morbidity or mortality. Examples include coronary artery disease, congestive heart failure, renal insufficiency, peripheral vascular disease, and stroke. Hypertension in the elderly poses an important problem for two reasons: (1) the proportion of elderly in the population is steadily increasing, and (2) the incidence of hypertension—and therefore the risk of cardiovascular disease—increases with age.[19,20] In the past there was a tendency to deemphasize elevated blood pressure readings in the elderly and to downplay the need for therapeutic intervention. Increasing blood pressure was viewed as part of the normal aging process, and blood

TABLE 1. Characteristics of Long-term Trials in Elderly Hypertensive Patients

						BP at Entry (mmHg)		BP Reduction (mmHg)*	
Study	Study Design	No. of Patients	Age (yr)	Males (%)	Duration (yr)	SBP Range	DBP Range	Control	Treatment
EWPHE, 1985	r, pc, db	840	≥ 60	30	4.7	160–239	90–119	16/11	21/7
Coope and Warrender, 1986	r, nb	884	60–79	31	4.4	≥ 170	≥ 105	10/10	16/10
STOP, 1991	r, pc, db	1627	70–84	37	2.1	180–230	90–120	9/6	22/9
SHEP, 1991	r, pc, sb	4736	> 60	43	4.5	160–219	< 90	15/6	11/4
MRC Working Group	r, pc, sb	4396	65–74	39	5.8	160–209	< 115	18/8	16/7

BP= blood pressure, SBP = systolic blood pressure, DBP = diastolic blood pressure, r = randomized, pc = placebo-controlled, db = double-blind, nb = nonblind, sb = single-blind, EWPHE = European Working Party Study on High Blood Pressure in the Elderly, STOP = Swedish Trial in Old Patients with Hypertension, SHEP = Systolic Hypertension in the Elderly Program, MRC = Medical Research Trial on Treatment of Hypertension in Older Adults.
* Differences between BP reduction in control and active treatment groups.
Adapted from Holzgreve H, Middeke M: Treatment of hypertension in the elderly. Drugs 46(Suppl 2): 24–31, 1993, with permission.

pressure measurements that would dictate treatment in young adults were often considered satisfactory or borderline in elderly patients. It was believed that the benefits of antihypertensive therapy might not be realized in patients with a limited life expectancy. It was also thought that elderly patients would be more susceptible to adverse drug reactions (e.g., orthostatic hypotension, cerebral hypoperfusion).[12, 21]

Several large, randomized, well-controlled trials have recently investigated the question of whether elderly hypertensives respond positively and receive benefits from pharmacologic intervention (Table 1). Examples include the European Working Party Study on High Blood Pressure in the Elderly (EWPHE), the Swedish Trial in Old Patients with Hypertension (STOP), the Medical Research Trial on Treatment of Hypertension in Older Adults (MRC II), and the Systolic Hypertension in the Elderly Program Study (SHEP).[22–25] The combined data from these trials demonstrate a reduction in both cerebrovascular and cardiovascular morbidity and mortality in treatment groups. A metaanalysis by Holzgreve[19] describes a 40% reduction in stroke incidence and 30% reduction in cardiovascular events with antihypertensive therapy.

The SHEP study was particularly interesting because it investigated whether antihypertensive therapy can decrease cardiovascular risk in elderly patients with isolated systolic hypertension (defined as systolic blood pressure 160–219 mmHg and diastolic pressure < 90 mm Hg), the most typical type of hypertension in the elderly.[25] In the United States, the prevalence of diastolic hypertension tends to plateau around the age of 55, whereas the prevalence of systolic hypertension continues to rise, even after 80 years of age. Before the results of SHEP were published, many physicians held the belief that isolated systolic hypertension, reflecting an increase in rigidity of the arterial system, was only an index of bad cardiovascular state and did not play a direct role in the development of cardiovascular complications.[26] This view persisted despite data from the Multiple Risk Factor Intervention Trial (MRFIT), which demonstrated that systolic blood pressure > 160 mmHg is a more significant risk factor, regardless of age, than diastolic pressure > 95 mmHg.[27]

The results of the SHEP trial showed conclusively that reduction or correction of hypertension strongly diminished the risk. Antihypertensive therapy was associated not only with a decrease in the number of cerebrovascular events but also with a 25% reduction in fatal and nonfatal coronary events and an even greater benefit in reducing episodes of congestive heart failure. The greatest benefit of therapy was seen in patients with severe underlying disease. Even patients over 80 years old benefit from treatment.[28,29]

CALCIUM ANTAGONISTS IN THE ELDERLY

Until recently, diuretics and beta blockers were the preferred drugs for first-line therapy of patients with hypertension. The Joint National Committee on Detection, Evaluation, and Treatment of High Blood Pressure recommended diuretics and beta blockers because of extensive and consistent evidence that these agents reduce both cardiovascular and cerebrovascular morbidity and mortality.[1,30] Angiotensin-converting enzyme (ACE) inhibitors, α-adrenergic antagonists, and calcium channel blockers are also commonly used for antihypertensive therapy. Many of these agents, however, are associated with side effects that may limit their therapeutic efficacy, especially among the elderly population, who are generally more susceptible to adverse drug reactions. The elderly also tend to have a higher incidence of autonomic dysfunction and are more prone to developing orthostatic hypotension than young hypertensive patients.[31,32] Diuretics may act to increase plasma lipid and uric acid concentrations while simultaneously reducing potassium levels. Beta blockers may increase serum cholesterol levels, blunt the hypoglycemic response, and increase fatigue.[33] Alpha blockers may precipitate orthostatic hypotension, and many ACE inhibitors produce a persistent cough. Therefore, attention has focused on calcium antagonists as potential first-line agents in the treatment of elderly patients with hypertension.

The calcium antagonists are a heterogeneous group of drugs with widely variable effects on heart muscle, sinus node function, atrioventricular conduction, peripheral blood vessels, and coronary circulation.[34–40] From a chemical standpoint, they can be divided into four main groups: phenylalkylamines (verapamil), benzothiazepines (diltiazem), dihydropyridines (nifedipine, nicardipine, nimodipine, amlodipine, felodipine, isradipine), and benzomidazolyl-substituted tetraline derivatives (mibefradil). Despite their heterogeneity, they all block the influx of calcium ions into the cells of vascular smooth muscle and myocardial tissue.[41]

Calcium ions play a fundamental role in the activation of cells and serve as the primary link between neurologic excitation and mechanical contraction of cardiac, smooth, and skeletal muscle. The most important characteristic of all calcium antagonists is their ability to inhibit selectively the inward flow of charge-bearing calcium ions when the calcium ion channels become permeable. The term *slow channel* is no longer used, because recently it was recognized that the calcium ion current develops faster than previously thought and that there are at least two types of calcium channels—L and T.[34, 42] The conventional calcium channel, which has been known to exist for a long time, is called the L-type channel. The function of the L-type channel is to admit the substantial amount of calcium ions required for initiation of contraction via calcium-induced calcium release from the sarcoplasmic reticulum. The L-type channel is blocked by all of the calcium antagonists, and its permeability is increased by catecholamines. The T-type channel appears to have more negative potentials than the L-type and probably plays an important role in the initial depolarization of sinus and atrioventricular nodal tissue. Specific blockers for the T-type channel are not yet available, but they are expected to inhibit the sinus and atrioventricular nodes profoundly.[43] Mibefradil is the first calcium antagonist to have selective blocking effects on the T-type as well as the L-type channel.

The observation that calcium antagonists are significantly more effective in inhibiting contraction in coronary and peripheral arterial smooth muscle than in cardiac and skeletal muscle is of great clinical importance. This differential effect is explained by the observation that arterial smooth muscle is more dependent on external calcium entry for contraction, whereas cardiac and skeletal muscle rely on a recirculating internal pool of calcium.[44] Because calcium antagonists are membrane-active drugs, they reduce the entry of calcium into cells and therefore exert a much greater effect on vascular wall contraction. This preferential effect allows calcium antagonists to dilate coronary and peripheral arteries in doses that do not severely affect myocardial contractility or have little, if any, effect on skeletal muscle.[34]

Based on the above mechanisms of action, the calcium antagonists appear well suited for use in elderly patients whose hypertensive profile is based on increasing arterial stiffness, decreased vascular compliance, and diastolic dysfunction secondary to atrial and ventricular stiffness.[45] Because they have multiple clinical applications, including treatment of angina pectoris and management of supraventricular arrhythmias, calcium antagonists also hold promise in the treatment of elderly hypertensive patients with comorbid cardiovascular conditions.

In general, calcium antagonists appear to be well tolerated by the elderly. Most adverse effects of the dihydropyridines are attributable to vasodilation; examples include headaches and postural hypotension. Postural hypotension is associated with an increased risk of dizziness and falls; thus, it is a serious concern for elderly patients. Peripheral edema also may result and be confused with symptomatic congestive heart failure.[46] Verapamil, which may be particularly useful in elderly patients with diastolic dysfunction, has been noted to increase constipation.[47] Age-related declines in renal or hepatic function alter drug disposition in elderly patients, mostly as a result of declines in first-pass metabolism. This decline decreases total body clearance and increases elimination half-life. Individualized dose adjustments or dosing schedules help to decrease adverse effects.[48] Further information about pharmacokinetics can be found in chapter 3.

Several studies have attempted to investigate the tolerance and efficacy of calcium antagonists in elderly patients. An extensive multicenter, open trial by Dubois et al. evaluated oral nicardipine in hypertensive patients.[49] Approximately 30,000 patients were enrolled (mean age = 64 years; 22% over the age of 74). The study showed significant blood pressure reduction in patients of all ages. Patients with comorbid conditions also showed some benefit: 80% of patients with clinical angina and 20% of patients with peripheral arterial disease became asymptomatic on nicardipine therapy. However, the full significance of these results is unclear, because the study lacked a control group. Nicardipine was also relatively well tolerated by elderly patients; only 11% withdrew from the study. Most of the adverse effects, including ankle edema and headaches, were attributed to vasodilation and resolved on discontinuation of the drug.

A small study by Forette et al. examined the efficacy and tolerance of nicardipine in a double-blind, placebo-controlled trial.[50,51] This study was confined to 31 elderly hypertensive patients (mean age = 84 years). As with the previous study, blood pressure was significantly reduced by nicardipine over 4 weeks of therapy. The average reduction was 36/15 mmHg (from 186/99 to 150/83) and was significant compared with placebo (2/5 mmHg reduction). Absorption of the drug was also noted by peak plasma levels drawn 1 hour after administration. Elderly patients showed higher plasma drug levels compared with younger subjects. The results were consistent with those obtained in a study by Buhler et al. of the efficacy and tolerance of verapamil.[52] More recent studies with felodipine ER also provide evidence of significant blood pressure reduction compared with placebo.[53,54] A study by Dunlay et al. showed significantly greater reduction in blood pressure of older patients receiving felodipine than of younger patients; this differential effect may be attributable to the higher mean plasma concentration of felodipine in elderly patients.[55,56] Further studies also demonstrated that felodipine was at least as effective as hydrochorothiazide, atenolol, and metoprolol as monotherapy and more effective than hydrochlorothiazide plus beta blocker in the elderly.[57–60]

Treating Elderly Hypertensives with Concomitant Disease

Calcium antagonists are well suited for treating elderly hypertensive patients with concomitant disease, especially coronary artery disease or heart failure due to diastolic dysfunction. In an elderly patient with both hypertension and ischemic heart disease (with or without angina pectoris), calcium antagonists are an appropriate therapy for both conditions.[60a] The combination of a calcium antagonist and a beta blocker is well tolerated and serves to offset transient increases in the neurohumoral reflex.[61]

Elderly patients with diastolic dysfunction have impaired myocardial relaxation, which leads to decreased ventricular filling. Impairment may be related to age or to myocardial ischemia from coronary disease.[62] Calcium antagonists may help to improve left ventricular diastolic function in such patients.[63]

Although calcium antagonists have been widely prescribed as first-line therapy for essential hypertension in the elderly, their overall effectiveness in reducing cardiovascular morbidity and mortality has not been fully demonstrated in large, long-term, randomized studies.[34] Retrospective or case-control studies have attempted to determine whether hypertensive patients treated with one class of antihypertensive agents have had more or fewer adverse events (i.e., myocardial infarcts or death) than patients treated with other drugs. Case-control studies attempt to balance the patients for other known risk factors, including age, sex, and blood pressure. These nonrandomized studies often have significant methodologic flaws that render them biased and undependable, even when obvious confounding variables are controlled.[64] Because the multiple factors associated with the prescription of a particular drug are difficult or impossible to control, case-control studies cannot test definitively whether a class of drugs has small-to-moderate risks or benefits.[65] Yet in the absence of randomized, prospective trials, they may provide the only source of data available.

One such retrospective case-control study, published in 1995 by Psaty et al., compared the medication used by 623 hypertensive patients (age range: 30–79 years) who had suffered a myocardial infarction with the medication used by 2,032 patients who had not suffered a myocardial infarction.[66] The data suggested an increased risk of myocardial infarction and mortality in hypertensive patients receiving short-acting calcium blockers compared with patients receiving diuretics and beta blockers. The overall adjusted risk of myocardial infarction for patients treated with calcium antagonists was 62% higher than for patients treated with diuretics and 57% higher than for patients treated with beta blockers.

The study by Psaty et al.[66] provides several possible mechanisms to account for the increased risk of myocardial infarction, including negative inotropic effects, proarrhythmic effects, proischemic effects from the coronary steal phenomenon, and a reflex increase in sympathetic activity that may produce ulceration and/or rupture of unstable plaque. As a result, the investigators proposed discontinuing the use of calcium antagonists as first-line antihypertensive agents.

A second study, published in 1995 by Pahor et al., dealt primarily with elderly hypertensive patients treated with short-acting calcium antagonists, ACE inhibitors, or beta blockers.[67] The study was a nonrandomized, prospective cohort analysis of 906 hypertensive patients over 71 years of age; the main endpoint was all-cause mortality over 4 years of follow-up. Pahor et al. provided more information about individual drugs by dividing the patients into catagories based on which hypertensive agent was used. The investigators also accounted for multiple confounding variables, including age, gender, high-density lipoprotein cholesterol, smoking, and comorbid conditions. They found that the risk of cardiac mortality was significantly higher with short-acting nifedipine than with any of the other agents. Compared with beta blockers, the relative risks for all-cause mortality were as follows: verapamil, 0.8 (not significant [NS]); diltiazem, 1.3 (NS); nifedipine, 1.7 ($p = 0.016$); and ACE inhibitors, 0.9 (NS).

Although the study by Pahor et al. was prospective, it was not randomized; therefore, the possiblity of bias still exists. Additional data obtained from a recent large, randomized, double-blind study have increased concern over the short-acting dihydropyridines.[68,69] The Multicenter Isradipine Diuretic Atherosclerosis Study (MIDAS) intended to evaluate the effects of isradipine, a short-acting dihydropyridine, on the course of carotid artery disease in hypertensive patients. Although the study did not achieve its stated goal, the results raised further concerns over the use of short-acting calcium antagonists in hypertensive patients. The incidence of vascular events (myocardial infarction, stroke, congestive heart failure, angina, and sudden death) was higher in the isradipine group than in the thiazide group (5.65% vs. 3.17%). MIDAS also found an increased incidence of angina pectoris and nonmajor vascular events and procedures in the isradipine-treated group.

RECENT PROSPECTIVE STUDIES

Is it possible to draw conclusions from currently available data about the clincial use of calcium antagonists in elderly hypertensive patients? Obviously the evidence is at best incomplete. It seems prudent to avoid short-acting dihydropyridine calcium antagonists in elderly hypertensive patients, especially those with a history of coronary disease.[70] Recommendations about the use of other calcium antagonists are not nearly as obvious. The studies cited above have assessed only the use of short-acting calcium blockers, which have not been approved for treatment of hypertension by the U.S. Food and Drug Administration. It also has been recognized that the longer-acting (or sustained-release) formulations of the same agents have different clinical effects.[71] Although the new agents were developed primarily to increase compliance by once-a-day dosing, their uniform plasma concentration has had dramatic physiologic effects. There appears to be less stimulation of the sympathetic nervous system, with consequent decreases in heart rate fluctuation and circulating catecholamine levels.[72,73]

The results of several prospective trials with amlodipine, a long-acting calcium antagonist, have supported this theory. The Prospective Randomized Amlodipine Survival Evaluation [PRAISE] trial,[74] which dealt primarily with the use of amlodipine in patients with heart failure, reported no adverse effect on patients with severe congestive heart failure and a favorable effect on patients with dilated cardiomyopathy. The Treatment of Mild Hypertension Study (TOMHS) trial[75] compared six antihypertensive interventions, including amlodipine. No evidence of adverse cardiac effects was reported, and amlodipine was the best tolerated of all drug treatments.

A recent case-control study by Alderman et al. further distinguishes cardiovascular outcomes with short-acting calcium antagonists from outcomes with the longer-acting formulations[76] (Table 2). Consistent with earlier studies, patients taking the short-acting calcium blockers were found to be at significantly increased risk for a first cardiovascular event compared with patients taking beta blockers. Yet the investigators found no increased cardiovascular risk in patients taking the long-acting calcium blockers (adjusted odds ratio = 0.76). Although these findings are new and highly promising, the study suffers from the same limitation of all retrospective analyses—potentially uncontrolled confounding variables.[76,77]

FUTURE STUDIES

In response to the need for more data related to the safety and efficacy of calcium antagonists, at least eight large-scale trials are planned or in progress (Table 3). Several of these trials are concerned solely with the use of calcium antagonists in the elderly.[78–84]

TABLE 2. Unadjusted Odds Ratios Between Cases and Controls by Drug Regimen at Index Date

Drug Class	Cases	Controls	Odds Ratio (95% CI)
Beta blocker monotherapy	26	18	1.00
All calcium antagonists*	91	71	0.89 (0.45–1.75)
Long-acting	69†	67	0.71 (0.36–1.42)
Short-acting	23†	4	3.98 (1.18–13.49)
Monotherapy			
Angiotensin-converting enzyme inhibitors	19	30	0.44 (0.19–1.01)
Diuretics	17	11	1.07 (0.41–2.82)
Other classes	5	5	0.69 (0.17–2.75)
Combinations without calcium antagonist‡	16	25	0.44 (0.19–1.05)
Withdrawn from medication	15	29	0.28 (0.11–0.76)

* Includes monotherapy and combination therapy.
† One patient was taking both long-acting and short-acting calcium antagonists.
‡ Excludes all monotherapy and all calcium antagonists
Adapted from Alderman MH, Cohen H, Roqué R, Madhavan S: Effect of long-acting and short-acting calcium antagonists on cardiovascular outcomes in hypertensive patients. Lancet 349:594–598, 1997, with permission.

The Antihypertensive and Lipid-Lowering Treatment to Prevent Heart Attack Trial (ALLHAT), sponsored by the National Heart, Lung, and Blood Institute, is one of the largest prospective, randomized studies ever undertaken. It is currently enrolling 40,000 patients over the age of 55 years. The goal of the study is to compare four antihypertensive interventions—long-acting calcium antagonist (amlodipine), ACE inhibitor (lisinopril), diuretic (chlorthalidone), and alpha blocker (doxazosin)—in terms of their ability to reduce coronary disease. One-half of the patients also will receive pravastatin to test the benefits of lowered cholesterol.

The Hypertension Optimal Treatment (HOT) study is a large international study based at the University of Göteborg in Sweden. HOT predicts that reduction of diastolic blood pressure with felodipine, a long-acting calcium blocker, may reduce cardiovascular events (myocardial infarction or stroke). An ACE inhibitor or beta blocker and a thiazide diuretic may be added to the initial regimen to reach the diastolic pressure goal. One-half of the patients also will receive 75 mg/day of aspirin.

The Systolic Hypertension in Europeans Study (SYST-EUR) is limited to patients 60 years of age and older with a resting systolic pressure of 160–219 mmHg and a diastolic pressure of <95 mmHg. Patients were randomized to nitrendipine or placebo. If additional titration is necessary, the active group will receive an ACE inhibitor, then a diuretic. These patients will be compared against the placebo group. The study seeks to evaluate the morbidity and mortality of stroke, myocardial infarction, and heart failure.

The Swedish Trial in Old Patients with Hypertension 2 (STOP 2) enrolled 6600 patients ranging in age from 70–84 years with a supine blood pressure of 180/105 mmHg or higher. The original STOP trial compared the effects of diuretics and beta blockers in elderly hypertensive patients in terms of cardiovascular morbidity and mortality. STOP 2 compares these two treatments against

TABLE 3. Ongoing Trials of Antihypertensive Therapy with Calcium Antagonists

Trial	Entry Criteria	Therapy	End-points
ALLHAT 5 years 40,000 pts	Age: > 55 yr BP: 140/90 or lower if treated Race: 55% African-American Other	1st: amlodipine, lisinopril, doxazosin, or chlorthalidone + pravastatin 2nd: reserpine, clonidine, atenolol, or hydralazine at physician's discretion	1st: fatal/nonfatal MI 2nd: morbidity, mortality, cost; benefits of cholesterol reduction
CONVINCE 5 years 15,000 pts	Age: ≥ 55 yr BP: systolic 140–190 or diastolic 90–100 or 175–100 if treated Other	1st: verapamil, atenolol, or HCTZ 2nd: HCTZ or beta blocker for achievement of BP goal 3rd: moexipril	1st: fatal/nonfatal MI, stroke; sudden cardiac death 2nd: cardiovascular disease, no. and time of deaths
HOT 2.5 years 18,000 pts	Age: 50–80 yr BP: diastolic 100–115	1st: felodipine + aspirin or placebo 2nd: captopril, enalapril, ramipril, atenolol, metoprolol, or propranolol, as needed	1st: fatal/nonfatal MI, stroke; achievement of BP goal 2nd: benefits of aspirin
INSIGHT 3 years 7,200 pts	Age: 55–80 yr BP: ≥ 150/95 (sitting) or systolic > 160 Other	1st: long-acting nifedipine or HCTZ-amiloride 2nd: atenolol or enalapril	1st: fatal/nonfatal MI, stroke; congestive heart failure 2nd: mortality
NORDIL 5 years 12,000 pts	Age: 50–69 yr BP: diastolic > 110 or > 100 (sitting) if treated or other risk factors present	1st: diltiazem, diuretic, or beta blocker 2nd: ACE inhibitor, beta blocker, or diuretic 3rd: ACE inhibitor, beta blocker, or diuretic	1st: fatal/nonfatal MI, stroke; sudden cardiac death 2nd: mortality, ischemic heart disease or attack, atrial fibrillation, congestive heart failure, renal disease
PREDICT 4–4.5 years 8,000–8,500 pts	Age: ≥ 55 yr BP: 90–109/14–179 Other	1st: diltiazem or chlorthalidone 2nd: open label therapy 3rd: open label therapy	1st: fatal/nonfatal cardiovascular disease
SISH 2 years 171 pts	Age: > 55 yr BP: systolic 140–159	1st: felodipine or placebo	1st: compare effects on EKG measurements of ventricular wall thickness and ventricular function
STOP 2 4 years 6,600 pts	Age: 70–84 yr BP: > 180/105 (sitting)	1st: felodipine, isradipine, atenolol, metoprolol, pindolol, enalapril, lisinopril, or HCTZ-amiloride	1st: fatal/nonfatal MI, stroke; sudden cardiac arrest 2nd: cardiovascular disease, cost, institutionalization, tolerability of therapy
SYST-EUR 5 years 3,000 pts	Age: ≥ 60 yr BP: systolic 160–219 or diastolic < 95 (sitting)	1st: nitrendipine or placebo 2nd: enalapril, HCTZ, or placebo	1st: fatal/nonfatal stroke 2nd: MI, congestive heart failure

pts = patients, BP = blood pressure (mmHg), MI = myocardial infarction, EKG = electrocardiographic, HCTZ = hydrochlorothiazide, ACE = angiotensin-converting enzyme. ALLHAT = Antihypertensive and Lipid-lowering Treatment to Prevent Heart Attack Trial, CONVINCE = Controlled Onset Verapamil Investigation of Cardiovascular Endpoints, HOT = Hypertension Optimal Treatment Study, INSIGHT = International Nifedipine Study Intervention as a Goal in Hypertension Treatment, NORDIL = Norwegian Diltiazem Intervention Study, PREDICT = Prospective Randomized Evaluation of Diltiazem CD Trial, SISH = Stage I Systolic Hypertension, STOP 2 = Swedish Trial in Old Patients with Hypertension 2, SYST-EUR = Systolic Hypertension in Europeans Study.

Adapted from Townsend RR, Kimmel SE: Calcium channel blockers and hypertension. 1: New trials. Hosp Pract 31:125–136, 1996.

treatment with a calcium blocker (felodipine or isradipine) or ACE inhibitor (enalapril or lisinopril). The primary end-points in STOP 2 are myocardial infarction, cardiac arrest, or stroke. The trial should be completed in 1998.

The Stage I Systolic Hypertension in the Elderly (SISH) is a pilot study that has enrolled elderly patients (over 60 years of age) with mild systolic hypertension (140–159 mmHg) and normal diastolic blood pressure. Felodipine is compared with placebo in an attempt to reduce systolic blood pressure from baseline by 10%. Although the study is too small to assess possible mortality differences, a comparison is being made between treatments in terms of effects on echocardiographic measurements of ventricular wall thickness and ventricular function. The results of the study should be available in late 1997.

CONCLUSION

Evidence from multiple studies indicates that medical treatment of the elderly patient with diastolic, systolic/diastolic, or isolated systolic hypertension will reduce morbidity and mortality. Calcium antagonists are well suited for use in elderly patients, whose elevated blood pressure is generally due to increased arterial stiffness and decreased vascular compliance. These agents are well tolerated in older people and produce relatively few adverse effects. In addition, they may be of substantial benefit in patients with concomitant conditions, particularly coronary artery disease (with or without angina pectoris) and diastolic cardiac dysfunction.

Final conclusions about the safety and efficacy of calcium antagonists must be deferred until the results of the current prospective studies are reported. Longer-acting calcium antagonists, which achieve stable plasma concentrations and decrease reflex stimulation of the sympathetic nervous system, should prove to be an excellent intervention in elderly hypertensive patients with stiff vasculature and increased peripheral vascular resistance and in those with concomitant coronary artery disease.

References

1. Joint National Committee on Detection, Evaluation and Treatment of High Blood Pressure: The Fifth Report of the Joint National Committee on Detection, Evaluation, and Treatment of High Blood Pressure (JNC-V). Arch Intern Med 153:154–183, 1992.
2. Folkow B: The pathophysiology of hypertension. Differences between young and elderly patients. Drugs 46(Suppl 2):3–7, 1993.
3. Folkow B: Sympathetic nervous control of blood pressure. Role in primary hypertension. Am J Hypertens 2:103S–111S, 1989.
4. Korner PL, Bobik A, Angus JA, et al: Resistance control in hypertension. J Hypertens 7(Suppl 4): S125–S134, 1989.
5. Yin FCP: The aging vasculature and its effects on the heart. In Weisfeldt ML (ed): The Aging Heart. New York, Raven Press, 1980, pp 137–214.
6. Wei JY: Use of calcium entry blockers in elderly patients. Special considerations. Circulation 80(Suppl 4):171–177, 1989.
7. Folkow B, Svanbork A: Physiology of cardiovascular aging. Physiol Rev 73:725–764, 1993.
8. Bynny RL: Hypertension in the elderly. In Laragh JH, Brenner BM (eds): Hypertension: Pathophysiology, Diagnosis and Management, vol. 2. New York, Raven Press, 1990, pp 1869–1888.
9. Wei JY: Cardiovascular anatomic and physiologic changes with age. Topics Geriatr Rehab 2:10–16, 1986.

10. Potter JF, Elahi d, Tobin JD, Andres R: The effect of age on the cardiothoracic ratio of man. J Am Geriatr Soc 30:404–409, 1982.
11. Epstein M: Aging and the kidney. J Am Soc Nephrol 7:1106–1122, 1996.
12. Black HR: Age-related issues in the treatment of hypertension. Am J Cardiol 72:10H–13H, 1993.
13. Lipsitz LA: Orthostatic hypotension in the elderly. N Engl J Med 321:952–956, 1989.
14. Messerli FH, Ventura HO, Glade LB, et al: Essential hypertension in the elderly: Haemodynamics, intravascular volume, plasma renin activity, and circulating catecholamine levels. Lancet ii:983–986, 1983.
15. Staessen J, Amery A, Fagard R: Isolated systolic hypertension in the elderly. J Hypertens 8:393–405, 1990.
16. Saltzberg S, Stroh JA, Frishman WH: Isolated systolic hypertension in the elderly: patho-physiology and treatment. Med Clinics North Am 72:523–547, 1988.
17. Anderson RJ, Reed G, Kirk LM: Therapeutic considerations for elderly hypertensives. Clin Ther 5:25–38, 1982.
18. LaPalio LR: Hypertension in the elderly. Am Fam Physician 52:1161–1165, 1995.
19. Holzgreve H, Middeke M: Treatment of hypertension in the elderly. Drugs 46(Suppl 2):24–31, 1993.
20. Gifford R, Moser M: Hypertension in the elderly: An update on treatment results and recommendations. Council Geriatr Cardiol (in press).
21. Aronow WS: Cardiovascular drug therapy in the elderly. In Frishman WH, Sonnenblick EH (eds): Cardiovascular Pharmacotherapeutics. New York, McGraw Hill, 1997, pp 1267–1281.
22. Amery A, Birkenhager W, Brixko P, et al: Morbidity and mortality results from the European Working Party on High Blood Pressure in the Elderly Trial. Lancet 1:1349–1354, 1985.
23. Dahlof B, Lindholm LH, Hansson L, et al: Morbidity and mortality in the Swedish Trial in Old Patients with Hypertension (STOP-Hypertension). Lancet 338:1281–1285, 1991.
24. MRC Working Party: Medical Research Council trial of treatment of hypertension in older adults: principal results. BMJ 304:405–412, 1992.
25. SHEP Cooperative Research Group: Prevention of stroke by antihypertensive drug treatment in older persons with isolated systolic hypertension. Final results of the Systolic Hypertension in the Elderly Program (SHEP). JAMA 255:3255–3264, 1991.
26. O'Rourke M: Arterial stiffness, systolic blood pressure, and logical treatment of arterial hypertension. Hypertens 15:339–347, 1990.
27. Stamler J, Neaton JD, Wentworth DN: Blood pressure (systolic and diastolic) and risk of fatal coronary heart disease. Hypertens 13(Suppl 5):2–12, 1989.
28. Menard J, Day M, Chatellier G, Laragh JH: Some lessons from Systolic Hypertension in the Elderly Program (SHEP). Am J Hypertens 5:325–330, 1992.
29. Lenfant C: High blood pressure. Some answers, new questions, continuing challenges. JAMA 275:1604–1606, 1996.
30. Kaplan NM, Gifford RW: Choice of initial therapy for hypertension. JAMA 275:1577–1580, 1996.
31. Alderman MH: Which antihypertensive drugs first—and why. JAMA 267:2786–2787, 1992.
32. Mancia G: Treatment of hypertension and prevention of cardiovascular morbidity. In Lichtlen PR, Reale A (eds): Adalat: A Comprehensive Review. Berlin, Springer-Verlag, 1991, pp 75–82.
33. Frishman WH: Alpha- and beta-adrenergic blocking drugs. In Frishman WH, Sonnenblick EH (eds): Cardiovascular Pharmacotherapeutics. New York, McGraw-Hill, 1997, pp 59–94.
34. Frishman WH: Calcium channel blockers. In, Frishman WH, Sonnenblick EH (eds): Cardiovascular Pharmacotherapeutics. New York, McGraw-Hill, 1997, pp 101–130.
35. Nayler WG: Pharmacological aspects of calcium antagonism. Short term and long term benefits. Drugs 46(Suppl 2):40–47, 1993.
36. Keefe D, Frishman WH: Clinical pharmacology of the calcium blocking drugs. In Packer M, Frishman WH (eds): Calcium Channel Antagonists in Cardiovascular Disease. Norwalk, CT, Appleton-Century-Crofts, 1984, pp 3–19.
37. Braunwald E: Mechanism of action of calcium-channel blocking agents. N Engl J Med 307:1618–1627, 1983.
38. Weiner DA: Calcium channel blockers. Med Clin North Am 72:83–115, 1988.
39. Frishman WH, Sonnenblick EH: Calcium channel blockers. In Schlant RC, Alexander RW (eds): Hurst's The Heart, 8th ed. New York, McGraw Hill, 1994, pp 1271–1290.
40. Frishman WH: Current status of calcium channel blockers. Curr Probl Cardiol 19:637–688, 1994.
41. Frishman WH (ed): Current Cardiovascular Drugs, 2nd ed. Philadelphia, Current Science, 1995, pp 129–148.
42. Opie LH, Frishman WH, Thadani U: Calcium channel antagonists. In Opie LH (ed): Drugs for the Heart, 4th ed. Philadelphia, W.B. Saunders, 1995, pp 50–83, 1995.

43. Mishra SK, Hermsmeyer K: Selective inhibition of T-type Ca^{2+} channels by RO 40-5967. Circ Res 7:144–148, 1994.
44. Erne P, Conen D, Kowski W, et al: Calcium antagonist induced vasodilation in peripheral, coronary, and cerebral vasculature as important factors in the treatment of elderly hypertensives. Eur Heart J 8(Suppl K):49–56, 1987.
45. Busse JC, Materson BJ: Geriatric hypertension: The growing use of calcium-channel blockers. Geriatrics 43:51–58, 1988.
46. Forette F, Bert P, Rigaud A: Are calcium antagonists the best option in elderly hypertensives? J Hypertens 12(Suppl 6):S19–S23, 1994.
47. Abernethy DR, Schwartz JB, Todd EL, et al: Verapamil pharmacodynamics and disposition in young and elderly hypertensive patients: Altered electrocardiographic and hypotensive responses. Ann Intern Med 105:329–336, 1986.
48. Schwartz JB, Abernethy DR: Responses to intravenous and oral diltiazem in elderly and younger patients with systemic hypertension. Am J Cardiol 59:1111–1117, 1987.
49. Dubois C, Blanchard D, Loria Y, Moreau M: Clinical trial of a new antihypertensive drug, nicardipine: Efficacy and tolerance in 29,104 patients. Curr Ther Res 42:727–736, 1987.
50. Forette F, Bellet M, Henry JF, et al: Effect of nicardipine in elderly hypertensive patients. Br J Pharmacol 20(Suppl 1):125S–129S, 1985.
51. Forette F, McClaran J, Hervy MP, et al: Nicardipine in elderly patients with hypertension: A review of experience in France. Am Heart J 117:256–261, 1989.
52. Bühler FRT, Hulthen UL, Kiowski W, et al: The place of the calcium antagonist verapamil in antihypertensive therapy. J Cardiovasc Pharmacol 4(Suppl 3):S350–S357, 1982.
53. Faulds D, Sorkin EM: Felodipine: A review of the pharmacology and therapeutic use of the extended release formulation in older patients. Drugs Agins 2:374–388, 1992.
54. Schneider J: Serum lipoproteins in hyperlipoproteinemic patients with essential hypertension treated with felodipine for two years. Curr Ther Res 50:118–126, 1991.
55. Dunlay MC, Lipschutz KH, Nelson EB, et al, for the Felodipine Study Group: A multicenter study of felodipine ER and placebo in elderly and young hypertensive patients [abstract]. Am J Hypertens 4:25A, 1991.
56. Todd PA, Faulds D: Felodipine: A reappraisal of the pharmacology and therapeutic use of its extended release formulation in cardiovascular disorders. Drugs 44:251–276, 1992.
57. Koenig W, Sund M, Binner L, et al: Comparison of once daily felodipine 10 mg ER and hydrochlorothiazide 25 mg in the treatment of mild to moderate hyper-tension. Eur J Clin Pharmacol 41:197–199, 1991.
58. Waite MA, Bone ME, Kubik MM, et al: A comparison of the efficacy and tolerability of felodipine ER and atenolol given as monotherapy in mild to moderate hypertension. Br J Clin Pharmacol 32:661–664, 1991.
59. Shapiro DA, Dunlay MC, Holmes GI, et al: Effect of age, race and degree of hypertension treated with felodipine, hydrochlorothiazide and atenolol [abstract]. Am J Hypertens 12:75A, 1989.
60. Weissel M, Stanek B, Flygt G: Felodipine is more effective than hydrochlorothiazide when added to a beta-blocker in treating elderly hypertensive patients. J Cardiovasc Pharmacol 15 (Suppl 4):S95–S98, 1990.

60a. Frishman WH, Michaelson MD: Use of calcium antagonists in patients with ischemic heart disease and systemic hypertension. Am J Cardiol 79:33–38, 1997.

61. Bassan MM: The combined use of calcium-channel blockers and beta blockers in the treatment of angina pectoris: an update. Pract Cardiol 13:51–54, 1987.
62. Setaro SF, Schulmon DS, Zaret BL, Soufer R: Congestive heart failure and intact systolic function: Improvement in clinical status and diastolic filling with verapamil. Clin Res 35:325A, 1987.
63. Manning WS, Hung G, Parker J, et al: Heart failure in the elderly: Diastolic dysfunction with preserved systolic function and normal coronary arteries [abstract]. J Am Coll Cardiol 11:160A, 1988.
64. Beever DG, Sleight P: Short acting dihydropyridine (vasodilating) calcium channel blockers for hypertension: is there a risk? Br Med J 312:1143–1145, 1996.
65. Buring JE, Glynn RJ, Hennekens CH: Calcium channel blockers and myocardial infarction. A hypothesis formulated but not yet tested. JAMA 274:654–655, 1995.
66. Psaty BM, Heckbert SR, Koepsell TD, et al: The risk of myocardial infarction associated with antihypertensive drug therapies. JAMA 274:620–625, 1995.
67. Pahor M, Guralnik JM, Corti MC, et al: Long-term survival and use of antihypertensive medications in older persons. J Am Geriatr Soc 43:1191–1197, 1995.
68. Borhani NO, Mercuri M, Borhani PA, et al: Final outcome results of the Multicenter Isradipine Diuretic Atherosclerosis Study (MIDAS): A randomized controlled trial. JAMA 276:785–791, 1996.

69. Chobanian AV: Calcium channel blockers. Lessons learned from MIDAS and other clinical trials. JAMA 276:829–830, 1996.
70. Fagan TC: Calcium antagonists and mortality. Another case of the need for clinical judgement. Arch Intern Med 155:2145, 1995.
71. Triggle DJ: Biochemical and pharmacologic differences among calcium channel antagonists: Clinical implications. In Epstein M (ed). Calcium Antagonists in Clinical Medicine. Philadelphia, Hanley & Belfus, 1992, pp 1–27.
72. Kiowski W, Erne P, Bertel O, Bolli P, Buhler S: Acute and chronic sympathetic reflex action and antihypertensive response to nifedipine. J Am Coll Cardiol 7:344–348, 1986.
73. Ruzicka M, Leenen FHH: Relevance of intermittent increases in sympathetic activity for adverse outcome on short-acting calcium antagonists. In Laragh JH, Brenner BM (eds): Hypertension: Pathophysiology, Diagnosis, and Management, 2nd ed. New York, Raven Press, 1995, pp 2815–2825.
74. Packer M, O'Connor CM, Ghali JK, et al: Effect of amlodipine on morbidity and mortality in severe chronic heart failure. Prospective Randomized Amlodipine Survival Evaluation Study Group. N Engl J Med 335:1104–1114, 1996.
75. Neaton JD, Grimm RH, Prineas RJ, et al: Treatment of Mild Hypertension Study. Final results. JAMA 270:713–724, 1993.
76. Alderman MH, Cohen H, Roqué R, Madhavan S: Effect of long-acting and short-acting calcium antagonists on cardiovascular outcomes in hypertensive patients. Lancet 349:594–598, 1997.
77. McMurray J, Murdoch D: Calcium antagonist controversy: The long and short of it. Lancet 349:585–586, 1997.
78. Townsend RR, Kimmel SE: Calcium channel blockers and hypertension. 1: New trials. Hosp Pract 31:125–136, 1996.
79. Davis BR, Cutler JA, Gordon DJ, et al: Rationale and design for the Antihypertensive and Lipid Lowering Treatment to Prevent Heart Attack Trial (ALLHAT). Am J Hypertens 9:342–360, 1996.
80. Elliot WJ: ALLHAT: The largest and most important clinical trial in hypertension ever done in the USA. Am J Hypertens 9:409–411, 1996.
81. The Hypertension Optimal Treatment Study (HOT Study). Blood Pressure 2:62–68, 1993.
82. Julius S: The Hypertension Optimal Treatment (HOT) Study in the United States. Am J Hypertens 9:41S–44S, 1996.
83. Amery A, Birkenhager W, Bulpitt CJ, et al: SYST-EUR: A multicenter trial on the treatment of isolated systolic hypertension in the elderly: Objectives, protocol, and organization. Aging 3:287–302, 1991.
84. Dahlof B, Hanson L, Lindholm LH, et al: STOP-Hypertension 2: A prospective intervention trial of newer versus older treatment alternatives in old patients with hypertension: Swedish Trial in Old Patients with Hypertension. Blood Pressure 2:136–141, 1993.

AYUB AKBARI M.D. / MARSHALL D. LINDHEIMER M.D., F.A.C.P., F.R.C.O.G.

18

Management of High Blood Pressure in Pregnancy

Hypertension is the most common medical complication of gestation, and the approach to this challenging clinical problem differs considerably from the approach in nonpregnant populations. First, the diagnostic spectrum of high blood pressure in pregnancy is wider because of two pregnancy-specific entities: preeclampsia, a complication associated with substantial maternal and fetal morbidity and mortality, and transient hypertension of pregnancy, a relatively more benign disorder. Second, there are two patients to deal with—the hypertensive mother and her unborn child.

This chapter focuses on managing the hypertensive disorders of pregnancy. It will review the striking alterations in hemodynamics that accompany normal gestation and highlight problems related to classification and diagnosis of disorders associated with hypertension in pregnancy. The main focus, however, is preeclampsia, the most potentially ominous of the common hypertensive disorders of pregnancy. The final section is devoted to the efficacy and safety of various antihypertensive drugs currently prescribed to pregnant women.

ALTERED HEMODYNAMICS AND VOLUME HOMEOSTASIS IN NORMAL PREGNANCY

Evaluation and treatment of high blood pressure in pregnant women require understanding of the cardiovascular changes which occur during gestation. Blood pressure decreases rapidly after conception; by midpregnancy, diastolic levels average 10 mmHg below nonpregnant values.[1–5] Blood pressure then increases slowly; near term it approaches nongravid levels and in the puerperium may even exceed such levels transiently.[6] Cardiac output (CO) increases by 30–50%; the increment begins soon after conception, peaks around gestational week 16, and is sustained at peak level until term.[7,8] Given the magnitude of the increase in CO, the decrease in blood pressure must be due to a striking decrement in peripheral vascular resistance—and indeed this is the case. *Thus normal pregnancy is a markedly vasodilated state.*

Knowledge of the cardiovascular changes in normal pregnancy enhances diagnostic skills. Although the definition of hypertension in gravidas remains diastolic pressure ≥ 90 and systolic pressure ≥ 140 mmHg, diastolic values of 75 mmHg in the second trimester and ≥ 85 mmHg in the third trimester or systolic levels ≥ 120 mmHg in midpregnancy and ≥ 130 mmHg in late pregnancy are suspect. Two large epidemiologic studies have demonstrated a significant

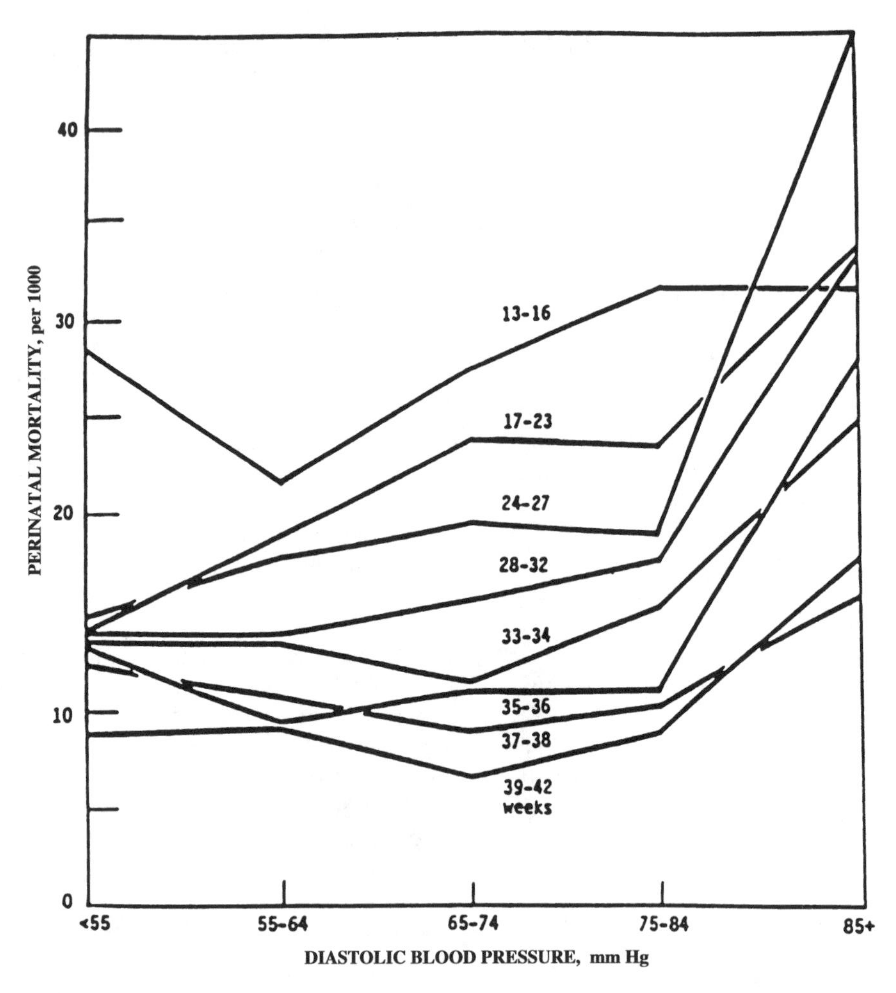

FIGURE 1. Trends in perinatal mortality rates as a function of diastolic pressure for gestational ages 13 weeks to term. Note the significant increase in perinatal mortality associated with pressures exceeding 84 mmHg. (From Friedman EA, Neff RK: Pregnancy and Hypertension. A Systematic Evaluation of Clinical Diagnostic Criteria. Littleton, MA, PSG Publishing, 1977, with permission.)

rise in fetal mortality when diastolic pressure at any stage of gestation is ≥ 85 mmHg[2] as well as significant increments in perinatal mortality and growth retardation when mean blood pressure exceeds 90 and 95 mmHg during second and third trimesters,[3] respectively (Fig. 1). One should also be aware that certain women with undiagnosed mild essential hypertension experience early in pregnancy a decrement in blood pressure that appears normal. They are then mislabeled as preeclamptic when frankly elevated values occur at term. An astute clinician cognizant of the fact that 124/78 mmHg at gestational week 16 may be abnormal can avoid such errors.

Review of mechanisms involved in maintenance of blood pressure during pregnancy, such as alterations in vasoconstricting and vasodilating hormones and autacoids, natriuretic peptides, and endothelium-produced factors, is beyond the scope of this chapter; the interested reader is referred elsewhere.[9–11] Of particular interest, however, are the marked increments in the production and circulating levels of renin, angiotensin, and numerous corticoids, especially aldosterone, deoxycorticosterone, and progesterone. There are also suggestions that increases in the production of vasodilating prostanoids, and more recently nitric oxide (NO) synthase, and NO-dependent and independent endothelium-derived relaxing factors (EDRFs) are responsible for the marked decrease in peripheral vascular resistance characteristic of pregnancy.[12–14]

Measurement of Blood Pressure

The literature about methods of measuring blood pressure during pregnancy is confusing. Controversies have included which posture is preferred (e.g., lateral recumbency vs. sitting) and which Korotkoff sound—K_4 (muffling) or K_5 (disappearance)—is a better measure of the diastolic level. Lower values with the patient in lateral recumbency merely reflect the difference in hydrostatic pressure when the cuff is positioned substantially above the left ventricle. In addition, the notion that the hyperdynamic circulation in pregnant women frequently leads to a large difference between K_4 and K_5, with the latter often approaching zero, is incorrect; K_5 is the sound closest to true diastolic pressure.[15–18] Thus, the approach to measurement is similar to that in nonpregnant populations; the patient should be seated and rest for > 5 minutes, and the cuff over the arm should be at the level of the heart. The role of ambulatory blood pressure monitoring in the detection of hypertensive disorders in pregnancy remains to be established.[19]

Volume Homeostasis

Pregnant women gain about 12.5 kg, mostly due to fluid accumulation; total body water increases by 6–9 L, of which 4–7 L is in the extracellular compartment.[10,20] Plasma volume rises approximately 50% and red blood cell mass 20–30%, resulting in dilutional anemia of pregnancy. Interstitial volume also increases, and *edema is a normal feature of pregnancy.*

Both renal plasma flow and glomerular filtration rate (GFR) increase markedly in pregnancy. The increase in GFR is reflected by decrements in both creatinine and urea nitrogen levels; the upper limits of normal are 0.8 mg/dl and 13 mg/dl, respectively.[20] *Thus values considered normal in nonpregnant populations may be abnormal in pregnant women.* Urinary protein excretion also rises; the upper limit of normal is 300 mg/24 hr (or twice that used at one's institution).

DETECTION AND CLASSIFICATION OF HYPERTENSION IN PREGNANCY

Many terms have been used to describe high blood pressure in pregnancy, including toxemia, pregnancy-induced or pregnancy-associated hypertension, gestosis, preeclampsia, and preeclamptic-toxemia, resulting in considerable confusion.This is especially true when the same term (e.g., pregnancy-induced hypertension [PIH]) is defined differently by various authors. Furthermore, classification schema that are often complex and overly detailed appear periodically around the globe. We prefer an approach developed by the American College of Obstetricians and Gynecologists Committee on Terminology in 1972 and endorsed most recently by the National High Blood Pressure Education Program (NHBPEP) in the United States.[21] This approach is concise and practical; it separates accurately the more benign from the more serious disorders. High blood pressure in pregnancy is divided into four categories: (1) chronic hypertension (both essential and secondary), (2) preeclampsia and eclampsia, (3) preeclampsia superimposed on chronic hypertension, and (4) transient hypertension of pregnancy.

Chronic Hypertension

The incidence of high blood pressure in pregnancy ranges between 7–10%, and about one-fourth of these cases are due to essential hypertension.[2,21,22] The diagnosis, most certain when a history of disease has been documented before pregnancy, is also highly probable when elevated pressure is detected before gestational week 20 or persists after the puerperium.[23] These pregnancies are usually uneventful if blood pressure elevations are mild and end-organ damage absent. As discussed below, however, affected mothers experience an increased incidence of superimposed preeclampsia, which is responsible for most of the morbidity.

Pregnancy in women with secondary causes of hypertension is unusual but does occur. In instances such as scleroderma, periarteritis nodosa, and Cushing's syndrome, patients and their fetuses do quite poorly.[24–27] Pheochromocytoma, which has a propensity to present during pregnancy, was once associated with substantial lethality. When detected, however, the disorder can be managed either surgically or pharmacologically, depending on the stage of gestation.[28,29] Alpha blockade has been used to control blood pressure until delivery, after which the tumor, if operable, is resected.[30] It is therefore prudent to screen for excessive catecholamines at the slightest suspicion. Renal angioplasty and stent placement for uncontrolled hypertension secondary to renal artery stenosis have been successful in pregnant women.[31,32]

Women with aldosterone-producing tumors have conceived. In some pregnant women the course of the disease has been similar to that in nongravid patients. Of interest, however, is that several patients have experienced amelioration of kaliuresis, normalization of serum plasma potassium levels, and even spontaneous reductions in blood pressure. These effects have been attributed to the pregnancy-induced rise in production of progesterone, a hormone that inhibits the action of aldosterone on the renal tubule.[33]

Preeclampsia and Eclampsia

Preeclampsia, especially when superimposed on chronic hypertension or renal disease (category 3), is the disorder most likely to endanger both mother and fetus.[2,10,11,23] *For this reason, when the underlying diagnosis is uncertain, it is prudent to treat the patient as if she has preeclampsia.* The disease occurs primarily in nulliparas and in the latter half of pregnancy (but there are exceptions), usually near term. The cardinal features of preeclampsia are hypertension and proteinuria, and

TABLE 1. Ominous Signs and Symptoms and Laboratory Abnormalities in Women with Preeclampsia

Preeclampsia is always potentially dangerous, but the indicators listed below may be particularly ominous.

Signs and symptoms

- Blood pressure ≥ 160 mmHg systolic or ≥ 110 mmHg diastolic
- Upper abdominal pain, especally epigastric and right upper quadrant pain
- Headache, visual disturbances, or other cerebral signs
- Cardiac decompensation (e.g., pulmonary edema); usually associated with underlying heart disease or chronic hypertension
- Retinal hemorrhages, exudates, or papilledema (extremely rare without other indicators of severity; almost always indicate underlying chronic hypertension)

Laboratory abnormalities

- Rapid increase in proteinuria, especially to ≥ 3 gm/24 hr, combined with rapidly decreasing serum albumin level
- Increasing serum creatinine levels, especially to > 2 mg/dl (177 μmol/L), unless known to be elevated previously
- Platelet count $10^5/mm^3$ or evidence of microangiopathic hemolytic anemia (e.g., schistocytes and/or increased lactate dehydrogenase or direct bilirubin levels)

The presence of intrauterine growth retardation and decreasing urine volumes also requires added vigilance.

many patients experience rapid weight gain and develop edema.[10,11,23] The patient also may manifest liver function and coagulation test abnormalities as well as thrombocytopenia. Of great concern is that the the disease may progress rapidly to a life-threatening convulsive phase, termed eclampsia, which is frequently preceded by premonitory symptoms and signs, including hyperreflexia, visual disturbances, severe headache, right upper quadrant or epigastric pain, and several laboratory changes[10,11,23,34] (Table 1). Eclampsia, however, also may occur suddenly and without warning in an asymptomatic patient whose blood pressure is but mildly elevated, which is one reason why preeclampsia, regardless of apparent severity, always presents a threat to mother and fetus.

One must be aware of a deceptively benign variant of preeclampsia in which the patient presents with minimal changes in blood pressure, borderline thrombocytopenia, modest increases in liver enzymes, and often no renal dysfunction. One is tempted to temporize. However, this form of preeclampsia may rapidly become life-threatening, developing in 24–48 hr into a florid syndrome characterized by hemolysis (of the microangiopathic hemolytic anemia variety with abundant schistocytes present on the peripheral blood smear), prominent changes in coagulation and liver function (platelet counts can plunge to $10^5/mm^3$ in 1 day, while transaminase and lactic dehydrogenase levels rise quickly, exceeding 1,000 IU within 24 hr). The variant is termed the **HELLP** syndrome (**h**emolysis, **e**levated **l**iver enzymes, **l**ow **p**latelet counts), and is associated with severe morbidity, unless the pregnancy is promptly terminated.[10,11,35] (See the discussion below of aggressive vs. conservative therapy when preeclampsia presents remote from term.)

Most cases of preeclampsia occur after midpregnancy or in the immediate puerperium, but exceptions exist. Preeclampsia presenting before the 20th gestational week has been described with hydatiform disease, nonimmune fetal hydrops disorder (including alpha thalassemia), abdominal pregnancy, and, even more rarely, chronic hypertension and renal disease.[10,11,23] At the other end of the spectrum is a disorder termed late postpartum eclampsia, which presents more than 48 hours after delivery and within the first 2 weeks of the puerperium.[10,11]

Transient Hypertension of Pregnancy

Some women develop mild-to-moderate hypertension in late gestation or the immediate puerperium; postpartum, blood pressure normalizes rapidly. Similar events often recur in subsequent gestations, but pregnancies are otherwise benign.[23] When such a scenario occurs late in the pregnancy of a nullipara, it may be difficult to distinguish transient hypertension of pregnancy from early preeclampsia; in this setting it is prudent to manage the patient as if she had preeclampsia.[21] Evidence also suggests that women with transient hypertension of pregnancy are destined to develop essential hypertension later in life; thus, yearly follow-ups to check blood pressure are appropriate for earlier detection of frank hypertension.[10,36]

There is another entity termed *late postpartum hypertension*, whose etiology may be comparable to that of transient hypertension of pregnancy. Patients experience a normotensive gestation, but their blood pressure rises about 2 weeks after delivery and remains elevated, usually at mild levels, for up to 6 months before normalizing.[10] *Late postpartum hypertension* is probably another harbinger of essential hypertension later in life. One must also be aware that women exposed to cocaine may present with syndromes that mimic both transient hypertension of pregnancy and preeclampsia.[37]

The above classification schema should prove clinically useful. At times, however, it may be difficult to distinguish among preeclampsia, chronic, secondary, or transient hypertension, and combinations of these entities by clinical criteria alone. Such dilemmas have been established years ago by clinical correlation studies in which renal biopsy was used to verify the diagnosis.[36] In one such study, a nephrologist and an obstetrician recorded their impression before the biopsy; they were completely correct in only 58% of patients (cited in ref. 36). This again underscores the prudence of treating all women with hypertensive disorders of pregnancy as if they had the more potentially morbid disorder (Table 2).

PATHOPHYSIOLOGY AND MANAGEMENT OF PREECLAMPSIA

Preeclampsia is more than hypertension; it is a systemic disorder.[11,38] This section explores the pathology and pathophysiology of the disease in terms of the major organ systems affected and discusses both management and attempts at prevention; the latter is based on a number of plausible hypotheses of what causes preeclampsia.

Hypertension and Vascular Reactivity

Hypertension in preeclampsia is due mainly to a reversal of the vasodilation characteristic of normal pregnancy, which is replaced by striking increases in

TABLE 2. Findings and Tests for Differentiating Preeclampsia (Pure or Superimposed) from Other Hypertensive Disorders

	Preeclampsia	Chronic Hypertension
Age	Extremes of age	Usually older (> 30 yr)
Parity	Most nulliparas	More often multiparous
Onset	Rare before gestational week 20	Pressure may be elevated ≤ 20 weeks
History	Negative	Positive; often hypertension in previous pregnancy
Physical examination	Fundoscopic evidence of vasospasm, cardiac status usually normal, CNS irritability (visual signs, headaches, hyperreflexia), upper abdominal or epigastric pain	End-organ findings when disease is long-standing, or pressure uncontrolled (i.e., arteriovenous nicking in fundi, cardiac hypertrophy), normal reflexes, normal abdominal examination
Complete blood count	Increased hematocrit suggesting hemoconcentration, decreasing platelet counts, schistocytes on smear	Normal (lupus erythematosus may mimic preeclamptic platelet changes)
Urinalysis and protein excretion	Often normal but may contain red blood cells and even casts; proteinuria usually present	Normal in essential hypertension, "active" when glomerular diseases exacerbate; minimal or no proteinuria in essential hypertension
Serum creatinine	Abnormal or rising values, especially when associated with oliguria, suggest severe preeclampsia	Normal in essential hypertension but may be increased in underlying glomerular disorders, which may exacerbate and mimic preeclampsia
Serum urate	Often elevated out of proportion to changes in creatinine levels	Usually normal
Liver function	Abnormal enzymes (transaminases, lactic dehydrogenase), increased bilirubin	Normal

peripheral vascular resistance. The increments are so marked that blood pressure rises despite decreases in cardiac output.[10,11,39] Normally the vasculature of normotensive gravidas manifests a decreased pressor responsiveness to a variety of vasoactive peptides and amines, especially angiotensin II (AII). In preeclampsia the vasculature becomes hyperresponsive to these hormones; the hyperresponsiveness to AII may occur months before the appearance of overt disease.[10,11,23] Hypertension is quite labile and may be accompanied by a reversal of the normal circadian blood pressure rhythms.[10,11]

The cause of altered vascular reactivity is obscure, but research during the past decade has focused on changes in the ratio of vasodilating and vasoconstricting prostanoids; some evidence suggests increments in the production of thromboxane.[10,13] Most recently investigators[13] have stressed the vasoconstricting potential of circulating pressor substances (e.g., AII and endothelin) if the adaptive increments in the production of NO synthase and NO-dependent or independent EDRF fail to occur or are inhibited.[10,11] They have searched for mechanisms that may damage the endothelium (the site of prostanoid, endothelin, and EDRF production) in preeclampsia, including numerous "bad" cytokines and insufficient antioxidant activity.[10,11,40,41] Other investigators have focused on gestational aberrations in calcitropic hormone, insulin, and magnesium metabolism.[11] As of 1998, however, all of these hypotheses remain unproved.

The Kidney and Volume Homeostasis

The characteristic renal lesion of preeclampsia, termed glomerular endotheliosis, is an enlarged ischemic glomerulus due to hypertrophy and swelling of the intracapillary cells. In fact, some investigators interpret the lesion as further evidence of an abnormal endothelium in all vessels of preeclamptic women.[36,42,43] Both glomerular filtration rate (GFR) and renal plasma flow are decreased; GFR falls about 25% in mild cases.[10,23] Thus, despite morphologic evidence of ischemia and obliteration of the urinary space, GFR in preeclamptic women is still above that in nonpregnant women, and this may not be appreciated if one is not aware of the upper limits of normal for serum creatinine in pregnancy (see above). Uric acid clearance, which normally rises in gestation, also decreases, often earlier and more than GFR; a circulating level ≥ 5.0 mg/dl (297 μmol/l) is abnormal in pregnancy.[10,20] Also of interest is a reversal of the normal hypercalciuria of pregnancy; hypocalciuria is difficult to explain by the small reductions in GFR, normally associated with preeclampsia.[11,44]

Abnormal proteinuria accompanies almost all cases of preeclampsia, and the diagnosis is suspect in its absence (although increased protein excretion may be a late feature).[23,36] The proteinuria may be minimal or severe and is nonselective.[10,36,45] In fact, preeclampsia is the major cause of nephrotic-range proteinuria in pregnancy.[36]

Renal sodium handling may be impaired in preeclampsia, but the degree of impairment varies considerably. In fact, some of the severest forms of the disease occur in the absence of edema (the "dry" preeclamptic patient), as underscored by a report in which 26 of 67 eclamptic women (women who convulsed) were edema-free.[46] Even in patients with marked edema, plasma volume is lower than in normal pregnancy, and there is evidence of hemoconcentration. The decreased intravascular volume represents, in part, extravasation of albumin into the interstitium, which increases oncotic pressure;[11,47] in fact, hypoalbuminemia may be present when proteinuria and liver dysfunction are minimal. Another feature of decreased volume is that central venous and pulmonary capillary wedge pressures tend to be low.[11,39] The alterations in intravascular volume, along with the decrements in placental perfusion and cardiac output, are a major reason to avoid diuretics in preeclamptic patients.

The cause of the impaired renal sodium excretion remains obscure. Filtered sodium, although decreased compared with normal gestation, is still higher than in the nongravid state.[10,11,23] More paradoxical is the finding that the renin-angiotensin-aldosterone system is suppressed, and circulating levels of natriuretic factor(s) are increased, despite the decrease in intravascular volume.[10,11]

Brain and Placenta

Eclampsia (the appearance of convulsions) remains a leading cause of maternal death, yet knowledge of its cause is limited. Thus controversies surround its treatment. Some authoritites suggest that eclampsia may relate to the coagulopathy associated with preeclampsia, noting fibrin deposition in the brain at autopsy.[48] Still others consider it a form of hypertensive encephalopathy,[49] a concept that seems difficult to accept because convulsion may occur suddenly in women with systolic levels below 140 mmHg. Nonetheless, vasoconstriction in preeclampsia at times may be selective, and investigations using ultrasonographic techniques suggest that severe cerebral vasospasm may occur even when peripheral vasoconstriction is less evident.[50,51] Reports using techniques such as cranial axial tomography and magnetic resonance vary, but many describe abnormalities, usually transient, that are consistent with localized hemorrhage or edema.[11,34,52–54] Unfortunately, because many of the studies were performed in patients receiving considerable fluid administration, some of the findings may be iatrogenic.

The best descriptions of gross and microscopic pathology are found in the 1973 monograph by Sheehan and Lynch, which describes the largest *autopsy* series in the literature.[42] Most of the necropsies were performed within 1 or 2 hours of death, thus eliminating most postmortem changes that may confound interpretation. The degree of hemorrhage and petechia varies from grossly visible to microscopic. The authors offer cogent arguments that the initiating events are vasoconstrictive and ischemic and that edema is a late or postpartum feature.

Uteroplacental blood flow decreases in preeclampsia, reflecting, in part, arterial lesions termed acute atherosis. The degree of placental infarction is higher in preeclamptic women than in normotensive gravidas.[11,42] The decrements in flow may relate to changes in the vascular resistance indices and evidence of reversal of blood flow when the patient is examined with Doppler ultrasound.[11,55]

There is also evidence of abnormal placentation in preeclampsia. In early pregnancy the uterine arteries are transformed from thick-walled, muscular vessels into sac-like, flaccid vessels that accommodate a tenfold increase in intrauterine blood flow.[11,56–60] This process, attributed to endovascular trophoblastic invasion, is virtually complete by the end of the second trimester. In women destined to develop preeclampsia, trophoblastic invasion of the spiral arteries is incomplete, and the arteries remain thick-walled and muscular. Recent research suggests that in preeclampsia the cytotrophoblast cells fail to mimic a vascular adhesion phenotype and to express adhesion molecules believed to be crucial to the process of vascular invasion.[61] These changes, some postulate, lead to placental hypoxemia and are the proximate cause of the disease.[40]

Other major systemic manifestations of preeclampsia, including coagulation problems and liver dysfunction (including the ominous HELLP syndrome), are discussed above. The pathologic changes in the liver, as well as rarer pathology in heart, lung, pituitary, and adrenal glands, are noted in the monograph by Sheehan and Lynch.[42]

Preeclampsia Superimposed on Another Disorder

Superimposed preeclampsia, especially when it occurs early in the second half of pregnancy, and/or on a background of preexisting hypertension, increases maternal and fetal risk substantially. However, as noted, its appearance may be difficult to distinguish clinically from an exacerbation of either chronic hypertension or an underlying renal disorder.[36] Because preeclampsia can have an explosive and ominous course, investigators continually search to develop tests which can predict imminent disease, or distinguish preeclampsia from the more benign transient hypertension of pregnancy, or here, to determine if superimposed preeclampsia has indeed occurred. There is also a search for markers that may provide clues to the cause of preeclampsia, as this would help devise strategies to prevent the disease or at least improve the management of this perilous disorder. In this respect, studies have focused on host of autacoids, circulating hormones, lipids, oxidant or antioxidant activity, markers of endothelial dysfunction, evidence of neutrophil activation, adhesion molecules, a host of cytokines, the isoelectric point of plasma albumin, urinary excretion of calcium, and so on.[10,11,40] Both the sensitivity and positive predictive value of virtually all these tests are either unknown or appear modest; as of 1998, none can be recommended. Table 2, however, describes a clinical approach which may help in the differentiation of preeclampsia from chronic or transient hypertension in pregnancy.

Can Preeclampsia Be Prevented?

Interventions to prevent preeclampsia or to abort its development at an early or preclinical stage would revolutionize prenatal care and save both maternal and fetal lives. Sodium restriction and/or administration of prophylactic diuretics, once a popular approach, was revived briefly in 1985 by a metaanalysis of randomized studies that had recruited about 7,000 women.[62] Analyses of the article, however, revealed no difference in the incidence of proteinuric hypertension; any seemingly positive effects were due to the ability of diuretics to reduce edema and/or provoke a modest decrease in blood pressure.

More recently, two approaches were advocated, each based on plausible hypotheses. The first, administration of low-dose aspirin (60–100 mg/day) after the 12th gestational week, was based on the hypothesis that hypertension and coagulation abnormalities in preeclampsia were due to an imbalance in the production of vasodilating and vasoconstricting prostaglandins. Low-dose aspirin inhibits platelet thromboxane synthesis but spares prostacyclin production. Initially, extremely favorable metaanalyses, primarily of small series, led physicians to prescribe low-dose aspirin prophylactically, especially to "high-risk" patients (e.g., pregnant women with chronic hypertension, renal disease, multiple gestations, or preeclampsia in a previous pregnancy, in whom the incidence of preeclampsia may approach 20%). This optimism proved premature. Several large, randomized, placebo-controlled, multicenter trials that included > 20,000 patients have failed to reveal any effects of aspirin on the incidence of preeclampsia. Of greater importance, aspirin prophylaxis had no effect on a variety of maternal and fetal adverse outcomes (reviewed in ref. 22).

An even more recent approach, calcium supplementation throughout gestation, is based on observations that preeclamptic women are hypocalciuric and that calciotropic hormones may play a role in the pathogenesis of preeclampsia. As in the case of aspirin, initial reports were extremely encouraging, and metaanalyses of early studies noted a significant reduction in preeclampsia (reviewed in ref. 22). However, a recent large, blinded, randomized, placebo controlled trial that included > 4,500 women failed to detect any effects of calcium supplementation on the incidence of preeclampsia.[63]

Are other trials on the horizon? Whereas enthusiasm for fish oil and its ingredient eicosapentanoic acid (surrogates for aspirin in their effects on prostaglandin metabolism) appears to have waned, investigators continue to contemplate new trials, again based on plausible hypotheses. Examples include dietary supplementation of L-arginine, antioxidants, and magnesium. Such trials, however, are quite expensive when designed appropriately. It seems more prudent to focus resources on basic research designed to determine the cause of preeclampsia before committing funds to further trials.

In certain populations, however, prophylactic strategies may improve outcomes. Some patients with a history of early-onset preeclampsia (≤ gestational week 34) harbor metabolic abnormalities or risk factors associated with vascular thrombosis. Screening such patients may reveal hyperhomocysteinemia, anticardiolipin antibodies, lupus anticoagulants, activated protein C resistance (factor V Leiden), or protein S deficiency.[64] The presence of any of these abnormalities suggests that a preeclampsia-like syndrome will recur in subsequent gestations and is an indication to consider prophylactic treatment with heparin and/or low-dose aspirin. Patients with hyperhomocysteinemia also may benefit from high-dose pyridoxine and folate.

MANAGEMENT

Because of its explosive nature, suspicion of preeclampsia is sufficient grounds to recommend hospitalization. Exceptionally and under carefully supervised circumstances, a well-informed patient may be managed by resting at home, and the utility of home nurse visits, as well as electronic fetal monitoring services, is under investigation. These aggressive approaches diminish the incidence of convulsions, minimize diagnostic error, and improve fetal outcome. Near term, induction of labor is the treatment of choice, but earlier in pregnancy one may carefully attempt temporization. Delivery is indicated, regardless of gestational age, if severe hypertension cannot be controlled within 24–48 hours or if the physician detects the ominous signs listed in Table 1 or signs of fetal jeopardy.

"Aggressive" vs. "Conservative" Management for Preeclampsia Remote from Term

Earlier studies,[65] as well as contemporaneous texts stress that preeclampsia remote from term warrants termination of pregnancy, due to the low rate of successful fetal outcomes, and more importantly, substantial risk to the mother's well-being. More recently, several articles have questioned this practice, suggesting an expanded role for "conservative" management with severe disease.[66–70] Of note, in many of the articles the appeal for a conservative approach appears more apparent than real. One group,[66,67,69] for instance, would temporize as long as blood pressure could be reasonably controlled, and in the absence of ominous signs and symptoms such as headache, blurred vision, epigastric pain, or laboratory evidence of coagulation or liver abnormalities, or indications of fetal jeopardy; precisely what we (see above) and others have recommended for decades when suggesting termination for severe disease. In this respect there is a recent trial,[66] where the investigators assigned patients for pregnancy termination or conservative management after labeling the disease "severe" based on a traditional classification pattern (e.g., level of blood pressure and degree of proteinuria), rather than by evolution of the clinical picture over a defined period of time. The positive results in the "conservative" arm are a cogent demonstration as well as a reminder that when managing women with hypertension in pregnancy we do not treat the diagnosis but evolving disease. However, there is another group recommending conservative management whose policies we disagree with.

Visser et al.[70] continue temporization even when patients develop early indications of the HELLP syndrome. Scrutiny of their results reveals several important points: (1) the study appears to have been inadequately controlled; (2) fetal loss was substantial, even though pregnancies up to gestational week 34 were included; and (3) the number of maternal bleeding episodes increased. Such increments in maternal complications appear too risky, and it is hoped that clinicians will continue to be vigilant in aggressively terminating gestations complicated by early-onset preeclampsia, using the clinical criteria noted above and by Sibai et al.,[66] who paradoxically entitled their criteria "conservative management." Finally, we note that all of the cited studies took place in tertiary care centers with maternal-fetal intensive care units—facilities that are not readily available to most practitioners.

Accelerated Hypertension and Imminent or Frank Eclampsia

Antihypertensive therapy is discussed in more detail below. When blood pressure rises rapidly near term or during delivery (especially with preeclampsia, whether pure or superimposed), parenteral therapy may be required (Table 3). The blood pressure level at which therapy should be initiated has been debated, but in 1990 a working group of NHBPEP indicated that diastolic levels ≥ 105 mmHg should be treated (although some current texts list the figure as ≥ 110 mmHg).[21] In certain circumstances, however, prompt treatment should be undertaken at lower levels. Examples include teen-age gravidas whose recent diastolic levels were ≤ 70 mmHg and patients with cerebral signs such as an excruciating headache, confusion, or somnolence.[71]

Previously the prevention and management of eclamptic convulsions were debated, at times acrimoniously.[21,22] American obstetricians had long considered parenteral magnesium sulfate the drug of choice because of its preventive and therapeutic value, whereas neurologists and internists condemned its efficacy and, along with European consultants, suggested diazepam or phenytoin. Some authorities even questioned whether any therapy other than blood pressure control alone was necessary for prophylaxis.[49,72] Two studies reported in 1995 radically altered such views. A randomized trial[73] of 1,680 eclamptic women demonstrated that intravenous magnesium sulfate was superior to both diazepam and phenytoin in preventing recurrent seizures. The second study,[74] scheduled to compare *prophylaxis* with magnesium sulfate and phenytoin in 4,500 patients, was stopped after only 2,123 women had been randomized because it appeared that seizures were occurring only in the phenytoin-treated group. Thus, one may conclude that magnesium sulfate is the drug of choice once a convulsion appears and that it is superior to phenytoin in preventing eclampsia. Neither trial, however, answered the question whether prophylactic therapy is necessary or whether blood pressure control alone will prevent preeclampsia.[75,76] Of note, in the prophylactic trial blood pressure was controlled

TABLE 3. Drug Therapy for Acute and Severe Hypertension in Pregnancy*

Drug	Dose and Route	Onset of Action	Adverse Effects†	Comments
Hydralazine (C)‡	5 mg IV or IM, then 5–10 mg every 20–40 min; or constant infusion of 0.5–10 mg/hr	IV: 10 min IM: 10–30 min	Headache, flushing, tachycardia, nausea, vomiting; possible increase in ventricular arrhythmia[110]	Drug of choice according to Working Group;[21] broad experience of safety and efficacy
Labetalol (C)	20 mg IV, then 20–80 mg every 20–30 min, up to 300 mg; or constant infusion of 1–2 mg/min to desired effect, then stop or reduce to 0.5 mg/min	5–10 min	Flushing, nausea, vomiting, tingling of scalp	Experience in pregnancy considerably less than with hydralazine
Nifedipine (C)	5–10 mg PO; repeat in 30 min if necessary, then 10–20 mg PO every 3–6 hr	10–15 min	Flushing, headache, tachycardia, nausea, inhibition of labor	May have synergistic interaction with magnesium sulfate; limited experience in pregnancy
Diazoxide (C)	30–50 mg IV every 5–15 min	2–5 min	Inhibition of labor; hyperglycemia, fluid retention with repeated doses	Doses of 150–300 mg may cause severe hypotension; may displace phenytoin from serum protein binding sites
Relatively contraindicated				
Sodium nitroprusside (C)§	0.5–10 μg/kg/min by constant IV infusion	Instantaneous	Cyanide toxicity, nausea, vomiting	Use only in critical care unit at low doses for briefest time feasible; may cause fetal cyanide toxicity

IV = intravenously, IM = intramuscularly, PO = orally.

* Indicated for acute elevation of Korotkoff phase V blood pressure > 105 mmHg; goal is gradual reduction to 90–100 mmHg.

† All agents may cause marked hypotension, especially in severe preeclampsia.

‡ (C) refers to Food and Drug Administration risk category.

§ We classify sodium nitroprusside in category D.

Modified from Barron WM: Hypertension. In Barron WM, Lindheimer MD (eds): Medical Disorders During Pregnancy, 2nd ed. St. Louis, Mosby, 1994, with permission.

similarly in each group, but only the women receiving phenytoin convulsed. One therefore may hypothesize that phenytoin, the quintessential anticonvulsant, is a central nervous system irritant in preeclamptic women.

Until new data are published, we continue to use magnesium sulfate both to prevent and to treat eclampsia. A loading dose of 4–6 gm (infused over 10 min; do not administer as a bolus) is followed by infusion of a sustaining solution (24 gm of magnesium sulfate in 1 liter of 5% dextrose in water) delivered at a rate of 1–2 gm/hr. The aim is to maintain plasma levels at 5–9 mg/dl (2.1–3.5 mM/L). Therapy should continue for 12–24 hours into the puerperium because one-third of patients with eclampsia have convulsions after childbirth.

Chronic Hypertension

Because most women with chronic hypertension during pregnancy have the essential variety, usually mild in nature, over 85% of their gestations are uncomplicated, and outcomes approach those of normotensive women. In a minority of cases, however, morbidity is increased, including abruption, acute renal failure, cardiac decompensation, and cerebral accidents in the mother, as well as growth retardation in the fetus. As noted, many of these poor outcomes are associated with superimposed preeclampsia. In addition, much of the morbidity seems to occur in gravidas over 30 years of age with poorly controlled blood pressure, sometimes of several years' duration, especially those with evidence of end-organ damage. Extremely obese women with chronic hypertension are at special risk for cardiac decompensation near term, especially if volume overloading occurs in labor. Such patients warrant an echocardiographic examination during pregnancy to evaluate ventricular compliance and performance.

Our approach to treatment of chronic hypertension follows the recommendations of the NHBPEP Working Group on Hypertension in Pregnancy, which have been incorporated into the fifth and sixth reports of the Joint National Committee for the Detection and Treatment of Hypertension.[77,78] These recommendations are summarized in Table 4. Antihypertensive medications prescribed to gravidas are reviewed in the final section. One should be aware, however, that few (if any) large, randomized, multicenter trials of safety and efficacy of antihypertensive medications in pregnant women have been performed. Currently, antihypertensive treatment is indicated when diastolic levels are ≥ 100 mmHg; treatment is recommended at lower levels if risk factors are present. We do not recommend treatment of mild hypertension during pregnancy, although it is advocated by some authoritites, but we stress the need for further studies of this important question.

ANTIHYPERTENSIVE DRUG THERAPY IN PREGNANCY

A physician contemplating the prescription of antihypertensive drugs to pregnant women must take into account the inadequacy of available research. Few large, multicenter, randomized trials have been performed with the sophistication required of drug studies in the late 1990s. Most investigations have been limited in scope, frequently performed at the request and with the support of the pharmacologic industry. Most published data reflect studies in which drug therapy was begun after midgestation, a time when virtually all risks of provoking congenital malformations have passed. In addition, there are no rigorous animal testing requirements, which must be developed and met before human trials are initiated. Examples include standardized means of evaluating effects of the drug on fetal ability to withstand hypoxic stress and a more complete analysis of morphologic and physiologic variables in neonates. The latter issue is highlighted by reports that administration of aminoglycosides to pregnant rats has harmful

TABLE 4. Drug Therapy for Chronic Hypertension in Pregnancy*†

Drug (FDA Class‡)	Daily Dose	Adverse Effects and Comments
Agent of choice		
Methyldopa (C)	500–3000 mg preferably in two divided doses	Drug of choice according to Working Group;[21] safety for mother and fetus (after 1st trimester) is well documented
Second-line agents		
Hydralazine (C)	50–300 mg in 2–4 divided doses	Few controlled trials but extensive experience with few serious adverse effects documented; several reports of neonatal thrombocytopenia
Labetalol (C)	200–1200 mg in 2–3 divided doses	Less experience than with methyldopa; efficacy and short-term safety appear equal to methyldopa
Beta-adrenergic inhibitors (C)	Dependent on specific agent used	May cause fetal bradycardia and impair fetal responses to hypoxia; risk of intrauterine growth retardation when begun in 1st or 2nd trimester[99]
Nifedipine (C)	30–120 mg of a slow-release preparation	Limited data; may inhibit labor; may have synergistic action with magnesium sulfate
Clonidine (C)	0.1–0.8 mg in 2 divided doses	Limited data
Prazosin (C)	1–30 mg in 2–3 divided doses	Limited data
Thiazide diuretics (C by manufacturer; D designation in reference 136)	Dependent on agent used	Most controlled studies in normotensive gravidas; few data in hypertensive gestation; implicated in volume depletion, electrolyte imbalance, pancreatitis, and thrombocytopenia
Contraindicated		
ACE inhibitors (D)§	Dependent on agent used	High rates of fetal loss in animals; in humans associated with many cases of oligohydramnios, fetopathy, intrauterine growth retardation, neonatal anuric renal failure, occasionally fatal; use only when absolutely necessary to preserve maternal well-being and when other agents are unsuccessful

ACE = angiotensin-converting enzyme.

* Note that safety during the first trimester has not been established for any antihypertensive agent.

† Drug therapy indicated for uncomplicated chronic hypertension when phase V diastolic blood pressure ≥ 100 mmHg. Treatment at lower levels may be indicated for patients with renal disease or diabetes mellitus.

‡ Food and Drug Administration Classification (see Table 5).

§ We classify ACE inhibitors in category X.

Modified from Barron WM: Hypertension. In Barron WM, Lindheimer MD (eds): Medical Disorders During Pregnancy, 2nd ed. St. Louis, Mosby, 1995, with permission.

tubular and glomerular effects on the fetal kidney, demonstrable only by sophisticated functional, histologic, and biochemical evaluations.[79] With these facts in mind, we review the current status of antihypertensive drug therapy during pregnancy in 1998.

Tables 3 and 4 summarize the major antihypertensive medications prescribed to pregnant women, and Table 5 summarizes the risk classification defined by the Food and Drug Administration (FDA), which is usually included with the drug's description in the Physician's Desk Reference (PDR). Few

TABLE 5. Food and Drug Administration Classification of Risk in Pregnancy

Category A
Controlled studies in women fail to demonstrate a risk to the fetus in the first trimester, and there is no evidence of risk in later trimesters. The possibility of fetal harm appears remote.

Category B
Either animal reproduction studies have not demonstrated a fetal risk but no controlled studies have been performed in pregnant women, or animal reproducton studies have shown an adverse effect (other than a decrease in fertility) that was not confirmed in controlled studies in women in the first trimester. There is no evidence of risk in later trimesters.

Category C
Either studies in animals have revealed adverse effects on the fetus (teratogenic or embryocidal effects or other) and there are no controlled studies in women, or studies in women and animals are not available. Drugs should be given only if the potential benefit justifies the potential risk to the fetus.

Category D
There is positive evidence of human fetal risk, but the benefits of use in pregnant women may be acceptable despite the risk (e.g., if the drug is needed in a life-threatening situation or for a serious disease for which safer drugs cannot be used or are ineffective). There is an appropriate statement in the "warnings" section of the labeling.

Category X
Studies in animals or humans have demonstrated fetal abnormalities, or there is evidence of fetal risk based on human experience, or both, and the risk of the use of the drug in pregnant women clearly outweighs any possible benefit. The drug is contraindicated in women who are or may become pregnant. There is an appropriate statement in the "contraindications" section of the labeling.

---

drugs are listed in category A (demonstrated safety in humans), highlighting the paucity of information in the field. Many physicians overuse drugs listed in category C (no human studies available), interpreting this classification as evidence of safety. The danger of such an approach is exemplified by the use of angiotensin-converting enzyme (ACE) inhibitors in pregnant women. Captopril and enalapril were originally listed in category C in the PDR. Despite reports of problems in animal models in the early 1980s, both drugs remained in category C until 1992. By then evidence that ACE inhibitors were associated with renal failure and death in the neonate had mounted. The manufacturer was required to include a "black box" label warning, and the drugs were changed to category D in the PDR. We had considered both drugs as category X (contraindicated in pregnant women) since 1985 and still do so.[80] Thus, we reiterate *drugs in FDA category C are not to be considered "safe." Rather such a classification should be interpreted as a warning that unknown adverse fetal effects may be present.*

Central Adrenergic Inhibitors

Methyldopa remains the drug of choice to initiate therapy for hypertension during gestation.[21] This status is based on its long history of effective use in gravidas as well as its prospective evaluation in several randomized trials, one of which included periodic evaluation of the offspring for 7.5 years.[81–85] Obstetricians are frequently criticized for using methyldopa on the grounds that it is less effective and more poorly tolerated than a host of newer drugs, many of which can be prescribed once daily. However, all studies comparing methyldopa with other agents have failed to establish their superiority, and the well-documented side-effect profile of methyldopa indicates that maternal side effects are generally mild and well tolerated.[86–88]

Clonidine is another $alpha_2$ agonist that has been prescribed to pregnant women,[89,90] but total experience with clonidine is quite limited compared with methyldopa. Of interest is a randomized trial that reported similar efficacy and safety for the two medications.[89] In a limited study of hospitalized patients, clonidine—alone or in combination with a placebo or hydralazine—was said to reduce the incidence of premature delivery.[90] Clonidine, however, should be avoided in early gestation because it is suspected of causing embryopathy (PDR, 1997). A controlled follow-up study of 22 neonates reported more problems with sleep disturbance in the clonidine-treated group.[91]

Vasodilators

Parenteral hydralazine is the drug of choice to treat acute severe hypertension, especially near term.[21] In the management of chronic hypertension, however, its use is primarily as a second line agent in conjunction with methyldopa or beta-adrenergic blockers (although some physicians now use labetalol or calcium antagonists in this role; see below). Data about the use of hydralazine as monotherapy are limited, although in one small study it was associated with increased frequencies of palpitations, dizziness, and headache.[92]

Beta-adrenergic Receptor Blockers

The safety and precise indications for prescribing beta blockers in pregnant women remain unclear despite an extensive literature[87,93–102] because of the heterogeneous nature of the studies. In many studies the drug was started rather late in gestation, which makes it hard to evaluate several concerns about its use, especially whether it causes fetal growth retardation.

The major concerns about beta-adrenergic blocking agents relate to the fetus. The drugs may alter the cardiovascular responses to hypoxia and cause bradycardia, respiratory depression, hypoglycemia, and growth retardation. Most of these concerns were due to animal studies and anecdotal reports in humans; only growth retardation seems to have been documented adequately.[99] In fact, most of the larger studies in humans have failed to detect adverse fetal effects.[93,94,97,103–105] Thus, beta-adrenergic blockers should not be used for long-term therapy of hypertension in pregnancy, unless methyldopa and hydralazine have failed to control blood pressure or cannot be administered to a given patient for some other reason.

Combined Alpha- and Beta-adrenergic Receptor Blockers

Labetalol is the drug within this group most extensively prescribed to pregnant women.[102,106–109] The parenteral form is used to treat severe hypertension; some believe that it controls pressure more efficiently and causes fewer side effects, such as headaches, flushing, and cardiac arrhythmias, than parenteral hydralazine.[102,110] Other studies find the two drugs similar, and hydralazine remains the drug of choice, following the dictum that the medication with the longest uneventful history in pregnancy should prove the safest.

The data about use of labetalol to treat chronic hypertension in pregnancy, limited to a few, albeit well-designed, randomized trials, have failed to establish its superiority over methyldopa,[107–109] and thorough follow-up studies of treated infants are yet to appear. Labetalol may be hepatotoxic in nonpregnant patients, and during pregnancy the liver may become even more vulnerable. Thus, we use labetalol only as a second-line agent.

Alpha-adrenergic Blockers

Data about the use of alpha-adrenergic blockers in pregnancy relate primarily to prazosin and are purely anecdotal and uncontrolled. Thus alpha-adrenergic blockers should not be prescribed to pregnant women.[111–113] The single exception is the rare case of pheochromocytoma, in which both prazosin and phenoxybenzamine have been used.[28–30,114]

Calcium Antagonists

Calcium antagonists, especially nifedipine, have been used to treat both acute and chronic hypertension in pregnancy (approximately 40 articles are reviewed in references 102 and 115–117). On theoretical grounds, calcium antagonists, which are direct muscle relaxants, should be useful in diseases such as preeclampsia, in which vascular constriction is severe. Fenakel et al.[118] found nifedipine significantly superior to hydralazine in a small group of hospitalized patients. In another trial, 100 mildly preeclamptic women treated with nifedipine had lower blood pressures than a similar number of patients allocated to bed rest alone.[119] Given the limited size of the study, however, outcomes were similar in both groups.

The remainder of the literature is ancedotal. When used to treat severe hypertension, the oral short-acting agents may provoke precipitous decreases in blood pressure,[120,121] especially when magnesium sulfate is infused. Magnesium sulfate also interferes with calcium-dependent excitation-contraction coupling, and case reports of severe hypotension and neuromuscular blockade have appeared when both drugs were used.[122,123] Thus, currently we do not use calcium antagonists when blood pressure rises suddenly near term.

Data about calcium antagonists and chronic hypertension in pregnancy are also limited.[102,115–117,124–126] We hesitate to use calcium antagonists in this setting because of the high incidence of superimposed preeclampsia, which may result in the precipitous need to start magnesium sulfate in a patient receiving long-acting calcium antagonists. Nimodipine, which is used by neurologists for treating patients with cerebral bleeding, is under study as a therapeutic agent in women with eclampsia.[127]

Angiotensin-converting Enzyme and Receptor Inhibitors

ACE inhibitors were discussed above as an example of why drugs originally listed as FDA category C should be viewed with caution. They have been linked to fetopathy and perhaps congenital anomalies and are associated with oliguria-anuria and fetal demise.[128–134] Since 1992, ACE inhibitors have been classified in category D (by us in category X), and the label includes a warning.

The manufacturers' approach to the newer angiotensin receptor inhibitors is similar. The policy is prudent despite the lack of supporting data.

Diuretics

In many trials diuretics have been prescribed to prevent preeclampsia (see above).[62] However, there are few data about their use in pregnant women with hypertension. In one small study, diuretics prevented most of the physiologic volume expansion associated with normal pregnancy.[135] Suboptimal plasma volume expansion in women with chronic hypertension has been associated with impaired fetal growth, but the data are far from persuasive.[10,136] Another problem with the use of diuretics is hyperuricemia, which may lead to a mistaken diagnosis of superimposed preeclampsia. Nevertheless, the NHBPEP Working Group,[21] recognizing the role of diuretics in populations with salt-sensitive blood pressure, especially those who develop rapid refractoriness to vasodilator therapy alone, noted that diuretics could be continued during gestation if prescribed before conception. However, they discouraged their use in women with preeclampsia. Thiazide diuretics seem to have a very safe record in pregnancy.[62] The more potent loop diuretics should be avoided, and furosemide is also embryotoxic (PDR, 1997).

Breastfeeding

Although many antihypertensive drugs can be detected in breast milk, sometimes in sufficient quantity to affect neonatal blood pressure, relevant data are unfortunately scarce. The reader is referred to the monograph by Briggs, Freeman and Yaffe,[137] recommendations of the Committee on Drugs of the American Academy of Pediatrics,[138] and brief reviews by White[139] and Bailey and Ito.[140] Many beta-blocking agents concentrate in breast milk in quantities sufficient to cause symptoms in the neonate, but propranolol seems to be an exception. On the other hand, the concentrations of ACE inhibitors, calcium antagonists, and diuretics are quite low, and mothers ingesting these agents can breast feed without concerns.[135–137,141]

References

1. MacGillivray I, Rose GA, Rowe B: Blood pressure survey in pregnancy. Clin Sci 37:395–407, 1969.
2. Friedman EA, Neff RK: Pregnancy and Hypertension. A Systematic Evaluation of Clinical Diagnostic Criteria. Littleton, MA, PSG Publishing, 1977.
3. Page EW, Christianson R: The impact of mean arterial pressure in the middle trimester upon the outcome of pregnancy. Am J Obstet Gynecol 125:740–746, 1976.
4. Wilson M, Morganti AA, Zervoudakis I, et al: Blood pressure, the renin-aldosterone system and sex steroids throughout normal pregnancy. Am J Med 68:97–104, 1980.
5. Capeless EL, Clapp JF: Cardiovascular changes in early phase of pregnancy. Am J Obstet Gynecol 161:1449–1453, 1989.
6. Walters BWJ, Thompson ME, Lee A, de Swiet M: Blood pressure in the puerperium. Clin Sci 71:589–594, 1986.
7. Robson SC, Hunter S, Boys RJ, Dunlop W: Serial study of factors influencing changes in cardiac output during human pregnancy. Am J Physiol 256:H1060–1065, 1989.
8. Poppas A, Shroff SG, Korcarz CE, et al: Serial assessment of the cardiovascular system in normal pregnancy. Circulation 95:2407–2415, 1997.

9. Rubin PC (ed): Handbook of Hypertension. Vol 10: Hypertension in Pregnancy. Amsterdam, Elsevier Science, 1988.
10. Lindheimer MD, Katz AI: Renal physiology and disease in pregnancy. In Seldin DW, Giebisch G (eds): The Kidney: Physiology and Pathophysiology, 2nd ed. New York, Raven Press, 1992, pp 3371–3431.
11. August P, Lindheimer MD. Pathophysiology of preeclampsia. In Laragh JH, Brenner BM (eds): Hypertension: Pathophysiology, Diagnosis, and Management, 2nd ed. New York, Raven Press, 1995, pp 2407–2426.
12. Fitzgerald DJ, Fitzgerald GA: Eicosanoids in the pathogenesis of preeclampsia. In Laragh JH, Brenner BM (eds): Hypertension: Pathophysiology and Management. New York, Raven Press, 1990, pp 1789–1804.
13. McGiff JG, Carrol MA: Magnesium, platelets and the vasculature in preeclampsia-eclampsia. Hypertens Pregnancy 13:217–226, 1994.
14. Sladek SM, Magness RR, Conrad KP: Nitric oxide and pregnancy. Am J Physiol 272:R441–463, 1997.
15. Johenning AR, Barron WM: Indirect blood pressure measurement in pregnancy: Korotkoff phase 4 vs phase 5. Am J Obstet Gynecol 167:573–580, 1992.
16. Blank SG, Helseth G, Pickering TG, et al: How should diastolic blood pressure be defined during pregnancy? Hypertension 24:234–240, 1994.
17. Shennan A, Gupta M, Halligan A, et al: Lack of reproducibility in pregnancy of Korotkoff phase 4 as measured by mercury sphygmomanometry. Lancet 347:139–142, 1996.
18. de Swiet M, Shennan A: Blood pressure measurement in pregnancy. Br J Obstet Gynaecol 103:862–863, 1996.
19. Halligan A, Shennan A, Thurston H, et al: Ambulatory blood pressure measurement in pregnancy: The current state of the art. Hypertens Pregnancy 14:1–16,1995.
20. Lindheimer MD, Katz AI: The normal and diseased kidney in pregnancy. In Schrier RW, Gottshalk CW(eds): Diseases of the Kidney, 6th ed. Boston, Little, Brown, 1977, pp 2063–2097.
21. National High Blood Pressure Education Program Working Group: Report on high blood pressure in pregnancy. Am J Obstet Gynecol 163:1689–1712, 1990.
22. Lindheimer MD: Preeclampsia-eclampsia 1996: Preventable? Have disputes on its treatment been resolved? Curr Opin Nephrol Hypertens 5:452–458, 1996.
23. Chesley LC: Hypertensive Disorders in Pregnancy. New York, Appleton-Century-Crofts, 1978.
24. Lamming GD, Symonds EM, Rubin PC: Pheochromocytoma in pregnancy: Still a cause of maternal death. Clin Exp Hypertens B:57–68, 1990.
25. Aron DC, Schnall AM, Sheeler LR: Cushings syndrome and pregnancy [see comments]. Am J Obstet Gynecol 162:244–252, 1990.
26. Ballou SP, Morley JJ, Kushner I: Pregnancy and systemic sclerosis. Arthritis Rheum 27:295–298, 1984.
27. Burkett G, Richards R: Periarteritis nodosa and pregnancy. Obstet Gynecol 59:252–254, 1982.
28. Molitch ME: Pituitary, thyroid, adrenal, and parathyroid disorders. In Barron WM, Lindheimer MD (eds): Medical Disorders During Pregnancy, 2nd ed. St. Louis, Mosby, 1995, pp 89–127.
29. Lyons CW, Colmorgen GH: Medical management of pheochromocytoma in pregnancy. Obstet Gynecol 72:450–451, 1988.
30. Greenberg M, Moawad AH, Weties BM, et al: Detection of an extraadrenal pheochromocytoma by magnetic resonance imaging during mid pregnancy. Radiology 161:475–476, 1986.
31. Easterling TR, Brateng D, Goldman ML, et al: Renal vascular hypertension during pregnancy. Obstet Gynecol 78:921–925, 1991.
32. Diego J, Guerra J, Pham C, Epstein M: Management of renovascular hypertension complicating pregnancy [abstract]. J Am Soc Nephrol 7:1549, 1996.
33. Lindheimer MD, Richardson DA, Ehrlich EM, Katz AI: Potassium homeostasis in pregnancy. J Reprod Med 32:517–532, 1987.
34. Sibai BM: Eclampsia. In Rubin PC (ed): Hand Book of Hypertension. Vol 10: Hypertension in Pregnancy. Amsterdam, Elsevier, 1988, pp 320–340.
35. Sibai BM, Justermann L, Velasco J: Current understanding of severe pre-eclampsia, pregnancy-associated haemolytic uremic syndrome, thrombotic thrombocytopenic purpura, haemolysis, elevated liver enzymes and low platelet syndrome, and postpartum acute renal failure: Different clinical syndromes or just different names. Curr Opin Nephrol Hypertens 4:346–355, 1994.
36. Fisher KA, Luger A, Spargo BH, Lindheimer MD: Hypertension in pregnancy: Clinical-pathological correlations and remote prognosis. Medicine 60:267–276, 1981.
37. Tawers CV, Picron RA, Nageotte MP, et al: Cocaine intoxication presenting as preeclampsia and eclampsia. Obstet Gynecol 81:545–547, 1993.

38. Roberts JM, Redman CWG: Pre-eclampsia: More than pregnancy-induced hypertension. Lancet 341:1447–1451, 1993.
39. Visser W, Wallenberg HCS: Central hemodynamic observation in untreated preeclamptic patients. Hypertension 17:1072–1077, 1991.
40. Conrad KP, Benyo DF: Placental cytokines and pathogenesis of preeclampsia. Am J Reprod Immunol 37:240– 249, 1997.
41. Ness RB, Roberts JM: Heterogenous causes constituting the single syndrome of preeclampsia. A hypothesis and its implications. Am J Obstet Gynecol 175:1365–1370, 1996.
42. Sheehan H, Lynch JB: Pathology of Toxemia of Pregnancy. London, Churchill, 1973.
43. Gaber L, Spargo BH, Lindheimer MD: Renal pathology in preeclampsia. Clin Obstet Gynaecol (Bailliére) 8:443–468, 1994.
44. Taufield PA, Ales K, Resnik L, et al: Hypocalciuria in preeclampsia. N Engl J Med 316:715–718, 1987.
45. Lindheimer MD, Katz AI: Renal Function and Disease in Pregnancy. Philadelphia, Lea & Febiger, 1977.
46. Sibai BM, McCubbin JH, Anderson GD, et al: Eclampsia: Observations from 67 recent cases. Obstet Gynecol 58:609–613, 1981.
47. Oian P, Maltau JM, Noddeland H, Fadnes HO: Transcapillary fluid balance in pre-eclampsia. Br J Obstet Gynaecol 93:235-239, 1986.
48. Mckay DG, Merrill SJ, Weiner AE, et al: The pathologic anatomy of eclampsia, bilateral renal cortical necrosis, pituitary necrosis, and other acute fatal complications of pregnancy, and its possible relationship to the generalized Shwartzman phenomenon. Am J Obstet Gynecol 66:507–539, 1953.
49. Donaldson JO: Neurology of Pregnancy, 2nd ed. Philadelphia, W.B. Saunders, 1989.
50. Belfort MA, Moise KJ: Effect of magnesium sulfate on maternal brain flow in preeclampsia. A randomized, placebo-controlled study. Am J Obstet Gynecol 167:661–666, 1992.
51. van den Veyver IB, Belfort MA, Rowe TF, Moise KJ: Cerebral vasospasm in eclampsia: Transcranial doppler ultrasound findings. J Matern Fet Med 3:9–13, 1994.
52. Dahmus MA, Barton JR, Sibai NM: Cerebral imaging in eclampsia: Magnetic resonance versus computed tomography. Am J Obstet Gynecol 167:935–941, 1993.
53. Moodley S, Bobat SM, Hoffman J, Bill PLA: Electroencephalogram and computed cerebral tomography findings in eclampsia. Br J Obstet Gynaecol 100:984–988, 1993.
54. Digre KB, Varner MW, Osborn AG, Crawford S: Cranial magnetic resonance imaging in severe preeclampsia vs eclampsia. Arch Neurol 50:399–406, 1993.
55. Cunningham FG, Macdonald PC, Gant NF, et al: Williams Obstetrics, 19th ed. Norwalk, CT, Appleton & Lange, 1993, pp 782–783.
56. Robertson WB, Brosens I, Dixon G: Maternal uterine vascular lesions in the hypertensive complications of pregnancy. In Lindheimer MD, Katz AI, Zuspan FP (eds): Hypertension in Pregnancy. New York, John Wiley, 1976, pp 115–129.
57. Fox H: The placenta in pregnancy hypertension. In Rubin PC (ed): Handbook of Hypertension. Vol 10: Hypertension in Pregnancy. New York, Elsevier Science, pp 16–37, 1988.
58. Khong TY, De Wolf F, Robertson WB, Brosens I: Inadequate maternal vascular response to placentation in pregnancies complicated by pre-eclampsia and by small for gestational age infants. Br J Obstet Gynaecol, 93:1049-1059, 1986.
59. Pijnenborg R, Anthony J, Davey DA, et al: Placental bed spiral arteries in the hypertensive disorders of pregnancy. Br J Obstet Gynaecol 98:648–655, 1991.
60. Zhou Y, Damsky CH, Chiu K, et al: Preeclampsia is associated with abnormal expression of adhesion molecules by invasive cytotrophoblasts. J Clin Invest 9:950–960, 1993.
61. Zhou Y, Damsky CH, Fisher SJ: Preeclampsia is associated with failure of human cytotrophoblasts to mimic a vascular adhesion phenotype. J Clin Invest 99:1–13, 1997.
62. Collins R, Yusuf S, Peto R: Overview of randomised trials of diuretics in pregnancy. BMJ 290:17–23, 1985.
63. Levine JR, Hauth JC, Curet LB, et al: Trial of calcium for prevention of preeclampsia. N Engl J Med 337:69–76, 1997.
64. Dekker GA, de Vries JI, Doelitzsch PM, et al: Underlying disorders associated with severe early-onset preeclampsia. Am J Obstet Gynecol 173:1042–1048, 1995.
65. Sibai BM, Taslimi M, Abdella TN, et al: Maternal and perinatal outcome of conservative management of severe preeclampsia in mid trimester. Am J Obstet Gynecol 156:1169–1173, 1987.
66. Sibai BM, Mercer BM, Schiff E, Friedman SA: Aggressive versus expectant management of severe preeclampsia at 28–32 weeks gestation: A randomized controlled trial. Am J Obstet Gynecol 171:818–822, 1994.

67. Schiff E, Friedman SA, Sibai BM: Conservative management of severe preeclampsia remote from term. Obstet Gynecol 84:626–630, 1994.
68. Visser W, van Pampus MG, Treffers PE, Wallenburg HCS: Perinatal results of hemodynamic and conservative temporizing treatment in severe preeclampsia. Eur J Obstet Gynecol Reprod Biol 53:175–181, 1994.
69. Chari R, Friedman SA, O'Brien JM, Sibai BM: Daily antenatal testing in women with severe preeclampsia. Am J Obstet Gynecol 173:1207–1210, 1995.
70. Visser W, Wallenburg HC: Temporizing management of severe pre-eclampsia with and without the HELLP syndrome. Br J Obstet Gynaecol 102:111–117, 1995.
71. Hinchey J, Chaves C, Appignani B, et al: A reversible posterior leukoencephalopathy syndrome. N Eng J Med 334:494–500, 1996.
72. Chua S, Redman CW: Are prophylactic anticonvulsants required in severe pre-eclampsia? Lancet 337:250–251, 1991.
73. Eclampsia Collaborative Group: Which anticonvulsant for women with eclampsia? Evidence from the Collaborative Eclampsia Trial. Lancet 345:1455–1463, 1995.
74. Lucas MJ, Leveno KJ, Cunningham FG: A comparison of magnesium sulfate with phenytoin for the prevention of eclampsia. N Eng J Med 333:201–205, 1995.
75. Moodley J, Moodley W: Prophylactic anticonvulsant therapy in the hypertensive crisis of pregnancy. The need for a large randomized trial. Hypertens Pregnancy 13:245–252, 1994.
76. Burrows RE, Burrows EA: The feasibility of a control population for randomized control trial of seizure prophylaxis in the hypertensive disorders of pregnancy. Am J Obstet Gynecol 173:929–935, 1995.
77. The Fifth Report of the Joint National Committee on Detection, Evaluation, and Treatment of High Blood Pressure. Arch Intern Med 153:154–183, 1993.
78. The Sixth Report of the Joint National Committee on Detection, Evaluation, and Treatment of High Blood Pressure. Arch Intern Med (in press). (Also published by the National Institutes of Health, Bethesda, MD).
79. Mallie JP, Coulon G, Billery C, et al: In utero aminoglycoside-induced nephrotoxicity in rat neonates. Kidney Int 33:36–44, 1988.
80. Lindheimer MD, Katz AI: Hypertension in pregnancy. N Eng J Med 313:675–680, 1985.
81. Redman CW: Fetal outcome in trial of antihypertensive treatment in pregnancy. Lancet 1:753–756, 1976.
82. Plouin PF, Breart G, Llado J, et al: A randomized comparison of early with conservative use of antihypertensive drugs in the management of pregnancy induced hypertension. Br J Obstet Gynaecol 97:134–141, 1990.
83. Kyle PM, Redman CWG: Comparative risk-benefit assessment of drugs used in the management of hypertension in pregnancy. Drug Safety 7:223–234, 1992.
84. Leather HM, Baker P, Humphreys DM, Chadd MA: A controlled trial of hypotensive agents in hypertension in pregnancy. Lancet 2:488–490, 1968.
85. Cockburn J, Moar VA, Ounsted M, Redman CW: Final report of study on hypertension during pregnancy. The effects of specific treatment on the growth and development of the children. Lancet 1:647–649, 1982.
86. Livingstone I, Craswell PW, Bevan EB, et al: Propranolol in pregnancy: Three-year prospective study. Clin Exp Hypertens B2:341–350, 1983.
87. Fidler J, Smith V, Fayers P, de Swiet M: Randomised controlled comparative study of methyldopa and oxprenolol in treatment of hypertension in pregnancy. BMJ 286:1927–1930, 1983.
88. Gallery EDM, Ross MR, Gyory AZ. Antihypertensive treatment in pregnancy: Analysis of different responses to oxprenolol and methyldopa. BMJ 291:563–566, 1985.
89. Horvath JS, Phippard A, Korda A, et al: Clonidine hydrochloride—A safe and effective antihypertensive agent in pregnancy. Obstet Gynecol 66:634–638, 1985.
90. Phippard AF, Fischer WE, Horvath JS, et al: Early blood pressure control improves pregnancy outcome in primigravid women with mild hypertension. Med J Aust 154:378–382, 1991.
91. Huisjes HJ, Hadders-Algra M, Touwen BCL: Is clonidine a behavorial teratogen in the human? Early Hum Dev 14:43–48, 1986.
92. Rosenfeld J, Bott-Kanner G, Boner G, et al: Treatment of hypertension during pregnancy with hydralazine monotherapy or with combined therapy with hydralazine and pindolol. Eur J Obstet Gynecol Reprod Biol 22:197–204, 1986.
93. Rubin PC, Butters L, Clark DM, et al: Placebo-controlled trial of atenolol in treatment of pregnancy-assocated hypertension. Lancet 1:431–434, 1983.
94. Bott-Kanner G, Hirsch M, Friedman S, et al: Antihypertensive therapy in the management of hypertension in pregnancy: A clinical double blind study of pindolol. Clin Exp Hypertens 11B:207–220, 1992.

95. Reynolds B, Butters L, Evans J, et al: First year of life after the use of atenolol in pregnancy-associated hypertension. Arch Dis Child 59:1061–1063, 1984.
96. Pruyn SC, Phelan JP, Buchanan GC: Long-term propranolol therapy in pregnancy: Maternal and fetal outcome. Am J Obstet Gynecol 135:485–489, 1979.
97. Wichman K, Ryden G, Karlberg B: A placebo controlled trial of metaprolol in the treatment of hypertension in pregnancy. Scand J Clin Lab Invest 44(169):90–95, 1984.
98. Gallery EDM, Saunders DM, Hunyor SN, Gyory AZ: Randomised comparison of methyldopa and oxprenolol for treatment of hypertension in pregnancy. BMJ 1:1591–1594, 1979.
99. Butters L, Kennedy S, Rubin PC: Atenolol in essential hypertension during pregnancy. BMJ 301:587-589, 1990.
100. Montan S, Ingemarsson I, Marsal K, Stoberg NO: Randomised controlled trial of atenolol and pindolol in human pregnancy: Effects on fetal hemodynamics. BMJ 304:946–949, 1992.
101. Marlettini MG, Crippa S, Morselli-Labate AM, et al: Randomised comparison of calcium antagonists and beta-blockers in the treatment of pregnancy induced hypertension. Curr Ther Res 48:684–694, 1990.
102. Sibai BM: Treatment of hypertension in pregnant women. N Eng J Med 335:257–265, 1996.
103. Hogstedt S, Lindeberg S, Axelsson O, et al: A prospective controlled trial of metoprolol-hydralazine treatment in hypertension during pregnancy. Acta Obstet Gynecol Scand 14:43–48, 1985.
104. Rubin PC, Butters L, Clark D, et al: Obstetric aspects of the use in pregnancy-associated hypertension of the beta-adrenoreceptor antagonist atenolol. Am J Obstet Gynecol 150:389–392, 1984.
105. Wichman K: Metoprolol in the treatment of mild to moderate hypertension in pregnancy: Effects on fetal heart activity. Clin Exp Hypertens Pregnancy B5:195–202, 1986.
106. Redman CWG: A controlled trial of the treatment of hypertension in pregnancy: Labetalol compared with methyldopa. In Riley A, Symonds EM (eds): The Investigation of Labetalol in the Management of Hypertension in Pregnancy. Amsterdam, Excerpta Medica, 1982, pp 101–110.
107. Michael CA, Potter JM: A comparison of labetalol with other antihypertensive drugs in the treatment of hypertensive disease of pregnancy. In Riley A, Symonds EM (eds). The Investigation of Labetalol in the Management of Hypertension in Pregnancy. Amsterdam, Excerpta Medica, 1982, pp 101–110.
108. Plouin PF, Breart G, Maillard F, et al: Comparison of antihypertensive efficacy and perinatal safety of labetalol and methyldopa in the treatment of hypertension in pregnancy: A randomised controlled trial. Br J Obstet Gynaecol 95:868–876, 1989.
109. Sibai BM, Mabie WC, Shamsa F, et al: A comparison of no medication versus methyldopa or labetalol in chronic hypertension during pregnancy. Am J Obstet Gynecol 162:960–967, 1990.
110. Bhorat IE, Naidoo DP, Rout CC, Moodley J: Malignant ventricular arrhythmias in eclampsia: A comparison of labetalol with hydralazine. Am J Obstet Gynecol 168:1292–1296, 1993.
111. Lube WF, Hodge JV: Combined alpha and beta-adrenoreceptor antagonism with prazosin and oxprenolol in control of severe hypertension in pregnancy. N Z Med J 94:169–172, 1981.
112. Rubin PC, Butters L, Low RA, Reid JL: Clinical pharmacological studies with prazosin during pregnancy complicated by hypertension. Br J Clin Pharmacol 16:543–547, 1983.
113. Dommisse J, Davey DA, Roos PJ: Prazosin and oxprenolol therapy in pregnancy hypertension. S Afr Med J 64:231–233, 1983.
114. Venuto R, Burstein P, Schneider R: Pheochromocytoma: Antepartum diagnosis and management with tumor resection in the puerperium. Am J Obstet Gynecol 150:431–432, 1984.
115. Childress CH, Katz VL: Nifedipine and its indications in obstetrics and gynecology. Obstet Gynecol 83:616–624, 1994.
116. Levin AC, Doering PL, Hatton RC: Use of nifedipine in the hypertensive diseases of pregnancy. Ann Pharmacol 28:1371–1378, 1994.
117. Lewis R, Sibai BM: The use of calcium-channel blockers in pregnancy. New Horizons 4:115–122, 1996.
118. Fenakel K, Fenakel G, Appelman Z, et al: Nifedipine in the treatment of severe preeclampsia. Obstet Gynecol 77:331–337, 1991.
119. Sibai BM, Barton JR, Akl S, et al: A randomized prospective comparison of nifedipine and bed rest versus bed rest alone in the management of preeclampsia remote from term. Am J Obstet Gynecol 167:879–884, 1992.
120. Impey L: Severe hypotension and fetal distress following sublingual administration of nifedipine to a patient with severe pregnancy induced hypertension at 33 weeks. Br J Obstet Gynaecol 100:959–961, 1993.

121. Waisman GD, Mayorga LM, Camera MI, et al: Magnesium plus nifedipine: Potentiation of hypotensive effect in preeclampsia? Am J Obstet Gynecol 159:308–309, 1988.
122. Snyder SW, Cardwell MS: Neuromuscular blockade with magnesium sulphate and nifedipine. Am J Obstet Gynecol 161:35–36, 1989.
123. Ben-Ami M, Giladi Y, Shalev E: The combination of magnesium sulphate and nifedipine: A cause of neuromuscular blockade. Br J Obstet Gynaecol 101:262–263, 1994.
124. Wide-Swensson DH, Ingemarsson I, Lunell NO, et al: Calcium channel blockade (isradipine) in treatment of hypertension in pregnancy: A randomized placebo-controlled study. Am J Obstet Gynecol 173:872–878, 1995.
125. Sibai BM: Diagnosis and management of chronic hypertension in pregnancy. Obstet Gynecol 78:451–461, 1991.
126. Magee LA, Schick B, Donnenfeld AE, et al: The safety of calcium channel blockers in human pregnancy: A prospective, multicenter cohort study. Am J Obstet Gynecol 174:823–828, 1996.
127. Belfort MA, Carpenter RJ Jr, Kirshon B, et al: The use of nimodipine in a patient with eclampsia: Color flow Doppler demonstration of retinal artery relaxation. Am J Obstet Gynecol 169:204–206, 1993.
128. Adverse drug reaction reporting systems. MMWR 46:240–242. 1997.
129. Rosa FW, Bosco LA, Graham CF, et al: Neonatal anuria with maternal angiotensin-converting enzyme inhibition. Obstet Gynecol 74:371–374, 1989.
130. Hanssens M, Keirse MJ, Vankelecom F, van Assche FA: Fetal and neonatal effects of treatment with angiotensin-converting enzyme inhibitors in pregnancy. Obstet Gynecol 78:128–135, 1991.
131. Piper JM, Ray WA, Rosa FW: Pregnancy outcome following exposure to angiotensin-converting enzyme inhibitors. Obstet Gynecol 80:429–432, 1992.
132. Barr M Jr, Cohen MM Jr: ACE-inhibitor fetopathy and hypocalvaria: The kidney–skull connection. Teratology 44:485–495, 1991.
133. Pryde PG, Sedman AB, Nugent CE, Barr M Jr: Angiotensin-converting enzyme inhibitor fetopathy. J Am Soc Nephrol 3:1575–1582, 1993.
134. Brent RL, Beckman DA: Angiotensin-converting enzyme inhibitors, an embryopathic class of drugs with unique properties: Information for clinical teratology counselors. Teratology 43:543–546, 1991.
135. Sibai BM, Grossman RA, Grossman HG: Effects of diuretics on plasma volume in pregnancies with long-term hypertension. Am J Obstet Gynecol 150:831–835, 1984.
136. Gallery EDM, Hunyor SN, Gyory AZ: Plasma volume contraction. A significant factor in both pregnancy-associated hypertension (pre-eclampsia) and chronic hypertension in pregnancy. Q J Med 192:593–602, 1979.
137. Briggs GG, Freeman RK, Yaffe SJ: A Reference Guide to Fetal and Neonatal Risk: Drugs in Pregnancy and Lactation, 4th ed. Baltimore, Williams & Wilkins, 1994.
138. Committee on Drugs of the American Academy of Pediatrics: The transfer of drugs and other chemicals into human breast milk. Pediatrics 84:924–936, 1989.
139. White W: Management of hypertension during lactation. Hypertension 6:297–300, 1984.
140. Bailey B, Ito S: Breast-feeding and maternal drug use. Pediatr Clin North Am 44:41–54, 1997.
141. Redman CWG, Kelly JG, Cooper WD: The excretion of enalapril and enalaprilat in human breast milk. Eur J Clin Pharmacol 38:99, 1990.

JERROLD H. LEVY M.D. / CATHERINE HURAUX M.D.
MARGARETA NORDLANDER Ph.D.

19

Treatment of Perioperative Hypertension

The treatment of perioperative hypertension differs considerably from that of chronic hypertension. During surgery acute changes in blood pressure are due to multiple factors, including rapid intravenous volume shifts and changes in sympathetic tone due to surgical stimulation, stress responses, or pain. Furthermore, because oral therapy is not possible, patients require parenteral therapy. Postoperative hypertension is primarily due to increased systemic vascular resistance[1] brought about by reflex changes in humoral factors, including increased levels of plasma renin and catecholamines;[2] stimulation of neural reflexes in the heart, coronary arteries, and great vessels;[3] and systemic hypothermia. Postoperative hypertension is a major problem after both cardiac and noncardiac surgery.[4–6] The incidence of postoperative hypertension ranges from 6–20% in various noncardiac surgical studies; it occurs more commonly in patients with preoperative hypertension related to the anesthetic regimen. Estafanous and Tarazi found an incidence of postoperative hypertension of 30–50% after coronary artery bypass surgery and 8–10% after valvular heart replacement.[7]

Potential approaches to perioperative hypertension with parenteral therapy include (1) alpha$_1$-adrenergic receptor blockade,[4] beta$_1$-adrenergic receptor blockade,[8,9] ganglionic blockade, and calcium channel blockade;[10–12] (2) stimulation of dopamine$_1$, central alpha$_2$-adrenergic receptors, or vascular guanylate cyclase and adenylate cyclase;[13] (3) inhibition of phosphodiesterase enzymes; and (4) inhibition of angiotensin-converting enzymes (Table 1). Inhibition of phosphodiesterase enzymes with the newer generation of type III (cyclic adenosine monophosphate- specific) drugs, such as milrinone, amrinone, and enoximone, is limited to pulmonary hypertension because these drugs have a potent inotropic effect. The use of prostaglandins is also reserved for pulmonary hypertension because they are potent platelet inhibitors. The major therapeutic approach to perioperative hypertension is directed at reversing systemic vasoconstriction with drugs that produce vasodilation. Because a spectrum of different drugs can be used to treat hypertension, the physiology of vascular responses and the different potential pharmacologic ways of treating perioperative hypertension are discussed.

PERIOPERATIVE HYPERTENSION

A recent study[14] pointed out that risk factors for hypertension in the postanesthesia care unit included increasing age, smoking, and renal disease. Patients

TABLE 1. Therapeutic Approaches to Perioperative Hypertension

- Alpha$_1$-adrenergic receptor blockade (phentolamine)
- Angiotensin-converting enzyme inhibition (enalaprilat)
- Beta-adrenergic blockade (esmolol, propranolol, metoprolol, atenolol)
- Calcium channel blockade (nicardipine, isradipine)
- Dopamine$_1$-receptor blockade (fenoldopam)
- Vascular guanylyl cyclase stimulation (nitrovasodilators: nitroprusside, nitroglycerin)
- Vascular adenylyl cyclase stimulation (pulmonary hypertension: prostacyclin, prostaglandin E$_1$)
- Phosphodiesterase inhibition (pulmonary hypertension: milrinone, amrinone, enoximone)

with a preoperative history of angina or inadequate ventilation during the postoperative period were also at increased risk for hypertension. Moreover, hypertension and tachycardia were found to be associated with long-term morbidity and mortality.

In cardiac surgical patients, blood pressure may be maintained at lower levels to avoid graft/suture line disruption by use of controlled hypotension. Currently, patients are being "fast-tracked" and rapidly weaned from mechanical ventilation after cardiac surgery. In addition, multiple mechanisms of mechanical manipulation of native and graft coronary arteries and the internal mammary artery (IMA) with suturing increase the risk of coronary artery or IMA spasm. Ventricular dysfunction is common even in patients with normal preoperative function because of stunning and reperfusion injury.

Hypertension after coronary artery bypass grafting has been reported in 30–80% of patients.[1,15–17,18–20] Augmentation of stress responses with elevation of plasma renin, angiotensin, epinephrine, and norepinephrine has been suggested as the pathophysiologic basis of the increased vascular resistance responsible for elevated arterial pressure.[1,18,22–24] Activation of neurogenic reflexes of the heart, coronary arteries, or great vessels also may be involved.[3,21,24–26]

Hypertension after cardiac surgery may lead to life-threatening complications, including hemorrhage, rupture of suture lines, cerebrovascular accidents, and myocardial ischemia.[8,15,24,27,28] The standard therapy for acute hypertension after coronary artery surgery has been sodium nitroprusside.[4,6,28–30] Nitroprusside has been associated with increased intrapulmonary shunting or myocardial ischemia.[4,8,28,31–34]

Many other pharmacologic agents also have been used to treat hypertension immediately after coronary artery bypass grafting, including nitroglycerin,[28,30,35–38] alpha-adrenoreceptor blockers,[4,39] beta-adrenoreceptor blockers,[8,27,40–42] angiotensin-converting enzyme inhibitors,[1,43] and 5-hydroxytryptamine antagonists.[39,44,45] Recently, calcium channel blockers, including nifedipine,[12,42,46–48] nicardipine,[38,49–54] isradipine,[11,55–61] diltiazem,[47] and verapamil[62] have been shown to reduce acute hypertension after coronary artery bypass grafting.

Cyclic Adenosine Monophosphate-Dependent Vasodilators

It is important to understand the mechanism of actions of the different vasodilators because certain drugs may have preferential effects on the pulmonary arterial, systemic arterial, or vascular capacitance bed. Cyclic adenosine-3′,5′-monophosphate (cyclic AMP) pathways are therapeutic approaches that have been applied primarily to pulmonary hypertension. Cyclic AMP may be increased by stimulating the $beta_2$-adrenergic receptor in vascular smooth muscle or inhibiting the breakdown of cyclic AMP by phosphodiesterase. Increasing cyclic AMP in vascular smooth muscle facilitates calcium uptake by intracellular storage sites, thus decreasing the amount of calcium available for contraction. The net effect of increasing calcium uptake is vascular smooth muscle relaxation and, hence, vasodilation. However, most catecholamines with $beta_2$-adrenergic activity (e.g., isoproterenol) and phosphodiesterase inhibitors (e.g., amrinone, milrinone, enoximone, aminophylline) have positive inotropic and other side effects that include tachycardia, glycogenolysis, and kaliuresis. Prostaglandin E_1 and prostacyclin also stimulate vascular adenylate cyclase independently of the $beta_2$-adrenergic receptor to produce pulmonary and systemic vasodilation. However, the prostaglandins are potent inhibitors of platelet and polymorphonuclear leukocyte aggregation and activation. Catecholamines with $beta_2$-adrenergic activity, phosphodiesterase inhibitors, and prostaglandin E_1 have been used to treat pulmonary hypertension and right ventricular failure. Prostacyclin has been approved for use in primary pulmonary hypertension and has been studied for treatment of hypertension after cardiac surgery.

Cyclic Guanosine Monophosphate and Nitrovasodilators

Drugs and endogenous mediators stimulate the vascular endothelium to release an endothelium-derived relaxing factor (EDRF or nitric oxide), which activates guanylate cyclase to generate cyclic guanosine monophosphate (cGMP). Nitrates and sodium nitroprusside, however, generate nitric oxide (NO) directly and independently of vascular endothelium. The mainstay of therapy for postoperative hypertension is sodium nitroprusside and, to a lesser extent, nitroglycerin. Both drugs generate NO to stimulate guanylate cyclase and to generate cyclic GMP in vascular smooth muscle. The active form of any nitrovasodilator is NO, in which the nitrogen exists in a +2 oxidation state. For any nitrovasodilator to be active, it must be first converted to NO. Nitroprusside is easily converted because the nitrogen is in a +3 oxidation state, with the NO molecule bound to the charged iron molecule in an unstable manner. Thus nitroprusside readily donates its NO moiety. Nitrogen molecules in nitroglycerin exist in a +5 oxidation state; thus, they must undergo several metabolic transformations before they are converted to an active molecule. Unlike nitroprusside, nitroglycerin does not produce coronary steal because the small intracoronary resistance vessels ($< 100\ \mu$) lack whatever metabolic transformation pathway is required to convert nitroglycerin into the active form of NO.

Nitroprusside is a widely used intravenous therapy for hypertension. Despite its clinical use, however, concerns remain about its potential toxicity (Table 2). Nitroprusside is a classic example of a drug that is pharmacokinetically friendly (because of its very short half-life) but pharmacodynamically unfriendly (because of the spectrum of potential toxic effects). In 1991, the Food and Drug Administration approved new labeling of nitroprusside that contained greater detail about the risk of cyanide toxicity, based on clinical use for almost 18 years.[63] Nitroprusside is composed of a ferrous ion complex with five

TABLE 2. Nitroprusside Therapy

- Potent venodilator and arterial vasodilator
- Cardiac output is often affected because of venodilation
- Volume replacement is often required for venodilation

cyanide moieties and a nitrosyl group. The molecule is 44% cyanide by weight and soluble in water. Nitroprusside interacts with oxyhemoglobin, dissociating and forming methemoglobin while releasing cyanide and NO.

Although both nitroprusside and nitroglycerin are effective vasodilators, they are also NO-based agents that produce venodilation, which contributes significantly to the labile hemodynamic state. Most clinicians have observed the tremendous fluctuations in blood pressure when sodium nitroprusside is initiated to treat hypertension in the immediate postoperative period. In hypertensive patients, intravenous volume administration is often used to allow infusion of nitroprusside because of the relative intravascular hypovolemia. Although nitroprusside is frequently used in the control of postoperative hypertension in other surgical interventions, it may contribute to myocardial ischemia through nonspecific coronary vasodilation and thus produce coronary steal.[6,9,33]

Beta-adrenergic Blockers

Beta-adrenergic blockers reduce heart rate and myocardial contractility, thereby decreasing cardiac output and reducing both diastolic and systolic blood pressure. Therefore, beta blockers should be considered for treatment of perioperative hypertension in hyperdynamic patients with high cardiac output and increased sympathomimetic tone. Because heart rate is a major determinant of myocardial blood supply, tachycardia must be treated, especially in patients with ischemic heart disease; beta blockers should be used as first-line therapeutic agents. $Beta_1$-adrenoreceptors are located in the cardiac and adipose tissue and $beta_2$-adrenoreceptors in various organs such as blood vessels. Stimulation of vascular $beta_2$-adrenoreceptors leads to vasodilation. Beta blockers also may have intrinsic sympathomimetic activity (ISA), membrane-stabilizing activities (MSAs), and vasodilating actions; they also may be relatively selective for $beta_1$-receptor blockade. The theoretical benefit of selective $beta_1$-blockade is that stimulation of $beta_2$-adrenoreceptors, leading to vasodilation, may counteract the unopposed alpha-adrenergic effects. However, the mechanisms by which beta blockers reduce blood pressure in clinical therapy of perioperative hypertension involve multiple effects on different pathways. Suppression of plasma renin activity and decrease in plasma norepinephrine concentration[64] are controversial theories.[65] Blockade of presynaptic beta-adrenoreceptors inhibits the release of norepinephrine and thereby reduces sympathetic activity in the heart, kidneys, and arterioles.[65] Beta blockers may have adverse effects on the metabolism of glucose and lipoproteins. Because epinephrine increases glycogenolysis and gluconeogenesis in the liver and skeletal muscle by stimulating $beta_2$ receptors, selective $beta_1$ blockers may be beneficial in diabetic patients. Finally, bronchial asthma, uncontrolled biventricular failure, and second- or third-degree atrioventricular block are absolute contraindications for beta blockade.

TABLE 3. Comparison of Selected Beta Blockers

Characteristic	Propranolol	Metoprolol	Atenolol	Esmolol	Labetalol
$Beta_1$ selectivity	0	+	+	+	0
ISA	0	0	0	0	0
MSA	0	±	0	+	±
Half-life	~ 4.5 hr	3–4 hr	6–7 hr	~ 9 min	~ 6 hr
Metabolism	Hepatic	Hepatic	Hepatic	Blood esterases	Hepatic
IV dose*	0.5–1 mg	0.5–1 mg		500 μg/kg for 1 min, then 50–200 μg/kg/min	2.5–10 mg

ISA = intrinsic sympathomimetic activity, MSA = membrane-stabilizing activity.
* Note: additional IV doses may be required.

Several beta blockers are used as antihypertensive agents in the perioperative period and may be administrated intravenously, including propranolol, metoprolol, atenolol, esmolol, and labetalol. Their characteristics are described in Table 3. Propranolol, a nonselective beta blocker, causes a decrease in heart rate and cardiac output and, at least initially, an increase in systemic vascular resistance. Metoprolol is a selective $beta_1$-adrenoreceptor blocker that has beneficial effects in the treatment of hypertension, pectoris angina, and supraventricular arrhythmia.[66] Abrahamsson and coworkers showed better distribution to the ischemic myocardium with metoprolol than with atenolol.[67] Atenolol is the most selective $beta_1$-adrenoreceptor blocker. In addition to lowering blood pressure, intravenous administration of atenolol reduces indices of isovolumic contraction and systolic function in patients with coronary artery disease.[68] Unfortunately, atenolol has a very long half-life and is not titratable to the rapidly changing environment that characterizes perioperative hypertension. Esmolol is an ultrashort-acting $beta_1$-selective blocker that is used commonly to treat acute hemodynamic changes during anesthesia and surgery. Esmolol is metabolized by erythrocyte cholinesterases that rapidly dissipate its clinical effects. In the emergency management of hypertension, esmolol effectively attenuates hemodynamic responses due to sympathetic activation or endogenous catecholamine release in the rapidly changing perioperative environment.[69,70] It can be administered as a bolus for immediate effects, but because of its short half-life, a concomitant infusion is needed to maintain appropriate therapeutic levels for any duration. Esmolol has been found to attenuate the hemodynamic responses to weaning from mechanical ventilation, awakening from anesthesia, and extubation.[71] Labetalol blocks both alpha- and beta-adrenoreceptors. Its administration is associated with little change in cardiac output and heart rate.[72] Sladen and coworkers reported the beneficial effect of intravenous labetalol for control of elevated blood pressure after coronary artery bypass.[73]

Angiotensin-converting Enzyme Inhibitors

The renin-angiotensin system is one of the key mechanisms involved in blood pressure regulation. Perioperative stress is associated with an increase in renin activity. The mechanism of action of angiotensin-converting enzyme (ACE) inhibitors is inhibition of a kininase in the vascular endothelium that is involved in the conversion of angiotensin I to angiotensin II, which increases blood pressure via direct stimulation of vascular smooth muscle receptors. Angiotensin II also stimulates secretion of aldosterone. One of the important considerations in the use of ACE inhibitors is their inhibitory effect on the breakdown of bradykinin, a potent releaser of NO from vascular endothelium. Other systems that may be involved in the effect of ACE inhibitors in lowering the blood pressure include plasma vasopressin level, prostaglandin pathway, and the kallikrein-kinin systems with release of bradykinin. However, the use of intravenous ACE inhibitors for hypertensive emergencies and perioperative hypertension is limited; acute renal failure has been reported in patients with renal artery stenosis. Enalaprilat, the hydrolysis product of enalapril, is currently the only agent available for intravenous infusion.[65] Enalaprilat is not titratable because of its length of action (5–15 minutes) and slow offset (12–24 hours) of action. Elimination half-life of enalaprilat is approximately 5 hours. Because contact activation with generation of activated Hageman factor (factor XIIa) during surgery, especially after extracorporeal circulation, may increase the generation of bradykinin, ACE inhibitors may contribute to hypotension after cardiac surgery. Because enalaprilat acts indirectly, ACE-inhibitors may not be the first choice for treating perioperative hypertension.

Calcium Channel Blockers

Calcium channel blockers (CCBs) belong to three chemically heterogeneous groups: phenylalkylamines (e.g., verapamil), benzothiazepines (e.g., diltiazem), and dihydropyridines (e.g., nifedipine). Any of these drugs may actively produce vasodilation, and all of them decrease calcium entry into or increase calcium mobilization out of vascular smooth muscle. The CCBs bind with high affinity to the L-type calcium channels and modulate their voltage-dependent calcium conductivity.[74] Calcium channels are thought to be membrane-spanning glycoproteins, with a water-filled pore, that function like ion-selective valves, allowing calcium to move from extracellular space to the cytosol.[75] The L-type channels are widely distributed in vascular and intestinal smooth muscle, endocrine tissue, myocytes, and cardiac pacemaker and conduction tissue. Although CCBs bind to L-type calcium channels present in almost all excitable cells, there are remarkable differences in the potency and efficacy of the various CCBs in various tissues.[75]

Phenylalkylamines, such as verapamil, are nearly equipotent in vascular smooth muscle and cardiac myocytes and conducting tissue. Thus they are vasodilators with additional cardiodepressant effects.[75–77] They also induce a negative chromotropic effect, resulting in reduced heart rate. Verapamil is commonly used for treatment of supraventricular tachycardia.[78] Because of its pronounced cardiac depressant effects, verapamil should not be used in patients with heart failure or impaired atrioventricular conduction or given in combination with beta blockers, which results in additional negative inotropic effects.

The pharmacodynamic pattern of benzothiazepines, represented by diltiazem, resembles that of the phenylalkylamines, but the benzothiazepines have slightly less cardiodepressant effects.[77] Diltiazem may be used for treatment of conduction disturbances because of its negative dromotropic effect.[78]

Because of the negative influence on cardiac contractility and conduction, phenylalkylamines and benzothiazepines should not be used for treatment of perioperative hypertension, especially in conjunction with cardiac surgery.

Dihydropyridines, which are predominantly dilators of peripheral resistance arteries, result in a more or less generalized vasodilatation that includes the renal, cerebral, intestinal, and coronary vascular beds. In doses that effectively reduce blood pressure, the dihydropyridines have little or no direct negative effect on cardiac contractility or conduction.[79] Their lack of negative chronotropic effects allows an initial reflex increase in heart rate, which vanishes during prolonged antihypertensive treatment.[80] Although the dihydropyridines may be considered more vasoselective than the phenylalkylamines and the benzothiazepines, the dihydropyridines also differ among themselves in this respect. Nifedipine and amlodipine are the least vasoselective of the dihydropyridines,[77, 79] whereas isradipine and felodipine are the most selective.[77] Nicardipine and nimodipine are intermediately selective[81] (Table 4).

CCBs do not relax venous smooth muscle, probably because of the relatively low density of L-type calcium channels in this vascular compartment. Therefore, CCBs, unlike nitrodilators, have little effect on cardiac filling pressure and preload. As a result, cardiac output is well maintained or increased when a CCB is given to reduce arterial pressure.

In addition to their hemodynamic effect, CCBs induce mild natriuresis, which may result in diuresis and/or prevention of the fluid retention seen with other vasodilators.[82] Vascular-selective dihydropyridines, such as nicardipine, also may reduce the incidence of ischemic episodes in patients who develop hypertension during cardiac surgery compared with sodium nitroprusside[49] and nitroglycerine.[83]

Additional factors should be considered in treating hypertension in cardiac surgical patients. After aortocoronary bypass grafting, patients have undergone direct mechanical manipulation of both native grafts and arteries as well as the internal mammary artery. Mechanical manipulation, suturing, and probing of native coronary arteries with metal probes may transiently damage vascular endothelium, setting up the potential risk for coronary or internal mammary spasm after cardiac surgery.[84,85] In addition, there are multiple opportunities for

TABLE 4. Selectivity for Vascular over Cardiac Effects and Pharmacokinetics of Calcium Channel Blockers

Calcium Channel Blocker	Vasoselectivity	Plasma Half-life (hr)	Plasma Clearance (ml/min/kg)
Verapamil	0	≈5	13
Diltiazem	+	≈5	12
Nifedipine	++	≈2	8
Isradipine	++++	≈8	10
Nicardipine	+++	< 1	7
Clevidipine	+++	< 0.1	150

thromboxane liberation after cardiopulmonary bypass and coronary revascularization, especially following protamine reversal of systemic heparinization.[86–89] All of these occurrences set up the potential for coronary spasm. Because calcium is the major intracellular mediator for vasoconstriction, calcium channel blockers may play an important role in the prophylaxis or therapy of coronary or internal mammary artery spasm after cardiac surgery; therefore, they are ideal drugs for treating hypertension in such patients.

Perioperative administration of CCBs has been limited by the unavailability of intravenous vascular-selective dihydropyridines. Clinical trials have indicated that isradipine may be a useful agent for treatment of hypertension after coronary artery bypass grafting.[11,56–61] Nicardipine, the first intravenous dihydropyridine available in the United States, offers an important therapeutic approach to perioperative hypertension.[10,49–54] However, isradipine has a long elimination time,[90] and the elimination time of nicardipine increases the longer the drug is infused,[91] which may be a disadvantage in the perioperative setting. Clevidipine, a new, ultrashort-acting, vascular-selective dihydropyridine, is presently in clinical phase II studies for control of blood pressure in cardiac surgical patients. Because of its rapid metabolism by ester hydrolysis by nonspecific tissue and blood esterases, clevidipine is rapidly eliminated from the blood and has a half-life of 1 minute.[92] Its blood pressure-reducing effect is rapid in onset, and at the end of drug infusion, blood pressure returns to baseline values within minutes. The hemodynamic profile of clevidipine[93] resembles that of the long-acting vasoselective dihydropyridines, felodipine and isradipine, but because of high plasma clearance (see Table 4), clevidipine allows control of rapid changes in blood pressure, which may occur during and after surgery.

Dopaminergic Receptor Agonists

Specific dopamine (DA) receptor agonists are a new class of agents under clinical investigation. Fenoldopam is a selective agonist for peripheral DA_1 receptors, which produce vasodilatation, increase renal perfusion, and enhance natriuresis.[94] Fenoldopam has a short duration of action; the elimination half-life is less than 10 minutes.[95] Fenoldopam undergoes extensive hepatic metabolism to form methyl, sulfate, and glucuronide conjugates, but it has no active or toxic metabolites. Aronson compared the systemic and renal vascular effects of hypotension obtained with fenoldopam and sodium nitroprusside in 10 anesthetized dogs. Renal blood flow was preserved with fenoldopam, whereas sodium nitroprusside caused redistribution of blood flow away from the kidneys during hypotension.[96]

Evidence from awake healthy patients suggests that hypotension following fenoldopam infusions may be relatively attenuated in normotensive patients because of an intact baroreceptor reflex that causes an increase in heart rate and sympathetic activity. These findings are characteristic of vasodilator therapy in healthy patients with intact sympathoadrenergic systems. The efficacy of fenoldopam infusion has been studied for treatment of postoperative hypertension by comparison with placebo. In doses of 0.1–1.5 mg/kg/min, fenoldopam reduced both systolic and diastolic blood pressure but also increased heart rate. There were no significant differences in plasma levels of epinephrine, norepinephrine, or dopamine in the fenoldopam-treated patients.[97] In another study of patients with acute severe hypertension, adults with supine diastolic blood pressure ≥ 120 mmHg were randomized to receive either fenoldopam or sodium nitroprusside therapy. Diastolic blood pressure was titrated to 95–110 mmHg or a maximal reduction of 40 mmHg for very high pressures. Infusions were maintained for at least 6 hours; then patients were weaned from intravenous therapy. The two antihypertensive agents were equivalent in controlling and maintaining diastolic blood pressure in terms of efficacy and acute adverse events. Because of its unique mechanism of action, fenoldopam may have advantages in selected subsets of patients. Further studies may be indicated in patient populations with pure hypertensive emergencies.[98]

SYSTEMIC VASCULAR RESISTANCE AS A GUIDE TO VASOACTIVE THERAPY

Systemic vascular resistance (SVR) is often used to guide vasodilator therapy in perioperative settings. It is calculated from measured values with the following equation:

(mean arterial pressure – central venous pressure) ÷ cardiac output = SVR

SVR calculated in this way may not be a true indicator of vascular responses, because we do not routinely index its value for body surface area. Three patients with identically measured hemodynamic parameters but different body weights and surface areas may have major differences in calculated SVR, although blood pressure and cardiac index are identical. Such discrepancies may add to the confusion of using SVR as a guide to intervention with vasoactive therapy in cases of systemic vasoconstriction. Drugs that cause peripheral vascular effects may demonstrate variable effects on hemodynamic function, depending on intravascular volume, left ventricular function, and extent of peripheral vasoconstriction. Using SVR instead of measured arterial blood pressure to guide vasodilator therapy may be misleading. Calculated SVR may increase when the patient is excessively vasodilated with sodium nitroprusside, because cardiac output may be dramatically reduced.

Given the limitations on the use of SVR, both systemic blood pressure and left ventricular filling pressures should be considered in treating perioperative hypertension. Afterload, as classically defined by studies of isolated myocardium, is proportional to systolic wall tension. By Laplace's law, left ventricular wall tension, in turn, is proportional to left ventricular pressure times radius of the left ventricle. Most commonly used vasodilators are NO-based and increase venous compliance, thereby lowering preload. However, one can often assess the extent of ventricular filling indirectly by observing the blood pressure response to positive-pressure ventilation or the effect of fluid challenge on cardiac output. The use of a dihydropyridine-derived CCBs with specific effects on the arterial vasculature, no significant effects on the venous capacitance bed or filling pressures, and favorable effects on the coronary arteries offers a specific therapeutic approach for systemic vasoconstriction. As the newer intravenous dihydropyridine-derived drugs become available, they will offer new options for treatment of perioperative hypertension.

CONCLUSION

Perioperative hypertension during cardiac or noncardiac surgery is a unique clinical problem characterized by systemic vasoconstriction, often with intravascular hypovolemia. It usually requires acute short-term intravenous therapy. Beta-adrenergic blockers are important first-line drugs for patients with hypertension and tachycardia, although beta blockers may have adverse side effects. Short-acting esmolol is a first-line beta blocker for perioperative use because of its titratability. The CCBs are important drugs with arterial vasodilating actions, and the new intravenous dihyropyridine compounds are especially promising because they have no negative inotropic effects and no effects on atrioventricular node conduction. Nicardipine is the first intravenous dihydropyridine CCB to become available for perioperative hypertension in the United States, and clevidipine is currently under investigation. Fenoldopam is a promising short-acting arterial-specific agent that stimulates dopamine$_1$ receptors and increases renal blood flow.

References

1. Roberts AJ, Niarchos AP, Subramanian VA, et al: Systemic hypertension associated with coronary artery bypass surgery: Predisposing factors, hemodynamic characteristics, humoral profile, and treatment. J Thorac Cardiovasc Surg 74:846–859, 1977.
2. Wallach R, Karp RB, Reves JG, et al: Pathogenesis of paroxysmal hypertension developing during and after coronary bypass surgery: A study of hemodynamic and humoral factors. Am J Cardiol 46:559–565, 1980.
3. Pratilas V, Pratila MG, Vlachakis ND, et al: Sympathetic nervous system tonicity and post coronary artery bypass hypertension. Acta Anesth Scand 24:69–73, 1980.
4. Roberts AJ, Niarchos AP, Subramanian VA, et al: Hypertension following coronary artery bypass surgery: Comparison of hemodynamic responses to nitroprusside, phentolamine, and converting enzyme inhibitor. Circulation 58(Suppl I):43–49, 1978.
5. Stinson EB, Holloway EL, Derby GC, et al: Control of myocardial performance early after open-heart operations by vasodilator treatment. J Thorac Cardiovasc Surg 73:523–529, 1977.
6. Fremes SE, Weisel RD, Baird RJ, et al: Effect of postoperative hypertension and its treatment. J Thorac Cardiovasc Surg 86:47–56, 1983.
7. Estafanous FG, Tarazi RC: Systemic arterial hypertension associated wtih cardiac surgery. Am J Cardiol 46:685–694, 1980.
8. Gray RJ, Bateman TM, Czer LSC, et al: Comparison of esmolol and nitroprusside for acute post-cardiac surgical hypertension. Am J Cardiol 59:887–891, 1987.
9. Mann T, Cohn PF, Holman BL, et al: Effect of nitroprusside on regional myocardial blood flow in coronary artery disease. Circulation 57:732–737, 1978.
10. David D, Dubois C, Loria Y: Comparison of nicardipine and sodium nitroprusside in the treatment of paroxysmal hypertension following aortocoronary bypass surgery. J Cardothorac Vasc Anesth 5:357–361, 1991.
11. Leslie J, Brister NW, Levy JH, et al: Treatment of post operative hypertension following coronary artery bypass surgery: Double-blind comparison of intravenous isradipine and sodium nitroprusside. Circulation 90:II256–II261,1994.
12. Hess W, Schulte-Sasse V, Tarnow J: Nifedipine versus nitroprusside for controlling hypertensive episodes during coronary artery bypass surgery. Eur Heart J 5:140–145, 1984.
13. Lucchesi BR: Role of calcium on excitation contraction in cardiac and vascular smooth muscle. 80:IV1–IV13, 1989.
14. Rose DK, Cohen MM, DeBoer, Math M: Cardiovascular events in the postanesthesia care unit, contribution of risk factors. Anesthesiology. 84:772–781, 1996.
15. Estafanous FG, Tarazi RC, Viljoen JF, et al: Systemic hypertension following myocardial revascularization. Am Heart J 85:732–738, 1973.
16. Viljoen JF, Estafanous FG, Tarazi RC: Acute hypertension immediately after coronary artery surgery. J Thorac Cardiovasc Surg 71:548–550, 1976.

17. Hoar PP, Hickey RF, Ullyot DJ: Systemic hypertension following myocardial revascularization: A method of treatment using epidural anesthesia. J Thorac Cardiovasc Surg 71:859–864, 1976.
18. Wallach R, Karp RB, Reves JG, et al: Pathogenesis of paroxysmal hypertension developing during and after coronary bypass surgery: a study of hemodynamic and humoral factors. Am J Cardiol 46:559–565, 1980.
19. Gal TJ, Cooperman LH: Hypertension in the immediate postoperative period. Br J Anesth 47:70–74, 1975.
20. Leslie JB: Incidence and aetiology of perioperative hypertension. Acta Anesthesiol Scand 37(Suppl 99):5–9, 1993.
21. Fouad FM, Estafanous FG, Tarazi RG: Hemodynamics of postmyocardial revascularization hypertension. Am J Cardiol 41:564–569, 1978.
22. Bailey DR, Miller ED, Kaplan JA, Rogers PW: The renin-angiotensin-aldosterone system during cardiac surgery with morphine-nitrous oxide anesthesia. Anesthesiology 42:538–544, 1975.
23. Taylor KM, Morton IJ, Brown JJ, et al: Hypertension and the renin-angiotensin system following open-heart surgery. J Thorac Cardiovasc Surg 74:840–845, 1977.
24. Hoar PF, Stone JG, Faltas AN, et al: Hemodynamic and adrenergic responses to anesthesia and operation for myocardial revascularization. J Thorac Cardiovasc Surg 80:242–248, 1980.
25. Peterson FD, Brown AM: Pressor reflexes produced by stimulation of efferent fibres in the cardiac sympathetic nerves of the cat. Circ Res 28:605–610, 1971.
26. James TN, Hageman GR, Urthaler F: Anatomic and physiologic considerations of a cardiogenic hypertensive reflex. Am J Cardiol 44:852–859, 1979.
27. Whelton PK, Flaherty JT, MacAllister NP, et al: Hypertension following coronary artery bypass surgery: Role of preoperative propranolol therapy. Hypertension 2:291–298, 1980.
28. Flaherty JT, Magee PA, Gardner TL, et al: Comparison of intravenous nitroglycerin and sodium nitroprusside for treatment of acute hypertension developing after coronary artery bypass surgery. Circulation 65:1072–1077, 1982.
29. Lappas DG, Lowenstein E, Waller J, et al: Hemodynamic effects of nitroprusside infusion during coronary artery operation in man. Circulation 54 (Suppl III):III4–III10, 1978.
30. Kaplan JA, Jones EL: Vasodilator therapy during coronary artery surgery. Comparison of nitroglycerin and nitroprusside. J Thorac Cardiovasc Surg 77:301–309, 1979.
31. D'Oliveira M, Sykes MK, Chakrabarti MK, et al: Depression of hypoxic pulmonary vasoconstriction by sodium nitroprusside and nitroglycerine. Br J Anesth 53:11–18, 1981.
32. Colley PS, Chenay FW, Hlastala MP: Ventilation-perfusion and gas exchange effects of sodium nitroprusside in dogs with normal and oedematous lungs. Anesthesiology 50:489– 895, 1979.
33. Becker LC: Conditions for vasodilator-induced coronary steal in experimental myocardial ischemia. Circulation 57:1103–1110, 1978.
34. Chiarello M, Gold HK, Leinbach RC, et al: Comparison between the effects of nitroprusside and nitroglycerin on ischemic injury during acute myocardial infarction. Circulation 54:766, 1976.
35. Tobias MA: Comparison of nitroprusside and nitroglycerin for controlling hypertension during coronary artery surgery. Br J Anesth 53:891–897, 1981.
36. Fremes SE, Weisele RD, Mickle DA, et al: A comparison of nitroglycerin and nitroprusside. I: Treatment of postoperative hypertension. Ann Thorac Surg 39:53–60, 1985.
37. Durkin MA, Thys D, Morris RB, et al: Control of perioperative hypertension during coronary artery surgery: A randomized double-blind study comparing isosorbide dinitrate and nitroglycerin. Eur Heart J 9(Suppl A):181–185, 1988.
38. Vecht RJ, Swanson KT, Nicolaides EP, et al: Comparison of intravenous nicardipine and nitroglycerine to control systemic hypertension after coronary artery bypass grafting. Am J Cardiol 64:19H–21H, 1989.
39. Mollhoff TH, Mulier JP, Lauwers P, Van Aken H: Ventilatory effects of ketanserine, urapidil and sodium nitroprusside following coronary artery bypass grafting. Anesthesiology 71(3A):A68, 1989.
40. Boudoulas H, Snyder GL, Lewis RP, et al: Safety and rationale for continuation of propranolol therapy during coronary bypass operations. Ann Thorac Surg 26:222–227, 1978.
41. Leslie JB, Kalayjian RW, Sirgo MA, et al: Intravenous labetolol for treatment of postoperative hypertension. Anesthesiology 67:413–416, 1987.
42. O'Leary G, Weisel RD, Young P, et al: Comparison of esmolol, nifedipine and nitroprusside for treatment of post ACB hypertension. Anesthesiology 71(3A):A194, 1989.
43. Niarchos AP, Roberts AJ, Case DB, et al: Hemodynamic characteristics of hypertension after coronary bypass surgery and effects of converting enzyme inhibitor. Am J Cardiol 43:586–593, 1979.
44. Van der Starre PJA, Harnick-de Weerd JE, Reneman RS: Nitroprusside and ketanserine in the treatment of postoperative hypertension following coronary artery bypass grafting: a haemodynamic and ventilatory comparison. J Hypertens 4(Suppl 1):S107–S110, 1986.

45. Lichtenthal PR, Rossi EC, Wade LD: Control of blood pressure after coronary artery bypass surgery with iv ketanserine: a placebo-controlled study. Anesthesiology 71(3A):A67, 1989.
46. Davis ME, Jones CJH, Feneck RO, Walesby RK: Intravenous nifedipine for control of hypertension in patients after coronary artery bypass graft surgery. J Cardiothorac Anesth 2:130–139, 1988.
47. Mullen JC, Miller Dr, Weisel RD, et al: Postoperative hypertension: a comparison of diltiazem, nifedipine and nitroprusside. J Thorac Cardiovasc Surg 96:122–132, 1988.
48. Iyer VS, Russel WJ: Nifedipine for postoperative blood pressure control following coronary artery vein grafts. Ann Royal Coll Surg Engl 68:73–75, 1986.
49. Van Wezel HB, Koolen JJ, Visser CA, et al: Antihypertensive and anti-ischemic effects of nicardipine and nitroprusside in patients undergoing coronary artery bypass grafting. Am J Cardiol 64:22H–27H, 1989.
50. Floyd J, Goldberg M, Halpern N, et al: Comparison of nicardipine and sodium nitroprusside for treatment of postoperative hypertension. Anesthesiology 71(3A):A66, 1989.
51. Dubois C, Lentdecker PH, David D, et al: Tolerance and efficacy of intravenous nicardipine on arterial hypertension during cardiac surgery. J Cardiovasc Pharmacol 12(Suppl 6):S198, 1988.
52. Casar G, Pool JL, Chelly JE, et al: Intravenous nicardipine for treatment of postoperative hypertension. Anesthesiology 67:A141, 1987.
53. Goldberg ME, Seltzer JL, Halpern N, et al: Nicardipine vs placebo for the treatment of postoperative hypertension. Anesthesiology 69:A39, 1988.
54. Perry SM, Smith D, Meacham MD, Wood M: Adrenergic response to nicardipine in the management of hypertension. Anesthesiology 69(3A):A17, 1988.
55. Lawrence CJ, Lestrade A, Chan E, De Lange S: Comparative study of isradipine and sodium nitroprusside in the control of hypertension in patients following coronary artery-bypass surgery. Acta Anesthesiol Scand 37(Suppl 99):53–55, 1993.
56. Lawrence CJ, Lestrade A, De Lange S: Isradipine, a calcium antagonist, in the control of hypertension following coronary artery bypass surgery. Am J Hypertens 4:207S–209S, 1991.
57. Marty J: Role of isradipine and other antihypertensive agents in the treatment of peri- and postoperative hypertension. Acta Anesthesiol Scand 37(Suppl 99):53–55, 1993.
58. Ruegg PC, Karmann U, Keller H: Management of perioperative hypertension using intravenous isradipine. Am J Hypertens 4:203S–206S, 1991.
59. Underwood SM, Davies SW, Feneck RO, Walesby RK: Comparison of isradipine with nitroprusside for control of blood pressure following myocardial revascularization hypertension: Effects on hemodynamics, cardiac metabolism, and coronary blood flow. J Cardiothorac Vasc Anesth 5:348–356, 1991.
60. Underwood SM, Feneck RO, Davies SW, Walesby RK: Use of isradipine in hypertension following coronary artery bypass surgery. Am J Med 86(Suppl 4A):81–87, 1989.
61. Brister NW, Barnette RE, Schartel SA, et al: Isradipine for treatment of acute hypertension after myocardial revascularization. Crit Care Med 19:334–338, 1991.
62. Casella ES, Blanck TTJ: Verapamil infusion for lowering blood pressure after cardiopulmonary bypass. J Cardiovasc Pharmacol 12(Suppl 6):S172, 1988.
63. Friederich JA, Buttterworth JF: Sodium nitroprusside: Twenty years and counting. Anesth Analg 81:152–162, 1995.
64. Man in't Velt AJ, Schalekamp MADH: Effects of 10 different beta-adrenoreceptor antagonists on hemodynamics, plasma renin activity and plasma norepinephrine in hypertension: The key role of vascular resistance changes in relation to partial agonist activity. J Cardiovasc Pharmacol 5(Suppl 1):S30, 1983.
65. Buris JF, Freis ED. Hypertensive emergencies. In Messerli FH (ed): Cardiovascular Drug Therapy, Messerli FH, Saunders, Philadelphia, W.B. Saunders, 1996, pp 148–160.
66. Benfield P, Clissold SP, Brogden RN: Metoprolol: An updated review of its pharmacodynamic and pharmacokinetic properties and therapeutic efficacy in hypertension, ischemic heart disease, and related cardiovascular disorders. Drugs 31:376, 1986.
67. Abrahamsson T, Ek B, Nerme V: The beta$_1$- and beta$_2$-adrenoreceptor affinity of atenolol and metoprolol: A receptor-binding study performed with differents radioligands in tissues from the rat, the guinea pig, and man. Biochem Pharmacol 37:203, 1988.
68. Thompson DS, Naqui N, Juul SM et al: Haemodynamic and metabolic effects of atenolol in patients with angina pectoris. Br Heart J 43:668, 1980.
69. Barbier GH, Shettigard UR, Appun DO: Clinical rationale use of an ultra-short acting betablocker: Esmolol. J Clin Pharmacol Ther 33:212–218, 1995.
70. Sintetos Al, Hulse J, Pritchett ELC: Pharmacokinetics and pharmacodynamics of esmolol administrated as an intravenous bolus. Clin Pharmacol Ther 41:112, 1987.

71. Fuhrman TM, Ewell Cl, Pippin WD, Weaver JM: Comparison of the efficacy of esmolol and alfentanil to attenuate the hemodynamic responses to emergence and extubation. J Clin Anesth 4:444–447, 1992.
72. Morel DR, Forster A, Suter PM: IV labetalol in the treatment of hypertension following coronary artery surgery. Br J Anaesth 54:1191, 1982.
73. Sladen RN, Klamerus KJ, Swalfford WJ, et al: Labetalol for the control of elevated blood pressure after coronary artery bypass grafting. J Cardiothorac Anesth 4:210, 1990.
74. Schwarz A, McKenna E, Vaghy PL: Receptors for calcium antagonists. Am J Cardiol 62:3G–6G, 1988.
75. Sun J, Triggle DJ: Calcium channel antagonists, cardiovascular selectivity of action. J Pharmacol Exp Ther 274:419–426, 1995.
76. Triggle DJ: Calcium channel drugs: Structure-function relationships and selectivity of action. J Cardiovasc Pharmacol 18(10):S1–S6, 1991.
77. Ljung B: Vascular selectivity of felodipine. Drugs 29(2):46–58, 1985.
78. Sung RJ, Elser B, McAllister RG Jr: Intravenous verapamil for termination of re-entrant supraventricular tachycardias: Intracardiac studies correlated with plasma verapamil concentrations. Ann Intern Med 93:682–689, 1980.
79. Nordlander M, Abrahamsson T, Åkerblom B, Thalén P: Vascular versus myocardial selectivity of dihydropyridine calcium antagonists as studied in vivo and in vitro. Pharmacol Toxicol 75:56–62, 1995.
80. Elvelin L, Jönsson L: The effect of dihydropyridine calcium antagonists on heart rate: Studies with felodipine. Curr Ther Res 55:736–746, 1994.
81. Perez-Vizcaino F, Tamargo J, Hof RP, Ruegg UT: Vascular selectivity of seven prototype calcium antagonists: A study at the single cell level. J Cardiovasc Pharmacol 22:768–775, 1993.
82. DiBona GF, Epstein M, Mann J, Nordlander M: Renal protection aspects of calcium antagonists. Kidney Int 41(36):1–118, 1992.
83. Apostolidou I, Despotis GJ, McCowley C, et al: Anti-ischemic effects of intravenous nicardipine and nitroglycerine following CABG surgery. Anaesth Analg 82:SCA43, 1996.
84. He G-W, Rosenfeldt FL, Buxton BF, Angus JA: Reactivity of human isolated internal mammary artery to constrictor and dilator agents. Implications for treatment of internal mammary artery spasm. Circulation 80:I141–I150, 1989.
85. Engleman RM, Loannis HR, Breyer RH, et al: Rebound vasospasm after coronary revascularization in association with calcium antagonist withdrawal. Ann Thorac Surg 37:469–472,1984.
86. Davies GC, Sobel M, Salzman EW: Elevated plasma fibrinopeptide A and thromboxane B_2 levels during cardiopulmonary bypass. Circulation 61:808–814, 1980.
87. Ylikorkala O, Saarela E, Viinikka L: Increased prostacyclin and thromboxane production in man during cardiopulmonary bypass. J Thorac Cardiovasc Surg 82:245–247, 1981.
88. Addonizio VP, Smith JB, Strauss JF, et al: Thromboxane synthesis and platelet secretion during cardiopulmonary bypass with bubble oxygenator. J Thorac Cardiovasc Surg 79:91– 96, 1980.
89. Teoh KH, Fremes SE, Weisel RD, et al: Cardiac release of prostacyclin and thromboxane A_2 during coronary revascularization. J Thorac Cardiovasc Surg 93:120–126, 1987.
90. Grossman E, Bren S, Messerli F: Isradipine in cardiovascular drug therapy. In Messerli FH (ed): Cardiovascular Drug Therapy. Philadelphia, W.B. Saunders, 1990, pp 1016–1024.
91. Cook E, Clifton GG, Vargas R, et al: Pharmacokinetics, pharmacodynamics and minimum effect of clinical dose of nicardipine. Clin Pharmacol Ther 47:706–718, 1990.
92. Ericsson H, Höglund L, Nordlander M, et al: The pharmacokinetics of clevidipine, an ultrashort acting calcium antagonist, during constant infusion in healthy volunteers. Clin Pharmacol Ther 1997 (in press).
93. Nordlander M, Björkman J-A, Regårdh C-G, Thalén P: Pharmacokinetic and hemodynamic effects of an ultra-short acting calcium antagonist. Br J Anaesth 76:12, 1996.
94. Van Zwieten PA: Cardiovascular receptors and drug therapy. In Messerli FH (ed): Cardiovascular Drug Therapy. Philadelphia, W.B. Saunders, 1996 pp 334–347.
95. Weber RR, McCoy CE, Fremiak JA, et al: Pharmacokinetics and pharmacodynamics of intravenous fenoldopam, a dopamine -1 receptor agonist in hypertensive patients. Br J Clin Pharmacol 25:17–22, 1988.
96. Aronson S, Goldberg LI, Glock D, et al: Effects of fenoldopam on renal blood flow and systemic hemodynamics during isoflurane anesthesia. J Cardiothorac Vasc Anesth 5:29–32, 1991.
97. Goldberg ME, Cantillo J, Nemiroff MS, et al: Fenoldopam infusion for treatment of postoperative hypertension. J Clin Anesth 5:386–391, 1993.
98. Panacek A, Bednarczyk EM, Dunbar LM, et al: Randomized, prospective trial of fenoldopam versus sodium nitroprussiate in the treatment of acute severe hypertension. Acad Emerg Med 2:959–965, 1995.

RONALD B. GOLDBERG M.D. / HERMES FLOREZ M.D.

20

Effects of Calcium Antagonists on Lipids

It has become increasingly apparent that hypertension is frequently associated with one or more other cardiovascular risk factors, such as truncal obesity, glucose intolerance, and dyslipidemia.[1] The recognition that hypertension needs to be viewed as one component of a multifactorial syndrome of increased cardiovascular risk has important consequences for its management. Despite the fact that treatment of hypertension has produced impressive reductions in the incidence of congestive heart failure, renal insufficiency, and stroke, the impact on the rate of coronary artery disease has been less than might be expected.[2] The reasons are not fully understood, but part of the explanation may lie with the fact that treatment directed primarily at blood pressure control without concern for other factors that contribute to coronary atherosclerosis may have limited protective value against coronary artery disease. Thus it has been suggested that the deleterious effects of beta blockers and diuretics on glucose tolerance and lipid metabolism may have contributed to their modest effects on coronary heart disease prevention compared with other hypertension-related end-organ events.[3]

As a result, modern management of hypertension requires that attention be directed at all associated cardiovascular risk factors. Lipid disorders, in particular, have attracted attention, because cholesterol-lowering therapy has been shown to produce significant effects in slowing the progression of coronary heart disease, whether patients are hypertensive or not.[4] The beneficial effect is highly correlated with the degree to which low-density lipoprotein cholesterol (LDL) is reduced[5] and is equal in magnitude to the extent to which an increase in LDL predicts an increased risk of coronary heart disease.[6] Elevation in triglyceride values appears to enhance the atherogenicity of LDL, as does a reduction in high-density lipoprotein cholesterol (HDL).

These developments are important not only to the preexisting lipid profile in patients with hypertension but also to the effects of antihypertensive agents on lipids and their metabolism. With the introduction of the newer antihypertensive drugs, more attention has been directed at their effects on circulating lipids and lipoproteins and their potential atherogenic properties. This chapter examines the effects of calcium channel blockers (CCBs) on lipids and their metabolism.

LIPID METABOLISM AND ATHEROGENESIS

Understanding the significance of an effect on lipids in the context of atherosclerosis requires an appreciation of the relationship between lipid transport and atherogenesis (Fig. 1).

Dietary lipid enters the circulation in chylomicrons, whereas endogenously synthesized lipids are released from the liver in very-low-density lipoprotein (VLDL). These triglyceride-rich lipoproteins undergo triglyceride hydrolysis by lipoprotein lipase located at the endothelial surface of most capillary networks, which are especially abundant in adipose tissue and muscle. This enzyme is inhibited by adrenergic stimulation and is the likely explanation for the rise in triglyceride levels documented with the use of beta blockers without intrinsic sympathomimetic activity as well as thiazide diuretics.[7] The resulting triglyceride-depleted, cholesterol-enriched remnant lipoproteins are normally taken up either completely (chylomicron remnants) or partially (VLDL remnants) by the liver via receptor-dependent processes. In populations ingesting relatively high cholesterol diets, fewer than 50% of the VLDL remnants are not taken up by the cholesterol-replete liver; they are converted via sinusoidal hepatic lipase to LDL. LDL provides cholesterol tissues in a tightly regulated manner via the LDL-receptor, which determines its rate of catabolism.

High-density lipoprotein (HDL) is assembled in the liver and intestine and acquires lipid and apoprotein from VLDL and chylomicrons during their

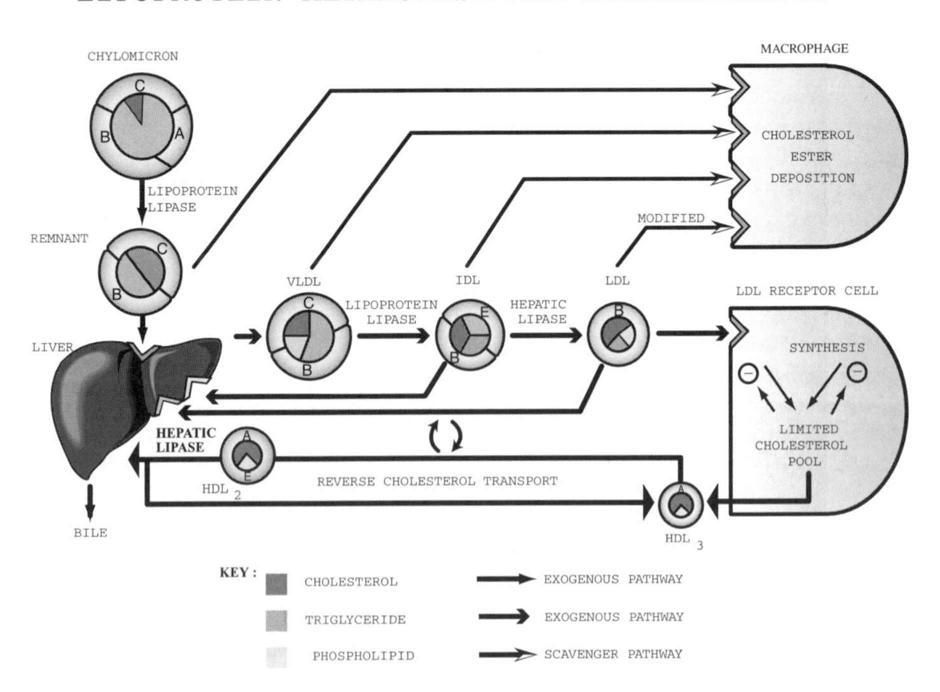

FIGURE 1. Lipoproteins and tissue cholesterol deposition.

metabolism by lipoprotein lipase. This process explains why beta blockers and thiazide diuretics, which reduce lipoprotein lipase activity, may lead to a modest reduction in HDL levels.[7] HDL is believed to mediate reverse cholesterol transport from cells to the other lipoproteins and the liver, thus helping to protect against cellular cholesterol overload.

Cholesterol-rich lipoproteins are thought to be atherogenic by virtue of their tendency to be taken up by arterial wall macrophages, thus contributing to foam cell formation. Although LDL may be the most important clinically measurable lipoprotein in this regard, evidence is accumulating that cholesterol-rich chylomicron remnants, which increase postprandially, and VLDL remnants may contribute significantly as well. Although this process is influenced by the absolute concentration of atherogenic particles, the extent to which these particles become peroxidized as a result of exposure to free radicals appears to be one of the critical requirements for their accumulation in macrophages. In addition, smaller LDL particles appear to be more susceptible to peroxidation and macrophage uptake than larger species. These new developments may explain the relationship between hypertriglyceridemia and atherogenesis: hypertriglyceridemic individuals have an increased postprandial accumulation of remnant particles and an increased frequency of small LDL particles. Hypertriglyceridemia is also associated with enhanced HDL catabolism, with lower HDL levels and perhaps a reduced capacity for reverse cholesterol transport.

EFFECTS OF CALCIUM CHANNEL BLOCKERS ON CIRCULATING LIPIDS AND LIPOPROTEINS

The level of circulating lipoproteins in part depends on cell vesicle secretory as well as receptor-mediated cellular uptake processes, which, in turn, are influenced by calcium-dependent mechanisms. Thus, it may be anticipated that chylomicron and VLDL secretion and remnant or LDL-receptor–dependent cell uptake may be influenced by CCBs. There have been no human studies of the effects of these agents on the kinetics of lipoprotein metabolism, principally because most studies of the effects of CCBs on steady-state plasma lipids and lipoproteins have detected little or no change in circulating levels.

Several investigators using animal models or cell culture systems, however, have demonstrated significant effects of CCBs on lipoprotein metabolism.[8–14] Despite differences between animals and humans and the limitations of extrapolating from tissue culture experiments, these reports raise the possibility that CCBs may have significant effects on lipoprotein metabolism despite the lack of significant changes in steady-state lipoprotein levels.

Animal and In Vitro Studies

Several reports indicate the possibility that CCBs may influence intestinal lipid absorption and secretion. Treatment of fat-fed rats with TA-3090, a benzothiazepine calcium antagonist, led to an increase in HDL cholesterol (HDL-C) without changes in triglycerides or LDL cholesterol (LDL-C), but with a significant decrease in postprandial chylomicronemia due to reduced intestinal triglyceride production.[8] In another feeding study using rabbits, the effects of nisoldipine and verapamil on the jejunal uptake of cholesterol and long-chain fatty acids were examined. Although no effect on cholesterol absorption was found, the two CCBs had significant and differing influences on the uptake of specific fatty acids.[9]

Experimental evidence also indicates that CCBs may affect VLDL metabolism. Using rat hepatocytes, verapamil was found to inhibit VLDL secretion directly, although the authors suggest that this effect was not due to the inhibition of calcium-dependent pathways.[10] In a hyperlipidemic rat model (JCR:LA-cp rats), diltiazem was also shown to reduce VLDL secretion. Of interest, although both female and male rats showed this effect, only male rats experienced a reduction in serum triglyceride levels, suggesting the presence of gender-specific modifiers of the CCB effect.[11] This study also raises the possibility that CCB effects on triglyceride levels may be more evident in hypertriglyceridemic states; as described above, normal male rats show no change in triglyceride levels in response to nifedipine or diltiazem.[8]

Evidence to support this contention was obtained in a rat model of chronic renal failure. Chronic renal failure is associated with hypertriglyceridemia due to reduced lipoprotein and hepatic lipase activities. In an inquiry focusing on the theory that secondary hyperparathyroidism associated with chronic renal failure increases the calcium burden in cells and that this increase may be responsible for the reduced lipase activities, the effects of verapamil on lipid metabolism were investigated in partially nephrectomized rats with and without parathyroidectomy.[12] Verapamil was shown to mimic the effects of parathyroidectomy: it prevented the fall in lipase activities and suppressed the rise in triglyceride levels resulting from chronic renal failure.

Evidence from cell culture systems suggests that CCBs may increase LDL uptake via the LDL-receptor. Stein et al. were the first to demonstrate this effect, using verapamil in isolated bovine aortic endothelial and smooth muscle cells as well as in human skin fibroblasts.[13] Impetus for the investigation flowed from the demonstration that Ca^{2+} is required for receptor-mediated endocytosis of LDL. Verapamil enhanced LDL uptake by increasing the number of LDL-receptors, an effect that requires protein synthesis. This effect also was found in both fibroblasts and human hepatoma cells with use of either verapamil or diltiazem but not with use of dihydropyridines.[14] If these effects occur in vivo, they may lead to an increased clearance of LDL through nonatherogenic pathways.

Lastly, although there have been no detailed experimental studies of HDL metabolism, CCBs have been shown to enhance HDL-mediated cell efflux of cholesterol, as discussed below.

In summary, although these results must be interpreted with caution in extrapolating the effects of CCBs to humans, animal and tissue culture experiments suggest the possibility that CCBs may have significant influences on triglyceride-rich lipoprotein and LDL metabolism. As discussed below, little evidence indicates that such influences have a significant bearing on steady-state plasma lipid and lipoprotein levels in humans. However, it is possible that compensatory mechanisms may minimize the effects of mild-to-moderate changes in lipoprotein metabolism on the lipid profile. In any event, each of the potential CCB effects described above tends to lower triglyceride and cholesterol levels, thus producing alterations regarded as beneficial from the standpoint of atherosclerosis (Table 1).

TABLE 1. Summary of the Effects of Calcium Channel Blockers on the Lipid Profile

Drug	Triglyceride (mg/dl)	Total Cholesterol (mg/dl)	LDL Cholesterol (mg/dl)	HDL Cholesterol (mg/dl)
Nifedipine	→	→	→ (↓)[19-21]	→ (↑)[25]
Verapamil	→	→ (↓, ↑)[22,23]	→	→
Diltiazem	→	→	→	→ (↑)[26]
Nicardipine	→	→	→	→
Nitrendipine	→	→	→	→ (↓)[27]
Nisoldipine	→	→	→	→ (↓)[28]
Isradipine	→	→	→	→ (↑ apo A-I)[29]
Amlodipine	→	→	→ (↓)[32]	→
Felodipine	→	→	→	→
Lacidipine	→	→	→	→

LDL = low-density lipoprotein, HDL = high-density lipoprotein, → = unchanged, ↓ = decreased, ↑ = increased, apo A-I = apolipoprotein A-I.

Human Studies

Since the original reports that diltiazem,[15] nifedipine,[16] and verapamil[17] had no effects on blood lipids, numerous studies essentially have confirmed these results. In a detailed and complete review of the literature, Trost and Weidman[18] analyzed a total of 43 published reports describing the measurement of lipids and lipoproteins in the course of studying diltiazem, nifedipine, and verapamil as well as the dihydropyridines, nitrendipine, nisoldipine, and nicardipine. As they point out, most studies lasted no longer than 6 weeks and therefore were easily influenced by acute fluctuations in body weight and nutrition of the relatively small numbers of participants (usually < 25).

Nevertheless, the data are mostly consistent in demonstrating no effect on levels of total, LDL, or HDL cholesterol or triglycerides. Three studies with nifedipine[19-21] and one with verapamil[22] reported small but significant decreases in total or LDL cholesterol, and two further studies, one with verapamil[23] and one with nitrendipine,[24] found small increases in total cholesterol. Two other studies, one using nifedipine[25] and the other using sustained-release diltiazem,[26] noted modest increases in HDL-C, whereas nitrendipine[27] and nisoldipine[28] were associated with small decreases in HDL-C. There seemed to be no effect trend related to type or mode of delivery of CCB, dose of agent, or age of subjects; thus there were no obvious explanations for these departures from the findings of most studies.

Since the publication of this detailed review, reports of investigations with the newer dihydropyridines—isradipine,[29-31] amlodipine,[32-34] felodipine,[35,36] and lacidipine[37]—have added further support to the conclusion that CCBs appear to be neutral with respect to their effects on the lipid profile. In addition, several long-term studies[34,35-38] have arrived at the same conclusion. A few studies have reported apolipoprotein measurements, particularly of HDL apo A-I and A-II, without detecting important effects.[29,30,39,40] One report documents that an 8-week course of nitrendipine enhanced fractional triglyceride transport in normolipidemic, hypertensive men.[41] However, fasting triglyceride levels were unchanged by the therapy in that study.[41] The novel CCB monatepil was recently demonstrated to reduce total and LDL cholesterol in comparison with nitrendipine, which had no effect, but this agent has significant $alpha_1$-blocking properties, which may explain its action.[42]

Several investigators have evaluated the effect of CCBs in type 2 diabetic patients with mild hypertriglyceridemia[35,43–45] and hypertensive hemodialyzed patients.[46] They found no effect. In type 1 diabetic patients with albuminuria but no hypertension, isradipine reduced VLDL cholesterol and triglyceride and increased HDL-C but did not affect albuminuria.[47] A long-term comparison of nifedipine and enalapril in a group of normolipidemic cigarette smokers and nonsmokers found that cigarette smoking significantly increased the tendency of nifedipine to raise the total cholesterol, LDL-C, and triglyceride and to lower the HDL-C noted in nonsmokers. Enalapril produced the opposite effects in nonsmokers, and this benefit was more pronounced in smokers. The authors suggested that increased sympathetic reactivity induced by smoking (which they measured) interacted with nifedipine to produce a deleterious effect on the lipid profile.[48] This interesting result merits further investigation.

Such results stand in clear contrast to the virtually unanimous findings of modest increases in triglycerides and a reduction in HDL-C levels reported with beta blockers and thiazide diuretics.[7] In many of the investigations that demonstrated no effect of CCBs on lipids, comparisons made in a parallel treatment arm with either a beta blocker or thiazide diuretic documented the familiar findings with both classes of drugs.[29,34,45,49–51] In addition, comparative studies with angiotensin-converting enzyme (ACE) inhibitors demonstrate that like CCBs, ACE inhibitors generally do not alter plasma lipids.[33,34,44,45,48] In the final analysis, these studies further emphasize the validity of the conclusion that CCBs produce no consistent alteration in the lipid profile. In comparison with antihypertensives such as beta blockers and thiazides, the lack of a deleterious effect on the fasting lipid profile represents an advantage. However, until more detailed kinetic studies are performed in both normolipidemic and dyslipidemic individuals, it seems premature to conclude that CCBs have no effect on the metabolism of circulating lipoproteins. Given the effects of CCBs on triglyceride metabolism noted in experimental studies,[8–12] studies of postprandial lipoproteins and LDL subtypes may be fruitful areas for future human research.

EFFECTS OF CALCIUM CHANNEL BLOCKERS ON CELLULAR LIPID METABOLISM

It has been known for many years that CCBs retard atherosclerotic plaque formation in experimental animals fed atherogenic diets despite the fact that serum lipids remain elevated.[52] This finding implies that CCBs reduce cholesterol accumulation in the arterial wall. The mechanism for this effect has been investigated in cell culture systems using monocyte-macrophage and smooth muscle cell lines.[53,54] Cholesterol accumulates pathologically mainly in cells of the monocyte-macrophage type via LDL-receptor–independent pathways mediated by scavenger receptors that do not downregulate when the cell is cholesterol-replete, thus providing a mechanism for cellular cholesterol overload.

Excessive uptake of cholesterol activates cholesterol esterification, which eventually leads to storage of cholesterol esters as lipid droplets. To bind to the scavenger receptor, LDL is believed to require chemical modification in the form of peroxidation, and oxidized LDL is a potent stimulator of monocyte-macrophage cholesterol ester accumulation and foam cell formation. Cholesterol-rich remnant particles (β-VLDL) are also able to induce foam cell formation in cultured cells.

As mentioned earlier, verapamil and diltiazem stimulate LDL uptake via the LDL-receptor, as does amlodipine.[52] This effect, however, does not lead to foam cell formation because of receptor downregulation. Nifedipine and verapamil but not diltiazem have been demonstrated to reduce cholesterol ester formation in macrophages treated with β-VLDL.[53] In addition, nifedipine has been shown to enhance cholesterol efflux from lipid-laden macrophages.[54] The realization that CCBs have antioxidant properties[55] has led to the investigation of their ability to prevent the in vitro oxidation of LDL. Several studies have documented that the dihydropyridines in particular have significant antioxidant effects in this regard.[56,57] In monkeys fed an atherogenic diet, amlodipine normalized the elevation of oxidized LDL in plasma and retarded atherosclerosis progression without reducing the raised LDL-C levels induced by the diet.[57] This newly discovered antioxidant activity of the CCBs represents a potentially important protective property against atherosclerosis and diseases characterized by increased free radical activity, such as diabetes. In summary, these investigations indicate that CCBs have important effects on cellular cholesterol metabolism that may reduce cholesterol accumulation in the arterial wall and retard formation of atherogenic plaque.

References

1. Haffner SM, Ferrannini E, Hazuda HP, et al: Clustering of cardiovascular risk factors in confirmed prehypertensive individuals. Hypertension 20:38–45, 1992.
2. MacMahon S, Peto R, Cutler J, et al: Blood pressure, stroke and coronary heart disease. Part I: Prolonged differences in blood pressure: Prospective, observational studies corrected for the regression dilution bias. Lancet 335:765–774, 1990.
3. Samelsson O, Wilmhelmsen L, Anderson OK, et al: Cardiovascular morbidity in relation to change in blood pressure and serum cholesterol levels in treated hypertension. JAMA 285:1768–1776, 1987.
4. Scandinavian Simvastatin Survival Study Group: Randomized trial of cholesterol lowering in 4444 patients with coronary heart disease: The Scandinavian Simvastatin Survival Study (4S). Lancet 344:1383–1389, 1994.
5. Stamler J, Wentworth D, Neaton JD: Is the relationship between serum cholesterol and risk of premature death from coronary heart disease continuous and graded? Findings in 356,222 primary screenees of the Multiple Risk Factor Intervention Trial (MRFIT). JAMA 256:2823– 2828, 1986.
6. Holme I: An analysis of randomized trials evaluating the effect of cholesterol reduction on total mortality and coronary heart disease incidence. Circulation 82:1916–1924, 1990.
7. Weidmann P, Uehlinger DE, Gerber A: Antihypertensive treatment and serum lipoproteins. J Hypertens 3:297–306, 1985.
8. Levy E, Smith L, Dumont L, et al: The effect of a new calcium channel blocker (TA 3090) on lipoprotein profile and intestinal lipid handling in rodents. Proc Soc Exp Biol Med 199:128–135, 1992.
9. Hyson DA, Thomson AB, Kappagoda CT: Differential and interactive effects of calcium channel blockers and cholesterol content of the diet on jejunal uptake of lipids in rabbits. Lipids 29: 281–287, 1994.
10. Nossen JO, Rustan AC, Drevon CA: Calcium antagonists inhibit secretion of very-low-density lipoprotein from culture rat hepatocytes. Biochem J 247:433–439, 1987.

11. Russell JC, Graham SE, Stewart B, et al: Sexual dimorphism in the metabolic response to the calcium channel antagonists, diltiazem and clentiazem, by hyperlipidemic JCR:LA-cp rats. Biochim Biophys Acta 1258:199–120, 1995.
12. Akmal M, Perkins S, Kasim SE, et al: Verapamil prevents chronic renal failure-induced abnormalities in lipid metabolism. Am J Kidney Dis 22:158–163, 1993.
13. Stein O, Leitersdorf E, Stein Y: Verapamil enhances receptor-mediated endocytosis of low density lipoproteins by aortic cells in culture. Arteriosclerosis 4:35–44, 1985.
14. Corsini A, Granata A, Fumagalli R, et al: Calcium antagonists and low density lipoprotein metabolism by human fibroblasts and human hepatoma cell line Hep G2. Pharmacol Res Commun 18:1–16, 1986.
15. Tsuyusaki T, Noro C, Yabata Y, et al: Clinical study on long-term oral administration of diltiazem hydrochloride (CRD-401). Jpn J Clin Exp Med 53:247, 1976 [in Japanese].
16. Delahaye JP, Touboul P, Gaspard P, et al: Le traitement de l'angine de poitrine de Prinzmetal par la nifedipine. Lyon Med 241:769–775, 1979.
17. Midtbo K, Hats O: Verapamil in the treatment of hypertension. Curr Ther Res 27:830–838, 1980.
18. Trost BN, Weidmann P: Effect of calcium antagonists on glucose homeostasis and serum lipids in non-diabetic and diabetic subjects: A review. J Hypertens 5(Suppl 4):S81–S104, 1987.
19. Franz IW, Wiewel D: Antihypertensive Wirkung von Nitredipin, Nifedipin und Acebutolol und deren Kombination auf den Ruhe- und Belastungsblutdruck bei Hochdruckkranken. Z Kardiol 74:111–116, 1985.
20. Landmark K: Antihypertensive and metabolic effects of long-term therapy with nifedipine slow-release tablets. J Cardiovasc Pharmacol 7:12–17, 1985.
21. Gonzalez Juanatey JR, Pose Reino A, Amaro Cedon A, et al: Valoracion de la eficacia antihipertensiva de la nifedipina en reposo y tras esfuerzo: Variacion de la actividad de renina plasmatica y de los lipidos plasmaticos durante el intervalo de tratamiento. Med Clin (Barc) 85:316–320, 1985.
22. Walldius G: Effect of verapamil on serum lipoproteins in patients with angina pectoris. Acta Scand Suppl 681:43–48, 1984.
23. Lehtonen A, Gordin A: Metabolic parameters after changing from hydrochlorothiazide to verapamil treatment in hypertension. Eur J Clin Pharmacol 27:153–157, 1984.
24. Fagan TC, Nelson EB, Lasseter KC, et al: Once- and twice-daily nitrendipine in patients with hypertension and noninsulin-dependent diabetes. Pharmacotherapy 6:128–136, 1986.
25. Vessby B, Abelin J, Finnson M, et al: Effects of nifedipine treatment on carbohydrate and lipoprotein metabolism. Curr Ther Res 33:1075–1081, 1983.
26. Pool PE, Seagren SC, Salel AF, et al: Effects of diltiazem on serum lipids, exercise performance and blood pressure: Randomized, double-blind, placebo-controlled evaluation for systemic hypertension. Am J Cardiol 56:86H–91H, 1985.
27. Francischetti EA, Oigman W, Fagundes VGA, et al: Long-term therapy with nitrendipine: Evaluation of its antihypertensive and metabolic effects. J Cardiovasc Pharmacol 9(Suppl 4):S107–S112, 1987.
28. Odigwe CO, McCulloch AJ, Williams DO, Tunbridge WMG: A trial of the calcium antagonist nisoldipine in hypertensive non–insulin-dependent diabetic patients. Diabetic Med 3:463–467, 1986.
29. Samuel P, Kirkendall W, Schaefer E, et al: Effects of isradipine, a new calcium antagonist, versus hydrochlorothiazide on serum lipids and apolipoproteins in patients with systemic hypertension. Am J Cardiol 62:1068–1071, 1988.
30. Lacourciere Y, Gape C, Brun D, et al: Beneficial effects of the calcium antagonist isradipine on apolipoproteins in hypertensive patients. Am J Hypertens 4:181S–184S, 1991.
31. McCarron DA, Weder AB, Egan BM, et al: Blood pressure and metabolic responses to moderate sodium restriction in isradipine-treated hypertensive patients. Am J Hypertens 10:68–76, 1997.
32. Lopez LM, Thorman AD, Mehta JL, et al: Effects of amlodipine on blood pressure, heart rate, catecholamines, lipids and responses to adrenergic stimulus. Am J Cardiol 66:1269–1271, 1990.
33. Omvik P, Thaulow E, Herland OB, et al: A double-blind, long-term, comparative study on quality of life, safety, and efficacy during treatment of amlodipine or enalapril in mild or moderate hypertensive patients: A multicenter study. J Cardiovasc Pharmacol 22(Suppl A):S13–S19, 1993.
34. Grimm RH Jr, Flack JM, Grandits GA, et al: Long-term effects on plasma lipids of diet and drugs to treat hypertension. Treatment of Mild Hypertension Study (TOMHS) Research Group. JAMA 275:1549–1556, 1996.
35. Shionoiri H, Takizawa T, Ohyama Y, et al: Felodipine therapy may not alter glucose and lipid metabolism in hypertensives. Felodipine Multicenter Prospective Study Group in Japan. Hypertension 23(Suppl 1):I215–I219, 1994.

36. Carroll J, Shamiss A, Zevin D, et al: Twenty-four-hour blood pressure monitoring during treatment with extended-release felodipine versus slow-release nifedipine: Cross-over study. J Cardiovasc Pharmacol 26:974–977, 1995.
37. Soro S, Ferrara A: Effect of lacidipine, a long-acting calcium antagonist, on hypertension and lipids: A 1 year follow-up. Eur J Clin Pharmacol 41:105–107, 1991.
38. Sawai K: Effects of long-term administration of diltiazem hydrochloride in hypertensive patients. Clin Ther 5:422–435, 1983.
39. Sasaki J, Arakawa K: Effects of short- and long-term administration of nifedipine on serum lipoprotein metabolism in patients with mild hypertension. Cardiovasc Drugs Ther 4:1033–1036, 1990.
40. Sasaki J, Kajiyama G, Kusukawa R: Effects of a new calcium channel blocker, MPC-1304, on blood pressure, serum lipoproteins and serum carbohydrate metabolism in patients with essential hypertension. Int J Clin Pharmacol Ther 6:366–370, 1995.
41. Marotta T, Ferrara LA, Pasanisi F, et al: Enhancement of exogenous triglyceride removal following calcium channel blockade. Artery 16:312–326, 1989.
42. Sasaki J, Ogihara T, Yokoyama M, et al: Comparative effects of monatepil, a novel calcium antagonist with α_1-adrenergic-blocking activity, and nitrendipine on lipoprotein and carbohydrate metabolism in patients with hypertension. Am J Hypertens 7:161S–166S, 1994.
43. Zanetti-Elshater F, Pingitore R, Beretta-Piccoli C, et al: Calcium antagonists for treatment of diabetes-associated hypertension. Metabolic and renal effects of amlodipine. Am J Hypertens 7:36–45, 1994.
44. Chan JC, Yeung VT, Leung DH, et al: The effect of enalapril and nifedipine on carbohydrate and lipid metabolism in NIDDM. Diabetes Care 17:859–862, 1994.
45. Giordano M, Matsuda M, Sanders L, et al: Effects of angiotensin-converting enzyme inhibitors, Ca^{++} channel antagonists, and α-adrenergic blockers on glucose and lipid metabolism in NIDDM patients with hypertension. Diabetes 44:665–671, 1995.
46. Riegel W, Horl WH, Heidland A: Long-term effects of nifedipine on carbohydrate and lipid metabolism in hypertensive hemodialyzed patients. Klin Wochenschr 64:1124–1130, 1986.
47. Norgaard K, Jensen T, Feldt-Rasmussen B: Effects of isradipine in Type I (insulin-dependent) diabetic patients with albuminuria and normal blood pressure. J Hum Hypertens 6:145–150, 1992.
48. Nazzaro P, Cicco G, Manzari M, et al: Lipids and cardiovascular reactivity changes in hypertensive cigarette smokers: Enalapril versus nifedipine treatment effects. Am J Med Sci 307 (Suppl 1):S150–S153, 1994.
49. Lacourciere Y, Poirer L, Lefebvre J, et al: Comparative effects of a new cardioselective betablocker nebivolol and nifedipine sustained release on 24-hour ambulatory blood pressure and plasma lipoproteins. J Clin Pharmacol 32:660–666, 1992.
50. Lijnen P, Van Hoof R, Amery A: Effects of celiprolol vs. nifedipine on serum lipoproteins in patients with mild to moderate hypertension. Cardiovasc Drugs Ther 8:509–513, 1994.
51. Saku K, Zhang B, Okamoto T, et al: Medium-term effects of betaxolol monotherapy and combination therapy with nitrendipine on lipoprotein and apolipoprotein metabolism in patients with mild to moderate essential hypertension. J Hum Hypertens 10:263–268, 1996.
52. Bernini F, Catapano AL, Corsini A, et al: Effects of calcium antagonists on lipids and atherosclerosis. Am J Cardiol 64:129I–134I, 1989.
53. Daugherty A, Ratert DL, Schonfeld G, et al: Inhibition of cholesterol ester deposition in macrophages by calcium entry blockers: An effect dissociated from calcium entry blockade. Br J Pharmacol 91:113–118, 1987.
54. Schmitz G, Robenek H, Bueck M, et al: Ca^{++} antagonists and ACAT inhibitors promote cholesterol efflux from macrophages by different mechanisms. Arteriosclerosis 8:46–56, 1988.
55. Lupo E, Locher R, Weisser B, et al: In vitro antioxidant activity of calcium antagonists against LDL oxidation compared with α-topopherol. Biochem Biophys Res Commun 203:1803–1808, 1994.
56. Napoli C, Chiariello M, Palumbo G, et al: Calcium-channel blockers inhibit human low-density lipoprotein oxidation by oxygen radicals. Cardiovasc Drugs Ther 10:417–424, 1996.
57. Kramsch DM, Sharma RC: Limits of lipid-lowering therapy: The benefits of amlodipine as an anti-atherosclerotic agent. J Hum Hypertens 9(Suppl 1):S3–S9, 1995.

ALEXANDER SCRIABINE M.D.

21

Calcium Antagonists and Central Nervous System Diseases

Calcium antagonists are primarily cardiovascular drugs, although their binding sites in the brain were discovered soon after their introduction in the therapy of angina pectoris and hypertension. The multiplicity of calcium channels in neurons and poor penetration of the blood–brain barrier by first-generation agents are probably responsible for the failure of early investigators to recognize the therapeutic potential of calcium antagonists for central nervous system (CNS) disorders. Even today the only application in CNS disorders approved by the Food and Drug Administration (FDA) is the use of nimodipine for prevention of neurologic deficits after subarachnoid hemorrhage. Calcium antagonists were studied experimentally in many models of CNS diseases with promising findings. Some of these findings were followed clinically, mostly with insufficient or inconclusive results. Anecdotal observations of the beneficial effects of calcium antagonists in certain CNS diseases were made in the clinic with no experimental support. This chapter summarizes the available information about the effects of calcium antagonists on the CNS in animals and humans and assesses their potential applications in neurology and psychiatry.

CALCIUM CHANNELS IN THE BRAIN

At least six types of voltage-dependent calcium channels were identified in neurons by electrophysiologic techniques.[1–8] They differ in conductance, activation and inactivation voltages, and sensitivity to various drugs or toxins capable of blockade or activation. The calcium channel types are identified by capital letters: L-, N-, T-, P-, Q-, and R-type channels (Table 1). Their distribution varies within the brain areas, and their location within neurons is not homogeneous. L-type channels are preferentially found in the neuronal cell bodies,[9] whereas N-type channels are located primarily in the presynaptic active zones. Each channel is a multiple subunit complex composed of the ion-conducting α_1 subunit and accessory α_2, β, and δ subunits. In the skeletal muscle an additional γ subunit is present. The α_1 subunit can function alone as a calcium channel; other subunits play only a modulatory role. In neurons five classes of calcium channel genes encode for the α_1 subunit; L-type currents appear to be generated by α_{1C} and α_{1D}, N-type by α_{1B}, and P-type by α_{1A}.[10] All identified subunits have now been purified, sequenced, cloned, and expressed. The molecular biology of calcium channels has recently been reviewed by Perez-Reyes and Schneider[11] as well as Spreyer et al.[12] According to these investigators, the expression of many

TABLE 1. Voltage-dependent Neuronal Calcium Channels and Their Antagonists

	Characteristics of Current			
Channel Type	Conductance PS	Activation Voltage	Inactivation	Antagonists
L	25	High	Slow	*Dihydropyridines:* nifedipine, nitrendipine, nisoldipine, nimodipine, isradipine, darodipine, amlodipine, felodipine, nicardipine, nilvadipine, cerebrocrast, and others *Phenylalkylamines:* verapamil, gallopamil, levemopamil, and others *Benzothiazepines:* diltiazem, clentiazem Others: flunarizine, cinnarizine, lidoflazine, lomerizine, lifarizine, nicergoline, R56865, SB 201825A, NNC 09-0026, and others
N	12–20	High	Moderate	GVIA, MVIIA (SNX-111), SNX-325
T	8	Low	Rapid	ω-Aga IIB, felodipine, niguldipine, flunarizine
P	10–12	Moderate	Very slow	ω-Aga IVA, daurisoline
Q	10–15	High	Moderate	MVIIC, Aga IVA
R	10–15	Moderate	Rapid	?

α_1 and β genes and the capability of a cell to control the composition of subunits contribute to the diversity of calcium channels.

ANTAGONISTS OF L-TYPE CALCIUM CHANNELS: CEREBROVASCULAR VS. NEURONAL EFFECTS

There are three major chemical classes of L-type calcium channel antagonists: dihydropyridines, phenylalkylamines, and benzothiazepines. In addition, many other drugs have been described as blocking L-type calcium channels. Each of the three classes has its distinct binding site at the calcium channel. Some of the blockers of L-type channels (e.g., felodipine, niguldipine) were found to block T-type channels as well. In vitro all L-type calcium channel antagonists block calcium currents in neurons, although their affinity for the neuronal channels varies greatly. In vivo antagonism of neuronal channels cannot be demonstrated with every calcium antagonist because of the limited ability of some drugs to enter the CNS or their higher selectivity for vascular or cardiac calcium channels. Peripheral vasodilatation tends to oppose the neuroprotective effect of calcium channel antagonists: pronounced lowering of arterial pressure produced at higher doses may lead to a decrease in cerebral blood flow and consequently enhance rather than reduce neuronal damage. The inverted U-shape dose–response curve for neuroprotective activity is, therefore, characteristic of L-type calcium channel antagonists. Some of the negative results with these drugs can be explained by the investigators' failure to establish a dose–response curve and to select an appropriate dose for clinical studies.

Cerebral vasodilator properties of antagonists of L-type calcium channels were recognized during the initial pharmacologic studies. The effects of nimodipine on the cerebral circulation were described by Kazda et al.,[13] whereas Towart[14] demonstrated its cerebrovascular selectivity. Hoffmeister et al. first described the CNS effects of nimodipine in conscious animals.[15] Their findings, as well as those of Cohen and Allen,[16] led to the clinical evaluation of nimodipine in the prevention of cerebrovascular spasms in patients with subarachnoid hemorrhage (SAH) (see below).

A reasonable assumption is that neuronal as well as cerebrovascular vasodilator effects of L-type calcium channel antagonists determine their effectiveness in the treatment of CNS diseases. Excessive uptake of calcium by neurons from the extracellular space and impaired cerebral blood flow are likely to be detrimental to normal brain function.

CENTRAL NERVOUS SYSTEM EFFECTS OF T-, N-, AND P-TYPE CALCIUM CHANNEL ANTAGONISTS

T-type calcium current is characterized by fast activation and inactivation kinetics and strong voltage dependence. It was described first by Llinás and Yarom[17] in neurons from the inferior olivary nucleus of guinea pigs and subsequently by Carbone and Lux in the dorsal root ganglion neurons of rats and chicks.[18] Some of the dihydropyridines with L-type channel-blocking activity (e.g., felodipine, niguldipine) also block T-type channels. Among other clinically used calcium antagonists, flunarizine was reported to block T-type calcium channels. Many experimental studies involving T-type currents used a peptide from funnel web spider venom, ω-Aga IIB.[19] This peptide is a potent but not selective antagonist of T-type currents.

The differences in the function of T-type and L-type channels are not yet completely understood, although T-type channels appear to control primarily spontaneous fluctuations of membrane potential. Such an oscillatory activity is involved in motor coordination but also facilitates convulsions; thus, the antagonists of T-type calcium channels may be indicated in the treatment of epilepsy. Other possible indications for T-type channel ligands are appetite control and sleep disturbances.

N-type calcium channels are activated by high voltage and involved primarily in the release of neurotransmitters (e.g., release of norepinephrine from sympathetic nerve endings). The first specific inhibitors of N-type calcium channels—peptides from the marine snail, *Conus geographus*—were discovered by Olivera et al.[20] They are known as ω-conotoxins: GVIA, MVIIA, and MVII. Subsequently many ω-conotoxins were synthesized at Neurex Corporation and carry SNX code numbers. SNX-111 is a synthetic MVIIA that was studied extensively as a neuroprotective agent.[21] Another relatively specific N-type calcium channel antagonist, SNX-325, was isolated from the toxin of a spider (*Sequestria florentina*).[22] SNX-325 blocks N-type channels selectively at nanomolar concentrations, whereas at micromolar levels it blocks most calcium currents. Koike et al.[23] described inhibition of N-type channels in isolated rat neurons by aniracetam and suggested that this effect may play a role in the drug's nootropic activity.

P-type calcium channels were discovered by Llinás et al. in 1992.[24] They are probably the most widely distributed calcium channels in the mammalian

CNS. P-type channels are characterized by high activation voltage with little or no inactivation. The specific and highly potent inhibitor of P-type channels is a 48-amino acid peptide, ω-Aga-IVA, which was found in a toxin from the funnel web spider (*Agelonopsis aperta*). In cerebellar Purkinje neurons ω-Aga-IVA blocked 90% of calcium current at the concentration of 20 nM.[24a]

Other spider venom peptides were found to be less specific antagonists of P-type calcium channels. Daurisoline, an alkaloid from a Chinese medicinal herb (*Menispermum dauricum*), was described as blocking P-type calcium channels.[25] The herb is used in traditional Chinese medicine for the treatment of epilepsy, hypertension, and asthma. Llinás et al. suggested that P-type channels are involved in neuronal degeneration.[24,26] It is conceivable, therefore, that antagonists of P-type calcium channels may be useful in the treatment of neurodegenerative diseases.

No specific inhibitors have been described for Q- or R-type calcium channels, although Q-type channels can be inhibited by ω-conotoxins MVIIC and Aga-IVA, the same peptides that inhibit N- and P-type channels, respectively.

EXPERIMENTAL AND CLINICAL STUDIES WITH L-TYPE CALCIUM CHANNEL ANTAGONISTS

Table 2 summarizes clinical studies of calcium antagonists in various CNS disorders.

Subarachnoid Hemorrhage

The first controlled clinical trial of calcium channel antagonists in the treatment of patients with SAH was conducted by Allen et al.[27] with nimodipine. Their rationale was based on experimental findings that in vitro nimodipine selectively antagonizes cerebrovascular spasms caused by serotonin and other vasoconstrictor substances and that prevention of vasospasm after SAH prevents or reduces neurologic deficits due to vascular spasms and/or blood clots. The trial was successful but failed to convince the FDA of the potential value of nimodipine. Two more large and successful clinical trials were performed in France[28] and the United Kingdom[29] before the FDA approved the use of nimodipine for improvement of neurologic outcome by reducing the incidence of ischemic deficits in patients with SAH from ruptured congenital aneurysms. Another calcium antagonist, nicardipine, also was found to have beneficial effects in the treatment of patients with SAH.[30] Although it is likely that other calcium antagonists can prevent neurologic deficits in patients with SAH, only nimodipine is approved in the United States for this indication.

Ischemic Stroke

The experimental basis for the use of L-type calcium antagonists in patients with ischemic stroke includes findings that calcium overload of ischemic neurons contributes to cell death and that many calcium antagonists are effective in various experimental stroke models.[31–33] The first clinical trial of oral nimodipine in patients with ischemic stroke was conducted by Gelmers et al.[34] It was highly

TABLE 2. Clinical Studies with Calcium Antagonists in Central Nervous System Disorders

Central Nervous System Disorders	Drugs	References
Subarachnoid hemorrhage (SAH)	Nimodipine	27–29
	Nicardipine	30, 49
Ischemic stroke	Nimodipine	34, 35
	Nicardipine	44
Head trauma	Nimodipine	47, 48
Global ischemia	Nimodipine	69, 70
Epilepsy	Flunarizine	82
	Nimodipine	83–86
Vascular dementia	Nimodipine	87
Primary degenerative and multiinfarct dementias	Nimodipine	88–90
Old-age dementia	Nimodipine	104–106
Alzheimer's dementia	Nimodipine	112, 113
Depression	Nimodipine	115
	Verapamil	116
Bipolar disease (manic depression)	Verapamil	117, 118
	Nimodipine	119–121
Panic attacks	Nimodipine	122, 123
	Verapamil	124
Migraine	Flunarizine	125–127
Cocaine addiction	Nimodipine	135
Hearing disorders, including tinnitus, Ménière's disease	Nimodipine	166, 167, 169, 170
Tourette's syndrome	Nifedipine	172, 173
	Verapamil	173
Essential tremor	Nimodipine	177

successful. Nimodipine, 30 mg every 6 hours orally, prevented neurologic deficits in the great majority of patients with ischemic stroke. No subsequent study produced such dramatic results. The main reason for the failure of other investigators to confirm the findings of Gelmers et al. became obvious in a meta-analysis of nine studies involving 3,700 patients.[35] The study by Gelmers et al. had the highest number of patients (50%) who received the drug within the first 12 hours. The metaanalysis clearly demonstrated that nimodipine was effective only if given within the first 12 hours after the stroke. Nimodipine was ineffective or even detrimental if started at 24 hours or longer after the stroke.

In experimental stroke models, many other L-type calcium channel antagonists, including nicardipine,[36] darodipine,[37] isradipine,[38] nilvadipine,[39] S-emopamil,[40] lomerizine,[41], lifarizine,[42] and clentiazem,[43] reduced brain infarct size. L-type calcium channel antagonists were effective in the experimental stroke models if given before or, in some studies, shortly after experimental stroke. The initial clinical studies with nicardipine were encouraging.[44] Recent data with the N-type calcium channel antagonist, SNX-111, suggest that the therapeutic window is substantially longer than with L-type channel antagonists and that

SNX-111 may be effective if given as late as 24 hours after the stroke.[21] Clinical studies with SNX-111 in patients with ischemic stroke are ongoing.

Head Trauma

Acute head trauma may lead to secondary ischemic brain injury by interfering with blood flow to brain tissue adjacent to the injured area. Because nimodipine, S-emopamil, and other calcium antagonists had beneficial effects in some animal models of brain injury,[45,46] Bailey et al.[47] evaluated the effects of nimodipine on outcome after head injury in humans. Nimodipine produced no substantial improvement in outcome in 350 patients with severe head injury, but some patients appeared to benefit slightly. The first trial was followed by a second multicenter trial involving 800 patients with severe head injury.[48] Again nimodipine was found to have no significant beneficial effects in most patients.

Initial clinical studies in patients with severe head injury suggested possible protection by SNX-111, an N-type calcium channel antagonist. Intravenous SNX-111 had beneficial effects in 4 of 7 patients with head trauma, and the investigators suggested that a double-blind study with SNX-111 is warranted in patients with head trauma.[48a]

Traumatic Subarachnoid Hemorrhage

Although the overall effects of nimodipine were not encouraging in the second trial in patients with head trauma, careful retrospective analysis of data suggested that nimodipine may have reduced neurologic deficits in a subgroup of patients who suffered SAH as a consequence of head trauma.

To evaluate further the usefulness of nimodipine in traumatic SAH, a prospective, randomized, double-blind, placebo-controlled trial, involving 123 patients, was conducted in 23 German neurosurgical clinics. The results were published by Harders et al.[49] The patients were treated with intravenous nimodipine followed by oral therapy for 3 weeks. Treatment was initiated within 12 hours after injury. Nimodipine significantly improved the outcome. Unfavorable outcome (death or severe disability) was observed in 46% of patients receiving placebo and only 25% of nimodipine-treated patients. The difference was highly significant statistically. On the basis of this study the FDA approved the use of nimodipine in patients with traumatic SAH.

Spinal Cord Injury

In cultured rat spinal cord neurons Bär et al.[50] found that depolarization-induced calcium entry inhibits neurite outgrowth and that nimodipine prevents this inhibition. In isolated spinal cord explants from mouse embryos, Tymianski and Tator demonstrated that nimodipine blocks the rise in $[Ca^{2+}]_i$ in response to potassium-induced depolarization.[51] This effect was demonstrable if nimodipine was applied before or during but not later than 10 minutes after application of high K^+. The authors concluded that the therapeutic window of nimodipine is probably too short for clinical application in the treatment of spinal cord injury.

Mixed results were obtained with nimodipine in in vivo models of spinal cord injury. In the initial study in rabbits, Faden et al.[52] failed to demonstrate any beneficial effect of nimodipine. Subsequently Fehlings et al.[52] found that nimodipine improved axonal function in rats with spinal cord trauma, but combination treatment with nimodipine and dextran was more effective than either substance alone. In baboons nimodipine, infused immediately after spinal cord injury, improved spinal cord blood flow and reduced spinal cord lesions.[54] In rabbits with spinal cord compression nimodipine infused with epinephrine improved motor deficits better than either treatment alone.[55] A dramatic prevention of neurologic deficits was observed with nimodipine in pigs subjected to aortic occlusion.[56] Haghighi et al.[57] failed to find any beneficial effects of nimodipine at hypotensive doses on axonal function in cats with spinal cord injury. Negative results with nimodipine in spinal injury models were reported by Holtz et al.[58] and Ceylan et al.[59] Nicardipine was also ineffective in a rat model of spinal cord injury.[60] Rhee et al.[61] reported recently that nimodipine alone failed to produce significant neurologic benefits in rats with spinal cord ischemia caused by thoracic aortic cross-clamping but decreased postischemic reperfusion hyperemia.

The general impression from the published experimental studies with nimodipine and other calcium antagonists is that they may produce beneficial effects in spinal cord injury if administered immediately after injury and at doses that cause no substantial fall in arterial pressure. Additional benefits may be obtained with simultaneous administration of dextran or a pressor agent to antagonize the hypotensive effect of calcium antagonists.

Global Ischemia

Experimental studies with calcium antagonists in animal models of global ischemia produced mixed results. Steen et al.[62,63] reported positive results with nimodipine in dogs and primates, whereas flunarizine was ineffective in the same laboratory.[64] In other studies flunarizine protected Mongolian gerbils[65] and rats[66] from global ischemic damage. Hypoxia is usually associated with an increase in brain levels of arachidonate and memory decline in rats; these effects were antagonized by nimodipine.[67] Nakamura et al.[68] found, however, that neither nimodipine nor nicardipine protected gerbils from delayed neuronal death after brief forebrain ischemia.

Global brain ischemia is encountered in patients resuscitated after cardiac arrest or a period of ventricular fibrillation. A drug that is neuroprotective in animal models of global ischemia may be expected to improve neurologic outcome in resuscitated patients. In the initial study of nimodipine in 230 patients with delayed resuscitation from out-of-hospital cardiac arrest, Roine and Kaste[69] found improved outcome and prolonged survival. In a subsequent larger trial no significant effect of nimodipine was observed.[70]

Brain Retraction Ischemia

During neurosurgery brain tissue often has to be retracted. This retraction causes ischemic damage and reduces cerebral blood flow as well as evoked potentials. In an attempt to reduce such damage with neuroprotective drugs, Andrews and Muto[71] developed an animal model in miniature swine. They studied the effects of mannitol and nimodipine, separately and together, in a group of 27 animals and found that both cerebral blood flow and evoked potential were better preserved during retraction ischemia in animals treated with a combination of mannitol and nimodipine than in controls or animals treated with either drug alone.[72] Nimodipine alone, as an infusion of 1.0 μg/kg/min, had little effect on normal ventilation but preserved evoked potentials and ameliorated alkalosis in hyperventilation. The authors recommend further studies and eventual use of the combination of mannitol and nimodipine as preventive treatment in surgery requiring brain retraction.

Epilepsy

L-type calcium channel antagonists of all three major classes have anticonvulsant activity in animal models. Most experimental studies used verapamil, flunarizine, or nimodipine.[73–78] According to Meyer et al.,[79] nimodipine protects rabbits from convulsions induced by direct electrical stimulation of the brain. In models of chronic epilepsy (amygdala-kindled seizures), flunarizine—but not nimodipine, verapamil, or nitrendipine—had anticonvulsant activity.[80] In another study verapamil and nimodipine reduced kindling rate and afterdischarge duration in adult and immature rats.[81]

In clinical studies, flunarizine[82] and nimodipine[83] had anticonvulsant activity in childhood lesional as well as adult epilepsy. Hans et al.[84] successfully treated drug-resistant posttraumatic epilepsy with nimodipine. Sasso and Mancia[85] studied the antiepileptic activity of nimodipine in a double-blind, placebo-controlled trial involving 32 patients with refractory epilepsy and found that nimodipine was superior to placebo in reducing the frequency and duration of seizures in 50% of patients. Meyer et al.,[86] however, were unable to demonstrate the effectiveness of nimodipine in patients with highly refractory seizure disorders.

In general, experimental and clinical evidence suggests that L- or T-type calcium channel antagonists can control some forms of epilepsy, even forms resistant to other drugs. Further clinical studies are required, however, to determine which calcium channel antagonists are most effective and in which forms of epilepsy.

Dementias

Vascular Dementia

The rationale for use of nimodipine in vascular dementia is based on its cerebral vasodilator effect and an assumption that the increase in cerebral blood flow may improve or slow the rate of deterioration of the mental state of patients with vascular dementia. In a double-blind, placebo-controlled study, Tobares et al. evaluated nimodipine in 65 patients with senile vascular dementia.[87] The drug was administered for 24 weeks. The results were evaluated by two different scales—the Crichton Geriatric Behavioral Rating Scale and the Performance Test of Activities of Daily Living. The study suggested cognitive improvement in nimodipine-treated patients, although beneficial effects were demonstrated only with the first scale.

Primary Degenerative and Multiinfarct Dementia

In a double-blind, placebo-controlled study, Fischof et al.[88,89] evaluated nimodipine in 130 patients, 68 of whom were diagnosed with primary degenerative dementia (PDD) and 62 with multiinfarct dementia (MID). After 12 weeks of treatment, nimodipine significantly improved cognitive functions as determined by either the Syndrome Short Test (SKT) or the Sandoz Clinical Assessment Geriatric Scale (SCAG). Nimodipine was effective in patients with either PDD or MID. Adan et al.[90] evaluated nimodipine in a double-blind, placebo-controlled study of 53 patients with either vascular dementia or MID. At the end of the study (86 days), 89% of patients receiving nimodipine, 30 mg 3 times/day, and 48% of patients receiving placebo improved (as assessed by a modified SCAG scale).

Old-age Dementia

The scientific basis for the evaluation of calcium antagonists in the treatment of old-age dementia includes the calcium hypothesis of brain aging proposed by Khachaturian.[91–93] According to this hypothesis, long-lasting disturbances in neuronal calcium homeostasis lead to impairment of brain function. This hypothesis is supported by many experimental studies, including Landfield's demonstration of calcium-dependent afterhyperpolarization in hippocampal neurons of aged rats,[94] findings by LeVere et al.[95,96] that nimodipine alleviates short-term memory dysfunction in old rats and nonhuman primates, and findings by Deyo et al.[97] and Straube et al.[98] that nimodipine facilitates acquisition of conditioned eyeblink responses in aging rabbits. Age-related abnormalities in cerebral microvessels, including membranous inclusions and microvascular deposits, were reduced by addition of nimodipine, 860 ppm, to the food of rats for 6 months. The behavioral effects of the drug were correlated with microvascular changes.[99] Nimodipine-induced improvement of motor deficits in aged rats, as demonstrated by Schuurman et al.,[100,101] provides additional justification for the use of nimodipine in aged humans. Yamamoto et al.[102] demonstrated that nicardipine and nilvadipine also ameliorate brain dysfunction in senescence-accelerated mice (SAM). They found that nimodipine attenuates age-related decrease in brain acetylcholine, serotonin, and dopamine levels and antagonizes an increase in monoamine oxidase-A activity.

One of the problems associated with the clinical trials of antidementia drugs in the elderly is the correct diagnosis of the type of brain dysfunction. A prerequisite for the acceptance of clinical results is proper identification of the disease. In Europe a popular diagnosis for elderly patients suffering from memory dysfunction is "chronic organic brain syndrome," a term that is not accepted by American psychiatrists. Many European psychiatrists differentiate between "organic" conditions, which are associated with anatomic evidence of brain lesions, and "functional" conditions, which have no such evidence. The best European results with nimodipine were obtained in elderly patients with memory dysfunction often attributed to "organic brain syndrome." Schmage et al.[104] summarized 11 double-blind European studies of nimodipine in such patients. Global improvements were assessed after 2–4 months of therapy. Excellent results were reported in 212 of 343 patients receiving nimodipine and only 63 of 347 patients receiving placebo.

Two multicenter studies involving a total of 755 elderly patients suffering from mental deterioration were conducted in Italy.[105] At an oral dose of 30 mg 3 times/day for 3 or 6 months, nimodipine improved cognitive functions as documented by psychobehavioral scales and psychometric tests. In another placebo-controlled study of 84 elderly patients, nimodipine was found to improve the symptoms of mental deterioration.[106] Although experimental design, diagnosis, or method of drug evaluation is open to criticism in these studies, the results should not be ignored. It is likely that one of the most promising indications for nimodipine or similar calcium antagonists is old-age dementia or memory dysfunction in the elderly.

Dementia Related to Human Immunodeficiency Virus

Infection with human immunodeficiency virus-1 (HIV-1) may be associated with neurologic manifestations involving cognition, movement, and sensation. These manifestations occur in one-third of adult patients with acquired immunodeficiency syndrome (AIDS) and are referred to as AIDS-related dementia.[107] The neuronal injury in AIDS-related dementia probably is caused indirectly by viral release of neurotoxic substances from macrophages or monocytes or by viral proteins (e.g., Gp 120). The neurotoxic substances that are likely to mediate such injury include eicosanoids, cytokines, and platelet-activating factor (PAF). These substances may increase intracellular calcium by increasing calcium currents in neurons through N-methyl-D-aspartate (NMDA) or other types of calcium channels. Because the increase in neuronal calcium is likely to be neurotoxic, it is logical to expect that NMDA- or L-type calcium antagonists may be neuroprotective in HIV-related neuronal injury. In vitro nimodipine and other calcium antagonists were shown to protect various neurons from Gp 120-induced toxicity.[108] No clinical studies of calcium antagonists in AIDS-related dementia have been published.

Alzheimer's Dementia

The idea that calcium antagonists may be useful in the treatment of Alzheimer's dementia is based on the assumptions that brain aging is a contributory factor and that abnormal calcium homeostasis facilitates neuronal death in brain aging.[93] If Ca^{2+} overload of ischemic neurons contributes to their death, prevention of such overload by calcium antagonists may prolong the life of diseased neurons and slow brain aging as well as the development of Alzheimer's disease. Other findings supporting the use of calcium antagonists in Alzheimer's disease include observations by Weis et al.[109] that nimodipine antagonizes β-amyloid-induced neurotoxicity in cultured cortical neurons and by Chisholm et al.[110] that nimodipine antagonizes elevation of intracellular free calcium caused by a toxic amyloid fragment (β-AP 25-35) in rat hippocampal neurons. There is also an age-independent loss of [^{3}H]nitrendipine-binding sites in patients with Alzheimer's disease.[111]

One of the initial trials of nimodipine in patients with Alzheimer's dementia was performed by Baumel et al.[112] In a double-blind, placebo-controlled study, nimodipine, at either 30 or 60 mg/day, was administered for 12 weeks. The results were questionable. Improvement was seen in only one variable (Clinical Global Impression [CGI]) and only with the low dose (30 mg). Subsequently, nimodipine was evaluated in two large multicenter, placebo-controlled trials involving a total of 1,648 patients with probable Alzheimer's disease in the United States and Canada.[113] A total of 1,377 patients completed the 26-week trials. Nimodipine was administered orally at higher doses than in previous trials (90 or 180 mg/day). The results indicated that primary outcome variables were not improved by nimodipine, although the retrospective analysis of patient subgroups indicated that in 497 more severely ill patients nimodipine at a dose of 180 mg reduced the rate of deterioration. This subgroup responded positively on the Buschke Selective Reminding (BSR) test, an index of memory storage. The overall results, however, were not sufficiently encouraging, and no additional clinical studies of nimodipine or other calcium antagonists have been conducted in patients with Alzheimer's disease.

Depression

Nimodipine was reported to be active in two animal models of depression: rat forced swimming and learned helplessness tests.[114] In 6 of 7 human patients, nimodipine decreased scores on the Hamilton Depression Scale.[115] Although verapamil appeared to have antidepressant activity in occasional patients, Höschl and Kozeny failed to confirm this effect in a double-blind, controlled study.[116]

Bipolar Disorders

Favorable results in bipolar disorders (manic depression) were reported with verapamil.[117,118] Verapamil appears to control mania and may be considered as an alternative to lithium. Beneficial effects of nimodipine were observed by Brunet et al.[119] in a small open study (6 manic depressive patients). Pazzaglia et al.[120] studied the effects of nimodipine in a controlled study of 12 patients with manic depression refractory to lithium and found that it reduced mood fluctuations. These findings suggested that the mechanism of action of nimodipine in manic depressive patients may differ from that of lithium and justified evaluation of the combination of nimodipine and lithium in manic depressive patients. According to Grunze et al.,[121] such a combination may be more effective than either drug alone and deserves further evaluation.

Panic Attacks

Panic attacks are defined as brief episodes of terror without obvious cause. The attacks are associated with a sympathetic crisis that includes tachycardia, dizziness, shortness of breath, and trembling. Such attacks recur, in some patients as often as a few times per week. The pathogenesis of panic attacks is still controversial, but a distinct abnormality in the temporal lobes of patients suffering from panic attacks has been described. The standard treatment is benzodiazepines, but nimodipine and verapamil were reported to produce beneficial effects.[122–124] The mechanism of action of calcium channel antagonists in patients with this disorder is not understood.

Migraine

Because the pathogenesis of migraine is still controversial, the rationale for the use of calcium antagonists as therapy cannot be precisely defined. Clinical trials of calcium antagonists were initiated on the assumptions that extracranial vasoconstriction causes migraine attacks and that calcium antagonists are effective vasodilators of cerebral as well as extracranial arteries. This theory is no longer considered likely. Migraine attacks are now thought to be of neural origins and appear to involve the release of serotonin.

Calcium antagonists have been studied extensively in the treatment of migraine. They are not effective for treatment of an acute attack but are effective as prophylactic agents to reduce the incidence of attacks. Clinical studies with flunarizine, nifedipine, nimodipine, and verapamil have been published. In numerous clinical studies, flunarizine, 10 mg/day, was consistently superior to placebo and equivalent to pizotifen.[125,126] According to Baumel,[127] flunarizine was more effective in migraine prophylaxis than other calcium antagonists and verapamil was more effective than nifedipine. These drugs seem to become more effective with continuous therapy and should be tried for at least 2 months. In his review of drug therapy for migraine,[128] Welch suggested that calcium antagonists may be beneficial by preventing vasoconstriction and release of 5-HT.

Drug and Alcohol Dependence

The binding sites on L-type calcium channels of dihydropyridines appear to be important in several aspects of drug and alcohol dependence. Calcium antagonists have been found to modulate drug or alcohol dependence in some relevant animal models.[129]

Isradipine and nimodipine were reported to suppress cocaine or morphine self-administration in drug-naive mice,[130] but Schindler et al.[131] failed to detect any effects of nimodipine, verapamil, or diltiazem on self-administration of cocaine in monkeys. Nimodipine blocked sensitization and conditioning of the motor stimulant effect of cocaine in rats[132] and decreased the sensitivity of mice and rats to the reinforcing effects of cocaine.[133] The lethality of cocaine in rats, however, was not reduced by either nifedipine or nimodipine.[134] Nimodipine had no significant effect on cocaine craving in 66 abstinent cocaine-dependent patients,[135] but the authors emphasized the need for further studies of nimodipine in the treatment of cocaine addiction and intoxication.

The density of dihydropyridine binding sites in rat brain increased after chronic morphine treatment, whereas repeated nimodipine administration attenuated morphine withdrawal signs in rats.[136,137] The naloxone-precipitated withdrawal signs after chronic treatment of rats with morphine were antagonized by either nimodipine or diltiazem.[138] Nifedipine, nimodipine, and nisoldopine enhanced the effects of morphine on 5-HT turnover while antagonizing its effects on DA turnover in rats.[139]

Physical dependence on barbital was reported to be enhanced by nifedipine[140] but suppressed by flunarizine.[141] The blockade of T-type calcium channels by flunarizine, but not nifedipine, was assumed to be responsbile for the observed difference in the effects of the two drugs. In mice nitrendipine, given chronically with barbital, reduced barbital withdrawal effects.[142]

Calcium antagonists clearly interact with ethanol in animals. Nifedipine, nitrendipine, nimodipine, and verapamil delayed or prevented the development of tolerance to ethanol in rats.[143–146] Nitrendipine and verapamil, but not felodipine, antagonized ethanol withdrawal hyperexcitability.[147–149] Surprisingly, diltiazem increased the severity of this phenomenon.[150] When ethanol-preferring rats were allowed free choice between ethanol (6 or 10%) and water, nifedipine, nimodipine, nitrendipine, verapamil, nicardipine, felodipine, and isradipine decreased ethanol intake, whereas diltiazem was ineffective.[151–154] In ethanol-preferring monkeys, Rezvani et al.[155] found that verapamil, but not diltiazem, reduces ethanol intake. Behavioral dysfunction of aged rats exposed to perinatal ethanol treatment was effectively antagonized by nimodipine.[156] The mechanism of interaction between calcium antagonists and ethanol is not precisely understood, but ethanol is known to increase dihydropyridine binding in the CNS.[157] Thus interaction of dihydropyridines with ethanol was demonstrated at the receptor level as well.

No convincing clinical evidence indicates that calcium antagonists are useful in the treatment of drug addiction or alcoholism, but the available animal data justify clinical trials.

Learning Disorders

Nimodipine reduced brain lesion-induced deficits in several higher-cognitive and sensorimotor-integrative tasks in rats.[158,159] It facilitated learning and increased excitability of hippocampal neurons in aging rabbits.[160] In aging monkeys nimodipine reduced age-related learning and performance deficits.[161] The optimal effects of nimodipine in primates were observed at drug blood levels ranging from 5–15 ng/ml. The dose–response relationship was an inverted U-shaped curve, and at higher levels the activity of nimodipine declined. To maintain therapeutically relevant blood levels of nimodipine in rats, Fanelli et al.[162] used subcutaneously implanted nimodipine pellets and found that nimodipine improved performance of reversal training on a spatial learning task in the Morris water maze. Levy et al.[163] also used implanted nimodipine pellets and concluded that nimodipine improved performance and learning of rats in the eight-arm radial maze. Slow release and steady blood levels are apparently important to the beneficial effects of nimodipine on performance and learning, as evidenced by the finding that single intraperitoneal injections of nimodipine prolonged acquisition of two-way avoidance behavior and impaired learning in mice.[164,165] The effects of calcium antagonists on learning or performance in humans have not been studied.

Hearing Disorders, Tinnitus, and Ménière's Disease

The initial observations of the effects of calcium antagonists on acute hearing loss and tinnitus were clinical. Handrock[166] studied the effects of nimodipine in 24 patients with sudden hearing loss; another group of 23 patients received pentofylline. The duration of treatment was 3 weeks. Hearing improved significantly in 92% of patients treated with nimodipine and 78% of patients treated with pentofylline. Theopold[167] reported an open study of 30 patients with inner ear diseases who were treated with nimodipine (30 mg 3 times/day) for 12 weeks. Of the 12 patients with sudden hearing loss, 6 experienced complete remission after nimodipine treatment. Three of 4 patients with tinnitus also had complete remission.

Jastreboff and Brennan[168] studied the effects of nimodipine on the auditory system in rats and guinea pigs. Nimodipine had a direct effect on cochlear microphonics and latencies of auditory nerve potentials. It also abolished salicylate-induced tinnitus.

In another clinical study of nimodipine, Davies et al.[169] treated 31 patients from a tinnitus clinic with nimodipine (30 mg 4 times/day) for 4 weeks. Symptoms improved substantially in 5 patients, whereas two reported worsening of symptoms. Lassen et al.[170] used nimodipine in 12 patients with Ménière's disease who had not responded to diuretics or vestibular suppressants. Nimodipine controlled vertigo in 7 of the 12. The authors recommended nimodipine as an alternative to the standard therapy of Ménière's disease, but the use of calcium antagonists for this indication is still controversial. In a more recent experimental study using guinea pigs, Van Benthem et al.[171] found that nimodipine at high intraperitoneal doses (15 mg/kg) may increase endolymph production—an undesirable effect in patients with Ménière's disease.

Tourette's Syndrome

Facial and vocal tics capable of progressing to generalized jerking movements are known as Gilles de la Tourette's syndrome. Goldstein[172] reported that nifedipine was an effective treatment in a 12-year-old boy. In another report nifedipine and verapamil, but not diltiazem, were active in two patients.[173] Many case reports but no large clinical trials with calcium antagonists have been published.

Diabetic Neuropathy

Robertson et al.[174] reported the effectiveness of nifedipine in the treatment of streptozotocin-induced diabetic neuropathy in rats. The treatment was preventive. Nifedipine, 40 mg/kg/day, was administered in food for 2 months. In control diabetic animals the sciatic nerve velocity was reduced by 23–28%, whereas nifedipine prevented this decrease without affecting nerve conduction velocity in normal animals. Its activity was attributed to a vasodilatory effect. Kappelle et al.[175] demonstrated a similar effect of nimodipine, 20 mg/kg, administered intraperitoneally every 48 hours for 10 weeks, and suggested that nimodipine may produce its effect by preventing calcium overload in neurons. The same investigators found subsequently[176] that nimodipine ameliorates diabetic neuropathy in spontaneously diabetic BioBreeding/Worcester rats and suggested three possible sites of action: directly on neurons, on vasa vasorum, and on nervi vasorum. No clinical effects of nimodipine on diabetic neuropathy in humans have been reported.

Essential Tremor and Parkinson's Disease

Biary et al.[177] evaluated the effects of nimodipine on essential tremor in a double-blind study of 16 patients. Improvement was seen in 8. The authors concluded that nimodipine is effective in some patients with essential tremor.

No clinical evidence indicates the effectiveness of calcium antagonists in the treatment of Parkinson's disease, but Kupsch et al.[178] reported that in mice nimodipine, delivered by implanted capsule, prevented neurotoxicity induced by 1-methyl-4-phenyl-1,2,3,6-tetrahydro-pyridine (MPTP) at the nigral level. Because MPTP is known to cause Parkinson's symptoms in animals and humans, nimodipine and similar calcium antagonists may be beneficial in the treatment of Parkinson's disease.

Neural Transplants

Because nimodipine was reported to increase cerebral blood flow under experimental and clinical conditions, Finger and Dunnett[179] investigated its effects on vascularization and growth of neural grafts from rat embryos implanted in adult rats with 6-OHDA lesions of the nigrostriatal pathway. They found that nimodipine, 15 mg/kg/day delivered by intubation for 2 weeks, increased the vascularization and volume of the grafts. The best effects of nimodipine were obtained under suboptimal transplant conditions. Possible benefits of nimodipine or other calcium antagonists in the survival of neural grafts should be evaluated in other experimental models.

Intractable Pain Syndrome

In experimental studies using a rat model of peripheral neuropathy, SNX-111 was found to block tactile allodynia.[180] The drug was tested clinically in patients with intractable, morphine-resistant pain. SNX-111 was administered by intrathecal infusion to a total of 31 patients. In 21 of 24 evaluable patients, partial or complete pain relief was achieved.[181] Controlled clinical studies with SNX-111 in patients with intractable pain are ongoing.

CALCIUM ANTAGONISTS IN COMBINATION WITH OTHER DRUGS

The future therapy of many CNS disorders is likely to involve "drug cocktails," either as fixed drug combinations or as simultaneous therapy with two or more drugs acting by different mechanisms. Successful clinical studies in patients with head trauma and SAH were reported with the combination of nimodipine and tirilazad.[182] The combined therapy reduced mortality and vasospasms and improved neurologic outcome at 6 months after SAH to a greater extent than nimodipine alone. The effectiveness of tirilazad was attributed to its ability to inhibit lipid peroxidation.

Uematsu et al.[183] described favorable interaction of nimodipine and dizocilpine in a feline model of stroke. The two drugs are expected to block calcium entry into neurons through two different types of channels (L-type and NMDA, respectively). Nimodipine was also reported to potentiate the analgesic effects of sufentanil and to antagonize the development of tolerance.[184]

Experimental results with combinations of calcium antagonists were highly promising. Nimodipine was shown to antagonize cisplatin neurotoxicity without interfering with its antitumor activity.[185]

CONCLUSION

Calcium antagonists and, more specifically, antagonists of L-type channels have been tested in many diseases of the CNS, including conditions for which no drug treatment or animal models are available. These trials opened new therapeutic opportunities, such as prevention of neurologic deficits in SAH with nimodipine and use of verapamil in lithium-resistant manic depression. The clinical trials with calcium antagonists in ischemic stroke did not lead to the approval of any drug for this indication but established the principle that drug therapy can reduce neurologic deficits in ischemic stroke and created an opportunity for introduction of new drugs.

As neuroprotective agents, calcium antagonists appear to be more successful in animal than in human clinical studies. Grotta[186] listed five reasons for the apparent differences: (1) better control of variables in animal models, (2) administration of drugs to animals before or immediately after injury, (3) difficulty in assessing clinical significance of positive results in animals, (4) statistical design of animal studies that facilitates detection of positive results, and (5) failure to publish negative animal findings.

The first three reasons are probably valid, whereas the validity of the last two is debatable. In general, the animal studies have reasonably good predictive value for the outcome of clinical studies, providing the results of animal experiments are properly considered in the design of clinical trials. If, for instance, the therapeutic window for a neuroprotective drug in stroke models is no longer than 6 hours after injury, there is no reason to assume that it will be different in humans.

In designing clinical protocols, investigators also should realize that some drugs, particularly neuroprotective agents, have inverted U-shaped dose–response curves. Clinical trials with one dose level of a drug and without determinations of blood level are at best uninformative and often misleading. Pharmacologists must provide clinical investigators with data about both dose–response and blood level–response relationships in various animal species. The optimal blood levels for the activity of many drugs are usually not substantially different in humans and animals. The failure to consider or interpret animal pharmacology properly is the most common reason for negative results in clinical trials.

Some of the possible indications for calcium antagonists (e.g., epilepsy, age-associated memory dysfunction, alcoholism, peripheral diabetic neuropathy) have not been studied sufficiently to permit definitive conclusions about potential effectiveness. Such indications are promising and should be explored further. Among the various current and potential indications for calcium antagonists in treatment of CNS diseases, their use in severe head trauma is probably the least promising. In general, it appears that calcium antagonists are most effective in prevention of ischemic damage rather than in treatment of existing, particularly severe brain damage. Brain damage develops more gradually in SAH than in ischemic stroke; thus, preventive treatment has a greater opportunity to benefit patients. Fisher et al.[187] recommended short-term prophylactic neuroprotection with calcium antagonists for patients undergoing vascular surgery and long-term neuroprotection for patients with transient ischemic attacks or atrial fibrillation.

References

1. Nowycky MC, Fox AP, Tsien RW: Three types of neuronal calcium channels with different calcium agonist sensitivity. Nature 316:440–443, 1985.
2. Choi DW: Calcium-mediated neurotoxicity: Relationship to specific channel types and role in ischemic damage. Trends Neurosci 11:465–469, 1988.

3. Bean BP: Classes of calcium channels in vertebrate cells. Annu Rev Physiol 51:367–384, 1989.
4. Catterall WA, de Jongh K, Rotman E, et al: Molecular properties of calcium channels in skeletal muscle and neurons. Ann NY Acad Sci 681:342–355, 1993.
5. Hofmann W, Biel M, Flockerzi V: Molecular basis for calcium channel diversity. Annu Rev Neurosci 17:399–418, 1994.
6. Bean BO, Mintz IM, Regan LJ, et al: Pharmacology of high-threshold calcium channels in rat neurons. Drugs Dev 2:61–70, 1993.
7. Fox AP, Scroggs RS, Mogul DJ, et al: Differential expression of dihydropyridine-sensitive calcium channels in different neurons. Drugs Dev 2:79–88, 1993.
8. Soong TW, Stea A, Hodson CD, et al: Structure and functional expression of a member of the low voltage-activated calcium channel family. Science 260:1133–1135, 1993.
9. Ahlijanian MK, Westenbroek RE, Catterall WA: Subunit structure and localization of dihydropyridine-sensitive calcium channels in mammalian brain, spinal cord and retina. Neuron 4:819–832, 1990.
10. Miljanich GP, Ramachandran R: Antagonists of neuronal calcium channels: Structure, function and therapeutic implications. Annu Rev Pharmacol Toxicol 35:707–734, 1995.
11. Perez-Reyes E, Schneider T: Molecular biology of calcium channels. Kidney Int 48:1111–1124, 1995.
12. Spreyer R, Franx JK, Eller A, et al: Molecular biology of calcium channels. In Busse W-D, Garthoff B, Seuter F (eds): Dihydropyridines. Berlin, Springer Verlag, 1993, pp 98–110.
13. Kazda S, Garthoff B, Krause HP, Schlossmann K: Cerebrovascular effects of the calcium antagonistic dihydropyridine derivative, nimodipine, in animal experiments. Arzneimittelforschung 32:331–333, 1982.
14. Toward R: The selective inhibition of serotonin-induced contractions of rabbit cerebrovascular smooth muscle by calcium antagonistic dihydropyridines. An investigation of the mechanism of action of nimodipine. Circ Res 48:650–657, 1981.
15. Hoffmeister F, Benz U, Heise A, et al: Behavioral effects of nimodipine in animals. Arzneimittelforschung 32:347–360, 1982.
16. Cohen RJ, Allen GS: Cerebral arterial spasm: The role of calcium in in vitro and in vivo analysis of treatment with nifedipine and nimodipine. In Wilkins RH (ed): Proceedings of the Second International Workshop on Vasospasm. Baltimore, Williams & Wilkins, 1980, pp 527–532.
17. Llinás R, Yarom Y: Electrophysiology of mammalian inferior olivary neurons in vitro. Different types of voltage dependent ionic conductances. J Physiol 315:549–567, 1981.
18. Carbone E, Lux HD: A low voltage activated, fully inactivating calcium channel in vertebrate sensory neurons. Nature 316:440–443, 1984.
19. Ertel EA, Cohen CJ, Leibowitz MD, et al: ω-Aga IIb: A high affinity peptide blocker of multiple types of voltage-gated channels. Biophys J 64:A117, 1993.
20. Olivera BM, Gray WR, Zeikus R, et al: Peptide neurotoxins from fish-hunting cone snails. Science 230:1338–1343, 1985.
21. Valentino K, Newcomb R, Gadbois T, et al: A selective N-type calcium channel antagonist protects against neuronal loss after global cerebral ischemia. Proc Natl Acad Sci USA 90:7894–7897, 1993.
22. Newcomb R, Palma A, Fox J, et al: SNX-325, a novel calcium antagonist from the spider *Sequestria florentina*. Biochemistry 34:8341–8347, 1995.
23. Koike H, Saito H, Matsuki N: Inhibitory effect of aniracetam on N-type current in acutely isolated rat neuronal cells. Jpn J Pharmacol 61:277–281, 1993.
24. Llinás R, Sugimori M, Hillman DE, Cherksey B: Distribution and functional significance of the P-type voltage-dependent channels in the mammalian central nervous system. Trends Neurosci 15:351–354, 1992.

24a. Bean BP, Mintz IM: Pharmacological classification of high-threshold calcium channels in rat neurons. In Busse W-D, Garthoff B, Seuter F (eds): Dihydropyridines. Berlin, Springer Verlag, 1993, pp 36–45.

25. Ming Ly Y, Frostl W, Dreesen J, Knöpfel T: P-type calcium channels are blocked by the alkaloid daurisoline. NeuroReport 5:1489–1492, 1994.
26. Llinás R, Sugimori M, Lin J-W, Cherksey B: Blocking and isolation of a calcium channel from neurons in mammals and cephalopods utilizing a toxin fraction (FTX) from funnel-web spider poison. Proc Natl Acad Sci USA 86:1689–1693, 1989.
27. Allen GS, Ahn HS, Preziosi TL, et al: Cerebral arterial spasm: A controlled trial of nimodipine in patients with subarachnoid hemorrhage. N Engl J Med 308:619–624, 1983.
28. Philippon J, Grob R, Dagreon F, et al: Prevention of vasospasm in subarachniod hemorrhage: A controlled study with nimodipine. Acta Neurochir 82:110–114, 1986.

29. Pickard JD, Murray GD, Illingworth R, et al: Effect of oral nimodipine on cerebral infarction and outcome after subarachnoid hemorrhage: British Aneurysm Nimodipine Trial BMJ 298:636–642, 1989.
30. Haley EC, Kassell NF, Torner JC, Kongable J: Nicardipine ameliorates angiographic vasospasm following subarachnoid hemorrhage [abstract]. Neurology 41(Suppl 1):Abs 346, 1991.
31. Gotoh O, Mohamed J, McCulloch J, et al: Nimodipine and the hemodynamic and histopathological consequences of middle cerebral artery occlusion in the rat. J Cerebr Blood Flow Metab 6:321–331, 1986.
32. Germano IM, Bartkowski HM, Cassel ME, Pitts LH: The therapeutic value of nimodipine in experimental cerebral ischemia: Neurological outcome and histologic findings. J Neurosurg 67:81–87, 1987.
33. Barone FC, Lysko PG, Price WJ, et al: SB 201823-A antagonizes calcium currents in central neurons and reduces the effects of focal ischemia in rats and mice. Stroke 26:1683–1690, 1995.
34. Gelmers HJ, Gorter K, deWeerdt CJ, Wiezer HJ: A controlled trial of nimodipine in acute ischemic stroke. N Engl J Med 318:203–207, 1988.
35. Mohr JP, Orgogozo JM, Harrison MJG, et al: Meta analysis of oral nimodipine trials in acute ischemic stroke. Cerebrovasc Dis 4:197–203, 1994.
36. Yamamoto M, Takenaka T: Neuroprotective action of nicardipine hydrochloride. CNS Drug Rev 1:91–106, 1995.
37. Sauter A, Rudin M: Prevention of stroke and brain damage with calcium antagonists in animals. Am J Hypertens 4:121S–127S, 1991.
38. Sauter A, Rudin M, Wiederholt KH, Holt RP: Cerebrovascular, biochemical and cytoprotective effects of isradipine. Am J Med 86(Suppl 4A):134–146, 1989.
39. Kawamura S, Yasui N, Shirasawa M, Fukasawa H: Effects of Ca^{2+} entry blocker (nilvadipine) on acute focal cerebral ischemia. Exp Brain Res 83:434–438, 1991.
40. Morikawa E, Ginsberg MD, Dietrich WD, et al: Postischemic (S)-emopamil therapy ameliorates focal ischemic brain injury in rats. Stroke 22:355–360, 1991.
41. Hara H, Morita T, Sukamoto T, Cutrer FM: Lomerizine (KB-2796), an new antimigraine drug. CNS Drug Rev 1:204–226, 1995.
42. Brown CM, Rush WR, Colquhoun HA: Lifarizine: A blocker of inactivated voltage-dependent sodium channels with cerebral neuroprotective action. CNS Drug Rev 1:149–167, 1995.
43. Kaminow L, Bevan J: Clentiazem reduces infarct size in rabbit middle cerebral artery occlusion. Stroke 22:242–246, 1991.
44. Patmore L, Whiting RL: Selective calcium entry blocking properties of nicardipine. Proceedings of the Ninth International Congress of Pharmacology, London, UK, July 29–August 3, 1984, p 880.
45. LeVere TE: Recovery of function after brain damage: The effects of nimodipine on the chronic behavioral deficit. Psychobiology 21:125–129, 1993.
46. Okiyama K, Rosenkrantz TS, Smith DH, et al: (S)-Emopamil attenuates acute reduction in regional cerebral blood flow following experimental brain injury. J Neurotrauma 11:83–95, 1995.
47. Bailey I, Bell A, Gray J, et al: A trial of the effects of nimodipine on outcome after head injury. Acta Neurochir 110:97–105, 1991.
48. The European Study Group on Nimodipine in Severe Head Injury: A multicenter trial of the efficacy of nimodipine on outcome after severe head injury. J Neurosurg 80:797–804, 1994.
48a. Demeyer I, Luther RR, Verhoeven FJS: SNX-111 in severe head injury: Preliminary results. Abstract presented at the Sixth International Symposium on Pharmacology of Cerebral Ischemia, Marburg, Germany, July 22–24, 1996.
49. Harders A, Kakarieka A, Braakman R, et al: Traumatic subarachnoid hemorrhage and its treatment with nimodipine. J Neurosurg 85:82–89, 1996.
50. Bär PR, Renkema GH, Veraart CM, et al: Nimodipine protects cultured spinal cord neurons from depolarization-induced inhibition of neurite growth. Cell Calcium 14:293–299, 1993.
51. Tymianski M, Tator CH: The direct effect of nimodipine on intracellular calcium in spinal neurons in explant culture. Drugs Dev 2:337–347, 1993.
52. Faden AI, Jacobs TP, Smith MT: Evaluation of the calcium channel antagonist nimodipine in experimental spinal cord ischemia. J Neurosurg 60:796–799, 1984.
53. Fehlings MG, Tator CH, Linden RD: The effect of nimodipine and dextran on axonal function and blood flow following experimental spinal cord injury. J Neurosurg 71:403–416, 1989.
54. Pointillart V, Gense D, Gross C, et al: Effects of nimodipine on posttraumatic spinal cord ischemia in baboons. J Neurotrauma 10:201–213, 1993.
55. Gambardella G, Collufio D, Caruso G, et al: Experimental incomplete spinal cord injury: Treatment with a combination of nimodipine and adrenaline. J Neurosurg Sci 39:67–74, 1995.

56. Schittek A, Bennink GBWE, Cooley DA, Langford LA: Spinal cord protection with intravenous nimodipine. J Thorac Cardiovasc Surg 104:1100–1105, 1992.
57. Haghighi SS, Chehrazi BB, Wagner FC: Effect of nimodipine-associated hypotension on recovery from acute spinal cord in cats. Surg Neurol 29:293–297, 1988.
58. Holtz A, Nystrom B, Herdin B: Spinal cord injury in rats: Inability of nimodipine or antineutrophil serum to improve spinal cord blood flow or neurologic status. Acta Neurol Scand 79:460–467, 1989.
59. Ceylan S, Ilbay K, Baykal S, et al: Treatment of acute spinal injuries: Comparison of thyrotropin-releasing hormone and nimodipine. Res Exp Med 192:23–33, 1992.
60. Black D, Markowitz RS, Kinkelstein SD, et al: Experimental spinal cord injury: Effect of a calcium antagonist (nicardipine). Neurosurgery 22:61–66, 1988.
61. Rhee RY, Gloviczki P, Cambria RA, et al: The effects of nimodipine on ischemic injury of the spinal cord during thoracic aortic clamping. Int Angiol 15:153–161, 1996.
62. Steen PA, Newberg LA, Milde JH, Michenfelder JD: Nimodipine improves cerebral blood flow and neurologic recovery after complete cerebral ischemia in the dog. J Cerebr Blood Flow Metab 3:38–43, 1983.
63. Steen PA, Gisvoid SE, Milde JH, et al: Nimodipine improves outcome when given after complete cerebral ischemia in primates. Anesthesiology 62:406–414, 1985.
64. Newberg LA, Steen PA, Milde JH, Michenfelder JD: Failure of flunarizine to improve cerebral blood flow or neurologic recovery in a canine model of complete cerebral ischemia. Stroke 15:666–671, 1984.
65. Bunnel OS, Louis TM, Saldanha RL, Kopelman AE: Protective action of calcium antagonists, flunarizine and nimodipine, on cerebral ischemia. Med Sci Res 15:1513–1514, 1987.
66. Deshpande JK, Wieloch T: Flunarizine, a calcium entry blocker, ameliorates ischemic brain damage in the rat. Anesthesiology 64:215–224, 1986.
67. Zupan G, Mršić Erakovic V, et al: The influence of nimodipine and amlodipine on the brain free arachidonic acid levels and behavior in hypoxia exposed rats. Can J Physiol Pharmacol 72:407, 1994.
68. Nakamura K, Hatakeyama T, Furuta S, Sakaki S: The role of early Ca^{2+} influx in the pathogenesis of delayed neuronal death after brief forebrain ischemia in gerbils. Brain Res 613:181–192, 1993.
69. Roine RO, Kaste M, Kinnunen P, et al: Nimodipine in out-of-hospital ventricular fibrillation: a randomized, double-blind, placebo-controlled clinical trial. In Scriabine A, Teasdale GM, Tettenborn D, Young W (eds): Nimodipine: Pharmacological and Clinical Results in Cerebral Ischemia. Berlin, Springer Verlag, 1991, pp 215–219.
70. Roine RO, Kaste M: European Resuscitation Nimodipine Study (ERNST) [abstract]. Stroke 27:169, 1996.
71. Andrews RJ, Muto RP: Retraction brain ischemia: Cerebral blood flow, evoked potentials, hypotension, and hyperventilation in a new animal model. Neurol Res 14:12–18, 1992.
72. Andrews RJ, Muto RP: Retraction brain ischemia: Mannitol plus nimodipine preserves both cerebral blood flow and evoked potentials during normoventilation and hyperventilation. Neurol Res 14:19–25, 1992.
73. Desai CK, Dickshit RB, Mansuri SM, Shah UH: Comparative evaluation of anticonvulsant activity of calcium channel blockers in experimental animals. Ind J Exp Biol 33:931–934, 1995.
74. Meyer FB, Anderson RE, Sundt TM, et al: Suppression of pentylenetetrazole induced seizures by oral administration of a dihydropyridine Ca^{2+} antagonist. Epilepsia 28:409–414, 1987.
75. Desmedt LKC, Niemeggeers CJE, Janssen PAJ: Anticonvulsive properties of cinnarizine and flunarizine in rats and mice. Arzneimittelforschung 15:1408–1413, 1975.
76. Dolin SI, Hunter AB, Halsey MJ, Little HJ: Anticonvulsant profile of the dihydropyridine calcium antagonists, nitrendipine and nimodipine. Eur J Pharmacol 152:19–27, 1988.
77. DeSarro GB, Meldrum BS, Nistico G: Anticonvulsant effects of some calcium entry blockers in DBA/2 mice. Br J Pharmacol 93:257–256, 1988.
78. Karpova MN, Pankov OYU, Kryzhanovskii GN, Glebov RN: Antiepileptic effects of nifedipine. Bull Exp Biol Med (Moscow) 112:260–262, 1991.
79. Meyer FB, Anderson RE, Sundt TM, Yaksh T: Inhibition of electrically induced seizures by a dihydropyridine calcium channel blocker. Brain Res 384:180–183, 1986.
80. Mack CM , Gilbert ME: An examination of the anticonvulsant properties of voltage-sensitive calcium channel inhibitors in amygdala kindled seizures. Psychopharmacology 106:365–369, 1992.
81. Wurpel JND, Iyer SN: Calcium channel blockers verapamil and nimodipine inhibit kindling in adult and immature rats. Epilepsia 35:443–449, 1994.
82. Overweg WJ, Binnie CD, Meijer JWA, et al: Double-blind placebo-controlled trial of flunarizine as add-on therapy in epilepsy. Epilepsia 25:217–222, 1984.

83. Pelliccia A, Sciaretta A, and Matricardi M: Nimodipine treatment for drug-resistant childhood epilepsy. Dev Med Child Neurol 32:1114–1116, 1990.
84. Hans P, Triffaux M, Bohomme V, et al: Control of drug-resistant epilepsy after head injury with intravenous nimodipine. Acta Anaesth Belg 45:175–178, 1994.
85. Sasso E, Mancia D: Putative antiepileptic activity of nimodipine. Epilepsia 36:S73, 1995.
86. Meyer FB, Cascino GD, Whisnant JP, et al: Nimodipine as an add-on for intractable epilepsy. Mayo Clin Proc 70:623–627, 1995.
87. Tobares N, Pedromingo A, Bigorra J: Nimodipine treatment improves cognitive functions in vascular dementia. In Bergener M, Reisberg B (eds): Diagnosis and Treatment of Senile Dementias. Berlin, Springer Verlag, 1989, pp 361–365.
88. Fischhof PK, Wagner G, Littschauer L, et al: Therapeutic results with nimodipine in primary degenerative dementia and multi-infarct dementia. In Bergener M, Reisberg B (eds): Diagnosis and Treatment of Senile Dementia. Berlin, Springer Verlag, 1989, pp 350–359.
89. Fischhof PK: Divergent neuroprotective effects of nimodipine in PDD and MID provide indirect evidence of disturbance in Ca^{2+} homeostasis in dementia. Methods Find Exp Clin Pharmacol 15:549–555, 1993.
90. Adan GO, Martin GA, Bello CJ, et al: Treatment of cerebrovascular insufficiency (multiinfarct dementia) with nimodipine in elderly patients. Invest Med Int 22:24–29, 1995.
91. Khachaturian ZS: Calcium hypothesis of Alzheimer's disease and brain aging. Ann NY Acad Sci 747:1–11, 1995.
92. de Jonge MC: The calcium hypothesis of brain aging: Its pharmacological rationale. Adv Neuropharmacol 1983, pp 227–240.
93. Smith T: The current status of the calcium hypothesis of brain aging and Alzheimer's disease. Report of a Heidelberg workshop. CNS Drug Rev 2(Suppl 1):1–45, 1996.
94. Landfield PW: Increased hippocampal Ca^{2+} channel activity in brain aging and dementia. Ann NY Acad Sci 747:351–364, 1995.
95. LeVere TE, Brugler T, Sandin M, Gray-Silva S: Recovery of function after brain damage. Facilitation by the calcium entry blocker, nimodipine. Behav Neurosci 103:561–565, 1989.
96. Jasmin S, Szelgia K, LeVere TE: The benefits of nimodipine for memory in aged animals and animals sustaining neocortical brain injury. Drugs Dev 2:263–273, 1993.
97. Deyo RA, Straube KT, Disterhoft JF: Nimodipine facilitates trace conditioning of the eyeblink response in aging rabbits. Science 243:809–811, 1989.
98. Straube KT, Deyo RA, Moyer JR, Disterhoft JF: Dietary nimodipine improves associative learning in aging rabbits. Neurobiol Aging 11:659–661, 1990.
99. de Yong GI, Traber J, Luiten PG: Formation of cerebrovascular abnormalities in the aging rat is delayed by chronic nimodipine application. Mech Ageing Dev 64:255–272, 1992.
100. Schuurman T, Klein H, Beneke M, Traber J: Nimodipine and motor deficits in the aged rat. Neurosci Res Comm 1:9–15, 1987.
101. Schuurman T, Traber J: Effects of nimodipine on behavior of old rats. In Traber J, Gispen WH (eds): Nimodipine and Central Nervous System Function: New Vistas. Stuttgart-New York, Schattauer, 1989, pp 195–208.
102. Yamamoto M, Suzuki M, Ozawa Y, et al: Effects of calcium channel blockers on impairment of brain function in senescence-accelerated mice. Arch Int Pharmacodyn 330:125–137, 1955.
103. Kabuto H, Yokoi I, Mori A, et al: Neurochemical changes related to aging in the senescence-accelerated mouse brain and the effect of chronic administration of nimodipine. Mech Ageing Dev 80:1–9, 1995.
104. Schmage N, Bergener M: Global rating, symptoms, behavior, and cognitive performance as indicators of efficacy in clinical studies with nimodipine in elderly patients with cognitive impairment syndromes. Int Psychogeriatr 4(Suppl 1):89–99, 1992.
105. Parnetti L, Senin U, Carosi M, et al: Mental deterioration in old age: Results of two multicenter, clinical trials with nimodipine. Clin Ther 15:394–406, 1993.
106. Canade V, Catapano F, Muraca L, Costentino A: Effects of nimodipine on mental deterioration in subjects with chronic cerebrovascular pathology. Minerva Med 82:111–114, 1991.
107. Lipton S: AIDS-related dementia and calcium homeostasis. Ann NY Acad Sci 747:205–224, 1994.
108. Lipton SA: Neuronal injury associated with HIV-1 and potential treatment with calcium-channel and NMDA antagonists. Dev Neurosci 16:145–151, 1994.
109. Weiss JH, Pike CJ, Cotman CW: Ca^{2+} channel blockers attenuate β-amyloid peptide toxicity to cortical neurons in culture. J Neurochem 62:372–375, 1994.
110. Chisholm LC, Hunnicut ES Jr, Davis JN: Neuronal $[Ca^{2+}]_i$ responses to different classes of calcium lowering agents. Drugs Dev 2:117–125, 1994.
111. Piggott MA, Candy JM, Perry RH: [^{3}H]nitrendipine binding in temporal cortex in Alzheimer's and Huntington's diseases. Brain Res 565:42–47, 1991.

112. Baumel B, Eisner LS, Karukin M, et al: Nimodipine in the treatment of Alzheimer's disease. In Bergener M, et al (eds): Diagnosis and Treatment of Senile Dementia. Berlin, Springer Verlag, 1989, pp 366–373.
113. Morich FJ, Bieber F, Lewis JM, et al: Nimodipine in the treatment of probable Alzheimer's disease. Clin Drug Invest 11:185–195, 1996.
114. de Vry J, de Beun R, Traber J: Dihydropyridine calcium channel inhibitors and psychiatric disorders: Preclinical data. Pharmacopsychiatry 26:208, 1993.
115. Walden J, Grunze H, Hellhammer D, et al: Treatment of depressive episodes with the calcium antagonist, nimodipine: Single case reports [abstract]. Neuropsychopharmacology 10(Suppl Pt 2):171–184, 1994.
116. Höschl C, Kozeny J: Verapamil in affective disorders. Biol Psychiatry 25:128–140, 1989.
117. Giannini AJ, Houser WL, Loiselle RH, Price WA: Antimanic effects of verapamil. Am J Psychiatry 139:502–504, 1984.
118. Dubovsky SL, Buzan R: The role of calcium channel blockers in the treatment of psychiatric disorders. CNS Drugs 4:47–57, 1995.
119. Brunet G, Cerlich B, Robert P, et al: Open trial of a calcium antagonist, nimodipine, in acute mania. Clin Neuropharmacol 13:224–228, 1990.
120. Pazzaglia PJ, Post RM, Ketter TA, et al: Preliminary controlled trial of nimodipine in ultrarapid cycling affective dysregulation. Psychiatry Res 49:257–272, 1993.
121. Grunze H, Walden J, Wolf R, Berger M: Combined treatment with lithium and nimodipine in a bipolar I manic syndrome. Prog Neuropsychopharmacol Biol Psychiatry 20:419–426, 1996.
122. Goldstein JA: Calcium channel blockers in the treatment of panic disorder. J Clin Psychiatry 46:546, 1985.
123. Klein E, Uhde TW: Controlled study of verapamil for treatment of panic disorder. Am J Psychiatry 145:431–434, 1988.
124. Gibbs DM: Hyperventilation-induced cerebral ischemia in panic disorder and the effect of nimodipine. Am J Psychiatry 149:1589–1591, 1992.
125. Mendenopoulos G, Manafi T, Logothetis J, Bostantjopoulou S: Flunarizine in prevention of classical migraine: A placebo controlled evaluation. Cephalalgia 5:31–37, 1985.
126. Rascol A, Montastruc JL, Rascol O: Flunarizine versus pizotifen: A double study in the prophylaxis of migraine. Headache 26:83–85, 1986.
127. Baumel B: Migraine: A pharmacological review with newer options and delivery modalities. Neurology 44(Suppl 3):S13–S17, 1994.
128. Welch KM: Drug therapy of migraine. N Engl J Med 329:1476–1483, 1993.
129. Little HJ: The role of calcium channels in drug dependence. Drug Alcohol Depend 38:173–194, 1995.
130. Kuzmin A, Zvartau E, Gessa GL, et al: Calcium antagonists isradipine and nimodipine suppress cocaine and morphine intravenous self-administration in drug-naive mice. Pharmacol Biochem Behav 41:497–500, 1992.
131. Schindler CW, Tella SR, Prada J, Goldberg SR: Calcium channel blockers antagonize some of cocaine's cardiovascular effects, but fail to alter cocaine's behavioral effects. J Pharmacol Exp Ther 272:791–198, 1995.
132. Reimer AR, Martin-Iverson MT: Nimodipine and haloperidol attenuate behavioral sensitization to cocaine but only nimodipine blocks the establishment of locomotion induced by cocaine. Psychopharmacology 113:404–410, 1994.
133. Kuzmin R, Semenova S, Ramsey NF, et al: Modulation of cocaine intravenous self-administration in drug naive animals by dihydropyridine Ca^{2+} channel modulators. Eur J Pharmacol 295:19–25, 1996.
134. Derlet RW, Tseng CC, Albertson TE: Cocaine toxicity and the calcium channel blockers nifedipine and nimodipine in rats. J Emerg Med 12:1–4, 1994.
135. Rosse RB, Alim TN, Fay-McCarthy M, et al: Nimodipine pharmacotherapeutic adjuvant therapy for inpatient treatment of cocaine dependence. Clin Neuropharmacol 17:348–358, 1994.
136. Ramkumar V, El-Fakahany EE: Prolonged morphine treatment increases rat brain DHP binding sites: Possible involvement in development of morphine dependence. Eur J Pharmacol 146:73–83, 1987.
137. Zharkovsky A, Totterman AM, Moisio J, Ahtee L: Concurrent nimodipine attenuates the withdrawal signs and the increase of cerebral dihydropyridine binding after chronic morphine treatment in rats. Naunyn-Schmiedeberg's Arch Pharmacol 347:483–486, 1993.
138. Colado MI, Alfaro MJ, Lopez F, et al: Effect of nimodipine, diltiazem and BAY K 8644 on the behavioral and neurochemical changes associated with naloxone-precipitated withdrawal in the rat: A comparison with clonidine. Gen Pharmacol 24:35–41, 1993.

139. Gaggi R, Roncada P, Gianni AM, Dall'Olio R: Interaction between dihydropyridine calcium antagonists and morphine on brain biogenic amines. Gen Pharmacol 25:923–929, 1994.
140. Suzuki T, Mizoguchi H, Motegi H, et al: Effects of nifedipine on physical dependence on barbital or diazepam in rats. J Toxicol Sci 20:415–425, 1995.
141. Suzuki T, Mizoguchi H, Noguchi H, et al: Effects of flunarizine and diltiazem on physical dependence on barbital in rats. Pharmacol Biochem Behav 45:703–712, 1993.
142. Rabbani M, Brown J, Butterworth AR, Little HJ: Dihydropyridine-sensitive calcium channels and barbiturate tolerance and withdrawal. Pharmacol Biochem Behav 47:675–680, 1994.
143. Wu PH, Pham T, Naranjo CA: Nifedipine delays the acquisition of alcohol tolerance. Eur J Pharmacol 139:233–236, 1987.
144. Little HJ, Dolin SJ: Lack of tolerance to ethanol after concurrent administration of nitrendipine. Br J Pharmacol 92:606P, 1987.
145. Smith JW, Wilson J, Little HJ: Co-administration of the calcium channel antagonist, nimodipine, decreases environmental-independent tolerance to ethanol. Br J Pharmacol 118(Suppl):40, 1996.
146. Sullivan JT: Effects of verapamil on the acquisition of ethanol tolerance. Life Sci 52:1295–1300, 1993.
147. Whittington MA, Little HJ: Nitrendipine, given during drinking, decreases the electrophysiological changes in the isolated hippocampal slice, seen during ethanol withdrawal. Br J Pharmacol 103:1677–1684, 1991.
148. Whittington MA, Dolin SJ, Patch TL, et al: Chronic dihydropyridine treatment can reverse the behavioural consequences of, and prevent addiction to, chronic alcohol treatment. Br J Pharmacol 103:1669–1676, 1991.
149. Watson WP, Cross A, Mysra A, et al: Felodipine shows separation between displacement of dihydropyridine binding and prevention of the ethanol withdrawal syndrome. Br J Pharmacol 112:1017–1024, 1994.
150. Watson WP, Little HJ: The calcium channel antagonist, diltiazem, increases the severity of ethanol withdrawal syndrome. Br J Pharmacol 107:208, 1992.
151. Resvani AH, Janowski DS: Decreased alcohol consumption by verapamil in alcohol preferring rats. Prog Neuropsychopharmacol Biol Psychiatry 14:623–631, 1990.
152. Fadda F, Garau B, Colombo J, Gessa GL: Isradipine and other calcium channel antagonists attenuate alcohol consumption in ethanol preferring rats. Alcohol Clin Exp Res 16:449–452, 1992.
153. Pucilowski O, Rezvani AH, Overstreet DH, Janowski DS: Calcium channel inhibitors attenuate consumption of ethanol, sucrose, and saccharin solution in rats. Behav Pharmacol 5:494–501, 1994.
154. de Beun R, Schneider R, Klein A, et al: Effects of nimodipine and other calcium channel antagonists in alcohol-referring AA rats. Alcohol 13:263–271, 1996.
155. Rezvani AH, Grady DR, Janowsky DS: Effect of calcium channel blockers on alcohol consumption in alcohol drinking monkeys. Alcohol 26:161–167, 1991.
156. Markel E, Felszeghy K, Luiten PG, Nyakas C: Beneficial effect of chronic nimodipine treatment on behavioral dysfunctions of aged rats exposed to perinatal ethanol treatment. Arch Gerontol Geriatr 21:75–88, 1995.
157. Guppy LJ, Littleton JM: Binding characteristics of the calcium channel antagonist [^{3}H] nitrendipine in tissues from ethanol dependent rats. Alcohol 29:283–293, 1994.
158. Jasmin S, Szeliga K, LeVere TE: The benefits of nimodipine for memory in aged animals and animals sustaining neocortical brain injury. Drugs Dev 2:263–273, 1993.
159. Finger S: Nimodipine and recovery from focal brain lesions. Drugs Dev 2:379–393, 1993.
160. Disterhoft JF, Thompson LT, Moyer JR, Kowalska M: Nimodipine facilitates learning and increases excitability of hippocampal neurons in aging rabbits. Drugs Dev 2:395–405, 1993.
161. Moss MB, Rosene DL: Therapeutic effects of nimodipine on related memory dysfunction in the monkey. Drugs Dev 2:249–261, 1993.
162. Fanelli RJ, McMonagle-Strucko K, Johnson DE, Janis RA: Behavioral and neurochemical effects of sustained release pellets of nimodipine. Drugs Dev 2:407–415, 1993.
163. Levy A, Kong RM, Stilman MJ, et al: Nimodipine improves spatial working memory and elevates hippocampal acetylcholine in young rats. Pharmacol Biochem Behav 39:781–786, 1991.
164. Nikolaev E, Kaczmarek L: Disruption of two-way active avoidance behavior produced by nimodipine. Pharmacol Biochem Behav 47:757–759, 1994.
165. Maurice T, Bayle J, Privat A: Learning impairment following acute administration of the calcium channel antagonist nimodipine in mice. Behav Pharmacol 6:167–175, 1995.
166. Handrock M: Sudden loss of hearing. In Betz E, Deck K, Hoffmeister F (eds): Nimodipine: Pharmacological and Clinical Properties. Stuttgart, Schattauer Verlag, 1985, pp 481–496.

167. Theopold HM: Nimodipine (Bay e 9736): A new concept in the treatment of inner ear diseases. Laryngol Rhinol Otol 64:609–613, 1985.
168. Jastreboff PJ, Brennan JF: Specific effects of nimodipine on the auditory system. Ann NY Acad Sci 522:716–718, 1988.
169. Davies E, Knox E, Donaldson I: The usefulness of nimodipine, an L-calcium channel antagonist, in the treatment of tinnitus. Br J Audiol 28:125–129, 1994.
170. Lassen LF, Hirsch BE, Kamerer DB: Use of nimodipine in the medical treatment of Ménière's disease. Am J Otol 17:577–580, 1996.
171. Benthem PPG, van Klis SFL, Albers FWJ, et al: The effect of nimodipine on cochlear potentials and Na^+/K^+-ATPase activity in normal and hydropic cochleas of the albino guinea pig. Hearing Res 77:9–18, 1994.
172. Goldstein JA: Nifedipine treatment of Tourette syndrome. J Clin Psychiatry 45:8, 1984.
173. Walsch TL, Lavenstein B, Licamele WL, et al: Calcium antagonists in the treatment of Tourette's disorder. Am J Psychiatry 143:1467–1468, 1986.
174. Robertson S, Cameron NE, Cotter MA: The effect of the calcium antagonist nifedipine on peripheral nerve function in the streptozotocin-diabetic rat. Diabetologia 35:1113–1117, 1992.
175. Kappelle AC, Bravenboer B, Traber J, et al: The Ca^{2+} antagonist nimodipine counteracts the onset of an experimental neuropathy in streptozotocin-induced, diabetic rats. Neurosci Res Community 10:95–104, 1992.
176. Kappelle AC, Biessels G, Bravenboer B, et al: Beneficial effects of the Ca^{2+} antagonist, nimodipine, on existing diabetic neuropathy in the BB/Wor rat. Br J Pharmacol 111:887–893, 1994.
177. Biary N, Bahou Y, Sofi MA, et al: The effect of nimodipine on essential tremor. Neurology 45:1523–1525, 1995.
178. Kupsch A, Gerlach M, Pupeter CG, et al: Pretreatment with nimodipine prevents MPTP-induced neurotoxicity at the nigral, but not at the striatal level in mice. NeuroReport 6:621–625, 1995.
179. Finger S, Dunnett SB: Nimodipine enhances growth and vascularization of neural grafts. Exp Neurol 104:1–9, 1989.
180. Bowersox SS, Valentino KL, Luther RR: Neuronal voltage-sensitive calcium channels. Drug News Perspect 7:261–268, 1994.
181. Brose WG, Pfeifer BL, Hasselbusch S, et al: Analgesia produced by SNX-111 in patients with morphine resistant pain. Abstracts of the Fifteenth Annual American Pain Society Meeting, Washington, DC, November 14–17, 1996.
182. Piek J: Clinical use of lazaroids in head injury and subarachnoid hemorrhage. Anesthetist 43(Suppl 1):103, 1994.
183. Uematsu D, Araki N, Greenberg JH, et al: Combined therapy with MK 801 and nimodipine for protection of ischemic brain damage. Neurology 41:88–94, 1991.
184. Hurle MA, Diaz A, Ruiz F, et al: Acute and chronic modulation of opioid effects by calcium channel blockers: A functional and biochemical study. Methods Find Exp Clin Pharmacol 16(Suppl 1):21, 1994.
185. Hamers FP, Van der Hoop RG, Steerenburg PA, et al: Putative neurotrophic factors in the protection of cisplatin-induced peripheral neuropathy in rats. Toxicol Appl Pharmacol 111:514–522, 1991.
186. Grotta J: Why do all drugs work in animals but none in stroke patients? 2. Neuroprotective therapy. J Intern Med 237:89–94, 1995.
187. Fisher M, Jones S, Sacco RL: Prophylactic neuroprotection for cerebral ischemia. Stroke 25:1075–1080, 1994.

KOICHI HAYASHI M.D., Ph.D. / TAKAO SARUTA M.D., Ph.D.
MURRAY EPSTEIN M.D., F.A.C.P.

22

Renal Hemodynamic Effects of Calcium Antagonists

Since the introduction of calcium antagonists three decades ago, attention has focused on their beneficial effects in the management of symptomatic coronary artery disease and their ability to lower blood pressure. Although the renal vasodilatory properties of calcium antagonists were described as early as 1962,[1] this seminal finding was virtually ignored. Only during the past two decades has there been a rediscovery of their beneficial effects on kidney function.[2–5] The purpose of this chapter is to discuss briefly the influence of calcium antagonists on renal hemodynamics. Chapter 24 reviews the current and potential therapeutic applications of this property in a wide array of acute renal insufficiency syndromes.

EXPERIMENTAL STUDIES IN ANIMALS

Effects of Calcium Antagonists on the Renal Microcirculation

We have recently reviewed the pharmacologic effects of calcium antagonists on renal hemodynamics.[2,3] Many of the initial studies assessing the effects of calcium antagonists on renal blood flow in intact animals yielded conflicting results.[2,5,7] To a large extent, the disparate responses can be attributed to differing experimental designs. Because calcium antagonists elicit vasodilation by disrupting the actions of vasoconstrictors, it seems logical that the renal response to calcium antagonists in any setting depends on the magnitude of basal renal vascular tone. Furthermore, calcium antagonists affect excitation-contraction coupling by modulating the function of specific types of calcium channels. The recruitment of these calcium antagonist-sensitive channels may vary with differing vasoconstrictors;[8,9] therefore, it also may be anticipated that the renal hemodynamic response to calcium antagonists depends on the nature of the factors determining basal renal vascular tone.

In brief, calcium antagonists do not affect the vasodilated-perfused kidney; however, they dramatically alter the response of the kidney to vasoconstrictor agents.[3,5] Under conditions of in vitro perfusion (e.g., perfusion pressures of 80–100 mmHg), the isolated rat kidney appears to possess little intrinsic vascular resistance. Accordingly, in the absence of exogenous vasoconstrictors, calcium antagonists exert no effect on renal perfusate flow (RPF) or glomerular filtration rate (GFR).[3,10,11] As we have detailed previously,[3] this attribute of the model facilitates pharmacologic investigation of the renal microvascular actions of vasoactive agents. Thus, in experimental settings in which renal vascular tone is established by a specific vasoconstrictor stimulus, the effects of calcium antagonists can be examined directly. For example, in the presence of norepinephrine (NE), calcium antagonists markedly augment GFR but produce only a modest improvement in renal plasma flow.[2,3,11]

As detailed in recent reviews,[2,3] organic calcium antagonists preferentially augment GFR during NE- and angiotensin II (AII)-induced vasoconstriction. We proposed that the selective augmentation of GFR by calcium antagonists thus may require the presence of an agonist that activates both afferent and efferent resistance vessels in the kidney but uses differing activating mechanisms in the two vessel types. Such agonists may elicit membrane depolarization and activate potential-dependent calcium channels (PDCs) in afferent, but not efferent arteriolar smooth muscle. Because organic calcium antagonists selectively inhibit the function of this calcium channel type, they increase GFR by preferentially attenuating afferent arteriolar tone.

If calcium antagonists augment GFR by preferentially attenuating afferent arteriolar vasoconstriction while preserving efferent arteriolar tone, the relative effects of calcium antagonists on GFR and RPF also may depend on the vasoconstrictor determining renal vascular tone (Table 1). Thus, in the presence of an agonist such as U44069, which preferentially increases afferent tone, calcium antagonists would increase both GFR and RPF to a similar extent. On the other hand, in the presence of an agonist having a greater effect on efferent arteriolar tone (i.e., AII), one would anticipate a preferential augmentation of GFR over RPF. Such a formulation is supported by our observations that calcium antagonists exert a preferential augmentation of GFR only during NE- or AII-induced renal vasoconstriction; this effect is most pronounced in AII-treated kidneys.

TABLE 1. Effects of Calcium Antagonists on Glomerular Filtration Rate (GFR) and Renal Perfusate Flow (RPF)

	Proposed Action on Renal Microvessels		Proposed Effect of:	
Agonist	Afferent Tone	Efferent Tone	Agonist Alone	Agonist + Calcium Antagonist
U44069 (Thromboxane)	↑	—	Preferential decrease in GFR	Parallel increase in both GFR and RPF
Norepinephrine	↑	↑	Decrease in both GFR and RPF	Preferential augmentation of GFR
Angiotensin II	↑	↑↑	Preferential decrease in RPF	Exaggerated preferential augmentation of GFR

Direct In Vitro and In Vivo Observations of the Renal Microcirculation

The predominant influence of calcium antagonists on GFR suggests that they antagonize preglomerular vasoconstriction. Direct assessment of the renal microvascular response is required, however, to determine their renal microvascular actions. Recent advances in renal physiology enable us to observe renal microvessels directly[12–21]: (1) micropuncture techniques, (2) isolated renal microvessels, (3) blood-perfused juxtamedullary nephrons, (4) in vivo hydronephrotic kidneys, and (5) isolated perfused hydronephrotic kidneys (Table 2).

Edwards et al.[12] initially developed the microdissected renal vessels to observe directly the renal microvascular responsiveness to various vasoactive agents. More recently, Ito et al.[14,15] succeeded in isolating the renal afferent and efferent arterioles, with an attached glomerulus and a thick ascending limb of Henle's loop. This preparation possesses both microvascular and tubular components and thus enables us to assess the net effects on the renal microvasculature.

Carmines et al.[16] developed an in vitro technique that allows direct visualization of the juxtamedullary nephron circulation. The kidney is isolated from the rat, and the papilla is folded back to expose the renal microvasculature of juxtamedullary nephrons. The perfusate consists of reconstituted blood (hematocrit, 33%), and renal microvessels can be observed at the juxtamedullary portion, with preserved perivascular structures adjoining the glomerulus.

Steinhausen et al.[18] introduced a novel technique for direct observation of the in vivo response of renal microvessels. They used the in vivo hydronephrotic rat kidney as a model for visualizing vascular responses within the renal microcirculation in situ. In their experiments, vasoactive agents are applied topically to the surface of the kidney. This approach circumvents systemic effects, allowing direct assessment of their renal actions in an intact and in vivo setting.

Finally, our laboratory developed a model of the isolated perfused hydronephrotic kidney that facilitates direct observation of the renal microvasculature under defined in vitro conditions.[19–21] This technique has been detailed previously. Briefly, unilateral hydronephrosis is induced in donor animals by unilateral ureteral ligation. After 8–10 weeks, renal tubular atrophy progresses to a stage that allows microscopic visualization of the renal microvessels. At this point, the hydronephrotic kidneys are excised and studied using a modification of the isolated perfused technique. The perfused kidney is placed on the stage of an inverted microscope modified to accommodate a heated chamber

TABLE 2. Methods for Assessing the Renal Microcirculation

	Methods	Authors	Characteristics
Indirect	Micropuncture		
	Laser Doppler		
Direct	Isolated microvessels	Edwards,[12] Yuan[13]	Without glomerulus
		Ito[14,15]	With glomerulus
	Juxtamedullary nephron	Carmines[16]	Blood-perfused
	Needle–charge-coupled device	Yamamoto[17]	Intravital
	Hydronephrotic kidney	Steinhausen[18]	In vivo
		Loutzenhiser[19–21]	In vitro

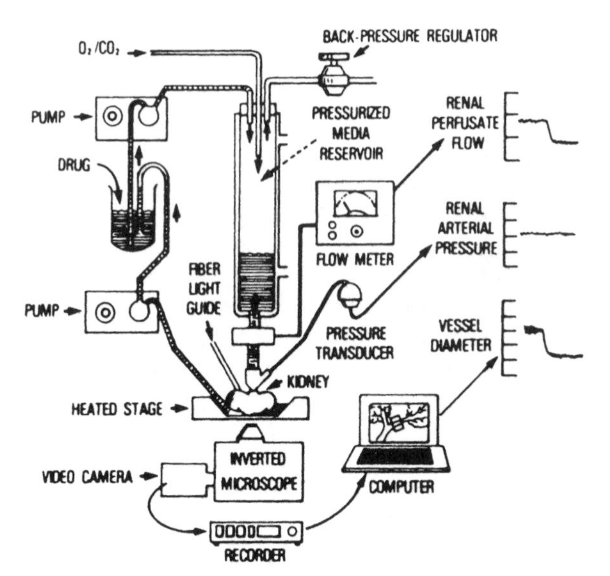

FIGURE 1. Apparatus used to study microvessels of isolated perfused hydronephrotic kidneys. Kidneys are perfused with artificial medium on the heated stage of an inverted microscope. Perfusate enters renal artery from a pressurized reservoir. Renal arterial pressure is kept constant by adjusting pressure within media reservoir. Video images of the microcirculation are transmitted to a microcomputer and vessel diameters are measured by automated software. (From Loutzenhiser R, Epstein M: Renal hemodynamic effects of calcium antagonists. In Epstein M, Loutzenhiser R (eds): Calcium Antagonists and the Kidney. Philadelphia, Hanley & Belfus, 1990, pp 33–74, with permission.)

equipped with a thin glass viewing port on the bottom surface (Fig. 1). Furthermore, this experimental model also allows us to measure changes in vessel diameter with an automated program custom-designed for determination of the mean distance between parallel edges of the selected microvessel.

Role of Potential-dependent Calcium Channels in Renal Microcirculation

Segmental Heterogeneity in Localization of Calcium Channels

Using the techniques described above, a number of studies demonstrated disparate renal microvascular actions of calcium antagonists. We used the isolated perfused hydronephrotic kidney model to characterize the distribution of potential-dependent calcium channels and their activity in the renal microcirculation.[20] Because membrane depolarization induced by high potassium (K) levels elicits preferential activation of potential-dependent calcium channels, this pharmacologic tool may clarify their role in mediating renal vascular tone. A medium high in potassium chloride (KCl; 30 mEq/L) causes marked constriction of the afferent arteriole but only a modest decrement in efferent arteriolar diameter (Fig. 2). Consequently, nifedipine reversed the KCl-induced afferent arteriolar constriction in a dose-dependent manner.

Carmines et al.[22] conducted an elegant study by directly assessing the intracellular calcium concentration ($[Ca^{2+}]_i$) of isolated rabbit glomeruli with attached afferent and efferent arterioles. They demonstrated that high K-induced depolarization elevated the $[Ca^{2+}]_i$ of the afferent arteriole from 150 ± 11 to 196 ± 12 nM, whereas $[Ca^{2+}]_i$ of the efferent arteriole was reduced from 188 ± 17 to 148 ± 13 nM, probably because of a decrease in electrochemical gradient driving force of Ca^{2+} leak into cells. They also demonstrated that the inhibition of potential-dependent calcium channels by nifedipine completely prevented the high K-induced rise in $[Ca^{2+}]_i$.

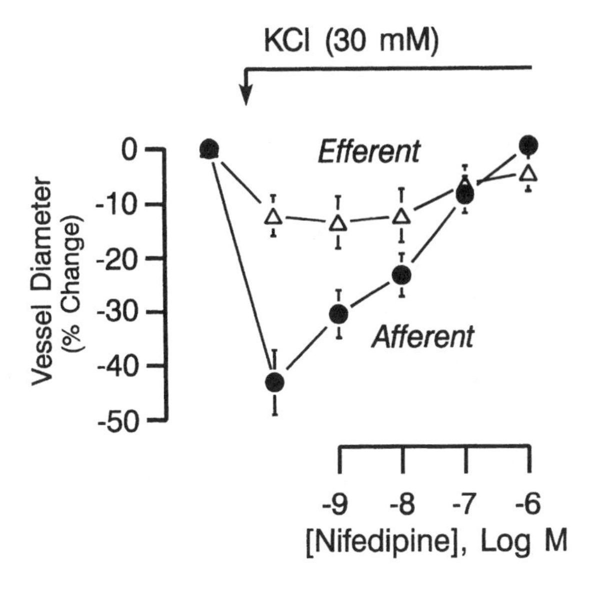

FIGURE 2. Effect of KCl-induced membrane depolarization and blockage of voltage-dependent calcium channels on renal microvessels. Afferent arterioles constricted and dilated prominently in response to KCl-induced membrane depolarization and blockade of calcium channels. In contrast, efferent arterioles responded only modestly to the agents. (Adapted with modification from Loutzenhiser R, Hayashi K, Epstein M: Divergent effects of KCl-induced depolarization on afferent and efferent arterioles. Am J Physiol 257:F561–F564, 1989.)

In addition to the indirect method of calcium channel activation by membrane depolarization, Steinhausen et al.[23] conducted an experiment in which calcium channels are activated directly by a calcium channel agonist, Bay K-8644. They demonstrated that Bay K-8644 caused a preferential constriction and vasomotion of the afferent arteriole, an observation qualitatively similar to that subtending KCl-induced activation of calcium channels.

In summary, from observations obtained under direct and indirect activation of potential-dependent calcium channels, it is reasonable to conclude that potential-dependent calcium channels predominate at the afferent arteriole. In contrast, these channels are sparse or functionally silent at the efferent arteriole. Such functional heterogeneity of the renal microvasculature greatly influences the actions of calcium antagonists on renal hemodynamics.

Effects of Calcium Antagonists on the Renal Microvasculature

Renal microvascular tone in vivo is influenced not only by vasoactive hormones, such as angiotensin II and norepinephrine, but also by myogenic and neural control. The observations in the isolated perfused normal kidney preparations suggest that angiotensin II and norepinephrine elicit both afferent and efferent arteriolar constrictions (see Table 1). To visualize directly the renal actions of calcium antagonists, we used the isolated perfused hydronephrotic kidney model and evaluated whether the underlying vascular tone determines the subsequent responses to calcium antagonists.[20,21,24–26] Nifedipine reverses the angiotensin II-induced constriction of the afferent arteriole in a dose-dependent manner (Fig. 3), whereas the efferent arteriolar constriction is refractory to nifedipine.[24] This vasoactive pattern is also observed with other calcium antagonists. Thus, angiotensin II-induced constriction of the afferent arteriole is reversed in a dose-dependent manner not only by dihydropyridine calcium antagonists (nifedipine, nicardipine, and amlodipine) but also by a benzothiazepine calcium antagonist (diltiazem), whereas the efferent arteriole is refractory to the vasodilator effects of both classes (Fig. 4). Similarly,

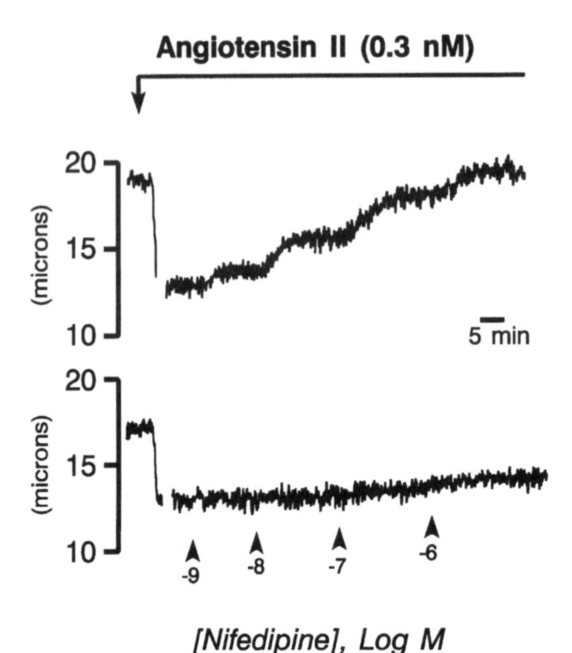

FIGURE 3. Representative tracings illustrating the effects of nifedipine on angiotensin II-induced constriction of renal microvessels. Nifedipine reversed the angiotensin II-induced constriction of an afferent arteriole in a dose-dependent manner but failed to inhibit the efferent arteriolar constriction. (From Hayashi K, Nagahama T, Oka K, et al: Disparate effects of calcium antagonists on renal microcirculation. Hypertens Res 19:31–36, 1996, with permission.)

nifedipine, nicardipine, and amlodipine inhibit the afferent arteriolar constriction during norepinephrine-induced constriction but have few effects on the efferent arteriole (Fig. 5).

Electrophysiologic assessment of the actions of angiotensin II and norepinephrine on the afferent arteriole indicates that both agents depolarize the membrane potential of the afferent arteriole.[27] These observations suggest that vasoconstrictors activate potential-dependent calcium channels at the afferent arteriole. A recent investigation from our laboratory demonstrates that angiotensin II activates chloride channels stimulated by Ca^{2+} released from the intracellular store, resulting in membrane depolarization.[28] In an analogous study, Harder et al.[29] demonstrated that elevations in perfusion pressure caused membrane depolarization and myogenic constriction of the canine interlobular artery. We also have demonstrated that increasing renal perfusion pressure elicited pressure-dependent constriction of the afferent, but not efferent arteriole (Fig. 6). The myogenic vasoconstrictor response is completely abolished by the calcium antagonists nifedipine[30] and diltiazem.[31] These findings suggest that potential-dependent calcium channels predominantly mediate activation mechanisms by angiotensin II, norepinephrine, and myogenic constriction of the afferent arteriole.

Role of Calcium Channels in Endothelin/Thromboxane Mimetic-induced Vasoconstriction

Calcium antagonists have a somewhat different effect on vasoconstriction of the afferent arteriole induced by endothelin and thromboxane agonists. Endothelin[21] and a thromboxane mimetic U44069[32] cause a marked afferent arteriolar constriction, and the subsequent addition of nifedipine inhibits the afferent arteriolar constriction in a dose-dependent manner. It should be noted, however,

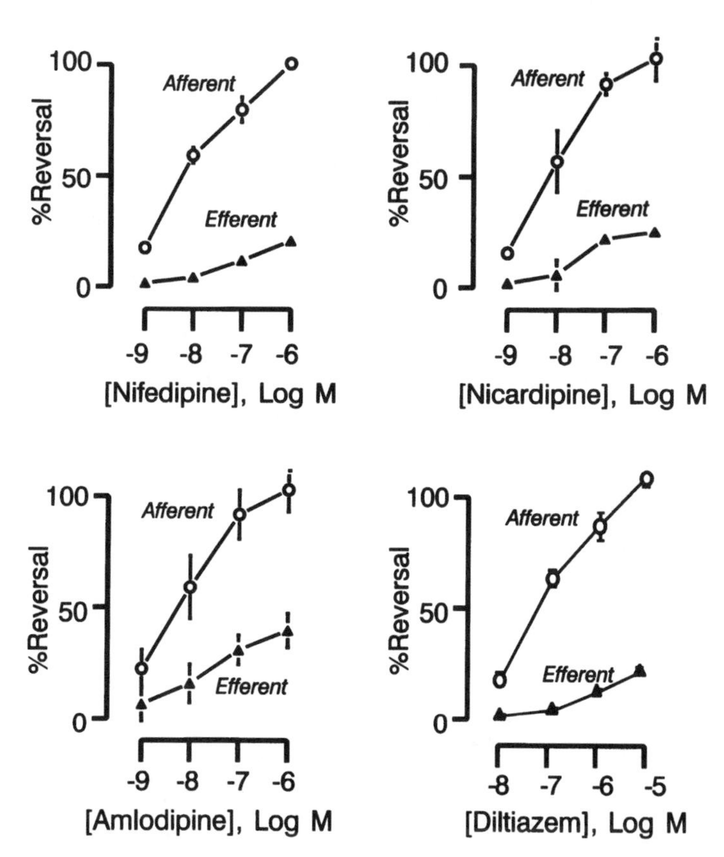

FIGURE 4. Effects of various calcium antagonists on angiotensin II-induced constriction of renal microvessels. Nifedipine, nicardipine, amlodipine, and diltiazem inhibited the angiotensin II-induced afferent arteriolar constriction in a dose-dependent manner. In contrast, efferent arteriolar constriction was resistant to calcium antagonists. (From Hayashi K, Nagahama T, Oka K, et al: Disparate effects of calcium antagonists on renal microcirculation. Hypertens Res 19:31–36, 1996, with permission.)

that nifedipine at a concentration of 10^{-6} M, which completely reversed the KCl-, angiotensin II-, and norepinephrine-induced afferent arteriolar constriction, failed to restore completely the endothelin–U44069-induced constriction. The divergent sensitivity to calcium antagonists under different vasoconstrictor stimuli may reflect the modulation of potential-dependent calcium channels by additional mechanisms. Our laboratory recently demonstrated that the response curve of balnidipine- and isradipine-induced dilation during endothelin-induced afferent arteriole constriction is shifted to higher concentrations than during KCl-induced constriction[33,34] (Fig. 7). Furthermore, this shift in the response curve is abolished by pretreatment with the protein kinase C inhibitor, staurosporine.[34] These observations suggest that the ability of calcium antagonists to inhibit afferent arteriolar constriction depends on the underlying vasoconstrictor stimuli.

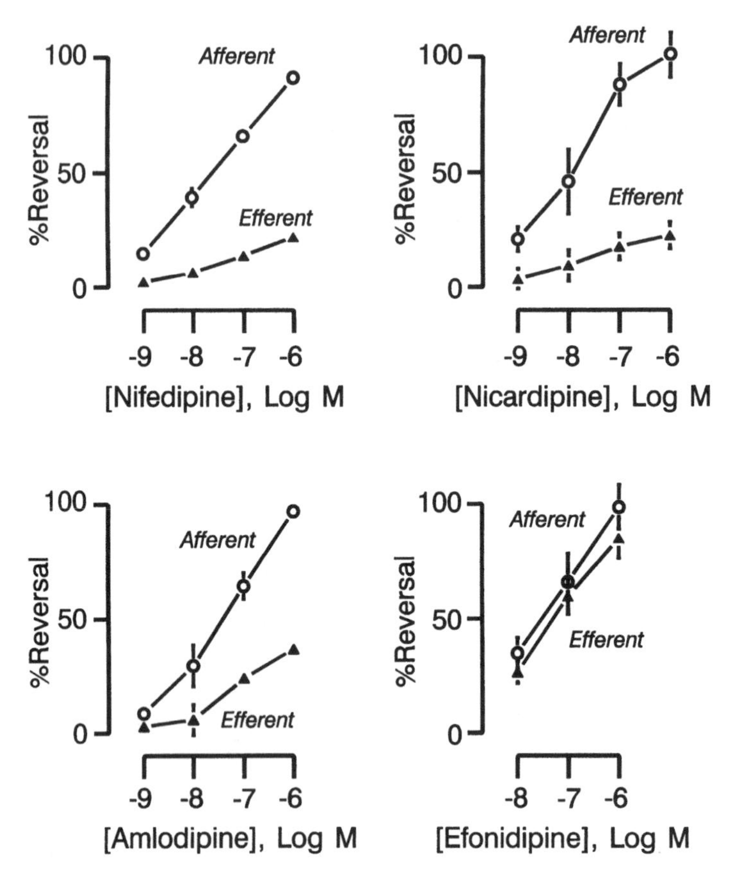

FIGURE 5. Effects of various calcium antagonists on norepinephrine-induced constriction of renal microvessels. Nifedipine, nicardipine, and amlodipine elicited predominantly afferent arteriolar vasodilation. In contrast, efonidipine dilated both afferent and efferent arterioles in a dose-dependent manner. (From Hayashi K, Nagahama T, Oka K, et al: Disparate effects of calcium antagonists on renal microcirculation. Hypertens Res 19:31–36, 1996, with permission.)

Renal Vasodilation Induced by Novel Calcium Antagonists

Recent developments in the structural modification of calcium antagonists afford various advantages to their clinical use. For example, amlodipine has a long half-life and causes less sympathetic nerve activation, whereas cilnidipine is reported to inhibit N-type calcium channels.[35] The additional characteristics of these calcium antagonists may influence the natural progression of renal disease.

From the standpoint of the renal microcirculation, recent investigations have revealed that, although traditional types of calcium antagonists preferentially dilate the afferent arteriole, some recently developed calcium antagonists markedly vasodilate the efferent arteriole as well. Tojo et al.[36] and our laboratory[26] have reported that manidipine elicits both afferent and efferent arteriolar

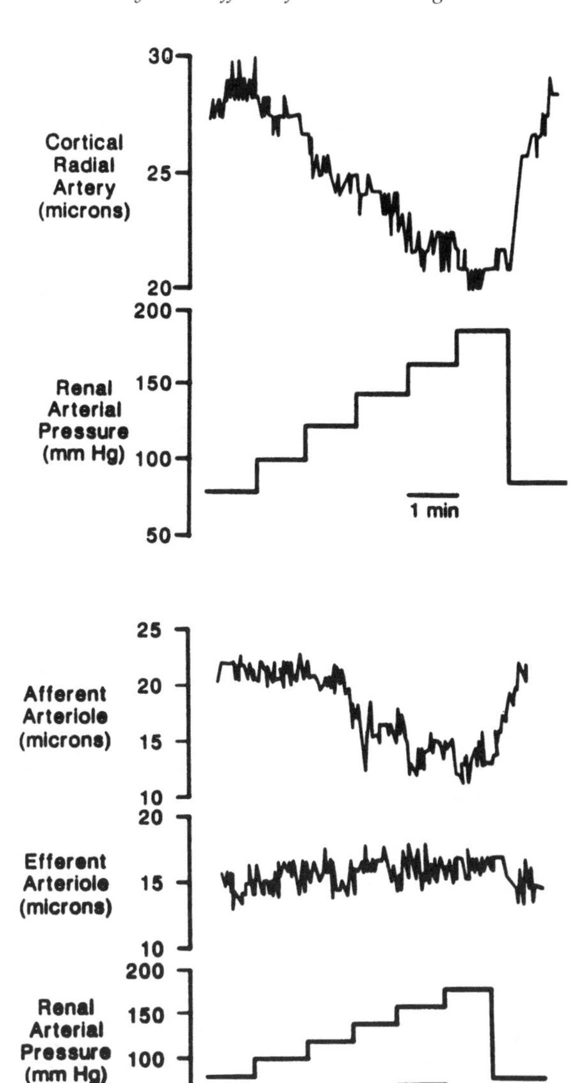

FIGURE 6. Pressure-induced changes in renal microvascular tone. Increasing renal perfusion pressure elicited pressure-dependent vasoconstriction of the afferent arteriole, whereas the efferent arteriole failed to constrict in response to pressure.

dilation in both in vivo and in vitro hydronephrotic kidney models, although the magnitude of the efferent arteriolar dilation is still lower than that of the afferent arteriolar dilation (Fig. 8). Furthermore, we recently demonstrated that a novel calcium antagonist, efonidipine, causes prominent efferent arteriolar dilation nearly identical in magnitude with the afferent arteriolar dilation (see Figs. 5 and 8).[24,26] In contrast, in the same preparation, nifedipine, nicardipine, and amlodipine cause predominantly afferent arteriolar dilation. Like dihydropyridine calcium antagonists, the benzothiazepine calcium antagonist, diltiazem, inhibits predominantly the afferent arteriolar constriction induced by angiotensin II (see Fig. 4) and a thromboxane mimetic, U44069.[32] Because these traditional calcium antagonists act on potential-dependent calcium channels, which predominate at the afferent arteriole, their effects on the

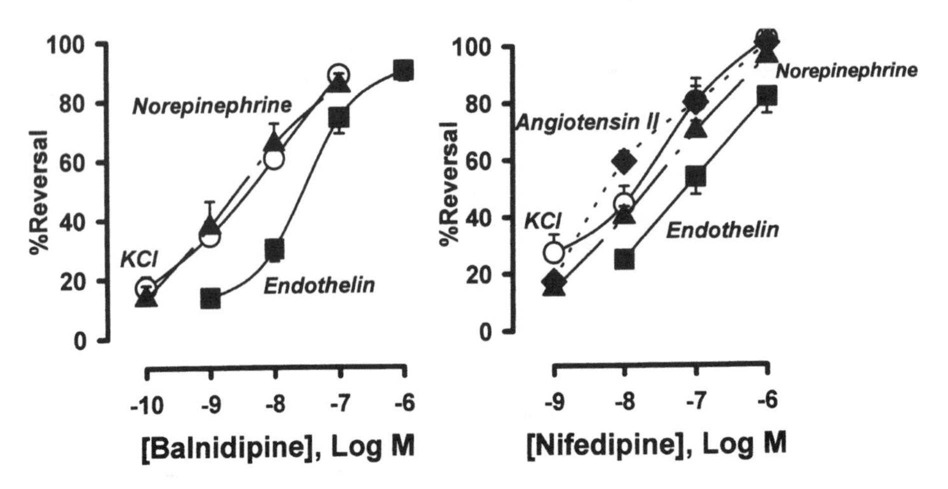

FIGURE 7. Reversal by calcium antagonists of afferent arteriolar constriction induced by angiotensin II, norepinephrine, KCl, and endothelin. Endothelin-induced constriction required higher concentrations of calcium antagonists to inhibit afferent arteriolar constriction than that induced by other vasoconstrictor stimuli.

efferent arteriole are most likely attributable to some of their additional properties rather than to class effects.

The novel calcium antagonists may exert salutary actions on renal hemodynamics. By acting on both afferent and efferent arterioles, they may reduce glomerular capillary pressure. It has been suggested that traditional calcium

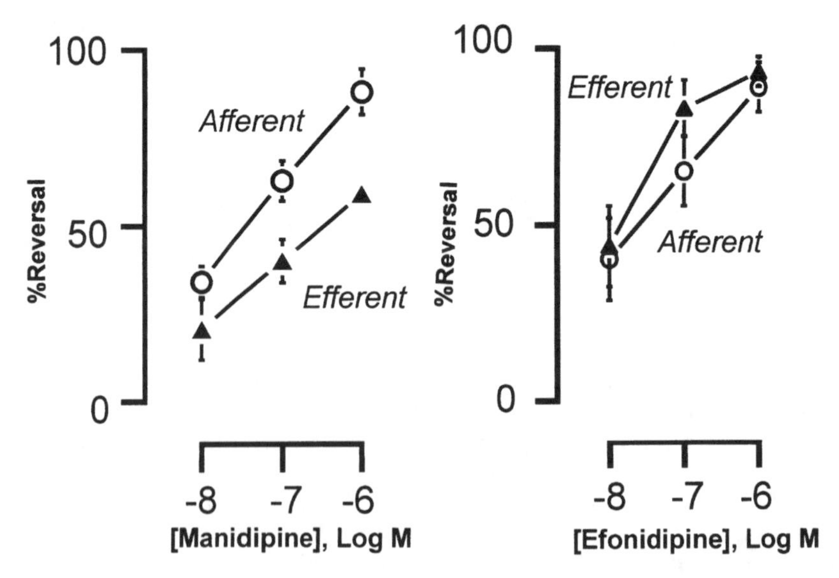

FIGURE 8. Reversal by manidipine and efonidipine of angiotensin II-induced constriction of afferent and efferent arterioles. Both manidipine and efonidipine elicited afferent and efferent arteriolar dilation, although efferent arteriolar dilation induced by manidipine was less than that induced by efonidipine. (Adapted from Saruta T, Kanno Y, Hayashi K, Konishi K: Antihypertensive agents and renal protection: Calcium channel blockers. Kidney Int 49(Suppl 55):S52–S56, 1996, with permission.)

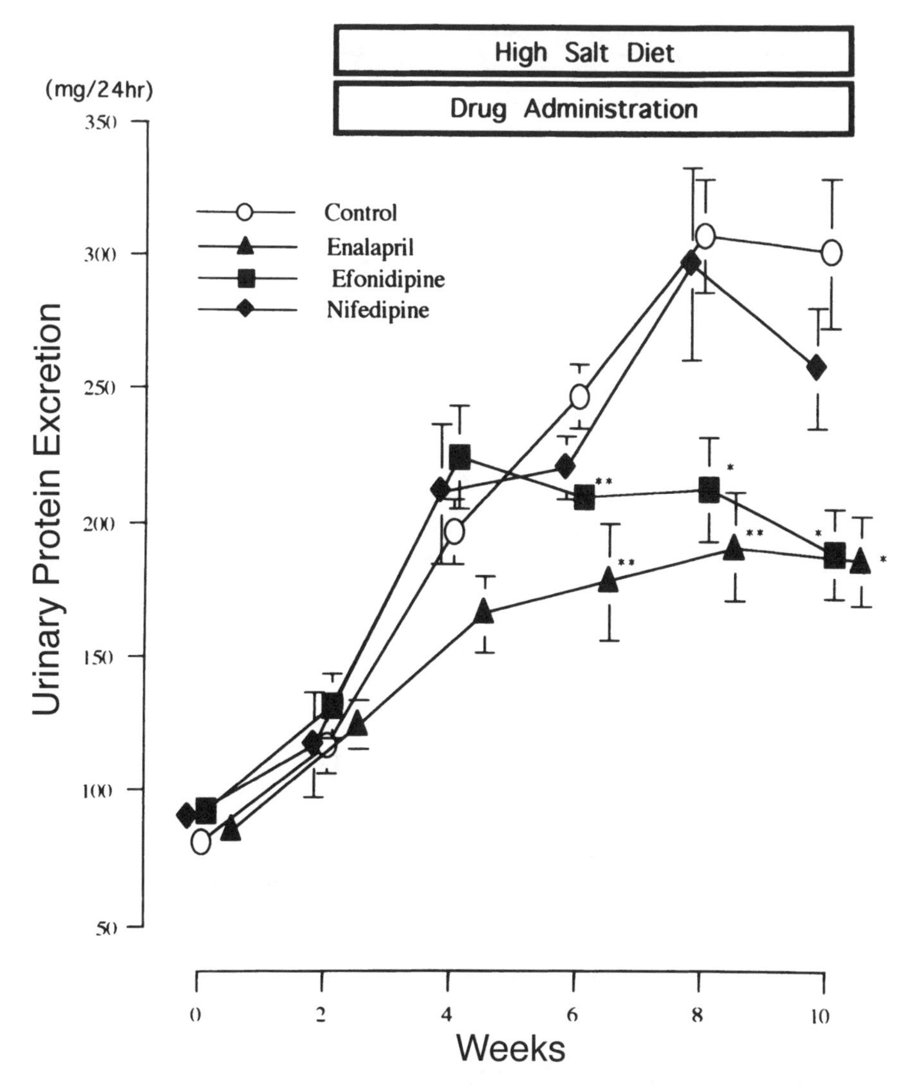

FIGURE 9. Effects of nifedipine, efonidipine, and enalapril on urinary protein excretion in 5/6 nephrectomized spontaneously hypertensive rats. Both enalapril and efonidipine inhibited the increases in proteinuria, whereas nifedipine failed to prevent the elevation in urinary protein excretion. (From Fujiwara et al, Clin Exp Hypertens, in press.)

antagonists, which mainly affect the afferent arteriole, may elevate glomerular capillary pressure. Because they also decrease systemic blood pressure, the net effects of calcium antagonists on glomerular hemodynamics are determined by the balance between the afferent arteriolar resistance and systemic blood pressure. In contrast to traditional calcium antagonists, which may elevate glomerular capillary pressure, efonidipine dilates both afferent and efferent arteriolar constriction. The vasodilator action of efonidipine on the efferent arteriole is noteworthy, because it is believed to reduce glomerular capillary pressure in concert with its blood pressure-lowering effect. Shudo et al.[37] recently reported

TABLE 3. Effects of Calcium Antagonists on Renal Microcirculation

	Authors	Ca Antagonist	Study	BP	R_A	R_E	P_{GC}	ΔP	Ref.
Normotensive									
Wistar rat	Pelayo	Verapamil	Acute	↘	↘	↔	↔	↔	38
	Ichikawa	Verapamil	Acute	↘	↘	↘	↔	↔	39
WKY rat	Isshiki	Diltiazem	Chronic	↓	↓	↔	↔		40
Hypertensive									
SHR	Isshiki	Diltiazem	Chronic	↓↓	↓↓	↓	↓		40
Remnant kidney	Yoshioka	Verapamil	Acute	↓	↘	↓	↓	↓	42
	Pelayo	Verapamil	Chronic	↔	↔	↔	↔	↓	43
	Anderson	Verapamil	Acute	↓	↔	↔	↓	↓	44
		Diltiazem	Acute	↓↓	↔	↔	↓	↓	44
	Dworkin	Nifedipine	Chronic	↓	↔	↑	↔	↔	45
SHR-UniNx	Dworkin	Nifedipine	Chronic	↓↓	↓	↔	↘	↓	49
	Dworkin	Amlodipine	Chronic	↓			↔	↔	50
DOCA-UniNx	Dworkin	Nifedipine	Chronic	↓	↓	↔	↔	↔	51
Diabetic									
DM rat-UniNx	Anderson	Nifedipine	Chronic	↘	↔	↔	↔	↔	47
DM dog-UniNx	Brown	TA-3090	Chronic	↘	↔	↔	↔	↔	48

WKY = Wistar-Kyoto rat, SHR = spontaneously hypertensive rat, UniNx = uninephrectomized, DOCA = deoxycorticosterone acetate, Ca = calcium, BP = blood pressure, R_A = afferent resistance, R_E = efferent resistance, P_{GC} = glomerular capillary pressure, ΔP = change in pressure.

that efonidipine acutely decreases proteinuria in spontaneously hypertensive rats, whereas systemic blood pressure is only partially reduced. Furthermore, we demonstrated that in salt-induced, subtotally nephrectomized spontaneously hypertensive rats, urinary protein excretion is reduced markedly to the same level as that observed in enalapril-treated rats (Fig. 9), whereas nifedipine reduced proteinuria only modestly. Systemic blood pressures were reduced to the same level in different treatment groups. Taken together, the acute and chronic effects of efonidipine on proteinuria suggest its beneficial effect on the glomerular capillary pressure, which may be mediated by renal microcirculatory effects on the efferent arteriole.

Role of Calcium Antagonists in the Progression of Renal Injury in Experimental Animals

The renal microvascular effects of traditional calcium antagonists suggest that they fail to correct, or may even cause, glomerular hypertension, although the actual glomerular capillary pressure depends on the balance between the renal perfusion pressure and the afferent arteriolar tone.

Numerous investigations have used micropuncture techniques to assess the influence of calcium antagonists on the response of the glomerular microcirculation to vasoconstrictor stimuli (Table 3). The response depends on numerous factors, including route of administration and types of experimental animals. Furthermore, if systemic blood pressure declines significantly, compensatory mechanisms are initiated that blunt the fall in renal resistance. Pelayo et al.[38] demonstrated that in normotensive rats the administration of verapamil in a modestly hypotensive dose caused a slight decrement in afferent arteriolar resistance but had no appreciable effect on efferent arteriolar resistance or glomerular capillary pressure. Furthermore, Ichikawa et al.[39] observed that a modestly hypotensive dose of verapamil minimally increases single nephron GFR, proportionately slightly reduces afferent and efferent arteriolar resistance, and does not affect glomerular capillary pressure. In addition, Isshiki et al.[40] demonstrated that in conscious Wistar-Kyoto rats, diltiazem diminished both systemic blood pressure and afferent arteriolar resistance but failed to reduce efferent arteriolar resistance or glomerular capillary pressure. Collectively these observations in normotensive animals suggest that calcium antagonists modestly reduce systemic blood pressure and cause a slight dilation of the afferent arteriole but do not affect glomerular capillary pressure. The effects on the efferent arteriole do not appear to favor any changes in arteriolar resistance.

In striking contrast, in vivo vasodilatory actions of calcium antagonists are profoundly unveiled when the kidney is preconstricted with renal nerve stimulation[38] and elevated systemic pressure.[30,31] Calcium antagonists markedly blunted the increments in nerve stimulation-induced arteriolar resistance (Fig. 10). Similarly, in spontaneously hypertensive rats, diltiazem elicited a marked decrease

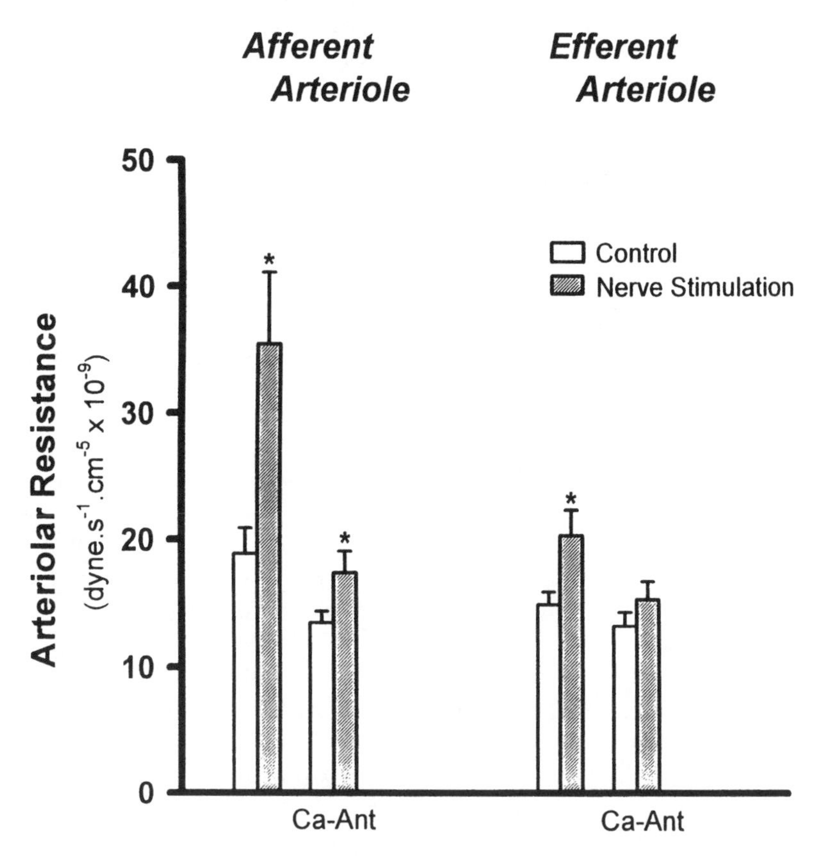

FIGURE 10. Effects of calcium antagonist verapamil on nerve stimulation-induced arteriolar tone. Verapamil markedly prevented the increase in afferent arteriolar resistance produced by nerve stimulation. * Indicates $p < 0.05$ vs. control. (Adapted with modification from Pelayo J: Modulation of renal adrenergic effector mechanisms by calcium entry blockers. Am J Physiol 252:F613–F620, 1987.)

in both systemic blood pressure and afferent arteriolar resistance but had no effect on the efferent arteriolar resistance.[40]

The remnant kidney model, considered to be an experimental analog of chronic renal failure, is characterized by an increase in plasma flow and glomerular filtration rate of single nephrons. These alterations are attributed to considerably reduced afferent arteriolar resistance.[41] In such renal microvascular settings, it is anticipated that calcium antagonists will exert no further vasodilatory action on the afferent arteriole. In fact, as illustrated in Table 3, most investigations reported that calcium antagonists did not alter afferent arteriolar resistance, whether administered acutely or chronically.[42–45] Glomerular capillary pressure thus may parallel the changes in systemic blood pressure, and the ability to normalize glomerular capillary pressure is more readily available when systemic blood pressure is decreased to the normal level.

Diabetic nephropathy manifests a diminished renal vascular tone, particularly at the afferent arteriole, and this renal microcirculatory alteration constitutes a pivotal determinant of the development of glomerular hypertension.[46] In experimental diabetes, Anderson et al.[47] and Brown et al.[48] demonstrated that calcium antagonists (nifedipine or TA-3090, respectively) reduced systemic blood pressure but had no effect on the afferent or efferent arteriolar tone, findings similar to those in remnant kidney models. In contrast to remnant kidneys, glomerular capillary pressure did not change. Although systemic blood pressure in diabetic animals was elevated compared with normal counterparts, the increments in systemic blood pressure are smaller than those in remnant kidney models. It follows, therefore, that the depressor actions of calcium antagonists are smaller than in remnant kidney models, and the modest reduction in blood pressure may not alter the glomerular capillary pressure. In kidneys in which the afferent arteriole is fully dilated, the effect of calcium antagonists on glomerular capillary pressure depends largely on systemic blood pressure.

In contrast to the subtotal nephrectomy model and diabetic kidneys, both heminephrectomized SHRs[49,50] and heminephrectomized DOCA-salt rats[51] manifest greater afferent arteriolar resistance (Table 4). Calcium antagonists

TABLE 4. Effects of Calcium Antagonists on Renal Microcirculation in Various Types of Renal Injury Models

		AP*	R_A†	R_E†	P_{GC}*	ΔP*	Ref.
5/6 Nx	Control	154 ± 6	1.20 ± 0.20	0.80 ± 0.10	68 ± 3	51 ± 3	44
	Diltiazem	102 ± 2¶	1.10 ± 0.20	0.70 ± 0.20	50 ± 2¶	35 ± 3¶	
SHR-UniNx	Control	172 ± 4	1.97 ± 0.26	1.10 ± 0.12	67 ± 2	55 ± 2	49
	Nifedipine	126 ± 3¶	1.00 ± 0.11¶	0.73 ± 0.06	60 ± 1	45 ± 1¶	
DOCA-UniNx	Control	139 ± 4	1.41 ± 0.19	0.94 ± 0.14	60 ± 1	49 ± 2	51
	Nifedipine	106 ± 2¶	0.83 ± 0.08¶	0.97 ± 0.08	62 ± 2	49 ± 1	
DM-UniNx	Control	118 ± 5	0.80 ± 0.10	0.70 ± 0.10	60 ± 1	47 ± 1	47
	Nifedipine	106 ± 2¶	0.60 ± 0.10	0.60 ± 0.10	60 ± 2	45 ± 2	

* Measured in mmHg.
† Measured in dyn/sec/$cm^{-5} \times 10^{10}$.
SHR = spontaneously hypertensive rat, Nx = nephrectomized, UniNx = uninephrectomized, DOCA = deoxycorticosterone acetate, R_A = afferent resistance, R_E = efferent resistance, P_{GC} = glomerular capillary pressure, ΔP = change in pressure, ¶ indicates $p < 0.05$ vs. control.

markedly reduce afferent arteriolar resistance, whereas efferent arteriolar resistance does not change appreciably. Collectively, these observations again support the formulation that the underlying vascular tone greatly influences the subsequent vasodilator actions of the calcium antagonist.

In summary, direct in vivo and in vitro observations in diverse experimental models indicate that calcium antagonists inhibit preglomerular vasoconstriction. In contrast, most studies suggest that the efferent arteriole appears to be refractory to their vasodilatory effects. Inferences from micropuncture studies are conflicting. Such divergent observations are most likely to be associated with the underlying vasoconstrictor tone in different experimental models; in remnant kidney models, heminephrectomized SHRs, and heminephrectomized DOCA rats, calcium antagonists apparently attenuate glomerular hypertension with variable effects on afferent and efferent arteriolar resistances.

RENAL FUNCTIONAL EFFECTS IN HUMANS

Studies in Normal Human Volunteers

In normal human volunteers, calcium antagonists have little effect on renal plasma flow or glomerular filtration rate. Leonetti et al.[52] reported that single doses of either nifedipine (10 mg orally) or verapamil (160 mg orally) failed to alter GFR in normal humans. Similarly, Wallia et al.[53] observed that nitrendipine (5–10 mg orally) had no effect on GFR or renal blood flow (RBF). Von Shaik et al.[54] also found that nicardipine (60 mg orally) had no effect on GFR in normal humans.

Studies in Hypertensive Patients

As in studies with experimental animals, hypertensive patients appear to be more sensitive to the renal hemodynamic effects of calcium antagonists than their normotensive counterparts. Over 20 years ago, Klutsch et al.[55] reported that nifedipine increased renal perfusion and GFR in a group of patients with essential hypertension. Subsequently, several reports have indicated that in contrast to the lack of effect in normal humans, hypertensive patients may exhibit an exaggerated renal hemodynamic response to calcium antagonists,[56] with sustained increases in GFR and RBF.[56] It has been suggested that these renal hemodynamic responses may not be sustained unless renal function was initially impaired. Such observations suggest that renal hemodynamic adaptations occur during chronic calcium antagonist therapy.

Several investigators have demonstrated a similar exaggerated vasodilatory response in normotensive offspring of hypertensive parents.[57,58] Blackshear et al.[57] reported that at least 50% of normotensive subjects with a family history of hypertension demonstrated an exaggerated renal vasodilatory response to diltiazem, suggesting an inherited abnormality of the renal vascular bed associated with hypertension. Montanari et al.[58] recently extended these observations. They investigated 9 young normotensive subjects with no family history of hypertension (F–) and 9 age-matched normotensive subjects with one parent with essential hypertension (F+). They determined effective renal plasma flow (ERPF) and GFR before and after administering a single 20-mg oral dose of nifedipine. Baseline renal function did not differ between the two groups. In contrast, nifedipine induced disparate renal hemodynamic and excretory responses. Whereas nifedipine did not alter ERPF of F– subjects, it produced an exaggerated renal vasodilator response in the F+ group, with a mean increase in ERPF of 31%. Concomitantly, GFR was unchanged. The studies of Blackshear et al.[57] and Montanari et al.[58] suggest an inherited trait associated with hypertension, manifested as an exaggerated response to calcium antagonists. Besides their utility in elucidating the pathogenesis of essential hypertension, these studies underscore the renal vasodilatory potential of calcium antagonists in patients with essential hypertension.

As detailed in a recent review,[3] the reported effects vary in hypertensive patients with impaired renal function. Some investigators have reported that the renal vasodilatory effects of calcium antagonists are blunted, whereas others have reported increases in GFR and RPF. These disparate effects may relate to the observation that nephron function is heterogeneous in chronic renal failure with some nephrons hyperfiltering, whereas others do not. Consequently, it is not altogether surprising that the renal hemodynamic responses to calcium antagonists may vary widely in patients with chronic renal failure. Furthermore, the blunted responses may reflect structural vascular damage and fixed renal vascular resistance in hypertensive patients with chronic renal failure.

Assessment of Intrarenal Hemodynamics Using Renal Function Curves

Because invasive approaches such as micropuncture cannot be performed in humans, several investigators have attempted to devise alternative approaches to approximate quantitatively intrarenal hemodynamics. Traditionally, such approaches have included Gomez's formulas[59] and, more recently, the application of the renal function curve (pressure-natriuresis relationship).[60,61]

Recently, Kimura et al.[62] applied renal function curves to a characterization of the effects of the dihydropyridine nicardipine on intrarenal hemodynamics in patients with essential hypertension. They estimated that nicardipine reduced R_A from 9,300 ± 900 to 7,400 ± 700 dyne/sec/cm^{-5} ($p < 0.01$), whereas no changes were noted in R_E. GFR, RPF, and RBF were not altered. Kimura et al.[62] proposed that neither RPF nor GFR increased despite afferent arteriolar dilatation because of a concomitant reduction in mean arterial pressure. Thus, studies using renal function curves in humans are in full accordance with experimental observations using several videomicroscopic techniques, including the isolated perfused hydronephrotic kidney,[2,3,5,21] the in vivo hydronephrotic kidney,[63] and the juxtamedullary nephron preparations.[64]

In summary, the renal hemodynamic response of calcium antagonists is a result of the interplay of numerous factors. Virtually all available in vitro studies suggest that calcium antagonists preferentially dilate the preglomerular circulation. In contrast, the renal vascular response in vivo in intact animals is not consistent and depends on the experimental setting and the magnitude of basal renal vascular tone.

It is possible that the renal microcirculatory effects of calcium antagonists differ in normal or hydronephrotic kidneys compared with remnant kidneys. Additional studies of the same calcium antagonists in different experimental models are necessary to resolve this issue.

CONCLUSION

Calcium antagonists exert several effects on renal hemodynamics and excretory function that contribute to their salutary or renal protective effects. Calcium antagonists preferentially attenuate afferent arteriolar vasoconstriction, with a concomitant reduction in systemic blood pressure. Because the net effects of calcium antagonists on the renal microcirculation depend on the balance between these hemodynamic factors, their ability to alter glomerular hemodynamics may vary depending on the underlying basal vascular tone observed in various renal diseases. Finally, recent development of new types of calcium antagonists may expand their usefulness in renal disorders.

References

1. Heidland A, Klutsch K, Obek A: Myogenbedingte vasodilatation bei nierenischamie. Munch Med Wochenschr 35:1636, 1962.
2. Loutzenhiser R, Epstein M: Effects of calcium antagonists on renal hemodynamics. Am J Physiol 249:F619–F629, 1985.
3. Loutzenhiser R, Epstein M: Renal hemodynamic effects of calcium antagonists. In Epstein M, Loutzenhiser R (eds): Calcium Antagonists and the Kidney. Philadelphia, Hanley & Belfus, 1990, pp 33–74.
4. Epstein M, Loutzenhiser R: Potential applicability of calcium antagonists as renal protective agents. In Epstein M, Loutzenhiser R (eds): Calcium Antagonists and the Kidney. Philadelphia, Hanley & Belfus, 1990, pp 275–298.
5. Loutzenhiser R, Epstein M: Renal microvascular actions of calcium antagonists. J Am Soc Nephrol 1:S3–S12, 1990.
6. Loutzenhiser R, Epstein M: Modification of the renal hemodynamic response to vasoconstrictors by calcium antagonists. Am J Neprhol 7(Suppl 1):7–16, 1987.
7. Dietz JR, Davis JO, Freeman RH, et al: Effects of intrarenal infusion of calcium entry blockers in anesthetized dogs. Hypertension 5:482–488, 1983.
8. Meisheri K, Hwang O, van Breemen C: Evidence for two separate Ca^{2+} pathways in smooth muscle plasmalemma. J Membr Biol 59:19–25, 1981.
9. Loutzenhiser R, Epstein M: Activation mechanisms of human renal artery: Effects of KCl, norepinephrine, and nitrendipine upon tension development and ^{45}Ca influx. Eur J Pharmacol 106:47–52, 1984.
10. Loutzenhiser R, Epstein M, Horton C, Sonke P: Reversal by the calcium antagonist nisoldipine of norepinephrine-induced reduction of GFR: Evidence for preferential antagonism of preglomearul vasoconstriction. J Pharm Exp Ther 232:382–387, 1985.
11. Loutzenhiser R, Horton C, Epstein M: Effects of diltiazem and manganese on renal hemodynamics. Studies in the isolated perfused rat kidney. Nephron 39:382–388, 1985.
12. Edwards RM: Segmental effects of norepinephrine and angiotensin II on isolated renal microvessels. Am J Physiol 244:F526–F534, 1983.
13. Yuan B, Robinette JB, Conger JD: Effects of angiotensin II and norepinephrine on isolated rat afferent and efferent arterioles. Am J Physiol 258:F741–F750, 1990.
14. Ito S, Carretero OA: An in vitro approach to the study of macula densa-mediated glomerular hemodynamics. Kidney Int 38:1206–1210, 1990.
15. Ito S, Ren Y: Evidence for the role of nitric oxide in macula densa control of glomerular hemodyanmics. J Clin Invest 92:1093–1098, 1993.
16. Carmines PK, Morrison KD, Navar LG: Angiotensin II effects on microvascular diameters of in vitro blood perfused juxtamedullary nephrons. Am J Physiol 251:F610–F618, 1986.
17. Yamamoto T, Tanaka H, Yoshiyuki J, et al: Visualization of intrarenal microvessels and evaluatino of the effect of angiotensin II by a needle probe microscope with a CCD camera. J Am Soc Nephrol 4:574, 1993.
18. Steinhausen M, Snoei H, Parekh N, et al: Hydronephrosis: A new method to visualize vas afferens, efferens, and glomerular network. Kidney Int 23:794–806, 1983.
19. Loutzenhiser R, Hayashi K, Epstein M: Atrial natriuretic peptide reverses afferent arteriolar vasoconstriction and potentiates efferent arteriolar vasoconstriction in the isolated perfused rat kidney. J Pharmacol Exp Ther 246:522–528, 1988.

20. Loutzenhiser R, Hayashi K, Epstein M: Divergent effects of KCl-induced depolarization on afferent and efferent arterioles. Am J Physiol 257:F561–F564, 1989.
21. Loutzenhiser R, Epstein M, Hayashi K, Horton C: Direct visualization of effects of endothelin on the renal microvasculature. Am J Physiol 258:F61–F68, 1990.
22. Carmines PK, Fower BC, Bell PD: Segmental distinct effects of depolarization on intracellular [Ca^{2+}] in renal arterioles. Am J Physiol 265:F677–F685, 1993.
23. Steinhausen M, Baehr M: Vasomotion and vasoconstriction induced by a Ca agonist in the split hydronephrotic kidney. Prog Appl Microcirc 14:25–29, 1989.
24. Hayashi K, Nagahama T, Oka K, et al: Disparate effects of calcium antagonists on renal microcirculation. Hypertens Res 19:31–36, 1996.
25. Saruta T, Hayashi K, Suzuki H: Renal effects of amlodipine. J Hum Hypertens 9(Suppl I):S11–S16, 1995.
26. Saruta T, Kanno Y, Hayashi K, Konishi K: Antihypertensive agents and renal protection: Calcium channel blockers. Kidney Int 49(Suppl 55):S52–S56, 1996.
27. Bührle CP, Nobiling R, Taugner R: Intracellular recordings from renin-positive cells of the afferent glomerular arteriole. Am J Physiol 249:F272–F281, 1985.
28. Takenaka T, Kanno Y, Kitamura Y, et al: Role of chloride channels in afferent arteriolar constriction. Kidney Int 50:864–872, 1996.
29. Harder DR, Gilbert R, Lomberd JH: Vascular muscle cell depolarization and activation in renal arteries on elevation of transmural pressure. Am J Physiol 253:F778–781, 1987.
30. Hayashi K, Epstein M, Loutzenhiser R: Pressure-induced vasoconstriction of renal microvessels in normotensive and hypertensive rats: Studies in the isolated perfused hydronephrotic kidney. Circ Res 65:1475–1484, 1989.
31. Loutzenhiser R, Epstein M, Horton C: Inhibition by diltiazem of pressure-induced afferent arteriolar vasoconstriction in the isolated perfused rat kidney. Am J Cardiol 59:A72–A75, 1987.
32. Hayashi K, Loutzenhiser R, Epstein M: Direct evidence that a thromboxane mimetic U44069 preferentially constricts the afferent arteriole. J Am Soc Nephrol 8:25–31, 1997.
33. Epstein M, Hayashi K, Loutzenhiser R: Renal microvascular effects of calcium antagonists on endothelin (ENDO)-norepinephrine (NE), and KCl-induced vasoconstriction. Presented at the Eleventh International Congress of Nephrology, Tokyo, Japan, 1990.
34. Takenaka T, Forster H, Epstein M: Protein kinase C and calcium channel activation as determinants of renal vasoconstriction by angiotensin II and endothelin. Circ Res 73:743–750, 1993.
35. Hosono M, Fujii S, Hiruma T, et al: Inhibitory effect of cilnidipine on vascular sympathetic neurotransmission and subsequent vasoconstriction in spontaneously hypertensive rats. Jpn J Pharmacol 69:127–134, 1995.
36. Tojo A, Kimura K, Matsuoka H, Sugimoto T: Effects of manidipine hydrochloride on the renal microcirculation in spontaneously hypertensive rats. J Cardiovasc Pharmacol 20:895–899, 1992.
37. Shudo C, Masuda Y, Sugita H, et al: Beneficial effect of efonidipine hydrochloride (NZ-105) on proteinuria in aged spontaneously hypertensive rats (SHR). Pharm Sci 1:333–335, 1995.
38. Pelayo J: Modulation of renal adrenergic effector mechanisms by calcium entry blockers. Am J Physiol 252:F613–F620, 1987.
39. Ichikawa I, Miele JF, Brenner BM: Reversal of renal cortical actions of angiotensin II by verapamil and manganese. Kidney Int 16:137–147, 1979.
40. Isshiki T, Amodeo C, Messerli FH, et al: Diltiazem maintains renal vasodilation without hyperfiltration in hypertension: Studies in essential hypertensive man and the spontaneously hypertensive rat. Cardiovasc Drug Ther 1:359–366, 1987.
41. Hostetter TH, Olson JL, Rennke HG, et al: Hyperfiltration in remnant nephrons: A potentially adverse response to renal ablation. Am J Physiol 241:F85–F93, 1981.
42. Yoshioka T, Shiraga H, Yoshida Y, et al: "Intact nephrons" as the primary origin of proteinuria in chronic renal disease: Study in the rat model of subtotal nephrectomy. J Clin Ivest 82:1614–1623, 1988.
43. Pelayo JC, Harris DCH, Shanley PF, et al: Glomerular hemodynamic adaptations in remnant nephrons: Effects of verapamil. Am J Physiol 254:F425–F431, 1988.
44. Anderson S: Renal hemodynamic effects of calcium antagonists in rats with reduced renal mass. Hypertension 17:288–295, 1991.
45. Dworkin LD, Benstein JA, Parker M, et al: Calcium antagonists and converting enzyme inhibitors reduce renal injury by different mechanisms. Kidney Int 43:808–814, 1993.
46. Hostetter TH, Rennke HG, Brenner BM: The case for intrarenal hypertension in the initiation and progression of diabetic and other glomerulopathies. Am J Med 72:375–380, 1982.
47. Anderson S, Rennke HG, Brenner BM: Nifedipine versus fosinopril in uninephrectomized diabetic rats. Kidney Int 41:891–897, 1992.

48. Brown SA, Walton CL, Crawford P, Bakris G: Long-term effects of antihypertensive regimens on renal hemodynamics and proteinuria. Kidney Int 43:1210–1218, 1993.
49. Dworkin LD, Feiner HD, Parker M, Tolbert E: Effects of nifedipine and enalapril on glomerular structure and function in uninephrectomized SHR. Kidney Int 39:1112–1117, 1991.
50. Dworkin LD, Tolbert E, Recht PA, et al: Effects of amlodipine on glomerular filtration, growth, and injury in experimental hypertension. Hypertension 27:245–250, 1996.
51. Dworkin LD, Levin RI, Benstein JA, et al: Effects of nifedipine and enalapril on glomerular injury in rats with deoxycorticosterone-salt hypertension. Am J Physiol 259:F598–F604, 1990.
52. Leonetti G, Cuspidi C, Sampieri L, et al: Comparison of cardiovascular, renal and humoral effects of acute administration of two calcium channel blockers in normotensive and hypertensive subjects. J Cardiovasc Pharmacol 4:S319–S324, 1982.
53. Wallia R, Greenberg A, Puschett JB: Renal hemodynamic and tubular transport effects of nitrendipine. J Lab Clin Med 105:498–503, 1985.
54. Van Schaik BAM, Van Nistelrooy AEJ, Geyskes GG: Antihypertensive and renal effects of nicardipine. J Lab Clin Med 104:498–503, 1985.
55. Klutsch K, Schmidt P, Grosswendt J: Der Einfluss von BAY A0140 auf die Nierenfunktion des Hypertonikers. Arztneimittelforschung 22:377–380, 1972.
56. Reams GP, Bauer JH: Acute and chronic effects of calcium antagonists on the essential hypertensive kidney. In Epstein M, Loutzenhiser R (eds): Calcium Antagonists and the Kidney. Philadelphia, Hanley & Belfus, 1990, pp 247–256.
57. Blackshear JL, Garnic D, Williams GH: Exaggerated renal vasodilator response to calcium entry blockade in first-degree relatives of essential hypertensive subjects. Hypertension 9:384–389, 1987.
58. Montanari A, Vallisa D, Ragni G, et al: Abnormal renal responses to calcium entry blockade in normotensive offspring of hypertensive parents. Hypertension 12:498–505, 1988.
59. Gomez DM: Evaluation of renal resistances, with special reference to changes in essential hypertension. J Clin Invest 30:1143–1155, 1951.
60. Guyton AC: Renal funtion curve—a key to understanding the pathogenesis of hypertension. Hypertension 10:1–6, 1987.
61. Kimura G, Saito F, Kojima S, et al: Renal function curve in patients with secondary forms of hypertension. Hypertension 10:11–15, 1987.
62. Kimura G, Deguchi F, Kojima S, et al: Effect of a calcium-entry blocker, nicardipine, on intrarenal hemodynamics in essential hypertension. Am J Kidney Dis 18:47–54, 1991.
63. Fleming JT, Parekh N, Steinhausen M: Calcium antagonists preferentially dilate preglomerular vessels of hydronephrotic kidney. Am J Physiol 253:F1157–F1163, 1987.
64. Carmines PK, Navar LG: Disparate effects of Ca channel blockers on afferent and efferent arteriolar responses to ANG II. Am J Physiol 256:F1015–F1020, 1989.

JOSE TUÑON M.D. / NIEVES TARIN M.D. / JESUS EGIDO M.D.

23

Effects of Calcium Antagonists on Vasoactive Hormones

Calcium channel blockers (CCBs) are a heterogeneous group of drugs in terms of their chemistry and pharmacodynamic effects, although they share a common mechanism of action. CCBs block the influx of calcium ions across cell membranes, including those of vascular smooth muscle and myocardial cells. Therefore, they are widely used as arterial vasodilators in the treatment of hypertension and coronary artery disease. Because of their mechanism of action, CCBs can modify several regulatory mechanisms of vasomotor tone and intravascular volume. This chapter reviews the effects of CCBs on the renin-angiotensin and endothelin systems. The physiopathologic aspects of both systems are briefly reviewed.

CALCIUM ANTAGONISTS AND THE RENIN-ANGIOTENSIN SYSTEM

General Aspects of the Renin-Angiotensin System

The renin-angiotensin system (RAS) is the main regulator of intravascular volume and systemic blood pressure. It was first described as an endocrine system, where the kidneys release renin, which converts the angiotensinogen produced by the liver into angiotensin I. The decapeptide angiotensin I is cleaved by angiotensin-converting enzyme (ACE), present mainly in the pulmonary vascular beds, to the octapeptide angiotensin II. The plasmatic RAS avoids excessive fall in blood pressure by vasoconstriction, renal fluid retention and, when hemorrhage is present, enhancement of the coagulation system. Angiotensin II exerts vasoconstricting effects both directly and through the production of endothelin. Furthermore, it increases the adrenergic drive through central[1] or peripheral[2–4] mechanisms. It also stimulates the production of aldosterone by the glomerulosa cells of the adrenal cortex, interferes with fibrinolysis, and has procoagulant effects. It increases platelet aggregation[5,6] and expression of plasminogen-activator inhibitor-1 (PAI-1).[7–9] Finally, ACE is responsible for the degradation of bradykinin, a hormone that produces vasodilation mediated by nitric oxide and prostacyclin release by the vascular endothelium.[10]

More recently a tissue RAS has been described.[11,12] Some investigators have suggested that plasmatic RAS is predominantly important for acute regulatory mechanisms, whereas the tissue RAS is involved mainly in chronic aspects of vascular regulation,[11,12] including the pathogenesis of atherosclerosis and renal disease.

Several studies have demonstrated the synthesis of RAS components in vascular walls,[13–18] where the limiting step in production of angiotensin II is tissue ACE.[15,19] In fact, ACE inhibitors reduce neointimal formation in response to vascular injury only when given during the time necessary to block tissue ACE.[20]

Angiotensin II Receptors and Signaling

Two major angiotensin II receptors have been cloned: AT_1 and AT_2.[21] Most of the known effects of angiotensin II in humans appear to be AT_1-mediated.[22] The AT_1 receptors belong to the superfamily of G protein-coupled receptors with seven transmembrane domains. They activate phospholypases C and D,[23] which stimulate the hydrolysis of phosphatidylinositol 4,5-biphosphate, yielding two second messengers, inositol 1,4,5-triphosphate (1,4,5-IP_3) and diacylglycerol. The 1,4,5-IP_3 binds specific receptors in the sarcoplasmic reticulum, releasing calcium to the cytosol. This process triggers the opening of channels in the cytoplasmic membrane, which allows the inflow of extracellular calcium by chemical gradient. Furthermore, diacylglycerol stimulates protein kinase C, which blocks the outflow channels of potassium, with subsequent cell depolarization and opening of voltage-activated calcium channels. Protein kinase C is also involved in contraction of vascular smooth muscle cells (VSMCs) and cell growth through *c-fos* induction.[23]

AT_2 receptors may belong to a unique class of seven transmembrane domain receptors, for which G protein coupling has not been demonstrated.[24] At present, however, their signaling mechanisms and role are not clear, although it has been speculated that they may cause signals that counterbalance those of AT_1[21]; for example, the growth-promoting effect of AT_1 receptors in coronary endothelial cells.[25]

Role of Tissue RAS in Atherosclerosis

Tissue RAS is involved in different aspects of atherosclerosis. In the initial stages it plays a role in endothelial dysfunction,[26] through increased production of angiotensin II and degradation of bradykinin,[27] and in uptake and peroxidation of low-density lipoprotein (LDL).[28] Tissue RAS is also involved in the progression of atherosclerotic plaque via VSMC proliferation induced by angiotensin II. Angiotensin II induces *c-myc* gene expression in cultured VSMCs[29] and DNA synthesis in VSMCs in rats.[30] In animal models, ACE expression is observed along with neointimal formation in VSMC after vascular injury,[19] and this process is reduced by ACE inhibitors[19,31–33] Finally, plaque rupture and thrombosis, the events that lead to acute coronary syndromes, are also related to tissue RAS. Angiotensin II increases expression of monocyte chemotactic protein-1 (MCP-1) in VSMCs through activation of nuclear factor κB (NF-κB).[34] This chemokine attracts macrophages to the atherosclerotic lesion,[34–36] where they release proteolytic enzymes that weaken the fibrous cap, enhancing the probability of plaque rupture[37–39] and secondary thrombosis. In addition, tissue RAS potentiates the thrombotic process by increasing platelet aggregation[5,6] and impairing fibrinolysis.[7–9]

Role of Tissue RAS in Renal Diseases

Tissue RAS is involved in chronic aspects of renal vascular regulation and in the development of some renal diseases[11,12] (Fig. 1). Angiotensin II is also synthetized locally by the kidneys independently of plasma levels.[40,41]

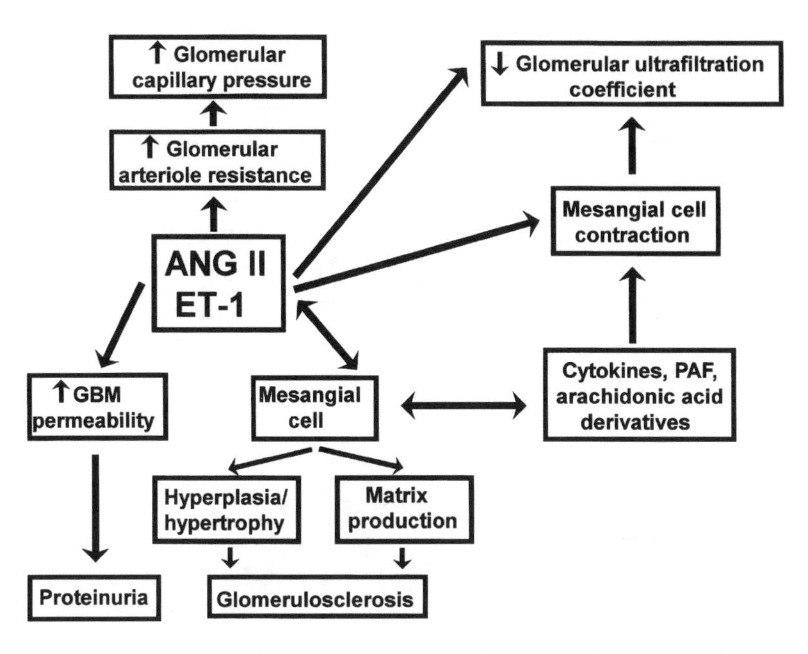

FIGURE 1. Glomerular actions of angiotensin II and endothelin-1. All the phenomena described in this figure have been demonstrated for angiotensin II. In contrast, some of them are only suspected for endothelin-1 (such as its actions on GBM permeability or mesangial cell hypertrophy) GBM-glomerular basement membrane. (From Egido J: Vasoactive hormones and renal sclerosis. Kidney Int 49:578–597, 1996, with permission.)

Expansion of the mesangium, due to proliferation or hypertrophy of mesangial cells and increased extracellular matrix, is the hallmark of a number of glomerular diseases.[42] Angiotensin II participates in mesangial cell growth (Fig. 2) and can induce either proliferation or hypertrophy[43–51] (Table 1). Most of our knowledge about angiotensin II comes from studies of its effects on VSMC. Under most conditions in cultured VSMC, angiotensin II induces both proliferation mediated by basic fibroblast growth factor (bFGF) and platelet-derived growth factor (PDGF) and an antiproliferative pathway mediated by transforming growth factor $beta_1$ ($TGF\beta_1$).[11,52] On balance, there is a hypertrophic response. Under some culture conditions, however, an imbalance between proliferative and antiproliferative response produces hyperplasia.[11] Although it is not known whether a similar process occurs in mesangial cells, angiotensin II induces cell proliferation in the presence of neutralizing antibodies to $TGF\beta_1$,[50] suggesting that $TGF\beta_1$ influences the actions of angiotensin II. A proliferative synergy takes place with other growth factors, such as epidermal growth factor (EGF).[53,54] Finally, angiotensin II directly stimulates mitogenesis, although the mechanisms are poorly defined. In rat kidneys, angiotensin II induces the expression of *c-fos* and *Egr-1* (early gene response-1)[55–57] and DNA synthesis.[58] However, unlike other growth factors, angiotensin II-stimulated DNA synthesis is delayed, probably because increased expression of new gene products is required before its initiation.[58]

Angiotensin II stimulates the synthesis of structural components of the extracellular matrix in mesangial and smooth muscle cells in vitro,[46,48–50,59–61] probably

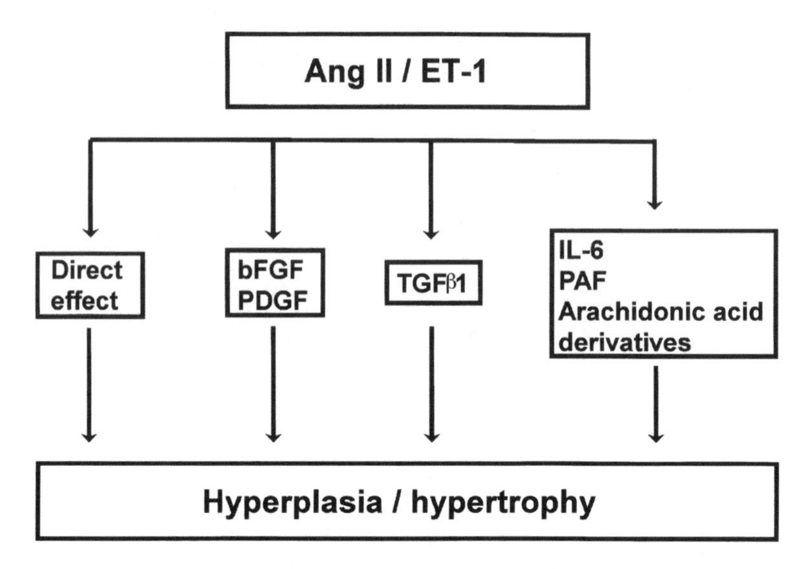

FIGURE 2. Angiotensin II and endothelin-1 as growth factors for mesangial cells. (From Egido J: Vasoactive hormones and renal sclerosis. Kidney Int 49:578–597, 1996, with permission.)

through its effects on cytokines and growth factors.[22] Among these, $TGF\beta_1$ has a major role,[48,62–66] acting through stimulation of the synthesis of matrix proteins and matrix protein receptors such as integrins, decreasing proteases, and increasing the activity of protease inhibitors.[62] Furthermore, $TGF\beta_1$ stimulates fibroblast chemotaxis and proliferation.[22] In fact, sustained expression of $TGF\beta_1$ contributes to the development of progressive renal fibrosis.[64] Moreover, $TGF\beta_1$ may lead to glomerulosclerosis, as demonstrated by using in vivo gene transfection.[65]

TABLE 1. Angiotensin II, Mesangial Cell Growth, and Matrix Proteins

	Origin of Cells	Cell Growth	Matrix
Fujiwara et al.[43]	Rat	Proliferation	Fibronectin
Homma et al.[44]	Rat	Hypertrophy	Collagen
Wolthuis et al.[45]	Rat	Proliferation	Fibronectin
Wolf et al.[46]	Murine cell line	Proliferation	Collagen I
Anderson et al.[47]	Mouse	Hypertrophy	
Ruiz-Ortega et al.[48]	Rat	Proliferation	Fibronectin Collagen IV
Ray et al.[49]	Human fetus	Proliferation	Fibronectin Biglycan
Kagami et al.[50]	Rat	Hypertrophy	Collagen I
Orth et al.[51]	Human adult	Proliferation Hypertrophy	

The conflicting results may be due partly to the different origin of the cells and variable culture conditions.
From Egido J: Vasoactive hormones and renal sclerosis. Kidney Int 49:578–597, 1996, with permission.

Angiotensin II increases the level of TGFβ_1 mRNA production of both active and latent TGFβ_1 in rat mesangial cells.[50] All of these data support the existence of a molecular cascade in which angiotensin II induces the synthesis and deposition of matrix components via TGFβ_1. This pathway is probably not unique to angiotensin II. Our work indicates that TGFβ_1 and matrix expression are induced by both platelet-activating factor (PAF)[67] and activation of Fc receptors[66] in mesangial cells.

Tissue RAS is also involved in renal interstitial fibrosis, and an increase of some of its components has been demonstrated in experimental models.[68] Angiotensin II infusion is associated with a monocyte/macrophage influx into the kidney, followed by an increase in the synthesis of fibronectin and interstitial collagens.[69] Furthermore, data from our laboratory support the hypothesis that angiotensin II directly stimulates renal fibroblasts, enhancing proliferation and synthesis of extracellular matrix proteins.[70] This observation agrees with previous findings in cardiac fibroblasts.[71–73]

Effects of Calcium Antagonists on the Renin-Angiotensin System

Calcium is an intracellular messenger used by the RAS and plays a central role in angiotensin II signaling. Thus, drugs that block the flow of calcium into the cells can interact with this system (Fig. 3). A discussion of the effects of CCBs on the levels of the different RAS components is followed by a review of their interactions with angiotensin II in cardiovascular and renal disease.

Effects on Renin

The hypotensive action of CCBs activates the sympathetic system, and especially the RAS, to limit their effects.[74–76] An increase in acute and chronic renin

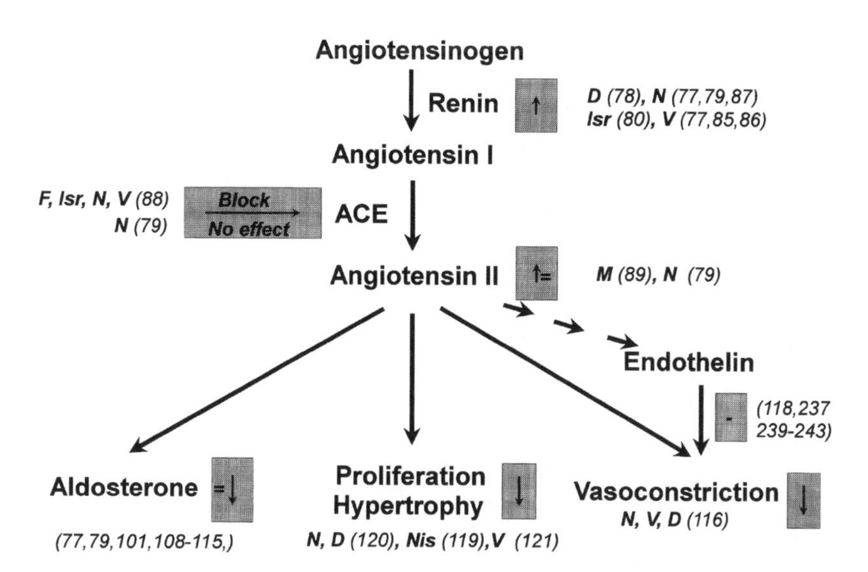

FIGURE 3. Effects of calcium blockers on the renin-angiotensin system. Arrows and signs inside the rectangles correspond to the actions of calcium antagonists and the numbers in parentheses to the references. ACE, angiotensin converting enzyme; D, diltiazem; F, felodipine; Isr, isradipine; L, lacidipine; M, manidipine; N, nifedipine; Nis, nisoldipine; V, verapamil; ↑, increase; ↓, decrease; =, no change.

levels, related to sodium load, has been described in patients receiving CCBs.[77–79] In rabbits, the increase in plasma renin induced by isradipine was more evident if the animals had been previously sodium-depleted with a diuretic.[80] This finding may explain in part why diuretics cause little additional decrease in blood pressure in patients treated with CCBs[81,82] as well as the fact that CCBs are the only antihypertensive agents whose action is enhanced by high sodium intake.[83] The mechanisms of CCB-induced increase in plasma renin levels have not been completely elucidated. The increase may be due in part to a reflex effect of blood pressure reduction because other vasodilators also raise plasma renin levels in rabbits.[84] In addition, unlike other vasodilators, CCBs specifically increase renin release in vitro through a direct action on the juxtaglomerular cells.[85–87]

Effects on Angiotensin-converting Enzyme

The effects of CCBs on ACE activity are controversial. Isradipine reduces ACE activity in healthy individuals and hypertensive patients,[88] and other CCBs have demonstrated a noncompetitive diminution of ACE activity in vitro.[88] However, in a different study, nifedipine did not affect ACE activity in healthy young subjects.[79] Further work is needed before definitive conclusions can be drawn.

Effects on Angiotensin II

Reports about the influence of CCBs on angiotensin II activity are conflicting. In one study, short-term oral administration of nifedipine increased angiotensin II plasma levels in healthy and hypertensive young people, although no effect was seen in older people.[79] Because ACE activity was unchanged, this finding was probably related to the increase in renin activity and plasma angiotensin I, the substrate of ACE. However, in another report, manidipine did not affect plasmatic angiotensin II when given orally for 7 days to diabetic adults.[89]

Effects on Aldosterone

CCBs inhibit the angiotensin II-induced increase in free cytosolic calcium and steroidogenesis[90–99] in glomerulosa cells by blocking calcium inflow through T- and L-type channels.[100] In some studies CCBs have been shown to reduce aldosterone levels in primary hyperaldosteronism—mainly in cases of idiopathic etiology—and have been proposed as a treatment.[98-104] However, these results have not been confirmed consistently[105–107] and require further investigation. Finally, in most reports of healthy and hypertensive patients, aldosterone levels are generally not affected,[77,79,108–114] although in some cases they are in fact reduced.[101,115] This finding contrasts with the effects of hydralazine, which increases plasma concentrations of both angiotensin II and aldosterone.[79]

Effects on Cardiovascular Responses to Angiotensin II

Nifedipine, verapamil, and diltiazem can reverse angiotensin II-induced vasoconstriction in humans.[116] This effect is due to a decrease in VSMC calcium inflow and facilitation of the vasodilator effects of nitric oxide.[117] Furthermore, the same agents inhibit the vasoconstricting action of endothelin, a peptide whose production is stimulated by angiotensin II.[118]

CCBs antagonize the changes in cell growth (hyperplasia or hypertrophy) induced by angiotensin II. In cultured cardiomyocites, nisoldipine inhibits the effects of angiotensin II on *c-fos* and *Egr-1* expression as well as total protein synthesis,[119] acting proximally on protein kinase C, probably through reduction of calcium influx. The same effects have been demonstrated in VSMCs, in which nifedipine and diltiazem inhibit PDGF and angiotensin II-induced DNA synthesis.[120] Furthermore, verapamil also blocks the basal and angiotensin II-stimulated DNA transcription in VSMCs,[121] an effect that may lead to a reduction in cellular hypertrophy. These data may explain the diminution in left ventricular hypertrophy achieved with CCBs in patients with systemic hypertension,[122–124] as evidence suggests that elevated blood pressure is not the only factor responsible for left ventricular hypertrophy.[122,125–127] In addition, the effects on VSMC proliferation agree in part with animal[128–130] and clinical studies,[131,132] which demonstrate that CCBs reduce the appearance of new atherosclerotic plaques. Nevertheless, CCBs had no overall effect on the progression or regression of coronary artery disease.[131,133]

Effects on Renal Disease

CCBs may have beneficial effects on kidney function. Unfortunately, most information about CCBs is related to their cardiovascular effects; few studies focus on their actions in renal disease. CCBs improve the renal vasoconstriction and toxicity elicited by cyclosporines[134] as well as acute renal failure after cadaveric renal transplantation[135] and radiocontrast administration.[136] Furthermore, they also may be renoprotective in essential hypertension and diabetes[137–139]; this effect is more evident with nondihydropyridine agents.[140]

Part of the beneficial effects of CCBs in renal disease derives from a reduction in arteriolar vasoconstriction. CCBs produce vasodilation in afferent arterioles with variable effects on the efferent arterioles.[141–143] These actions are probably related in part to the ability of CCBs to reverse angiotensin II-induced vasoconstriction, as explained above. However, not all of the benefits of CCBs at the renal level are due to their effects on vascular tone. Although the greater vasodilation of afferent arterioles in hypertension may increase glomerular capillary pressure and accelerate glomerulosclerosis,[144] this effect has not been observed in clinical studies.[137–139] Therefore, CCBs may have additional renoprotective effects.[144,145] Given the similarities between glomerular mesangial cells and VSMCs, CCBs probably also reduce angiotensin II-induced hypertrophy and proliferation in renal cells. However, additional studies are needed to determine the exact mechanisms of the renoprotective effects of CCBs.

Therapeutic Implications

CCBs have multiple interactions with the RAS, which must be taken into account from a therapeutic perspective. For example, the increase in renin response to CCBs after diuretic-induced salt depletion suggests that this combination is probably not the best to improve blood pressure control.[81,82]

Of special interest is the inhibition of the vasoconstriction, cellular hypertrophy, and cellular proliferation induced by angiotensin II. These actions make CCBs complementary to ACE inhibitors in lowering blood pressure and preventing cardiovascular complications. In regard to coronary atherosclerosis, however, the prospects are different. Despite reduction in the appearance of new coronary artery lesions, CCBs have not shown marked benefits in terms of infarction and mortality rates.[146] Cholesterol-lowering drugs achieve a greater reduction in plaque progression and, of greater importance, in plaque rupture and thrombosis, events that lead to acute coronary syndromes. Thus, the role of CCBs in the treatment of coronary artery disease is mainly related to their anti-ischemic effects, although beta blockers are usually preferred in the absence of specific contraindications.[146]

The possible role of CCBs in renal protection is discussed in chapter 13.

CALCIUM ANTAGONISTS AND ENDOTHELINS

General Aspects of Endothelins

Endothelins (ETs) are vasoactive 21-amino acid peptides synthetized by the endothelium and many other tissues.[147] Although the slow regulation of their biosynthesis and secretion resembles that of inflammatory cytokines, their mechanisms of action are similar to other vasoconstrictors.

Three isoforms of ET have been characterized: ET-1, ET-2, and ET-3.[147–151] The mature forms are produced from prepropolypeptides (preproendothelins) of about 200 residues, which are encoded by three different genes.[152] Preproendothelins are processed by a furin-like processing protease[153] into biologically inactive intermediates, called big endothelins-1, -2, or -3. Big ETs are converted to ETs via cleavage at the common *Trp*21 residue by endothelin-converting enzyme.[154,155]

The three endothelins are distributed differently throughout the body. ET-1 is the only isoform detected in vascular endothelial cells. Preproendothelin-1 is constitutively expressed in cultured endothelial cells, the intima of blood vessels,[147,149] brain, kidney, lung, and other tissues.[156] ET-2 and ET-3 are expressed in the brain, kidney, adrenal glands, and intestines.[157,158]

Regulation of Endothelin Production

The production of ETs is regulated at the level of mRNA transcription. Once synthetized, they are secreted via the constitutive pathway without further regulation.[147,159–162] The expression of preproendothelin is stimulated by substances derived from platelets, coagulation products, cytokines, and vasopressor hormones, including angiotensin II.[147,149,160–163] However, feed-back mechanisms avoid excessive ET production. Agonists such as thrombin stimulate, in addition to ET, nitric oxide (NO) production, which exerts a negative feedback on ET synthesis through a mechanism dependent on cyclic guanosine monophosphate (cGMP).[161,164] Similar effects occur with angiotensin II.[165] In addition, VSMCs and platelets can produce factors that inhibit ET production.[166]

Mechanism of Action

Endothelins act on two structurally and functionally distinct subtypes of the G protein-coupled superfamily of receptors with seven transmembrane domains. These subtypes, called ET_A and ET_B,[167–170] have different isopeptide selectivities. Whereas the ET_B receptors have equipotent affinity for the three isoforms of ETs,[170,171] the order of affinity for the ET_A subtype is ET-1 > ET-2 >> ET-3.[169,172] Both receptors have different tissue distribution, although there is partial overlap.

ET_A receptors are present in VSMCs and mesangial cells. Activation of these receptors by ETs induces vasoconstriction through an increase of intracellular calcium.[147,173–183] However, the precise mechanism of signal transduction has not been completely elucidated.[184] In brief, after binding to cell membrane, ETs activate phospholipase C, leading to the formation of inositol triphosphate (IP_3) and diacylglycerol.[185–189] IP_3 promotes a rise in cytosolic calcium derived from intracellular stores, and diacylglycerol stimulates protein kinase C, which is involved in VSMC contraction. Furthermore, ETs stimulate nonselective cation channels, causing depolarization and subsequent activation of voltage-dependent L-type calcium channels.[174,190] Thus, ET-induced rise in intracellular calcium is due to both mobilization of intracellular stores and influx of extracellular calcium, but the contribution of these two pathways has not been precisely delineated.[191,192] Finally, low levels of ET-1 are also involved in processes of cellular differentiation, hypertrophy, and proliferation.[193–195]

ET_B receptors have been classically involved in vasodilation. They are present in endothelial cells and mediate the production of nitric oxide and prostacyclin.[168] This action explains the transient vasodilation that precedes the pressor effect of ETs.[147,174,196–198] However, it is well known that ET_B receptors may act either as vasodilators or vasoconstrictors, depending on the species and location in the body.[199–206] Moreover, ET_A receptors can also induce prostacyclin synthesis in rats.[207] Thus it seems that ET receptors exert a variety of sometimes opposite effects, depending on the species and location in the body.

Role of Endothelins

Like the RAS, ETs participate in renal vascular regulation (Fig. 1), and their levels are increased in acute renal failure.[208] They have contractile effects on renal arterioles[177,209,210]—mediated primarily by ET_B receptors—and reduce renal blood flow, glomerular filtration rate,[177,209–211] urine volume, and sodium excretion.[212] Furthermore, ETs may be related to renal disease by acting as mitogens on mesangial cells[187] via ET_A receptors[213] and protein kinase C activation[214] (Fig. 2). They also potentiate the action of growth factors that bind to tyrosine kinases, such as bFGF and epidermal growth factor (EGF).[215] Like angiotensin II, ET-1 regulates VSMC mitogenesis through a mechanism that involves $TGF\beta_1$ synthesis. Endothelins are also suspected to regulate extracellular matrix remodeling,[22] and they have been shown to increase the expression of fibronectin and collagen IV through a mechanism that involves protein kinase C activation and $TGF\beta_1$.[48,60]

Data about ET plasma levels in hypertensive patients are conflicting. Levels have been reported to be either elevated[216–218] or normal.[219–221] However, local production of ET may vary within the body,[209] and plasma levels may not reflect the real activity of the ET system.[222] Given their vasoconstrictor effects, ETs can also play a role in hypertension.[223] Although vascular reactivity to ETs is not clearly increased,[224–227] their potentiating actions on other vasoconstrictor peptides are augmented in hypertension.[196]

Increased levels of circulating ETs have been noted in atherosclerosis[228,229] and acute myocardial infarction.[230] ETs may play a role in both entities, because they have proliferative effects on VSMCs[149,187,195] and increase monocyte chemotaxis in vitro.[231] Furthermore, ETs are produced by VSMCs in vitro[232,233] and in atherosclerotic plaques[229] as well as in cultured endothelial cells stimulated with oxidized low-density-lipoproteins.[234] In addition, ET-1 may be implicated in the pathophysiology of myocardial ischemia and reperfusion, in both of which an increase of ET-1 binding site density is found.[235]

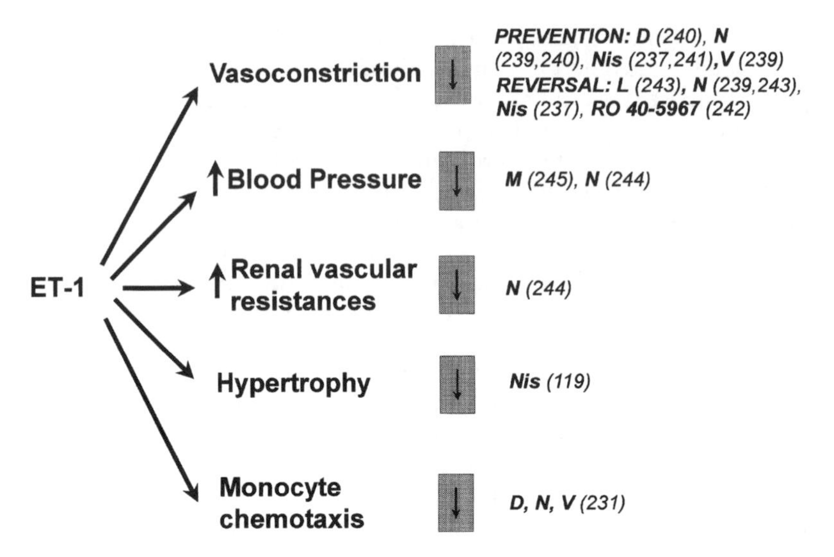

FIGURE 4. Effects of calcium blockers on endothelin-1 actions. Arrows inside the rectangles correspond to the actions of calcium antagonists and the numbers in parentheses to the references. ACE, angiotensin converting enzyme; D, diltiazem; F, felodipine; Isr, isradipine; L, lacidipine; M, manidipine; N, nifedipine; Nis, nisoldipine; V, verapamil; ↑, increase; ↓, decrease; =, no change.

Effects of Calcium Antagonists on Endothelins

ETs share common signaling mechanisms with the RAS. Again, calcium plays a critical role in these events. Multiple interactions of CCBs with ETs have been described and may explain their pharmacologic actions (Fig. 4).

Effects on Endothelin Production

Dihydropyridines decrease expression of ET-1 in the heart of spontaneously hypertensive rats[236] as well as plasma[89] and vascular levels of ET-1 in humans,[237] probably by inhibition of its synthesis.[237] It is not known whether these actions are due to calcium entry blockade or, in contrast, whether they are nonspecific, since calcium ionophores do not consistently increase ET production.[237,238]

Effects on Hemodynamic Actions

Pretreatment with CCBs prevents or reduces ET-induced vasoconstriction in human arteries in vivo[239] and in vitro.[237,240,241] CCBs also unmask the vasodilator effect of ET-1. The vasodilation observed with coinfusion of verapamil or nifedipine with ET-1 is greater than that achieved with the CCB alone.[239] In addition, CCBs relax arteries precontracted with ET-1 in vitro[237,242,243] and in vivo.[239] However, vascular relaxation is complete only when low doses of ET-1 are used. This observation suggests that the mechanism of contraction in response to low concentrations of ET-1 relies mainly on calcium influx and, with high concentrations, on the release of intracellular calcium.[237] Thus, a great part of the hypotensive effects of CCBs is probably due to the antagonism of ET-induced calcium inflow. However, the response to CCBs varies with vessel location,[191] and in some vessels CCBs have no effect on ET-induced vasoconstriction.[192] This finding suggests that the role of calcium influx in the response to ET-1 varies with the location of the vessel[191,192] and perhaps is related to the distribution of ET_A and ET_B receptors.

Effects on Left Ventricular Hypertrophy

Nisoldipine inhibits *c-fos* and *Egr-1* expression triggered by ET-1 as well as total protein synthesis in rat cardiomyocytes in vitro.[119] The precise mechanism is unknown. This effect adds to the above-mentioned inhibition of angiotensin II-induced protein synthesis and helps to explain the reduction in left ventricular hypertrophy achieved by CCBs in systemic hypertension.[122–124]

Effects on the Renal Actions

Given intravenously, CCBs reverse ET-1-induced increases in renal vascular resistance[244–246] and blood pressure[244,247] and decreases in renal blood flow and glomerular filtration rates.[246,247] These effects are secondary to their antagonism of ET-induced vasoconstriction.

Effects on Coronary Artery Disease

Nifedipine, diltiazem, and verapamil reduce ET-1 induced monocyte chemotaxis in vitro.[231] This effect may be of benefit in atherosclerosis, a process in which monocytes migrate into the vascular wall and contribute to the progression of disease. The reduction of monocyte chemotaxis may account for the above-mentioned decrease in the appearance of new atherosclerotic plaques induced by CCBs. In addition, amlodipine suppresses the increase in ET-1 binding sites induced by ischemia and reperfusion in the heart.[235] Thus, part of the benefit achieved by CCBs in coronary artery disease may be mediated through their antagonism of ETs. However, further studies are needed to confirm this hypothesis.

Therapeutic Implications

Knowledge of ETs is less extensive than knowledge of the renin-angiotensin system. Existing data help to explain how CCBs work in clinical practice. For example, the vasodilator effect for which they are used in hypertension and coronary artery disease is achieved in part by antagonism of ET-induced vasoconstriction. Moreover, these interactions may play a role in clinical aspects such as left ventricular hypertrophy, progression of atherosclerosis, and cardiac ischemia and reperfusion. However, much work remains to be done to elucidate the roles and mechanisms of action of ETs and to determine to what extent the beneficial effects of CCBs depend on their interactions with these peptides.

References

1. Sholkens BA, Jung W, Rascher W, et al: Brain angiotensin II stimulates release of pituitary hormones, plasma catecholamines and increases blood pressure in dogs. Clin Sci 59 (Suppl 6):53s-56s, 1980.
2. Hilgers KF, Veelken R, Rupprecht G, et al:Angiotensin II facilitates sympathetic transmission in rat hind limb circulation. Hypertension 21:322–328, 1993.
3. Lokhandwala MF, Amelang E, Buckley JP: Facilitation of cardiac sympathetic function by angiotensin II: Role of presynaptic angiotensin receptors. Eur J Pharmacol 52:405–409, 1978.
4. Draper AJ, Meghji S, Redfern PH: Enhanced presynaptic facilitation of vascular adrenergic neurotransmission in spontaneously hypertensive rats. J Auton Pharmacol 9:103–111, 1989.
5. Swartz SL, Moore TJ: Effect of angiotensin II on collagen-induced platelet aggregation in normotensive subjects. Thromb Haemost 63:87–90, 1989.

6. Someya N, Morotomi Y, Kodama K, et al: Suppressive effects of captopril on platelet aggregation in essential hypertension. J Cardiovasc Pharmacol 6:840–843, 1984.
7. Rdiker PM, Gaboury CL, Conlin PR, et al: Stimulation of plasminogen activator inhibitor in vivo by infusion of angiotensin II: Evidence of a potential interaction between the renin-angiotensin system and fibrinolytic function. Circulation 87:1969–1973, 1993.
8. Vaughan DE, Lazos SA, Tong K: Angiotensin II regulates the expression of plasminogen activator inhibitor in cultured endothelial cells. J Clin Invest 95:995–1001, 1995.
9. Hamdan AD, Quist WC, Gagne JB, Feener EP: Angiotensin-converting enzyme inhibition suppresses plasminogen activator inhibitor-1 expression in the neointima of balloon-injured rat aorta. Circulation 93:1073–1078, 1996.
10. Dzau VJ, Chobanian A:. Renin-angiotensin system and atherosclerotic vascular disease. In Fuster V, Ross R, Topol EJ (eds): Atherosclerosis and Coronary Artery Disease, 1st ed. Philadelphia, Lippincott-Raven, 1996, pp 237–242.
11. Dzau VJ, Mukoyama M, Pratt RE: Molecular biology of angiotensin receptors: Target for drug research? J Hypertens 12(Suppl 2):S1–S5, 1994.
12. Mulrow PJ: The intrarenal renin-angiotensin system. Curr Opin Nephrol Hypertens 1:41–44, 1993.
13. Lindpainter K, Ganten D: The cardiac renin-angiotensin system: an appraisal of present and experimental and clinical evidence. Circ Res 68:905–921, 1991.
14. Hirsch AT, Talsness CE, Schunkert H, et al: Tissue specific activation of cardiac angiotensin converting enzyme in experimental heart failure. Circ Res 69:475–482, 1991.
15. Schunkert H, Dzau VJ, Tang SS, et al: Increased rat cardiac angiotensin converting enzyme activity and mRNA expression in pressure overload left ventricular hypertrophy: Effects on coronary resistance, contractility and relaxation. J Clin Invest 86:1913–1920, 1990.
16. Finckh M, Hellmann W, Ganten D, et al: Enhanced cardiac angiotensinogen gene expression and angiotensin converting enzyme activity in tachypacing-induced heart failure in rats. Basic Res Cardiol 86:303–316, 1991.
17. Lindpainter K, Lu W, Niedermajer N, et al: Selective activation of cardiac angiotensinogen gene expression in postinfarction ventricular remodeling in the rat. J Mol Cell Cardiol 25:133–143, 1993.
18. Naftilan AJ, Zuo WM, Inglefinger J, et al: Localization and differential regulation of angiotensinogen mRNA expression in the vessel wall. J Clin Invest 87:1300–1311, 1991.
19. Rakugi H, Kim DK, Krieger JE, et al: Induction of angiotensin converting enzyme in the neointima after vascular injury. Possible role in restenosis. J Clin Invest 93:339–346, 1994.
20. Dzau VJ, Pratt R: Tissue renin-angiotensin in experimental restenosis after vascular injury: evidence of local activation. J Cardiovasc Pharmacol 20(Suppl B):S28–S32, 1992.
21. Dzau VJ: Circulating versus local renin-angiotensin-system in cardiovascular homeostasis. Circulation 77(Suppl 19):1–4, 1988.
22. Egido J: Vasoactive hormones and renal sclerosis. Kidney Int 49:578–597, 1996.
23. Sadoshima J, Izumo S: Signal transduction pathways of angiotensin II-induced *c-fos* gene expression in cardiac myocites *in vitro*. Circ Res 73:424–438, 1993.
24. Mukoyama M, Nakajima M, Horiuchi M, et al: Expression cloning of type 2 angiotensin II receptor reveals a unique class of seven-transmembrane receptors. J Biol Chem 268:24539–24542, 1993.
25. Stoll M, Steckelings UM, Paul M, et al: The angiotensin AT_2 receptor mediates inhibition of cells proliferation in coronary endothelial cells. J Clin Invest 95:651–657, 1995.
26. Mancini GBJ, Henry GC, Macaya C, et al: Angiotensin-converting enzyme inhibition with quinapril improves endothelial vasomotor dysfunction in patients with coronary artery disease. The TREND (Trial on reversing endothelial dysfunction) study. Circulation 94:258–265, 1996.
27. Vaughan DE, Pfeffer MA: Ventricular remodeling following myocardial infarction and angiotensin-converting enzime inhibitors. In Fuster V, Ross R, Topol EJ (eds): Atherosclerosis and Coronary Artery Disease, 1st ed. Philadelphia, Lippincott-Raven, 1996, pp 1193–1204.
28. Keidar S, Kaplan M, Aviram M: Angiotensin II-modified LDL is taken up by macrophages via the scavenger receptors leading to cellular cholesterol accumulation. Arterioscler Thromb 16:97–105, 1996.
29. Naftilan AJ, Pratt RE, Dzau VJ: Induction of platelet-derived growth factor A-chain and c-*myc* gene expressions by angiotensin II in cultured rat vascular smooth muscle cells. J Clin Invest 83:1419–1424, 1989.
30. Daemen MAJP, Lombardi DM, Bosman FT, Schwartz SM. Angiotensin II induces smooth muscle cell proliferation in the normal and injured rat arterial wall. Circ Res 68:450–456, 1991.
31. Powell JS, Clozel JP, Müller RKM, et al: Inhibitors of angiotensin-converting enzyme prevent myointimal proliferation after vascular injury. Science 245:186–188, 1989.

32. Clozel JP, Muller RK, Roux S, et al: Influence of the status of the renin-angiotensin system on the effect of cilazapril on neointima formation after vascular injury in rats. Circulation 88:1222–1227, 1993.
33. Bell L, Madri JA: Influence of the angiotensin system on endothelial and smooth muscle cell migration. Am J Pathol 137:7–12, 1990.
34. Hernández-Presa M, Bustos C, Ortego M, et al: Angiotensin converting enzyme inhibition prevents arterial NF-κB activation, MCP-1 expression and macrophage infiltration in a rabbit model of early accelerated atherosclerosis. Circulation 95:1532–1541, 1997.
35. Neiken NA, Coughlin SR, Gordon D, Wilcox JN: Monocyte chemoattractant protein-1 in human atheromatous plaques. J Clin Invest 88:1121–1127, 1991.
36. Yla-Herttuala, Lipton BA, Rosenfeld ME, et al: Expression of monocyte chemoattractant protein-1 in macrophage rich areas of human and rabbit atherosclerotic lesions. Proc Natl Acad Sci U S A 88:5252–5256, 1991.
37. Henney AM, Wakeley PR, Davies MJ, et al: Localization of stromelysin gene expression in atherosclerotic plaques by in situ hybridazation. Proc Natl Acad Sci U S A 88:8154–8158, 1991.
38. Matrisian LM: The matrix-degrading metalloproteinases. Bioessays 14:455–463, 1992.
39. Galis ZS, Sukhova GK, Lark MW, Libby P: Increased expression of matrix-metalloproteinases and matrix-degrading activity in vulnerable regions of human atherosclerotic plaques. J Clin Invest 94:2493–2503, 1994.
40. Seikaly MG, Arant BS, Seney FD: Endogenous angiotensin concentrations in specific intrarenal fluid compartments of the rat. J Clin Invest 86:1352–1357, 1990.
41. Paul M, Wagner J, Dzau VJ: Gene expression of the renin-angiotensin system in human tissues. Quantitative analysis by polymerase chain reaction. J Clin Invest 91:2058–2064, 1993.
42. Marx M, Sterzel B Sorokin L: Renal matrix and adhesion in injury and inflammation. Curr Opin Nephrol Hypertens 2:527–535, 1993.
43. Fujiwara Y, Takama T, Shin S, et al: Angiotensin II (AII) stimulates mesangial cell growth through pfosphoinositide (PI) cascade [abstract]. Proc Am Soc Nephrol 24A, 1988.
44. Homma T, Hoover RL, Ichikawa I, Harris RC: Angiotensin II (AII) induces hypertrophy and stimulates collagen production in cultured rat mesangial cell (MC) [abstract]. Clin Res 38:358, 1990.
45. Wolthuis A, Boes A, Rodeman HP, Grond J: Vasoactive agents affect growth and protein synthesis of cultured rat mesangial cells. Kidney Int 41:124–131, 1992.
46. Wolf G, Haberstroh U, Neilson EG: Angiotensin II stimulates the proliferation and biosynthesis of type I collagen in cultured mesangial cells. Am J Pathol 140:95–107, 1992.
47. Anderson PW, Do YS, Hsueh WA: Angiotensin II causes mesangial cell hypertrophy. Hypertension 21:29–35, 1993.
48. Ruiz-Ortega M, Gómez-Garre D, Palacios I, et al: Effect of angiotensin II and endothelin on matrix protein synthesis by cultured mesangial cells. A complicated network of interactions [abstract]. J Am Soc Nephrol 4:664, 1993.
49. Ray PE, Bruggeman LA, Horikoshi S, et al: Angiotensin II stimulates human fetal mesangial cell proliferation and fibronectin biosynthesis by binding to AT_1 receptors. Kidney Int 45:177–184, 1994.
50. Kagami S, Border WA, Miller DE, Noble NA: Angiotensin II stimulates extracellular matrix protein synthesis through induction of transforming growth factor-β expression in rat glomerular mesangial cells. J Clin Invest 93:2431–2437, 1994.
51. Orth S, Weinreich T, Bonisch S, Weih M, Ritz E. Angiotensin II induces hypertrophy and hyperplasia in adult human mesangial cells. Exp Nephrol 3:23–33, 1995.
52. Gibbons GH, Pratt RE, Dzau VJ: Vascular smooth muscle cell hypertrophy vs hyperplasia. Autocrine transforming growth factor-beta 1 expression determines growth response to angiotensin II. J Clin Invest 90:456–461, 1992.
53. Bagby SP, Kirk EA, Mitchell LH, et al: Proliferative synergy of ANG and EGF in porcine aortic vascular muscle cells. Am J Physiol 265:F239–F249, 1993.
54. Saltis J, Agrotis A, Bobik A: Transforming growth factor-β_1 enhances the proliferative effects of epidermal growth factor on vascular smooth muscle from the spontaneously hypertensive rat. J Hypertens S6:S184–S185, 1991.
55. Rupprecht HD, Dann P, Sukhatme VP, et al: Effect of vasoactive agents on induction of Egr-1 in rat mesangial cells: correlation with mitogenicity. Am J Physiol 263:F623–F636, 1992.
56. Rosenberg ME, Hostetter TH, Kren S, Chmielewski D: In vivo effects of angiotensin II and norepinephrine on early growth response genes in the rat kidney. Kidney Int 43:601–609, 1993.
57. Neyses LK, Nouskas J, Luyken J, et al: Induction of immediate-early genes by angiotensin II and endothelin-1 in adult rat cardiomyocytes. J Hypertens 11:927–934, 1993.

58. Weber H, Taylor DS, Molloy CJ: Angiotensin II induces delayed mitogenesis and cellular proliferation in rat aortic smooth muscle cells. J Clin Invest 93:788–798, 1994.
59. Wolf G, Neilson EG: Angiotensin II as a renal growth factor. J Am Soc Nephrol 3:1531–1540, 1993.
60. Ruiz-Ortega M, Gómez-Garré D, Alcazar R, et al: Involvement of angiotensin II and endothelin in matrix protein production and renal sclerosis. J Hypertens 12:S51–S52, 1994.
61. Simon G, Abraham G, Altman S: Stimulation of vascular glycosaminoglycan synthesis by subpressor doses of angiotensin II in rats. Hypertension 23(Suppl I):148–151, 1994.
62. Border WA, Ruoslahti E: Transforming growth factor β in disease: The dark side of tissue repair. J Clin Invest 90:1–7, 1992.
63. Okuda S, Langiuino LR, Ruoslahti E, Border WA: Elevated expression of transforming growth factor β and proteoglycan production in experimental glomerulonephritis. J Clin Invest 86:453–462, 1990.
64. Yamamoto T, Noble NA, Miller DE, Border WA: Sustained expression of TGF-β_1 underlies development of progressive kidney fibrosis. Kidney Int 45:916–927, 1994.
65. Isaka Y, Fujiwara Y, Ueda N, et al: Glomerulosclerosis induced by in vivo transfection of transforming growth factor β or platelet-derived growth factor gene into the rat kidney. J Clin Invest 92:2597–2601, 1993.
66. López-Armada MGJ, Gómez-Guerrero C, Egido J: Immune complexes stimulate the expression and synthesis of matrix proteins in cultured rat and human mesangial cells. Role of transforming growth factor β [abstract]. J Am Soc Nephrol 25:812, 1994.
67. Ruiz-Ortega M, Largo R, Bustos C, et al: Platelet activating factor (PAF) stimulates the expression and synthesis of extracellular matrix proteins in cultured renal cells [abstract]. J Am Soc Nephrol 683:25, 1994.
68. Rosenberg ME, Smith LJ, Correa-Rotter R, Hostetter TH: Paradox of the renin-angiotensin system in chronic renal disease. Kidney Int 45:403–410, 1994.
69. Johnson RJ, Alpers CE, Yoshimura A, et al: Renal injury from angiotensin II-medicated hypertension. Hypertension 19:464–474, 1992.
70. Ruiz-Ortega M, González E, Egido J: Molecular characterization of angiotensin II effects on renal interstitial fibroblasts [abstract]. J Am Nephrol Soc 6:909, 1995.
71. Villarreal FJ, Kim NN, Ungab GD, et al: Identification of functional angiotensin II receptors on rat cardiac fibroblasts. Circulation 88:2849–2861, 1993.
72. Schorb W, Booz GW, Dostal DE, et al: Angiotensin II is mitogenic in neonatal rat cardiac fibroblasts. Circ Res 72:1245–1254, 1993.
73. Crabos M, Roth M, Hahn AWA, Erne P: Characterization of angiotensin II receptors in cultured adult rat cardiac fibroblasts. J Clin Invest 93:2372–2378, 1994.
74. Hof RP, Hof-Miyashita A: Different peripheral vasodilator effects of isradipine in sodium-loaded and sodium-depleted rabbits. Gen Pharmacol 19:243–247, 1988.
75. Nievelstein HMNW, VanEssen H, Tijssen CM, et al: Systemic and regional hemodynamic actions of calcium entry blockers in conscious spontaneously hypertensive rats. Eur J Pharmacol 113:187–198, 1985.
76. Hof RP, Hof A: The renin-angiotensin system modulates the peripheral vascular effects of the calcium antagonist isradipine in anesthetized rabbits. J Cardiovasc Pharmacol 12:233–238, 1988.
77. Muiesan G, Agabiti-Rosei E, Castellano M, et al: Antihypertensive and humoral effects of verapamil and nifedipine in essential hypertension. J Cardiovasc Pharmacol 4:S325–S329, 1982.
78. Chaffman M, Brodgen RN: Diltiazem: A review of its pharmacologic properties and therapeutic efficacy. Drugs 29:387–454, 1985.
79. Himaratsu K, Yamagishi F, Kubota T, Yamada T: Acute effects of the calcium antagonist, nifedipine, on blood pressure, pulse rate, and the renin-angiotensin-aldosterone system in patients with essential hypertension. Am Heart J 104:1346–1350, 1982.
80. Hof RP, Umemura K, Evenou JP, et al: Effects of isradipine on plasma renin activity in sodium-loaded and -depleted conscious rabbits. J Cardiovasc Pharmacol 19:503–507, 1992.
81. MacGregor GA, Markandu ND, Smith SJ, Sagnella GA: Does nifedipine reveal a functional abnormality of arteriolar smooth muscle cell in essential hypertension? The effect of altering sodium balance. J Cardiovasc Pharmacol 7(Suppl 6):S178–S181, 1985.
82. Salvetti A, Magagna A, Innocenti P, et al: The combination of chlorthalidone with nifedipine does not exert an additive antihypertensive effect in essential hypertensives: a cross-over multicenter study. J Cardiovasc Pharmacol 17:332–335, 1991.
83. Bellini G, Battilana G, Puppis E, et al: Renal responses to acute nifedipine administration in normotensive and hypertensive patients during normal and low sodium intake. Curr Ther Res 35:974–981, 1984.

84. Hof RP, Evenou JP, Hof-Miyashita A: Similar increase in circulating renin after equihypotensive doses of nitroprusside, dihydralazine or isradipine in conscious rabbits. Eur J Pharmacol 136:251–254, 1987.
85. Craven PA, DeRubertis FR: Ca^{2+}-dependent modulation of renin release from isolated glomeruli: Apparent independence from alterations in cGMP. Metabolism 34:651–657, 1985.
86. Churchill PC: Second messengers in renin secretion. Am J Physiol 249:F175–F184, 1985.
87. Marre M, Misumi J, Raemsch KD, et al: Diuretic and natriuretic effects of nifedipine on isolated perfused rat kidneys. J Pharmacol Exp Ther 223:263–270, 1982.
88. Casarini DE, Carmona AK, Plavnik FL, et al: Calcium channel blockers as inhibitors of angiotensin I-converting enzyme. Hypertension 26(Pt 2):1145–1148, 1995.
89. Hirakata H, Iino K, Ishida I, et al: Effects of a new calcium antagonist, manidipine, on the renal hemodynamics and the vasoactive humoral factors in patients with diabetes mellitus. Blood Press Suppl 3:124–129, 1992.
90. Millar JA, Struthers AD: Calcium antagonists and hormone release. Clin Sci 66:249–255, 1984.
91. Shima S, Kawashima Y, Hirai M: Studies on cyclic nucleotides in the adrenal gland.VIII. Effects of angiotensin on adenosine 3´, 5´-monophosphate and steroidogenesis in the adrenal cortex. Endocrinology 103:1361–1367, 1978.
92. Fakunding JL, Chow R, Catt KJ: The role of calcium in the stimulation of aldosterone production by adrenocorticotropin, angiotensin II, and potassium in isolated glomerulosa cells. Endocrinology 105:329–333, 1979.
93. Fakunding JL, Catt KJ: Dependence of aldosteron stimulation in adrenal glomerulosa cells on calcium uptake: effects of lanthanum and verapamil. Endocrinology 107:1345–1353, 1980.
94. Schiffrin EL, Lis M, Gutkowska J, Genest J: Role of Ca^{2+} in response of adrenal glomerulosa cells to angiotensin II, ACTH, K^+, and ouabain. Am J Physiol 241:E42–E46, 1981.
95. Foster R, Lobo MV, Rasmussen H, Marusic ET: Calcium: Its role in the mechanism of action of angiotensin II and potassium in aldosterone production. Endocrinology 109:2196–2201, 1981.
96. Foster R, Rasmussen H: Angiotensin-mediated calcium efflux from adrenal glomerulosa cells. Am J Physiol 245:E281–E287, 1983.
97. Kojima K, Kojima I, Rasmussen H: Dihydropyridine calcium agonist and antagonist effects on aldosterone secretion. Am J Physiol 247:E645–E650, 1984.
98. Capponi AM, Lew PD, Jornot L, Vallotton MB. Correlation between cytosolic free Ca^{++} and aldosterone production in bovine adrenal glomerulosa cells. J Biol Chem 259:8863–8869, 1984.
99. Kojima I, Kojima K, Kreutter D, Rasmussen HR: The temporal integration of the aldosterone secretory response to angiotensin occurs via two intracellular pathways. J Biol Chem 259:1448–1457, 1984.
100. Python CP, Rossier MF, Vallotton MB, Capponi AM: Peripheral-type benzodiazepines inhibit calcium channels and aldosterone production in adrenal glomerulosa cells. Endocrinology 132:1489–1496, 1993.
101. Yokoyama T, Shimamoto K, Iimura O: Mechanism of inhibition of aldosterone secretion by a Ca^{2+} channel blocker in patients with essential hypertension and patients with primary aldosteronism. Nippon Naibunpi Gakkai Zasshi 71:1059–1074, 1995.
102. Nadler JL, Hsueh W, Horton R: Therapeutic effect of calcium channel blockade in primary aldosteronism. J Clin Endocrinol Metab 60:896–899, 1985.
103. Opocher G, Rocco S, Murgia A, Mantero F: Effect of verapamil on aldosterone secretion in primary aldosteronism. J Endocrinol Invest 10:491–494, 1987.
104. Bravo EL, Fouad FM, Tarazi RC: Calcium channel blockade with nifedipine in primary aldosteronism. Hypertension 8(Suppl I):I-191, 1986.
105. Carpene G, Rocco S, Opocher G, Mantero F: Acute and chronic effect of nifedipine in primary aldosteronism. Clin Exp Hypertens A 11:1263–1272, 1989.
106. Stimpel M, Ivens K, Wanbach G, Kaufmann W: Are calcium antagonists helpful in the management of primary aldosteronism? J Cardiovasc Pharmacol 12(Suppl 6):S131–S134, 1988.
107. Bursztyn M, Grossman E, Rosenthal T: The absence of long-term therapeutic effect of calcium blockade in the primary aldosteronism of adrenal adenomas. Am J Hypertens 1:88S–90S, 1988.
108. Mohanty PK, Sowers JT, McNamara C, et al: Effects of diltiazem on hormonal and hemodynamic responses to lower body negative pressure and tilt in patients with mild to moderate systemic hypertension. Am J Cardiol 56:28H–33H, 1985.
109. Pedersen OL, Mikkelsen E, Christensen NJ, et al: Effect of nifedipine on plasma renin, aldosterone and catecholamines in arterial hypertension. Eur J Clin Pharmacol 15:235–240, 1979.
110. Marone C, Luisoli S, Bomio F, et al: Body sodium-blood volume state, aldosterone and cardiovascular responsiveness after calcium entry blockade with nifedipine. Kidney Int 28:658–665, 1985.

111. Van Schaik BAM, Van Nistelrooy AEJ, Geyskes GG: Antihypertensive and renal effects of nicardipine. Br J Clin Pharmacol 18:57–63, 1984.
112. Van Schaik BAM, Hene RJ, Geyskes GG: Influence of nicardipine on blood pressure, renal function and plasma aldosterone in normotensive volunteers. Br J Clin Pharmacol 20:88S–94S, 1985.
113. Guthrie GP, McAllister RG, Kotchen TA: Effects of intravenous and oral verapamil upon pressure and adrenal steroidogenic responses in normal man. J Clin Endocrinol Metab 57:339–343, 1983.
114. Frohlich ED. Hemodynamic effects of calcium entry blocking agents in normal and hypertensive rats and man. Am J Cardiol 56:21H–27H, 1985.
115. Shamiss A, Peleg E, Rosenthal T, Ezra D: The role of atrial natriuretic peptide in the diuretic effect of Ca^{2+} entry blockers. Eur J Pharmacol 233:113–117, 1993.
116. Andrawis NS, Craft N, Abernethy DR: Calcium antagonists block angiotensin II-mediated vasoconstriction in humans: comparison with their effect on phenilephrine-induced vasoconstriction. J Pharmacol Exp Ther 261:879–884, 1992.
117. Lüscher TF, Wenzel RR, Moreau P, Takase H: Vascular protective effects of ACE inhibitors and calcium antagonists: theoretical basis for a combination therapy in hypertension and other cardiovascular diseases. Cardiovasc Drugs Ther 9(Suppl 3):509–523, 1995.
118. Lüscher TF, Yang Z, Kiowski W, et al: Endothelin-induced vasoconstriction and calcium antagonists. J Hum Hypertens 6 (Suppl 2):S3–S8, 1992.
119. Grohé C, Nouskas J, Vetter H, Neyses L: Effects of nisoldipine on endothelin-1 and angiotensin II- induced immediate/early gene expression and protein synthesis in adult rat ventricular cardiomyocites. J Cardiovasc Pharmacol 24:13–16, 1994.
120. Ko YD, Sachinidis A, Graack GH, et al: Inhibition of angiotensin II and platelet-derived growth factor-induced vascular smooth muscle cell proliferation by calcium entry blockers. Clin Investigator 70:113–117, 1992.
121. Andrawis NS, Abernethy DR: Verapamil blocks basal and angiotensin II-induced RNA synthesis of rat aortic vascular smooth muscle cells. Biochem Biophys Res Commun 183:767–773, 1992.
122. Messerli FH: Antihypertensive therapy—Going to the heart of the matter. Circulation 81:1128–1135, 1990.
123. Amodeo C, Kobrin I, Ventura HO, et al: Immediate and short-term haemodynamic effects of diltiazem in patients with hypertension. Circulation 73:108–113, 1986.
124. Hansson L, Dahlöf B: Calcium antagonists in the treatment of hypertension: State of the art. J Cardiovasc Pharmacol 15(Suppl4):S71–S75, 1990.
125. Frohlich ED: Left ventricular hypertrophy, cardiac diseases and hypertension: Recent experiences. J Am Coll Cardiol 14:1587–1594, 1989.
126. Cody RJ: Regression of left ventricular hypertrophy in resistant hypertension. J Am Coll Cardiol 16:143–144, 1990.
127. Liebson P: Clinical studies of drug reversal of hypertensive left ventricular hypertrophy. Am J Hypertens 3:512–517, 1990.
128. Henry PD, Bentley KI: Suppresion of atherogenesis in cholesterol fed rabbits treated with nifedipine. J Clin Invest 68:1366–1369, 1981.
129. Jackson CL, Bush RC, Bowyer DE: Mechanism of antiatherogenic action of calcium antagonists. Atherosclerosis 80:17–26, 1989.
130. Triggle DJ: Calcium antagonists in atherosclerosis: a review and commentary. Cardiovasc Drug Rev 6:320–335, 1989.
131. Lichtlen PR, Hugenholtz PG, Rafflenbeul W, et al: Retardation of angiographic progression of coronary artery disease by nifedipine. Results of the International Nifedipine Trial on Antiatherosclerotic Therapy (INTACT). Lancet 335:1109–1113, 1990.
132. Loaldi A, Polese A, Montorsi P: Comparison of nifedipine, propranolol and isosorbide dinitrate on angiographic progression and regression of coronary arterial narrowings in angina pectoris. Am J Cardiol 63:433–439, 1989.
133. Waters D, Lespérance J, Francetich M, et al: A controlled clinical trial to assess the effect of a calcium channel blocker on the progression of coronary atherosclerosis. Circulation 82:1940–1953, 1990.
134. Dawidson I, Rooth P, Lu C, et al: Verapamil improves the outcome after cadaver renal transplantation. J Am Soc Nephrol 2:983–990, 1991.
135. Wagner K, Albrecht S, Neumayer H: Prevention of posttransplant acute tubular necrosis by calcium antagonist diltiazem: A prospective randomized study. Am J Nephrol 7:287–291, 1987.
136. Russo D, Testa A, Della Volpe L, Sansone G: Randomized prospective study on renal effects in two different contrast media in humans. Nephron 55:254–257, 1990.
137. Kasiske BL, Kalil RS, Ma JZ, et al: Effect of anti-hypertensive therapy on the kidney in patients with diabetes: a meta-regression analysis. Ann Intern Med 118:129–138, 1993.

138. Slataper R, Vicknair N, Sadler R, Bakris GL: Comparative effects of different antihypertensive treatment on progression of diabetic renal disease. Arch Intern Med 153:973–980, 1993.
139. Zucchelli P, Zuccala A, Borghi M, et al: Long-term comparison between captopril and nifedipine in the progression of renal insufficiency. Kidney Int 42:452–458, 1992.
140. Kilaru P, Bakris GL: Microalbuminuria and progressive renal disease. J Hum Hypertens 8:809–817, 1994.
141. Yoshioka T, Shiraga H, Yoshida Y, et al: "Intact nephrons" as the primary origin of proteinuria in chronic renal disease: Study in the rat model of subtotal nephrectomy. J Clin Invest 82:1614–1623, 1988.
142. Anderson S: Renal hemodynamic effects of calcium antagonists in rats with reduced renal mass. Hypertension 17:288–295, 1991.
143. Isshiki T, Amodeo C, Messerli FH, et al: Diltiazem maintains renal vasodilation without hyperfiltration in hypertension: Studies in essential hypertension man and the spontaneously hypertensive rat. Cardiovasc Drugs Ther 1:359–366, 1987.
144. Epstein M: Calcium antagonists and renal protection. Arch Intern Med 152:1573–1584, 1992.
145. Messerli FH, Aepfelbacher FC: Cardiac effects of calcium antagonists in hypertension. In Messerli FH (ed): Cardiovascular Drug Therapy, 2nd ed. Philadelphia, W.B. Saunders, 1996, pp 908–915.
146. Moss AJ: Role of calcium channel blockers in postinfarction survival. In Fuster V, Ross R, Topol EJ (eds): Atherosclerosis and Coronary Artery Disease, 1st ed. Philadelphia, Lippincott-Raven, 1996, pp 1215–1222.
147. Yanagisawa M, Kurihara H, Kimura S, et al: A novel potent vasoconstrictor peptide produced by vascular endothelial cells. Nature 332:411–415, 1988.
148. Yanagisawa M, Inouf A, Ishikawa T, et al: Primary structure, synthesis and biological activity of rat endothelin, an endothelium derived vasoconstrictor peptide. Proc Natl Acad Sci U S A 85:6964–6967, 1988.
149. Yanagisawa M, Masaki T. Biochemistry and molecular biology of endothelins. Trends Pharmacol Sci 10:374–378, 1989.
150. Inoue A, Yanagisawa M, Kimura S, et al: The human endothelin family: Three structurally and pharmacologically distinct isopeptides predicted by three separate genes. Proc Natl Acad Sci U S A 86:2863–2867, 1989.
151. Huggins JP, Pelton JT, Miller RC: The structure and specificity of endothelin receptors: their importance in physiology and medicine. Pharmacol Ther 59:55–123, 1993.
152. Inoue A, Yanagisawa M, Takuwa Y, et al: The human prepro-endothelin-1 gene. J Biol Chem 264:14954–14959, 1990.
153. Laporte S, Denault JB, D'Orleans-Juste P, Leduc R: Presence of furin mRNA in cultured bovine endothelial cells and possible involvement of furin in the processing of the endothelin precursor. J Cardiovasc Pharmacol 22(Suppl 8):S7–S10, 1993.
154. Kimura S, Kasuya Y, Sawamura T, et al: Structure-activity relationships of endothelin: Importance of the C-terminal moiety. Biochem Biophys Res Commun 156:1182–1186, 1988.
155. Opgenorth TJ, Wu-Wong JR, Shiosaki K: Endothelin-converting enzymes. FASEB J 6:2653–2659, 1992.
156. Sakurai T, Yanagisawa M, Inoue A, et al: cDNA cloning, sequence analysis and tissue distribution of rat preproendothelin-1 mRNA. Biochem Biophys Res Commun 175:44–47, 1994.
157. Bloch KD, Eddy RL, Shows TB, Quertermous T: cDNA cloning and chromosomal assignment of the gene encoding endothelin 3. J Biol Chem 264:18156–18161, 1989.
158. Ohkubo S, Ogi K, Hosoya M, et al: Specific expression of human endothelin-2 (ET-2) gene in a renal adenocarcinoma cell line. Molecular cloning of cDNA encoding the precursor of ET-2 and its characterization. FEBS Lett 274:136–140, 1990.
159. Yanagisawa M: The endothelin system. A new target for therapeutic intervention. Circulation 89:1320–1322, 1994.
160. Schini VB, Hendrickson H, Heublein DM, et al: Thrombin enhaces the release of endothelin from cultured porcine aortic endothelial cells. Eur J Pharmacol 165:333–334, 1989.
161. Boulanger C, Lüscher TF: Relase of endothelin from the porcine aorta. Inhibition by endothelium-derived nitric oxide. J Clin Invest 85:587–590, 1990.
162. Kohno M, Yasunari K, Murakawa K, et al: Relase of immunoreactive endothelin from porcine aortic strips. Hypertension 15:718–723, 1990.
163. Yoshizumi M, Karihara H, Morita T, et al: Interleukin 1 increases the production of endothelin-1 by cultured endothelial cells. Biochem Biophys Res Commun 166:324–329, 1990.
164. Lüscher TF, Oemar BS, Boulanger CM, Hahn WA: Molecular and cellular biology of endothelin and its receptors (Pt 1). J Hypertens 11:7–11, 1993.
165. Lüscher TF, Boulanger CM, Dohi Y, Yang Z: Endothelium-derived contracting factors. Hypertension 19:117–130, 1992.

166. Stewart DJ, Langleben D, Cernacek P, Cianflone K: Endothelin release is inhibited by coculture of endothelial cells with cells of vascular media. Am J Physiol 259:H1928–H1932, 1990.
167. Arai H, Hori S, Aramori I, et al: Cloning and expression of cDNA encoding an endothelin receptor. Nature 348:730–732, 1990.
168. Sakurai T, Yanagisawa M, Takuwa Y, et al: Cloning of a cDNA encoding a non-isopeptide-selective subtype of the endothelin receptor. Nature 348:732–735, 1990.
169. Hosoda K, Nakao K, Hiroshi A, et al: Cloning and expression of human endothelin 1 receptor cDNA. FEBS Lett 287:23–27, 1991.
170. Sakamoto A, Yanagisawa M, Sakurai T, et al: Cloning and expression of human cDNA for the ETB endothelin receptor. Biochem Biophys Res Commun 178:656–663, 1991.
171. Ogawa Y, Nakao K, Arai H, et al: Molecular cloning of a non-isopeptide-selective human endothelin receptor. Biochem Biophys Res Commun 178:248–255, 1991.
172. Lin HY, Kaji EH, Winkel GK, et al: Cloning and functional expression of a vascular smooth muscle endothelin 1 receptor. Proc Natl Acad Sci U S A 88:3185–3189, 1991.
173. Wallnofer A, Weir S, Ruegg U, Cauvin C: The mechanism of action of endothelin-1 as compared with other agonists in vascular smooth muscle. J Cardiovasc Pharmacol 13(Suppl 5):23–31, 1989.
174. Kiowski W, Lüscher TF, Linder L, Büler FR: Endothelin-1 induced vasoconstriction in humans. Reversal by calcium channel blockade but not by nitrovasodilators or endothelium-derived relaxing factor. Circulation 83:469–475, 1991.
175. Miller VM, Komori K, Burnett JJ, Vanhoutte PM: Differential sensitivity to endothelin in canine arteries and veins. Am J Physiol 257:H1127–H1131, 1989.
176. Lüscher TF, Yang Z, Tschudi M, et al: Interaction between endothelin-1 and endothelin-derived relaxing factor in human arteries and veins. Circ Res 66:1088–1094, 1990.
177. Miller WL, Redfield MM, Burnett JJ: Integrated cardiac, renal and endocrine actions of endothelin. J Clin Invest 83:317–320, 1989.
178. Brain SD, Crossman DC, Buckley TL, Williams TJ: Endothelin-1: Demostration of potent effects on the microcirculation of humans and other species. J Cardiovasc Pharmacol 13(Suppl 5): 147–149, 1989.
179. Chabrier PE, Auguet M, Roubert P, et al: Vascular mechanism of action of endothelin-1: Effect of calcium antagonists. J Cardiovasc Pharmacol 13(Suppl 5):218–219, 1989.
180. Clark JG, Larkin SW, Benjamin N, et al: Endothelin-1 is a potent vasoconstrictor in dog peripheral vasculature in vivo. J Cardiovasc Pharmacol 13(Suppl 5):211–212, 1989.
181. D'Orleans JP, Finet M, De Nucci G, Vane JR: Pharmacology of endothelin-1 in isolated vessels: effect of nicardipine, methylene blue, hemoglobin and gossypol. J Cardiovasc Pharmacol 13(Suppl 5):19–22, 1989.
182. Goetz KL, Wang BC, Madwed JB, et al: Cardiovascular, renal, and endocrine responses to intravenous endothelin in conscious dogs. Am J Physiol 255:R1064–R1068, 1988.
183. Walder CE, Thomas GR, Thiemermann C, Vane JR: The hemodynamic effects of endothelin-1 in the pithed rat. J Cardiovasc Pharmacol 13(Suppl 5):93–97, 1989.
184. Pollock DM, Keith TL, Highsmith R:. Endothelin receptors and calcium signaling. FASEB J 9:1196–1204, 1995.
185. Marsden PA, Danthuluri NR, Brenner BM: Endothelin action on vascular smooth muscle involves inositol triphosphate and calcium mobilization. Biochem Biophys Res Commun 158:86–93, 1989.
186. Resink TJ, Scott-Burden T, Buhler FR: Endothelin stimulates phospholipase C in cultured vascular smooth muscle cells. Biochem Biophys Res Commun 157:1360–1368, 1988.
187. Simonson MS, Wann S, Mene P, et al: Endothelin stimulates phospholipase C. Na^+/H^+ exchange, c-fos expression, and mitogenesis in rat mesangial cells. J Clin Invest 83:708–712, 1989.
188. Segiura M, Inagami T, Hare GM, Johns JA: Endothelin action: inhibition by a protein kinase C inhibitor and involvement of phosphoinositols. Biochem Biophys Res Commun 158:170–176, 1989.
189. Van Renterghem C, Vigne P, Barhanin J, et al: Molecular mechanism of action of the vasoconstrictor peptide endothelin. Biochem Biophys Res Commun 157:977–985, 1988.
190. Goto K, Kasuya K, Matsuki N, et al: Endothelin activates the dihydropyridine-sensitive, voltage-dependent Ca^{2+} channel in vascular smooth muscle. Proc Natl Acad Sci U S A 86:3915–3918, 1989.
191. Cauvin C, Lukeman S, Cameron J, et al: Differences in norepinephrine activation and diltiazem inhibition of Ca^{2+} channels in isolated rabbit aorta and mesenteric resistance vessels. Circ Res 56:822–828, 1985.
192. Sunman W, Martin G, Hair WM, et al: Effect of calcium antagonists on endothelin-induced contraction of isolated human resistance arteries: differences related to site of origin. J Hum Hypertens 7:189–191, 1993.
193. Ito H, Hirata Y, Hiroe M, et al: Endothelin-1 induces hypertrophy with enhanced expression of muscle-specific genes in cultured neonatal rat cardiomyocites. Circ Res 69:209–215, 1991.

194. Hahn AWA, Resink TJ, Kern F, et al: Effects of endothelin-1 on vascular smooth muscle cell phenotypic differentiation. J Cardiovasc Pharmacol 20(Suppl 12):33–36, 1992.
195. Dubin D, Pratt RE, Cooke JP, Dzau VJ: Endothelin, a potent vasoconstrictor, is a vascular smooth muscle mitogen. J Vasc Med Biol 1:13–16, 1989.
196. Dohi Y, Lüscher TFL: Endothelin in hypertensive resistance arteries. Intraluminal and extraluminal dysfunction. Hypertension 18:543–549, 1991.
197. Wagner TD, De Nucci G, Vane JR: Rat endothelin is a vasodilator in the isolated perfused mesentery of the rat. Eur J Pharmacol 159:325–326, 1989.
198. Wright CE, Fozard JR: Regional vasodilation is a prominent feature of the haemodynamic response to endothelin in anaesthetized, spontaneously hypertensive rats. Eur J Pharmacol 155:201–203, 1988.
199. Télémaque S, Gratton JP, Claing A, D'Orléans-Juste P: Endothelin-1 induces vasoconstriction and prostacyclin release via the activation of endothelin ETA receptors in the perfused rabbit kidney. Eur J Pharmacol 237:275–281, 1993.
200. Wellings RP, Warnwe TD, Thiemermann C, et al: Vasoconstriction in the rat induced by endothelin-1 is blocked by PD145065. J Cardiovasc Pharmacol 22(Suppl 8):S103–S106, 1993.
201. Pollock DM, Opgenorth TJ: Evidence for endothelin-induced renal vasoconstriction independent of ETA receptor activation. Am J Physiol 264:R222–R226, 1993.
202. Karet FE, Kuc RE, Davenport AP: Novel ligands BQ123 and BQ3020 characterize endothelin receptor subtypes ETA and ETB in human kidney. Kidney Int 44:36–42, 1993.
203. Douglas SA, Vickery-Clark LM, Ohlstein EH: Endothelin-1 does not mediate hypoxic vasoconstriction in canine isolated blood vessels: effect of BQ123. Br J Pharmacol 108:418–421, 1993.
204. Hay DWP, Luttmann MA, Hubbard WC, Undem BJ: Endothelin receptor subtypes in human and guinea-pig pulmonary tissues. Br J Pharmacol 110:1175–1183, 1993.
205. Ihara M, Saeki T, Fukurodo T, et al: A novel radioligand 125-I BQ-3020 selective for endothelin (ETB) receptors. Life Sci 51:47–52, 1992.
206. White DG, Cannon TR, Garratt H: Endothelin ETa and ETb receptors mediate vascular smooth-muscle contraction. J Cardiovasc Pharmacol 22(Suppl 8):S144–S148, 1993.
207. Wright HM, Malik KU: Prostacyclin synthesis elicited by endothelin-1 in rat aorta is mediated by an ETA receptor via influx of calcium and is independent of protein kinase C. Hypertension 26 (Pt 2):1035–1040, 1995.
208. Firth JD, Ratcliffe PJ, Raine AE, Ledingham JG: Endothelin: An important factor in acute renal failure? Lancet ii:1179–1182, 1988.
209. Lüscher TF, Bock AH, Yang Z, Diederich D: Endothelium-derived relaxing and contracting factors: perspectives in nephrology. Kidney Int 39:575–590, 1991.
210. Badr KF, Murray JJ, Breyer MD, et al: Mesangial cell, glomerular and renal vascular responses to endothelin in the rat kidney. Elucidation of signal transduction pathways. J Clin Invest 83:336–342, 1989.
211. Reference deleted.
212. King AJ, Brenner BM, Anderson S: Endothelin: A potent renal and systemic vasoconstrictor peptide. Am J Physiol 256:C1101–C1107, 1989.
213. Gomez-Garre D, Ruiz-Ortega M, Ortego M, et al: ACE inhibition decreases endothelin-induced fibronectin synthesis in cultured mesangial cells [abstract]. Nephrol Dial Transplant 9:892., 1994.
214. Simonson MS: Endothelins: Multifactorial renal peptides. Physiol Rev 73:375–411, 1993.
215. Clavell AL, Burnett JC: Physiologic and pathophysiologic roles of endothelin in the kidney. Curr Opin Nephrol Hypertens 3:66–72, 1994.
216. Shichiri M, Hirata Y, Ando K, et al: Plasma endothelin levels in hypertension and chronic renal failure. Hypertension 15:493–496, 1990.
217. Khono M, Yasunari K, Murakawa K, et al: Plasma immunoreactive endothelin in essential hypertension. Am J Med 88:614–618, 1990.
218. Saito Y, Nakao K, Mukoyama M, et al: Application of monoclonal antibodies for endothelin to hypertensive research. Hypertension 15:734–738, 1990.
219. Miyauchi T, Yanagisawa M, Suzuki N, et al: Venous plasma concentrations of endothelin in normal and hypertensive subjects [abstract]. Circulation 80(Suppl II):II2280, 1989.
220. Davenport AP, Ashby MJ, Eaton P, et al: A sensitive radioimmunoassay measuring endothelin-like immunoreactivity in human plasma: Comparison of levels in patients with essential hypertension and normotensive control subjects. Clin Sci 78:261–264, 1990.
221. Schiffrin EL, Thibault G: Plasma endothelin in human esential hypertension. Am J Hypertens 4:303–308, 1991.
222. Lüscher TF, Oemar BS, Boulanger CM, Hahn AWA: Molecular and cellular biology of endothelin and its receptors (Pt 2). J Hypertens 11:121–126, 1993.

223. Lüscher TF, Seo B, Bühler B: Potential role of endothelin in hypertension. Hypertension 21:752–757, 1993.
224. Miyauchi T, Ishikawa T, Tomore Y, et al: Characteristics of pressor response to endothelin in spontaneously hypertensive and Wistar-Kyoto rats. Hypertension 14:427–431, 1989.
225. Deiderich D, Yang Z, Bühler FR, Lüscher TF: Endothelium-derived relaxing factor and endothelin in resistance arteries of hypertensive rats [abstract]. J Vasc Med Biol 1(3):167, 1989.
226. Tomobe Y, Miyauchi T, Saito A, et al: Effects of endothelin on the renal artery from spontaneously hypertensive and Wistar-Kyoto rats. Eur J Pharmacol 152:373–374, 1988.
227. Criscione L, Nellis P, Riniker B, et al: Reactivity and sensitivity of mesenteric vascular beds and aortic rings of spontaneously hypertensive rats to endothelin: effects of calcium entry blockers. Br J Pharmacol 100:31–36, 1990.
228. Yasuda M, Kohono M, Tahara A, et al: Circulating immunoreactive endothelin in ischaemic heart disease. Am Heart J 119:801–806, 1990.
229. Lerman A, Edwards BS, Hallett JW, et al: Circulating and tissue endothelin immunoreactivity in advanced atherosclerosis. N Engl J Med 325:997–1001, 1991.
230. Miyauchi T, Yanagisawa M, Tomizawa T, Masaki T: Increased plasma concentrations of endothelin-1 and big endothelin-1 in acute myocardial infarction. Lancet ii:53–54, 1989.
231. Achmad TH, Rao GS: Chemotaxis of human blood monocytes toward endothelin-1 and the influence of calcium channel blockers. Biochem Biophys Res Commun 189:994–1000, 1992.
232. Hahn AWA, Resink TJ, Scott-Burden T, et al: Stimulation of endothelin mRNA and secretion in rat vascular smooth muscle cells: A novel autocrine function. Cell Regulation 1:649–659, 1990.
233. Resink TJ, Hahn AW, Scott-Burden T, et al: Inducible endothelin mRNA expression and peptide secretion in cultured human vascular smooth muscle cells. Biochem Biophys Res Commun 168:1303–1310, 1990.
234. Boulanger CM, Tanner FC, Bea M, et al: Oxidized low-density lipoproteins induce mRNA expression and release of endothelin from the human and porcine endothelium. Circ Res 70:1191–1197, 1992.
235. Nayler WG, Ou RC, Gu XH, Casley DJ: Effect of amlodipine pretreatment on ischaemia-reperfusion-induced increase in cardiac endothelin-1 binding site density. J Cardiovasc Pharmacol 20:416–420, 1992.
236. Feron O, Salomone S, Godfraind T: Blood pressure-independent inhibition by lacidipine of endothelin-1-related cardiac hypertrophy in salt-loaded, stroke-prone spontaneously hypertensive rats. J Cardiovasc Pharmacol 26 (Suppl 3): S459–S461, 1995.
237. Liu JJ, Casley D, Wojta J, et al: Effects of calcium- and ETA-receptor antagonists on endothelin-induced vasoconstriction and levels of endothelin in the human internal mammary artery. Clin Exp Pharmacol P 21:49–57, 1994.
238. Yanagisawa M, Kurihara H, Kimura S, et al: A novel potent vasoconstrictor peptide produced by vascular endothelial cells. Nature 332:411–415, 1988.
239. Kiowski W, Linder L, Erne P: Vascular effects of endothelin-1 in humans and influence of calcium channel blockade. J Hypertens Suppl 12(Suppl 1):S21–S26, 1994.
240. Fried G, Liu YA: Effects of endothelin, calcium channel blockade and EDRF inhibition on the contractility of human uteroplacental arteries. Acta Physiol Scand 151:477–484, 1994.
241. Balligand JL, Godfraind T: Effect of nisoldipine on contractions evoked by endothelin-1 in human isolated distal and proximal coronary arteries and veins. J Cardiovasc Pharmacol 24:618–625, 1994.
242. Boulanger CM, Nakashima M, Olmos L, et al: Effects of the Ca^{2+} antagonist RO 40-5967 on endothelium-dependent responses of isolated arteries. J Cardiovasc Pharmacol 23:869–876, 1994.
243. Meyer P, Lang MG, Flammer J, Luscher TF: Effects of calcium channel blockers on the response to endothelin-1, bradykinin and sodium nitroprusside in porcine ciliary arteries. Exp Eye Res 60:505–510, 1995.
244. Kaasjager KA, van Rijn HJ, Koomans HA, Rabelink TJ: Interactions of nifedipine with the renovascular effects of endothelin in humans. J Pharmacol Exp Ther 275:306–311, 1995.
245. Takenaka T, Epstein M, Forster H, et al: Attenuation of endothelin effects by a chloride channel inhibitor, indanyloxiacetic acid. Am J Physiol 262:F799–F806, 1992.
246. Loutzenhiser R, Epstein M, Hayashi K, Horton C: Direct visualization of effects of endothelin on the renal microvasculature. Am J Physiol 258:F61–F68, 1990.
247. Takahashi K, Katoh T, Fukunaga M, Badr KF: Studies on the glomerular microcirculatory actions of manidipine and its modulation of the systemic and renal effects of endothelin. Am Heart J 125(Pt 2):609–619, 1993.

MURRAY EPSTEIN M.D., F.A.C.P.

24

Calcium Antagonists and the Kidney: Implications for Renal Protection

Since the introduction of calcium antagonists three decades ago, attention has focused on their beneficial effects in the management of symptomatic coronary artery disease and their ability to lower blood pressure. Although the renal vasodilatory properties of calcium antagonists were described as early as 1962,[1] this seminal finding was virtually ignored. Only during the past two decades has this important effect been rediscovered, with an increasing awareness of the beneficial effects on kidney function.[2–6] This review briefly considers the constellation of effects of calcium antagonists on renal function and their current and potential therapeutic applications. The beneficial effects of calcium antagonists not only protect against a wide array of acute renal insufficiency syndromes but also have an emerging role in retarding progressive renal disease.

EFFECTS OF CALCIUM ANTAGONISTS ON THE RENAL MICROCIRCULATION

The pharmacologic effects of calcium antagonists on renal hemodynamics[2,3] and renal microcirculation are reviewed in chapter 23. In summary, direct in vivo and in vitro observations in diverse experimental models indicate that in the acute setting calcium antagonists antagonize preglomerular vasoconstriction. Collectively, studies from several laboratories indicate that calcium antagonists reverse afferent arteriolar vasoconstriction induced by widely divergent stimuli, including putative mediators of deranged renal hemodynamics such as endothelin. Such observations suggest that the activation of potential dependent calcium channels (PDCs) is a final common mechanism of afferent arteriolar vasoconstriction by diverse agonists. In contrast, the efferent arteriole is highly refractory to the vasodilatory effects of calcium antagonists, indicating remarkable intraorgan heterogeneity of mechanisms that activate smooth muscle within the renal microcirculation. Inferences from micropuncture studies are conflicting, but most studies in remnant kidney models suggest that calcium antagonists attenuate glomerular hypertension with variable effects on afferent arteriolar resistance (R_A) and efferent arteriolar resistance (R_E).

RENAL FUNCTIONAL EFFECTS IN HUMANS

Studies in Normal Humans

As discussed in chapter 23, in normal human volunteers calcium antagonists have little effect on renal plasma flow or glomerular filtration rate. In contrast, hypertensive patients appear to be more sensitive to the renal hemodynamic effects of calcium antagonists than their normotensive counterparts.[7] Several investigators have demonstrated an exaggerated vasodilatory response in normotensive offspring of hypertensive parents.[8,9] Blackshear et al.[8] reported that at least 50% of normotensive subjects with a family history of hypertension demonstrated an exaggerated renal vasodilatory response to diltiazem, suggesting that hypertension is associated with an inherited abnormality of the renal vascular bed. Subsequently, Montanari et al.[9] extended and verified these findings.

Additional Postulated Mechanisms Mediating Renal Protection

Aside from their renal microcirculatory effects, calcium antagonists have additional properties that may contribute to their ability to afford renal protection under diverse experimental conditions and perhaps in clinical disorders.[2,4] The more prominent mechanisms postulated to mediate the renal protective actions of calcium antagonists include their ability to lessen injury by retarding renal growth,[10,11] to attenuate mesangial entrapment of macromolecules,[12,13] to countervail or attenuate the mitogenic effects of diverse cytokines and growth factors, including platelet-derived growth factor (PDGF) and platelet-activating factor (PAF),[12] and to act as free radical scavengers[12] (Table 1).

The results of several studies suggest that calcium antagonists can protect the kidney after reduction of renal mass. Dworkin and associates postulated that in great part glomerular injury depends not only on the pressure developed within the glomerular capillary, but also on tension in the vessel wall.[14] Tension appears to be influenced equally by glomerular pressure (P_{GC}) and vessel radius (R_{GC}). Thus, if the glomerular capillary radius increases when kidneys hypertrophy, then wall tension rises from both a hemodynamic and a structural basis. Conversely, therapies that prevent hypertrophy may decrease tension by reducing R_{GC}. Consistent with this hypothesis, Dworkin et al.[15] reported that R_{GC} was significantly increased in rats 8 weeks after one and two-thirds nephrectomy. Administration of nifedipine was associated with a reduction in R_{GC} sufficient to cause a decline in tension similar in magnitude to that produced by agents such as angiotensin-converting enzyme (ACE) inhibitors that retard injury by reducing P_{GC}.

TABLE 1. Known and Postulated Mechanisms Mediating the Renoprotective Actions of Calcium Antagonists

1. Reduction in systemic blood pressure	5. Amelioration of uremic nephrocalcinosis
2. Reduction of renal hypertrophy	6. Attenuation of mitogenic effects of growth factors
3. Modulation of mesangial traffic of macromolecules	7. Possible blockade of pressure-induced calcium entry
4. Reduction in metabolic activity of remnant kidneys	8. Decreased free radical formation

Modified from Epstein M, Loutzenhiser R: Potential applicability of calcium antagonists as renal protective agents. In Epstein M, Loutzenhiser R (eds): Calcium Antagonists and the Kidney. Philadelphia, Hanley & Belfus, 1990, p 289.

In addition, calcium antagonists may attenuate mesangial entrapment of macromolecules. Several lines of evidence suggest that angiotensin II may affect renal function by mechanisms other than those that are hemodynamically mediated. Thus, angiotensin II has been demonstrated to influence the transport of blood-borne macromolecules into the mesangium.[16,17] In an elegant study, Raij and Keane[13] demonstrated that angiotensin II, when given to rats in subpressor doses, both increases the uptake and decreases the disappearance rate of macromolecules such as radiolabeled IgG in the mesangium. Entrapment of macromolecules in the mesangium may lead to mesangial injury due to stimulation of local inflammation and ultimately to expansion of mesangial cell and/or matrix with subsequent progression to glomerular sclerosis.[13] The proposal that calcium antagonists counteract the mesangial effects of angiotensin II suggests an additional mechanism whereby they may attenuate or retard the development of glomerulosclerosis. Additional studies are required to substantiate these proposals and to ascertain whether such effects occur with clinically relevant doses of calcium antagonists.

Additional salutary effects may relate to the ability of calcium antagonists to modulate and countervail the promotion of mesangial hyperplasia and hypertrophy by diverse growth factors and cytokines.[18,19] Mesangial cells generate PAF, which induces platelet aggregation as well as chemotaxis and activation of leukocytes. Mesangial cells also release PDGF in response to certain injuries. The release of PDGF is further enhanced in the presence of thrombin, suggesting an additional link between the coagulation system and glomerular injury.[20] In a preliminary report, Shultz and Raij noted that calcium antagonists inhibit the mitogenic effect of PDGF and thrombin on mesangial cells.[21] In addition, calcium antagonists inhibit thrombin-induced stimulation of production of PAF by endothelial cells,[22] suggesting another mechanism whereby they may attenuate glomerular injury induced by PAF (see chapter 23).

The renoprotective action of calcium antagonists may be attributable in part to their putative interaction with toxic free radicals. Several lines of evidence indicate that when stimulated by certain cytokines or the terminal components of complement, mesangial cells can generate oxygen free radicals. Since these forms of reactive oxygen species are mediators of tissue injury and are known to cause glomerular injury in various models of renal ischemia and inflammation,[23–26] they may further propagate glomerular damage. The proposed role of calcium antagonists as free-radical scavengers may contribute to their ability to counter or attenuate glomerular damage.

An additional mechanism by which calcium antagonists may exert their protective effects includes amelioration of mitochondrial calcium overload, which results in mitochondrial malfunction and eventual cell death.[27,28] Indeed, calcium antagonists have been reported to increase survival of ischemic tubular cells in culture.[29]

NATRIURETIC EFFECTS OF CALCIUM ANTAGONISTS

A large body of evidence indicates that calcium antagonists are natriuretic. Numerous investigations have evaluated the direct effect of calcium antagonists administered to experimental animals.[7,30–33] An increase in the rate of urinary sodium excretion ($U_{Na}V$) was a prominent feature in all reports, whereas GFR increased in only one-half of the studies, indicating that the natriuresis can be dissociated from increases in filtered sodium load. Increases in sodium excretion were observed despite decreases in blood pressure, which ordinarily favor increased sodium reabsorption by reducing renal blood flow. Thus, contrary to earlier proposals, the natriuretic effects of calcium antagonists cannot be attributed solely to hemodynamic effects.

Over a dozen investigations have assessed the acute natriuretic effects of calcium antagonists in humans with or without hypertension.[30,34] Renal blood flow (RBF) and GFR were reported to increase in less than one-half of the subjects. Nevertheless, an increase in $U_{Na}V$ was observed in hypertensive subjects in almost all studies and in normal subjects in the majority of studies. Natriuresis was demonstrable without concomitant changes in RBF or GFR, indicating that the natriuretic responses may occur independently of intrarenal hemodynamic effects.

There are fewer studies of the long-term effects of calcium antagonists on renal sodium handling.[35,36] One of the more informative long-term studies was conducted by Luft et al,[36] who evaluated 8 normal and 8 mildly hypertensive subjects with a consistent sodium intake of 150 mmol/day while housed in a clinical research center. The study included a 7-day placebo period and a 7-day period during which the subjects received the dihydropyridine, nitrendipine, 20 mg/day. The administration of nitrendipine was associated with a decrease in blood pressure in hypertensive subjects and a 1-kg weight loss in both groups. Both normal and hypertensive subjects experienced a natriuresis for several days. The total negative sodium balance amounted to about 150 mmol in both normal and hypertensive subjects, which is not dissimilar to values observed when thiazide diuretics are used in the treatment of patients with essential hypertension.[37]

The attainment of a new steady state of sodium balance after the initial days of treatment with calcium antagonists confounds attempts to demonstrate whether the natriuretic effect persists. Thus determination of basal sodium excretory rates may not be definitive. In an attempt to circumvent this problem, several investigators used the natriuretic response to saline-induced volume expansion as an index of natriuretic potential. Ruilope et al.[38] investigated long-term renal sodium handling by using sequential challenges with an intravenous sodium load. Eight patients with mild-to-moderate essential hypertension were studied before and after 1, 8, and 24 weeks of treatment with nitrendipine. Patients were placed on a 120-mEq sodium diet 5 days before initiation of the studies. The natriuretic response to saline administration was markedly augmented at 1, 8, and 24 weeks of calcium antagonist therapy compared with the initial study before administration of nitrendipine. Collectively, these data indicate that the augmented natriuretic response of nitrendipine to saline-induced volume expansion persists during 6 months of therapy.

Although the above observations are consistent with the formulation that the natriuretic potential of nitrendipine is sustained, several limitations of the study design confound interpretation of the results. As we have discussed previously,[39–41] rapid volume expansion with exogenous solutions such as saline has a number of drawbacks related to its lack of specificity. For example, saline infusion increases the volume of all fluid compartments and induces concomitant alterations in plasma composition.

Studies from our laboratory over the past 30 years have circumvented many of these methodologic problems by applying a newly developed investigative tool—the model of head-out water immersion—to the assessment of renal sodium and water homeostasis in both normal humans and diverse edematous states.[39–44] Earlier studies from our laboratory demonstrated that this maneuver produces a prompt, sustained diuresis and natriuresis[43,44] and markedly suppresses levels of plasma renin activity (PRA), plasma aldosterone (PA), and arginine vasopressin (AVP).[40,41]

We have used the immersion model to assess the long-term natriuretic potential of a dihydropyridine calcium antagonist (isradipine) in 5 patients with essential hypertension.[34] In essence, we assessed whether isradipine was capable of inducing a sustained state of increased natriuretic potential, as suggested by the studies of Ruilope.[38] Patients underwent an initial immersion study before treatment. Isradipine was started at 2.5 mg orally twice daily, and the dosage was doubled weekly until diastolic blood pressure was controlled at levels of < 95 mmHg. Figure 1 depicts a representative study in patient A. The initial study was conducted before initiation of antihypertensive therapy, at which time his blood pressure was 140/110 mm Hg. The initial immersion study before therapy resulted in a natriuresis with the $U_{Na}V$ increasing threefold from a prestudy level of 86 to a level of 300 μEq/min during hour 3 of immersion. Cumulative sodium excretion was 40.4 mEq/3 hours. The first immersion study during isradipine therapy was carried out

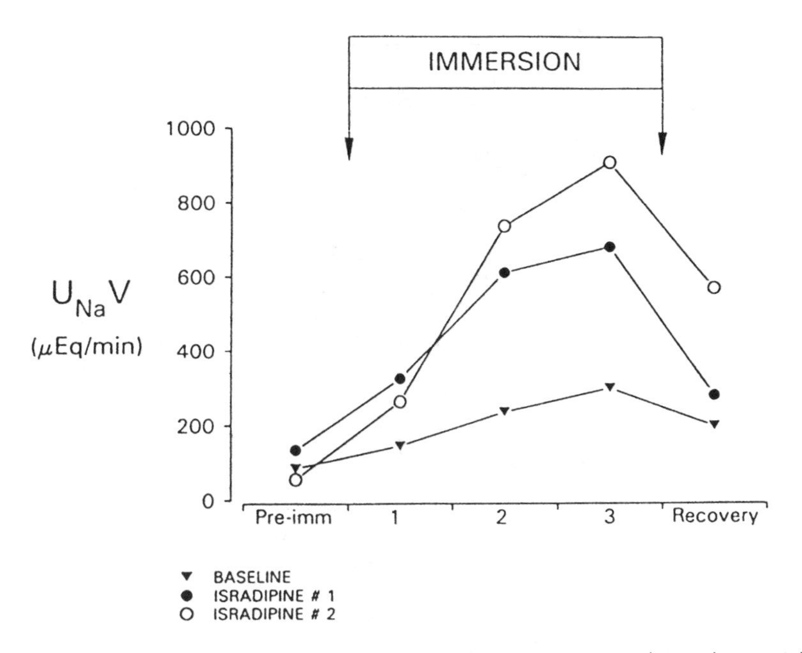

FIGURE 1. Changes in the rate of sodium excretion ($U_{Na}V$) in patient A during three serial immersion studies. The initial study *(closed triangle)* was conducted without isradipine. The latter two studies were carried out 28 *(closed circles)* and 84 *(open circles)* days after initiation of isradipine. It is readily apparent that the natriuretic effects of immersion are magnified by isradipine and that these effects persist for at least 84 days. (From Epstein M, De Micheli AG, Forster H: Natriuretic effects of calcium antagonists in humans: A review of experimental evidence and clinical data. Cardiovasc Drugs Rev 9:399–413, 1991, with permission.)

28 days after initiation of treatment. The blood pressure at this time was 123/91 mmHg. The natriuresis was markedly enhanced, with $U_{Na}V$ increasing from 134 μEq/min to a peak of 681 μEq/min during the third hour of immersion. Cumulative sodium excretion was 97.1 mEq/3 hours, more than twice the value during the initial immersion study.

Eight weeks later (on the 84th day of isradipine therapy), when his blood pressure was 116/92 mmHg, the patient underwent a repeat immersion study. Immersion induced a 15-fold increase in sodium excretion (from a preimmersion level of 56 to a peak of 910 μEq/min during hour 3).

Fractional excretion of sodium (FE_{Na}) mirrored the changes in $U_{Na}V$ in patient A (Fig. 2). During the initial study before isradipine treatment, FE_{Na} increased from 0.47 to a peak of 1.31% during hour 3. The natriuresis was markedly augmented during isradipine therapy; during the initial isradipine study FE_{Na} increased from 0.73 to 3.19%. During the final study, FE_{Na} increased even further to 3.93%.

Figure 3 summarizes the natriuretic effects for the 5 patients who completed both immersion studies before and after isradipine therapy. The peak in natriuretic response during the initial immersion study (without drug therapy) is compared with the peak during the immersion study after control of blood pressure with isradipine. The titration period necessary to obtain blood pressure control (diastolic blood pressure < 95 mmHg) varied. Consequently, at the time of the second immersion study, the duration of isradipine therapy ranged from

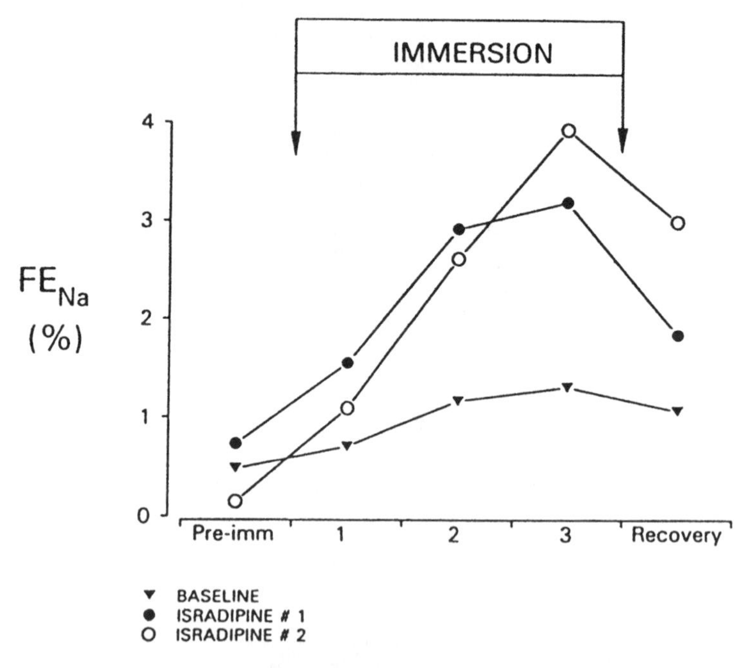

FIGURE 2. Changes in fractional excretion of sodium in patient A during three serial immersion studies. The initial study was conducted without isradipine. The latter two studies were carried out 28 *(closed circles)* and 84 *(open circles)* days after initiation of isradipine. (From Epstein M, De Micheli AG, Forster H: Natriuretic effects of calcium antagonists in humans: A review of experimental evidence and clinical data. Cardiovasc Drugs Rev 9:399–413, 1991, with permission.)

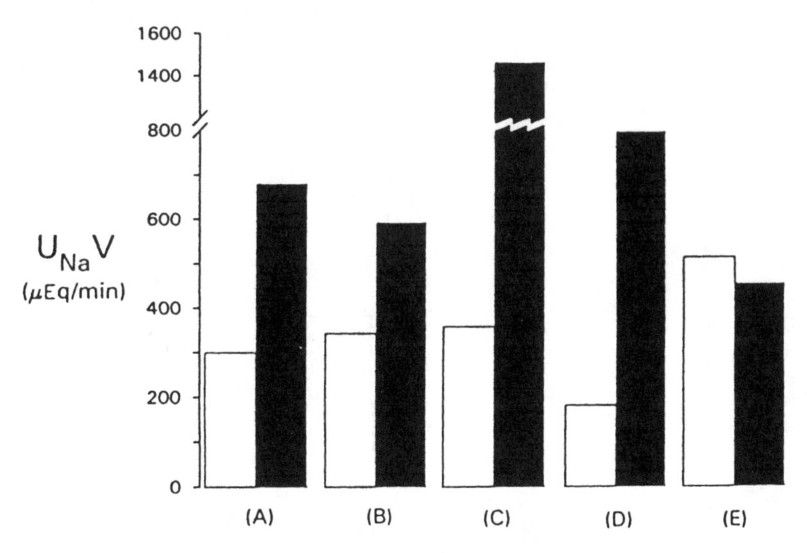

FIGURE 3. Peak natriuretic responses of five patients with essential hypertension undergoing immersion before *(open bars)* and after control of blood pressure with isradipine *(closed bars)*. (From Epstein M, De Micheli AG, Forster H: Natriuretic effects of calcium antagonists in humans: A review of experimental evidence and clinical data. Cardiovasc Drugs Rev 9:399–413, 1991, with permission.)

9 (patient B) to 30 days (patient C). Four of the 5 patients manifested a striking immersion-induced natriuresis while on isradipine, exceeding the natriuresis observed before isradipine therapy. Because basal sodium excretion during the preimmersion hour tended to be higher in the second (isradipine) immersion than in the initial immersion studies, the natriuretic responses were compared using delta $U_{Na}V$ (peak $U_{Na}V$ – basal $U_{Na}V$) as an index of the natriuresis. The pattern of natriuretic responses was indistinguishable from that in Figure 3. Patient E manifested similar increments in $U_{Na}V$ during the two studies (an increase in $\Delta\ U_{Na}V$ of 374 μEq/min during immersion without isradipine compared with 362 μEq/min with isradipine).

Fractional excretion of sodium (FE_{Na}) mirrored the changes in $U_{Na}V$ (Fig. 4). During the initial immersion study with isradipine treatment, FE_{Na} increased from 0.49 ± 0.16 to a peak of 1.74 ± 0.57% during hour 3 ($p < 0.05$). The natriuresis was markedly augmented by isradipine therapy, with FE_{Na} increasing from 0.66 ± 0.11 to 3.64 ± 0.58% ($p < 0.05$). In addition, cumulative sodium excretion averaged 50 ± 8 mEq/3 hours during the initial immersion study vs. 113 ± 24 mEq/3 hours during the isradipine immersion, mirroring the peak changes in $U_{Na}V$. The natriuresis was accompanied by a marked kaliuretic response with a greater than twofold increase in urinary potassium excretion (U_KV). The kaliuretic responses during immersion with and without isradipine were virtually identical (peak U_KV of 123 ± 15 vs. 125 ± 31 μEq/min, respectively).

Urinary flow rates tended to mirror the changes for $U_{Na}V$. Four of the patients had a greater immersion-induced diuresis while taking isradipine. Indeed, in two patients (C and D) the diuretic response on isradipine was twofold greater than during immersion without drug.

Collectively our studies confirm previous suggestions that calcium antagonists exert long-term effects on sodium homeostasis.[38,45–48] Although basal urinary

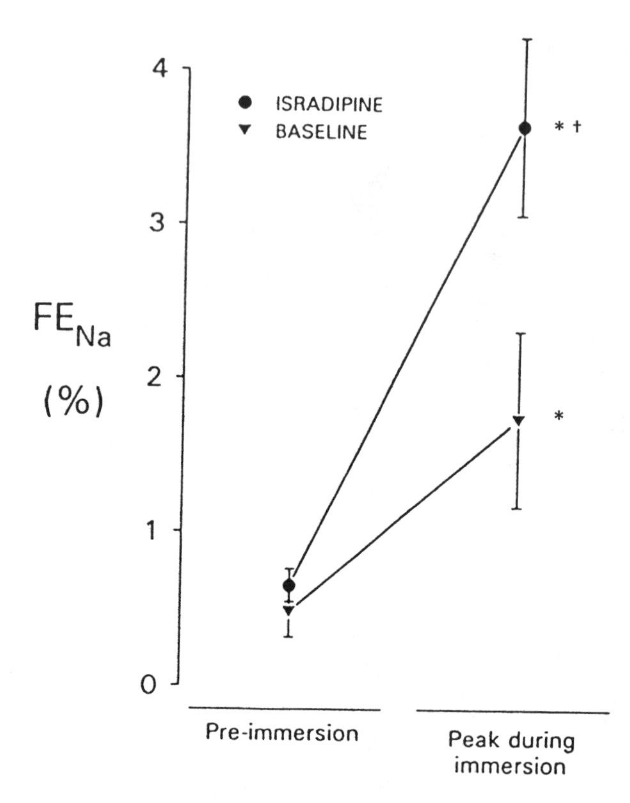

FIGURE 4. Changes in fractional excretion of sodium (FE_{Na}) in five patients with essential hypertension undergoing immersion before and after control of blood pressure with isradipine. The augmentation of FE_{Na} during isradipine therapy exceeded the increment induced by immersion in the absence of isradipine administration. Results are mean ± SE (*$p < 0.05$ compared with pre-immersion; †$p < 0.05$ for peak with vs. without isradipine). (From Epstein M, De Micheli AG, Forster H: Natriuretic effects of calcium antagonists in humans: A review of experimental evidence and clinical data. Cardiovasc Drugs Rev 9:399–413, 1991, with permission.)

sodium excretion was not increased by isradipine, we demonstrated that the natriuretic and diuretic responses to immersion-induced volume expansion are magnified after periods of isradipine therapy exceeding 9–30 days. The natriuretic potential of isradipine is further underscored by the fact that natriuresis occurred under conditions that favor sodium retention: systemic blood pressure was reduced, and renal perfusion pressure may have been lessened. In summary, most studies of the long-term effects of calcium antagonists indicate that elevated blood pressure can be reduced without impairing renal hemodynamics or renal excretory function.

THERAPEUTIC APPLICATIONS OF CALCIUM ANTAGONISTS IN HYPERTENSION AND RENAL DISEASE

Based on a consideration of the pharmacologic effects of the calcium antagonists on renal hemodynamics, it is tempting to consider current usage and to speculate about potential applications in clinical medicine.

Management of Essential Hypertension

The striking effects of calcium antagonists on renal hemodynamics and renal sodium handling support their use in the treatment of hypertension.[4] Medications, such as hydralazine, that directly reduce peripheral vascular resistance

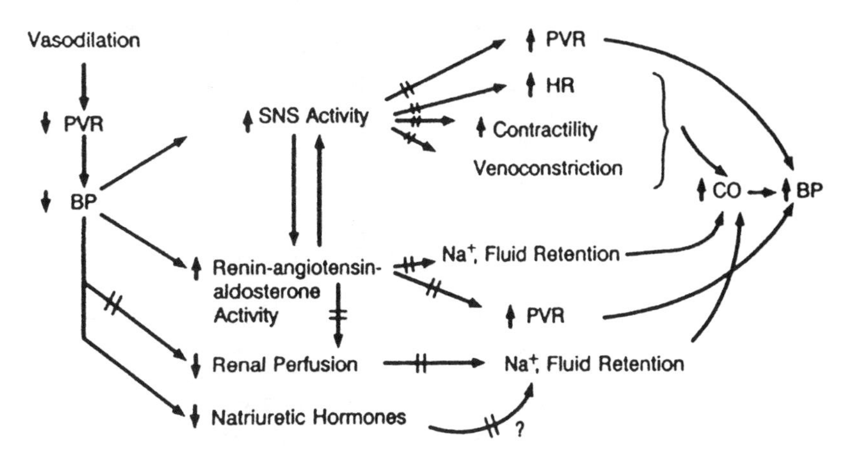

FIGURE 5. A summary of the mechanisms whereby calcium antagonists attenuate the expected adaptive changes in peripheral vascular resistance (PVR), heart rate, cardiac output, and extracellular fluid volume that otherwise would lead eventually to a reduction in the initial blood pressure-lowering response. // symbols indicate countervailing mechanisms that are attenuated by calcium antagonists. (From Epstein M, Loutzenhiser R (eds): Calcium Antagonsists and the Kidney. Philadelphia, Hanley & Belfus, 1990, p 277, with permission.)

have been used in antihypertensive therapy for many years, but their effectiveness is limited by reactive stimulation of renal and hormonal responses that counteract their antihypertensive actions[4] (Fig. 5) and by the induction of sodium retention. The consequent volume expansion results in pseudotolerance to the antihypertensive effects of hydralazine.

In contrast to direct-acting vasodilators such as hydralazine and minoxidil, calcium antagonists attenuate the expected adaptive changes in peripheral vascular resistance (PVR), heart rate, cardiac output, and extracellular fluid volume that lead to a reduction in the initial blood pressure-lowering response of direct vasodilators. As examples, calcium antagonists interfere with angiotensin II- and alpha adrenergic-mediated vasoconstriction. They also countervail the sodium-retaining effects of decreased renal perfusion[3,4] and possibly decrease levels of natriuretic hormones (indicated by the symbol //in Fig. 5).

Prophylaxis Against Acute Renal Failure

Aside from the role of calcium antagonists in treating hypertension, their salutary effects on renal hemodynamics, in concert with their effects on cellular calcium metabolism, suggest a future role in managing certain types of acute renal insufficiency.[4,49,50] Table 2 summarizes several such examples. Possibilities include use of their ability to augment renal perfusion in clinical settings in which renal hemodynamics are compromised. Before considering such clinical applications, it is relevant to review briefly the effects of calcium antagonists in experimental models of acute renal failure (ARF).

Experimental Acute Renal Failure

Over 70% of ARF in humans results from hypoxic injury to the kidney. Intrarenal hemodynamic and metabolic events cannot be accurately assessed in humans

TABLE 2. Current and Potential Applications of Calcium Antagonists for Prophylaxis Against Acute Renal Insufficiency

1. Transplantation
   - Organ preservation during harvesting of kidneys
   - Improved graft function and fewer episodes of rejection
2. Amelioration of renal insufficiency due to:
   - Radiocontrast agents
   - Cyclosporine nephrotoxicity
   - Amphotericin nephrotoxicity
   - Aminoglycoside nephrotoxicity ?
   - Chemotherapy (cisplatin) ?

with ARF. Therefore, various experimental models of renal hypoperfusion or complete interruption of renal blood supply have been used in an attempt to obtain more direct insight into the pathophysiologic events and consequences of hypoxia. Examples include norepinephrine-induced ARF and renal failure resulting from clamping of the renal artery. None of these models, however, truly mimics human ARF.

The role of calcium antagonists as protective agents against ARF has been assessed in a number of experimental models, including not only the "clamping" model and norepinephrine infusion, but also glycerol-induced ARF, gentamicin-induced renal failure, and ischemia induced by radiocontrast agents.

Models of Postischemic Acute Renal Failure. An experimental model that has attracted much investigative attention is postischemic ARF induced by vascular clamping. As detailed in comprehensive reviews,[51,52] the available data conflict with regard to the effects of calcium antagonists in this setting. Malis et al.[53] found no protection in a rat model of ischemia induced by a 40-minute renal artery clamping, regardless of whether verapamil was administered before or after treatment, intrarenally or intravenously, or in vivo or in isolated perfused kidneys. Only when verapamil was administered before 40 minutes of norepinephrine-induced ischemia was a protective effect demonstrated.[53] In contrast, Goldfarb et al.[54] found partial protection in rats (improved creatinine clearance at 24 hours) when verapamil was infused intravenously for 15 minutes before and during 70 minutes of ischemia, but not when it was infused immediately after ischemia.

Bock et al.[55] reexamined the protective effects of verapamil in renal ischemia induced by vascular clamping in rats. Verapamil was administered in doses of 0.1 and 1.0 mg/kg immediately before clamping the renal artery for 60 minutes. Compared with results in controls, both doses of verapamil failed to afford protection in terms of urine flow, plasma creatinine, creatinine clearance, or histologic examination 24 and 48 hours after ischemia. The authors concluded that they were unable to confirm earlier reports of a protective effect of verapamil in the rat model of ischemic renal failure.

The above-cited differences may be explained by the use of varying doses of calcium antagonists, different preischemic infusion periods (some of which lasted only 20 minutes), and various periods (some lasting only 15 minutes) of postischemic infusion.[53] None of these experiments evaluated the long-term application of calcium antagonists.

Hertle and Garthoff[56] examined the effects of nisoldipine on renal function after contralateral nephrectomy and 60 minutes of normothermic ischemia. Nisoldipine (300 ppm) was incorporated into a standard diet as well as administered 1 hour prior to ischemia (10 mg/kg orally). The investigators assessed the effects of nisoldipine on survival, serum urea, serum creatinine, urine volume, and creatinine clearance. Treatment with nisoldipine resulted in the survival of all animals (compared with 67% in the untreated group) and improved immediate and long-term renal function (14 days). In contrast, nisoldipine was not effective when administered only in the postischemic period.

Most investigations of the effects of calcium antagonists on experimental ARF have been carried out in anesthetized dogs or rats. To circumvent the confounding influence of anesthesia, Wagner et al.[57] investigated the effects of calcium antagonist treatment in a model of postischemic ARF in conscious dogs monitored by an implanted electromagnetic flow probe. The investigators reported that when diltiazem was administered before and after ischemia, both GFR and RBF were preserved. In contrast, when diltiazem was administered only after induction of ischemia, RBF was improved, but the effect on GFR was nonsignificant. The authors attributed the protective effect of diltiazem to a reversal of the vasoconstriction as well as a protective effect on the ischemic tubular cells.

Thalen et al.[58] used a model of renal artery occlusion to assess the protective effects of the dihydropyridine felodipine. In an initial set of experiments, they observed that after 30 or 60 minutes of renal artery occlusion, the subsequent recovery of renal function, as assessed by endogenous creatinine clearance, was greater in rats treated with felodipine (45 nmol/kg IV) during the occlusion period than in a corresponding group of rats treated with vehicle. The survival rate after 60 minutes of occlusion was 11% in the control rats but increased to 70% in the felodipine-treated rats.[58]

In a second series of experiments, acute renal damage was evaluated by assessing the extent of erythrocyte trapping in the kidney after 30 minutes of reperfusion following 60 minutes of renal artery occlusion.[58] Previous investigators have demonstrated that the degree of erythrocyte trapping, especially in the inner stripe, correlates with the severity of renal functional damage after ischemia, as assessed by GFR, urine flow rate, urinary concentrating ability, and RBF.[59,60] Consequently, Thalen et al.[58] considered differences of erythrocyte trapping as an index of the degree of protection afforded by felodipine. They observed that felodipine administration (45 nmol/kg) during the occlusion period attenuated renal damage compared with the damage sustained by the control animals, which received only the vehicle for felodipine. Kidney weight and hematocrit were also better maintained in the felodipine-treated rats. Of interest, renal damage was also reduced by the t-butyl analog of felodipine, H186/86, which is devoid of vasodilatory effects. Collectively, these results demonstrate that treatment with the calcium antagonist felodipine protects the kidney from ischemic/reperfusion injuries. They suggest that such tissue protection is not related to hemodynamic effects alone, because the hemodynamically inactive dihydropyridine H186/86 also reduced the extent of renal damage. The investigators speculated that the renal protective effects may be attributable in part to an additional antiperoxidant or scavenger-like effect inherent in the dihydropyridine molecule.[58]

Glycerol-induced Acute Renal Failure. Another experimental model used to assess the renal protective effects of calcium antagonists is glycerol-induced ARF. Lee et al.[61] compared the effects of diltiazem with the effects of an ACE inhibitor on the natural history of glycerol-induced ARF in rats. Animals pretreated with diltiazem for 3 days before the administration of glycerol developed a less severe renal failure syndrome than either the captopril-treated or control group. Treatment decreased the extent of tubular cell necrosis and was associated with rapid histologic and functional recovery.

Passive Heymann Nephritis. Passive Heymann nephritis (PHN) is an experimental model of in situ immune complex disease. Arai et al.[61a] studied the effects of manidipine administration in rats with PHN and demonstrated that manidipine treatment attenuated the increase in urinary albumin excretion and fractional clearance of albumin in PHN. They suggested that one of the mechanisms by which manidipine reduces proteinuria in PHN might be attributable to the decreased accumulation of lipid peroxidation products in the renal cortex.[61a]

Aminoglycoside Nephrotoxicity. In experimental aminoglycoside-induced ARF, binding of the aminoglycoside to the luminal brush border membrane of the proximal tubule has been postulated to alter the composition of membrane phospholipid, to increase permeability to calcium, and to impair mitochondrial function.[62] Because of such increased permeability to calcium, calcium antagonists might be expected to preserve cellular integrity and to protect against aminoglycoside-induced nephrotoxicity, as suggested by Eliahou et al.[63] Lee et al.[64,65] evaluated the protective effect of a calcium antagonist on gentamicin nephrotoxicity in the rat. Administration of gentamicin for 12 days caused a substantial decrease in both GFR and renal plasma flow (RPF). Concurrent treatment with the calcium antagonist nitrendipine promoted a significant increase in GFR but did not measurably influence renal blood flow. Urinary excretion of N-acetyl-glucosaminidase (NAG) and beta-glucosidase was measured to assess direct tubular cell toxicity resulting from administration of aminoglycoside. A progressive rise in the excretion of both enzymes began on day 3 and continued throughout the study. The administration of nitrendipine in conjunction with gentamicin almost completely abrogated the increased enzymuria. Although recognizing the obvious limitations of an isolated study in an animal model, the authors recommended that additional studies be carried out to evaluate the possible protective role of calcium antagonists in aminoglycoside nephrotoxicity.

Watson et al.[66] investigated the effects of verapamil on the development of renal insufficiency and calcium accumulation in renal tissue after administration of aminoglycoside. Sprague-Dawley rats were given gentamicin (120 mg/kg body weight/day) by daily subcutaneous injection for either 6 or 9 days. Subgroups of animals received verapamil added to the drinking water at a concentration of 10 mg/dl, beginning 2 days before the start of gentamicin administration. Control groups were treated with sham injections of dextrose/water for 6 or 9 days; for two of these groups, verapamil was added to drinking water. The investigators reported that the degree of functional damage and accumulation of cortical tissue calcium after 6 or 9 days of administration of gentamicin (120 mg/kg body weight/day) was not significantly different in rats whose drinking water contained verapamil (10 mg/dl) and control animals. The accumulation of tissue calcium correlated with the degree of reduction of creatinine clearance and probably reflected the extent of lethal tubular cell injury.

Additional studies are required to evaluate further the effects of calcium antagonists in experimental aminoglycoside nephrotoxicity and to resolve discrepant findings.

Protection Against Chronic Amphotericin B Nephrotoxicity. Amphotericin B is an antifungal antibiotic that currently is the agent of choice for certain otherwise uniformly fatal systemic fungal infections.[67,68] Unfortunately, chronic administration of amphotericin in humans has been associated with persistent decreases in GFR and RPF.[69] In rats, acute infusion of amphotericin also induces marked derangements in renal hemodynamics, consisting of intense renal vasoconstriction and diminished GFR.[70,71] Tolins and Raij observed that adverse renal hemodynamic effects persist during chronic daily administration of amphotericin in rats.[72] They also demonstrated that calcium channel blockade preceding acute amphotericin administration completely prevented renal vasoconstriction and blunted the fall in GFR induced by amphotericin.[70]

The investigators extended their observations to the chronic setting.[73] They hypothesized that during daily administration of amphotericin cotreatment with a calcium channel blocker may prevent amphotericin nephrotoxicity. Rats were given diltiazem (45 mg/kg 1 hour before and 1 hour after amphotericin) or vehicle by gastric tube and amphotericin B (5 mg/kg/day IP) for 10 days. Cotreatment with diltiazem ameliorated the amphotericin-induced rise in creatinine, fall in GFR rate, and fall in RPF. Additional studies are required to confirm and extend these encouraging preliminary findings.

Protection Against Radiocontrast-induced Acute Renal Failure. Several lines of evidence have demonstrated that the intrarenal administration of radiocontrast medium prolongs the vasoconstrictive response and reduces GFR.[74,75] In a preliminary study, Bakris and Burnett[76] investigated the effects of calcium entry modulation on the renal hemodynamic response to radiocontrast agents. They demonstrated that injection of radiocontrast solution (diatrizoate meglumine) in anesthetized dogs reduced mean GFR by 42% (Fig. 6). In contrast, this decrement was attenuated by the infusion of either verapamil (50 μg/min) or diltiazem (40 μg/min). Similarly, chelation with ethylene glycol tetraacetic acid (EGTA) (10 μmoles/min) attenuated the renal vasoconstriction. These hemodynamic responses were achieved at intrarenal doses that did not alter systemic arterial pressure or baseline RBF. The investigators concluded that calcium antagonism with several different agents or chelation of calcium with EGTA attenuates both the magnitude and duration of radiocontrast-mediated intrarenal vasoconstriction.[76] These preliminary studies suggest that calcium entry constitutes an important mediator in the vasoconstrictive phase that attends administration of radiocontrast medium.

Oliet et al.[77] extended these observations to humans. In a preliminary communication,[77] they reported the results of a small, randomized, double-blind study of the protective effect of nifedipine in patients undergoing renal arteriography. Patients treated with placebo showed a blunted natriuresis and diuresis and an increase in excretion of NAG. In contrast, patients treated with nifedipine sublingually did not develop these alterations in renal function.[77]

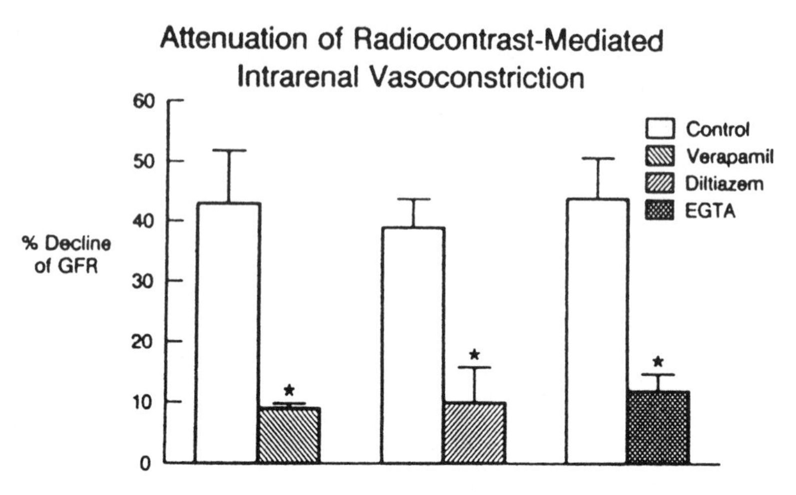

FIGURE 6. Effects of calcium entry modulation on glomerular filtration rate (GFR) in anesthetized dogs receiving radiocontrast solution (diatrizoate meglumine). Injection of radiocontrast reduced mean GFR by 42%. This decrement was attenuated by infusion of verapamil, diltiazem, or EGTA. * = $p < 0.05$ compared with control. (From data of Bakris GL, Burnett JC: A role for calcium in radiocontrast-induced reductions in renal hemodynamics. Kidney Int 27:465–468, 1985.)

The authors interpreted their data as suggesting that a calcium antagonist may ameliorate renal damage associated with radiocontrast agents.

Two full-length reports of randomized, prospective studies have extended these preliminary findings. Neumayer et al.[78] carried out a prospective, randomized, double-blind study assessing the effects of 3 days of nitrendipine treatment on radiocontrast-induced nephrotoxicity. They investigated the protective effects of pretreatment with a calcium antagonist and demonstrated that it confers protection against subsequent administration of radiocontrast media. Nineteen patients were relegated to the control group and 16 to the calcium antagonist group. All patients received a nonionic contrast medium, either intravenously or intraarterially, for the following radiographic examinations: computed tomography, renal arteriography, angiography of the lower extremities, or venography. Patients in the treatment group received nitrendipine, 20-mg orally starting 1 day before examination and continuing for 2 days thereafter. GFR was determined by inulin clearance on the day before, the day of, and 2 days following radiographic examination. In addition, urinary excretion of total protein was quantitated, and the following urinary enzyme excretory rates were determined: alanine-aminopeptidase (AAP), gammaglutamyltranspeptidase (gamma-GT), and N-acetyl-β-glucosamidase (β-NAG). These enzymes are of renal tubular cell origin and are considered by some investigators to be indicative of tubular dysfunction.[79]

In the treatment group, nitrendipine administration preserved GFR.[78] In contrast, control patients manifested a significant (27%) reduction in GFR on day 2 after injection of contrast medium ($p \geq 0.01$). Moreover, the increase in enzymuria of three different renal enzymes (gamma-GT, AAP, and β-NAG) and urinary protein excretion was significantly ameliorated by nitrendipine. The investigators interpreted their data as indicating that prophylactic administration of calcium antagonists ameliorates radiocontrast-induced renal dysfunction.

Russo et al.[80] investigated the effects of nifedipine on renal function in patients undergoing intravenous pyelography. Thirty male patients were randomly allocated to one of three experimental groups: (1) a high-osmolality radiocontrast agent (diatrizoate meglumine); (2) a low-osmolality medium (iopamidol); and (3) nifedipine plus a high-osmolality medium. The GFR and RPF were determined by means of simultaneous clearances of inulin and para-aminohippurate (PAH) before and at 30, 60, and 120 minutes after radiocontrast administration. Patients receiving diatrizoate without calcium antagonists showed a progressive decrease in GFR and an initial decrement in RPF. In contrast, GFR did not decrease in patients receiving diatrizoate plus nifedipine pretreatment. RPF did not decrease and indeed increased significantly at 30 and 60 minutes after radiocontrast administration. The authors concluded that treatment with a calcium antagonist prevented the anticipated decrements in renal hemodynamics induced by hyperosmolar radiocontrast medium.

In summary, the available data suggest that calcium antagonists protect, at least acutely, against the development of acute renal insufficiency in patients receiving radiocontrast agents. Additional controlled, prospective, randomized trials are warranted to substantiate and extend these findings. In an analogous manner, calcium antagonists have been demonstrated to exert salutary effects in protecting against other clinical syndromes of ARF.

ROLE IN TRANSPLANT-ASSOCIATED ACUTE RENAL FAILURE

Various investigators have clearly demonstrated that the prophylactic administration of calcium antagonists to donor kidneys ameliorates posttransplantation delayed graft function and renal insufficiency. Because the human kidney cannot sustain warm ischemic periods longer than 30–60 minutes without irreversible renal injury, renal hypothermia has been used widely to decrease the metabolic demands of the tissue and to diminish renal injury. Harvested human cadaver kidneys, flushed with a cold physiologic solution mimicking intracellular fluid, can be stored for up to 72 hours and still maintain adequate function after transplantation. Despite this attempt to minimize renal ischemia with hypothermia, a significant number of cold-stored cadaver kidneys undergo a period of reversible acute tubular necrosis (ATN) after transplantation. During the past decade, numerous investigations have been undertaken to study enhancement of hypothermic protection with the addition of calcium antagonists.

In a prospective, randomized trial, Wagner et al.[81] evaluated the influence of diltiazem on the development of ATN after cadaveric kidney transplantation. Diltiazem was added to Eurocollins solution perfusate (20 mg/L) at the time of donor nephrectomy. The graft recipient received a preoperative bolus injection of diltiazem, which was followed by maintenance diltiazem therapy. In the control group ($n = 22$), 9 patients (41%) developed ATN compared with 2 patients (10%) in the diltiazem group ($p < 0.05$). In the control group, 3.5 ± 0.4 hemodialyses per patient were necessary compared with 0.6 ± 0.2 in the diltiazem group ($p < 0.005$) (Fig. 7).

Neumayer and Wagner[82] carried out two prospective, randomized studies that confirmed their earlier findings and delineated the importance of donor pretreatment. They investigated the effects of differing treatment regimens of

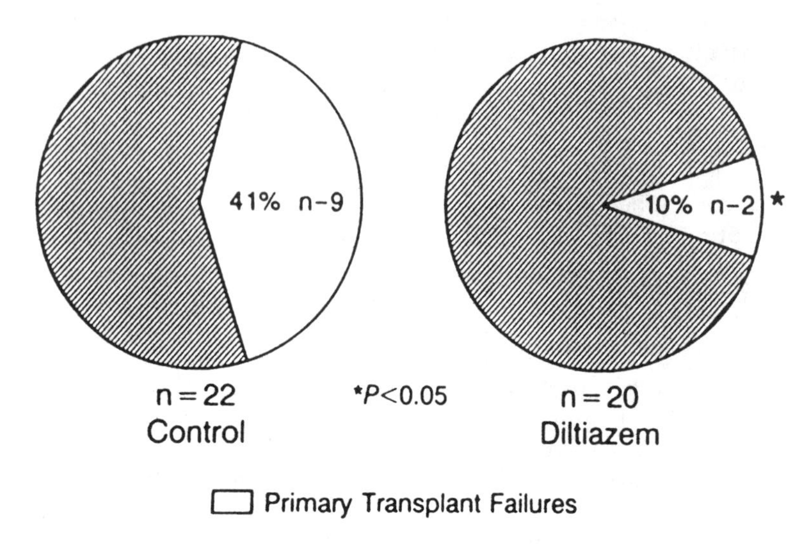

FIGURE 7. Effects of calcium antagonists on subsequent development of posttransplant acute renal insufficiency in 42 transplant recipients. Nine of 22 patients who did not receive diltiazem developed renal insufficiency *(left panel)*. Only 2 of the 20 patients treated with diltiazem developed renal insufficiency *(right panel)*. (Redrawn from data of Neumayer HH, Wagner K: Prevention of delayed graft function in cadaver kidney transplants by dilatiazem: Outcome of two prospective, randomized clinical trials. J Cardiovasc Pharmacol 10:S170–S177, 1987.)

diltiazem. In study 1, diltiazem was added to the Eurocollins solution used to perfuse the donor kidney, and treatment was continued in the recipients after transplantation. Patients in study 2 received a similar regimen but without donor pretreatment. The authors interpreted their data as indicating that combined treatment of donor and recipient with diltiazem lowers the incidence of delayed graft function in recipients treated with cyclosporine. In contrast to the salutary effects of combined treatment of donor and recipient (study 1), the authors concluded that posttransplant treatment with calcium antagonists of the recipient alone was less effective in protecting against postgraft failure.

Other investigators have confirmed the protective effects of calcium antagonists in patients undergoing renal transplantation. Frei et al.[83] randomized 110 recipients of cadaver kidneys to either a control group ($n = 56$) or a group receiving kidney grafts pretreated with diltiazem ($n = 54$). The treatment group received intravenous diltiazem 2 hours before surgery, followed by a continuous infusion up to 72 hours after surgery. Diltiazem was then continued orally until day 30, and if the patient required antihypertensive therapy, diltiazem was continued throughout the study period of 1 year. The investigators observed that initial graft nonfunction was significantly reduced by diltiazem (13% vs. 34%, $p < 0.01$). They could not, however, demonstrate a long-term effect on graft function.

Dawidson et al.[84,85] conducted several randomized clinical trials of the effects of calcium antagonists in conferring renal protection in cadaver kidney transplant recipients treated with cyclosporine. Their initial study[85] randomized 40 patients to evaluate prevention of cyclosporine nephrotoxicity with verapamil. Treatment with verapamil was initiated on day 3 before cyclosporine treatment was started and continued for 10 days. Duplex Doppler flow measurements were obtained daily, approximately 6 hours after the oral administration of cyclosporine. Verapamil promptly enhanced blood flow velocity, especially in patients in whom low flows were encountered initially. This finding was observed on the initial day of verapamil administration or on day 3 postoperatively (i.e., even before cyclosporine had been started).

Dawidson et al. extended these observations with an additional calcium antagonist, nifedipine. They retrospectively compared the clinical course of patients (n = 17) who received calcium antagonists (verapamil and nifedipine) during the first year after transplantation vs. patients (n = 24) who received no calcium antagonists.[86] They noted a lower incidence of both first and second rejection and improved graft function at 1 year after transplantation in the calcium antagonists-treated group. These results suggest that the short-term improvements in both graft function and rejection episodes with calcium antagonists seen in the earlier studies are maintained for the first year after transplantation if the drugs are continued. In addition, this study demonstrates that nifedipine, like verapamil and diltiazem, has beneficial effects.

In a later study, Dawidson et al.[87] expanded their observations with a prospective study of verapamil in renal transplant recipients. They randomized 59 patients to receive either verapamil for 14 days or no calcium antagonist therapy, along with conventional four-drug immunosuppression. Actuarial graft survival at 1 year was statistically different in the two groups. The verapamil-treated patients had a 93% rate of 1-year graft survival compared with 72% in the non–calcium antagonist-treated patients. These extended observations mirror the short-term observations of the earlier prospective study by Dawidson et al.[85]

The effects of calcium antagonist therapy on delayed graft function and graft survival are reviewed in chapter 25.

Protection Against Cyclosporine Nephrotoxicity

Cyclosporine A (CyA) is the immunosuppressant of choice in human organ transplantation because it improves graft survival and is not associated with myelosuppression. Unfortunately, CyA therapy may lead to a wide spectrum of nephrotoxic complications.[88,89] Nephrotoxicity remains the major hurdle in the wider application of this important drug.

Despite much study, the mechanisms by which cyclosporine injures the kidney remain incompletely defined. The nephrotoxicity of CyA was initially considered to be primarily a tubulotoxic effect, but recent studies suggest an important hemodynamic component. For example, cardiac allograft recipients receiving CyA have a twofold elevation in renal vascular resistance compared with those receiving azathioprine.[90] The results of several studies suggest that the renal vasoconstriction is reversible, with rapid improvement in renal function when CyA dosage is reduced.[88] In renal allograft recipients whose medication is changed from CyA to azathioprine, RBF (as assessed by the clearance of I^{131} orthoiodohippurate) increased, suggesting that the CyA-associated rise in renal vascular resistance was reversible even after 12 months of therapy.[91]

Within the past several years, an increasing body of evidence indicates that calcium antagonists exert protective actions against cyclosporine nephrotoxicity.[84,86,87,92] Although the exact mechanisms whereby calcium antagonists ameliorate cyclosporine nephrotoxicity are not completely defined, it is conceivable that they act by countervailing the effects of thromboxane and/or endothelin.[93,94] Cyclosporine augments thromboxane A_2, thereby inducing renal vasoconstriction.[93] Several studies from our laboratory demonstrate that calcium antagonists reverse both thromboxane-mediated and endothelin-mediated afferent arteriolar vasoconstriction.[95,96] These findings provide a scientific framework for anticipating a protective effect of calcium antagonists on CyA-induced renal vasoconstriction.

An additional mechanism whereby calcium antagonists confer protection may relate to their ability to attenuate endothelial dysfunction. We have recently examined the vasodilatory effects of acetylcholine (ACH) on norepinephrine (NE)-induced microvascular constriction in isolated perfused hydronephrotic rat kidneys pretreated with CyA.[97] The vasodilatory responses evoked by ACH were blunted in CyA rats. In control and CyA rats, the addition of nitro-L-arginine abolished afferent arteriolar (AA) and efferent arteriolar (EA) vasodilation induced by ACH, suggesting that the sustained vasodilatory responses of AA and EA to ACH depends mainly on nitric oxide synthesis. Our data indicate that ACH-induced vasodilation is decreased in renal microvessels of CyA rats, suggesting that nitric oxide synthesis is reduced in the renal vasculature of CyA rats. This reduction may contribute to the pathogenesis of CyA nephrotoxicity.

The findings to date about the protective effects of calcium antagonists against cyclosporine nephrotoxicity are encouraging (see chapter 25). Additional studies are required to confirm and extend preliminary observations that the reversal of cyclosporine-induced vasoconstriction is accompanied by an augmentation of GFR. Furthermore, we await confirmation of preliminary indications that chemically dissimilar calcium antagonists are equally protective and clearer delineation of the precise dosages of calcium antagonists required to confer such protection.

Interaction of Calcium Antagonists and Cyclosporine

The administration of calcium antagonists in the setting of concomitant cyclosporine therapy has additional important clinical implications. The calcium antagonists diltiazem, verapamil, and nicardipine have been shown to increase blood levels of cyclosporine in humans and consequently to reduce cyclosporine dose requirements.[98–104] This effect is related to the ability of the calcium antagonists to interfere with the cytochrome P450 enzyme system in the liver, which is critical to the degradation of cyclosporine.[104] Of interest, this interaction is not a class effect. Whereas most calcium antagonists increase CyA blood levels,[100,104] nifedipine does not share a similar capability to depress cyclosporine metabolism, largely because it is not concentrated in the liver to the same extent as other calcium antagonists.[104]

Although selected calcium antagonists augment blood cyclosporine levels, this augmentation does not necessarily increase nephrotoxicity.[84,98] From a theoretical standpoint, it has been argued that calcium antagonists (not nifedipine) may be advantageous in cyclosporine-treated patients. Thus, one may maintain or improve cyclosporine-based immunosuppression without increasing the dose of cyclosporine and still avoid cyclosporine-induced renal dysfunction (see chapter 25 for additional discussion).

Calcium Antagonists as Adjuvant Therapy in Cancer Chemotherapy

Calcium antagonists interact with a specific membrane protein (P-glycoprotein) that transports drugs, including cyclosporine and some cytotoxic drugs, across the cell membrane. This effect is the basis of the proposed role of calcium antagonists as adjuvant therapy in cancer chemotherapy.[105] At present little information about this application is available.

ROLE IN CHRONIC RENAL FAILURE

Effects in Experimental Chronic Renal Failure

During the past decade, major investigative interest has focused on the determinants of the progression of chronic renal disease. Chronic renal insufficiency, once established, tends to progress to end-stage renal failure. The mechanisms underlying the progression of renal disease have remained incompletely defined. Numerous studies have attempted to characterize the deleterious intrarenal processes that result from the initial insult and persist after its cause has disappeared.[106–110] In light of studies suggesting that calcium antagonists exert a protective effect in diverse models of experimental ARF, it is interesting to consider their effects in models of experimental chronic renal failure and clinical progressive renal insufficiency.

In animal experiments, the loss of functional renal mass is followed by a wide range of adaptive morphologic and functional changes. Examples include glomerular and tubular hypertrophy, increases in intraglomerular capillary pressure (glomerular hypertension), increased mesangial traffic of macromolecules, and proximal tubular hypermetabolism. Such changes have been implicated in the development of progressive renal scarring and insufficiency. Furthermore, the late deposition of calcium within scarred animal and human kidneys is also thought to be detrimental to the underlying scarring process.

Although their renal microcirculatory effects should not favor an attenuation of glomerular hypertension, calcium antagonists have additional properties that may contribute to their ability to afford renal protection under diverse experimental conditions and perhaps in clinical disorders.[2,4,10–29] Some of the more prominent mechanisms postulated to mediate the renal protective actions of calcium antagonists are listed in Table 1 and discussed on pages 2–3.

Several investigators have studied the effects of chronic administration of calcium antagonists in experimental chronic renal failure.[106–113] Goligorsky et al.[106] investigated the effect of chronic verapamil administration in the remnant kidney rat model. They examined calcium metabolism of the renal cortex and concomitantly assessed the morphologic criteria of nephrocalcinosis, abnormal tubular cell Ca^{2+} kinetics, and mitochondrial and morphologic changes in tubular basement membrane. At this early stage, however, changes in renal function had not yet occurred. Goligorsky et al. concluded that chronic verapamil administration ameliorates uremic nephrocalcinosis.

Harris et al.[107] investigated the effects of long-term administration of a calcium antagonist on the progression of experimental chronic renal failure in rats. The studied indices included degree of renal functional deterioration, extent of histologic damage and nephrocalcinosis, and cumulative survival. The investigators concluded that verapamil protects against renal dysfunction, histologic damage, nephrocalcinosis, and myocardial calcification and improves survival in the remnant model of chronic renal disease.[107]

Although Harris et al.[107] attributed the protective effects of verapamil to a reduction in the accumulation of calcium in renal tissue, their results did not exclude the possibility that verapamil also affords protection by affecting glomerular dynamics in remnant nephrons or that other mechanisms affect the course of progressive renal disease.[108–110] They extended their studies to the remnant kidney model by using micropuncture techniques to assess the effects of chronic administration of verapamil on glomerular capillary dynamics in rats with reduced renal mass.[115]

The data about calcium antagonists and progression of chronic renal disease are not consistent. As an example, Jackson and Johnston[112] compared the effects of enalapril and felodipine in Sprague Dawley rats after subtotal nephrectomy. Six weeks after surgery, plasma creatinine concentration was lower in the enalapril-treated group than in the felodipine-treated group. The glomerulosclerosis index (obtained from a histologic score) was reduced by enalapril but not by felodipine. The authors interpreted these results as suggesting that ACE inhibitors may have specific intrarenal effects that retard the deterioration of renal function that accompanies reduction of renal mass.

In a recent editorial, Bidani and Griffin[113] summarized the results of many of these trials and concluded that the key to ostensible discrepancies may lie in the interplay between the loss of intrarenal autoregulation, which occurs with the administration of calcium antagonists, and the magnitude of the reduction in blood pressure. Calcium antagonists decrease afferent arteriolar tone, allowing greater transmission of systemic blood pressure to the glomerular capillaries. In turn, this effect may be counterbalanced by the effect of calcium antagonists in lowering systemic blood pressure; thus, there is no net change or even reduction in the intraglomerular pressure.

Finally, recent studies have suggested that several of the newer calcium antagonists may differ in their renal microcirculatory profile and may be more favorable to the kidney. Tojo et al.[115a] characterized the effects of manidipine hydrochloride on renal microcirculation in SHR rats using classic micropuncture techniques. After two months of manidipine administration there were no changes in SNGFR, but SNG plasma flow was increased. Glomerular transcapillary hydraulic pressure difference (ΔP), which is the parameter most often associated with the development of glomerulosclerosis, was reduced significantly. Both afferent and efferent (R_A and R_E) arteriolar resistance were reduced. Consequently, manidipine might be a desirable antihypertensive agent for patients with renal disease because of its renal microcirculatory profile that could retard progression of renal disease.

In summary, several (but not all) recent studies suggest that calcium antagonists may protect against the deterioration of renal function when renal mass has been reduced either by surgical ablation of renal tissue or by primary renal disease. Additional studies are required to standardize the therapeutic regimens, including intervals between induction of the renal disease and initiation of calcium antagonist treatment.

Clinical Experience in Chronic Renal Failure

Despite these provocative observations, only recently have investigators attempted to extrapolate these findings to the clinical arena.[116–123]

In 1994, we reviewed the literature critically to ascertain whether treatment with calcium antagonists modifies progression of nondiabetic chronic renal disease in humans.[123] At that time, only a few long-term studies (i.e., 12-month or longer studies) were available, and they suggested that calcium antagonists may attenuate the decline in renal function of patients with chronic renal failure. The majority of available studies in humans, however, were nonrandomized, of too short duration, or confounded by investigative difficulties precluding definite conclusions as to whether calcium antagonists have renoprotective effects.

In a retrospective study, Brazy and Fitzwilliam[117] determined that calcium antagonists slowed the decline in renal function in patients with various chronic renal diseases. Calcium antagonists were found to be superior to other commonly used antihypertensive agents, despite equivalent reductions in blood pressure. Other investigators have used proteinuria as a surrogate endpoint for assessing progression and have failed to observe a beneficial effect of calcium antagonists on proteinuria in hypertensive patients with chronic renal failure.[119] In view of the emerging importance of calcium antagonists in the antihypertensive armamentarium, prospective randomized studies are required to delineate their effects on the progression of renal insufficiency.

This important clinical question has recently been addressed by several randomized prospective trials that documented a renal protective effect with dihydropyridine calcium antagonists.[120–122] Zucchelli et al.[120] reported the results of a prospective, randomized, controlled trial in which they compared the effects of the ACE inhibitor captopril and nifedipine on both hypertension and progression of renal insufficiency. During the 1-year prerandomization period, patients were treated with various standard antihypertensive regimens, that is, combinations that included beta blockers, furosemide, clonidine, and hydralazine. Subsequently, 121 hypertensive patients with chronic renal failure were randomly allocated to either captopril or slow-release nifedipine treatment for 3 years. The progression rate of renal insufficiency, assessed as 1/serum creatinine vs. time, creatinine clearance vs. time, and ^{99m}TC diethylenetriamine-penta-acetic acid clearance vs. time, was attenuated to a similar degree in both treatment groups. The investigators proposed that their data were "consistent with the hypothesis that both calcium antagonists and ACE inhibitors possess a renoprotective effect."

Various factors related to the experimental design and the conduct of the study, however, confound the certainty of such a conclusion. First, because the mean arterial pressure was higher during standard antihypertensive therapy, the better control of blood pressure may have contributed to the retardation of the progression rate of renal insufficiency by ACE inhibitors or calcium antagonists. Furthermore, a low-protein diet was instituted concomitantly with the initiation of the "standard antihypertensive therapy" phase of the study; observations from the MDRD study suggest that such an intervention is associated initially with an accelerated decline in GFR. Thus, dietary protein restriction may have exaggerated the decline in renal function during the initial year of study, thereby overestimating the differences between the standard therapy phase and the findings after randomization. Finally, the interpretation of the renal survival curves is fraught with difficulty. Of the 121 patients who underwent randomization at the end of the initial year, only 37 in the captopril group and 32 in the nifedipine group completed the 3-year study. In the third year, 11 of the remaining 46 patients in the nifedipine group and only 2 of the remaining 44 patients in the captopril group reached end-stage renal disease ($p < 0.005$). The reduction in statistical power may account for the absence of a difference in overall renal survival between the two groups. The resultant diminution in the number of patients during the final year of study confounds interpretation of the renal survival curves. Despite these reservations, the clinical course of the patients in group II (nifedipine) suggests that calcium antagonists may be equally protective and mandates additional long-term prospective studies that circumvent such concerns.

Velussi et al.[121] compared the effects of cilazapril and amlodipine on GFR and albumin excretion in hypertensive patients with non–insulin-dependent diabetes mellitus (NIDDM). Twenty-six patients were normoalbuminuric and 18 were microalbuminuric. GFR was measured by plasma clearance of ^{51}Cr EDTA at baseline and every 6-12 months during 3-year follow-up. As shown in Figure 8, the GFR decline (mean ± SE) per year in the normoalbuminuric patients during cilazapril therapy was similar to that observed during amlodipine

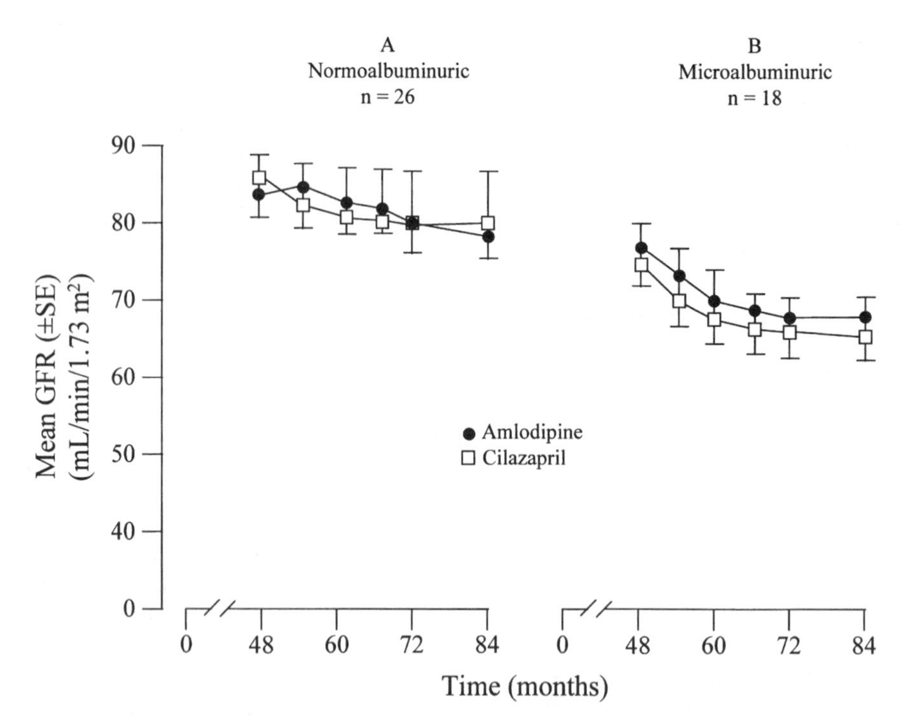

FIGURE 8. Comparison of the effects of cilazapril versus amlodipine on GFR and albumin excretion in hypertensive NIDDM patients studied for 3 years. Twenty-six patients were normoalbuminuric and 18 were microalbuminuric. The rate of decline of GFR measured by ^{51}Cr EDTA did not differ in the two groups. (Adapted from Velussi M, Brocco E, Frigato F, et al: Effects of cilazapril and amlodipine on kidney function in hypertensive NIDDM patients. Diabetes 45:216–222, 1996, with permission.)

therapy. Similarly, the GFR decline per year in the microalbuminuric patients during cilazapril therapy was similar to that observed with amlodipine. Cilazapril and amlodipine lowered the albumin excretion rate (AER) to a similar extent in normoalbuminuric and microalbuminuric patients.

In a carefully conducted recent study, Rossing et al.[122] compared the effects of the long-acting dihydropyridine nisoldipine with those of the ACE inhibitor lisinopril on proteinuria and decline in GFR in 49 hypertensive patients with insulin-dependent diabetes mellitus (IDDM). In a recent interim report of the results at the end of the first year of study, the authors reported a striking dissociation between antiproteinuric effects and effects on GFR. Albuminuria was reduced by 47% in the lisinopril group vs. no decrement in the nisoldipine group. In marked contrast, the decline in GFR appeared to be less steep in the nisoldipine group compared with the decline in the lisinopril group. These observations clearly demonstrate that a dihydropyridine, presumably acting through mechanisms independent of renal microcirculatory effects, is renoprotective. In summary, the available results are consistent with the hypothesis that ACE inhibitors and calcium antagonists act in a complementary manner to attenuate the progression of chronic renal failure.

The relatively few studies available for review are consistent with the assertion that ACE inhibitors and probably calcium antagonists attenuate the progression of chronic renal failure.

Do Chemically Dissimilar Calcium Antagonists Differ in Their Renal Effects?

In addition to the question of whether calcium antagonists as a class are renoprotective, interest has focused on the question of whether there are differences within the class (e.g., dihydropyridine versus nondihydropyridine calcium antagonists.)

Of interest, studies conducted in patients with diabetes mellitus have served as the clinical forum for recent suggestions that chemically dissimilar calcium antagonists may have different renal effects. Evidence from two recent meta-analyses suggests that nondihydropyridine calcium antagonists may have a greater renoprotective effect than dihydropyridines. De Zeeuw[131] conducted a metaanalysis of studies to ascertain whether ACE inhibitors differ from other antihypertensive agents in efficacy in lowering proteinuria. Included were 41 studies consisting of 1124 diabetic and nondiabetic patients, of whom 558 had nondiabetic renal disease. The authors concluded that ACE inhibitors conferred an antiproteinuric effect beyond that attributable to their blood pressure lowering effect. Of interest, differences in effect were found among the calcium antagonists; nifedipine had the least effect.

Weidmann et al.,[132] in an extension of their previous metaanalyses of antihypertensive therapy in diabetic nephropathy, reevaluated published studies in diabetic patients with microalbuminuria or clinical proteinuria treated for > 4 weeks with conventional therapy (diuretics, beta blockers, or both), ACE inhibitors, nifedipine, or other calcium antagonists. Greater decreases in proteinuria were observed in patients given ACE inhibitors than in patients given conventional therapy; proteinuria was unchanged with nifedipine treatment. Why nifedipine did not exert an antiproteinuric effect and why the rate of decline in GFR in the nifedipine-treated group exceeded the rate in the other groups were not established. Clearly, rigorous prospective, comparative studies are needed to corroborate differences among calcium antagonists and the mechanisms mediating such differences.

More recently Maki et al.[133] conducted a metaanalysis of investigations with follow-up times of at least 6 months to assess the effects of different antihypertensive agents on proteinuria and GFR in patients with kidney disease. In contrast to their earlier metaanalysis, they included studies in diabetic and nondiabetic patients with renal disease. When the results of 14 randomized controlled trials were examined, ACE inhibitors caused a greater decrease in proteinuria, improvement in GFR, and decline in mean arterial pressure compared with controls. In a multivariate analysis of controlled and uncontrolled trials, each 10 mm Hg reduction in blood pressure decreased proteinuria, but ACE inhibitors and nondihydropyridine calcium antagonists were associated with additional declines in proteinuria that were independent of blood pressure changes and diabetes. Each 10-mmHg reduction in blood pressure caused a relative improvement in glomerular filtration rate (0.18 ml/min per month), but among diabetic patients there was a tendency for dihydropyridine calcium antagonists to cause a relative reduction in GFR (-0.68 ml/min per month).

Maki et al.[133] emphasized that the results of their retrospective analysis should be interpreted with caution and proposed that direct long-term comparison studies are needed to determine whether there are class differences between ACE inhibitors and calcium antagonists, and whether there are specific differences within the classes (e.g., dihydropyridine versus nondihydropyridine calcium angatonists).

Only one investigative group has demonstrated such findings in randomized prospective studies. Lash and Bakris[134] have recently conducted a 4-year follow-up study in patients with nephropathy from type II diabetes. They observed that therapy with lisinopril and verapamil reduces nephrotic-range proteinuria. Of interest, the combination of the two agents induced a greater effect than either agent alone.[134] This benefit occurred in the absence of additional antihypertensive effects of the combination.

Are the Renal Vasodilatory Effects of Calcium Antagonists Detrimental?

The demonstrated ability of calcium antagonists to dilate preferentially the afferent arteriole has raised questions as to whether they may be detrimental in the long term. Theoretical considerations suggest that vasoactive agents that preferentially reduce the resistance of the afferent arteriole should favor an increase in P_{GC}. Whether this increase actually occurs in either normal humans or patients with chronic renal failure (in analogy with the remnant rat kidney model) has not been established. At least one factor, systemic blood pressure, must be taken into account. Even in the face of afferent arteriolar vasodilatation, the P_{GC} depends on the net effect of systemic and renal microvascular influences. Thus, a concomitant reduction in MAP would result in a negligible increase in P_{GC}.[124]

Despite fears of a theoretical increase in adverse renal effects, preliminary reports indicate that long-term (> 4 years) calcium antagonist therapy in patients with chronic renal failure is associated with well-maintained renal function.

Clearly, additional studies are required to determine the mechanisms whereby calcium antagonists fail to accelerate renal damage despite the potential for persistent glomerular capillary hypertension. As discussed previously, it is becoming increasingly apparent that several factors in addition to elevated hydrostatic pressure may mediate progressive glomerular injury.[113] An attractive hypothesis, therefore, is that the potential adverse effects of increased P_{GC} are countered by concomitant salutary effects of calcium antagonists, as enumerated elsewhere (see chapter 23).

ROLE IN DIABETIC NEPHROPATHY

On the basis of several lines of clinical and experimental evidence, ACE inhibitors have been advocated as advantageous in patients with diabetic nephropathy.[125–130] On the other hand, investigation of the benefit of calcium antagonists in this setting is at an earlier stage.

Most available clinical trials have assessed the effects of ACE inhibitors in patients with IDDM. The largest and most compelling of these trials was conducted by the Diabetes Collaborative Study Group,[127] which reported the results of a trial designed to determine whether the ACE inhibitor captopril is more effective in slowing the progression of diabetic nephropathy than agents that act primarily by reducing blood pressure. This randomized, controlled trial compared captopril with placebo in patients with IDDM who had urinary protein excretion > 500 mg/day and serum creatinine concentrations > 2.5 mg/dl. The primary end point was doubling of the baseline serum creatinine concentrations. Serum creatinine concentrations doubled in 25 patients in the captopril group and 43 patients in the conventional therapy group. Captopril therapy was associated with a 50% reduction in risk for the combined end points of death, dialysis, and transplantation. This reduction was independent of the small disparity in blood pressure between the groups.

Several studies also have documented a renoprotective effect of dihydropyridines. Velussi et al.[121] compared the effects of cilazapril and amlodipine on GFR and AER in normoalbuminuric and microalbuminuric hypertensive patients with NIDDM. As detailed above, they showed that both drugs caused a similar decline of GFR and AER in the two patient groups (see Fig. 8). The comparison of nisoldipine and lisinopril by Rossing et al.[122] (see page 455) in patients with IDDM clearly supports the hypothesis that ACE inhibitors and calcium antagonists act in a complementary manner to attenuate the progression of diabetic nephropathy. The other available studies, although relatively few, are consistent with this theory.

As detailed in a recent editorial,[135] a number of recent long-term prospective trials have suggested that calcium antagonists may also be efficacious in retarding progression in diabetic nephropathy. The reader must consider a number of crucial issues when reviewing the growing literature related to diabetic nephropathy. It is unclear whether the antiproteinuric effects of ACE inhibitors or calcium antagonists are related to intrarenal hemodynamic effects, changes in glomerular permselectivity, or both. The extent to which a decrease in proteinuria constitutes a surrogate endpoint for renal protection remains to be established. Clearly, additional prospective studies are required to determine the long-term effect of calcium antagonists and, to a lesser extent, of ACE inhibitors on the natural course of decline of GFR (and, one hopes, on the progression of anatomic abnormalities) in diabetes mellitus (see chapters 14 and 15).

EFFECTS OF CALCIUM ANTAGONISTS ON RENAL FUNCTION IN PATIENTS WITH RENOVASCULAR HYPERTENSION

Theoretically calcium antagonists may influence renal function in patients with renovascular hypertension. Within the past several years, increasing attention has been devoted to the potential adverse effects of ACE inhibitors on renal function in susceptible patients. Specifically, it has been shown that ACE inhibitors may induce acute reversible renal insufficiency in patients with bilateral renal artery stenosis, stenosis of a solitary kidney, or class III or IV congestive heart failure.[136–137] It is believed that ACE inhibitors mediate renal insufficiency by attenuating angiotensin II-mediated efferent tone. Whereas ACE inhibitors and angiotensin II antagonists block both renal afferent and efferent arteriolar actions of angiotensin II,[138] calcium antagonists reverse only the afferent effects.[4–6] These observations provide a theoretical framework favoring the use of calcium antagonists in patients with renovascular hypertension.

Ribstein et al.[139] compared the effects of nifedipine and captopril in 6 patients with bilateral renal artery stenosis and 4 patients with renal artery stenosis in a solitary kidney (Fig. 9). Patients were challenged with an acute

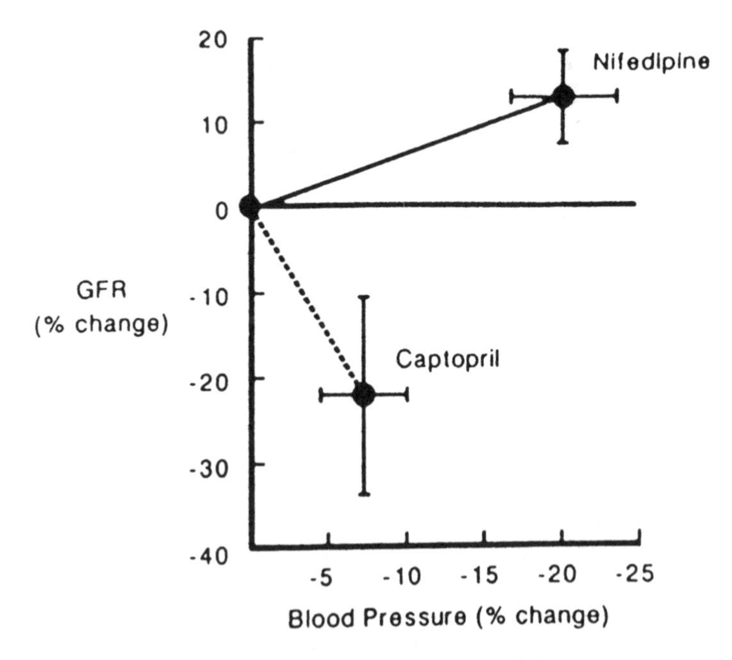

FIGURE 9. Comparison of the acute effects of captopril and nifedipine on GFR in patients with bilateral renal artery stensois (n = 6) or with renal arterial stenosis of a solitary kidney (n = 4). In contrast to the reduction of GFR induced by captopril, GFR tended to increase with nifedipine. (Redrawn from data of Ribstein J, Mourad G, Mimran A: Contrasting acute effects of captopril and nifedipine on renal function in renovascular hypertension. Am J Hypertens 1:239–244, 1988.)

administration of nifedipine, 20 mg, and subsequently with the acute administration of captopril, 50 mg. Nifedipine induced a decline in MAP of 19 ± 4%. The GFR did not decrease but rather increased slightly (+13 ± 6%). In contrast, acute administration of captopril induced a fall in GFR (-23 ± 11%) despite a lesser decrement in MAP. The authors proposed that the relatively beneficial effect of calcium antagonists on GFR was attributable to preferential afferent arteriolar vasodilation. Regardless of mechanism, these preliminary findings suggest that patients with renal artery stenosis are not at risk for developing acute renal insufficiency when treated with calcium antagonists.

Miyamori et al.[140] compared the effects of captopril and nifedipine on split renal function in 6 patients with renovascular hypertension secondary to unilateral renal artery stenosis. Patients received either captopril or nifedipine. The GFR in the stenotic kidney decreased markedly during captopril administration (from 24 ± 6 SE to 11 ± 2 ml/min, $p < 0.01$), whereas it decreased only slightly during nifedipine administration (to 19 ± 5 ml/min). Effective RPF increased in the nonstenotic kidneys in response to both drugs.

In summary, because of their preferential vasodilatory effect on the preglomerular arterioles, calcium antagonists may be preferred as antihypertensive agents in patients with renovascular hypertension.

CONCLUSION

Calcium antagonists exert several effects on renal hemodynamics and excretory function that probably contribute to their renoprotective effects:

Role as Prophylactic Agents Against the Development of Acute Renal Failure

1. Calcium antagonists preferentially attenuate afferent arteriolar vasoconstriction, suggesting that they may benefit patients at increased risk for developing acute renal failure. Specifically, several studies indicate that calcium antagonists protect against the development of acute renal failure immediately after cadaveric renal transplantation.
2. Several recent studies have indicated that pretreatment with calcium antagonists attenuates the development of enzymuria or acute azotemia after administration of radiocontrast agents.
3. Calcium antagonists mitigate acute cyclosporine-induced renal vasoconstriction (and presumably nephrotoxicity).

Additional clinical investigation is required to delineate other settings in which calcium antagonists may have an acute renoprotective effect, such as in amphotericin and aminoglycoside nephrotoxicity.

Role in Retarding Progression of Chronic Renal Failure

In addition to acute renal prophylaxis, it is becoming increasingly apparent that calcium antagonists also may retard the progression of chronic renal disease. Although their renal microcirculatory effects should not favor an attenuation of glomerular hypertension, calcium antagonists have additional properties that may contribute to their ability to provide renal protection. These putative mechanisms include the ability to retard renal growth, possibly to lessen the mesangial entrapment of macromolecules, and to mitigate the mitogenic effects of diverse growth and inflammatory mediators. Several clinical trials have demonstrated a unique renal protective effect of ACE inhibitors over and above that attributable to lowering of blood pressure. Of interest, several recent prospective, randomized trials have suggested that calcium antagonists and ACE inhibitors may be equally protective. Collectively, the available results are consistent with the formulation that ACE inhibitors and probably calcium antagonists attenuate the progression of chronic renal failure.

Acknowledgments

The author is indebted to Elsa V. Reina for her expert preparation of the manuscript. Portions of this chapter are adapted with permission from Epstein M, Loutzenhiser R (eds): Calcium Antagonists and the Kidney. Philadelphia, Hanley & Belfus, 1990.

References

1. Heidland A, Klutsch K, Obek A: Myogenbedingte vasodilatation bei nierenischamie. Munch Med Wochenschr 35:1636, 1962.
2. Loutzenhiser R, Epstein M: Effects of calcium antagonists on renal hemodynamics. Am J Physiol 249:F619–F629, 1985.
3. Loutzenhiser R, Epstein M: Renal hemodynamic effects of calcium antagonists. In Epstein M, Loutzenhiser R (eds): Calcium Antagonists and the Kidney. Philadelphia, Hanley & Belfus, 1990, pp 33–74.
4. Epstein M, Loutzenhiser R: Potential applicability of calcium antagonists as renal protective agents. In Epstein M, Loutzenhiser R(eds): Calcium Antagonists and the Kidney. Philadelphia, Hanley & Belfus, 1990, pp 275–298.
5. Loutzenhiser R, Epstein M: Renal microvascular actions of calcium antagonists. J Am Soc Nephrol 1:S3–S12, 1990.
6. Epstein M: Calcium antagonists and renal protection: Current status and future perspectives. Arch Intern Med 152:1573–1584, 1992.
7. Reams GP, Bauer JH: Acute and chronic effects of calcium antagonists on the essential hypertensive kidney. In Epstein M, Loutzenhiser R (eds): Calcium Antagonists and the Kidney. Philadelphia, Hanley & Belfus, 1990, pp 247–256.
8. Blackshear JL, Garnic D, Williams GH, Harrington DP, Hollenberg, NK: Exaggerated renal vasodilator response to calcium entry blockade in first-degree relatives of essential hypertensive subjects. Hypertension 9:384–389, 1987.
9. Montanari A, Vallisa D, Ragni G, et al: Abnormal renal responses to calcium entry blockade in normotensive offspring of hypertensive parents. Hypertension 12:498–505, 1988.
10. Dworkin LD: Effects of calcium channel blockers on experimental glomerular injury. J Am Soc Nephrol l:S21–S27, 1990.
11. Dworkin LD: Impact of calcium entry blockers on glomerular injury in experimental hypertension. Cardiovasc Drugs Ther 4:1325–1330, 1990.
12. Sweeney C, Shultz P, Raij L: Interactions of the endothelium and mesangium in glomerular injury. J Am Soc Nephrol 1:S13–S20, 1990.
13. Raij L, Keane W: Glomerular mesangium: Its function and relationship to angiotensin II. Am J Med 79(Suppl 36):24–30, 1985.
14. Dworkin LD, Benstein JA: Antihypertensive agents, glomerular hemodynamics and glomerular injury. In Epstein M, Loutzenhiser R (eds): Calcium Antagonists and the Kidney. Philadelphia, Hanley & Belfus, 1990, pp 155–176.
15. Dworkin LD, Levin RI, Benstein JA, et al: Effects of nifedipine and enalapril on glomerular injury in rats with deoxycorticosterone-salt hypertension. Am J Physiol 259:F598–F604, 1990.
16. Schor N, Ichikawa I, Brenner BM: Mechanisms of action of various hormones and vasoactive substances on glomerular ultrafiltration in the rat. Kidney Int 20:242–251, 1981.
17. Keane WJ, Raij L: Relationship among altered glomerular barrier permselectivity, angiotensin II and mesangial uptake of macromolecules. Lab Invest 52:599–604, 1985.

18. Sweeney CJ, Raij L: Interactions of the endothelium and mesangium in glomerular injury: Potential role of calcium antagonists. In Epstein M (ed): Calcium Antagonists in Clinical Medicine, 1st ed. Philadelphia, Hanley & Belfus, 1992, pp 413–426.
19. Egido J: Vasoactive hormones and renal sclerosis. Kidney Int 49:578–597, 1996.
20. Shultz PJ, Knauss T, Mene P, Abboud HE: Mitogenic signals for thrombin in mesangial cells: Regulation of phospholipase C and PDGF genes. Am J Physiol 257:F366–F374, 1989.
21. Shultz P, Raij L: Inhibition of human mesangial cell proliferation by calcium blockers. Hypertension 15(Suppl I):76–80, 1990.
22. Tolins JP, Melemed A, Sulciner D, et al: Calcium channel blockade inhibits platelet activating factor production by human umbilical vein endothelial cells. Lipids 26:1218–1222, 1991.
23. Weiss SJ, Young L, LoBuglio AF, et al: Role of hydrogen peroxide in neutrophil-mediated destruction of cultured endothelial cells. J Clin Invest 68:714–721, 1981.
24. Diamond JR, Bonventre JV, Karnovsky MJ: A role for oxygen free radicals in amino nucleoside nephrosis. Kidney Int 29:478–483, 1986.
25. Paller MS, Hoidal JR, Ferris TE: Oxygen free radicals in ischemic acute renal failure in the rat. J Clin Invest 74:1156–1164, 1984.
26. Shah SV: Role of reactive oxygen metabolites in experimental glomerular disease. Kidney Int 35:1093–1106, 1989.
27. Burke TJ, Singh H, Schrier RW: Calcium handling by renal tubules during oxygen deprivation injury to the kidney prior to reoxygenation. J Cardiovasc Drugs Ther 4:1319–1324, 1990.
28. Schrier RW, Burke TJ: Calcium channel-blockers in experimental and human acute renal failure. Adv Nephrol 17:287–300, 1988.
29. Schwertschlag U, Schrier RW, Wilson P: Beneficial effects of calcium channel blockers and calmodulin binding drugs on in vitro renal cell anoxia. J Pharmacol Exp Ther 238:119–124, 1986.
30. Luft FC, Weinberger MH: Calcium antagonists and renal sodium homeostasis. In Epstein M, Loutzenhiser R (eds): Calcium Antagonists and the Kidney. Philadelphia, Hanley & Belfus, 1990, pp 203–212.
31. Roy MW, Guthrie GP, Holladay FP, Kotchen TA: Effects of verapamil on renin and aldosterone in the dog and rat. Am J Physiol 245:E410–E416, 1983.
32. Bell AJ, Linder A: Effects of verapamil and nifedipine on renal function and hemodynamics in the dog. Renal Physiol 7:329–343, 1984.
33. DiBona GF, Sawin LL: Renal tubular site of action of felodipine. J Pharmacol Exp Ther 228: 420–424, 1984.
34. Epstein M, De Micheli AG, Forster H: Natriuretic effects of calcium antagonists in humans: A review of experimental evidence and clinical data. Cardiovasc Drug Rev 9:399–413, 1991.
35. Thananopavarn C, Golub H, Eggena P, et al: Renal effects of nitrendipine monotherapy in essential hypertension. J Cardiovasc Pharmacol 6(Suppl):125–128, 1985.
36. Luft FC, Aronoff GR, Sloan RS, et al: Calcium channel blockade with nitrendipine: effects on sodium homeostasis, the renin-angiotensin system, and the sympathetic nervous system in humans. Hypertension 7:438–442, 1985.
37. Freis ED, Wanko A, Wilson IM, Parish AE: Chlorothiazide in hypertensive and normotensive patients. Ann NY Acad Sci 71:450–455, 1958.
38. Ruilope LM, Miranda B, Rodicio JL: Characterization of the long-term natriuretic effects of calcium antagonists. In Epstein M, Loutzenhiser R (eds): Calcium Antagonists and the Kidney. Philadelphia, Hanley & Belfus 1990, pp 225–232.
39. Epstein M: Renal effects of head-out water immersion in man: Implications for an understanding of volume homeostasis. Physiol Rev 58:529–581, 1978.
40. Epstein M, Pins DS, Sancho J, Haber E: Suppression of plasma renin and plasma aldosterone during water immersion in normal man. J Clin Endocrinol Metab 41:618–625, 1975.
41. Epstein M, Preston S, Weitzman RE: Iso-osmotic central blood volume expansion suppresses plasma arginine vasopressin in normal man. J Clin Endocrinol Metab 52:256–262, 1981.
42. Epstein M: Renal effects of head-out water immersion in humans: A 15 year update. Physiol Rev 72:563–619, 1992.
43. Epstein M, Duncan D, Fishman LM: Characterization of the natriuresis caused in normal man by immersion in water. Clin Sci 43:275–287, 1972.
44. Epstein M, Katsikas JL, Duncan DC: Role of mineralocorticoids in the natriuresis of water immersion in normal man. Circ Res 32:228–236, 1973.
45. Krishna GG, Riley LJ Jr, Deuter G, et al: Natriuretic effect of calcium-channel blockers in hypertensives. Am J Kidney Dis 5:566–572, 1991.
46. Luft FC, Aronoff GR, Fineberg NS, Weinberger MH: Effects of oral calcium, potassium, digoxin, and nifedipine on natriuresis in normals humans. Am J Hypertens 2:14–19, 1989.

47. MacGregor GA, Prevahouse JB, Cappuccio FP, Markandu ND: Nifedipine, diuretics and sodium balance. J Hypertens 5(Suppl 4):S127–S131, 1987.
48. Ribstein J, Codis P, duCailar G, Mimran A: Effects of captopril and nitrendipine on the response to acute volume expansion in essential hypertension. J Hypertens 6(Suppl 4):S473– S475, 1988.
49. Epstein M: Calcium antagonists and the kidney: Implications for renal protection. Kidney Int 41 (Suppl 36):S66–S72, 1992.
50. Loutzenhiser R, Epstein M: Calcium antagonists and the kidney. Am J Hypertens 2:154S–161S, 1989.
51. Puschett JB: Calcium antagonists and renal ischemia. In Epstein M, Loutzenhiser R (eds): Calcium Antagonists and the Kidney. Philadelphia, Hanley & Belfus, 1990, pp 177–185.
52. Michael U, Lee SM: The role of calcium antagonists in nephrotoxic models of renal failure. In Epstein M (ed): Calcium Antagonists and the Kidney. Philadelphia, Hanley & Belfus, 1990, pp 187–201.
53. Malis CD, Cheung JY, Leaf A, et al: Effects of verapamil in models of ischemic acute renal failure in rats.Am J Physiol 243:S735–S742, 1983.
54. Goldfarb D, Iaina A, Serban I, et al: Beneficial effects of verapamil in ischemic acute renal failure in the rat. Proc Soc Exp Biol Med 172:389–392, 1983.
55. Bock AH, Brunner FP, Torhorst J, Thiel G: Failure of verapamil to protect from ischaemic renal damage. Nephron 57:299–305, 1991.
56. Hertle L, Garthoff B: Calcium channel blocker nisoldipine limits ischemic damage in rat kidney. J Urol 134:1251–1254, 1985.
57. Wagner K, Schultze G, Molzahn M, Neumayer HH: The influence of long-term infusion of the calcium antagonist diltiazem on postischemic acute renal failure in conscious dogs. Klin Wochenschr 64:135–140, 1986.
58. Thalen PG, Nordlander MIL, Sohtell MEH, Svensson LET: Attenuation of renal ischaemic injury by felodipine. NaunynSchmiedebergs Arch Pharmacol 343:411–417, 1991.
59. Jacobsson J, Odlind B, Tufveson G, Wahlberg J: Effects of cold ischemia and reperfusion on trapping of erythrocytes in the rat kidney. Transplant Int 1:75–79, 1988.
60. Mason J, Torhorst J, Welsch J: Role of the medullary perfusion defect in the pathogenesis of ischemic renal failure. Kidney Int 26:283–293, 1984.
61. Lee SM, Hillman BJ, Clark RL, Michael UF: The effects of diltiazem and captoril on glycerol-induced acute renal failure in the rat: Functional, pathologic, and microangiographic studies. Invest Radiol 20:961–970, 1985.
61a. Arai T, Kobayashi S, Nakajima T, Hishida A: Reduction of proteinuria by a calcium antagonist, manidipine, in rats with passive Heymann nephritis. Int Congr Ser Excerta Med, Diuretic IV:647–650, 1993.
62. Mergner WJ, Smith MW, Sahaphong S, et al: Studies on the pathogenesis of ischemic cell injury. VI: Accumulation of calcium by isolated mitochondria in ischemic rat kidney cortex. Virchows Arch (Cell Pathol) 26:1–16, 1977.
63. Eliahou H, Iaina A, Serban S: Verapamil's beneficial effect and cyclic nucleotides in gentamicin-induced acute renal failure in rats. Proceeding of the IXth International Congress of Nephrology, 1984, p 323A.
64. Lee SM, Pattison ME, Michael UF: Nitrendipine protects against aminoglycoside nephrotoxicity in the rat. J Cardiovasc Pharmacol 9(Suppl 1):S65–S69, 1987.
65. Lee SM, Michael UF: The protective effect of nitrendipine on gentamicin acute renal failure in rats. Exp Mol Pathol 43:107–114, 1985.
66. Watson AJ, Gimenez LF, Klassen DK, et al: Calcium channel blockade in experimental aminoglycoside nephrotoxicity. J Clin Pharmacol 27:625–627, 1987.
67. Utz JP: Chemotherapy of the systemic mycoses. Med Clin North Am 66:221–234, 1982.
68. Utz JP, Bennett JE, Brandiss MD, et al: Amphotericin B toxicity. Ann Intern Med 61:334–354, 1964.
69. Douglas JB, Healy JK: Nephrotoxic effects of amphotericin B, including renal tubular acidosis. Am J Med 46:154–162, 1969.
70. Tolins JP, Raij L: Adverse effects of amphotericin B administration on renal hemodynamics in the rat. Neurohumoral mechanisms and influence of calcium channel blockade. J Pharmacol Exp Ther 245:594–599, 1988.
71. Cheng JT, Witty RT, Robinson RR, Yager WE: Amphotericin B and nephrotoxicity: Increased renal resistance and tubule permeability. Kidney Int 22:626–633, 1982.
72. Tolins JP, Raij L: Chronic amphotericin B nephrotoxicity in the rat: Protective effect of prophylactic salt loading. Am J Kid Dis ll:313–317, 1988.
73. Tolins JP, Raij L: Chronic amphotericin B nephrotoxicity in the rat: Protective effect of calcium channel blockade. J Am Soc Nephrol 2:98–102, 1991.

74. Talner LB, Davidson AJ: Renal hemodynamic effects of contrast media. Invest Radiol 1968;2:310–317, 1968.
75. Caldicott WJH, Hollenberg NK, Abrams HS: Characteristics of response of renal vascular bed to contrast media. Invest Radiol 15:539–547, 1970.
76. Bakris GL, Burnett JC: A role for calcium in radiocontrast-induced reductions in renal hemodynamics. Kidney Int 27:465–468, 1985.
77. Oliet A, Lumbreras C, Mateo S, et al: Calcium channel blockade minimizes the renal toxicity of radiocontrast agents. J Cardiovasc Pharmacol 12(Suppl 6):S164, 1988.
78. Neumayer HH, Junge W, Kufner A, Wenning A: Prevention of radiocontrast media-induced nephrotoxicity by the calcium channel blocker nitrendipine: A prospective randomized clinical trial. Nephrol Dial Transplant 4:1030–1036, 1989.
79. Kunin CM, Chesney RW, Craig WA, et al: Enzymuria as a marker of renal injury and disease: Studies of N-acetyl-β-glucosaminidase in the general population and in patients with renal disease. Pediatrics 62:751–760, 1978.
80. Russo D, Testa A, Della Volpe L, Sansone G: Randomized prospective study on renal effects of two different contrast media in humans: Protective role of calcium channel blocker. Nephron 55:254–257, 1990.
81. Wagner K, Albrecht S, Neumayer H: Prevention of posttransplant acute tubular necrosis by the calcium antagonist diltiazem: A prospective randomized study. Am J Nephrol 7:287–291, 1987.
82. Neumayer HH, Wagner K: Prevention of delayed graft function in cadaver kidney transplants by diltiazem: Outcome of two prospective, randomized clinical trials. J Cardiovasc Pharmacol l0:S170–S177, 1987.
83. Frei U, Margreiter R, Harms A, et al: Preoperative graft reperfusion with a calcium antagonist improves initial function: Preliminary results of a prospective randomized trial in 110 kidney recipients. Transplant Proc 19:3539–3541, 1987.
84. Dawidson I, Rooth P: Effects of calcium antagonists in ameliorating cyclosporine A nephrotoxicity and post-transplant ATN. In Epstein M, Loutzenhiser R (eds): Calcium Antagonists and the Kidney. Philadelphia, Hanley & Belfus, 1990, pp 233–246.
85. Dawidson I, Rooth P, Fry WR, et al: Prevention of acute cyclosporine A-induced renal blood flow inhibition and improved immunusupression with verapamil. Transplantation 48:575–580, 1989.
86. Palmer BF, Dawidson J, Sagalowsky A, et al: Improved outcome of cadaveric renal transplantation due to calcium channel blockers. Transplantation 52:640–645, 1991.
87. Dawidson I, Rooth P, Lu C, et al: Verapamil improves the outcome after cadaver renal transplantation. J Am Soc Nephrol 2:983–990, 1991.
88. Myers BD: Cyclosporine nephrotoxicity. Kidney Int 30:964–974, 1986.
89. Bennett WM, DeMattos A, Meyer MM, et al: Chronic cyclosporine nephropathy: The Achilles' heel of immunosuppressive therapy. Kidney Int 50:1089–1100, 1996.
90. Myers BD, Ross J, Newton L, et al: Cyclosporine-associated chronic nephropathy. N Engl J Med 311:699–705, 1984.
91. Curtis JJ, Luke RG, Dubovsky E, et al: Cyclosporine in therapeutic doses increases renal allograft vascular resistance, Lancet 2:477, 1986.
92. Rooth P, Dawidson I, Diller K, Taljedal I-B: Beneficial effects of calcium antagonist pretreatment and albumin infusion on cyclosporine A-induced impairment of kidney microcirculation in mice. Transplant Proc l9:3602–3605, 1987.
93. Perico N, Benigni A, Zoa C, Remuzzi G: Functional significance of exaggerated renal thromboxane A2 synthesis induced by cyclosporine A. Am J Physiol 251:F581–F587, 1986.
94. Kon V, Sugiura M, Inagami T, et al: Role of endothelin in cyclosporine induced glomerular dysfunction. Kidney Int 37:1487–1491, 1990.
95. Loutzenhiser R, Epstein M, Hayashi K, Horton C: Direct visualization of effects of endothelin on the renal microvasculature. Am J Physiol 258:F61–F68, 1990.
96. Loutzenhiser R, Epstein M, Horton C, Sonke P. Reversal of renal and smooth muscle actions of the thromboxane mimetic U-44069 by diltiazem. Am J Physiol 250:F619–F626, 1986.
97. Takenaka T, Hashimoto Y, Epstein M. Diminished acetylcholine-induced vasodilation in renal microvessels of cyclosporine-treated rats. J Am Soc Nephrol 3:42–50, 1992.
98. Neumayer H, Wagner K: Prevention of delayed graft function in cadaver kidney transplants by diltiazem. Lancet 2:1355–1356, 1985.
99. Brockmoller J, Neumayer HH, Wagner K, et al: Pharmacokinetic interaction between cyclosporine and diltiazem. Europ J Clin Pharmacol 38:237–242, 1990.
100. Weir MR: Calcium channel blockers in organ transplantation: Important new therapeutic modalities. J Am Soc Nephrol l:S28–S38, 1990.

101. Maggio TG, Bartels DW: Increased cyclosporine blood concentrations due to verapamil administration. Drug Intell Clin Pharm 22:705–707, 1988.
102. Kohlhaw K, Wongeit K, Frei U, et al: Effect of the calcium channel blocker diltiazem on cyclosporine blood levels and dose requirements. Transplant Proc 20(Suppl 2):572–574, 1988.
103. Cantarovich M, Hiesse C, Lockiec F, et al: Confirmation of the interaction between cyclosporine and the calcium channel blocker nicardipine in renal transplant patients. Clin Nephrol 28:190–193, 1987.
104. Henricsson S, Lindholm A: Inhibition of cyclosporine metabolism by other drugs in vitro. Transplant Proc 20(Suppl 2):569–571, 1988.
105. Ince P, Elliot K, Appleton DA, et al: Modulation by verapamil of vincristine pharmacokinetics and sensitivity to metaphase arrest of the normal rat colon in organ culture. Biochem Pharmacol 41:1217–1225, 1991.
106. Goligorsky MS, Chaimovits C, Rapoport J, et al: Calcium metabolism in uremic nephrocalcinosis: preventive effect of verapamil. Kidney Int 27:774–779, 1985.
107. Harris DCH, Hammond WS, Burke TJ, Schrier RW: Verapamil protects against progression of experimental chronic renal failure. Kidney Int 31:41–46, 1987.
108. Klahr S, Schreiner G, Ichikawa I: The progression of renal disease. N Engl J Med 318:1657–1666, 1988.
109. Brenner BM, Meyer TW, Hostetter TH: Dietary protein intake and the progressive nature of kidney disease: The role of hemodynamically mediated injury in the pathogenesis of progressive glomerular sclerosis in aging, renal ablation and intrinsic renal disease. N Engl J Med 307:652–659, 1982.
110. Hostetter TH, Olson JL, Rennke HG, et al: Hyperfiltration in remnant nephrons: a potentially adverse response to renal ablation. Am J Physiol 241:F85–F93, 1981.
111. Eliahou HE, Cohen D, Herzog D, et al: The control of hypertension and its effect on renal function in rat remnant kidney. Nephrol Dial Transplant 3:38–44, 1988.
112. Jackson B, Johnson CI. The contribution of systemic hypertension to progression of chronic renal failure in the rat remnant kidney: Effect of treatment with an angiotensin converting enzyme inhibitor or a calcium inhibitor. J Hypertens 6:495–501, 1988.
113. Bidani A, Griffen KA: Calcium channel blockers and renal protection: Is there an optimal dose? [editorial]. J Clin Lab Med 125:553–555, 1995.
114. Brunner F P, Thiel G, Hermle M, et al: Long term enalapril and verapamil in rats with reduced renal mass. Kidney Int 36:969–977, 1989.
115. Pelayo JC, Harris DCH, Shanley PF, et al: Glomerular hemodynamic adaptations in remnant nephrons: effects of verapamil. Am J Physiol 254:F425–F431, 1988.
115a. Tojo A, Kimura K, Matsuoka H, Sugimoto T: Effects of manidipine hydrochloride on the renal microcirculation in spontaneously hypertensive rats. J Cardiovasc Pharmacol 20:895–899, 1992.
116. Eliahou HE, Cohen D, Hellberg B, et al: Effect of the calcium channel blocker nisoldipine on the progression of chronic renal failure in man. Am J Nephrol 8:285–290, 1988.
117. Brazy PC, Fitzwilliam JF: Progressive renal disease: role of race and antihypertensive medications. Kidney Int 37:1113–1119, 1990.
118. Herlitz H: Long-term effects of felodipine in patients with reduced renal function. Kidney Int 41(Suppl 36):S110–S113, 1992.
119. Reams G, Lau A, Knaus V, Bauer JH: The effect of nifedipine GITS on renal function in hypertensive patients with renal insufficiency. J Clin Pharmacol 31:468–472, 1991.
120. Zucchelli P, Zuccala A, Borghi M, et al: Long-term comparison between captopril and nifedipine in the progression of renal insufficiency. Kidney Int 42:452–458, 1992.
121. Velussi M, Brocco E, Frigato F, et al: Effects of cilazapril and amlodipine on kidney function in hypertensive NIDDM patients. Diabetes 45:216–222, 1996.
122. Rossing P, Tarnow L, Boelskifter S, et al: Differences between nisoldipine and lisinopril on glomerular filtration rates and albuminuria in hypertensive IDDM patients with diabetic nephropathy during the first year of treatment. Diabetes 46:481–487, 1997.
123. ter Wee P, De Micheli AG, Epstein M: Effects of calcium antagonists on renal hemodynamics and progression of non-diabetic chronic renal disease. Arch Intern Med 154:1185–1202, 1994.
124. Anderson S: Renal hemodynamic effects of calcium antagonists in rats with reduced renal mass. Hypertension l7:288–295, 1991.
125. Keane WF, Anderson S, Aurell M, et al: Angiotensin converting enzyme inhibitors and progressive renal insufficiency. Ann Intern Med 111:503–516, 1989.
126. Noth RH, Krolewski AS, Kaysen GA, et al: Diabetic nephropathy. Hemodynamic basis and implications for disease management. Ann Intern Med 110:795–813, 1989.
127. Lewis EJ, Hunsicker LG, Bain RP, Rohde RD, for the Collaborative Study Group: The effect of angiotensin-converting enzyme therapy on diabetic nephropathy. N Engl J Med 329:1456–1462, 1993.

128. Hostetter TH, Rennke HG, Brenner BM: The case for intrarenal hypertension in the initiation and progression of diabetic glomerulopathies. Am J Med 72:375–380, 1982.
129. Zatz R, Dunn BR, Meyer TW, et al: Prevention of diabetic glomerulopathy by pharmacological amelioration of glomerular capillary hypertension. J Clin Invest 77:1925–1930, 1986.
130. Myers BD, Meyer TW: Angiotensin-converting enzyme inhibitors in the prevention of experimental diabetic glomerulopathy. Am J Kidney Dis 13:20–24, 1989.
131. De Zeeuw D: Meta-analysis of non-diabetic disease blood pressure trials [abstract]. Proceedings of the XIII International Congress on Nephrology, Madrid, Spain, July 1995, p 65.
132. Weidmann P, Schneider M, Boehlen L: Effects of different antihypertensive drugs in diabetic nephropathy: a meta-analysis [abstract]. Proceedings of the XIII International Congress on Nephrology, Madrid, Spain, July 1995, p 66.
133. Maki DD, Ma JZ, Louis TA, Kasiske BL: Long-term effects of antihypertensive agents on proteinuria and renal function. Arch Intern Med 155:1073–1080, 1995.
134. Lash JP, Bakris GL: Effects of ACE inhibitors and calcium antagonists alone or combined on progression of diabetic nephropathy. Nephrol Dial Transplant 10(Suppl 9):56–62, 1995.
135. Epstein M: Calcium antagonists and diabetic nephropathy. Arch Intern Med 151:2361–2364, 1991.
136. Packer M, Lee WH, Medina N, et al: Influence of diabetes mellitus on changes in left ventricular performance and renal function produced by converting enzyme inhibition in patients with severe chronic heart failure. Am J Med 82:1119–1126, 1987.
137. Hricik DE, Dunn MJ: Angiotensin-converting enzyme inhibitor-induced renal failure: Causes, consequences, and diagnostic uses. J Am Soc Nephrol 1:845–858, 1990.
138. Loutzenhiser R, Epstein M, Hayashi K, et al: Characterization of the renal microvascular effects of angiotensin II antagonists, DuP 753: studies in isolated perfused hydronephrotic kidneys. Am J Hypertens 4:309S–314S, 1991.
139. Ribstein J, Mourad G, Mimran A: Contrasting acute effects of captopril and nifedipine on renal function in renovascular hypertension. Am J Hypertens 1:239–244, 1988.
140. Miyamori I, Yasuhara S, Matsubara T, et al: Comparative effects of captopril and nifedipine on split renal function in renovascular hypertension. Am J Hypertens 1:359–363, 1988.

MATTHEW R. WEIR M.D.

25

The Clinical Utility of Calcium Antagonists in Renal Transplant Recipients

Calcium antagonists are a widely used therapeutic class of medications in patients with hypertension and angina pectoris.[1] They have emerged as commonly used drugs in organ transplant recipients.[2,3] This chapter describes the clinical benefits of calcium antagonists in renal transplant recipients and focuses on some of the controversy concerning their use as adjuvant immunomodulatory agents or agents to diminish allograft dysfunction due to either reperfusion injury or exposure to nephrotoxic agents. Adverse events associated with calcium antagonists are also discussed and put into perspective along with their important therapeutic advantages.

Calcium antagonists, although diverse in chemical structure, exert their biologic activity primarily through inhibition of transmembrane cellular influx of calcium.[4–20] Because changes in intracellular calcium content regulate a broad variety of cellular biochemical processes, including gene expression, it is not surprising that calcium antagonists significantly affect many different cellular activities, which may have clinical importance in organ transplant recipients. Among these effects are the activation and function of lymphocytes,[4–9] regulation of vascular smooth muscle contraction,[10–12] platelet activation,[13–16] and cell membrane stabilization.[17–20]

Prior reviews of the clinical benefits of calcium antagonists in renal transplant recipients have described their efficacy for the treatment of high blood pressure and their ability to diminish posttransplant allograft dysfunction.[1,2] These observations are well established (Table 1). Prior reviews also have suggested that calcium antagonists may boost the effects of conventional immunosuppression and even diminish the rate of rejection.[1,2] This issue remains controversial and is discussed in light of recent reviews of clinical trials. Other clinical data indicate that calcium antagonists may attenuate some of the nephrotoxic effects of cyclosporine.[21–24] In addition, this chapter examines the theoretical possibility that calcium antagonist therapy may attenuate interstitial fibrotic changes in the kidney in response to cyclosporine administration. Some of the adverse effects of chronic calcium antagonist therapy are discussed, including leg edema and gingival hypertrophy, along with important observations about drug interactions with cyclosporine levels.

HYPERTENSION IN ORGAN TRANSPLANT RECIPIENTS: EFFICACY OF CALCIUM ANTAGONIST THERAPY

Hypertension is common after transplantation and occurs in approximately 90% of cyclosporine-treated patients.[25–28] The pathogenesis of hypertension is

TABLE 1. Therapeutic Benefits of Calcium Antagonists in Organ Transplant Recipients

Proven	Unproven
Antihypertensive activity	Augment immunosuppression
Diminish renal allograft nonfunction	Attenuate vascular remodeling and restructuring
Acutely attenuate cyclosporine-mediated reduction in renal function	Chronically reduce cyclosporine-mediated renal injury

multifactorial. Cyclosporine therapy enhances sodium and water retention by the kidney, which results in a low renin salt-sensitive form of essential hypertension.[27,28] In addition, there is evidence of both intrarenal and systemic vasoconstriction.[21,24,27–30] Sodium retention is likely due, in part, to preglomerular vasoconstriction from cyclosporine with an associated reduction in renal plasma flow and a resultant perception by the kidney of diminished effective arterial blood volume. Intrarenal neurohumoral systems, such as the renin-angiotensin-aldosterone system, respond to diminished volume and through a variety of mechanisms increase renal sodium and water retention.[30] Cyclosporine also may have direct antinatriuretic effects independent of its vasoconstrictive properties. These observations probably explain why dietary salt restriction and pharmacotherapies that enhance natriuresis exhibit antihypertensive effects in cyclosporine-treated renal transplant recipients[31–36] (Fig. 1).

Systemic vasoconstriction is also evident as a result of cyclosporine therapy (Fig. 2). Experimental studies demonstrate that cyclosporine inhibits vascular

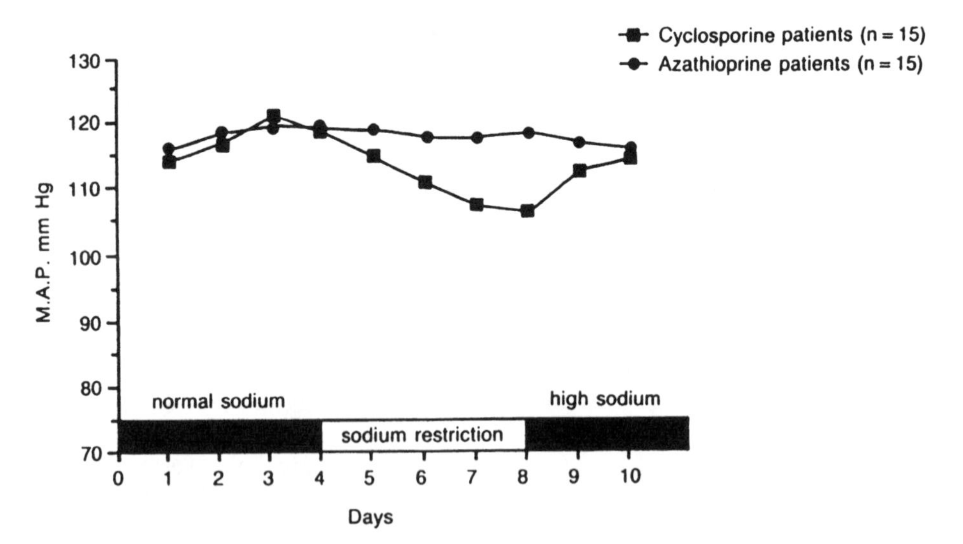

FIGURE 1. Posttransplant hypertension. Mean arterial pressures of 15 cyclosporine-treated and 15 azathioprine-treated renal transplant recipients. The patients had similar blood pressures on the normal sodium (150 mEq) diet. With sodium restriction, the cyclosporine group had a significant ($p < 0.01$) decrease in blood pressure compared with either the baseline pressure values or the pressure values of the azathioprine group, which did not decrease with sodium restriction. (From Curtis JJ, Luke RG, Jones D, et al: Hypertension in cyclosporine-treated renal transplant recipients is sodium dependent. Am J Med 85:134–138, 1988, with permission.)

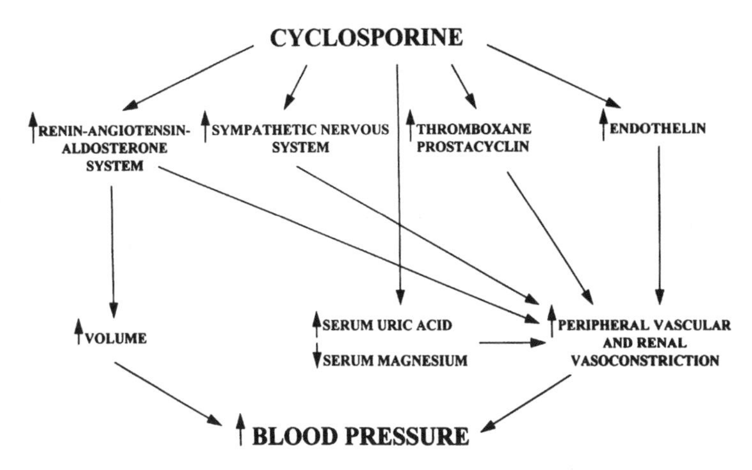

FIGURE 2. The various pathways by which cyclosporine therapy may result in an increase in blood pressure.

relaxation responses and may also directly activate other neurohumoral systems, such as the renin-angiotensin system or sympathetic nervous system.[26–28,37–39] Cyclosporine therapy also may shift the balance toward production of vasoconstrictive over vasodilatory arachidonic acid metabolites and enhance endothelin production.[40,41] Hypomagnesemia or hyperuricemia as a result of cyclosporine therapy or related to intrinsic renal dysfunction or use of diuretics also may facilitate blood pressure elevation.[42,43] Consequently, there are numerous pathways whereby cyclosporine therapy may induce elevation in blood pressure in addition to other predisposing conditions such as impaired graft function or hypertension due to disease of the native kidneys.

Calcium antagonists possess different mechanisms of action for reducing blood pressure— mechanisms that may be ideal in cyclosporine-treated renal transplant recipients.[32,44–48] They enhance sodium and water excretion by the kidney through a direct effect on the renal tubule and an increase in both renal plasma flow and glomerular filtration rate due to selective afferent arteriolar dilatation.[32] Calcium antagonists also directly antagonize the vasoconstrictor effects of several of the putative cyclosporine-stimulated neurohumoral factors, such as angiotensin II, norepinephrine, thromboxane, and endothelin.[44–48]

In clinical trials of patients with essential hypertension, calcium antagonists have demonstrated potent antihypertensive activity, even in low renin salt-sensitive patients ingesting an ad libitum salt diet.[33,34,49] Their antihypertensive effects are preserved despite greater dietary salt intake, and some provocative studies have suggested that blood pressure reduction is even greater with greater dietary salt intake. This finding may be related to a higher baseline blood pressure at the onset of calcium antagonist therapy.[34] Clinical trials comparing the efficacy of calcium antagonists with other therapies in hypertensive cyclosporine-treated renal transplant recipients have illustrated their efficacy not only in controlling blood pressure but also in maintaining glomerular filtration rate and renal plasma flow.[35] Some studies have compared the short-term effects of calcium antagonists and angiotensin-converting enzyme (ACE) inhibitors in both cyclosporine- and non–cyclosporine-treated renal transplant recipients. Mimran and colleagues[50] noted that captopril was more effective in non–cyclosporine-treated patients, whereas its effects were blunted in the cyclosporine-treated patients, consistent with the renal sodium retention and reduced plasma renin activity associated with cyclosporine-induced hypertension. In contrast, other investigators have reported the potent efficacy of calcium antagonists in organ transplant recipients despite cyclosporine therapy.[35,36]

However, there are some clinical differences in the acute and chronic effects of antihypertensive drugs in cyclosporine-treated patients. In a short-term crossover study using captopril or nifedipine for 48 hours, Curtis et al. observed that, despite similar reductions in blood pressure with each drug, renal plasma flow and glomerular filtration rate were unchanged with the calcium antagonist, yet decreased with the ACE inhibitor[35] (Fig. 3). However, in a long-term comparative study of lisinopril and nifedipine in cyclosporine-treated renal transplant recipients, Mourad and colleagues[51] noted that after 12 and 30 months of follow-up, both antihypertensive agents resulted in equal long-term reductions in blood pressure and renal vascular resistance with no significant changes in either glomerular filtration rate or urinary albumin excretion. However, all patients had stable and well-preserved graft function. More extensive experience involving patients with graft dysfunction and proteinuria is required to determine whether one class has an advantage over another in terms of deterioration of graft function and blood pressure control.

Some clinicians prefer the use of calcium antagonists, particularly in the posttransplant period, simply because, unlike ACE inhibitors, they do not induce functional renal insufficiency, which may complicate posttransplant management because of concerns over other causes of graft dysfunction such as volume depletion, rejection, or urologic abnormalities. In addition, calcium antagonists do not interfere with the recovery of hematocrit after transplant, whereas in some patients ACE inhibitors may interfere with erythropoiesis.[52]

Calcium antagonists may be more effective than other therapies in preserving diurnal variation in blood pressure (the normal nocturnal reduction should approximate 10% of the day-time mean) in cyclosporine-treated transplant recipients.[53] This effect may have considerable clinical significance, because some investigators have suggested that preservation of diurnal variation in blood pressure may be important in reducing the vascular risk associated with hypertension.

Clinical evidence also suggests that urate clearance is higher and serum uric acid is lower with calcium antagonist therapy in cyclosporine-treated renal transplant recipients.[54] This finding is another potential benefit because cyclosporine enhances renal urate reabsorption with resultant hyperuricemia and greater risk for gout.[55]

One of the disadvantages of calcium antagonist therapy is that pedal edema may occur. Although usually a modest problem, in some patients, particularly in women, edema may be a significant cosmetic and discomforting problem. It is not related to volume overload and probably results from a capillary leak syndrome. No studies demonstrate that either thiazide-type or loop diuretics attenuate the edema. Caution is urged to avoid overzealous efforts to diurese transplant patients with pedal edema due to calcium antagonists; such efforts may result in prerenal azotemia and electrolyte depletion. In addition, overzealous diuresis may offset the beneficial uricosuric effect of calcium antagonists.

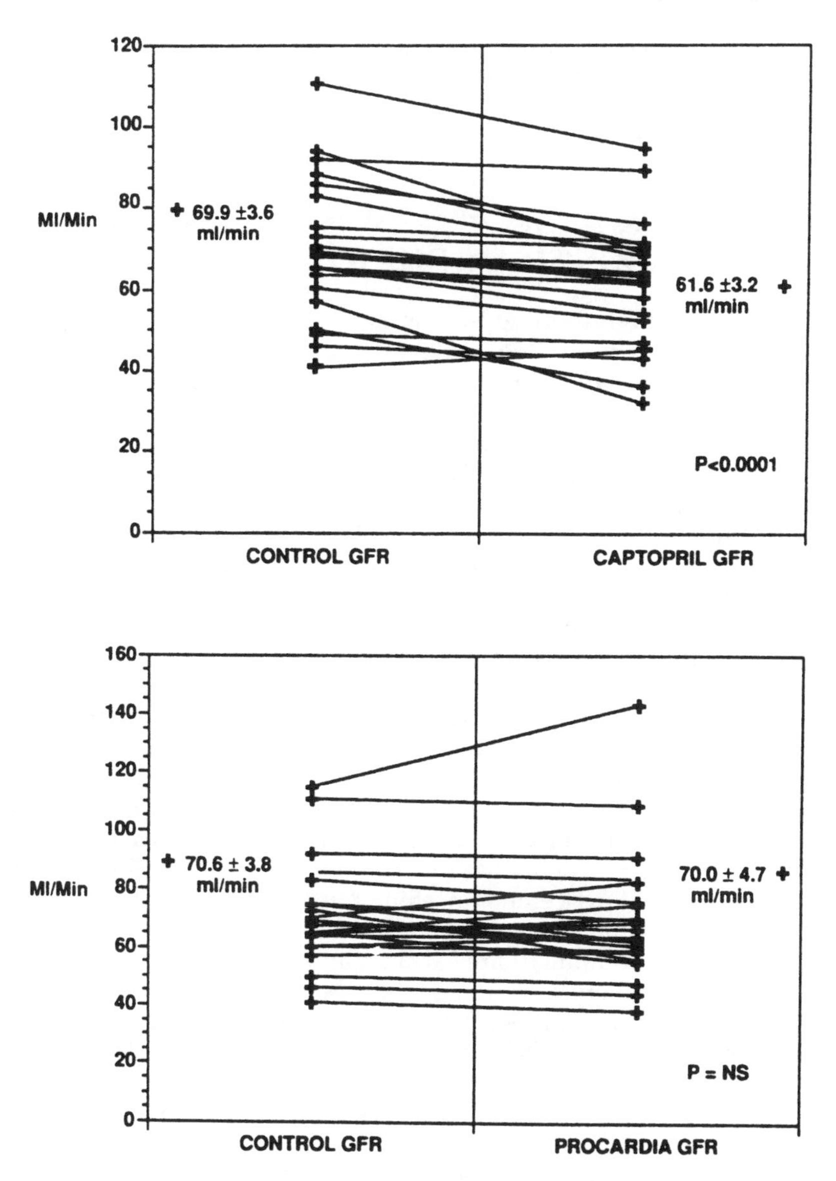

FIGURE 3. Eighteen patients in a crossover study and 10 additional patients (5 taking captopril and 5 taking nifedipine) who were studied in a parallel study are combined in these line graphs depicting the GFR before and after therapy with the antihypertensive medication captopril or nifedipine. Twenty-two of the 23 patients given captopril had a decrease in GFR. Mean ± SE is shown for each group before and after treatment. (From Curtis JJ, Laskow DA, Jones PA, et al: Captopril induced fall in glomerular filtration rate in cyclosporine-treated hypertensive patients. J Am Soc Nephrol 3:1570–1574, 1993, with permission.)

Either a reduction in dose or discontinuation of the drug may be indicated if the edema is debilitating.

Clinical experience with calcium antagonists supports their use, particularly in cyclosporine-treated renal transplant recipients, because they are effective in controlling blood pressure despite variation in dietary salt consumption without adverse effects on either glomerular filtration rate or renal plasma flow. Consequently, stringent dietary salt restriction, which is not only impractical but may be potentially hazardous in selected patients, is not necessary.

POSTTRANSPLANT ISCHEMIC RENAL INJURY: EFFECT OF CALCIUM ANTAGONISTS ON DELAYED GRAFT FUNCTION

Delayed graft function resulting from ischemic renal injury is an important clinical problem. It is associated with inferior allograft and patient survival rates compared with grafts without early dysfunction.[56] Ischemic renal injury poses a problem in organ transplantation, because sharing of organs to improve tissue matching requires longer cold ischemia time. Scientific evidence demonstrates that cytosolic accumulation of calcium may be a critical factor in initiating, propagating, and maintaining cellular ischemic injury.[57] In fact, cellular accumulation of calcium may be correlated with increasing severity of histologic damage in experimental models of renal ischemia.[57]

Normal physiologic conditions are characterized by a large gradient between extracellular and intracellular calcium concentrations, with a much lower level of calcium within the cell and large amounts of calcium in extracellular fluid. Ischemia may alter transcellular calcium distribution through several different pathways (Fig. 4). The net result is accumulation of calcium within the cytosol and mitochondria. Unfortunately, greater cytosolic and mitochondrial accumulation of calcium interferes with synthesis of adenosine triphosphate (ATP) and thus may limit the ability of the cell to export and sequester calcium.[57] Greater cytosolic calcium accumulation enhances vascular hyperresponsiveness to circulating vasoconstrictors as well as to renal nerve stimulation.[57] Diminished ATP synthesis also allows cellular swelling due to inhibition of the sodium-potassium-ATPase pump with subsequent risk for cell lysis.[57]

Cellular calcium accumulation is also an important factor in reperfusion injury because the delivery of oxygen, in the presence of xanthine oxidase, may lead to oxygen free radical formation, which is toxic to cell membranes and renders them more permeable to calcium.[57] Consequently, ischemic changes of the cell, for various reasons, result in cellular accumulation of calcium, which may initiate, propagate, and perpetuate ischemic damage and result in allograft nonfunction.

Because calcium antagonists experimentally diminish the transmembrane flux of calcium into cells, their ability to attenuate ischemic renal injury has been studied extensively. Malis et al.[58] have demonstrated that verapamil may protect against norepinephrine-induced renal failure in rats by preventing complete cessation of renal blood flow. Similarly, Burke et al.[59] have demonstrated that calcium antagonists reduce ischemic renal injury even when administered after total cessation of blood flow for 40 minutes. Some investigators have suggested that the loss of autoregulation of renal blood flow and the resultant increase in vascular responsiveness to circulating vasopressors in ischemic models may be related to elevated cellular calcium concentration[57] Consequently, blocking cellular calcium accumulation makes scientific sense for preventing ischemic cellular injury.

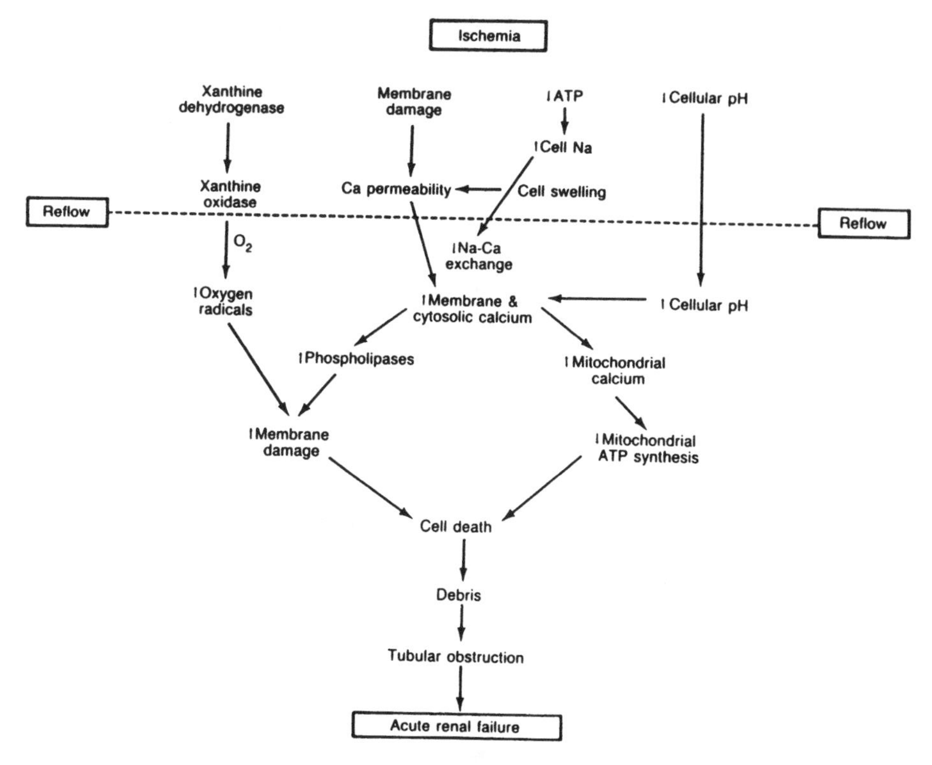

FIGURE 4. Multiple possible pathways whereby ischemia and reperfusion lead to graft injury. (From Schrier RW, Arnold PE, Van Putten NJ, et al: Cellular calcium in ischemic acute renal failure: Role of calcium entry blockers. Kidney Int 32:313–321, 1987, with permission.)

The encouraging results in experimental models have led to a number of clinical trials designed to assess whether calcium antagonists effectively reduce the incidence of immediate graft dysfunction and thus improve overall graft survival rates. The first such studies were conducted in 1987 by Neumeyer and Wagner at the Klinikum Steiglitz in Berlin.[60] The investigators treated the donor kidneys as well as the recipients with diltiazem. A significant reduction in the incidence of allograft dysfunction was associated with an improvement in renal blood flow and glomerular filtration rate within the first week after transplantation in calcium antagonist-treated patients.

Various other clinical studies have been conducted since then. Ladefoged and Andersen[61] analyzed nine different reports[62–69] describing the effect of calcium antagonists on the incidence of delayed graft function. The rate of delayed graft function in the control groups showed considerable variation (15–81%). Needless to say, the effect of a specific therapy to improve graft function would be more evident in a program with a higher incidence of delayed graft function. In five of the nine studies, treatment with calcium antagonists showed a significant beneficial effect (Fig. 5). The odds ratio for each of the nine studies illustrates overlapping confidence intervals and significant heterogeneity of results. However, a highly significant ($p < 0.0001$) reduction in the rate of delayed graft function was observed if the studies are combined. The overall

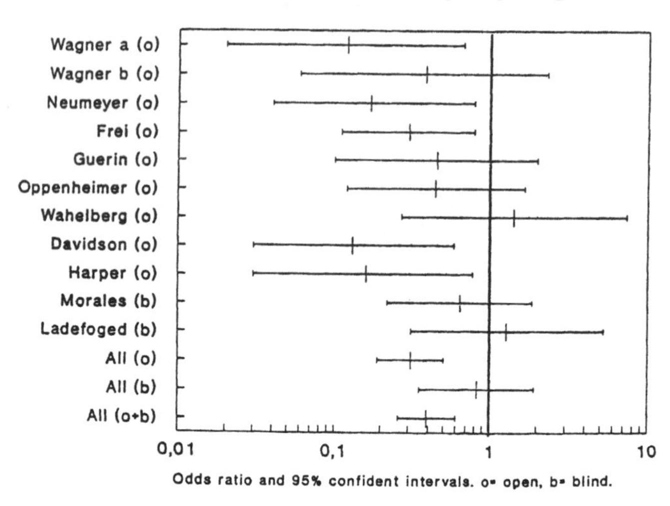

FIGURE 5. The odds ratios for 9 different studies reporting the impact of calcium antagonist therapy on the incidence of delayed graft function. (From Ladefoged SD, Andersen CB: Calcium channel blockers in kidney transplantation. Clin Transpl 8:128–133, 1994, with permission.)

benefit was a 21.8% reduction in the rate of delayed graft function (confidence intervals: 13.6–31.1%).

Since the Ladefoged analysis was published, two other reports have examined the influence of calcium antagonists on graft function. Ponticelli et al.[36] found no difference in the prevalence of acute tubular necrosis at any point between patients treated with nifedipine and patients who did not receive a calcium antagonist. Vasquez and Pollak[70] evaluated 65 cyclosporine-treated cadaveric renal allograft recipients with posttransplant hypertension. Thirty-two of the patients received chronic calcium antagonist therapy, whereas 33 did not in the 1-year follow-up period. The investigators noted that only patients who received chronic administration of a calcium antagonist before transplantation showed evidence of improved renal allograft function at 3 and 12 months after transplant. The likely explanation is lessened ischemic insult. Although this study is small, it emphasizes an important observation. Clinical trials that demonstrate a beneficial effect on early renal function used calcium antagonists in the patients or donor kidney either before transplant or in the intra- and/or perioperative period. When calcium antagonists are administered later in the posttransplant period, there is less evidence of benefit.

Studies that report no beneficial effect of calcium antagonists on delayed graft function either have a low delayed graft function rate or started calcium antagonist therapy more than 72 hours after transplantation. For example, in the prospective study of Pirsch et al.,[71] despite a low delayed graft function rate at their center and late addition of calcium antagonist therapy after conversion from antilymphocyte sera to cyclosporine therapy, there was still a trend toward reduction in the incidence in graft dysfunction from 18% to 9% (p = NS).

In summary, despite their small size and varying study designs and forms of induction immunosuppression, clinical trials demonstrate that calcium antagonist therapy before transplant, during the procedure, or in the immediate perioperative or postoperative period decreases the rate of allograft dysfunction. This effect is most apparent at centers with a higher incidence of delayed graft function.

Several critical questions remain unanswered by available data. What dose of calcium antagonist is needed to demonstrate benefit, and how long should therapy be continued? Which is more important, treatment of the donor or recipient? Or is treatment of both the optimal strategy? What are the differences among the calcium antagonists? Most clinical trials have been conducted with the nondihydropyridines, diltiazem and verapamil. Available clinical data suggest that all calcium antagonists exhibit similar beneficial effects; future trials are needed to distinguish among them.

Another unresolved and important question is whether perioperative or intraoperative administration of calcium antagonists will reduce the requirement for antilymphocyte antibody induction therapy. Nephrotoxic immunosuppressants such as cyclosporine and tacrolimus may be given with greater impunity in the immediate postoperative period if ischemic renal injury can be attenuated with calcium antagonist therapy. This approach may result in significant cost-savings. In addition, preliminary studies suggest that calcium antagonist therapy, in the intraoperative or perioperative period, also may improve preservation and function of pancreas, lung, and liver allografts.[72–74]

AMELIORATION OF DRUG NEPHROTOXICITY BY CALCIUM ANTAGONISTS

Both cyclosporine and tacrolimus possess potent immunosuppressant properties, but their use is limited by nephrotoxicity. The two drugs share similarities and dissimilarities in terms of their nephrotoxic effects. The exact nephrotoxic mechanisms of each drug remain poorly described. The numerous theories include intense intrarenal vasoconstriction, leading to ischemic injury and fibrotic changes; a direct toxic effect on renal tubular epithelial cells; arteriolopathic injury due to platelet activation; and/or elaboration of soluble mediators of fibrosis, such as transforming growth factor β.[75,76] The following discussion focuses primarily on the interrelationship between calcium antagonists and cyclosporine-mediated renal injury; few data assess the effect of calcium antagonists on tacrolimus-mediated renal injury. Moreover, clinical experience with cyclosporine and calcium antagonists in transplantation is more extensive.

Cellular uptake of calcium is probably involved in the vasoconstrictive activities of cyclosporine, at least as one of its secondary effectors. For example, it has been demonstrated that cyclosporine enhances vasopressin-induced calcium mobilization and contraction in mesangial cells.[77] Therefore, antagonists of cellular calcium uptake may provide a suitable approach to attenuate the pathophysiologic basis of vasoconstriction in the presence of cyclosporine. The ability of calcium antagonists to attenuate vasoconstriction related to cyclosporine has been demonstrated in both experimental and clinical studies. Barros et al.[78] and Sulemanlar et al.[79] demonstrated that the calcium antagonist verapamil improved renal perfusion and function in cyclosporine-treated rats compared with placebo. Similarly, Rooth et al.[21] demonstrated improved flow in the subcapsular renal microcirculation in mice, if they

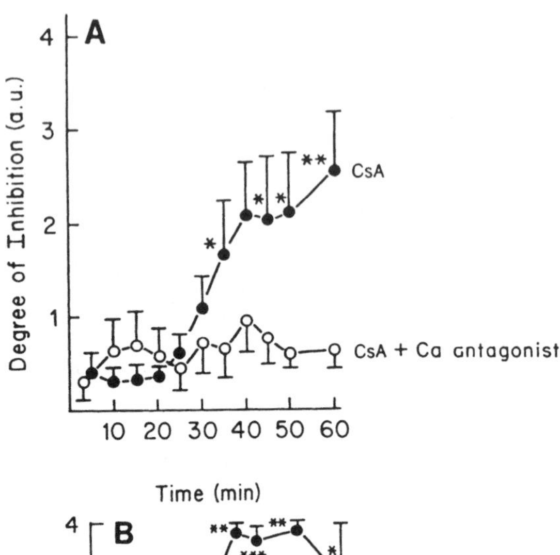

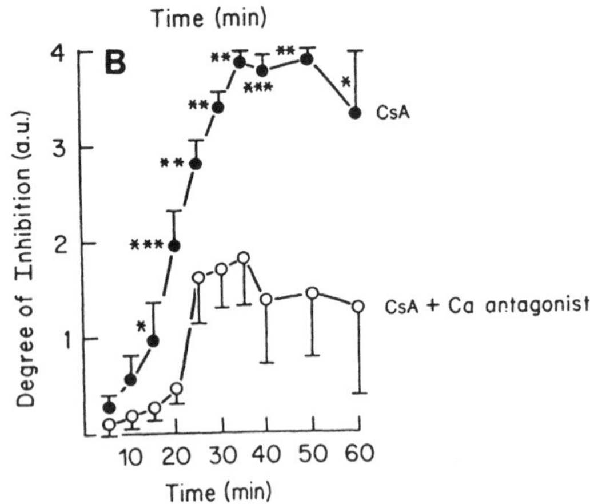

FIGURE 6. Pretreatment with the calcium antagonist verapamil (0.35 mg/kg) markedly improved the kidney subcapsular blood flow at different infusion rates of cyclosporine. *A,* At low infusion rates (0.18–0.22 mg/kg/min; N = 10), inhibition was completely prevented. *B,* At higher infusion rates (0.9–2.4 mg/kg/min; N = 13), the effect was partial (mean ± SEM). (From Rooth P, Dawidson I, Dillen K, Taljedal IB: Protection against cyclosporine-induced impairment of renal microcirculation by verapamil in mice. Transplantation 45:433–437, 1988, with permission.)

were pretreated with the calcium antagonist verapamil before infusion of cyclosporine (Fig. 6).

Similarly, human studies have demonstrated a benefit of pretreatment with calcium antagonists to prevent cyclosporine-mediated vasoconstriction. For example, Weir et al.[22] studied the acute effects of intravenous cyclosporine on renal hemodynamics in healthy humans. Pretreatment with verapamil minimized cyclosporine-induced reduction in renal plasma flow, although no significant protective effect on glomerular filtration rate was noted. In a study of human renal allograft recipients, Dawidson et al.[23] observed that verapamil attenuated cyclosporine-mediated reduction in renal parenchymal diastolic blood flow measured by Doppler in the first week after renal transplantation. In a more sophisticated study, Ruggenenti et al.[24] sequentially measured hourly glomerular filtration rates in renal allograft recipients after an oral dose of cyclosporine. Pretreatment with the calcium antagonist lacidipine completely prevented the fall in glomerular filtration rate that occurred within 2 hours of peak cyclosporine concentration after a single oral dose (Fig. 7). Thus, both experimental and human evidence support the concept that calcium antagonists limit acute cyclosporine-mediated vasoconstriction within the kidney. This observation

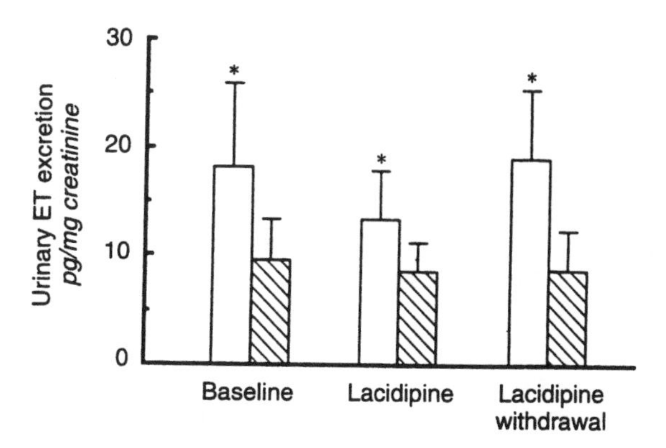

FIGURE 7. Simultaneous monitoring of blood cyclosporine (CsA) concentration and glomerular filtration rate (GFR) in a representative patient before (*A*, baseline), during lacidipine treatment (*B*, lacidipine), and 7 days after lacidipine withdrawal (*C*, lacidipine withdrawal). (From Ruggenenti P, Perico N, Mosconi L, et al: Calcium channel blockers protect transplant patients from cyclosporine-induced daily renal hypoperfusion. Kidney Int 43:706–711, 1993, with permission.)

may have clinical significance because chronic intrarenal vasoconstriction may be one of the properties of cyclosporine that leads to the development of progressive interstitial fibrotic changes. Studies are needed to demonstrate that calcium antagonists chronically interfere with cyclosporine-mediated intrarenal vasoconstriction in human organ transplant recipients.

Calcium antagonists may possess other properties that limit cyclosporine-mediated renal injury. Some investigators have suggested that calcium antagonists alter cyclosporine metabolism to create less nephrotoxic metabolites. Kunzendorf et al.[80] demonstrated that diltiazem increases the serum concentration of the M17 metabolite of cyclosporine, which may be less nephrotoxic yet possesses similar immunosuppressive properties as other more toxic metabolites (Fig. 8). Nagineni et al.[81] demonstrated that calcium antagonists are capable of limiting proximal tubular cell uptake of cyclosporine, which may reduce renal injury. However, other investigators have noted that calcium antagonists

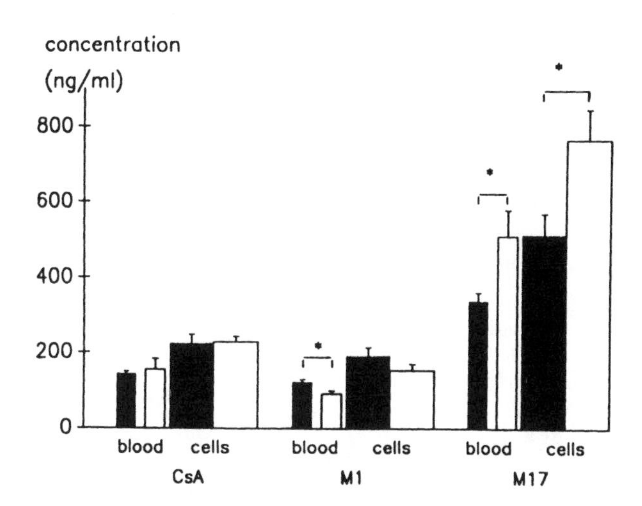

FIGURE 8. Concentrations of cyclosporine and cyclosporine metabolites M1 and M17, in whole blood and packed cells, separated at 37°C in kidney graft recipients treated with (□) or without (■) diltiazem. Results are expressed as mean ± SEM (*$p < 0.05$). (From Kunzendorf FW, Walz G, Brockmoeller J, et al: Effect of diltiazem upon metabolism and immunosuppressive action of cyclosporine in kidney graft recipients. Transplantation 52:280–284, 1991, with permission.)

do not alter the tissue distribution of cyclosporine.[82,83] Despite the scant evidence from these limited experimental clinical studies, the possibility remains that alterations in cyclosporine metabolism, particularly those induced by the nondihydropyridine calcium antagonists, may lead to the formation of metabolites that are less nephrotoxic or possibly to adjustments in the transcellular distribution of metabolites so that the risk for renal injury is decreased. Clearly, more clinical studies are needed.

The ability of calcium antagonists to reduce platelet aggregation in response to either epinephrine or adenosine diphosphate (ADP) has been demonstrated in vitro.[13–15] It is possible that calcium antagonists limit platelet-mediated vascular injury, which may be involved in the vasculopathic process that occurs with chronic rejection. Calcium antagonists have been demonstrated to reduce progressive neointimal hyperplasia at the venous anastomosis in polytetrafluoroethylene arteriovenous hemodialysis grafts.[84] This benefit may be related to inhibition of platelet aggregation and/or other trophins or mitogens of vascular smooth muscle.

Recent experimental evidence has also focused on the immunoregulatory and fibrogenic activities of cyclosporine based on its ability to stimulate expression of transforming growth factor β (Fig. 9). Cyclosporine-associated renal dysfunction and hypertension may be related to cyclosporine-induced overexpression of transforming growth factor β_1.[76] This hormone enhances extracellular matrix accumulation and increases endothelin production by vascular smooth muscle cells.[85,86] Consequently, it may represent a mechanistic link between the development of hypertension and interstitial fibrosis related to prolonged cyclosporine usage. Calcium antagonists inhibit the stimulatory effect of transforming growth factor β_1 on collagen synthesis by human mesangial cells in culture.[87] Of great interest, in vitro experimental studies used calcium antagonists in therapeutic concentrations customarily achieved in humans (Fig. 10). Also of interest was the observation that structurally different calcium antagonists caused significant and dose-dependent inhibition of human mesangial cell mitogenesis and protein synthesis and consequently growth.

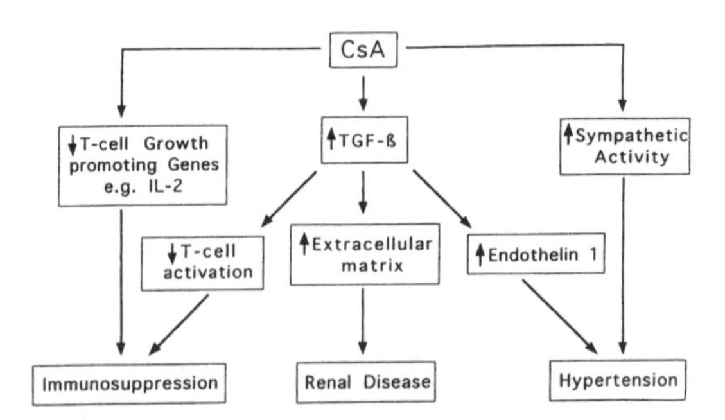

FIGURE 9. Potential sequelae of cyclosporine-mediated augmentation of TGF-β expression. In this schema of events, the cyclosporine-associated increase in TGF-β expression is hypothesized to contribute to the following: immunosuppression, renal interstitial fibrosis, and hypertension. In this paradigm, TGF-β represents the mechanistic link for the clinically desirable (immunosuppression) and deleterious (hypertension, fibrosis) consequences of cyclosporine use. (From Khanna A, Baogui L, Stenzel KH, Suthanthiran M: Regulation of new DNA synthesis in mammalian cells by cyclosporine. Demonstration of a transforming growth factor β-dependent mechanism of inhibition of cell growth. Transplantation 57:577–582, 1994, with permission.)

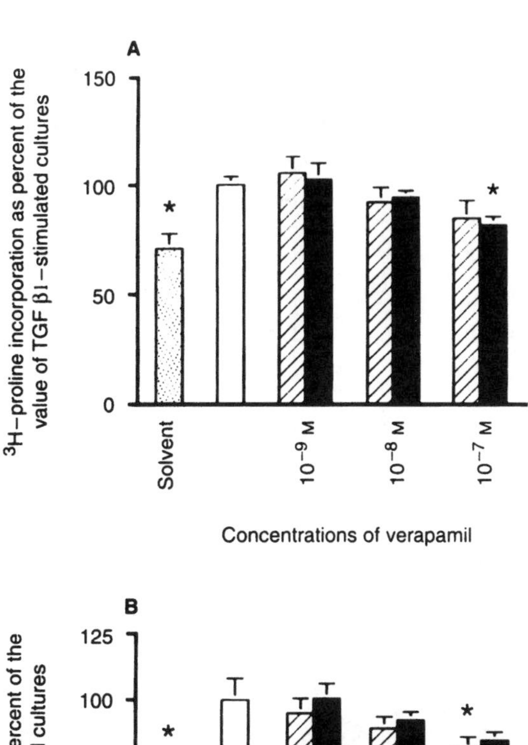

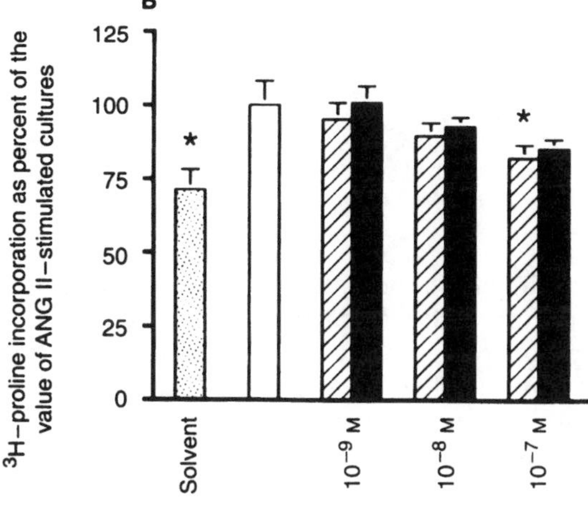

FIGURE 10. Effect of different concentrations of (R)-verapamil and (S)-verapamil on ^{3}H-proline incorporation into the collagen fraction (cpm = counts per minute) of human mesangial cell cultures stimulated with TGF β1 *(A)* and Ang II *(B)* (24 hr of incubation). Symbols in *A* are (□) TGF β1 2 ng/ml + solvent, (striped box) TGF β1 2 ng/ml + (R)-verapamil, and (■) TGF β1 2 ng/ml + (S)-verapamil. Symbols in *B* are (□) Ang II 10^{-9} M + solvent, (striped box) Ang II 10^{-9} M + (R)-verapamil, and (■) Ang II 10^{-9} M + (S)-verapamil. ^{3}H-proline incorporation values of stimulated cultures are given as 100%. Each bar is the mean ± SD of six parallel cultures of one experiment. Results were confirmed in 5 independent experiments. *$P < 0.04$ versus stimulated cultures. (From Orth SR, Nohiling R, Bonisch S, Ritz E: Inhibitory effect of calcium channel blockers on human mesangial cell growth: Evidence for actions independent of L-type Ca^{++} channels. Kidney Int 49:868–879, 1996, with permission.)

This effect was observed despite the presence of cytokines such as platelet-derived growth factor and epidermal growth factor and even in the presence of pressor agonists such as angiotensin II and endothelin 1.

To support the concept that calcium antagonists may have in vivo properties that alleviate progressive vasculopathy, Schroeder et al.[88] randomized cardiac allograft recipients to receive conventional cyclosporine-based immunosuppression either with or without the calcium antagonist diltiazem. The diltiazem-treated group had stabilization of coronary artery luminal diameter as measured by quantitative coronary angiography after 1 year, whereas the group that did not receive diltiazem developed progressive luminal diameter obliteration (Fig. 11). These observations were clinically significant, because the dilitazem-treated group exhibited a trend toward reduced incidence of

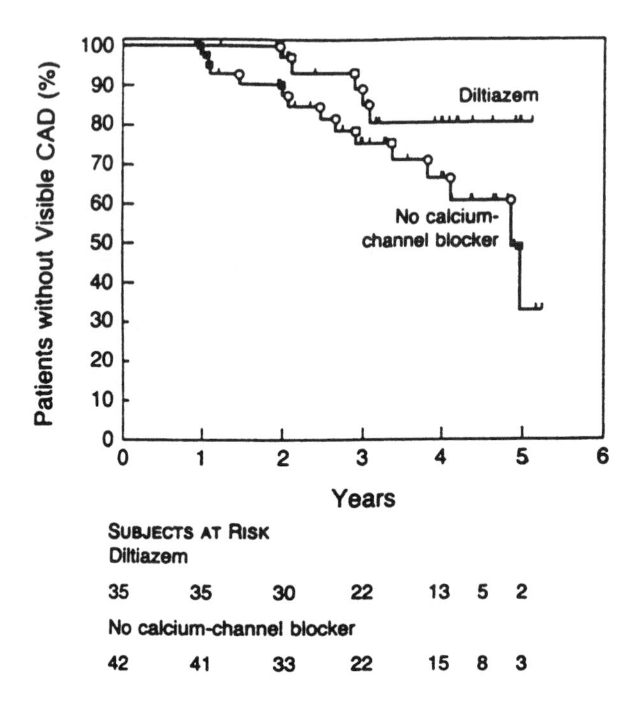

FIGURE 11. The proportion of patients free of visible evidence of coronary artery disease (CAD) after transplantation. (From Schroeder JS, Gao SZ, Alderman EL, et al: A preliminary study of diltiazem in the prevention of coronary artery disease in heart-transplant recipients. N Engl J Med 328:164–170, 1993, with permission.)

cardiovascular events and death. It is likely that calcium antagonists, either through an effect on the platelet or growth-promoting properties of cyclosporine or perhaps even by diminishing chronic immunogenic damage of the vascular wall, were involved in the beneficial effects on posttransplant coronary artery disease. Similar effects are likely to be observed within the vascular tree of the kidney and other allografts with prolonged follow-up.

In summary, calcium antagonists may limit nephrotoxicity in response to cyclosporine or other immunosuppressant nephrotoxins in various ways. Also important, however, is improved blood pressure control in addition to limitation of chronic endothelial damage as a result of various growth-stimulating perturbations. Although no long-term data about calcium antagonists and renal function involve large numbers of patients studied in a prospective, randomized fashion, data from experimental studies and the clinical effects in transplant coronary artery disease indicate potential benefit.

POTENTIAL IMMUNOMODULATORY PROPERTIES OF CALCIUM ANTAGONISTS

Experimental evidence supports the concept that increased intracellular calcium is critical for many events in lymphocyte activation and function.[4–9] However, the relative importance of the intracellular shifts of calcium pools vs.

transmembrane flux of calcium has not been completely defined. Preliminary investigations demonstrated that various chemically different calcium antagonists inhibit several aspects of both the afferent and efferent limbs of immunity in a concentration-dependent fashion.[89–91] Although the drug concentrations that may have this inhibitory effect do not result in cytotoxicity, they are substantially higher than what can be achieved in vivo with chronic dosing. Therefore, careful evaluation of the concentrations used in in vitro experiments vs. the concentrations achievable with in vivo dosing is necessary. For example, Weir[92] and others[91] have demonstrated the in vitro immunosuppressive properties of the calcium antagonist verapamil. When verapamil was used with cyclosporine at a concentration ≥ 5 mmol, mitogen-induced proliferative responses of human peripheral blood mononuclear cells were reduced significantly compared with either drug used alone.[92] In fact, 50% less cyclosporine was required to cause the same degree of immunosuppression with verapamil than when cyclosporine was used alone. In clinical practice, chronic dosing of verapamil usually results in serum levels of 0.2–0.8 mmol. However, because calcium antagonists are potently tissue-bound, their in vivo effects may exceed their in vitro effects for a given serum concentration.

In addition, calcium antagonists, particularly those that alter cyclosporine metabolism, may lead to the formation of metabolites with greater tissue penetration or greater production of a more immunosuppressive metabolite that further interferes with lymphocyte activation and function. Some experimental evidence supports this concept,[80] whereas other evidence does not.[82,83]

The antimitogenic properties of calcium antagonists, albeit with higher in vitro concentrations than can be achieved clinically, are of interest because their in vitro inhibitory effects in lymphocytes and human mesangial cells are independent of L-type calcium channel inhibition.[87,93] This finding suggests that calcium antagonists may possess other antiproliferative properties, perhaps related to inhibition of precursor molecule cellular incorporation[93,94] or inhibition of potassium efflux, which is known to be important in the activation and function of lymphocytes.[95,96] Consequently, some of the potential immunomodulatory properties of calcium antagonists may be unrelated to their influence on calcium conductance through L-type channels. To support these observations, in vitro experimental studies assessing the response of lymphocytes and mesangial cells to various mitogens demonstrate that the isomeric forms of verapamil (R^+,S^-) exhibit similar antiproliferative effects,[87,93] even though the S^- isomer has an inhibitory effect on blocking the calcium channel and R^+ isomer has no effect at all.

A recent report by Hailer et al.[94] demonstrated that cytokine-induced expression of adhesion molecules on human umbilical vein endothelial cells was not affected by the presence of calcium antagonists. These findings support the concept that the immunomodulating properties of calcium antagonists are independent of adhesion molecule expression.

In vivo experimental studies have yielded conflicting results in regard to the immunosuppressant properties of calcium antagonists. Foegh et al.[95] reported that in a rat heterotopic cardiac allograft model only high doses of nifedipine or verapamil significantly increased mean survival time compared with azathioprine. Corteza et al.[96] reported that nifedipine induced significant suppression of delayed-type hypersensitivity to antigen in mice, whereas verapamil and diltiazem were found to be ineffective. Their results contrast with the report by

Dumont et al.,[97] who found that oral treatment with diltiazem in canine heterotopic heart transplant recipients resulted in complete abrogation of mitogen-stimulated lymphocyte proliferation, suggesting that long-term exposure of immunocompetent cells to calcium antagonists plays a role in the expression of their immunomodulatory properties. Others[98] have reported that benzothiazepine-like calcium antagonists (diltiazem and clentiazem) may alter the metabolism of cyclosporine to achieve either greater tissue penetration of the metabolites or greater production of a more immunosuppressive metabolite that further interferes with lymphocyte activation and function. As opposed to phenylalkylamines or dihydropyridines, benzothiazepines resulted in significant immunosuppression in a rat heterotopic cardiac allograft model when used alone; they also interacted beneficially with cyclosporine to prolong allograft survival.

Clinical trials in humans have also yielded conflicting results. Chitwood et al.[99] found no effect of calcium antagonists on lymphocyte function in humans. On the other hand, in their study of 20 kidney allograft recipients treated with cyclosporine and one of three different calcium antagonists, Carozzi and colleagues[100] found subtle but consistent measurable suppression of serum levels of interleukin 1β, interleukin 2, interleukin 6, and tumor necrosis factor alpha (TNFα).

Data about the effect of calcium antagonists on rejection incidence in patients receiving cyclosporine-based immunosuppression are limited. Ladefoged and Andersen[61] identified nine studies[62–67,101,102] describing the effect of calcium antagonists on the rate of rejection in renal transplant recipients. A significant reduction was reported in two studies, whereas no reduction was observed in seven. Figure 12 depicts the odds ratio for risk of rejection in each study. An average reduction of 6.8% (confidence interval: –2.7% to –16.3%) was observed if all of the studies were averaged together ($p = 0.38$).The studies by Wagner and Neumayer,[62] which demonstrated a reduction in the incidence of rejection, demonstrated evidence that calcium antagonist-treated patients had higher cyclosporine levels, which may explain their observations. The studies have major problems: (1) almost all were open-label and non-randomized and (2) duration of follow-up was limited (usually < 6 months). Thus, based on currently available experimental and human in vivo data, little evidence supports the concept that calcium antagonists have immunomodulatory properties in cyclosporine-treated renal transplant recipients.

INTERACTION OF CALCIUM ANTAGONISTS WITH CYCLOSPORINE LEVELS

Calcium antagonists can alter cyclosporine metabolism by inhibiting the hepatic cytochrome P450 system (Table 2). This effect has been best described with verapamil, diltiazem, and nicardipine.[103–108] Of interest, the three drugs are structurally different calcium antagonists. In contrast, nifedipine, isradipine, and felodipine, all members of the dihydropyridine class of calcium antagonists, do not alter cyclosporine levels.[109–111] Whether amlodipine, which is also a dihydropyridine calcium antagonist, intereferes with cyclosporine metabolism is controversial. Pesavento et al.[109] demonstrated that amlodipine led to a 40% increase in cyclosporine levels, and Vanderchaaf et al.[112] reported that amlodipine resulted in a 23% increase in cyclosporine levels. These data contrast with the results of

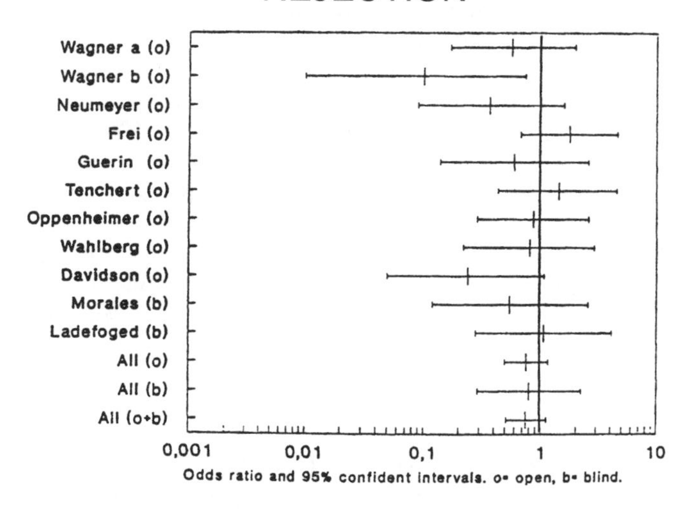

FIGURE 12. The odds ratios for 11 different studies reporting the impact of calcium antagonist therapy on the incidence of rejection (From Ladefoged SD, Andersen CB: Calcium channel blockers in kidney transplantation. Clin Transpl 8:128–133, 1994, with permission.)

a study by Toupance et al.,[113] who reported no difference in cyclosporine levels in patients treated with amlopidine for 4 weeks. However, in the Pesavento and Vanderchaaf studies, the transplant patients received amlodipine for a longer period, and the investigators theorized that, because of the long half-life of amlopidine (> 45 hours), the longer duration of their studies may have allowed the inhibitory effect on cyclosporine metabolism to manifest. Pesavento et al.[109] theorized that the effect of amlopidine in increasing cyclosporine levels may be related to its ability not only to bind to the dihydropyridine ring binding site but also to interact with the verapamil and diltiazem binding sites.

In addition, another dihydropyridine calcium antagonist, nicardipine, substantially elevates cyclosporine levels by as much as 250–370%.[106,107] This observation also indicates that not all members of each chemically separate calcium antagonist class are the same.

No published reports demonstrate that calcium antagonists affect tacrolimus levels. However, because tacrolimus is a macrolide drug that affects hepatic cytochrome P450 function, it is assumed that it may react with calcium antagonists much like cyclosporine.

TABLE 2. Interactions Between Calcium Antagonists and Cyclosporine

Increases Cyclosporine Level		No Effect on Cyclosporine Level	
Class	**Drug**	**Class**	**Drug**
Dihydropyridine	Amlodipine	Dihydropyridine	Felodipine
Benzothiazepine	Diltiazem	Dihydropyridine	Isradipine
Dihydropyridine	Nicardipine	Dihydropyridine	Nifedipine
Phenylalkylamine	Verapamil	Dihydropyridine	Nitrendipine

As previously discussed, alteration in cyclosporine metabolism may not be detrimental. This property may be important in reducing overall cost of immunosuppressant therapy. In addition, altered metabolism of cyclosporine may result in either less nephrotoxic metabolites or perhaps even more immunosuppressive metabolites, which may be helpful in organ transplantation. However, clinical data to support these observations are limited, and potential benefits must be weighed against the likelihood of a more complicated immunosuppressive regimen with greater potential for mistakes if drug doses are adjusted and levels are not carefully followed.

One interesting report[114] demonstrates that diltiazem, but not verapamil or nifedipine, can decrease the clearance of intravenously administered cyclosporine. The etiology of this effect is not well explained, but it may be related to hepatic metabolism. This interaction needs to be carefully monitored in patients placed on intravenous cyclosporine.

GINGIVAL HYPERTROPHY: THE INTERRELATIONSHIP BETWEEN CYCLOSPORINE AND CALCIUM ANTAGONISTS

Gingival hyperplasia is a common side effect of chronic cyclosporine therapy.[115–117] Two recent clinical trials evaluated whether concomitant calcium antagonist therapy worsens this side effect. Bokenkamp et al.[116] analyzed the gingival status of 106 children transplanted at their center and noted a significantly higher degree of gingival overgrowth in children receiving a combination of cyclosporine and nifedipine than in children receiving cyclosporine or nifedipine alone. After elimination of calcium antagonists from the antihypertensive regimen and improved dental care with chlorhexidine gel, the investigators noted a reduction in the degree of gingival hyperplasia. In another recent study, King et al.[117] performed a cross-sectional investigation in 66 renal transplant recipients who received either cyclosporine alone (n = 18), cyclosporine and nifedipine (n = 15), or cyclosporine and diltiazem (n = 12) and a control group treated with azathioprine (n = 21). The hyperplastic index of gingival hypertrophy in the first three groups was substantially greater than in the control group. The investigators concluded that cyclosporine, either alone or in combination with a calcium antagonist, induced a significant degree of gingival enlargement compared with controls.

Although the two studies do not prove that calcium antagonists worsen cyclosporine-mediated gingival hypertrophy, they certainly suggest that when gingival hypertrophy occurs, one of the therapeutic approaches should be consideration of another antihypertensive medication in addition to proper dental hygiene.

INFLUENCE OF CALCIUM ANTAGONISTS ON GRAFT AND PATIENT SURVIVAL

Calcium antagonists are effective antihypertensive agents and may provide benefit in terms of cyclosporine-mediated nephrotoxicity and allograft dysfunction in the immediate posttransplant period. They may possess beneficial immunomodulatory effects and inhibit some of the mitogenic processes that

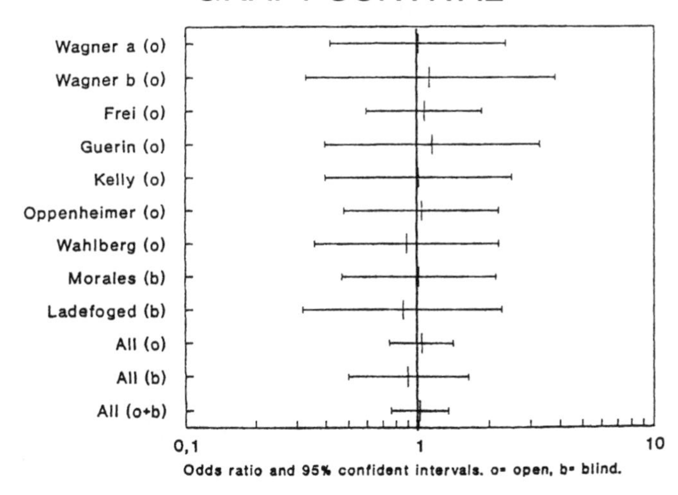

FIGURE 13. The odds ratios for 9 different studies reporting the impact of calcium antagonist therapy on graft survival (From Ladefoged SD, Andersen CB: Calcium channel blockers in kidney transplantation. Clin Transpl 8:128–133, 1994, with permission.)

lead to vascular remodeling and restructuring and progressive vascular obliteration. Such effects may be related to either nonimmunologic or immunologic causes of progressive graft dysfunction. Consequently, calcium antagonists may improve renal allograft survival rates.

Unfortunately, limited data prospectively evaluate large numbers of renal transplant recipients, with and without calcium antagonists, over a sufficient period to assess overall impact. The recent review by Ladefoged and Andersen[61] evaluated graft survival in all published studies of calcium antagonists in transplantation. Only seven studies[62,64–67,118] were reported. The odds ratios for the seven studies are depicted in Figure 13. No evidence of a statistically significant benefit was observed in any of the studies ($p = 0.92$). However, when Ladefoged and Andersen evaluated the effect of calcium antagonists on graft function, they found a total of 12 studies[62–69,101,102,118] that used serum creatinine as an indicator. A significant improvement was noted in 5 of the 12 studies. In seven additional studies[62,63,65,68,69,102] graft function was assessed with radioisotopic tracer techniques or inulin. A significant improvement was reported in 5 of the 7 studies. These results suggest that calcium antagonists may have a beneficial effect. Complicating this analysis, however, were the different chemical classes of calcium antagonists used, the different immunosuppressive regimens, the small study sizes and limited duration of follow-up (usually 1 year or less), and the lack of a prospective, randomized, controlled design. The longest study reported in the literature was the 5-year study by Ponticelli et al.,[36] who assessed long-term renal function in hypertensive cyclosporine-treated renal transplant recipients who received nifedipine ($n = 52$) and 58 controls who received no calcium antagonist. They noted no difference in mean plasma creatinine level at either 1 or 5 years of follow-up in either of the two groups, suggesting no beneficial effects of calcium antagonists on renal function.

However, this study was small and retrospective, with too many confounding variables to make any significant statement about whether treatment with calcium antagonists had an overall effect on long-term renal function.

CONCLUSIONS

There is no question that calcium antagonists represent an important therapeutic consideration in cyclosporine-treated renal allograft recipients. The major benefits are the ability to control blood pressure without compromise of renal function and attenuation of delayed graft function, as supported by available evidence.[60,68,69,119–123] The second benefit will be most appreciated in centers with a high rate of delayed graft function. Because attenuation of delayed graft function has been associated with improvement in renal allograft survival rates, it is likely that calcium antagonists have a beneficial effect on graft survival in such patients. However, prospective, randomized clinical trials, using double-blind techniques that include treatment of the donor graft with calcium antagonists before or during harvesting and perfusion and possibly during reperfusion or intraoperatively, need to be conducted to assess long-term benefit. A second approach is to assess long-term control of blood pressure, proteinuria, and renal function in patients treated with and without calcium antagonists in order to determine whether better blood pressure control and possible attenuation of cyclosporine-mediated fibrotic changes will improve preservation of renal function.

Additional studies need to be performed to evaluate the benefits of calcium antagonists in high-risk recipients, such as older patients, regrafted patients, and patients with diabetes who receive a kidney transplant. An interesting small study by Weinrauch et al.[124] demonstrated that diabetic kidney transplant recipients who received calcium antagonists (n = 36) survived longer than those who did not (n = 32) because of decreased need for dialytic therapy and lower incidence of septic death and multiple septic episodes.

Finally, clinical trials also need to focus on the different chemical classes of calcium antagonists, because one class may have advantages over the others. Such research may be important in terms of side effects and effect on blood pressure, proteinuria, and even cyclosporine metabolism because of their relevance to overall immunosuppression and toxicity.

Acknowledgment

The author acknowledges the expert secretarial assistance of Valerie Heisler.

References

1. Buhler FR: Calcium antagonists. In Laragh JH, Brenner BM (eds): Hypertension: Pathophsiology, Diagnosis, and Management. New York: Raven, 1990, pp 2169–2179.
2. Weir MR: Therapeutic benefits of calcium channel blockers in cyclosporine-treated organ transplant recipients: Blood pressure control and immunosuppression. Am J Med 90(Suppl 5A):32S–36S, 1991.
3. Weir MR, Suthanthiran M: Supplementation of immunosuppressive regimens with calcium channel blockers: Rationale and clinical efficacy. Clin Immunother 2:458–467, 1994.
4. Whitney RB, Sutherland RM: Enhanced uptake of calcium by transforming lymphocytes. Cell Immunol 5:137–147, 1972.

5. Parker CW: Correlation between mitogenicity and stimulation of calcium uptake in human lymphocytes. Biochem Biophys Res Commun 61:1180–1186, 1974.
6. Dubois JH, Crumpton MJ: Extracellular Ca^{++} requirement for initiation of lymphocyte growth. Biochem Soc Trans 8:721–722, 1980.
7. Greene WC, Parker CW: Calcium and lymphocyte activation. Cell Immunol 25:74–89, 1976.
8. Diamantstein T, Ulmer A: The control of immune response in vitro by Ca^{++}. Immunology 28:121–125, 1975.
9. Freedman MH: Early biochemical events in lymphocyte activation. Cell Immunol 44:290–313, 1979.
10. Laher I, Hwa JJ, Bevan JA: Calcium and vascular myogenic tone. In Paoletti R, Vanhoute P (eds): Calcium Antagonists: Pharmacology and Clinical Research. Proc Annu NY Acad Sci 552:216–225, 1988.
11. Scheid CR, Fay FS: Transmembrane ^{45}Ca fluxes in isolated smooth muscle cells: Basal Ca^{2+} fluxes. Am J Physiol 246:C422–C430, 1984.
12. Khalil RA, van Breeman C: Sustained contraction of vascular smooth muscle: Calcium influx or C-kinase activation? J Pharmacol Exp Ther 244:537–542, 1988.
13. Mehta JL: Influence of calcium-channel blockers on platelet function and arachidonic acid metabolism. Am J Cardiol 55:158B–164B, 1985.
14. Addonizio VP, Fisher CA, Strauss JF, et al: Effects of verapamil and diltiazem on human platelet function. Am J Physiol 250:366–371, 1986.
15. Mehta J, Mehta P, Ostrowski N: Calcium blocker diltiazem inhibits platelet activation and stimulates vascular prostacyclin synthesis. Am J Med Sci 291:20–24, 1986.
16. Jelenska MM, Kopec M: Platelet-mediated collagen and fibrin retraction: Effect of prostaglandins, cyclic AMP, calcium antagonists, and N-ethylmaleimide. Thromb Res 30:499–509, 1983.
17. Trump BF, Berezesky IK, Laiho KV, et al: The role of calcium in cell injury: A review. Scanning Electron Microsc 2:437–462, 1980.
18. Arnold PE, Van Putten VJ, Lumlertgul D, et al: Adenine nucleotide metabolism and mitochondrial Ca^{2+} transport following renal ischemia. Am J Physiol 250:F357–F363, 1986.
19. Chien KR, Abrams J, Serroni A, et al: Accelerated phospholipid degradation and associated membrane dysfunction in irreversible, ischemic liver cell injury. J Biol Chem 253:4809–4817, 1978.
20. McCord JM: Oxygen-derived free radicals in post-ischemic tissue injury. N Engl J Med 312:159–163, 1985.
21. Rooth P, Dawidson I, Dillen K, Taljedal IB: Protection against cyclosporine-induced impairment of renal microcirculation by verapamil in mice. Transplantation 45:433–437, 1988.
22. Weir MR, Klassen DK, Shen SY, et al: Acute effects of intravenous cyclosporine on blood pressure, renal hemodynamics, and urine prostaglandin production of healthy humans. Transplantation 49:41–47, 1990.
23. Dawidson I, Rooth P, Fry WR, et al: Prevention of acute cyclosporine-induced renal blood flow inhibition and improved immunosuppression with verapamil. Transplantation 48:575–580, 1989.
24. Ruggenenti P, Perico N, Mosconi L, et al: Calcium channel blockers protect transplant patients from cyclosporine-induced daily renal hypoperfusion. Kidney Int 43:706–711, 1993.
25. Chapman JR, Mareen R, Arias M, et al: Hypertension after renal transplantation: a comparison of cyclosporine and conventional immunosuppression. Transplantation 43:860–864, 1987.
26. Bellet M, Cabrol C, Sassano P, et al: Systemic hypertension after cardiac transplantation: Effect of cyclosporine on the renin-angiotensin system. Am J Cardiol 56:927–931, 1985.
27. Bennett WM, Porter GA: Cyclosporine associated hypertension. Am J Med 85:131–133, 1988.
28. Murray BM, Paller MS, Ferris TE: Effect of cyclosporine administration on renal hemodynamics in conscious rat. Kidney Int 28:767–774, 1985.
29. Curtis JJ, Luke RG, Dubousky E, et al: Cyclosporine in therapeutic doses increases renal allograft vascular resistance. Lancet 2:477–479, 1986.
30. Luke RG: Mechanism of cyclosporine-induced hypertension. Am J Hypertens 4:468–471, 1991.
31. Curtis JJ, Luke RG, Jones D, et al: Hypertension in cyclosporine-treated renal transplant recipients is sodium dependent. Am J Med 85:134–138, 1988.
32. Romero JC, Raij L, Granger GP, et al: Multiple effects of calcium channel blockers on renal function in hypertension. Hypertenson 10:140–151, 1987.
33. MacGregor GA, Cappuccio FP, Markandu ND: Sodium intake, high blood pressure, and calcium channel blockers. Am J Med 82(Suppl 3B):16–22, 1987.
34. Nicholson JP, Resnick JM, Laragh JH: The antihypertensive effect of verapamil at extremes of dietary sodium intake. Ann Intern Med 107:329–334, 1987.

35. Curtis JJ, Laskow DA, Jones PA, et al: Captopril induced fall in glomerular filtration rate in cyclosporine-treated hypertensive patients. J Am Soc Nephrol 3:1570–1574, 1993.
36. Ponticelli C, Montagnino G, Aroldi A, et al: Hypertension after renal transplantation. Am J Kidney Dis 21(Suppl 2):73–78, 1993.
37. Xue H, Bukoski RD, McCarron DA, et al: Induction of contraction in isolated rat aorta by cyclosporine. Transplantation 43:715–718, 1987.
38. Murray BM, Paller MS: Beneficial effects of renal denervation and prazosin on GFR and renal blood flow after cyclosporine in rats. Clin Nephrol 25:S37–S39, 1986.
39. Scherrer V, Vissing SF, Morgan BJ, et al: Cyclosporine-induced sympathetic activation and hypertension after heart transplantation. N Engl J Med 323:693–699, 1990.
40. Coffman TM, Carr DR, Yarger WE, et al: Evidence that renal prostaglandin thromboxane production is stimulated in chronic cyclosporine nephrotoxicity. Transplantation 43:282–285, 1986.
41. Kon V, Sugiura M, Inagami T, et al: Cyclosporine (Cy) causes endothelin-dependent acute renal failure [abstract] 37:486(A), 1990.
42. June CH, Thompson CB, Kennedy MS, et al: Correlation of hypomagnesemia with the onset of cyclosporine-associated hypertension in marrow transplant patients. Transplantation 41:47–51, 1986.
43. June CH, Thompson CB, Kennedy MS, et al: Profound hypomagnesemia and renal magnesium wasting associated with the use of cyclosporine for marrow transplantation. Transplantation 39:620–624, 1985.
44. Loutzenhiser R, Epstein M: Effects of calcium antagonists on renal hemodynamics. Am J Physiol 249:F619–F626, 1985.
45. Goldberg JP, Schrier RW: Effects of calcium membrane blockers on in vivo vasoconstrictive properties of norepinephrine, angiotensin II, vasopressin. Miner Electrolyte Metab 10:178–183, 1984.
46. Loutzenhiser R, Epstein M: The renal hemodynamic effects of calcium antagonists. In Epstein M, Loutzenhiser R (eds): Calcium Antagonists and the Kidney. Philadelphia, Hanley & Belfus, 1990, pp 33–73.
47. Loutzenhiser R, Epstein M, Horton C: Reversal of renal and smooth muscle actions of the thromboxane mimetic U-44069 diltiazem. Am J Physiol 19:F619–F626, 1986.
48. Ichikawa I, Miele JF, Brenner BM, et al: Reversal of renal cortical actions of angiotensin II by verapamil and manganese. Kidney Int 16:137–147, 1979.
49. Saunders E, Weir MR, Kong BW, et al: A comparison of the efficacy and safety of a beta-blocker, a calcium channel blocker, and a converting enzyme inhibitor in hypertensive blacks. Arch Intern Med 150:1707–1713, 1990.
50. Mimran A, Mourad G, Ribstein J: The renin-angiotensin system and renal function in kidney transplantation. Kidney Int 38(Suppl 30):S114–S117, 1990.
51. Mourad G, Ribstein J, Mimran A: Converting-enzyme inhibitor versus calcium antagonists in cyclosporine-treated renal transplants. Kidney Int 43:419-425, 1993.
52. Julian BA, Gaston RS, Barker CV, et al: Erythropoiesis after withdrawal of enalapril in post-transplant erythrocytosis. Kidney Int 46:1397–1403, 1994.
53. Schwietzer GKW, Hartmann A, Kober G, Jungmann E, Stratmann D, Kaltenbach M, Schoeppe W: Chronic angiotensin-converting enzyme inhibition may improve sodium excretion in cardiac transplant hypertension. Transplantation 59:999–1004, 1995.
54. Sennesael JJ, Lamote JG, Violet I, et al: Divergent effects of calcium channel and angiotensin-converting enzyme blockade on glomerular tubular function in cyclosporine-treated renal allograft recipients. Am J Kidney Dis 27:701-708, 1996.
55. Hsiao-Yi L, Rocher LL, McQuillan MA, et al: Cyclosporine- induced hyperuricemia and gout. N Engl J Med 321:287–292, 1989.
56. The Canadian Multicentre Transplant Study Group: A randomized clinical trial of cyclosporine in cadaveric renal transplantation. N Engl J Med 314:1219–1225, 1986.
57. Schrier RW, Arnold PE, Van Putten NJ, et al: Cellular calcium in ischemic acute renal failure: Role of calcium entry blockers. Kidney Int 32:313–321, 1987.
58. Malis CD, Cheung JY, Leaf AL, et al: Effects of verapamil in models of ischemic acute renal failure in the rat. Am J Physiol 245:F735–F742, 1983.
59. Burke TJ, Arnold PE, Gordon JA, et al: Protective effect of intrarenal calcium membrane blockers before or after renal ischemia. Functional, morphological, and mitochondrial studies. J Clin Invest 74:1830–1841, 1984.
60. Neumayer HH, Wagner K: Influence of the calcium antagonist diltiazem on delayed graft function in cadaver kidney transplantation: Results of a six month follow-up. Transplant Proc 19:1353–1357, 1987.

61. Ladefoged SD, Andersen CB: Calcium channel blockers in kidney transplantation. Clin Transpl 8:128–133, 1994.
62. Wagner K, Neumayer HH: Calciumantagonismus und akutes Nirenversagen. Studien zur Pathogenese des akuten Transplantatversagens nach Nierentransplantation und seiner Prävention durch den Calciumantagonisten Diltiazem. Stuttgart, Schattauer, 1988, pp 53–79.
63. Neumayer HH, Schriber M, Wagner K: Prevention of delayed graft function by diltiazem and iloprost. Transplant Proc 21:1221–1224, 1989.
64. Frei U, Margreiter R, Harms A, et al: Peroperative graft reperfusion with a calcium antagonist improves initial function: Preliminary results of a prospective randomized trial in 110 kidney recipients. Transplant Proc 19:3539–3541, 1987.
65. Guerin C, Berthoux P, Broyet C, Berthoux F: Effects du diltiazem sur la pression arterielle et la function renale du transplante renal sous ciclosporine A. Arch Mal Cœur 82:1223–1227, 1989.
66. Oppenheimer F, Alcaraz A, Manalich M, et al: Influence of the calcium blocker diltiazem on the prevention of acute renal failure after renal transplantation. Transplant Proc 24:50–51, 1992.
67. Wahlberg J, Hanas E, Bergstrom C, et al: Diltiazem treatment with reduced dose of cyclosporine in renal transplant recipients. Transplant Proc 24:311–312, 1992.
68. Dawidson I, Rooth P, Alway C, et al: Verapamil prevents posttransplant delayed function and cyclosporine A nephrotoxicity. Transplant Proc 22:1379–1380, 1990.
69. Harper SJ, Moorhouse J, Veitch PS, et al: Nifedipine improves immediate, and 6- and 12-month graft function in cyclosporine A (CyA) treated renal allograft recipients. Transplant Int 5 (Suppl 1):S69–S72, 1992.
70. Vasquez EM, Pollak R: Effect of calcium channel blockers on graft outcome in cyclosporine-treated renal allograft recipients. Transplantation 60:885–887, 1995.
71. Pirsch JD, D'Alessandro AM, Roecker FB, et al: A controlled, double-blind, randomized trial of verapamil and cyclosporine in cadaveric renal transplant patients. Am J Kidney Dis 21:189–195, 1993.
72. Grewal HP, Garland L, Novak K,et al: Risk factors for postimplantation pancreatitis and pancreatic thrombosis in pancreas transplant recipients. Transplantation 56:609–612, 1993.
73. Swoboda L, Clancy DE, Donnibriuk MA, Rieder-Nelissen CM: The influence of verapamil on lung preservation. A study on rabbit lungs with a reperfusion model allowing physiological loading. Thorac Cardiovasc Surg 41:85–92, 1993.
74. Chavez-Cantaya RE, Pino DeSola G, Ramirez-Romero P, et al: Ischemia and reperfusion injury of the rat liver: the role of nemodipine. J Surg Res 60:199–206, 1996.
75. Van Buren DH, Burke JF, Lewis RM: Renal function in patients receiving long-term cyclosporine therapy. J Am Soc Nephrol 4:S17–S22, 1994.
76. Khanna A, Baogui L, Stenzel KH, Suthanthiran M: Regulation of new DNA synthesis in mammalian cells by cyclosporine. Demonstration of a transforming growth factor β-dependent mechanism of inhibition of cell growth. Transplantation 57:577–582, 1994.
77. Mayer-Lehnert H, Schrier RW: Potential mechanism of cyclosporine A induced vascular smooth muscle contraction. Hypertension 13:352–360, 1989.
78. Burros EJG, Boim MA, Ajzen H, et al: Glomerular hemodynamics and hormonal participation in cyclosporine nephrotoxicity. Kidney Int 32:19–25, 1987.
79. Sulemanlar G, Lien YH, Shapiro JL, et al: Efficacy of verapamil in preventing chronic cyclosporine nephrotoxicity in the rat. J Am Soc Nephrol (in press).
80. Kunzendorf U, Walz G, Brockmoeller J, et al: Effect of diltiazem upon metabolism and immunosuppressive action of cyclosporine in kidney graft recipients. Transplantation 52:280–284, 1991.
81. Nagineni CN, Misra BC, Lee DBN, et al: Cyclosporine A–calcium channels interaction: A possible mechanism for nephrotoxicity. Transplant Proc 19:1358–1362, 1987.
82. Scoble JE, Senior JCM, Chan P, et al: In vitro cyclosporine toxicity: The effect of verapamil. Transplantation 47:647–650, 1979.
83. McMillen MA, Baumgarten WK, Schaefer HC, et al: The effect of verapamil on cellular uptake organ distribution and pharmacology of cyclosporine. Transplantation 44:395–401, 1987.
84. Taber TE, Maikranz PS, Haag BW, et al: Maintenance of adequate hemodialysis access. Prevention of neointimal hyperplasia. ASAIO Trans 41:842–846, 1995.
85. Kovacs EJ: Fibrogenic cytokines: the role of immune mediators in the development of scar tissue [Review]. Immunol Today 12:17–23, 1991.
86. Kurihara H, Yoshizumi M, Sugiyama T, et al: Transforming growth factor-β stimulates the expression of endothelin mRNA by vascular endothelial cells. Biochem Biophys Res Commun 159: 1435–1440, 1989.

87. Orth SR, Nohiling R, Bonisch S, Ritz E: Inhibitory effect of calcium channel blockers on human mesangial cell growth: Evidence for actions independent of L-type Ca^{++} channels. Kidney Int 49:868–879, 1996.
88. Schroeder JS, Gao SZ, Alderman EL, Hunt SA, et al: A preliminary study of diltiazem in the prevention of coronary artery disease in heart-transplant recipients. N Engl J Med 328:164–170, 1993.
89. Birx DL, Berger M, Fleisher TA: The interference of T cell activation by calcium channel blocking agents. J Immunol 133:2904–2909, 1984.
90. Weir MR, Peppler R, Gomolka D, et al: Additive inhibition of afferent and efferent immunological responses of human peripheral blood mononuclear cells by verapamil and cyclosporine. Transplantation 51:851–857, 1991.
91. McMillen MA, Lewis T, Jaffe BM, et al: Verapamil inhibition of lymphocyte proliferation and function in vitro. J Surg Res 39:76–80, 1985.
92. Weir MR, Peppler R, Gomolka D, Handwerger BS: Additive effect of cyclosporine and verapamil may occur through different mechanisms that may be dependent or independent on the slow calcium channel. Transplant Proc 21:866–870, 1989.
93. Weir MR, Gomolka D, Peppler R, Handwerger BS: Mechanisms responsible for inhibition of lymphocyte activation by agents which block membrane calcium or potassium channels. Transplant Proc 25:605–609, 1993.
94. Hailer NP, Blaheta RA, Harder S, et al: Modulation of adhesion molecule expression on endothelial cells by verapamil and other Ca^{++} channel blockers. Immunobiology 191:38–51, 1994.
95. Foegh ML, Khirabadi RS, Ramwell PW: Prolongation of rat cardiac allograft survival by Ca^{++} blockers. Transplantation 40:211–212, 1985.
96. Corteza Q, Shen S, Revie D, Chretien P: Effects of calcium channel blockers on in vivo cellular immunity in mice. Transplantation 47:339–342, 1989.
97. Dumont L, Libersan D, Chen HF, et al: Immunosuppressive activity of diltiazem-like calcium antagonists in experimental heart transplantation [abstract]. Can J Cardiol 7:(Suppl A):125A, 1991.
98. Dumont L, Chen H, Dalozo P, et al: Immunosuppressive properties of the benzothiazepine calcium antagonists diltiazem and clentiazem, with and without cyclosporine, in heterotopic rat heart transplantation. Transplantation 56:181–184, 1993.
99. Chitwood KK, Heim-Duthoy KL: Immunosuppressive properties of calcium channel blockers [Review]. Pharmacotherapy 13:447–454, 1993.
100. Carozzi S, Nasini MG, Pietrucci A, et al: Immunosuppressive effects of different calcium channel blockers in human kidney allografts. Transplantation Proc 27:1054–1057, 1995.
101. Tenschert W, Harfmann P, Meyr-Moldenhauser WH, et al: Kidney protective effect of diltiazem after renal transplantation with long cold ischemia time and triple-drug immunosuppression. Transplant Proc 23:1334–1335, 1991.
102. Dawidson I, Rooth P, Fry WR, et al: Prevention of acute cyclosporine-induced renal blood flow inhibition and improved immunosuppression with verapamil. Transplantation 48:575–580, 1989.
103. Pochet JM, Pirson Y: Cyclosporine-diltiazem interaction (Letter). Lancet 1:979, 1986.
104. Lindholm A, Henrieson S: Verapamil inhibits cyclosporine metabolism. Lancet 1:1962–1963, 1987.
105. Renton KW: Inhibition of hepatic microsomal drug metabolism by the calcium channel blockers diltiazem and verapamil. Biochem Pharmacol 34:2549–2553, 1985.
106. Bourbigot B, Guiserix J, Airiau J, et al: Nicardipine increases cyclosporine blood levels [Letter]. Lancet 1:1447, 1986.
107. Cantarovich M, Hiesse C, Lockiec F, et al: Confirmation of the interaction between cyclosporine and the calcium channel blocker nicardipine in renal transplant patients. Clin Nephrol 28:190–193, 1987.
108. Kronbach T, Fischer V, Meyer U: Cyclosporine metabolism in human liver: Identification of a cytochrome P-450 III gene family as the major cyclosporine-metabolizing enzyme explains interactions of cyclosporine with other drugs. Clin Pharmacol Ther 43:630–635, 1988.
109. Pesavento TE, Jones PA, Julian BA, Curtis JJ: Amlodipine increases cyclosporine levels in hypertensive renal transplant patients: Results of a prospective study. J Am Soc Nephrol 7:831–835, 1996.
110. Endresen L, Bergan S, Holdaas H, et al: Lack of effect of the calcium antagonist isradipine on cyclosporine pharmacokinetics in renal transplant patients. Ther Drug Monit 13:490–495, 1991.
111. Cohen DJ, Teng SN, Appel GB: Influence of oral felodipine on serum cyclosporine concentrations. Clinical Transplantation 8:541–545, 1994.

112. Van der Schaaf M, Hene RJ, Floor M, et al: Hypertension after renal transplantation calcium channel or converting enzyme blockade? Hypertension 25:77–81, 1995.
113. Toupance O, Lavaud S, Canivet E, et al: Antihypertensive effect of amlodipine and lack of effect with cyclosporine metabolism in renal transplant recipients. Hypertension 24:297–300, 1994.
114. Sketris IS, Methot ME, Nicol D, et al: Effect of calcium channel blockers on cyclosporine clearance and use in renal transplant patients. Ann Pharmacother 28:1227–1231, 1994.
115. Nell A, Riegler B, Ulm C, et al: Stimulation of platelet mitogen-induced prostaglandin I 2 synthesis in peridontal tissue of cyclosporine A treated patients. Wien Klin Wochenschr 107:278–282, 1995.
116. Bokenkamp A, Bohnhorst B, Beier C, et al: Nifedipine aggravates cyclosporine A-induced gingival hyperplasia. Pediatr Nephrol 8:181–185, 1994.
117. King GN, Fullinfau R, Huggins TJ, et al: Gingival hyperplasia in renal allograft recipients receiving cyclosporin-A and calcium antagonists. J Clin Periodontol 20:286–293, 1993.
118. Kelly JJ, Walker RG, D'Apice AJF, Kincaid-Smith P: A prospective study of the effect of diltiazem in renal allograft recipients receiving cyclosporin A: Preliminary results. Transplant Proc 22:2127–2128, 1990.
119. Neumayer HH, Kunzendorf U, Schreiber M: Protective effects of calcium antagonists in human renal transplantation. Kidney Int 41(Suppl 36):S87–S93, 1992.
120. Dawidson I, Rooth P, Lu C, et al: Verapamil improves the outcome after cadaveric renal transplantation. J Am Soc Nephrol 2:983–990, 1991.
121. Palmer BF, Dawidson I, Sagalowsky A, et al: Improved outcome of cadaveric renal transplantation due to calcium channel blockers. Transplantation 52:640–645, 1991.
122. Chrysostomou A, Walker RG, Russ GR, et al: Diltiazem in renal allograft recipients receiving cyclosporine. Transplantation 55:300–304, 1993.
123. Suthanthiran M, Haschemeyer RH, Riggio RR, et al: Excellent outcome with a calcium channel blocker-supplemented immunosuppressive regimen in cadaveric renal transplantation. A potential strategy to avoid antibody induction protocols. Transplantation 55:1008–1013, 1993.
124. Weinrauch LA, D'Elia JA, Gleason RE, et al: Role of calcium channel blockers in diabetic renal transplant patients: Preliminary observations on protection from sepsis. Clin Nephrol 44:185–192, 1995.

MURRAY EPSTEIN M.D., F.A.C.P. / BERNARD WAEBER M.D.

26

Fixed-dose Combination Therapy with Calcium Antagonists

The principle of polypharmacy has gained almost universal acceptance for patients with certain conditions, such as angina, who do not respond to a single agent. Moreover, the treatment of hypertension often includes the use of combination therapy (addition of one drug with a complementary mechanism of action to another drug) to enhance the blood pressure-lowering effect of a single agent. Although this practice is common in medicine, the concept of polypharmacy with fixed-dose combination therapy has been slow to gain widespread acceptance.

The pendulum has swung back and forth regarding the advisability of fixed-dose combination antihypertensive agents vs. stepwise addition of the separate components.[1] Historically, the combination of reserpine with hydrochlorothiazide and hydralazine, Ser-Ap-Es, was the first fixed-dose combination approved for use in the United States.[2] The historical worldwide evolution of fixed-dose combinations for treatment of hypertension is summarized in Table 1.

Until recently, few studies were designed specifically to assess whether there is an advantage to fixed-dose combinations. Fortunately, the recent approval by the Food and Drug Administration (FDA) of four fixed-dose combinations of an angiotensin-converting enzyme (ACE) inhibitor and a calcium antagonist has once again focused attention on this issue. Because several additional fixed-dose combinations are awaiting regulatory approval, clinicians will be challenged to examine their approach to combination therapy. Consequently, this is an appropriate time to examine critically the role of fixed-dose agents in the antihypertensive armamentarium. This chapter focuses primarily on the role of the most recently developed fixed-dose combinations, i.e., those containing a calcium channel blocker (CCB) coadministered with either an ACE inhibitor or a beta blocker.

MECHANISMS OF ANTIHYPERTENSIVE ACTION

From among the vast armamentarium of antihypertensive medications, CCBs and ACE inhibitors have emerged over the past several years as the most commonly used agents.[3] Their increased use compared with diuretics and beta blockers is due to numerous factors, including a favorable efficacy-tolerability profile and good marketing efforts. Although both CCBs and ACE inhibitors reduce

TABLE 1. Historical Evolution of Fixed-dose Combination Agents

1960s	1980s
Reserpine/thiazide diuretic	Angiotensin-converting enzyme (ACE) inhibitor/ thiazide diuretic
Reserpine/hydralazine/thiazide diuretic	
Methyldopa/thiazide diuretic	1990s
1970s	Low-dose beta blocker/thiazide diuretic
Thiazide diuretic/potassium-sparing diuretic	Calcium antagonist/beta blocker
Thiazide diuretic/spironolactone	Calcium antagonist/ACE inhibitor
Beta blocker/thiazide diuretic	Angiotensin II antagonist/thiazide diuretic
Clonidine/thiazide diuretic	

arterial pressure, they work by different mechanisms. CCBs inhibit calcium influx into muscle cells,[4–6] whereas ACE inhibitors both reduce synthesis of angiotensin II and lead to an accumulation of bradykinin, a peptide that induces vasodilation by stimulating the release of nitric oxide (NO) from the endothelium.[7,8]

The major hemodynamic abnormality in patients with essential hypertension is an increase in peripheral vascular resistance. Considerable evidence suggests that the elevation in peripheral vascular resistance is mediated, in part, by abnormal transmembrane flux of calcium.[9] This mechanism becomes evident when blockade of calcium-mediated electromechanical coupling in contractile tissues produces arteriolar vasodilation in hypertensive patients but has no effect in normotensive people.[10]

Activation of the renin-angiotensin-aldosterone system is also likely to contribute to the genesis of hypertension, in part because of the potent vasoconstrictive qualities of angiotensin II and the sodium retention induced by aldosterone. Angiotensin II also enhances the release of norepinephrine by sympathetic nerve terminals, which may represent another important pressor mechanism.[11] The local synthesis of angiotensin II in the vasculature may be an additional factor that increases blood pressure.[12] This mechanism may account for the fact that ACE inhibitors reduce blood pressure even when plasma renin is normal or low. Consequently, it is logical to inhibit the renin-angiotensin-aldosterone system with pharmacologic therapy.

The mechanisms of action of beta blockers are still not fully elucidated and may involve a centrally mediated decrease in sympathetic tone, presynaptic inhibition of norepinephrine release, and suppression of renin secretion.[13]

Because CCBs, ACE inhibitors, and beta blockers reduce arterial pressure by different mechanisms, their combination may increase overall antihypertensive efficacy. This is particularly true for combinations of CCBs with ACE inhibitors or beta blockers.

Table 2 summarizes the potential advantages of fixed-dose antihypertensive combination medications.

RATIONALE FOR COMBINATION OF CALCIUM CHANNEL BLOCKERS WITH ACE INHIBITORS AND BETA BLOCKERS

Enhanced Antihypertensive Efficacy

The rationale for prescribing fixed-dose combinations of antihypertensive agents relates in part to the concept that antihypertensive efficacy may be

TABLE 2. Potential Advantages of Fixed-dose Antihypertensive Combination Medications

1. Simplicity of use and convenience for patient and physician
2. Simple titration process (of the combination per se)
3. Improved compliance, with possibly enhanced efficacy
4. Potentiation of antihypertensive effects
   - Additive or synergistic effect
   - Permitting of full blood pressure-lowering effect in patients tending to have less than full response to one component. Examples: beta blocker + diuretic in elderly, angiotensin-converting enzyme inhibition + diuretic in elderly.
5. Reduction in side effects by allowing lower dosage of one or both components. Example: less thiazide-induced hypokalemia with captopril/thiazide combination.
6. Offsetting of undesirable side effects. Example: obviation of calcium antagonist-induced edema by addition of angiotensin-converting enzyme inhibitor.
7. Cost of fixed-dose combinations is usually less than cost of the constituents prescribed separately.

Adapted from Epstein M, Oster JR: Hypertension: Practical Management. Miami, Battersea, 1988, p 128, with permission.

enhanced when two classes of agents are coadministered. First, there is increasing awareness of heterogeneity in responsiveness to treatment.[14] Crossover studies demonstrate that patients who respond to one class of drug do not necessarily respond to a different class.[15,16] Consequently, good blood pressure control can be achieved in a larger proportion of patients by use of low doses of two drugs that act on different physiologic systems.[16]

Second, combination therapy serves to countervail the counterregulatory mechanisms that are triggered whenever pharmacologic intervention is initiated and that act to limit the efficacy of the antihypertensive medication.[1] For example, selective arterial vasodilators, such as hydralazine or minoxidil, stimulate both the sympathetic nervous system and the renin-angiotensin system.[17] These agents induce profound sodium avidity by the kidney and hence volume expansion. Of interest, the CCB-mediated vasodilation provides variable effects on sympathetic activation and paradoxically natriuretic and diuretic responses.[18–20] Short-acting dihydropyridine (nifedipine-like) CCBs induce a reflex increase in sympathetic neuronal tone much like other vasodilators.[18] Longer-acting dihydropyridine CCBs also may increase sympathetic tone, especially at initiation of treatment, albeit not to the same extent as the short-acting agents.[18] Conversely, nondihydropyridine CCBs, such as verapamil or diltiazem, have no effect on autonomic function.[21]

The activation of the sympathetic nervous system secondary to calcium channel blockade with a dihydropyridine limits the fall in blood pressure. This phenomenon is reflected by the observation that the acute decrease in blood pressure induced by nifedipine is inversely correlated in hypertensive patients with the simultaneous increase in heart rate and is associated with a stimulation of renin secretion.[22] It therefore seems appealing to add a medication that buffers the counterregulatory responses of the sympathetic nervous system and the renin-angiotensin system, such as a beta blocker[13] or an ACE inhibitor.[23]

Enhanced Tolerability

For initial therapy in most hypertensive patients, full doses of multiple drugs in combination are not routinely prescribed because of inability to titrate each of the constituents and to separate individual side effects; in addition, patients may be exposed to superfluous therapy. However, if low doses of two antihypertensive agents with different modes of action are combined, dose-dependent adverse effects are minimized because smaller doses are used to achieve control.

This concept has been nicely demonstrated by Fagan (Fig. 1).[24] With a low dose of drug A, only a partial therapeutic effect is obtained and adverse effects (A′) are minimal. If the dose is raised to B, the greater effect will be accompanied by more adverse effects (B′). However, if a low dose of another drug is added, with minimal side effects, the extra benefit will be obtained without more adverse effects, which remain at A′.

Low-dose combination therapy was validated recently by multifactorial trials:[25,26] low doses of bisoprolol fumarate (an ultraselective $beta_1$ blocker) and hydrochlorothiazide in combination decreased systolic and diastolic blood pressure with few adverse effects (Fig. 2). Because efficacy was achieved with no more side effects than placebo, the FDA approved this low-dose combination as first-line therapy.[27]

Another relevant concept for consideration of fixed combination therapy is enhanced tolerability. In brief, this notion suggests that one drug of a fixed combination can antagonize some of the adverse effects of the second drug.[28]

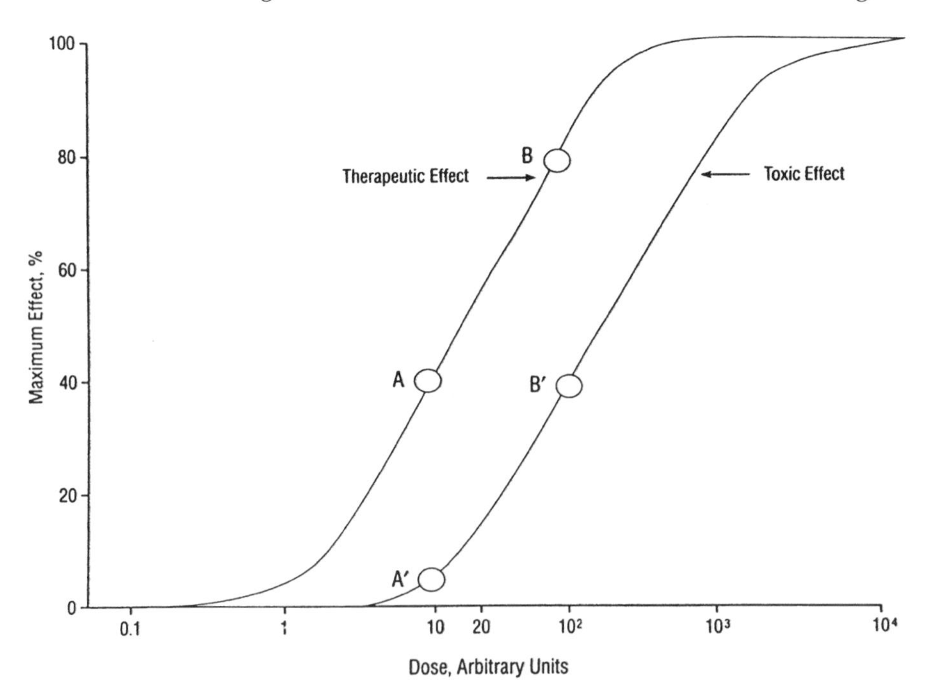

FIGURE 1. Linear dose-response curves showing theoretical therapeutic and toxic effects. The horizontal axis is a logarithmic scale with arbitrary dose units. The vertical axis is a linear scale showing the maximal possible response. (Adapted from Fagan TC: Remembering the lessons of basic pharmacology. Arch Intern Med 154:1430–1431, 1994.)

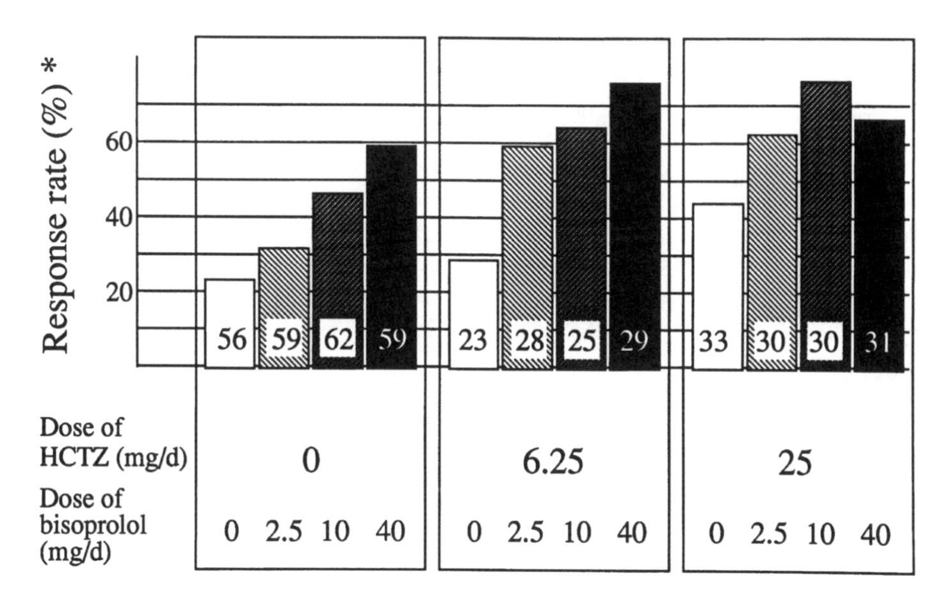

FIGURE 2. Response rate after a 12-week treatment with various doses of bisoprolol and hydrochlorothiazide (HCTZ) given in combination. (Adapted from Frishman WH, Bryzinski BS, Coulson LR, et al: A multifactorial trial design to assess combination therapy in hypertension. Treatment with bisoprolol and hydrochlorothiazide. Arch Intern Med 154:1461–1468, 1994.)

Ideally, the side-effect profile of the components of a combination should be mutually counteractive or at least nonadditive; certainly they must not be additive. Good examples of potential counteractive effects include the attenuation of thiazide-induced urinary losses of potassium and magnesium by potassium-sparing diuretics and the reduction of hydralazine-induced tachycardia with reserpine. Others include the mitigation by beta blockers or captopril of diuretic-induced hypokalemia, prevention of propranolol-induced reduction in cardiac output with hydralazine, and prevention of the dihydropyridine-induced heart rate acceleration by beta blockers and ACE inhibitors.

Improved Compliance

Compliance is another important consideration in evaluating a role for fixed-dose combination therapy. As detailed elsewhere, despite the availability of many newer antihypertensive agents, hypertensive patients continue to remain at higher risk of premature death than the general population.[29] The persistence of morbidity and mortality may be accounted for by the frequent failure to achieve adequate blood pressure control.[30] For instance, the Nutrition and Health Examination Survey in the United States found that nearly 75% of treated hypertensive patients failed to attain a target blood pressure of less than 140 mmHg systolic and less than 90 mmHg diastolic.[31,32] The extent of the problem is also revealed by a recent survey indicating that experience in Europe is similar. The Cardiomonitor Study is an independent survey of approximately 1500 physicians performed biannually in several European countries by an independent healthcare market research company. Of the 11,613 treated hypertensive patients included in a recent survey, only 37% had reached the target diastolic blood pressure set by the doctor.[33] These examples demonstrate that blood pressure continues to be inadequately controlled in the majority of treated hypertensive patients. Such poor control persists even though detailed management guidelines for hypertension have been extensively discussed and the conclusions have been widely publicized.[34] Of interest, in the Cardiomonitor survey by far the most common cause for insufficient blood pressure control advanced by the doctors was poor compliance (70 %). The point of view of patients, however, was quite different: 80% claimed that they took the prescribed treatment every day.[33]

Clearly we must narrow the gap between what the guidelines say and what we achieve in clinical practice if we are to improve the quality of care for hypertensive patients. Although several initiatives may serve to bridge this gap,[35] a pivotal approach is the use of agents with more acceptable side-effect profiles, thereby enhancing patient compliance. One such initiative is the use of the newer, low-dose fixed combinations of antihypertensive drugs. In addition to an enhanced tolerability profile, these regimens can reduce the number of tablets that the patient is required to take. Having to take fewer tablets should lessen the likelihood that the patient may confuse medications and should reduce the number of treatment failures that result from missed doses. Some evidence indicates that the compliance rate among patients receiving a CCB is the highest with once-daily administration, followed by a twice-daily, thrice-daily, and four-times-daily administration.[36] Hence, because of its convenience for the patient, once-daily dosing is expected to enhance compliance.[37] This advantage generally applies to fixed-dose combinations.

Dosing Flexibility

A frequently voiced objection to fixed-dose combinations is the lack of flexibility in dosing. This objection does not appear to be valid. If needed, the possibility remains of doubling the dose of the combined antihypertensive agents by prescribing 2 tablets instead of 1 tablet of the fixed-dose combination. Of primary importance, finding the most suitable dosage for a given patient on a trial-and-error basis is difficult. It is tempting, of course, to increase the doses in the absence of satisfactory blood pressure control. This strategy, however, often results in an increased incidence of adverse reactions.

Cost of Drugs

Another substantive consideration that directly affects the success of therapeutic interventions is the cost of drugs. Drug cost is of great importance in preventing target-organ disease. The cost of medication is one of the major reasons for poor adherence to antihypertensive medication, which results in less than optimal blood pressure control. An interesting attribute of fixed-dose combinations is the reduced cost in many instances. The cost of fixed-dose combinations is generally less than the cost of the constituents prescribed separately, in part because of the requirements of government health reimbursement systems. In the United States, the cost of a fixed combination of benazepril and amlodipine

(Lotrel) is less than the cost of the two component drugs purchased separately.[38] Similarly, the current plans are to price the fixed-dose combination of trandolapril and verapamil at a cost less than the cost of the two components purchased separately.

Enhancement of Favorable Effects on Target Organs

Several lines of evidence suggest that combinations of different agents, especially CCBs and ACE inhibitors, may have additive effects not only on blood pressure but also on preservation of target organs such as the heart and kidney. CCBs and ACE inhibitors may act in a complementary fashion either to favor regression of left ventricular hypertrophy[39] or to retard the progression of renal disease in patients with glomerular disease.[40] The combination of a beta blocker with a dihydropyridine CCB also seems to offer some advantage in the management of hypertension and may be particularly beneficial for the prevention and treatment of ischemic heart disease.[41,42]

COMBINATIONS OF ANTIHYPERTENSIVE AGENTS

The availability of five main groups of antihypertensive agents (diuretics, beta blockers, ACE inhibitors, CCBs, $alpha_1$ blockers) provides ten possible combinations of two drugs. Several recently approved combinations (including fixed-dose combinations of CCBs with ACE inhibitors, ACE inhibitors or beta blockers with diuretics, and dihydropyridine CCBs with beta blockers) have been investigated.[1,43] Our focus is CCB-containing combinations.

Beta Blocker and Calcium Antagonist

Beta blockers combine well with dihydropyridine calcium antagonists because of their complementary modes of action,[41,42] which enhance not only antihypertensive efficacy but also tolerability. Thus, beta blockers prevent the baroreceptor reflex-mediated increase in heart rate observed in some patients during calcium entry blockade. On the other hand, the dihydropyridine-induced vasodilatation may attenuate the unwanted effects of beta blockade on the peripheral circulation.

The first fixed-dose combination of a beta blocker and calcium antagonist available in a number of European countries contained atenolol (50 mg) and a sustained-release formulation of the dihydropyridine CCB, nifedipine (20 mg). Control of blood pressure was superior to that achieved by either component alone. As anticipated, heart rate was significantly decreased by atenolol and the combination compared with nifedipine alone.[44,45] The vasodilatation-induced side-effects, especially flushes and hot sweats, were significantly less frequent during atenolol and combination therapy than during treatment with nifedipine alone.

Another well-studied fixed-dose combination of a beta blocker and calcium antagonist consists of felodipine (5 or 10 mg), a dihydropyridine, and metoprolol (50 or 100 mg), a cardioselective beta blocker. Recent studies have demonstrated that adequate blood pressure reduction can be achieved in a greater

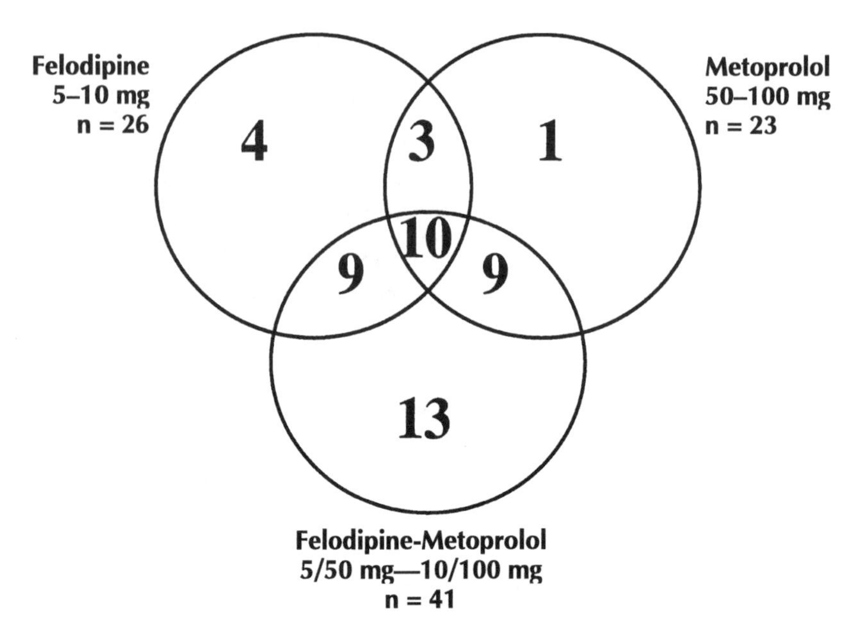

FIGURE 3. Number of patients whose diastolic blood pressure normalized (≤ 90 mmHg) after 12 weeks of treatment with felodipine and metoprolol, given alone or in combination. The overlap between circles shows the patients who were good responders to two or more of the treatments. (Adapted from Dahlöf BL, Jönsson O, Borgholst G, et al: Improved antihypertensive efficacy of the felodipine-metoprolol extended-release tablet compared with each drug alone. Blood Pressure 2 (Suppl 1):37–45, 1993.)

percentage of patients using this combination than with either component alone.[16,46] In a 12-week, double-blind, randomized, three-way crossover multicenter study, the efficacy and tolerability of the fixed combination administered once daily were compared with the efficacy and tolerability of either drug alone.[16] Figure 3 discloses the number of patients with normalized diastolic blood pressure (≤ 90 mmHg) during the 12-week treatments. Felodipine (F) plus metoprolol (M), 5 mg F + 50 mg M or 10 mg F + 100 mg M, normalized diastolic blood pressure in 41 of 58 patients, in contrast with the 26 patients whose blood pressure was controlled with felodipine alone (5–10 mg) and the 23 patients controlled with metoprolol alone (50–100 mg). Of interest was the large heterogeneity in blood pressure responses. Some patients responded to whatever treatment was used, whereas others responded during administration of two or even one drug only. All three treatments were well tolerated.

It is well established that isolated systolic hypertension represents a major risk factor in the elderly and that pharmacologic treatment is associated with a clear-cut reduction in the incidence of cardiovascular complications.[47,48] In a recent trial, 28 patients with isolated systolic hypertension (systolic blood pressure ≥ 160 mmHg and diastolic blood pressure < 95 mmHg), aged 60 years or older, were randomized according to a double-blind crossover design to 6-week treatment periods with either felodipine (5 mg/day), metoprolol (50 mg/day), or the combination of both (5 mg felodipine + 50 mg metoprolol/day), with stepwise upward titration, if necessary, to 20 mg felodipine and 200 mg metoprolol.[49] At the end of the treatment phase, systolic blood pressure 2 hours after dosing was < 140 mmHg in 70 % of patients on combination therapy, whereas the corresponding values for felodipine and metoprolol were 45% and 24%, respectively. Despite the additive effect of the combined treatment on blood pressure, no increase in adverse effects was observed.

Antihypertensive therapy may have an important effect on quality of life. In a recently performed double-blind trial, 947 hypertensive patients were randomized to 3 months of treatment with a fixed combination of felodipine (5 mg) and metoprolol (50 mg), enalapril (10 mg), or placebo.[50] The study drugs were given once a day, and their dosage could be doubled if needed. The reduction of blood pressure was significantly better with the combination than with enalapril and placebo, but there was no difference in well-being, as assessed by validated self-administered questionnaires, among the three treatments.

In contrast to dihydropyridines, the combination of a beta blocker and a phenylalkylamine such as verapamil should be avoided, because the risk of atrioventricular conduction abnormalities or cardiodepression is increased. For the same reason, caution must also be exerted when considering the combination of a beta blocker with a benzothiazepine such as diltiazem. The new CCB mibefradil may be combined safely with a beta blocker. Mibefradil blocks not only L-type calcium channels (like the other CCBs available so far) but also T-type calcium channels and has no effect on atrioventricular conduction and cardiac contractility[51] (see chapter 10).

Calcium Antagonist and ACE Inhibitor

It is of interest to consider how the principles favoring fixed-dose combination therapy may apply to combinations of a calcium antagonist and an ACE inhibitor.

Antihypertensive Efficacy

Several studies of the free combination of calcium antagonists with ACE inhibitors in patients with essential hypertension have indicated that the antihypertensive efficacy of the components is additive.[52–58] Of note, the coadministration of a CCB (felodipine, 5–20 mg/day) and an ACE inhibitor (enalapril, 5–20 mg/day) is more effective in lowering blood pressure in elderly patients with isolated systolic hypertension than either agent alone.[59]

Experience with fixed-dose combinations of a CCB and ACE inhibitor has accumulated mainly with verapamil and trandolapril.[60–62] In a double-blind, placebo-controlled trial, 746 hypertensive patients were randomized to 6 weeks of treatment with either placebo, verapamil SR (sustained-release) monotherapy (120, 180, or 240 mg), trandolapril monotherapy (0.5, 2, or 8 mg), or verapamil SR/trandolapril combinations (0.5/120, 0.5/180, 0.5/240, 2/120, 2/180, 2/240, 8/120, 8/180, or 8/240 mg).[61] Figure 4 illustrates the blood pressure reductions in a surface analysis. Obviously, in terms of antihypertensive efficacy, combinations of small doses of the two agents were advantageous.The percentage of patients with adverse reactions, on the other hand, was similar for monotherapy and combination therapy.

An important issue is the comparative antihypertensive efficacy of CCB/ACE inhibitor combinations and other types of combinations. Diuretics are generally considered the drugs of choice to add to ACE inhibitors for the treatment

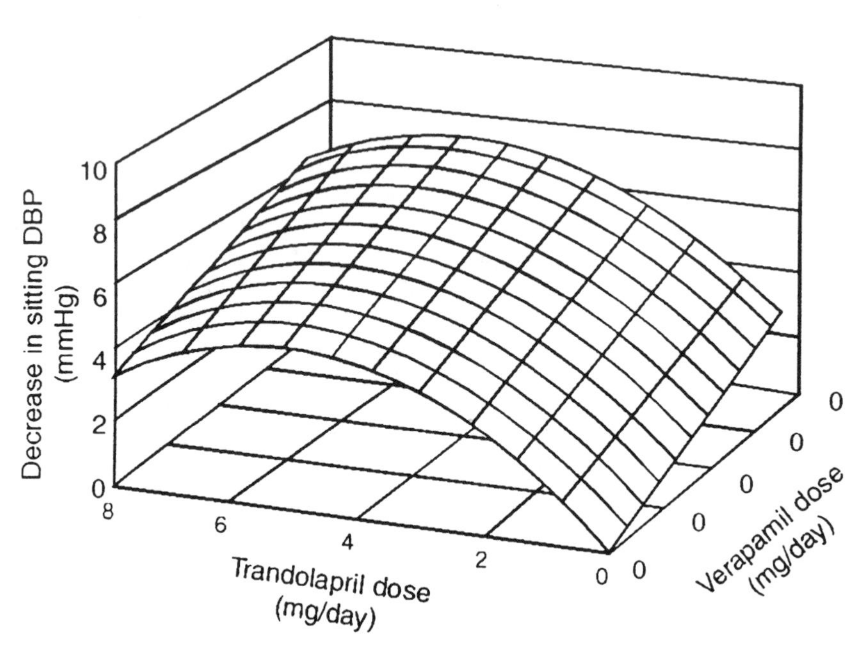

FIGURE 4. Monotherapy and combination therapy with different doses of trandolapril and verapamil SR in hypertension. (Adapted from De Quattro V, Lee D, and the Trandolapril Study Group: Fixed-dose combination therapy with trandolapril and verapamil SR is effective in primary hypertension. Am J Hypertens 10 (Pt 2):138–145, 1997.)

of hypertension. In fact, however, the combination of a CCB with an ACE inhibitor is as effective as the combination of a thiazide with an ACE inhibitor, and the incidence of side effects is the same.[58,60] Also of note, the coadministration of a CCB and ACE inhibitor is equally effective in terms of lowering blood pressure as the combination of a beta blocker and diuretic.[60] In one study, 1647 hypertensive patients received the dihydropyridine isradipine, 2.5 mg twice daily, for 4 weeks.[63] Nonresponders (diastolic blood pressure > 90 mmHg) were randomly allocated to a 4-week treatment with either placebo, the beta blocker bopindolol (0.5 or 1 mg/day), the diuretic metolazone (1.25 or 2.5 mg/day), the ACE inhibitor enalapril (10 or 20 mg/day), or isradipine, 5 mg twice daily. Figure 5 shows the changes in systolic and diastolic blood pressure induced by the various treatments. Regardless of dosage, most drugs led to a comparable reduction in blood pressure; exceptions were the lower dosage of bopindolol and the high dose of isradipine, which were less effective in lowering systolic blood pressure.

The safety profile of CCBs, especially dihydropyridines, may be improved by simultaneous ACE inhibition. For example, the incidence of ankle edema due to CCB treatment decreases when an ACE inhibitor is coadministered.[64] This effect is largely due to the venodilating capacity of the ACE inhibitors. Of greater importance, ACE inhibition blunts the reflex increase in sympathetic nerve activity mediated by dihydropyridine (Fig. 6). In the single-blind study by Gennari et al., 24 hypertensive patients with a diastolic blood pressure of 95–115 mmHg were randomized to receive for 4 weeks either captopril, 50 mg

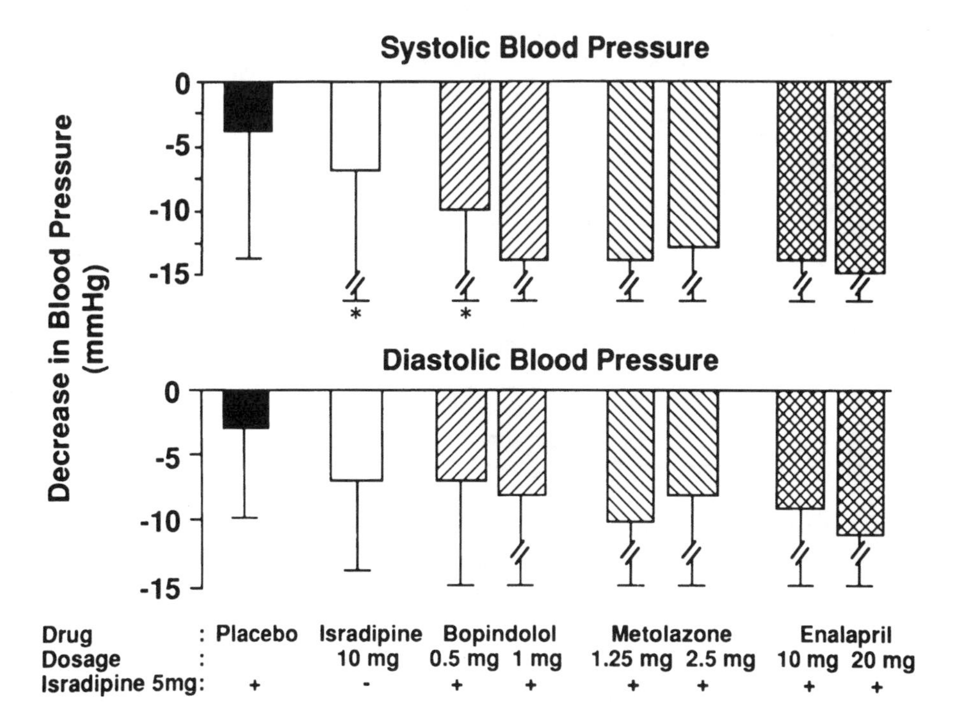

FIGURE 5. Changes in blood pressure induced by 4 weeks of treatment with either isradipine alone, 2.5 mg twice a day, or isradipine, 2.5 mg once a day, in combination with placebo, pindolol, metolazone or enalapril. All patients were still hypertensive after 4 weeks of treatment with isradipine, 2.5 mg once a day. (Adapted from Lüscher TF, Waeber B: Efficacy and safety of various combination therapies based on a calcium antagonist in essential hypertension: Results of a placebo-controlled randomized trial. J Cardiovasc Pharmacol 21:305–309, 1993.)

twice daily, or nitrendipine, 20 mg once daily.[54] For the following 4 weeks one-half of the patients were treated with placebo and one-half with nitrendipine, 10 mg/day, plus captopril, 25 mg twice daily. Panel A of Figure 6 summarizes the patients treated initially with captopril. The ACE inhibitor produced a significant fall in diastolic blood pressure with no concomitant change in heart rate. The patients randomized to placebo returned to pretreatment blood pressure values, whereas the association of captopril and nitrendipine led to an additional fall in blood pressure, which was significant when compared with the ACE inhibitor alone, and heart rate remained unchanged. Panel B of Figure 6 depicts the patients treated first with nitrendipine. This calcium antagonist caused a significant blood pressure decrease that was associated with significant heart rate acceleration. Baseline values were reached again by patients allocated to placebo during the last 4 weeks of the study. In the remaining patients treated by the combination of nitrendipine, 10 mg/day, and captopril, 50 mg/day, a further significant fall in blood pressure was observed in comparison with the nitrendipine monotherapy phase. During combination therapy, however, there was no evidence of a reflex increase in heart rate. Finally, one may expect reductions in CCB-associated constipation (verapamil) and flushing and headache (dihydropyridines), because the dose of the CCBs can be kept low when they are used in combination with ACE inhibitors.

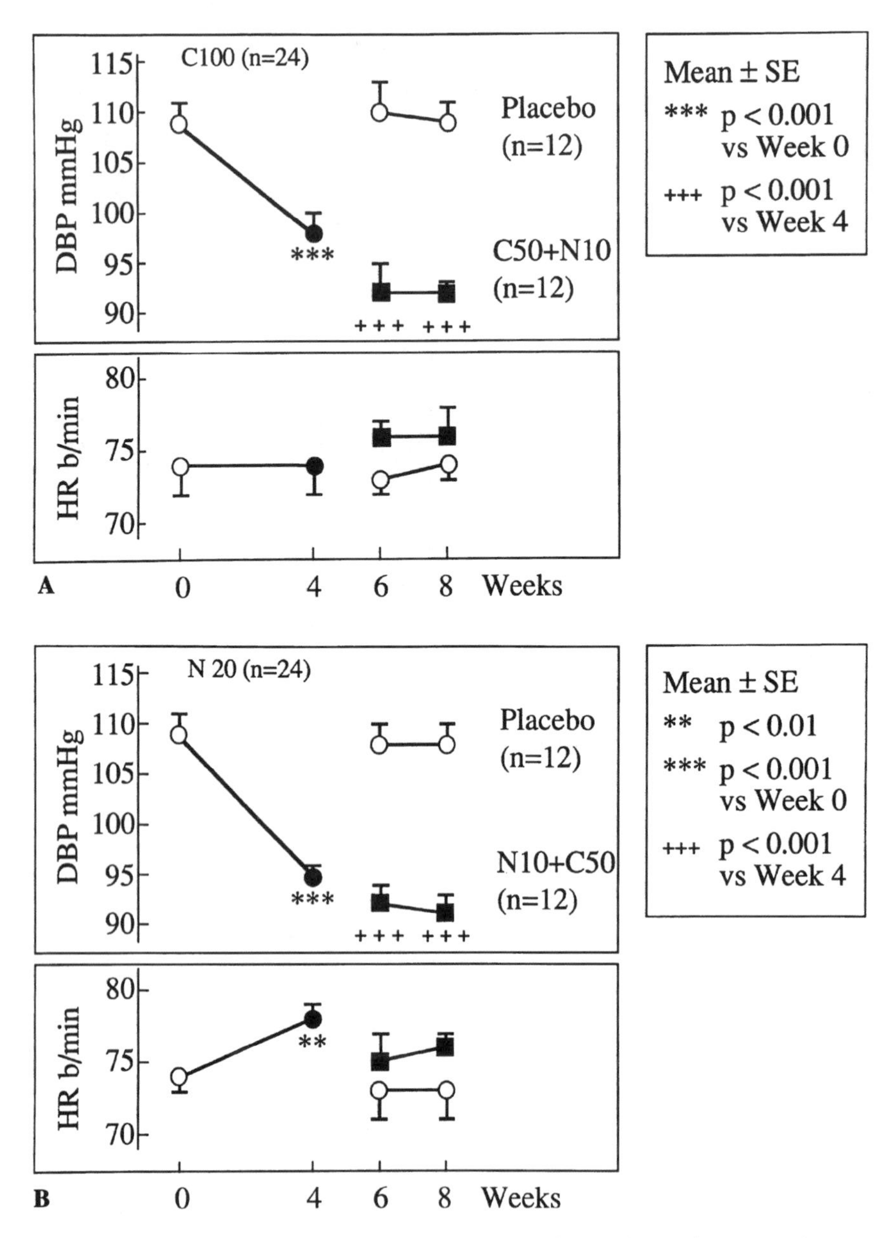

FIGURE 6. Blood pressure and heart rate response to captopril (C) and nitrendipine (N) administered alone or in combination. *A,* Patients initially treated with captopril. *B,* Patients initially treated with nitrendipine. (Adapted from Gennari C, Nami R, Pavese G, et al: Calcium-channel blockade (nitrendipine) in combination with ACE inhibition (captopril) in the treatment of mild to moderate hypertension. Drugs Ther 3:319–325, 1989.)

Collectively, these and other considerations have culminated in the development of several CCB/ACE inhibitor combinations. Four fixed-dose calcium antagonist/ACE inhibitor combination agents are now available for antihypertensive therapy in the United States (Table 3). Lotrel, a fixed-dose

TABLE 3. Fixed-dose Combination Agents for Treatment of Hypertension in the United States

Brand Name	Formulation	Components	Doses	Approved
Lotrel	Capsule	Amlodipine Benazepril	Amlodipine 2.5 mg, benazepril 10 mg Amlodipine 5 mg, benazepril 10 mg Amlodipine 5 mg, benazepril 20 mg	1995
Tarka	Sustained-release capsule	Verapamil SR Trandolapril	Verapamil HCL SR 120 mg, trandolapril 0.5 mg Verapamil HCL SR 180 mg, trandolapril 1 mg Verapamil HCL SR 180 mg, trandolapril 2 mg	Nov. 1996
Lexxel	Extended-release tablet	Enalapril Felodipine	Enalapril 5 mg, felodipine 5 mg	Dec. 1996
Teczem	Tablet	Diltiazem Enalapril	Diltiazem 180 mg, enalapril 5 mg	Oct. 1996

combination of amlodipine besylate and benazepril HCl, is available as a capsule containing three doses of amlodipine and benazepril. Lexxel is available as a tablet containing a core of extended-release felodipine, which is coated with enalapril. Tarka is a fixed-dose sustained-release capsule containing three doses of verapamil SR and trandolapril. Teczem has a core of extended-release diltiazem malate, which is coated with enalapril.

Beneficial Effects on Target Organ Damage

A theoretical framework supports the concept of greater amelioration of disease in target organs with CCB/ACE inhibitor combination therapy than would be expected from the antihypertensive strength of either component alone.[65,66] Although this notion is somewhat speculative, it constitutes a provocative and appealing aspect of antihypertensive therapy.

Left Ventricular Hypertrophy. The benefits of both ACE inhibitors and calcium antagonists on reversal of ventricular hypertrophy (LVH) are well established.[59,60] As a class, CCBs may be slightly less potent in producing regression of LVH than ACE inhibitors. Of interest, several studies have suggested that nondihydropyridine calcium antagonists have a greater effect than dihydropyridine agents, independently of treatment duration and quality of blood pressure control. Theoretically, this difference may be explained by some degree of stimulation of the sympathetic nervous system during dihydropyridine-induced vasodilatation, with an ensuing increase in exposure of the heart to the trophic effects of norepinephrine. It remains unknown, however, whether the combination of a CCB with an ACE inhibitor ought to be particularly efficacious in preventing and/or regressing LVH.

Preservation of Renal Function. The potential renoprotective effects of CCB/ACE inhibitor combination therapy on surrogate end-points of renal disease merit consideration. Ample evidence suggests that ACE inhibitors are particularly effective in reducing microalbuminuria and proteinuria in patients with essential hypertension as well as in insulin-dependent and non–insulin-dependent diabetics.[40] This beneficial effect is mainly due to a reduction in intraglomerular pressure resulting from preferential dilation of the efferent arteriole. The ACE inhibition-induced reduction in proteinuria is associated with slowed progression to end-stage renal failure. The effect of CCBs on microalbuminuria and proteinuria is less clear,[20] particularly in the the case of dihydropyridine CCBs. For example, unlike other dihydropyridines, nifedipine may increase rather than decrease urinary albumin excretion. Nondihydropyridine calcium antagonists, on the other hand, usually reduce proteinuria. Combining a CCB, especially diltiazem or verapamil, and an ACE inhibitor may be more effective in reducing microalbuminuria and albuminuria than either agent alone, as suggested by a study of diabetic patients in whom the combination of diltiazem and lisinopril was superior to monotherapy with either agent in reducing proteinuria.[67] The level of blood pressure achieved during treatment was not a confounding factor because the protocol was designed to titrate antihypertensive therapy to achieve similar blood pressure control in the three groups. A recently completed, 5-year study in non–insulin-dependent diabetics extended these observations using verapamil and lisinopril.[68] The combined treatment slowed the progression of renal disease to a greater extent than either agent alone. The additive effect on proteinuria reduction was also noted (see chapters 14 and 15 for detailed discussion).

Although these preliminary findings are provocative, carefully controlled, randomized prospective studies of long duration are needed to establish the precise effects of a combination of CCBs and ACE-inhibitors on the progression of both diabetic nephropathy and chronic renal failure. Such trials should assess a number of variables, including the optimal dose of both drugs and to what level blood pressure should be lowered.

At present one such study has been initiated to compare ACE inhibitor combined with calcium antagonist therapy vs. either agent alone in patients with nondiabetic renal disease. This large, randomized prospective study was initiated to study renal protection in nondiabetic renal disease. The European Multicenter Study on Progression in Nondiabetic Renal Disease (NEPHROS) will compare the renoprotective effects of ramipril, felodipine, or the combination of ramipril and felodipine (M. Aurell, personal communication). Two hundred patients with nondiabetic renal disease will be recruited in the United Kingdom, France, Germany, Israel, and Sweden. A second study will soon be initiated to compare the renoprotective effects of trandolapril, verapamil, or the combination of trandolapril plus verapamil.

Metabolic Effects

The combination of a CCB and an ACE inhibitor has the advantage of neutrality in regard to glucose and lipid metabolism, as demonstrated repeatedly in diabetic patients.[69] This observation is important because insulin release involves calcium. Consequently, when high doses of CCBs are used alone to reduce arterial pressure, a worsening of preexisting hyperglycemia has been reported.

Electrolyte Imbalance

Another bothersome effect of ACE inhibitors, especially in patients with preexisting renal insufficiency, is hyperkalemia. Frequently, control of hypertension in patients with chronic renal insufficiency necessitates more potent classes of drugs such as CCBs and ACE inhibitors. Patients who tend to develop hyperkalemia when treated with ACE inhibitors may benefit from the accessibility of CCBs, which do not increase and in fact may reduce the rate of increase in plasma potassium.[70] Moreover, because lower doses of ACE inhibitors are generally required with combination therapy, hyperkalemia has not been reported as a major side effect in previously completed studies of such groups.

Edema

Edema is the most common side effect reported with CCBs and usually relates to arteriolar vasodilation rather than to decreased urinary sodium excretion. The incidence of edema is markedly reduced when the CCB is combined with an ACE inhibitor. Although studies are relatively sparse, a large body of anecdotal evidence indicates that the addition of an ACE inhibitor to patients receiving CCBs attenuates or reverses the peripheral edema. Of interest, Salvetti et al.[64] reported that ankle edema associated with nifedipine therapy disappeared in three of four patients after addition of captopril.

The proposal that edema is attenuated by the fixed-dose combination has recently been validated. Gradman et al.[71] recently investigated the dose-response relationships and safety of enalapril and felodipine ER given alone and in combination in 707 patients. After 4 weeks of placebo run-in, patients were randomized to placebo, felodipine (2.5, 5, 10 mg), enalapril (5, 20 mg), or enalapril plus felodipine combinations for 8 weeks, according to a double-blind, 3×4 factorial-design protocol. The combination of enalapril and felodipine was associated with less peripheral edema than felodipine alone (4% vs. 11% for felodipine alone). Data submitted for the new drug application for the amlodipine-benazepril combination indicate that the incidence of edema in female patients decreased from 9.1 % with amlodipine alone to 3.2 % with the combination of amlodipine and benazepril (unpublished data). Presumably the improvement is attributable to venodilation by ACE inhibitors, which allows correction of the relative outflow obstruction caused by the massive increase in arterial flow from vasodilation of the arterial tree.[72]

UNRESOLVED ISSUES AND FUTURE CHALLENGES

Presently, we are at an early stage of the evolution of the fixed-dose combination therapy. Although the theoretic construct is promising, additional experience is needed and a number of questions must be addressed:

1. How does one decide that a fixed-dose combination is suitable for a patient?
2. Should fixed-dose combinations be first-line therapy or used only when single agents fail?
3. Is there a pharmacoeconomic advantage to going directly to fixed-dose combinations rather than trying monotherapy first? For example, does eliminating stepwise adjustment of drugs and dosages (and associated physician evaluation time) confer some economic advantage to either the patient or the physician?
4. What are the criteria for selecting separate components of a fixed-dose combination? Possibilities include wide therapeutic indices (wide margins of

safety) and match-up of pharmacokinetic properties so that one agent of the pair does not accumulate more than the other.

Evolving Role of Fixed-dose Combination Agents for Initial Monotherapy

It is reasonable to consider the role of fixed-dose combination agents for initial monotherapy. Prudence clearly dictates that physicians should avoid initial fixed-dose combination therapy in patients who may be sensitive to the individual components (e.g., beta blockers for patients with asthma and diuretics for patients with sulfur skin allergy). However, for appropriate patients, convenience, enhanced compliance, and cost savings are predicted benefits that assume increasing importance. Currently, the only FDA-approved combination available on the market for initial therapy is bisoprolol/6.25 HCTZ. Based on the attributes reviewed in this chapter, as more experience accrues, it is probable that more combinations will become available because the concept of improved efficacy without an increase in adverse effects is clinically compelling.

CONCLUSION

The most common rationale for combining antihypertensive drugs is enhanced efficacy in terms of reducing blood pressure. Enhancing tolerability and reducing or limiting side effects assume increased relevance with the focus on improving patient compliance. An additional exciting and novel concept is the possibility of enhancing salutary effects on target organs by combination therapy—over and above the effect expected from the fall in arterial pressure alone. This approach may become a desirable therapeutic goal, provided that such benefits are documented rigorously by carefully conducted morbidity and mortality studies.

References

1. Oster JR, Epstein M: Fixed-dose combination medications for the treatment of hypertension: A critical review. J Clin Hypertens 3:278–293, 1987.
2. Ausubel H, Levine ML: Treatment of hypertension with Ser-Ap-Es. Curr Ther Clin Exp 9:29–35, 1967.
3. Manolio TA, Cutler JA, Furberg CD, et al: Trends in pharmacologic management of hypertension in the United States. Arch Intern Med 155:829–837, 1995.
4. Epstein M: Calcium antagonists in the management of hypertension. In Epstein M (ed): Calcium Antagonists in Clinical Medicine. Philadelphia, Hanley & Belfus, 1992, pp 213– 230.
5. Bakris GL: Calcium abnormalities and the diabetic, hypertensive patient: Implications of renal preservation. In Epstein M (ed): Calcium Antagonists in Clinical Medicine. Philadelphia, Hanley & Belfus, 1992, pp 367–388.
6. Ram JL, Standley PR, Sowers JR: Calcium function in vascular smooth muscle and its relationship to hypertension. In Epstein M (ed): Calcium Antagonists in Clinical Medicine. Philadelphia, Hanley & Belfus, 1992, pp 29–48.
7. Waeber B, Nussberger J, Brunner HR: Angiotensin-converting enzyme inhibitors. In Laragh JH, Brenner BM (eds): Hypertension: Pathophysiology, Diagnosis, and Management. New York, Raven Press, 1995, pp 2861–2875.
8. Bönner G: Do kinins play a significant role in the antihypertensive and cardioprotective effects of angiotensin I-converting enzyme inhibitors? In Laragh JH, Brenner BM (eds): Hypertension: Pathophysiology, Diagnosis, and Management. New York, Raven Press, 1995, pp 2877–2893.

9. Blaustein MP: Sodium ions, calcium ions, blood pressure regulation and hypertension: A reassessment of a hypothesis. Am J Physiol 232:165–173, 1977.
10. Pedersen OL, Christensen NJ, Ramsch KD: Comparison of acute effects of nifedipine in normotensive and hypertensive man. J Cardiovasc Pharmacol 2:357–366, 1980.
11. Zimmerman BG, Sybert EG, Wong PC: Interaction between sympathetic and renin-angiotensin system. J Hypertens 2:581–588, 1984.
12. MacFadyen RJ, Lees KR, Reid JL: Tissue and plasma angiotensin converting enzyme and the response to ACE inhibitor drugs. Br J Clin Pharmacol 31:1–13, 1991.
13. Prichard BNC, Cruickshank JM: Beta blockade in hypertension: Past, present and future. In Laragh JH, Brenner BM (eds): Hypertension: Pathophysiology, Diagnosis, and Management. New York, Raven Press, 1995, pp 2827–2859.
14. Sever P: The heterogeneity of hypertension: Why doesn't every patient respond to every antihypertensive drug ? J Hum Hypertens 9 (Suppl 2):33–36, 1995.
15. Attwood S, Bird R, Burch K, et al: Within-patient correlation between the antihypertensive effects of atenolol, lisinopril and nifedipine. J Hypertens 12:1053–1060, 1994.
16. Dahlöf B, Jönsson L, Borgholst O, et al: Improved antihypertensive efficacy of the felodipine-metoprolol extended-release tablet compared with each drug alone. Blood Pressure 2 (Suppl 1): 37–45, 1993.
17. Koch-Weser J: Drug therapy: Hydralazine. N Engl J Med 295:320–323, 1976.
18. Ruzicka M, Leenen FHH: Relevance of intermittent increases in sympathetic activity for adverse outcome on short-acting calcium antagonists. In Laragh JH, Brenner BM (eds): Hypertension: Pathophysiology, Diagnosis, and Management. New York, Raven Press, 1995, pp 2815– 2825.
19. Epstein M: Calcium antagonists should continue to be used for first line treatment of hypertension. Arch Int Med 155:2150–2156, 1995.
20. Zanchi A, Brunner HR, Waeber B, Burnier M: Renal haemodynamic and protective effects of calcium antagonists in hypertension. J Hypertens 13(Pt 1):1363–1375, 1995.
21. Kailasam MT, Parmer RJ, Cervenka JH, et al: Divergent effects of dihydropyridine and phenylalkylamine calcium channel antagonist classes on autonomic function in human hypertension. Hypertension 26:143–149, 1995.
22. Kiowski W, Erne P, Bertel O, et al: Acute and chronic sympathetic reflex activation and antihypertensive response to nifedipine. J Am Coll Cardiol 7:344–348, 1986.
23. Ménard J, Bellet M: Calcium antagonists-ACE inhibitors combination therapy: Objectives and methodology of clinical development. J Cardiovasc Pharmacol 21(Suppl 2):49–54, 1993.
24. Fagan TC: Remembering the lessons of basic pharmacology. Arch Intern Med 154:1430–1431, 1994.
25. Frishman WH, Bryzinski BS, Coulson LR, et al: A multifactorial trial design to assess combination therapy in hypertension. Treatment with bisoprolol and hydrochlorothiazide. Arch Intern Med 154:1461–1468, 1994.
26. Prisant ML, Weir MR, Papademetriou V, et al: Low-dose drug combination therapy: An alternative first-line approach to hypertension treatment. Am Heart J 130:359–366, 1995.
27. Fenichel RR, Lipicky RJ: Combination products as first-line pharmacotherapy. Arch Intern Med 154:1429–1430, 1994.
28. Messerli FH: Combination therapy in hypertension. J Hum Hypertens 6(Suppl. 2):19–21, 1992.
29. Lindholm L, Ejlertsson G, Schersten B: High risk of cerebrovascular morbidity in well-treated male hypertensives: A retrospective study of 40–59-year-old hypertensives in a Swedish primary care district. Acta Med Scand 216:251–259, 1984.
30. Isles CG, Walker LM, Beevers GDI: Mortality in patients of the Glasgow Blood Pressure Clinic. J Hypertens 4:141–156, 1986.
31. Burt VL, Whelton P, Roccella EJ: Prevalence of hypertension in the U.S. adult population. Results from the Third National Health and Nutrition Examination Survey, 1988–1991. Hypertension 25:305–313, 1995.
32. Frohlich ED: There's good news and not so good news [editorial]. Hypertension 25:303–304, 1995.
33. Hosie J, Wiklund IK: Managing hypertension in general practice: Can we do better? J Hum Hypertens 9(Suppl 2):15–18, 1995.
34. Swales JD: Management guidelines for hypertension: Is anyone taking notice? J Hum Hypertens 9(Suppl 2):9–13, 1995.
35. Ménard J, Chatellier G: Limiting factors in the control of blood pressure: Why is there a gap between theory and practice? J Hum Hypertens 9(Suppl 2):19–23, 1995.
36. Farmer KC, Jacobs EW, Philipps CR: Long-term patient compliance with prescribed regimens of calcium channel blockers. Clin Ther 16:316–326, 1995.
37. Rudd P: Clinicians and patients with hypertension: Unsettled issues about compliance. Am Heart J 130:572–579, 1995.

38. Cost to pharmacist for 1 year's/day's treatment, according to January 1996 Average Wholesale Price (AWP) listed in Medi-Span® Prescription Pricing Guide.
39. Messerli FH, Michalewicz L: Cardiac effects of combination therapy. Am J Hypertens 10 (Pt 2):146–152, 1997.
40. Ritz ES, Orth SR, Strzelczyk P: Angiotensin converting enzyme inhibitors, calcium channel blockers, and their combination in glomerular disease. J Hypertens 15(Suppl 2):21–26, 1997.
41. Nayler WG: The potential for added benefits with beta-blockers and calcium antagonists in treating cardiovascular disorders. Drugs 35(Suppl 4):1–8, 1988.
42. Sever P: Combination therapy with calcium-entry blockers and beta-adrenoceptor antagonists in hypertension. Drugs Ther 3:327–332, 1989.
43. Waeber B, Brunner HR: Combination antihypertensive therapy: Does it have a role in rational therapy? Am J Hypertens 10(Pt 2):131–137, 1997.
44. Stanley NN, Thirkettle JL, Varma MPS, et al: Efficacy and tolerability of atenolol, nifedipine and their combination in the management of hypertension. Drugs 35(Suppl 4):29–35, 1988.
45. Anderton JL, Vallance BD, Stanley NN, et al: Atenolol and sustained release nifedipine alone and in combination in hypertension. A randomized, double-blind, cross-over study. Drugs 35(Suppl 4):22–26, 1988.
46. Dahlöf B, Hosie J, for the Swedish/United Kingdom Study Group: Antihypertensive efficacy and tolerability of a fixed combination of metoprolol and felodipine in comparison with the individual substances in monotherapy. J Cardiovasc Pharmacol 16:910–916, 1990.
47. Staessen J, Amery A, Fagard R: Isolated systolic hypertension in the elderly. J Hypertens 8:393–405, 1990
48. SHEP Cooperative Research Group: Prevention of stroke by antihypertensive drug treatment in older persons with isolated systolic hypertension. JAMA 265:3255–3264, 1991.
49. Wing LM, Russell AE, Tonkin AL, et al: Felodipine, metoprolol and their combination compared with placebo in isolated systolic hypertension in the elderly. Blood Pressure 3:82–89, 1994.
50. Detry JM, Dahlöf B, Garcia-Puig J, et al: Quality of life and blood pressure control: Results from a randomized, double-blind, placebo-controlled clinical trial with a felodipine-metoprolol combination or enalapril. J Hypertens 15 (Suppl 4):37, 1997.
51. Lüscher T, Clozel JP, Noll G: Pharmacology of the calcium antagonist mibefradil. J Hypertens 15(Suppl 3):11–18, 1997.
52. Singer DRJ, Markandu ND, Shore AC, McGregor GA: Captopril and nifedipine in combination for moderate to severe essential hypertension. Hypertension 9:629–633, 1987.
53. Morgan T, Anderson A, Hopper J: Enalapril and nifedipine in essential hypertension: Synergism of the hypotensive effects in combination. Clin Exp Hypertens 10:779–789, 1988.
54. Gennari C, Nami R, Pavese G, et al: Calcium-channel blockade (nitrendipine) in combination with ACE inhibition (captopril) in the treatment of mild to moderate hypertension. Drugs Ther 3:319–325, 1989.
55. Anderson A, Morgan T: Interaction of enalapril with sodium restriction, diuretics and slow-channel calcium-blocking drugs. Nephron 55(Suppl 1):70–72, 1990.
56. Fitscha P, Meisner W, Hitzenberger G: Antihypertensive effects of isradipine and captopril as monotherapy or in combination. Am J Hypertens 4:151s–153s, 1991.
57. Maclean D: Combination therapy with amlodipine and captopril for resistant systemic hypertension. Am J Cardiol 73:55A–58A, 1994.
58. Letellier P, Overlack A, Agnes E, Desche P, and the Multicentre Study Group on Treatment Association with Perindopril: Perindopril plus nifedipine versus perindopril plus hydrochlorothiazide in mild to severe hypertension: a double-blind multicentre study. J Hum Hypertens 8:145–149, 1994.
59. Wing LM, Russell AE, Tonkin AL, et al: Mono- and combination therapy with felodipine or enalapril in elderly patients with systolic hypertension. Blood Pressure 3:90–96, 1994.
60. De Leeuw PW, Kroon AA: Fixed low-dose combination of an angiotensin converting enzyme inhibitor and a calcium channel blocker drug in the treatment of essential hypertension. J Hypertens 15(Suppl 2):39–42, 1997.
61. De Quattro V, Lee D, and the Trandolapril Study Group: Fixed-dose combination therapy with trandolapril and verapamil SR is effective in primary hypertension. Am J Hypertens 10 (Pt 2): 138–145, 1997.
62. Veratran Study Group: Effects of verapamil SR, trandolapril, and their fixed combination on 24-h blood pressure. The Veratran Study. Am J Hypertens. 10:492–499, 1997.
63. Lüscher TF, Waeber B: Efficacy and safety of various combination therapies based on a calcium antagonist in essential hypertension: Results of a placebo-controlled randomized trial. J Cardiovasc Pharmacol 21:305–309, 1993.

64. Salvetti A, Innocenti PF, Lardella M, et al: Captopril and nifedipine interactions in the treatment of essential hypertensives: A crossover study. J Hypertens 5 (Suppl 4):139–142, 1987.
65. Cruickshank JM, Lewis J, Moore V, Dodd C: Reversibility of left ventricular hypertrophy by differing types of antihypertensive therapy. J Hum Hypertens 6:85–90, 1992.
66. Dahlöf B, Pennert K, Hansson L: Reversal of left ventricular hypertrophy in hypertensive patients: A meta-analysis of 109 treatment studies. Am J Hypertens 5:95–110, 1992.
67. Bakris GL: Effects of diltiazem or lisinopril on massive proteinuria associated with diabetes mellitus. Ann Intern Med 112:707–708, 1990.
68. Bakris GL, Copley JB, Vicknair N, Leurgans S: Effects of nondihydropyridine calcium antagonists on progression of nephropathy resulting from noninsulin dependent diabetes: A five year follow-up prospective study. J Invest Med 43:440A, 1995.
69. Teuscher A, Weidmann PU: Requirements for antihypertensive therapy in diabetic patients: Metabolic aspects. J Hypertens 15(Suppl 2):67–75, 1997.
70. Solomon R, Dubey A: Diltiazem enhances potassium disposal in subjects with end-stage renal disease. Am J Kidney Dis 14:420–426, 1992.
71. Gradman AH, Cutler NR, Davis PJ, et al, for the Enalapril/Felodipine ER Factorial Study Group: Combined enalapril and felodipine extended release (ER) for systemic hypertension. Am J Cardiol 79:431–435, 1997.
72. Gustafsson D, Grände P-O, Borgström P, Lindberg L: Effects of calcium antagonists on myogenic and neurogenic control of resistance and capacitance vessels in cat skeletal muscle. J Cardiovasc Pharmacol 12:413–422, 1988.

WILLIAM J. ELLIOTT M.D., Ph.D.

27

The Costs of Treating Hypertension

Since the early 1970s the trend to spend more resources on medical care has been growing worldwide. Part of the reason has been a concomitant decrease in expenditures on armaments and other defense-related budgetary items, but another important factor is the growing number of adults who do not die in war but survive to suffer a natural death, usually attended by a physician or other medical care provider. Perhaps the United States is best example of such changes. The U.S. Office of Management and Budget estimates that, in 1969, expenditures for defense were 42.2% of the budget, with 7.5% allocated for medical care, pension and disability benefits (excluding Social Security), and medical research. Allocations for the same items in 1995 included 16.2% for defense and 19.3% for health-related expenditures. The changes in total dollar amounts are even more staggering: defense allocations were $82.5 billion in 1969 and grew to $272 billion in 1995 (a 330% increase), whereas medically related expenditures ballooned from $14.4 billion in 1969 to $322 billion in 1995 (a 2,263% increase). Such exponential growth leads policymakers to seek ways to limit further expenditures in this fastest-growing segment of the budget.

It is probably only natural that hypertension became the first target (if not simply one of the first several targets) for restriction of further growth of medical expenditures.[1] Hypertension is a common condition in nearly all developed countries and currently ranks first among reasons for people in the United States to visit a physician.[2] In addition, both government and pharmaceutical companies have a spent a great deal of money on hypertension. The results are widespread public awareness, sound estimates of prevalence in the general population, development of effective medications, and solid knowledge about the effectiveness of treatment in reducing stroke, myocardial infarction (MI), and even deaths from cardiovascular disease.

ESTIMATES OF THE TOTAL COST OF TREATING HYPERTENSION

It is difficult, if not impossible, to estimate the exact cost of hypertension and its sequelae. The cost varies according to country, time at which the estimate is made, economic factors (including current inflation rate), and perspective of the estimator. For example, 1996 estimates of the cost of hypertension in the United States, derived from the American Heart Association, put the total cost at $20.6 billion,[3] but this estimate does not include the cost of dialysis, transplantation, or treatment of hypertension due to chronic renal impairment, because these aspects lie within the purview of the National Kidney Foundation.

If one includes the estimated $7–10 billion for the End Stage Renal Disease (ESRD) Program (most of which goes to dialysis centers and medical personnel) and the $2–4 billion for the costs of renal transplantation related to hypertension (which is only 37% of the total, based on the moving average of patients reaching ESRD from 1985–1995), another $9–14 billion must be added to the estimate, bringing the total cost of hypertension in the United States to at least $32 billion. Even this estimate appears moderately low, because the per patient cost averages only $610 annually.

It is somewhat more interesting to look at how various authoritative bodies break down the total cost of hypertension treatment. The Joint National Committee on Detection, Evaluation and Treatment of High Blood Pressure has pointed out that "the cost of medications can amount to 70–80% of the total expenditures to treat hypertension."[2] This estimate is derived from data gathered in developing nations, in which hospitals, emergency departments, and nursing homes are uncommon (if they exist at all); most medical care is donated by missionaries. In contrast, the American Heart Association's estimates (based on a 1988 study by the Health Care Financing Administration and projected forward with inflation rates pertinent to each sector of medical care) suggest that only 20% of U.S. expenditures for treatment of hypertension is used to purchase pharmaceuticals[3,4] (Fig. 1). In fact, it is likely that, if we treated hypertension more efficiently, more than the current 21% of Americans with hypertension could achieve blood pressures < 140/90 mm Hg,[5] which would significantly reduce the nearly $97 billion currently spent on hospital and nursing home charges for cardiovascular disease alone.[3] This huge amount is currently a target for fierce cost-cutting, from rapid "rule-out MI" protocols, which obviate the need for hospitalization, to streamlined protocols for in-hospital drug use and early discharge after myocardial infarction.

Although the above data allocate a percentage of the total expenditure to "lost output" (i.e., the amount of money that the patient would have contributed to

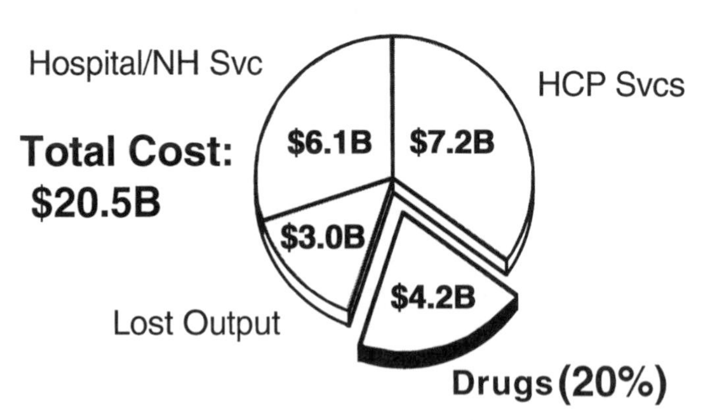

FIGURE 1. Estimates from the American Heart Association of how the $20.6 billion spent on hypertension in the United States is divided. Purchase of antihypertensive drugs accounts for only 20% of the total. HCP Svcs = health care provider services, which is the amount paid for physicians and nurses to see patients. Lost output refers to the amount of money diverted from the gross domestic product by transportation expenses to and from the physician's office and the amount of useful work that would have been done if the patient did not need to interrupt the usual productive day for interacting with the health care provider. (Adapted from Heart and Stroke Facts: 1996 Statistical Supplement. Dallas, TX, American Heart Association. 1996, p 23.)

the economy had it not been necessary to spend time at the physician's office, at the pharmacy, and in transit), several other hidden costs of hypertension are not usually considered. These costs add to the already high and very real costs of the sequelae of hypertension, currently estimated at about $17,500 per nonfatal myocardial infarction, $15,000 per nonfatal stroke, $22,000 per year for nursing home care, $50,000 per year for dialysis, and $55,000 per year after renal transplant. These hidden costs include the cost of evaluation of adverse effects of antihypertensive medication, the cost of time spent by physicians and patients in addressing laboratory studies and other perceived adverse effects of treatment, administration of health care benefit plans (currently thought to add about 20% to the cost of care in the United States, but probably less in other countries), and indirect costs that result from adverse events (including heart attack, stroke, and hospitalization).

Probably the biggest indirect cost to the health care system in the United States, which is found to a much smaller extent in other countries, is the cost of the tort system, which is used to compensate people who suffer adverse outcomes that can be attributed to a specific source (e.g., medication, physician). The population-based (and epidemiologically sound) recommendations of the New Zealand Consensus Committee for treatment of hypertension may work well in New Zealand but probably would have major medicolegal consequences in the United States. Based on local pharmacoeconomic calculations[6,7] and other considerations, New Zealand's Expert Panel on Hypertension recommends treatment only if the absolute risk of an adverse event (e.g., stroke, myocardial infarction, death) exceeds 10% in the next 8 years.[8] Lifestyle modifications without pharmacologic therapy are recommended for all patients with sufficiently mild hypertension and insufficient additional risk factors to raise the absolute risk above 10%. Although this strategy appropriately targets high-risk individuals, it also allows about a 1% incidence of adverse events in people who are "denied" drug treatment. If only 10% of this number were successful in litigation (which in the United States typically starts at an award of $3 million), it would be far more cost-effective simply to treat all patients and presumably successfully defend the remainder of the plaintiff lawsuits by claiming that "all that could be done medically was done, consonant with good local medical practice." In addition to the potential legal costs of a system that denies drug treatment to people with a low absolute risk[9] (if implemented in the United States, which has more lawyers per capita than any other country), there are the human costs of pain, suffering, and mental and emotional distress for patients who suffer events and their families. Perhaps if we could put a monetary value on this "utility," we might paraphrase the words of Oscar Wilde: "The only thing worse than the cost of treating hypertension is the cost of not treating hypertension." This statement, however, has not been verified by formal cost-effectiveness analyses, which, unlike juries, do not affix monetary amounts to human pain and suffering.

Another area in which few data exist but which is important in considering what to do to control the cost of hypertension is the dichotomy between the acquisition costs for drug therapy and the total cost of care for hypertension. The acquisition costs for antihypertensive medications have been estimated for many locales and at different times but vary unpredictably. Many American physicians incorrectly assume that the Joint National Committee's Fifth Report "preferred" diuretics or beta blockers as first-line therapy because many drugs

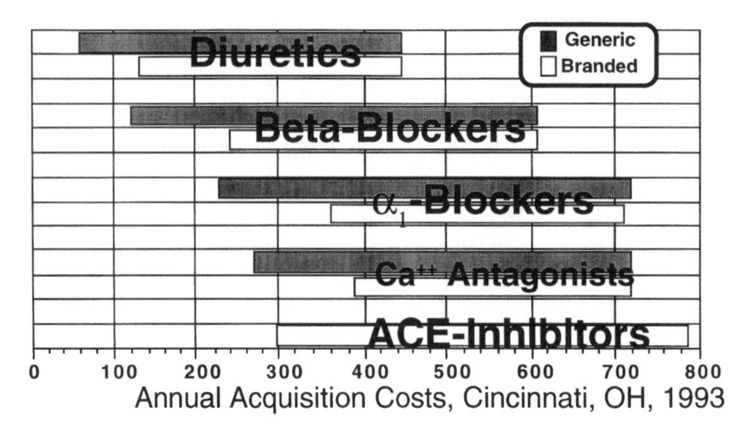

FIGURE 2. Range of retail cost for acquiring antihypertensive medications for each of the commonly prescribed drug classes in Cincinnati in 1993. Note that no generic angiotensin converting enzyme inhibitor was available when data were collected. (Adapted from Reif MC, Carter VL: The retail cost of antihypertensive therapy: Physician and patient as educated consumers. Am J Hypertens 7:571–575, 1994.)

in these classes are available generically and, in general, are cheaper to purchase than the alternative classes of medication (calcium antagonists, angiotensin-converting enzyme [ACE] inhibitors, $alpha_1$ blockers, and alpha-beta blockers). A large survey of Cincinnati pharmacies revealed that these conclusions are correct for the single cheapest drug within each pharmacologic class, but there was great overlap across agents when the range from the least expensive to the most expensive was examined[10] (see Fig. 2). The importance of examining not only the initial acquisition costs but the total cost of care across classes of medication has not yet been widely appreciated, and relevant data are rare. One survey of a large clinic in Omaha indicated that, after adding the costs of other pharmaceuticals, laboratory monitoring, additional visits for titration, and evaluation of adverse experiences attributed to the initially chosen medication, there were no significant differences in total cost of care across the five classes of antihypertensive medication.[11] Similar conclusions have been reached in preliminary studies of hypertensive patients controlled for 1 year on monotherapy at a large Chicago clinic.[12] The dichotomy between the relatively low cost of treating hypertension in uncomplicated patients and the high cost of caring for patients with end-organ disease is perhaps best illustrated in the John Deere experience, in which 4% of hypertensive patients incurred 56% of total costs (DJ Wilson, M.D., personal communication, 1996).

REVIEW OF RESULTS OF COST-EFFECTIVENESS CALCULATIONS

The conclusions of the first cost-effectiveness publication for hypertension treatment have withstood the test of two decades of revision and recalculation. In 1976 and 1977, Weinstein and Stason published several seminal papers that primarily targeted strategies for screening. They concluded that treatment was more cost-effective for higher-risk individuals and that if lower-risk patients (e.g., those with lower levels of initial blood pressure) were treated, the increase in cost was dramatic, but effectiveness rose only marginally. This inverse relationship between cost per life-year saved and absolute risk for a given population has been demonstrated many times since, with more desirable cost-effectiveness results obtained in older vs. younger patients, men vs. women, and higher vs. lower levels of pretreatment blood pressures.[7,13–17] Subsequent analyses of measuring blood pressure have also demonstrated that, in general, screening the general population for hypertension costs (rather than saves) money but is nonetheless recommended for American physicians.[18,19] Newer refinements in cost-effectiveness calculations (including methods for calculating marginal costs,[20] cost-utility analyses incorporating quality-of-life measurements,[16,21] and multivariate regression of cost data[22]) have been applied only recently to antihypertensive therapies.

Cost-effectiveness calculations are largely driven by reductions in nonfatal event rates. Although medical students are typically taught that improving mortality rates may be as important as (if not more important than) reducing morbidity, economic analyses are improved mainly in treatments to prevent expensive events that the patient survives. A treatment with optimal economic properties, therefore, is a treatment that allows affected patients to return to work at 100% of capacity (if they are to survive) but at the same time sharply reduces survival in patients who are unable to return to work and ultimately require more health care expenditures in the future for the same disease. The definition of what constitutes a "good treatment" depends somewhat on whether the evaluator is a physician or an economist. The dichotomy between economic and medical viewpoints is perhaps best understood by hospital administrators, who have learned that "the best hospital admission is a dead admission," because it results in full diagnosis-related group (DRG) payment but requires little hospital care or expense.

One of the earliest cost-effectiveness calculations comparing various classes of antihypertensive drug therapy was performed for young (35–64-year-old) patients requiring 20 years of drug therapy, using the Coronary Heart Disease Policy Model. Across a wide range of initial assumptions, propranolol had the lowest cost per life-year saved (at $10,900), followed by hydrochlorothiazide ($16,400). Costs per life-year saved for nifedipine and captopril were $61,900 and $72,100, respectively. This paper has been frequently used as a standard of comparison for the cost-effectiveness of other preventive modalities. Recent efforts to compare classes of antihypertensive medications have provided different conclusions. For instance, an alpha blocker has been claimed to have a higher cost-effectiveness result than a beta blocker (primarily because of the putative beneficial effects on serum lipids[23]).

Perhaps because of a highly-developed system of socialized medicine, the population of Sweden has been the basis for many cost-effectiveness calculations.[15,17,23–26] The results of one of the best studies are shown in Table 1.[27] These estimates are based on a computer model of event rates extrapolated from the Framingham Heart Study, the metaanalysis by Collins et al.,[28] and cost of treatment with diuretics or beta blockers of 900 Swedish kronor per year (about $112 U.S., which is much lower than recent estimates of the annual cost of care for hypertension in the United States[3,10]). Among the many variations in results of sensitivity analyses, Jönsson has recalculated the cost of a life-year saved if treatment were begun with calcium antagonists or ACE inhibitors (estimated

TABLE 1. Cost for One Additional Year of Life for Hypertension Treated with Diuretics or Beta Blockers as First-line Treatments*

	Age < 45 Years		Age 45–69 Years		Age ≥ 70 Years	
Diastolic BP (mmHG)	**Men**	**Women**	**Men**	**Women**	**Men**	**Women**
90–94	$118,375	$313,250	$8,500	$26,875	$3,125	$2,625
95–99	$97,500	$236,750	$4,250	$16,625	$1,750	$875
100–104	$79,500	$173,500	$8,500	$7,375	$375	—†
≥ 105	$55,000	$93,250	—	—	—	—

* Based on the reductions in stroke and myocardial infarction calculated from metaanalyses of clinical trials against placebo, translated into 1992 U.S. dollars.
† — = for this subgroup, antihypertensive drug therapy saves money.
Adapted from Jönsson BG: Cost-benefit of treating hypertension. J Hypertens (Suppl 12): S65–S75, 1994.

cost of 2000 Swedish kronor annually, again a great deal lower than estimates of recent costs in the United States[10,29–31]), with the assumption that these drugs may achieve all of the reduction in myocardial infarction predicted from epidemiologic studies.[28] The additional (marginal) costs of a life-year saved (over and above the cost of diuretic or beta-blocker treatment) for patients of various ages and initial blood pressures are shown in Table 2.[24] Whereas it was cost-saving (in the long term) to use diuretics or beta blockers to treat older patients with moderately elevated blood pressures (denoted by "—" in Table 1), there are no patient subgroups for which treatment results in overall cost-savings when the newer classes of antihypertensive agents are used, despite the presumed achievement of "expected efficacy" in reduction of myocardial infarction.

Results such as these have led some countries and some medical care systems to mandate initial therapy with the less expensive (and presumably more cost-effective) drugs. In Belgium and South Africa,[32] therapy for all hypertensive patients begins with a diuretic; only if problems are encountered are other drugs allowed. In Canada, recent discussions have also led to the conclusion that initial therapy should start with one of the less expensive drugs,[33] although the debate

TABLE 2. Additional Cost per Year of Life Saved for Hypertension Treated with Angiotensin-converting Enzyme Inhibitors or Calcium Antagonists as Routine First-line Treatments*

	Age < 45 Years		Age 45–69 Years		Age ≥ 70 Years	
Diastolic BP (mmHG)	**Men**	**Women**	**Men**	**Women**	**Men**	**Women**
90–94	$114,875	$779,750	$17,625	$80,875	$13,125	$21,375
95–99	$97,000	$702,750	$14,375	$74,000	$11,500	$19,875
100–104	$81,375	$628,625	$9,875	$68,875	$10,375	$18,375
≥ 105	$60,750	$536,000	$6,750	$62,000	$8,375	$16,500

* Over and above the costs of Table 1 for diuretics or beta blockers; calculated for optimal reductions in stroke and myocardial infarction, as expected from epidemiologic studies (i.e., better than the metaanalyses of clinical trials against placebo), translated into 1992 U.S. dollars.
Adapted from Jönsson BG: Cost-benefit of treating hypertension. J Hypertens (Suppl 12): S65–S75, 1994.

continues.[34] The World Health Organization and U.S. Joint National Committee have arrived at the same conclusion but without such detailed discussions of or reference to cost-effectiveness calculations.[2,35] In countries where few resources are available for allocation to antihypertensive therapy, there is great concern that even the less expensive therapies are too costly for the national exchequer.[36]

One of the more popular substudies in many modern clinical trials is the pharmacoeconomic evaluation of prospectively gathered data, which may provide a sound estimate of cost-effectiveness. This technique has been used in the Captopril Cooperative Study Group (for type I diabetics with renal impairment),[37] the Hypertension Optimal Treatment study,[38] and the STOP-Hypertension Trial,[15] among others. One major impetus for these studies is the requirement in some countries (e.g., Canada, Australia) for an evidence-based pharmacoeconomic analysis, which accompanies the request for marketing approval from the government. Products with a high cost per life-year saved (typically > $U.S. 100,000) generally do not receive unrestricted approval.

There have been a few calculations of the cost-effectiveness of government-based efforts to raise public awareness (and eventually treatment) of hypertension. Such calculations have generally been more favorable than calculations based on specific drug therapies. In some locales (e.g., North Karelia in Finland[39]), incremental costs of treating patients with newly discovered hypertension are borne by the socialized medical care system (thus incurring little marginal cost); in others (e.g., the United States), public awareness has been heightened by various methods (especially public service announcements) that incur little cost except for development, initial recording, and distribution.[40] The widely recommended implementation of lifestyle modification for hypertension therapy has not been well or thoroughly evaluated by formal cost-effectiveness calculations,[41–43] but there is concern that individual patient education by dieticians and the implementation of exercise programs by trained personnel may be more costly (and perhaps even less effective over the long term) than drug treatment.[35,41]

RESULTS OF ADMINISTRATIVE ATTEMPTS TO IMPROVE THE COST-EFFECTIVENESS OF HYPERTENSION TREATMENT

As might be expected, based on the fact that hypertension therapy is not cost-saving overall for most patients, health care administrators have attempted to improve cost-effectiveness by both curbing expenditures and eliminating costly treatment to low-risk patients. The current treatment mandate in Belgium and South Africa and the discussion document promulgated by the New Zealand Consensus Conference (both discussed above) are examples of the second strategy. Perhaps more interesting are the long-term results of systemwide limits on the availability of chronic medical therapies, the best example of which may have occurred in New Hampshire in 1988.[44] When state-supported payments to recipients of Medicaid were limited to three prescriptions per month (the statewide average at the time), statewide prescription drug costs decreased by 35% (as expected by the health care planners), but this decrease was accompanied by a 20% increase in use of hospital resources and a 220% increase in admissions to nursing homes. One might suspect that many of the hospital and nursing home admissions were for patients whose chronic medical problems (including hypertension) worsened when they had insufficient chronic medications paid for by the state. When the cap on prescription drugs was withdrawn (after 11 months), expenditures for medications and hospitalization rates returned toward baseline levels, but few patients who entered nursing homes were discharged. Thus the experiment probably resulted in higher actual costs during and after the limit on prescriptions than before its institution.

Although hypertension has not been a focus for the New Hampshire Medicaid recipients, a formal cost calculation has been made for patients with psychiatric problems.[45] When the state limited coverage to three prescriptions per month per Medicaid enrollee, the cost of various drugs used to treat psychiatric diseases decreased 15–49% (depending on the type of medication, e.g., antidepressants or antipsychotics.). Unfortunately, the side effect of this cost savings was an even bigger increase in utilization of community mental health centers, day hospitals, and emergency mental health services (20–40%). All of these services declined toward baseline utilization rates after the cap on pharmaceuticals was rescinded. After accounting for all excessive utilization of health care services, it was determined that overall the per-patient costs increased by $1530 during the time that the limit on prescription drugs was in place; this was 17 times the per-patient cost savings realized from reductions in drug prescribing and dispensing.

Data about control of hypertension after such governmental attempts to limit health care expenditures are available from the state of California, which withdrew benefits for 270,000 MediCal beneficiaries in 1982. One year later, blood pressure control among 182 former enrollees deteriorated significantly, and the percentage of patients with diastolic blood pressures over 100 mmHg more than tripled (from 3% to 19%).[46] Evidence from the Rand Health Insurance Experiment also suggests that free health care improves blood pressure control.[47] One-half of 3958 relatively healthy people less than 65 years old were randomized to free health care; the other half had to bear part of the cost. After 3–5 years, more cases of hypertension had been diagnosed among recipients of free care (presumably because the number of physician visits was 33% higher), and their blood pressures were 2–4 mmHg lower (on average).[48] This clinically insignificant lowering of blood pressure in a population translates to an anticipated long-term 14–28% reduction in stroke and 4-10% reduction in myocardial infarction.[28,49] It thus appears that, in some settings, restricting access to needed chronic medical therapy can be "penny wise and pound foolish" and actually lead to long-term increases in cost rather than sustained short-term reductions, as intended.

A more successful statewide program to limit use of expensive medication was implementation of telephone-triaged prior authorization for nonsteroidal antiinflammatory drugs (NSAIDs) in Tennessee.[50] Healthcare authorities noted that in 1988 Tennessee spent $99 million per year on NSAIDs, most of which was paid for newer, brand-name drugs that were heavily promoted by industry without a great deal of evidence that one was better than another. Beginning in November 1989, physicians wishing to prescribe a brand-name NSAID for a Tennessee Medicaid recipient were required to obtain permission via telephone before such a prescription would be honored, much like the system for preauthorization hospital admissions implemented across the United States. In the year following implementation of this policy, total NSAID prescriptions were reduced by 19%, and expenditures for NSAIDs dropped by 53%. For the estimated $75,000 per annum required to administer the program, a savings of $12.4 million was realized during the first two years alone. Unlike other programs to limit availability of medications,[44,45] no major increases in medical, emergency department, or hospital services were noted, nor was there a major increase in state expenditures for alternate drugs (e.g., acetaminophen). The effects of the additional administrative burden of this program on the stress levels and blood pressures of physicians, office staff, and patients have not yet been examined. Although this system is no longer used in Tennessee (where the states pays only for generic ibuprofen), the extension of the telephone-triaged prior authorization system for limiting expensive medications is currently under study for specific antihypertensive agents in several locales.[51,52]

Another method of administrative control of drug costs that is becoming more widespread is based at the pharmacy.[52] Some states have laws that grant the pharmacist the privilege of "therapeutic substitution," which is based on the principle of dispensing a less expensive drug rather than the drug prescribed by the physician if the two products have been designated "therapeutically equivalent" by a panel of local experts. Although this practice is currently restricted to only a few of the 50 states, many of the mail-order pharmacies have been established in these states, presumably to take advantage of cost savings for nonresidents whose prescriptions are mailed to states where therapeutic substitution is allowed.

Another method that has had a more mixed reception is the practice of influencing the pharmacist to request a change for a prescribed drug product, sometimes with a telephone call to the physician. This method initially had a poor reputation, because a certain pharmaceutical company paid a bounty for each patient switched (after telephoning the physician for approval) to their less expensive product. Ongoing "research," however, shows that much of a pharmacy's budget can be saved when the pharmacy manager implements a systemwide program requiring each staff pharmacist to contact a physician who has prescribed the more expensive product and to "educate" the physician about the availability of a cheaper alternative with presumed equivalent therapeutic effects.[53,54] These "studies" often have been funded by the company with the less expensive alternative, which makes conflict of interest a concern. In addition, the results generally focus only on pharmacy costs and ignore additional patient visits to the physician to accomplish dose-titration of the new drug with the lower acquisition cost.

SIMPLE PHYSICIAN-DEPENDENT SUGGESTIONS FOR IMPROVING COST-EFFECTIVENESS IN TREATING HYPERTENSION

Physicians have several options to improve the cost-effectiveness of treating hypertension, in good conscience and with little risk:[55–57]

1. Be sure that the patient truly has hypertension before launching an expensive treatment program, particularly in patients with "borderline" and lower stages of hypertension. A large number of such patients must be treated

to prevent one adverse event.[58] The current approach in the United States[2] (also recommended by the World Health Organization[35]) is to minimize initial testing (including measurement of blood pressure on several occasions in the physician's' office), although some have suggested that diagnostic approaches that initally are more costly eventually may be proven cost-effective.[59–63]

2. Prescribe the less expensive drugs whenever they are appropriate and feasible.[30,31,35,57] Acquisition costs of ACE inhibitors, for example, vary significantly, even among brand-name formulations. Examination of a recent pricelist indicates a statistically significant relationship between average wholesale price[64,65] and time on the market[4] (see Fig. 3). Although this relationship was not found for all classes of antihypertensive drugs (e.g., calcium antagonists), among the several available formulations of older calcium antagonists (e.g., verapamil, diltiazem, nifedipine), it was valid.[66] Perhaps for this reason consumers and healthcare insurers should support the development of new drugs.

3. Maximize compliance with medications.[34,67,68] It has been estimated that an accurate assessment of compliance at each visit adds only 90 seconds to the clinician's contact time with the patient.[69] When the patient is not compliant, the medical care and pharmacy cost that preceded noncompliance are wasted, and the patient is at higher risk for adverse sequelae of hypertension because blood pressure is not lowered as expected. Thus, cost is increased, and effectiveness is lowered when compliance with medications is suboptimal.

Other methods of limiting the cost of treating hypertension may have a negative effect on independent pharmacists, who, unfortunately, are already an endangered species of health care professional:

1. In prescribing maintenance therapy (e.g., antihypertensive agents that have been effective and well-tolerated after a few months), dispensing a large number of pills per visit to the pharmacy (e.g., 100-day supply) saves about

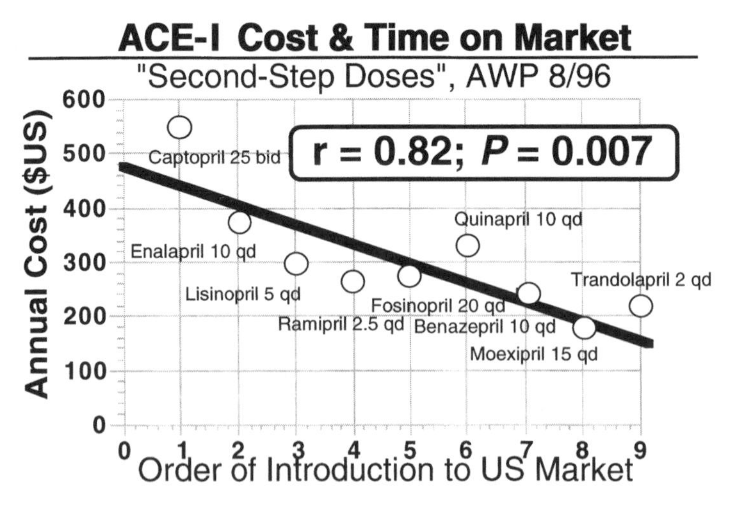

FIGURE 3. Relationship between the average wholesale price (AWP) of angiotensin-converting enzyme inhibitors and the order of introduction into the U.S. market. Regression analysis suggests that the marketers of the twelfth ACE inhibitor will have to pay the patients to take it, because the acquisition cost is projected to be less than zero dollars annually. (Adapted from Medispan Prescribing Pricing Guide. Indianapolis, IN, Medispan, 1996.)

30% (compared with three 30-day supplies) of pharmacy time and expense as well as patient time and effort.

2. Many enterprising and dexterous patients prefer to have a prescription for a pill that contains twice their usual dose of medication. They bisect the pill, consuming only half at each administration. Because the costs of medications do not typically increase geometrically as the dose is increased (and the costs of a notable few—quinapril, benazepril, moexipril, trandolapril, nisoldipine, and terazosin—do not increase at all), this approach can be a substantial money-saver. If, however, mistakes are made in cutting pills or taking the appropriate dose, this approach may increase cost, because safety and efficacy may be threatened.

3. Some patients who require more than a single drug may benefit financially from prescription of "combination pills" (containing a dose of each medication), if the doses are well chosen and convenient to take.[70]

4. Many patients are eligible for programs that allow them to obtain prescription medications without out-of-pocket cost. The medications are intended either as samples (generally frowned on by industry) or as assistance to people of lesser financial means.[71] If the patient lives in South Carolina (which has a system known as "Communi-Care") or other locales where special programs have been negotiated between the state and pharmaceutical companies, a call to a toll-free number will provide the name and location of a nearby pharmacy that provides antihypertensive medication directly from a major manufacturer at no cost to the patient.

CONCLUSION

For most patients with hypertension, treatment is likely to cost money overall. This cost can be a major impediment to the financial success and even survival of an already burgeoning health care system. Efforts to restrict access to effective medications have not always been cost-saving in the long term, but more emphasis is likely to be placed on newer administrative efforts to make prescription of expensive medications more difficult (if not impossible). Outpatient prescription drugs for hypertension are currently favored targets for cost-containment, with recommendations from many authorities to use more diuretics or beta blockers. More data should be accrued about differences (if any) in total cost of care for hypertension across drug classes before administrative decisions to limit prescribing are implemented. Several simple methods of improving the cost-effectiveness of hypertension treatment include establishing a proper diagnosis, enhancing compliance, prescribing less expensive medications and combination pills when appropriate, and taking advantage of indigent care programs from pharmaceutical companies when available.

References

1. Swales JD: Economics and the treatment of hypertension. J Hypertens 13:1357–1361, 1995.
2. Joint National Committee: The fifth report of the Joint National Committee on Detection, Evaluation and Treatment of High Blood Pressure. Arch Intern Med 153:154–186, 1993.
3. Heart and Stroke Facts: 1996 Statistical Supplement. Dallas, TX, American Heart Association. 1996, p 23.

4. Elliott WJ: The costs of treating hypertension: What are the long-term realities of cost containment and pharmacoeconomics? Postgrad Med 99:241–248, 1996.
5. Burt VL, Cutler JA, Higgins M, et al: Trends in the prevalence, awareness, treatment, and control of hypertension in the adult US population. Data from the health examination surveys, 1960 to 1991. Hypertension 26:60–69, 1995.
6. Kawachi I, Malcolm LA: Treating mild to moderate hypertension: Cost-effectiveness and policy implications. J Cardiovasc Pharmacol 16(Suppl 7):S126–S128, 1990.
7. Kawachi I, Malcolm LA: The cost-effectiveness of treating mild-to-moderate hypertension: A reappraisal. J Hypertens 9:199–208, 1991.
8. Jackson R, Barham P, Bills J, et al: Management of raised blood pressure in New Zealand: A discussion document. BMJ 307:107–110, 1993.
9. Alderman MH: Blood pressure management: Individualized treatment based on absolute risk and the potential for benefit. Ann Intern Med 119:329–335, 1993.
10. Reif MC, Carter VL: The retail cost of antihypertensive therapy: Physician and patient as educated consumers. Am J Hypertens 7:571–575, 1994.
11. Hilleman DE, Mohiuddin SM, Lucas BD Jr, et al: Cost-minimization analysis of initial antihypertensive therapy in patients with mild-to-moderate essential diastolic hypertension. Clin Ther 16:88–102, 1994.
12. Elliott WJ: Costs associated with changing antihypertensive drug monotherapy: "Preferred" vs. "alternative" therapy [abstract]. Am J Hypertens 8:80A, 1995.
13. Kawachi I: Epidemiology of stroke. Importance of preventive pharmacological strategies in elderly patients and associated costs. Drugs Aging 5:288–299, 1994.
14. Lindholm LH, Johannesson M: Cost-benefit aspects of treatment of hypertension in the elderly. Blood Pressure 3:11–14, 1995.
15. Johannesson M, Dahlöf B, Lindholm LH, et al: The cost-effectiveness of treating hypertension in elderly people: An analysis of the Swedish Trial in Old Patients with Hypertension (STOP-Hypertension). J Intern Med 234:317–323, 1993.
16. Bulpitt CJ, Fletcher AE: Cost-effectiveness of the treatment of hypertension. Clin Exp Hypertens 15:131–146, 1993.
17. Johannesson M: The impact of age on the cost-effectiveness of hypertension treatments: An analysis of randomized drug trials. Med Decis Making 14:236–244, 1994.
18. Littenberg B, Garber AM, Sox HC Jr: Screening for hypertension. Ann Intern Med 112:192–202, 1990.
19. Littenberg B: A practice guideline revisited: Screening for hypertension. Ann Intern Med 122:937–939, 1995.
20. Torgerson DJ, Spencer A: Marginal costs and benefits. BMJ 312:35–36, 1996.
21. Nease RF Jr, Owens DK: A method for estimating the cost-effectiveness of incorporating patients' preferences into practice guidelines. Med Decis Making 14:382–392, 1994.
22. Shepard DS, Stason WB, Perry HM Jr, et al: Multivariate cost-effectiveness analysis: An application to optimizing ambulatory care for hypertension. Inquiry 32:320–331, 1995.
23. Lindgren B: The cost-benefit approach to pricing new medicines: Doxazosin versus beta-blocker treatment in Sweden. Am Heart J 119:748–753, 1990.
24. Johannesson M: The cost-effectiveness of the switch towards more expensive antihypertensive drugs. Health Policy 28:1–13, 1994.
25. Johannesson M: The cost-effectiveness of hypertension treatment in Sweden. Pharmacoeconomics 7:242–250, 1995.
26. Johannesson M, Agewall S, Hartford M, et al: The cost-effectiveness of a cardiovascular multiple-risk-factor intervention programme in treated hypertensive men. J Intern Med 237:19–26, 1995.
27. Jönsson BG: Cost-benefit of treating hypertension. J Hypertens Suppl 12:S65–S75, 1994.
28. Collins R, Peto R, MacMahon S, et al: Blood pressure, stroke, and coronary heart disease. Part 2: Short-term reductions in blood pressure: Overview of randomised drug trials in their epidemiological context. Lancet 335:827–838, 1990.
29. Manolio TA, Cutler JA, Furberg CD, et al: Trends in pharmacologic management of hypertension in the United States. Arch Intern Med 155:829–837, 1995.
30. Psaty BM, Koepsell TD, Yanez ND, et al: Temporal patterns of antihypertensive medication use among older adults, 1989 through 1992. An effect of the major clinical trials on clinical practice? JAMA 273:1436–1438, 1995.
31. Kaplan NM, Gifford RW Jr: Choice of initial therapy for hypertension. JAMA 275:1577–1580, 1996.
32. Hypertension Society of South Africa: Guidelines for the management of hypertension at primary health care level [endorsed by the Medical Association of South Africa and the Medical Research Council]. S Afr Med J 85:1321–1325, 1995.

33. Cottrell K: Experts discuss cost-effective treatment of hypertension during world conference in Canada. Can Med Assoc J 153:1332–1335, 1995.
34. Schueler K. Cost-effectiveness issues in hypertension control. Can J Public Health 85(Suppl 2): S54–S56, 1994.
35. WHO Expert Committee: Hypertension control. World Health Organ Tech Rep Ser 862:1–83, 1996.
36. Arroyo P, Herrera J, Fernandez V: The therapeutic approach to the control of hypertension. Its impact on health policy. Clin Exp Hypertens 17:1121–1126, 1995.
37. Rodby RA, Firth LM, Lewis EJ, for the Collaborative Study Group: An economic analysis of captopril in the treatment of diabetic nephropathy. Diabetes Care 19:1051–1061, 1996.
38. The HOT Study Group: The Hypertension Optimal Treatment Study (The HOT Study). Blood Pressure 2:62–68, 1993.
39. Nissinen A, Tuomilehto J, Kottke TE, Puska P: Cost-effectiveness of the North Karelia hypertension program 1972–77. Med Care 24:767–780, 1984.
40. Roccella EJ, Lenfant C: Considerations regarding the cost and effectiveness of public and patient education programmes. J Hum Hypertens 6:463–467, 1992.
41. Johannesson M, Fagerberg B: A health-economic comparison of diet and drug treatments in obese men with mild hypertension. J Hypertens 10:1063–1070, 1992.
42. Cutler JA: Combinations of lifestyle modification and drug treatment in management of mild-moderate hypertension: A review of randomized clinical trials. Clin Exp Hypertens 15:1193–1204, 1993.
43. Public health focus: Physical activity and the prevention of coronary heart disease. MMWR 42:669–672, 1993.
44. Soumerai SB, Ross-Degnan D, Avorn J, et al: Effects of Medicaid drug-payment limits on admission to hospitals and nursing homes. N Engl J Med 325:1072–1077, 1991.
45. Soumerai SB, McLaughtlin TH, Ross-Degnan D, et al: Effects of a limit on Medicaid drug-reimbursement benefits on the use of psychotropic agents and acute mental health services by patients with schizophrenia. N Engl J Med 331:600–605, 1994.
46. Lurie N, Ward NB, Shapiro MF, et al: Termination of MediCal benefits: A follow-up study one year later. N Engl J Med 314:1266–1268, 1986.
47. Brook RH, Ware JE Jr, Rogers WH, et al: Does free health care improve adults' health? N Engl J Med 309:1426–1434, 1983.
48. Keeler EB, Brook RH, Goldberg GA, et al: How free care reduced hypertension in the health care experiment. JAMA 254:1926–1931, 1985.
49. Rose G: Strategy of prevention: Lessons learned from cardiovascular disease. BMJ 282:1847–1851, 1981.
50. Smalley WE, Griffin MR, Fought RL, et al: Effect of a prior-authorization requirement on the use of nonsteroidal antiinflammatory drugs by Medicaid patients. N Engl J Med 332:1612–1617, 1995.
51. Aucott JN, Pelecanos E, Dombrowski R, et al: Implementation of local guidelines for cost-effective management of hypertension. A trial of the firm system. J Gen Intern Med 11:139–146, 1996.
52. Jameson J, VanNoord G, Vanderwoud K: The impact of a pharmacotherapy consultation on the cost and outcome of medical therapy. J Fam Pract 41:469–472, 1995.
53. McDonough KP, Weaver RH, Viall GD: Enalapril to lisinopril: Economic impact of a voluntary angiotensin-converting enzyme-inhibitor substitution program in a staff-model health maintenance organization. Ann Pharmacother 26:399–404, 1992.
54. Simons WR, Rizzo JA, Stoddard M, Smith ME: The costs and effects of switching calcium channel blockers: evidence from Medicaid claims data. Clin Ther 17:154–173, 1995.
55. Stason WB: Opportunities for improving the cost-effectiveness of antihypertensive treatment. Am J Med 81(Suppl 6C):45–49, 1986.
56. Stason WB: Opportunities to improve the cost-effectiveness of treatment for hypertension. Hypertension 18(Suppl 3):I-161–I-166, 1991.
57. World Hypertension League: Economics of hypertension control. Bull World Health Organ 73:417–424, 1995.
58. Insua JT, Sacks HS, Lau TS, et al: Drug treatment of hypertension in the elderly: A meta-analysis. Ann Intern Med 121:355–362, 1994.
59. Krakoff LR, Schechter C, Fahs M, Andre M: Ambulatory blood pressure monitoring: Is it cost-effective? J Hypertens 9(Suppl):S28–S30, 1991.
60. Krakoff LR: Ambulatory blood pressure monitoring can improve cost-effective management of hypertension. Am J Hypertens 6(Suppl):220S–224S, 1993.
61. Yarows SA, Khoury S, Sowers JR: Cost effectiveness of 24-hour ambulatory blood pressure monitoring in evaluation and treatment of essential hypertension. Am J Hypertens 7:464–468, 1994.

62. Imai Y: Clinical significance and cost-effectiveness of 24-hour ambulatory blood pressure monitoring. Tohoku J Exp Med 176:1–15, 1995.
63. Pierdomenico SD, Mezzetti A, Lapenna D, et al: "White-coat" hypertension in patients with newly diagnosed hypertension: Evaluation of prevalence by ambulatory monitoring and impact on cost of health care. Eur Heart J 16:692–697, 1995.
64. Moexipril: Another ACE Inhibitor for Hypertension. Medical Lett Drugs Ther 37:75–76, 1995.
65. Medispan Prescribing Pricing Guide. Indianapolis, IN, Medispan, 1996.
66. Elliott WJ: Pharmacoeconomics of antihypertensive drugs: A reason for consumers to support drug development? [abstract]. Clin Pharmacol Ther 101:208, 1997.
67. Elliott WJ: Compliance strategies. Curr Opin Nephrol Hypertens 3:271–278, 1994.
68. Bittar N: Maintaining long-term control of blood pressure: The role of improved compliance. Clin Cardiol 18(6 Suppl 3):III12–III16, 1995.
69. Fishman T: The 90-Second Intervention: A patient compliance mediated technique to improve and control hypertension. Public Health Rep 110:173–178, 1995.
70. Kaplan NM: Implications for cost-effectiveness. Combination therapy for systemic hypertension [editorial]. Am J Cardiol 76:595–597, 1995.
71. Report of the Task Force on the Availability of Cardiovascular Drugs to the Medically Indigent. Circulation 85:850–860, 1992.

FRANS H. H. LEENEN M.D., Ph.D., F.R.C.P.C.

28

Dihydropyridine Calcium Antagonists and Sympathetic Activity: Relevance to Cardiovascular Morbidity and Mortality

The 1,4-dihydropyridines have several properties that theoretically make them attractive for treatment of hypertension and prevention as well as management of coronary artery disease (CAD). However, they also have actions that are potentially detrimental, leading, for example, to worsening of angina or heart failure. Such negative properties may be intrinsic to the class or specific for a given compound or formulation. This chapter briefly reviews the subclasses of dihydropyridines and outlines evidence that although dihydropyridines can decrease sympathetic activity, rapid lowering of blood pressure after dosing may override this effect, resulting in intermittent increases in sympathetic activity. The final discussion focuses on the concept that different subclasses may affect outcome differently, depending on the pattern of blood pressure control and changes in sympathetic tone.

SUBCLASSES OF DIHYDROPYRIDINES

Dihydropyridines decrease blood pressure primarily by arterial vasodilation.[1] This response strongly correlates with plasma drug concentration.[2–4] Thus, although the primary hemodynamic effects are rather similar for the whole class, differences in pharmacokinetic profile—even for different formulations of the same dihydropyridine—have a pronounced effect on the pattern of hemodynamic changes. Based on pharmacokinetics, the dihydropyridines can be separated into three distinct subclasses: rapid-, short-acting agents, controlled or slow-release formulations of short-acting agents, and slow-onset, long-acting agents (Table 1).

Plasma concentrations of rapid-, short-acting agents increase quickly and often markedly after dosing, leading to a rapid decrease in blood pressure. Plasma concentrations also diminish quickly because of the short elimination half-life, with a return of blood pressure to predosing levels. Frequent administration of relatively low doses, therefore, provides more sustained blood pressure control than higher or less frequent doses (e.g., nifedipine capsule, 10 mg, 1 every 6 hr vs. 2 every 12 hr).

Newer formulations of the above subclass show a decreased rate of absorption and, therefore, a more gradual and less marked increase in plasma concentrations. Speed of onset and duration of antihypertensive effect are clearly

TABLE 1. Subclasses of Dihydropyridines According to Pharmacokinetics

Rapid-, short-acting agents (e.g., nifedipine capsules, felodipine tablets)
Absorption is rapid, but plasma levels (and effects) also disappear rapidly.

Controlled or slow-release formulations of short-acting agents (e.g., nifedipine GITS, felodipine ER)
Absorption is more gradual and prolonged, resutling in more stable hemodynamic effects over a 24-hour dosing interval.

Slow-onset, long-acting agents (e.g., amlodipine, lacidipine)
These agents have a long elimination half-life. Hemodynamic effects develop gradually over days or weeks, and during maintenance therapy hemodynamic and blood pressure effects are sustained and stable.

From Leenen FHH: Clinical relevance of 24-hour blood pressure control by 1,4-dihydropyridines. Am J Hypertens 9(Suppl.):97s–104s, 1996, with permission.

different. For example, the slower and more steady rate of absorption of nifedipine from nifedipine gastrointestinal therapeutic system (GITS) results in fairly stable plasma concentrations and stable antihypertensive effect over 24 hours during maintenance treatment.[5,6] However, not all once-daily, slow-release formulations demonstrate this consistency. For example, with once-daily dosing the extended-release (ER) felodipine tablet or sustained-release nifedipine (Adalat CC) exhibit a clear peak to trough variation in plasma drug concentration,[4,7] which may result in fluctuations in the extent of antihypertensive effect over 24 hours.[4,8]

The intrinsically long-acting agents (e.g., amlodipine) are absorbed fairly slowly. Because of their elimination half-life of 2–3 days,[9] they reach steady-state plasma concentrations after 1–2 weeks. The antihypertensive effect is, therefore, rather minimal at initial dosing and gradually increases over subsequent weeks.[10] This subclass, therefore, causes an antihypertensive effect with gradual onset but sustained duration over a 24-hour dosing interval during maintenance treatment.[11]

Thus, the rapid-, short-acting dihydropyridines differ from the slow-onset, long-acting agents not only in duration of action but also in pattern of hemodynamic responses—i.e., rapid on-and-off vs. stable. During maintenance therapy stable effects are sustained by slow-onset, long-acting dihydropyridines such as amlodipine as well as some of the slow-release formulations such as nifedipine GITS. Differences between the two subclasses emerge during short periods of noncompliance. Thus, consistent with the short elimination half-life of nifedipine, after one missed dose of nifedipine GITS the antihypertensive effect diminishes over the next 12–24 hours.[6] In contrast, the antihypertensive effect of amlodipine decreases gradually, and after 2 missed doses most of the antihypertensive effect is still present.[11] Because short periods of noncompliance are rather common in many hypertensive patients,[12,13] extended therapeutic coverage contributes to stable hemodynamic effects during chronic treatment.

Figures 1 and 2 provide examples of blood pressure response for the three subclasses: rapid and marked lowering of blood pressure by nifedipine capsule; sustained antihypertensive effect throughout the day by nifedipine GITS, which clearly diminishes over the day after a missed dose; and sustained antihypertensive effect by amlodipine persisting for at least 2 days.

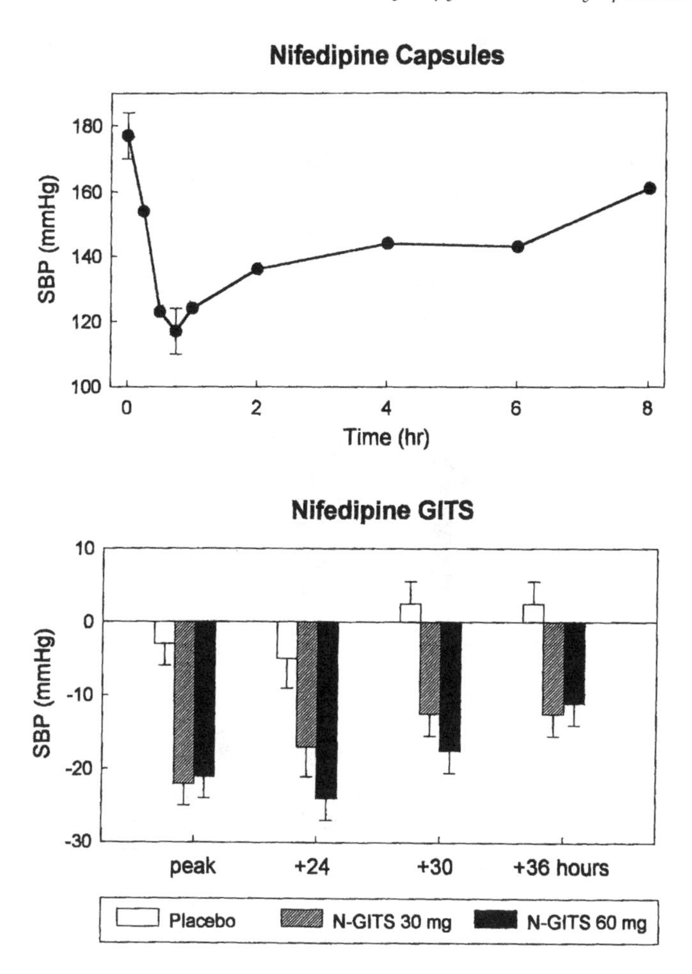

FIGURE 1. Systolic blood pressure responses to nifedipine capsule *(top panel)* and nifedipine GITS *(bottom panel)* in hypertensive patients. Two nifedipine 10-mg capsules were administered at t = 0 to 6 hypertensive patients taking concomitant beta-blocker treatment.(Data from Myers MG, Raemsch KD: Comparative pharmacokinetics and antihypertensive effects of the nifedipine tablet and capsule. J Cardiovasc Pharmacol 10(Suppl 10):S76–S78, 1987, with permission.) The nifedipine GITS (N-GITS) data were obtained at the end of 4 weeks of double-blind treatment with placebo (n = 20), nifedipine GITS, 30 mg/day (n = 16), or nifedipine GITS, 60 mg/day (n = 20) with ambulatory blood pressure monitoring up to 36 hours after dosing. Values are absolute values (means ± SEM, top panel) or changes in blood pressure (means ± SEM, bottom panel). (Figure redrawn from Zanchetti A: Trough and peak effects of a single daily dose of nifedipine gastrointestinal therapeutic system (GITS) as assessed by ambulatory blood pressure monitoring. J Hypertens 12(Suppl 5):S23-S27, 1994, with permission.)

DIHYDROPYRIDINES AND SYMPATHETIC ACTIVITY

Direct Effects of Dihydropyridines on Sympathetic Activity

The dihydropyridines may affect sympathetic tone by mechanisms both related and unrelated to their hemodynamic effects. Intrinsic effects, not related to hemodynamic effects, include inhibition of energy-dependent catecholamine uptake into storage vesicles,[14,15] induction of leakage of amines from their vesicles,[15] and

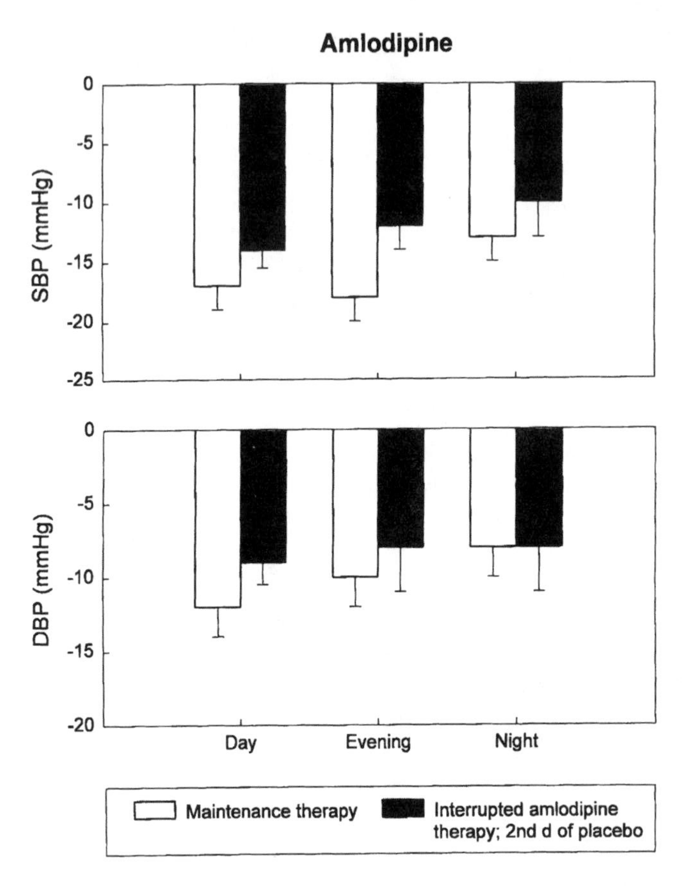

FIGURE 2. Persistence of the antihypertensive effect of amlodipine during the second day of placebo treatment (i.e., up to 72 hours after administration of active drug), which interrupted active therapy with amlodipine for 9–10 weeks. Values represent mean (± SEM) decreases from values at the end of the placebo run-in period, obtained by 24-hour ambulatory blood pressure monitoring. Placebo day blood pressure was 158 ± 3/102 ± 2 mmHg. Amlodipine dose, 5 mg (n = 7) and 10 mg (n = 13). All decreases in blood pressure were significant ($p < 0.05$), and reductions with maintenance therapy did not differ significantly from reductions during the second day of placebo. (From Leenen FHH, Fourney A, Notman G, Tanner J: Persistence of anti-hypertensive effect after 'missed doses' of calcium antagonist with long (amlopidine) vs short (diltiazem) elimination half-life. Br J Clin Pharmacol 41:83–88, 1996, with permission.)

activation of the metabolic degradation of catecholamines via the catalysing enzyme, catechol-O-methyltransferase. These effects may be more prominent with lipophilic dihydropyridines; for example compared with felodipine, amlodipine has only minor effects on catecholamine stores.[15] Peripheral effects on sympathetic nerves may lead, to variable degrees, to increased release and reduced reuptake of catecholamines by dihydropyridines. As reviewed by Eikenburg and Lokhandwala,[16] neurotransmitter overflow elicited by electrical stimulation of sympathetic nerves is relatively resistant to inhibition by calcium antagonists. In fact, some studies have reported that electrically stimulated release of neurotransmitters can be enhanced. However, to what extent these findings in vitro also occur in vivo in therapeutic concentrations is still unclear.

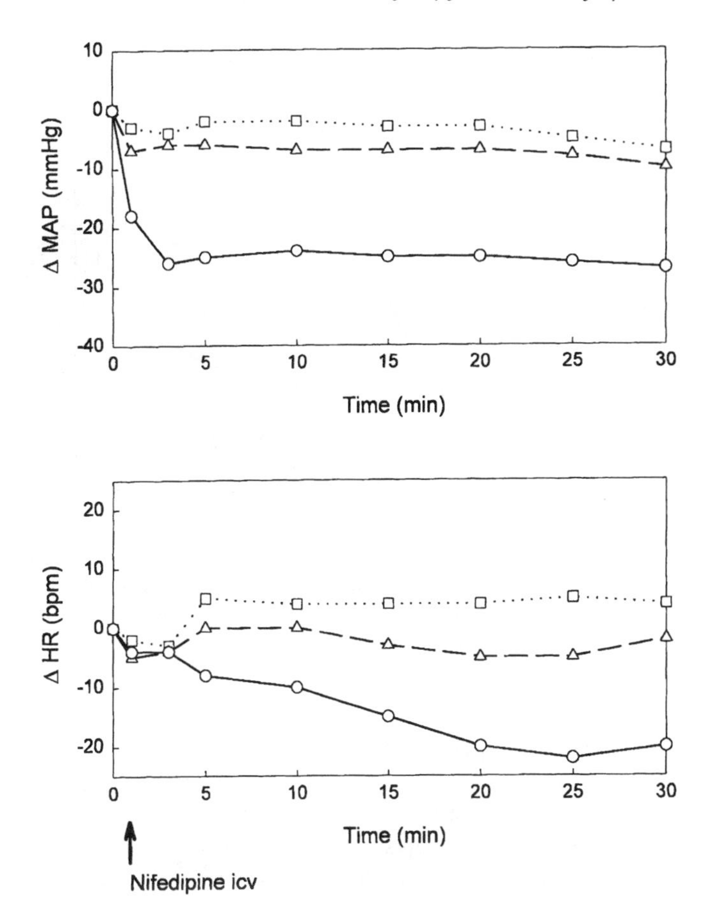

FIGURE 3. Changes in mean arterial pressure (MAP) and heart rate (HR) after administration of nifedipine into the lateral ventricle of the brain of spontaneously hypertensive rats. Values are means ± SEM (n = 5–6). □ · · · □ = vehicle, △ - - △ = nifedipine, 5 μg/kg, ○—○ = nifedipine, 50 μg/kg. * $p < 0.05$ vs. vehicle. (Data redrawn from Laurent S, Girerd X, Tsoukaris-Kupfer D, et al: Opposite central cardiovascular effects of nifedipine and BAY k 8644 in anesthetized rats. Hypertension 9:132–138, 1987, with permission.)

In addition to the above peripheral effects, dihydropyridines also have effects in the central nervous system (CNS), which lead to decreases in blood pressure and heart rate, presumably by decreasing sympathetic outflow. For example, administration of nifedipine into the lateral ventricle of the brain of spontaneously hypertensive rats leads to dose-related decreases in blood pressure and heart rate (Fig 3), which can be prevented by pretreatment with a ganglion blocker. The ganglion blocker does not prevent the hypotensive response to intravenously administered nifedipine.[17] At least in part, central effects may relate to excitation of neurons in the nucleus tractus solitarius of the brainstem (either directly or indirectly), resulting in withdrawal of sympathetic tone and decreases in blood pressure and heart rate.[18]

The central sympathoinhibitory effect of dihydropyridines appears to be therapeutically relevant. Thus, chronic treatment of spontaneously hypertensive rats with nisoldipine (administered in food) normalized the elevated cardiac

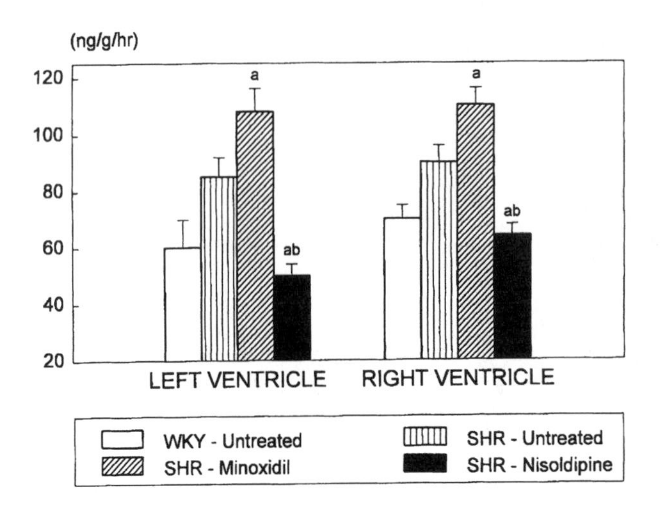

FIGURE 4. Left and right ventricular norepinephrine turnover rates in untreated Wistar-Kyoto (WKY) rats and spontaneously hypertensive rats (SHR) treated with either minoxidil (120 mg/L drinking water) or nisoldipine (2 mg/gm food). Bars represent means ± SEM. [a]$p < 0.05$ vs. untreated SHR, [b]$p < 0.05$ vs. untreated WKY rats. (Figure redrawn from Tsoporis J, Fields N, Leenen FHH: Contrasting effects of calcium antagonists vs. arterial vasodilators on cardiac anatomy and sympathetic activity in spontaneously hypertensive rats. J Cardiovasc Pharmacol 17(Suppl 2):S166–S168, 1991, with permission.)

sympathetic activity, as assessed by norepinephrine turnover rates. In contrast, the arterial vasodilator minoxidil further increased sympathetic activity[19] (Fig 4). Similarly, chronic treatment with amlodipine or manidipine significantly lowered levels of plasma norepinephrine in spontaneously hypertensive rats, whereas hydralazine caused a further increase.[20] Clearly, long-term treatment with dihydropyridines can lower sympathetic activity in hypertensive animals.

Baroreflex-mediated Sympathetic Hyperactivity

Rapid decreases in blood pressure result in deactivation of arterial baroreceptors and, as a consequence, increased sympathetic neural activity and withdrawal of parasympathetic tone. In normotensive or hypertensive rats, sympathetic activity (as assessed, for example, by plasma norepinephrine or renal sympathetic nerve activity) increases in parallel with the decrease in blood pressure caused by a dihydropyridine.[21,22] Similar findings have been reported for hypertensive humans.[23–26] Indeed, after acute administration of a fast-acting dihydropyridine, the extent of the increase in sympathetic activity may be similar to that observed after administration of classical arterial vasodilators, such as hydralazine.[24]

Even at acute administration, however, the extent of the increase in plasma norepinephrine and heart rate also depends on both the rate and extent of the decrease in blood pressure. As indicated in Table 2, the nifedipine capsule causes larger and faster increases in plasma norepinephrine and heart rate than the nifedipine slow-release tablet, reflecting the faster rate but similar degree of decrease in blood pressure with the capsule. This difference in response is

TABLE 2. Effects of Nifedipine Administered Orally as Capsule or Slow-release Tablet on Plasma Norepinephrine and Heart Rate in Healthy Subjects

	Capsule (2 × 10 mg)	Slow-release Tablet (20 mg)
Plasma norepinephrine (pg/ml)		
Baseline	152 ± 47	191 ± 64
Time 1*	326 ± 96§	260 ± 121
Time 2†	330 ± 129§	278 ± 100∞
Changes in heart rate (beats/min)		
Time max‡	30 min	240 min
Increase	+ 16 ± 9§	+ 7 ± 5

Values are means ± SEM (n = 6).
* Time 1: 45 and 90 minutes after nifedipine, capsule and slow-release (SR) tablet, respectively.
† Time 2: 120 and 240 minutes after nifedipine, capsule and SR tablet, respectively.
‡ Time max: time of maximal increase in heart rate vs. baseline.
Systolic blood pressure (BP) did not change, and diastolic BP decreased by 5–10 mmHg, without marked differences between the formulations.
§ $p < 0.01$, ∞ $p < 0.05$.
All data derived from Kleinbloesem CH, van Brummelen P, van de Linde JA, et al: Nifedipine: Kinetics and dynamics in healthy subjects. Clin Pharmacol Ther 35:742–749, 1984, with permission.

consistent with resetting of the baroreflex with the slower decrease in blood pressure.

During chronic treatment central inhibitory effects on sympathetic tone may develop (see above), and the arterial baroreflex is reset to a lower blood pressure.[27] Arterial baroreflex sensitivity for control of heart rate has been reported either to remain unchanged[28,29] or to increase somewhat.[27,30] During chronic treatment, related to the putative sympathoinhibitory effect and resetting of the baroreflex, one may anticipate less or no increase in sympathetic activity for the same lower level of blood pressure as found after acute treatment. However, after each dose of a fast-absorbed, fast-acting dihydropyridine the blood pressure again shows a rapid fall from the predosing level. This rapid fall again may lead to an arterial baroreflex-mediated increase in sympathetic activity. The higher the dose and the longer the dosing interval, the larger and faster the extent of the fall in blood pressure and, therefore, presumably the larger the increase in sympathetic activity. If this concept is correct, one can expect changes in sympathetic activity depending on the hemodynamic profile—i.e., intermittent increases in sympathetic activity shortly after dosing during chronic treatment with rapid-acting dihydropyridines but no increases (and possibly decreases) during chronic treatment with slow-onset, long-acting dihydropyridines.

Effects of Rapid-, Short-acting Dihydropyridines on Sympathetic Activity

As outlined above, the first-dose effects of rapid-, short-acting dihydropyridines on sympathetic activity were recognized shortly after the introduction of dihydropyridines in clinical care.[24,31] Subsequent studies suggested that this increase in sympathetic activity disappears during chronic treatment, and this belief was commonly held during the 1980s.[25,32] However, some of the early studies reported

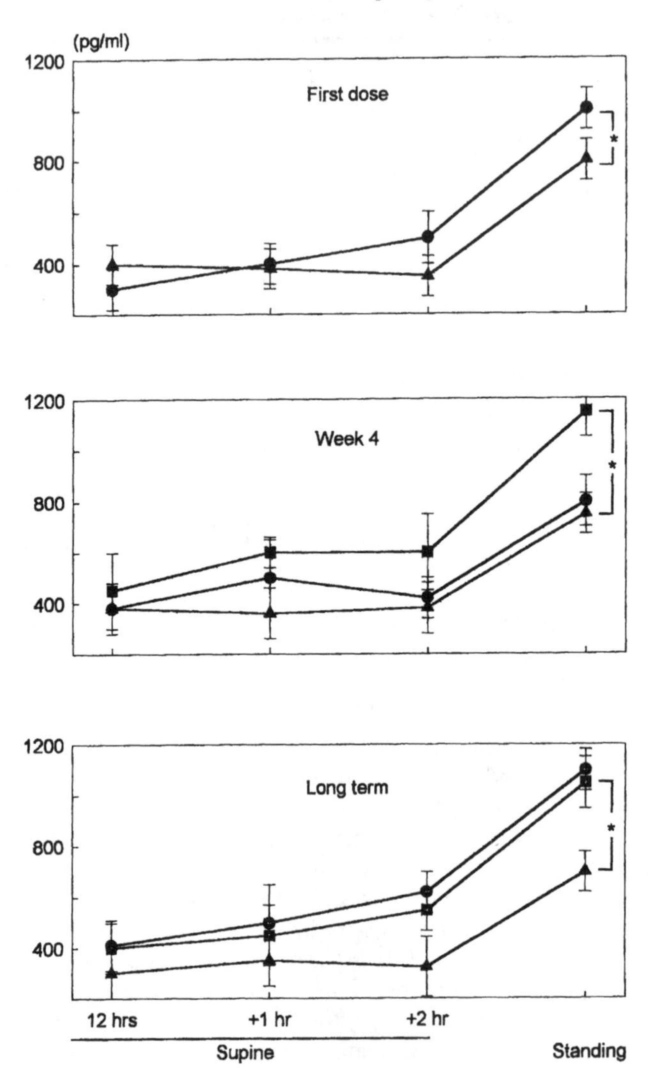

FIGURE 5. First-dose, short-term (4 weeks), and chronic (6 and 12 months) effects of felodipine tablets on plasma norepinephrine. Values represent mean ± SEM (for number of patients, see Leenen FHH, Holliwell DL, 1992). The acute and short-term studies were double-blind, placebo-controlled; the long-term follow-up was an open study with most patients taking 10–20 mg twice daily. First dose and week 4: ▲—▲, placebo; ●—●, felodipine, 5mg; ■—■, felodipine, 10 mg. Long term: ▲—▲, placebo; ●—●, 6 months; ■—■, 12 months. (Figure redrawn from Leenen FHH, Holliwell DL: Antihypertensive effect of felodipine associated with persistent sympathetic activation and minimal regression of left ventricular hypertrophy. Am J Cardiol 69:639–645, 1992, with permission.)

persistent increases during chronic treatment.[33–36] Several factors contribute to these divergent results, such as timing of blood sampling relative to dosing (i.e., shortly after dosing versus later), sampling with supine resting subjects vs. active subjects, level of dosing, and dosing interval (Fig. 5 and Table 3). The

TABLE 3. Percent Change from Baseline in Plasma Norepinephrine with Nitrendipine*

	Nitrendipine, 10 mg		Nitrendipine, 20 mg	
	Supine	Standing	Supine	Standing
Plasma norepinephrine (%)	+5 ± 30	+33 ± 29†	+23 ± 12†‡	+65 ± 25†‡

* Measured 2 hours after dosing during chronic treatment. Values represent mean ± SEM (n = 8).
† $p < 0.05$ vs. baseline.
‡ $p < 0.05$, 20 vs. 10 mg response.
From El-Beheiry H, Ruedy N, Wall RA, et al: The acute and chronic effects of nitrendipine on plasma catecholamines in hypertensive patients. Can J Cardiol 9:41–46, 1993, with permission.

first dose of felodipine tablet, 5 mg, caused a significant increase in plasma norepinephrine in both recumbent and upright positions (see Fig. 5). After 4 weeks of therapy, only a minor increase in plasma norepinephrine was noted 12 hours after dosing. However, the next dose of felodipine again significantly increased plasma norepinephrine, compared with both placebo and end of dosing. This increase was minor for the group receiving 5 mg twice daily but prominent for the group receiving 10 mg twice daily. The larger dose also caused larger decreases in blood pressure. This pattern persisted for the next 12 months: minor increases at end of dosing but clear increases after dosing, particularly in standing position.[23] Table 3 shows dose-response findings for nitrendipine: increases were larger ($p < 0.05$) after the 20-mg dose than after the 10-mg dose, and the extent of the increase was more pronounced in standing vs supine position for both doses.[26] Grossman et al.[37] reported that 12 hours after dosing isradipine (average dose: 5 mg twice daily) affected plasma norepinephrine only to a minor extent but caused a major increase in response to hand grip (relative to pretreatment response).

Also illustrative is a study by Savonitto et al.,[38] who used felodipine tablet, 5 mg twice daily (Fig. 6):

> After 7 days of treatment blood pressure values 12 h after drug administration were similar to those achieved acutely at the time of maximal drug effect, while heart rate returned to pretreatment levels . . . . When the next dose of felodipine was administered, a further hypotensive effect was obtained and, although the blood pressure reduction was smaller than that obtained acutely, reflex tachycardia was exactly the same as that following the first administration. Thus, reflex tachycardia was not reduced, but reset to lower blood pressure levels, or, perhaps, considering the smaller blood pressure reduction involved, was even amplified.

Consistent with this pattern of heart rate, plasma noreprinephrine showed a similar increase at day 7 and day 1 (Table 4). Such an enhanced response relative to the fall in blood pressure is consistent with an enhanced dihydropyridine-induced gain of the baroreflex, as reported in some studies.[27,30]

Kiowski et al.[25] noted no increase in plasma norepinephrine during chronic treatment but used a slow-release tablet of nifedipine, administered 3 times daily. This dosing schedule may have resulted in a fairly stable decrease in blood pressure and therefore no sympathetic activation. Bruun et al.[32] also used nifedipine tablets and reported an increase in supine plasma norepinephrine after 2 weeks but not at 6 and 12 weeks. However, the dosing schedule was not stated, nor the time after dosing when the samples were obtained.

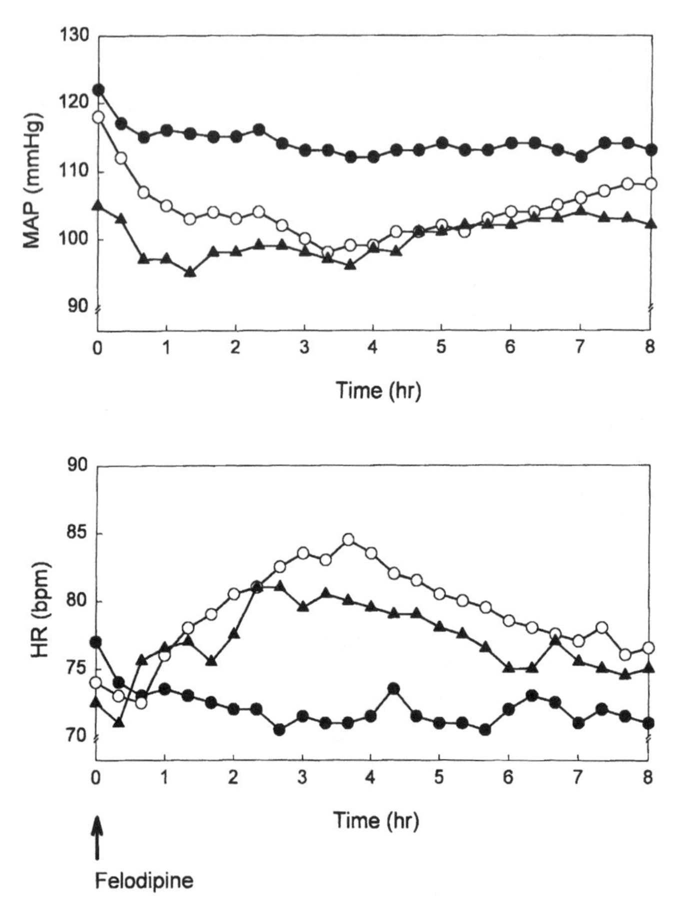

FIGURE 6. Eight-hour profile of mean blood pressure and heart rate after placebo; after first dose of felodipine tablet, 5 mg; and after 7 days, 5 mg twice daily. Mean values for 11 patients with mild hypertension. ●—●, day 1, placebo; ○—○, day 1, first dose of felodipine, 5 mg; ▲—▲, day 7, after 6 days of felodipine, 5 mg twice daily. (Redrawn from Savonitto S, Bevilacqua M, Chebat E, et al: β-adrenergic receptors and reflex tachycardia after single and repeated felodipine administration in essential hypertension. J Cardiovasc Pharmacol 17:970–975, 1991, with permission.)

Effects of Once-daily, Slow-release Formulations on Sympathetic Activity

Whereas nifedipine GITS causes stable plasma concentrations and antihypertensive effects over a 24-hour dosing interval, felodipine ER and Adalat CC (another slow-release formulation of nifedipine) exhibit clear peak to trough variations in plasma concentrations and in antihypertensive effect. Studies of sympathetic activity appear to follow the same divergent pattern. Thus, clear (60–70%) increases in arterial norepinephrine, particularly norepinephrine spill-over rate, were caused by intravenous administration of felodipine. Similar increases were noted after 8 weeks of felodipine ER, 10 mg once daily, 2–3 hours after dosing.[39] Koenig et al.[40] reported a significant (+44%) increase in plasma norepinephrine in 28 hypertensive patients after 2 weeks of treatment

TABLE 4. Plasma Catecholamines Before and 2 Hours After Felopidine Tablet

	Day 1		Day 7	
	Before Dose	**2 Hr After Dose**	**Before Dose**	**2 Hr After Dose**
Plasma norepinephrine (pg/ml)	304 ± 126	487 ± 178*	356 ± 145†	535 ± 168*
Plasma epinephrine (pg/ml)	28 ± 14	42 ± 25	30 ± 12	31 ± 11

Values are mean ± SD (n = 11).
* $p < 0.01$ difference within days (before vs. after dose).
† $p < 0.05$ vs. predose values between days.
From Savonitto S, Bevilacqua M, Chebat E, et al: β-adrenergic receptors and reflex tachycardia after single and repeated felodipine administration in essential hypertension. J Cardiovasc Pharmacol 17:970–975, 1991, with permission.

with felodipine ER, 10 mg/day but at 24 hours after dosing in sitting position. In both studies, no persistent increases in heart rate were noted, consistent with a dissociation of baroreflex control of sympathetic activity and heart rate, probably by changes in vagal control of heart rate.

Somewhat in contrast to the findings in the above studies, Kailasam et al.[41] reported that the first dose of felodipine ER, 5 mg, caused a significant (20%) increase in plasma norepinephrine at 4 hours after dosing in supine position but only a minor, nonsignificant increase after 4 weeks of dosing with 5–10 mg/day. However, timing of blood sampling relative to chronic dosing was not clear. Moreover, after the same 4 weeks of follow-up, treatment with verapamil was associated with a 33% decrease in plasma norepinephrine.[41] This finding may reflect a true sympathoinhibitory effect of verapamil or a shift in baseline over the follow-up period. Thus, lower dose, supine position, and possible shift in baseline may have caused the smaller effects noted in this study.

Consistent with a sustained antihypertensive effect over 24 hours for nifedipine GITS, no changes in plasma norepinephrine were reported after 4 weeks of treatment by Halperin et al.[42] or after 2, 6, and 12 months of treatment by Phillips et al.[43] In the second study, 16 patients with severe hypertension were followed for 1 year. Despite a high dose of nifedipine GITS (median: 120 mg/day), plasma norepinephrine tended to decrease over the 1 year of follow up, primarily in patients receiving nifedipine monotherapy.

Effects of Slow-onset, Long-acting Dihydropyridines on Sympathetic Activity

Dihydropyridines such as amlodipine cause an antihypertensive effect that gradually develops over 2–3 weeks,[10,44] and only minor, if any at all, dosing-related fluctuations in blood pressure are apparent during chronic treatment (see Fig. 2). Arterial baroreflex-induced increases in sympathetic activity, therefore, are rather unlikely during acute or chronic treatment with amlodipine. Instead, the putative intrinsic effects of dihydropyridines on sympathetic activity may become apparent. In a recent study,[45] chronic treatment of 17 patients with mild-to-moderate hypertension with amlodipine, 5–10 mg/day, lowered the average daily blood pressure by 21 ± 3 /13 ± 2 mmHg with no change in heart

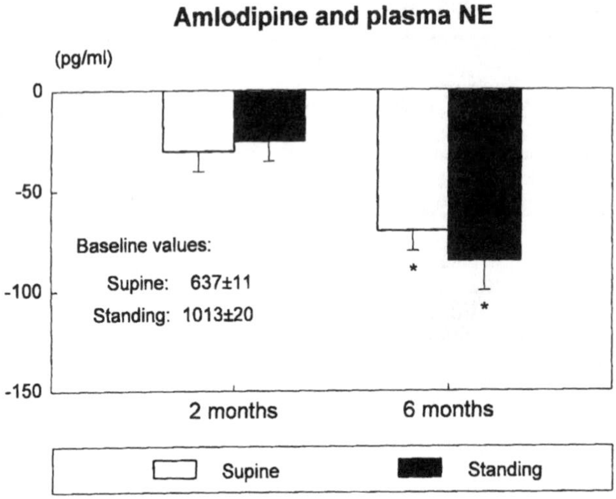

FIGURE 7. Changes in plasma norepinephrine (NE) during chronic treatment with amlodipine in hypertensive patients. Values are mean (± SEM) changes from baseline (n = 17). *$p < 0.05$ vs. baseline. (From Leenen FHH, Fourney A: Comparison of the effects of amlodipine and diltiazem on 24-hour blood pressure, plasma catecholamines, and left ventricular mass. Am J Cardiol 78:203–207, 1996, with permission.)

rate (-1 ± 2 beats/min). As shown in Figure 7, plasma norepinephrine showed minor decreases after 2 months and significant decreases in both supine and standing positions after 6 months of therapy about 4–6 hours after dosing. However, in the absence of a parallel, placebo-treated control group, it is not possible to ascertain whether these decreases reflect a time effect or a true effect of the calcium-antagonist.

In another study of amlodipine, Abernethy et al.[9] reported no changes in plasma norepinephrine in 13 young and 15 older hypertensive patients after 2 weeks on amlodipine, 2.5 mg/day, and after 12 weeks (median doses: 7 and 9 mg/day, respectively) at 24 hours after dosing.

Summary

Extensive evidence suggests that dihydropyridines, on the one hand, can lower sympathetic activity by central mechanisms (based on animal studies) and, on the other hand, can markedly increase sympathetic activity (in the case of rapid- and short-acting agents in both animals and humans). Such increases appear to persist with once-daily, slow-release formulations that cause rapid decreases in blood pressure after dosing but not with slow-release formulations or intrinsically long-acting agents that result in persistent, stable blood pressure control. However, surprisingly few studies have evaluated the effects of once-daily dihydropyridines on sympathetic activity. No study to date has compared directly two different dihydropyridines (e.g.. felodipine ER vs. nifedipine GITS) to evaluate the actual extent of differences in effects on sympathetic activity. No study to date has evaluated the chronic effects of nifedipine GITS or amlodipine vs. placebo to determine whether these agents do not increase sympathetic activity (or perhaps cause a decrease), taking into account changes in baseline over time by including a placebo group or placebo crossover.

From a methodologic point of view, one must consider that dihydropyridines also may increase norepinephrine clearance from the plasma in relation to an increase in cardiac output.[39,41] The degree of sympathetic activity may then be greater than suggested by venous levels of plasma norepinephrine.

RELEVANCE OF SYMPATHETIC HYPERACTIVITY TO OUTCOME

Koenig et al.[40] concluded that "the persistently increased sympathetic activity, indicated by elevated plasma norepinephrine levels, after 2 weeks of treatment with felodipine ER may not be clinically relevant as assessed by an essentially unchanged heart rate." Lindquist et al.[39] concluded from their study of felodipine ER that "the therapeutic implications of these findings are unknown, but sympathetic activity may influence the vascular adaptation to antihypertensive treatment."

In contrast, Julius[46] considers sympathetic hyperactivity a major coronary risk factor for hypertensive patients. The reasons are multifactorial and include both acute and chronic mechanisms. Acute mechanisms due to increased cardiac sympathetic tone include (1) induction or potentiation of cardiac arrhythmias[47,48] via stimulation of either alpha or beta receptors and (2) increased cardiac work and myocardial oxygen demand due to positive inotropic and chronotropic effects, with increased preload[49] and afterload. Such effects may cause or exacerbate clinical manifestations of myocardial ischemia, particularly in patients with coronary artery disease (CAD). These clinical manifestations include worsening of angina or congestive heart failure, myocardial infarction (MI), and sudden death.[47] Other semiacute effects include activation of platelets, which leads to enhanced coronary or cerebral thrombosis.

More chronic effects include development or maintenance of arterial and cardiac hypertrophy via both direct and indirect effects (such as stimulation of the renin-angiotensin system and increased hemodynamic load). Through several mechanisms cardiovascular hypertrophy adversely affects outcome.[50] Chronically increased sympathetic tone also may affect mechanisms leading to enhanced development of atherosclerosis. Clearly sympathetic hyperactivity, whether acute, chronic, or intermittent, cannot be considered as clinically irrelevant and may affect outcome via many mechanisms.

Because of these and other potentially adverse properties, effective blood pressure control or management of angina does not necessarily translate into improved outcome in assessing all-cause or specifically cardiac-related morbidity and mortality.[51,52] The balance of positive vs. negative properties for all-cause or cardiac outcome also may vary depending on the disease treated (e.g., congestive heart failure, hypertension, mild stable angina, unstable angina, or suspected MI). Factors such as age also may play a major role. For example, rapid lowering of blood pressure in a young hypertensive patient with well-functioning baroreflexes and appropriate cerebral auto-regulation may cause minor dizziness. In contrast, the same dose in an older hypertensive woman with osteoporosis may lead to more marked dizziness and possibly a fall, hip-fracture, and further negative consequences. Clearly outcome studies are pivotal before one can be confident about the appropriateness of a given class of drugs for a specific indication in a given patient.

DIHYDROPYRIDINE SUBCLASSES AND CARDIAC OUTCOME

This chapter focuses on cardiac outcome in patients with coronary artery disease (CAD) or hypertension for the subclasses of dihydropyridines outlined in Table 1. Rapid-, short-acting formulations, slow-release formulations, and slow-onset, long-acting agents share the intrinsic properties of dihydropyridines. One cannot conclude a priori that these properties contribute only to positive outcome and not to negative outcome. On the other hand, the three subclasses show marked differences in a number of aspects, several of which clearly can affect outcome. The balance of positive and negative influences is therefore likely to differ with the three subclasses but also may differ with the condition treated and perhaps with patient characteristics such as age.

Coronary Artery Disease

Rapid-, Short-acting Dihydropyridines

Early in their use, it became apparent that rapid-, short-acting dihydropyridines exacerbate angina in a significant subgroup of patients with CAD.[53,54] A similar proischemic effect[55] has been described for nifedipine capsules, nisoldipine, and nicardipine. Rapid lowering of perfusion pressure and reflex increase in sympathetic activity may cause such an effect.[55] In addition, coronary steal may contribute. Egstrep and Anderson[56] compared nifedipine (10–20 mg 3 times/day) with metoprolol in patients with stable angina , who were divided into two subgroups according to presence or absence of collateral circulation. Whereas metoprolol reduced total and silent ischemia in both subgroups, nifedipine reduced ischemic episodes in patients with poor or no collateral flow but significantly increased both total and silent ischemia in the 12 patients with good collateral flow. Heart rate at onset of ischemia was similar with placebo and nifedipine. The authors suggested that nifedipine may cause a coronary steal phenomenon in patients with significant coronary collateral flow. Actual flow, however, was not measured. This omission is relevant, because dihydropyridines can actually increase flow to ischemic myocardium by dilating collateral vessels.[57] A possible relationship between time after dosing and hypotension that affects collateral flow and thereby increases ischemic episodes, as described by Boden et al.,[54] was not evaluated.

Such a proischemic effect may lead not only to worsening of angina but, if severe enough, also to myocardial (re)infarction or death in patients with stable CAD. No actual outcome studies have been reported for rapid-, short-acting dihydropyridines in patients with stable CAD. Efficacy studies are suggestive for a trend toward negative outcome in patients with angina. For example, Thadani et al.[58] reported two sudden deaths and four cases of development of unstable angina in 137 patients with stable angina randomized to different doses of nidoldipine twice daily for 4 weeks compared with no instances of death or unstable angina in the placebo group (n = 48). Similarly, in a double-blind, crossover study of 46 patients with stable angina, four developed unstable angina and two others developed non-–Q-wave infarctions while taking nicardipine.[59]

Chronic studies to evaluate effects on progression of CAD also provide evidence for concern. In the International Nifedipine Trial on Antiatherosclerotic Therapy (INTACT) in patients with stable CAD, 8 cardiac deaths occurred during 2-year follow-up among 173 patients taking nifedipine capsules (up to 20 mg 4 times/day) compared with two cardiac deaths among 175 patients taking placebo ($p = 0.115$). The incidence of nonfatal MIs (8 vs. 7) did not differ significantly.[60] In the 2-year study by Waters et al.,[61] 14 MIs occurred among the 192 patients taking nicardipine compared with 8 MIs in the 191 patients taking placebo. Altogether, the studies in patients with stable CAD suggest that rapid-, short-acting dihydropyridines have a negative effect on cardiac outcome.

Several studies have evaluated the hypothesis that nifedipine capsules may prevent or delay coronary events in patients with unstable angina pectoris or suspected MI. None of these studies provided evidence for a positive effect on outcome. On the contrary, large studies such as the HINT or SPRINT 2 were discontinued prematurely because of concern about adverse outcome. In SPRINT 2, the review committee overseeing the study recommended discontinuation after 43 deaths (6.3%) in the nifedipine group compared with 29 deaths (4.3%) in the placebo group ($p = 0.10$ for a two-sided test).

The first metaanalysis[62] of the effects of dihydropyridines on outcome in patients with (suspected) MI or unstable angina included 21 acute and long-term trials. The second metaanalysis[63] added studies of outcome in stable angina. The actual data are shown in Table 5.

Both individual studies and metaanalyses may fall short of conventional significance, and one may question the validity of combining results of studies performed in patients with stable vs. unstable CAD.[64] Nonetheless, evidence in both patient populations points toward a negative effect on outcome with rapid-, short-acting dihydropyridines. Taking into account also the high incidence of

TABLE 5. Effects of Dihydropyridines on Outcomes in Acute Myocardial Infarction or Unstable Angina and in Combined Stable and Unstable Coronary Artery Disease

	Treatment			
	Active	Control	Odds Ratio	95% CI
Acute MI or unstable angina				
Mortality	365/4731	330/4733	1.13	0.97–1.32
(Re)infarction	124/3645	111/3680	1.14	0.68–1.92
Stable and Unstable CAD*				
Mortality	379/5137	335/5135	1.16	0.99–1.35
(Re)infarction	138/3838	119/3871	1.19	0.92–1.53

* Adverse outcome in patients taking dihydropyridines is significant ($p < 0.02$) when end-points are combined.

CI = confidence interval, MI = myocardial infarction, CAD = coronary artery disease.

Data from Held PH, Yusuf S, Furberg CD: Calcium channel blockers in acute myocardial infarction and unstable angina: An overview. BMJ 299:1187–1190, 1989; and Yusuf S, Held PH, Furburg CD: Update of effects of calcium antagonists in myocardial infarction or angina in light of the Second Danish Verapamil Infarction Trial (DAVIT-II) and other recent studies. Am J Cardiol 67:1295–1297, 1991, with permission.

side effects and problems with compliance for dosing 3 or 4 times/day as well as the availability of alternative treatments, rapid-, short-acting dihydropyridines are clearly not appropriate for use in patients with either stable or unstable CAD.

Long-acting Dihydropyridines

Various double-blind, placebo-controlled studies (for references, see Ruzicka and Leenen[65]), ranging in duration up to ½ year, have evaluated the effects of nifedipine GITS and amlodipine in patients with stable chronic angina. In contrast to the high frequency of worsening angina with nifedipine capsules,[53] all of these studies reported only improvements in angina. Of equal importance, none of the studies, which included 5–600 patients, reported deaths, myocardial infarctions, or development of unstable angina during active treatment. Thus it appears that when long-acting dihydropyridines are used for symptomatic relief in patients with stable CAD, the antiangina responses are predictable and may not be associated with adverse outcome compared with placebo.

Comparison with beta blockers also shows no evidence of adverse effects of long-acting dihydropyridines. For example, in a double-blind study comparing the effects of nifedipine retard tablets (20 mg twice daily; n = 121) with metoprolol (n =128) for up to 10 weeks of treatment, 7 cardiovascular events occurred in patients taking metoprolol vs. 3 in patients taking nifedipine retard.[66] Results from the Total Ischemic Burden European Trial (TIBET) show that atenolol (50 mg twice daily) and nifedipine SR (20–40 mg twice daily) cause similar improvements in ischemic parameters in patients with mild chronic stable angina.[67] Figure 8 shows the Kaplan-Meier survival curves: during 2-year follow-up 12.8% of patients taking atenolol (n = 226) vs. 11.2% of patients taking nifedipine SR (n = 232) had a hard end-point. For hard and soft end-points combined, the rates were 20.8% and 19.8%, respectively.[68] Thus, the safety profile of the longer-acting dihydropyridines to date is clearly not toward an adverse outcome, and an actual outcome study (even for an intermediate long-acting dihydropyridine) in patients with stable CAD is consistent with this conclusion. At present, no study with sufficiently large samples to detect differences in outcome of 10–20% have compared nifedipine GITS and amlodipine with placebo or beta blocker. In addition, no studies so far compared outcome in patients developing unstable angina or MI during treatment with a long-acting dihydropyridine or other agent (e.g., beta blocker). This scenario is not uncommon. Should the dihydropyridine be stopped or continued? In addition, it is also not known how nifedipine GITS or amlodipine will perform in patients with unstable angina or suspected MI. Will the outcome indeed be different from the outcome with nifedipine capsules?

Hypertension

No studies addressing the effects of dihydropyridines on cardiac outcome in hypertensive patients have been reported (see Addendum). In the absence of such evidence, the effects of dihydropyridines on an intermediate end-point—left ventricular hypertrophy (LVH)—are discussed, along with findings from relevant nonoutcome studies. LVH is recognized as a pathophysiologic substrate for the development of heart failure, ventricular arrhythmias, myocardial

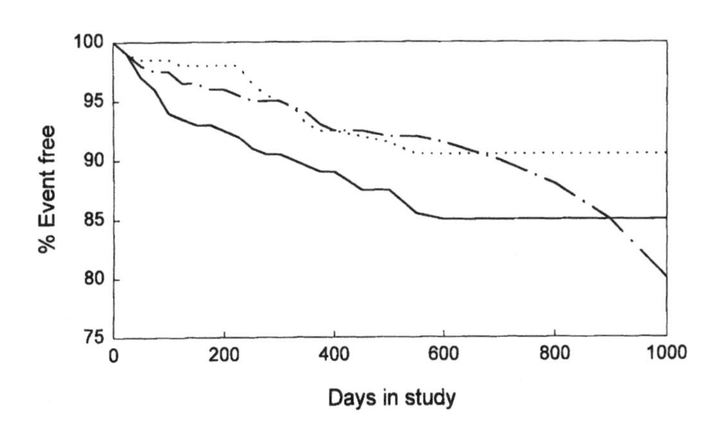

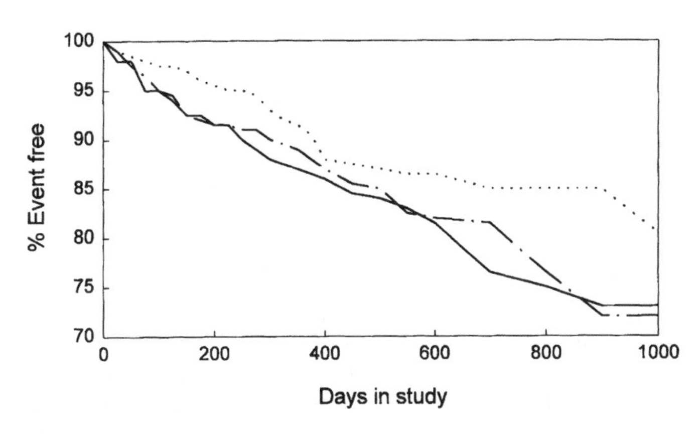

FIGURE 8. Kaplan-Meier survival curves showing percent of event-free patients with mild chronic stable angina by days for hard end-points (top panel) and hard and soft end-points (bottom panel) in the TIBET study. — = atenolol (n = 226), – · – = nifedipine (n = 232), and · · · · = atenolol + nifedipine (n = 224). (Redrawn from Dargie HJ, Ford I, Fox KM, on behalf of the TIBET study group: Total Ischemic Burden European Trial (TIBET). Effects of ischemia and treatment with atenolol, nifedipine SRS and their combination on outcome in patients with chronic stable angina. Eur Heart J 17:104–112, 1996, with permission.)

ischemia, and sudden death.[50] Persistence or progression of LVH despite antihypertensive therapy may reflect activation of mechanisms negatively affecting the cardiovascular system. The effects of the different subclasses of dihydropyridines on cardiac hypertrophy in hypertensive patients were recently reviewed.[69]

Rapid-, Short-acting Dihydropyridines

Left Ventricular Hypertrophy Studies of rapid-, short-acting dihydropyridines performed in the 1980s showed variable and mainly negative results for changes in left ventricular mass during chronic treatment. In the overview by

Cruickshank et al.[70] of regression of LVH with differing types of antihypertensive therapy, the short-acting dihyrdopyridines, nifedipine and nitrendipine, were significantly less effective ($p < 0.05$) in decreasing LV mass index compared with ACE-inhibitors or α-methyldopa and tended to be less effective ($p < 0.10$) than nonhydropyridine calcium antagonists (verapamil and diltiazem). Decreases in blood pressure were similar for the different classes. Even after correction for differences in duration of treatment, dihydropyridines remained less effective.

Coronary Events The Multicentre Isradipine Diuretic Atherosclerosis Study (MIDAS) was designed to evaluate the effects of short-acting isradipine ($n = 442$) vs. hydrochlorothiazide ($n = 441$) on progression of atherosclerosis in the carotid arteries of hypertensive patients.[71] Both treatments lowered diastolic blood pressure to the same extent, but the diuretic was more effective in lowering systolic blood pressure (by 3–4 mmHg; $p = 0.002$). During 3-year follow-up, all-cause mortality was the same (8 deaths with isradipine vs. 9 deaths with diuretic). The incidence of major vascular events was higher with isradipine than with the diuretic (25 vs. 14, respectively; $p = 0.07$), mainly because of an increased incidence of "angina pectoris objectively documented in hospital" (11 cases vs. 3 cases, respectively; $p = 0.03$). This end-point, however, is rather soft and difficult to interpret. Moreover, even if the difference is relevant, it can also be explained by differences in blood pressure control. On the other hand, fewer patients receiving isradipine developed cancer during the 3-year follow-up (13 vs. 20 with diuretic).

In a case-control study by Psaty et al.,[72] higher doses of short-acting calcium antagonists were associated with an increased risk of myocardial infarction compared with either diuretics or beta blockers. In a cohort study by Pahor et al.[73] in older hypertensive patients, short-acting nifedipine was associated with decreased survival, which again was more pronounced at higher doses. However, as stated by Pahor et al.[73] and in many commentaries and editorials,[74,75] the observational nature of case-control or cohort studies involves many potential flaws and limitations. The results of observational studies are often in contrast with the results of prospective randomized trials.[76] For example, the use of reserpine for treatment of hypertension was discredited by three case-comparison studies, published in the same issue of the Lancet in 1974, that suggested an association between reserpine and breast cancer. Subsequently, large studies found no such association. Fraser[77] states that "the myth of this association has been described as a classic error of study design."

Long-acting Dihydropyridines

Left Ventricular Hypertrophy In the analysis by Cruickshank et al.,[70] the short-acting dihydropyridines were clearly less effective than ACE inhibitors in decreasing left ventricular mass. Several more recent studies directly compared the more long-acting dihydropyridines with ACE inhibitors. Similar decreases in left ventricular mass were noted after 24 weeks of treatment with slow-release nifedipine ($n = 14$) or perindopril ($n = 16$) by Schulte et al.,[78] after 1 year of treatment with isradipine ($n = 16$) or lisinopril ($n = 16$) by Bielen et al.,[79] and after 6 months of treatment with amlodipine ($n = 12$) or enalapril ($n = 12$) by Agabiti-Rosei et al.[80] Combining data from four such studies, Fagard et al.[81] reported similar percentages of decrease in systolic blood pressure, left ventricular wall thickness, and left ventricular mass for ACE-inhibitors and dihydropyridines.

Coronary Events The Shanghai Trial of Nifedipine in the Elderly (STONE) was performed in a Chinese population with a low incidence of coronary events.[82] The risk for strokes was clearly decreased by a nifedipine slow-release tablet given twice daily compared with placebo. This trial does not provide relevant information about dihydropyridines and coronary events in terms of their effectiveness in decreasing risk for strokes compared with diuretics or beta blockers.

Two case-control studies evaluated the association between different classes of antihypertensive agents and risk of myocardial infarction. In the Norwegian study performed by Aursnes et al.[83] in 1991–1993, treatment with diuretics or beta blockers had little effect on the incidence of MI compared with no treatment. In contrast, long-term treatment with α_1-blockers, ACE inhibitors, or calcium antagonists (73% dihydropyridines, mainly longer-acting agents) was associated with a significant reduction. Jick et al.[84] reported a case-control study from the United Kingdom General Practice research database of the relation between different antihypertensive drug therapies and MI in patients with no known clinical or laboratory risk factors for MI other than hypertension during the period 1993–1994. The matched relative risk (RR) estimate comparing users of longer-acting dihydropyridines with users of beta blockers was 0.6 (95% 0.1, 0.2–1.4); for shorter-acting dihydropyridines it was 1.2 (95% 0.1, 0.7–2.1). Neither study provides evidence of adverse coronary outcome with the use of longer-acting dihydropyridines. However, the limitations of case-control studies, stated above, also apply to these studies.

Large, randomized clinical trials to compare the effects of amlodipine (ALLHAT study[85]) and nifedipine GITS (INSIGHT study[86]) on cardiovascular morbidity and mortality with the effects of standard diuretic treatment are currently underway.

SYMPATHETIC ACTIVITY: DETERMINANT OF OUTCOME WITH USE OF DIHYDROPYRIDINES

The previous section outlined the cardiac outcome in patients with CAD or hypertension treated with short- vs. longer-acting dihydropyridines. It appears that for intermediate end-points, regression of LVH and control of angina, longer-acting dihydropyridines, in contrast with short-acting dihydropyridines, do not adversely affect outcome and are similarly effective in decreasing left ventricular mass compared with ACE inhibitors and in controlling angina compared with beta blockers. Some evidence suggests a similar contrast for coronary events outcome.

Can sympathetic hyperactivity explain most of the negative outcome associated with short-acting vs. longer-acting dihydropyridines? Several lines of evidence suggest that intermittent increases in sympathetic activity by short-acting dihydropyridines contribute significantly to adverse outcome in patients with CAD:

1. In patients with chronic stable angina, the addition of nifedipine capsule, 20 mg 3 times/day, to beta blocker treatment caused none of the adverse

effects[57,87,88] reported for nifedipine capsule alone. Hypotension-induced angina may still occur 20–30 minutes after dosing, although it may disappear after the dose is lowered.[54]

2. In patients with unstable angina, no adverse responses were reported when nifedipine capsule was added to beta blocker therapy.[89] This study showed that during 4-month follow-up addition of nifedipine capsule (up to 20 mg 4 times/day; n = 68) to conventional treatment is beneficial in patients with angina at rest compared with addition of placebo (n = 70). Benefit (defined as probability of no failure of medical therapy) was substantial, with no evidence of excessive mortality or MI in the nifedipine group.

3. In the HINT study, the increased risk of developing MI, noted in patients with unstable angina pectoris randomized to nifedipine capsule, did not occur in patients randomized to combined therapy with nifedipine capsule and metoprolol or in patients on maintenance treatment with beta blocker randomized to nifedipine capsule.[90]

4. In patients with chronic stable angina pectoris (n = 207), nifedipine GITS decreased anginal episodes and total number of ischemic events. The extent of these improvements was similar for nifedipine GITS alone or in addition to beta blocker therapy.[91] This finding is consistent with the absence of sympathetic activation, which may worsen angina, by nifedipine GITS.

The overall results of these studies suggest that sympathetic activation may explain to a large extent differences in cardiac outcome with use of short-acting vs. longer-acting dihydropyridines. Other possible mechanisms for negative outcome, such as hypotension, obviously may be contributing factors. Clinical trials comparing outcome with short-acting vs. longer-acting dihydropyridines are unlikely. However, the subclass of longer-acting dihydropyridines is not homogeneous. Compounds such as amlodipine provide persistent hemodynamic effects in both compliant and partially compliant patients,[69] whereas slow-release formulations such as nifedipine GITS provide satisfactory persistent control only in compliant patients.[6] Furthermore, felodipine ER, taken once daily, may not provide persistent control even in compliant patients and may still cause sympathetic activation. Thus, one may hypothesize the following order of effectiveness in terms of positive outcome: amlodipine > nifedipine GITS > felodipine ER. Direct comparisons are obviously needed to substantiate this hypothesis.

CONCLUSION

Intermittent as opposed to gradual and sustained hemodynamic effects lead to substantial differences among the three subclasses of dihydropyridines. Sympathetic activation represents one of the major differences, and sympathetic hyperactivity may lead to various negative cardiovascular effects, which affect outcome both acutely or over the long term. Rapid- , short-acting dihydropyridines cause intermittent sympathetic hyperactivity and in several clinical settings may lead to an adverse outcome. Long-acting dihydropyridines causing sustained hemodynamic effects appear not to cause sympathetic hyperactivity (in fact, they may even decrease it), and improved outcome (relative to short-acting agents) has been established in terms of side-effect profile,[92] LVH, and control of angina in patients with stable CAD. Proper outcome studies of actual coronary events are still lacking in terms of safety (i.e., negative outcome in the control of symptoms such as angina) and prevention of adverse events in the treatment of hypertension. Considering the number of ongoing clinical trials that address the effects on outcome of new classes of antihypertensive drugs, it is clear that lowering of blood pressure per se is no longer considered an appropriate surrogate end-point for clinical outcome. As a logical consequence of this major shift in paradigm, untested classes should not be considered on equal footing with the proven classes (e.g., in hypertension, diuretics, and beta blockers). Pending the results of ongoing trials, one cannot assume equivalency of outcome, as suggested by Epstein,[93] Messerli,[94] and Chalmers.[95] Long-acting dihydropyridines are promising and are clearly different from the short-acting forms, but more outcome data are required before one can be confident that long-acting dihydropyridines do not share some of the adverse events associated with the rapid-, short-acting forms. In the meantime they should remain second- or third-line agents (i.e., after diuretics and/or beta blockers).

ADDENDUM

Since the submission of this manuscript, several studies have been published on the subject of cardiac outcomes during treatment with dihydropyridines. In the DEFIANT-II study,[96] 542 patients were randomized 7–10 days post-MI to either placebo or nisoldipine coat-core once daily for 6 months, and 60–70% of the patients were also taking β-blockers. After 6 months of treatment a positive trend for reduction of mortality (1 death in nisoldipine group vs. 7 in placebo group; $p < 0.07$) and for the combined end-point of mortality and adverse cardiac events (204 in nisoldipine group vs. 222 in placebo group; $p = 0.09$) was found.

In another case-control study, Alderman et al.[97] evaluated the effects of long- vs. short-acting calcium antagonists on cardiovascular outcomes in hypertensive patients. Compared with patients receiving β-blocker monotherapy, patients treated with long-acting calcium antagonists ($n = 136$) had an adjusted odds ratio of 0.76 (95% CI = 0.41–1.43), but patients receiving short-acting calcium antagonists ($n = 27$) were at a significantly increased risk of adverse outcomes (adjusted odds ratio 3.88; 95% CI = 1.15–13.11; $p = 0.029$). The actual calcium antagonists used in this study were not specified.

Finally, the results of the SYST-EUR trial were reported at the recent meeting of the European Society of Hypertension.[98] In this study, 4,695 elderly patients with isolated systolic hypertension were randomized to either placebo or dihydropyridine nitrendipine once daily. A decrease in systolic blood pressure by 10.7 mmHg in the treatment group vs. placebo group over a follow-up period of two years was associated with a 42% reduction in the primary clinical end-point of (fatal or nonfatal) stroke, and a significant (31%) reduction in cardiovascular events. Also reported in the nitrendipine group were decreased incidence of bleeding and cancer.

This chapter outlined the differences between the long- and short-acting forms of dihydropyridines. Consistent with this concept, results of these three studies provide evidence for the safety of long-acting dihydropyridines following myocardial infarction and for improved outcomes in hypertensive patients.

Acknowledgments

The author gratefully acknowledges the excellent secretarial assistance of Ms. Sandy Boehmer. The author is supported by a career investigatorship from the Heart and Stroke Foundation of Ontario. The author's research discussed in this chapter was supported by operating grants from the Heart and Stroke Foundation of Ontario and several pharmaceutical companies. This chapter represents an update of concepts previously published by the author.[65, 69, 92]

References

1. Struyker-Boudier HAJ, Smits JFM, De Mey JGR: The pharmacology of calcium antagonists: A review. J Cardiovasc Pharmacol 15(Supp 14):S1–S10, 1990.
2. Kleinbloesem CH, van Brummelen P, van de Linde JA, et al: Nifedipine: Kinetics and dynamics in healthy subjects. Clin Pharmacol Ther 35:742–749, 1984.
3. Edgar B, Colste P, Haglund K, Regardh CG: Pharmacokinetics and haemodynamic effects of felodipine as monotherapy in hypertensive patients. Clin Invest Med 10:388–394, 1987.
4. Saseen JJ, Carter BL, Brown TER, et al: Comparison of nifedipine alone and with diltiazem or verapamil in hypertension. Hypertension 28:109–114, 1996.
5. Middlemost SJ, Sack M, Davis J, et al: Effects of long-acting nifedipine on casual office blood pressure measurements, 24-hour ambulatory blood pressure profiles, exercise parameters and left ventricular mass and function in black patients with mild to moderate systemic hypertension. Am J Cardiol 70:474–478, 1992.
6. Zanchetti A: Trough and peak effects of a single daily dose of nifedipine gastrointestinal therapeutic system (GITS) as assessed by ambulatory blood pressure monitoring. J Hypertens 12(Suppl 5):S23-S27, 1994.
7. Bailey DG, Arnold JMO, Bend JR, et al: Grapefruit juice-felodipine interaction: Reproducibility and characterization with the extended release drug formulation. Br J Clin Pharmacol 40:135–140, 1995.
8. Myers MG, Leenen FHH, Tanner J: Differential effects of felodipine and nifedipine on 24-hour blood pressure and left ventricular mass. Am J Hypertens 8:712–718, 1995.
9. Abernethy DR, Gutkowska J, Winterbottom LM: Effects of amlodipine, a long-acting dihydropyridine calcium antagonist in aging hypertension: Pharmacodynamics in relation to disposition. Clin Pharmacol Ther 48:76–86, 1990.
10. Donnelly R, Meredith PA, Miller SHK, et al: Pharmacodynamic modeling of the antihypertensive response to amlodipine. Clin Pharmacol Ther 54:303–310, 1993.
11. Leenen FHH, Fourney A, Notman G, Tanner J: Persistence of anti-hypertensive effect after 'missed doses' of calcium antagonist with long (amlodipine) vs short (diltiazem) elimination half-life. Br J Clin Pharmacol 41:83–88, 1996.
12. Mallion JM, Dutrey-Dupagne C, Vaur L, et al: Benefits of electronic pillboxes in evaluating treatment compliance of patients with mild to moderate hypertension. J Hypertens 14:137–144, 1996.
13. Leenen FHH, Wilson TW, Bolli P, et al: Patterns of compliance with once vs. twice daily antihypertensive drug therapy in primary care. Can J Cardiol 1997 [submitted].
14. Tachikawa E, Takahashi S, Shimizu C, et al: Effects of nicardipine and other Ca^{2+} antagonists on catecholamine transport into chromaffin granule membrane vesicles. Res Commun Chem Pathol Pharmacol 45:305–308, 1984.
15. Terland O, Gronberg M, Flatmark T: The effect of calcium channel blockers on the H^+-ATPase and bioenergetics of catecholamine storage vesicles. Eur J Pharmacol 207:37–41, 1991.
16. Eikenburg DC, Lokhandwala MF: Calcium antagonists and sympathetic neuroeffector function. J Auton Pharmac 6:237–255, 1986.
17. Laurent S, Girerd X, Tsoukaris-Kupfer D, et al: Opposite central cardiovascular effects of nifedipine and BAY K 8644 in anesthetized rats. Hypertension 9:132–138, 1987.
18. Higuchi S, Takeshita A, Ito N, et al: Arterial pressure and heart rate responses to calcium channel blockers administered in the brainstem in rats. Cir Res 57:244–251, 1985.
19. Tsoporis J, Fields N, Leenen FHH: Contrasting effects of calcium antagonists vs. arterial vasodilators on cardiac anatomy and sympathetic activity in spontaneously hypertensive rats. J Cardiovasc Pharmacol 17(Suppl 2):S166–S168, 1991.
20. Kushiro T, Asagami T, Takahashi A, et al: Effects of long-acting calcium channel blocker (CCB) and hydralazine on insulin sensitivity in spontaneously hypertensive rats (SHR). Am J Hypertens 8:72A, 1995.

21. Takishita S, Muratani H, Kawazoe N, et al: Acute effects of manidipine on renal blood flow and sympathetic nerve activity in conscious, spontaneously hypertensive rats. Blood Pressure 1(Suppl 3):53–59, 1992.
22. Imai K, Higashidate S, Prados PR, et al: Relation between blood pressure and plasma catecholamine concentration after administration of calcium antagonists to rats. Biol Pharm Bull 17:907–910, 1994.
23. Leenen FHH, Holliwell DL: Antihypertensive effect of felodipine associated with persistent sympathetic activation and minimal regression of left ventricular hypertrophy. Am J Cardiol 69:639–645, 1992.
24. Murphy MB, Scriven AM, Brown MJ, et al: The effects of nifedipine and hydralazine induced hypotension on sympathetic activity. Eur J Clin Pharmacol 23:479–482, 1982.
25. Kiowski W, Bertel O, Erne P, et al: Hemodynamic and reflex responses to acute and chronic antihypertensive therapy with the calcium entry blocker nifedipine. Hypertension 5(Suppl I):70–74, 1983.
26. El-Beheiry H, Ruedy N, Wall RA, et al: The acute and chronic effects of nitrendipine on plasma catecholamines in hypertensive patients. Can J Cardiol 9:41–46, 1993.
27. Smith SA, Mace PJE, Littler WA: Felodipine, blood pressure, and cardiovascular reflexes in hypertensive humans. Hypertension 8:1172–1178, 1986.
28. Littler WA, Young MA: The effect of nicardipine on blood pressure, its variability and reflex cardiac control. Br J Clin Pharmacol 20:115S–119S, 1985.
29. Kiowski W, Erne P, Bertel O, et al: Acute and chronic sympathetic reflex activation and antihypertensive response to nifedipine. J Am Coll Cardiol 7:344–348, 1986.
30. Littler WA, Watson RDS, Stallard TJ, McLeay RAB: The effect of nifedipine on arterial pressure and reflex cardiac control. Postgrad Med J 59(Suppl 2):109–113, 1983.
31. Lederballe Pedersen O, Christensen NJ, Raemsch KD: Comparison of acute effects of nifedipine in normotensive and hypertensive man. J Cardiovasc Pharmacol 2:357–366, 1980.
32. Bruun NE, Ibsen H, Nielsen F, et al: Lack of effect of nifedipine on counterregulatory mechanisms in essential hypertension. Hypertension 8:655–661, 1986.
33. Muiesan G, Agabiti-Rosei E, Castellano M, et al: Antihypertensive and humoral effects of verapamil and nifedipine in essential hypertension. J Cardiovasc Pharmacol 4(Suppl 3):S325–S329, 1982.
34. Katzman PL, Hulthen UL, Hokfelt B: The effect of 8 weeks treatment with the calcium antagonist felodipine on blood pressure, heart rate, working capacity, plasma renin activity, plasma angiotensin II, urinary catecholamines and aldosterone in patients with essential hypertension. Br J Clin Pharmacol 21:633–640, 1986.
35. Katzman PL, Hulthen L, Hokfelt B: Catecholamines, renin-angiotensin-aldosterone, and cardiovascular response during exercise following acute and long-term calcium antagonism with felodipine in essential hypertension. J Cardiovasc Pharmacol 10:439–444, 1987.
36. Katzman PL, Hulthen UL, Hokfelt B: Effects of the calcium antagonist felodipine on the sympathetic and renin-angiotensin-aldosterone systems in essential hypertension. Acta Med Scand 223:125–131, 1988.
37. Grossman E, Messerli FH, Oren S, et al: Disparate cardiovascular response to stress tests during isradipine and fosinopril therapy. Am J Cardiol 72:574–579, 1993.
38. Savonitto S, Bevilacqua M, Chebat E, et al: β-adrenergic receptors and reflex tachycardia after single and repeated felodipine administration in essential hypertension. J Cardiovasc Pharmacol 17:970–975, 1991.
39. Lindqvist M, Kahan T, Melcher A, Hjemdahl P: Acute and chronic calcium antagonist treatment elevates sympathetic activity in primary hypertension. Hypertension 24:287–296, 1994.
40. Koenig W, Binner L, Gabrielsen F, et al: Catecholamines and the renin-angiotensin-aldosterone system during treatment with felodipine ER or hydrochlorothiazide in essential hypertension. J Cardiovasc Pharmacol 18:349–353, 1991.
41. Kailasam MT, Parmer RJ, Cervenka JH, et al: Divergent effects of dihydropyridine and phenylalkylamine calcium channel antagonist classes on autonomic function in human hypertension. Hypertension 26:143–149, 1995.
42. Halperin AK, Icenogle MV, Kapsner CO, et al: A comparison of the effects of nifedipine and verapamil on exercise performance in patients with mild to moderate hypertension. Am J Hypertens 6:1025–1032, 1993.
43. Phillips RA, Ardeijan M, Shimabukuro S, et al: Normalization of left ventricular mass and associated changes in neurohormones and atrial natriuretic peptide after 1 year of sustained nifedipine therapy for severe hypertension. J Am Coll Cardiol 17:1595–1602, 1991.
44. Bainbridge AD, Herlihy O, Meredith PA, Elliot HL: A comparative assessment of amlodipine and felodipine ER: Pharmacokinetic and pharmacodynamic indices. Eur J Clin Pharmacol 45:425–430, 1993.

45. Leenen FHH, Fourney A: Comparison of the effects of amlodipine and diltiazem on 24- hour blood pressure, plasma catecholamines, and left ventricular mass. Am J Cardiol 78:203–207, 1996.
46. Julius S: Sympathetic hyperactivity and coronary risk in hypertension. Hypertension 21:886–893, 1993.
47. Cohn JN: Sympathetic nervous system activity and the heart. Am J Hypertens 2:353S-356S, 1989.
48. Baron HV, Lesh MD: Autonomic nervous system and sudden cardiac death. J Am Coll Cardiol 27:1053–1060, 1996.
49. Leenen FHH, Reeves RA: The beta-receptor-mediated increase in venous return in humans. Can J Physiol Pharmacol 65:1658–1665, 1987.
50. Levy D, Garrison RJ, Savage DD, et al: Prognostic implications of echocardiographically determined left ventricular mass in the Framingham study. N Engl J Med 322:1561–1566, 1990.
51. Furberg CD, Psaty BM, Meyer JV: Nifedipine. Dose-related increase in mortality in patients with coronary heart disease. Circulation 92:1326–1331, 1995.
52. Alderman MH: More news about calcium antagonists. Am J Hypertension 9:710–712, 1996.
53. Stone PH, Muller JE, Turi ZG, et al: Efficacy of nifedipine therapy in patients with refractory angina pectoris: Significance of the presence of coronary vasospasm. Am Heart J 106:644–652, 1983.
54. Boden WE, Korr KS, Bough EW: Nifedipine-induced hypotension and myocardial ischemia in refractory angina pectoris. JAMA 253:1131–1135, 1985.
55. Waters D: Proschemic complications of dihydropyridine calcium channel blockers. Circulation 84:2598–2600, 1991.
56. Egstrup K, Andersen Jr PE: Transient myocardial ischemia during nifedipine therapy in stable angina pectoris, and its relation to coronary collateral flow and comparison with metoprolol. Am J Cardiol 71:177–183, 1993.
57. Packer M: Combined beta-adrenergic and calcium-entry blockade in angina pectoris. N Engl J Med 320:709–718, 1989.
58. Thadani U, Zellner SR, Glasser S, et al: Double-blind, dose-response, placebo-controlled multicenter study of nisoldipine: A new second-generation calcium channel blocker in angina pectoris. Circulation 84:2398–2408, 1991.
59. Gheorghiade M, Weiner DA, Chakko S, et al: Monotherapy of stable angina with nicardipine hydrochloride: double-blind, placebo-controlled, randomized trial. Eur Heart J 10:695–701, 1989.
60. Lichtlen PR, Hugenholtz PG, Rafflenbeul W, et al: Retardation of angiographic progression of coronary artery disease by nifedipine. Results of the International Nifedipine Trial on Antiatherosclerotic Therapy (INTACT). Lancet 335:1109–1113, 1990.
61. Waters D, Lesperance J, Francetich M, et al: A controlled clinical trial to assess the effect of a calcium channel blocker on the progression of coronary atherosclerosis. Circulation 82:1940–1953, 1990.
62. Held PH, Yusuf S, Furberg CD: Calcium channel blockers in acute myocardial infarction and unstable angina: An overview. BMJ 299:1187–1190, 1989.
63. Yusuf S, Held PH, Furburg CD: Update of effects of calcium antagonists in myocardial infarction or angina in light of the Second Danish Verapamil Infarction Trial (DAVIT-II) and other recent studies. Am J Cardiol 67:1295–1297, 1991.
64. Opie LH, Messerli FH: Nifedipine and mortality. Grave defects in the dossier. Circulation 92:1068–1073, 1995.
65. Ruzicka M, Leenen FHH: Relevance of 24-hour blood pressure profile and sympathetic activity for outcome on short- versus long-acting 1,4-dihydropyridines. Am J Hypertens 9:86–94, 1996.
66. Savonitto S, Ardissino D, Egstrup K, et al: Combination therapy with metoprolol and nifedipine versus monotherapy in patients with stable angina pectoris. Results of the International Multicenter Angina Exercise (IMAGE) study. J Am Coll Cardiol 27:311–316, 1996.
67. Fox KM, Mulcahy D, Findlay I, et al, on behalf of the TIBET study group: The Total Ischemic Burden European Trial (TIBET). Effects of atenolol, nifedipine SR and their combination on the exercise test and the total ischemic burden in 608 patients with stable angina. Eur Heart J 17:96–103, 1996.
68. Dargie HJ, Ford I, Fox KM, on behalf of the TIBET study group: Total Ischemic Burden European Trial (TIBET). Effects of ischemia and treatment with atenolol, nifedipine SRS and their combination on outcome in patients with chronic stable angina. Eur Heart J 17:104–112, 1996.
69. Leenen FHH: Regression of cardiac hypertrophy by 1,4 dihydropyridines in hypertensive patients. J Hypertens 1997 [in press].

70. Cruickshank JM, Lewis J, Moore V, Dodd C: Reversibility of left ventricular hypertrophy by differing types of antihypertensive therapy. J Hum Hypertens 6:85–90, 1992.
71. Borhani NO, Mercuri M, Borhani PA, et al: Final outcome results of the Multicenter Isradipine Diuretic Atherosclerosis Study (MIDAS). A randomized controlled trial. JAMA 276:785–791, 1996.
72. Psaty BM, Heckbert SR, Koepsell TD, et al: The risk of myocardial infarction associated with antihypertensive drug therapies. JAMA 274:620–625, 1995.
73. Pahor M, Guralnik JM, Corti M, et al: Long-term survival and use of antihypertensive medications in older persons. J Am Ger Soc 43:1–7, 1995.
74. Buring JE, Glynn RJ, Hennekens CH: Calcium channel blockers and myocardial infarction. A hypothesis formulated but not yet tested [editorial]. JAMA 274:654–655, 1995.
75. Mancia G, van Zwieten PA: How safe are calcium antagonists in hypertension and coronary heart disease? J Hypertens 14:13–17, 1996.
76. Mayes LC, Horwitz RI, Feinstein A: A collection of 56 topics with contradictory results in case-control research. Int J Epidemiol 17:680–685, 1988.
77. Fraser HS: Reserpine: A tragic victim of myths, marketing, and fashionable prescribing. Clin Pharmacol Ther 60:368–373, 1996.
78. Schulte K, Meyer-Sabellek M, Liederwald K, et al: Relation of regression of left ventricular hypertrophy to changes in ambulatory blood pressure after long-term therapy with perindopril versus nifedipine. Am J Cardiol 70:468–473, 1992.
79. Bielen EC, Fagard RH, Lijnen PJ, et al: Comparison of the effects of isradipine and lisinopril on left ventricular structure and function in essential hypertension. Am J Cardiol 69:1200–1206, 1992.
80. Agabiti-Rosei ML, Muiesan ML, Rizzoni D, et al: Cardiovascular structural changes and calcium antagonist therapy in patients with hypertension. J Cardiovasc Pharmacol 24(Suppl A): S37–S43, 1994.
81. Fagard R, Lijnen P, Staessen J, Thijs L, Amery A: Mechanical and other factors relating to left ventricular hypertrophy. Blood Pressure 3(Suppl 1):5–10, 1994.
82. Gong L, Zhang W, Zhu Y, et al: Shanghai Trial of Nifedipine in the Elderly (STONE). J Hypertens 14:1237–1245, 1996.
83. Aursnes I, Litleskare I, Froyland H, Abdelnoor M: Association between various drugs used for hypertension and risk of acute myocardial infarction. Blood Pressure 4:157–163, 1995.
84. Jick H, Dervy LE, Gurewich V, Vasilakis C: The risk of myocardial infarction associated with antihypertensive drug treatment in persons with uncomplicated essential hypertension. Pharmacology 16:321–326., 1996
85. Davis BR, Cutler JA, Gordon DJ, et al, for the ALLHAT: Rationale and Design for the Antihypertensive and Lipid Lowering Treatment to Prevent Heart Attack Trial (ALLHAT). Am J Hypertens 9:342–360, 1996.
86. Brown MJ, Castaigne A, Ruilope LM, et al: INSIGHT: International Nifedipine GITS Study Intervention as a Goal in Hypertension Treatment. J Hum Hypertens 10(Suppl 3):S157–S160, 1996.
87. Johnston DL, Lesoway R, Humen DP, Kostuk WJ: Clinical and hemodynamic evaluation of propranolol in combination with verapamil, nifedipine and diltiazem in exertional angina pectoris: A placebo controlled, double blind, randomized, crossover study. Am J Cardiol 55:680–687, 1985.
88. Findlay IN, MacLeod K, Ford M, et al: Treatment of angina pectoris with nifedipine and atenolol: efficacy and effect on cardiac function. Br Heart J 55:240–245, 1986.
89. Gerstenblith G, Ouyang P, Achuff SC, et al: Nifedipine in unstable angina. A double-blind, randomized trial. N Engl J Med 306:885–889, 1982.
90. The Holland Interuniversity Nifedipine/Metoprolol Trial (HINT) Research Group: Early treatment of unstable angina in the coronary care unit: A randomized, double blind, placebo controlled comparison of recurrent ischemia in patients treated with nifedipine or metoprolol or both. Br Heart J 56:400–413, 1986.
91. Parmley WW, Nesto RW, Singh BN, et al: The N-CAP Study Group. Attenuation of the circadian patterns of myocardial ischemia with nifedipine GITS in patients with chronic stable angina. J Am Coll Cardiol 19:1380–1389, 1992.
92. Leenen FHH: Clinical relevance of 24-hour blood pressure control by 1,4-dihydropyridines. Am J Hypertens 9(Suppl.):97s–104s, 1996.
93. Epstein M: Calcium antagonists: Still appropriate as first line antihypertensive agents. Am J Hyper 9:110–121, 1996.
94. Messerli FH: What, if anything, is controversial about calcium antagonists? Am J Hypertens 9:177S–181S, 1996.

95. Chalmers J: Treatment guidelines in hypertension: Current limitations and future solutions. J Hypertens 14(Suppl 4):S3–S8, 1996.
96. The DEFIANT-II Research Group: Doppler flow and echocardiography in functional cardiac insufficiency: Assessment of nisoldipine therapy. Results of the DEFIANT-II Study. Eur Heart J 18:31–40, 1997.
97. Alderman MH, Cohen H, Roque R, Madhaven S: Effect of long-acting and short-acting calcium antagonists on cardiovascular outcomes in hypertensive patients. Lancet 349:594–598, 1997.
98. Husten L: Calcium antagonists "not guilty." Lancet 349:1818, 1997.

MURRAY EPSTEIN, M.D., F.A.C.P.

29

Safety of Calcium Antagonists as Antihypertensive Agents: An Update

Since their introduction more than 25 years ago, calcium antagonists have emerged as one of the most attractive and widely used classes of antihypertensive agents. Of the 20–25 million patients receiving medication for hypertension in the United States, about one-fourth are taking calcium antagonists. Their wide appeal is attributable to several features, including efficacy, beneficial characteristics such as metabolic neutrality, and relatively few, nuisance-type side effects.[1,2] In addition, recent investigations have focused on their possible protective effects on target organs, such as the heart and kidney,[3,4] further enhancing their appeal (see chapters 5, 14, 15, and 24).

Despite these attributes, several retrospective analyses have suggested that calcium antagonists may be detrimental and may promote adverse cardiovascular events. More recently, Pahor et al.[5,6] proposed that calcium antagonists increase the risk of cancer by interfering with apoptosis.

The first metaanalysis of the effects of 1,4-dihydropyridines on outcome included 21 clinical trials and suggested an adverse trend in mortality and subsequent development of myocardial infarction.[7] Psaty et al. reported the findings of an observational, case-control study of patients who had suffered a myocardial infarction (cases) and patients who had not (controls) with respect to prior use of various antihypertensive drugs.[8] They concluded that hypertensive patients receiving calcium antagonists had a significantly greater risk of myocardial infarction compared with patients receiving beta blockers (relative risk [RR] = 1.63, 95% confidence interval [CI] = 1.23–2.16). On the basis of this and other retrospective analyses, Furberg and Psaty proposed that the use of calcium antagonists as first-line antihypertensive agents should be discontinued.[8]

Concerns raised about the study include the fact that the retrospective, population-based, case-control design used by Psaty et al. precluded "randomization of treatment allocation." In her recent book entitled Science on Trial,[9] which deals with the breast implant controversy, Angell succinctly summarizes the pitfalls in epidemiologic cohort studies:

> Probably the chief difficulty in epidemiologic studies is choosing groups of people who are alike in every way except for the exposure in question (in cohort studies) or the disease in question (case-control studies). Yet this is essential. Otherwise, some other difference between the groups might account for the results and badly mislead everyone. Other differences between groups that may confuse the results are termed "confounding variables." For example, cigarette smokers are more likely to drink alcohol than are nonsmokers. So when an epidemiologic study shows a link between

cigarette smoking and a disease, it is necessary to determine whether the real association is with smoking or whether it might possibly be with drinking (the confounding variable in this case). It could be the combination—or even some other factor that might be different between smokers and nonsmokers.

Although there are statistical methods for neutralizing confounding variables, they are not perfect, and they are of no use whatsoever unless the confounding variables are known and measured.

Recently Stason et al.[9a] performed a metaanalysis of published randomized, controlled trials to assess the safety of nifedipine in hypertensive patients and to compare rates of adverse cardiovascular events in patients receiving nifedipine (as monotherapy or in combination with other agents) and patients receiving other active drugs. Outcomes of interest were definitive cardiovascular events (death, nonfatal myocardial infarction, nonfatal stroke, revascularization procedure during the study period) and episodes of new-onset or increased angina. Review of 1,880 citations revealed 98 randomized, controlled clinical trials that met protocol criteria. Fourteen events occurred in 5,198 exposures to nifedipine (0.27%) and 24 events in 5,402 exposures (0.44%) to other active controls. Unadjusted odds ratios for nifedipine vs. controls were 0.49% (95% CI = 0.22–1.09) for definitive cardiovascular events and 0.61% (95% CI = 0.31–1.17) for all events (cardiovascular events plus increased angina). The odds ratio for nifedipine monotherapy (sustained- or extended-release form in 91% of exposures) was not significantly higher for definitive cardiovascular events (1.40; 95% CI = 0.49–4.03) or all events (1.39; 95% CI = 0.59–3.32). The odds ratio for nifedipine in combination with another agent was significantly lower for definitive cardiovascular events (0.09; 95% CI = 0.01–0.66) and all events (0.15; 95% CI = 0.03–0.65). Differences in odds ratios for nifedipine monotherapy and combined therapy were statistically significant ($p = 0.02$ for definitive events; $p = 0.001$ for all events). The authors concluded that their metaanalysis supports the safety of sustained- and extended-release nifedipine in the treatment of mild or moderate hypertension when it is used in combination with other drugs.[9a]

Although metaanalyses and observational studies clearly have limits,[10,11] Psaty et al.[12] raised an important question that deserves consideration—whether calcium antagonists, as a group, promote adverse cardiovascular events. Furthermore, media reporting of the presentation triggered concern among users of calcium antagonists and even among patients taking other antihypertensive drugs. This reaction underlines the importance and relevance of critically considering the issues raised by Psaty et al.

I have previously debated Furberg on these issues and have suggested that the allegations are not relevant to the calcium antagonists in current usage.[13,14] The merits of Furberg and Psaty's recent retrospective analysis and the constraints of the experimental design have been discussed extensively.[10] This chapter, therefore, focuses on a number of key issues, including (1) the markedly disparate effects of differing calcium antagonist formulations and their clinical implications, which are clearly relevant to this controversy; (2) the putative mechanisms proposed by Furberg to account for enhanced morbidity; and (3) data derived from recent prospective studies that are inconsistent with Furberg's findings.

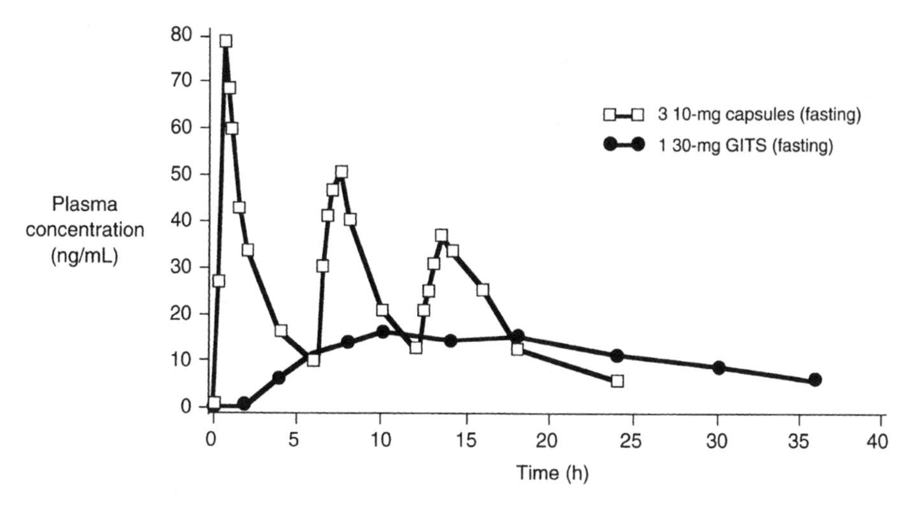

FIGURE 1. Comparison of the differing pharmacokinetic profiles of a similar dose of nifedipine administered either as a capsule or as the gastrointestinal therapeutic system (GITS) formulation. Repetitive administration of the nifedipine capsule results in rapid attainment of peak concentration (within 1 hour) with rapid decline. In contrast, plasma drug concentration of nifedipine GITS reaches a plateau slowly with maintenance of relatively constant drug levels for an extended duration.

DIFFERENT FORMULATIONS OF THE SAME CALCIUM ANTAGONIST CAN PRODUCE MARKEDLY DIFFERENT PHARMACOKINETIC AND PHARMACODYNAMIC EFFECTS

To group all of the markedly heterogeneous calcium antagonists together as if they represent uniform and identical agents is erroneous. The fact that calcium antagonists are heterogeneous and consist of chemically dissimilar agents is well established.[15,16] Less appreciated, perhaps, is the fact that different formulations of the same chemical moiety can produce markedly different hemodynamic and neurohormonal effects.[17–19] The earliest calcium antagonists were short-acting. Subsequently, the drug delivery systems for the short-acting agents were modified to provide more fully and consistently maintained calcium antagonist activity.

After administration of a nifedipine capsule, plasma drug concentration peaks rapidly at high levels within 1 hour and then falls rapidly (Fig. 1). In contrast, plasma drug concentration of nifedipine gastrointestinal therapeutic system (GITS), which uses an osmotic pump system to deliver the drug in a steady infusion over 24 hours,[20] attains a plateau slowly, peaking at approximately 6 hours, and maintains a relatively constant drug level over 24 hours. Such new slow-release formulations were developed primarily to produce a sustained 24-hour therapeutic effect. An added benefit of equal importance, albeit not widely appreciated, is that uniform plasma concentrations do not provoke activation of the renin-angiotensin and sympathetic nervous systems.[17]

The rate of drug delivery into the systemic circulation has profound effects on the hemodynamic and neurohumoral responses to a dihydropyridine calcium antagonist. In a study of intravenously administered nifedipine, Kleinbloesem et al.[18] showed clearly that the rate of drug delivery determined the pattern of response (Fig. 2A). They compared the effects of a rapid bolus dose and slow infusion on systemic hemodynamics. The rapid attainment of an effective plasma concentration via intravenous bolus dose (and exponential infusion) caused no appreciable fall in blood pressure because of an associated increase in adrenergic activity and a marked increase in heart rate (and presumably in cardiac output) (Fig. 2B). In contrast, an identical plasma concentration attained gradually over several hours by slow intravenous infusion resulted in no adrenergic response and no increase in heart rate but a significant decrease in blood pressure (see Fig. 2B). These studies emphasize the importance of the *rate* of attainment of plasma levels in determining adrenergic and cardioacceleratory response.

During chronic treatment with dihydropyridines, major fluctuations in blood pressure (rapid onset and offset of antihypertensive effects) during the dosing interval may persist for short-acting drugs and formulations.[19,21,22] In contrast, slow-release formulations of otherwise rapidly absorbed dihydropyridines achieve a more gradual and sustained antihypertensive effect. The results of Frohlich et al.[23] are representative of several recent studies that highlight the different effects of alternative formulations of calcium antagonists on intermittent increases in sympathetic activity. The investigators compared the effects of different formulations of nifedipine on the neurohormonal and plasma renin activity response at trough. They administered either nifedipine GITS, once daily, or nifedipine capsule, 3 times/day, to 10 patients with mild-to-moderate hypertension. Both formulations reduced mean arterial pressure and total peripheral resistance to a similar extent. At trough, plasma noradrenaline levels of patients receiving nifedipine GITS were virtually identical to those of patients receiving placebo (375 ± 60 vs. 387 ± 71 pg/ml). In contrast, trough noradrenaline levels tended to be substantially higher in patients receiving nifedipine capsules 3 times/day, although this adrenergic activation did not attain statistical significance.

More recently, Rosseau et al.[24] compared the effects of nisoldipine coat core (CC), an extended-release formulation, with an immediate-release formulation of nisoldipine. They observed no effect on heart rate or neurohormonal levels after administration of nisoldipine CC and suggested that this was due to the more favorable pharmacokinetic profile that avoids sudden vasodilatation and hypotension.

Demonstrations of Sustained Vasodilatory Efficacy of the Newer Long-acting Calcium Antagonists

As detailed previously, the disadvantage of a multiple daily dosing requirement for the older short-acting calcium antagonists has led to the development of long-acting agents with either intrinsically long half-lives and durations of action (e.g., amlodipine) or long-acting formulations of established drugs (e.g,. nifedipine GITS).[15,25,26] Although the longer durations of action have been established by pharmacokinetic studies, the clinically relevant issue is whether the newer slow-release agents are able to interfere in a sustained manner with the pressor responses to adrenergic and nonadrenergic vasoconstrictor stimuli.[27–29] A recent study by Ueda et al.[29] is of interest. The investigators compared the duration and consistency of action of amlodipine and nifedipine GITS by

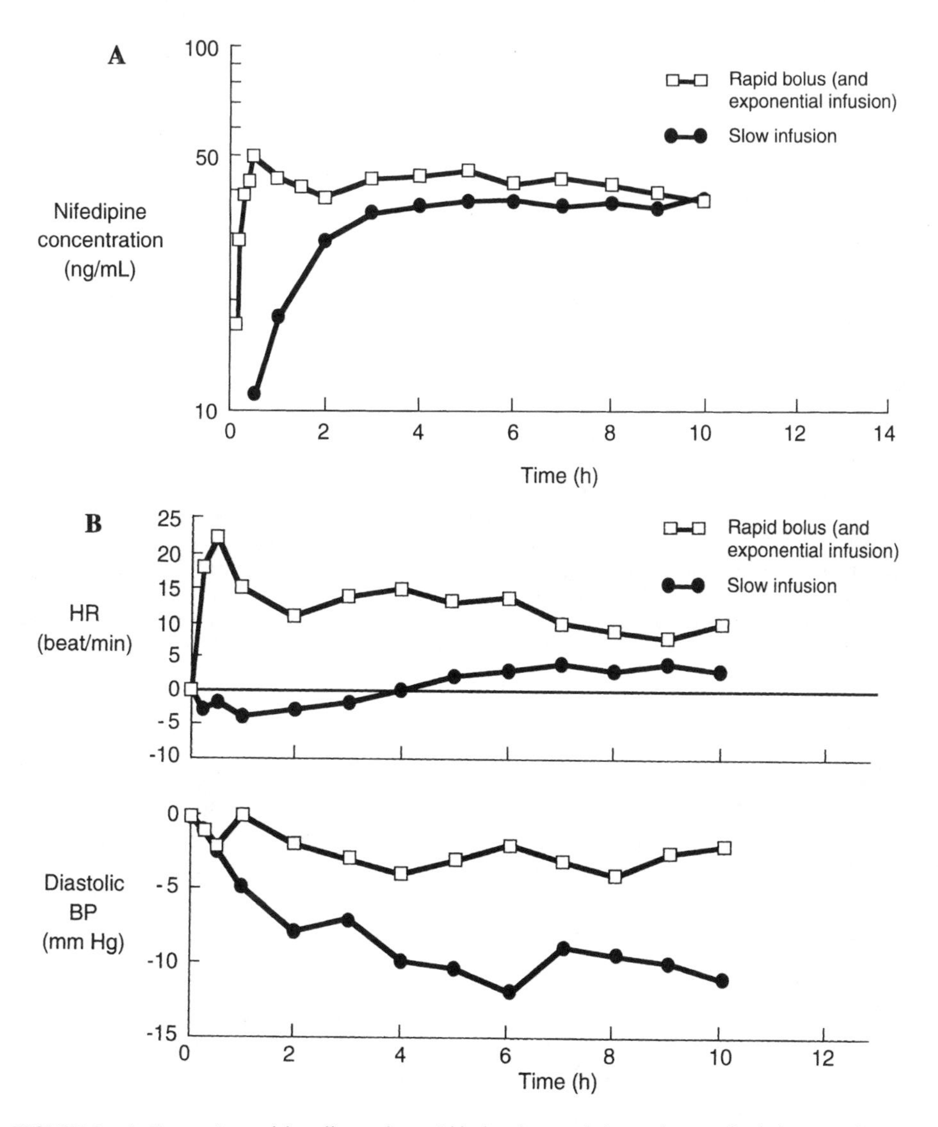

FIGURE 2. *A,* Comparison of the effects of a rapid bolus dose and slow infusion of nifedipine on blood pressure and heart rate. A rapid intravenous bolus dose (and exponential infusion) rapidly attained an effective level of nifedipine. In contrast, a progressive incremental intravenous infusion attained an identical plasma nifedipine level more gradually. Results are mean ± standard deviation. *B,* The rapid attainment of an effective plasma concentration via a rapid intravenous bolus dose (and exponential infusion) caused no appreciable fall in blood pressure. In contrast, if an identical plasma nifedipine concentration is attained gradually, the decrease in blood pressure is significant, but there is no increase in heart rate. Results are mean ± standard deviation. (Adapted from Kleinbloesem CH, van Brummelen P, Danlöf M, et al: Rate of increase in the plasma concentration of nifedipine as a major determinant of its hemodynamic effects in humans. Clin Pharmacol Ther 41:26–30, 1987, with permission.)

determining the duration of the attenuation of the pressor responses to noradrenaline and angiotensin II. Both drugs significantly attenuated the pressor responses to both vasoconstrictor substances at 24 hours after dosing, confirming their 24-hour duration of action.

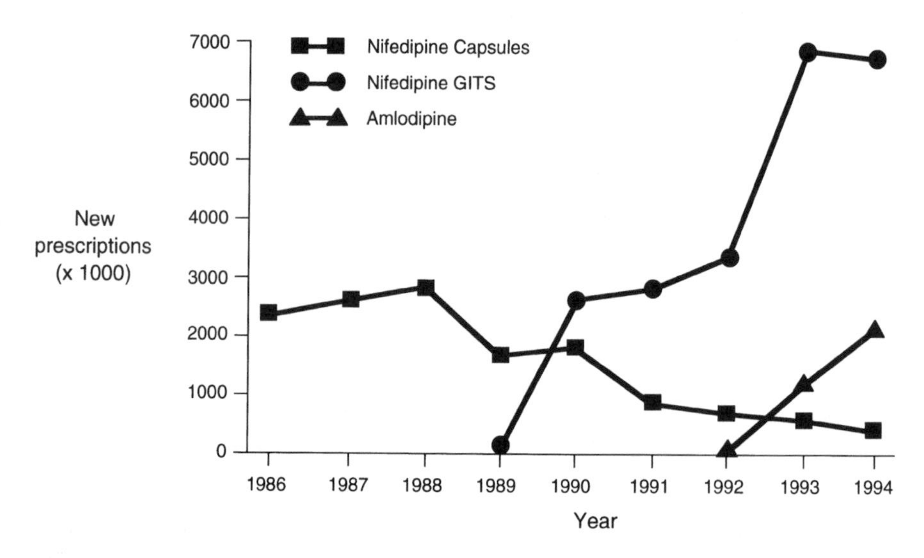

FIGURE 3. Introduction and growth of new prescriptions of different formulations of nifedipine since 1986. Nifedipine (Procardia®) was available in the United States only as a short-acting formulation in the late 1980s. The slow-release gastrointestinal therapeutic system (GITS) formulation was not introduced until 1989, and its prescription and usage did not increase appreciably until 1991–1992. (Data from IMS NPA audit.)

There are numerous examples of the importance of the slow onset and longer duration of action of calcium antagonists in clinical practice. A case in point is nisoldipine. The short-acting formulation of nisoldipine was not antianginal,[30] whereas the long-acting core-coat form achieved positive results.[31]

Relevance of Different Formulations to Recent Retrospective Reviews

The different pharmacodynamic profiles of the rapidly absorbed agents, rapidly acting formulations, and the more recently introduced slow-release, slow-acting formulations and their disparate cardioacceleratory and neurohormonal effects clearly have clinical relevance. Nifedipine was available in the United States only as a short-acting formulation in the late 1980s (Fig. 3). The slow-release GITS formulation was not introduced until 1989, and its prescription and usage did not increase appreciably until 1991–1992.

The data for the benzothiazepine, diltiazem, are similar. The only diltiazem formulation available in the late 1980s was the tablet. Diltiazem slow-release (SR) was not introduced until 1988, and its usage did not peak until 1991–1992. The more recent slow-acting formulation, diltiazem CD, was not introduced until 1991. Consequently, we must assume that studies encompassing the period that began in 1986 must have included many patients using the more rapid-release formulations.

Psaty's study[12] encompassed the period from 1986 to 1993 when the slow-release formulations were not readily available or, in some instances, had not yet been introduced. Indeed, the trial routinely used three short-acting prototype calcium antagonists—nifedipine, verapamil and diltiazem, two of which (nifedipine and diltiazem) are not indicated for the treatment of hypertension by the Food and Drug Administration. Consequently, it is reasonable to assume that the short-acting formulations were the agents observed by Psaty et al.[12] to cause wide fluctuations in drug concentrations and vasodilatory effects during apparent steady-state treatment, with resultant cardioacceleration and recurring sympathetic activation.[17]

WHY HUMAN STUDIES ON THE REGRESSION OF ATHEROSCLEROSIS ARE NEGATIVE

The previous discussion of the differences between the rapidly absorbed, short-acting formulations and the more recently introduced longer-acting formulations may account, at least in part, for the negative results to date of human studies of the regression of atherosclerosis. None of the drugs administered so far has provided sustained blood levels over the entire 24-hour period. All of the calcium antagonist formulations used in regression studies to date have been shown to evoke reactive cardioacceleration during apparent steady-state conditions.[17] Consequently, a reasonable hypothesis to account for the increased adverse vascular events in regression studies is that medications provoking a counter-regulatory response with cardioacceleration may exacerbate ischemia in patients with basal myocardial ischemia. Conversely, agents producing true steady-state levels (i.e., a GITS-like preparation, amlodipine, or lacidipine) do not evoke such unwanted counter-regulatory responses and therefore may not promote increased adverse vascular events.

To validate this hypothesis, regression using agents that produce true steady-state levels needs to be assessed in large prospective studies. One such study has been initiated recently—the Prospective Randomized Evaluation of the Vascular Effects of Norvasc Trial (PREVENT), a prospective 3-year, double-blind, placebo-controlled trial to evaluate regression. It will compare the effect of Norvasc (amlodipine) with that of placebo on the development and progression of atherosclerotic lesions in coronary and carotid arteries in patients with coronary artery disease. Another regression study currently in progress is the European Lacidipine Study on Atherosclerosis (ELSA).[32] This 4-year European study is similar to the Multicenter Isradipine Diuretic Atherosclerosis Study (MIDAS)[33] but uses lacidipine to assess the efficacy of calcium antagonists in retarding the progression of atherosclerotic lesions. Lacidipine, like amlodipine, possesses intrinsic pharmacokinetic properties that result in relatively stable plasma drug concentrations.

Blood Pressure Variability as a Determinant of End-organ Damage

Aside from the theoretical construct detailed above that favors a reduction in coronary risk, additional considerations suggest that the more smoothly sustained activity of the newer agents, such as amlodipine, may be of additional benefit in the treatment of hypertension. Variability in blood pressure, independent of the absolute level, correlates with end-organ damage in hypertensive patients. Both cross-sectional[34] and longitudinal data[35] suggest that end-organ damage in hypertension is produced by both average daily blood

TABLE 1. Proposed Mechanisms for Increased Relative Risk of Cardiovascular Events with Calcium Antagonists

1. Negative inotropic effects
2. Proarrhythmic effects
3. Proischemic effects (from coronary steal)
4. Prohemorrhagic effects

pressure values and magnitude of daily variations in blood pressure. Consequently, drugs that are short acting and associated with major fluctuations in blood pressure may fail to attenuate end-organ damage despite ostensible blood pressure control.

PROPOSED MECHANISMS OF THE ADVERSE CARDIOVASCULAR EVENTS ATTRIBUTED TO CALCIUM ANTAGONISTS

Furberg and Psaty have proposed that several mechanisms account for the increased incidence of adverse events putatively attributed to calcium antagonists[8,12] (Table 1). This section considers whether the suggested mechanisms are supported by a rigorous scientific database.

Negative inotropic effects. In isolated myocardial preparations, all calcium antagonists have been shown to exert negative inotropic effects.[2] Nevertheless, a number of caveats are in order. Calcium antagonists vary in their effects on measurements of in vivo left ventricular function depending on a number of factors: left ventricular function before treatment; whether ischemia or hypertension is alleviated concomitantly; dose used; and extent of arterial dilatation. The negative inotropic effects of calcium antagonists are dose-dependent and, with the relatively low doses that are used clinically, may not produce important adverse effects on cardiac performance. Finally, the intrinsic negative inotropic properties of calcium antagonists are greatly modified by a baroreceptor-mediated reflex augmentation of beta-adrenergic tone consequent to vasodilation.[2]

Proarrhythmic effect. Psaty et al.[12] proposed that a proarrhythmic effect may account for the increased mortality observed in patients treated with calcium antagonists. The cellular electrophysiologic basis of arrhythmia formation in the heart is detailed in a recent review[36] (see also chapter 8). It is apparent that calcium overload in myocardial cells is an important factor in the genesis of diverse serious arrhythmias, especially rate and rhythm disorders associated with myocardial ischemia, infarction, and reperfusion. These conditions result in significant elevations in cellular calcium levels in regions of poor or no perfusion. Calcium antagonists block the entry of calcium into heart cells, and they have been shown to be effective in preventing experimentally induced ischemic and reperfusion arrhythmias.[36]

Theoretically, there are circumstances in which dosage with certain calcium antagonists may lead to myocardial calcium overload. For example, as detailed in a recent review by Chakko and Bassett, it is possible that a $beta_1$-mediated increase in calcium entry may disturb myocardial calcium regulation, especially in diseased hearts in which cellular calcium regulation is already compromised (see chapter 8). Dosage with a short-acting calcium antagonist may cause sufficient recurrent vasodilation several times a day to be proarrhythmic. Despite such considerations, a review of the available data fails to support the notion that calcium antagonists are proarrhythmic.

Rigorous prospective data about proarrhythmic and antiarrhythmic effects of calcium antagonists are relatively sparse. The proarrhythmic mechanism proposed by Psaty et al., however, seems unlikely in view of the extensive experimental work demonstrating that calcium antagonists have an antiarrhythmic effect on both ischemic and reperfusion arrhythmias.[36] For example, a recent Finnish study of 155 consecutive patients resuscitated after out-of-hospital ventricular fibrillation demonstrated a clear antiarrhythmic effect of the dihydropyridine calcium antagonist nimodipine. Ventricular fibrillation persisted in 1 of 75 nimodipine-treated patients compared with 12 of 80 control subjects ($p < 0.01$).[37] Furthermore, ventricular fibrillation recurred in 1 patient receiving nimodipine and 12 patients receiving placebo, a clear benefit in favor of the calcium antagonist ($p = 0.006$). In summary, although excessive concentrations of various calcium antagonists may conceivably cause varying degrees of atrioventricular block, no evidence suggests that the newer extended-release agents are proarrhythmic.[38]

Chapters 7–9 review in depth the subject of calcium antagonists and arrhythmias.

Proischemic effect. Several reports have suggested that apart from their beneficial antianginal effects in patients with angina pectoris, dihydropyridine calcium antagonists also may promote proischemic complications.[17,39] In their metaanalysis, Psaty and Furberg related the proischemic effect to "coronary steal."[8] The steal phenomenon presumably induces myocardial ischemia by reducing coronary resistance and directing flow from poststenotic underperfused areas to normally perfused myocardium.[40,41] Highlighted in a recent editorial, the evidence to date supporting the concept that nifedipine causes coronary steal in humans is not rigorous.[42] A number of examples are inconsistent with this notion. Malacoff et al.[43] measured regional myocardial blood flow in patients with coronary artery disease with ^{133}xenon and observed that nifedipine improved regional myocardial blood flow to ischemic segments.

An important study of myocardial perfusion in patients with variant angina during exercise was reported by Kugiyama et al.[44] They observed that nifedipine improved exercise duration and reduced the size of the perfusion defect on thallium imaging; propranolol shortened exercise duration and actually increased the size of the perfusion defect compared with placebo. Other studies also failed to show that nifedipine causes a coronary steal phenomenon when coronary perfusion was assessed;[45] most suggest that it has a favorable effect on coronary perfusion in humans.[46] In addition, studies have shown that nifedipine improves left ventricular dysfunction associated with ischemia in humans, a finding that argues against coronary steal.[47] Thus, the concept that nifedipine promotes coronary steal in humans and is proischemic remains speculative.

A subsequent study in patients with chronic stable angina by Parmley et al.[48] disclosed that nifedipine GITS reduced the weekly number of anginal episodes from 5.7 to 1.8 ($p = 0.0001$) and the number of ischemic events (assessed by ambulatory electrocardiographic monitoring) from 7.3 to 4.0 ($p = 0.0001$). The drug, either alone or in combination with a beta blocker, reduced ischemia over a 48-hour period. Specifically, Parmley et al. observed no proischemic effect (worsening of ischemia on ambulatory monitoring) with nifedipine GITS. Such studies clearly call into question generalizations that calcium antagonists, *as a class*, are proischemic.

Data from Recent Prospective Studies

To date, much of the controversy has centered on the limitations of meta-analyses and retrospective studies. Although such analyses are helpful and suggestive, they do not substitute for randomized, prospective studies with hard endpoints. Within the past several years, four prospective studies in well-defined patient populations have become available; the Prospective Randomized Amlodipine Survival Evaluation (PRAISE),[49] the Shanghai Trial of Nifedipine in the Elderly (STONE),[50] the Vasodilator-Heart Failure Trial (V-HeFT III),[51] and the Doppler Flow and Echocardiography in Functional Cardiac Insufficiency: Assessment of Nisoldipine Therapy (DEFIANT II).[52] All four studies provide data that are relevant to the present controversy. Both PRAISE and VHeFT provided mortality data about patients with severe congestive heart failure. DEFIANT II provided data in the setting of postmyocardial infarction. All studies failed to reveal an increased mortality in patients treated with calcium antagonists.

The STONE study was a single-blind, placebo-controlled trial of 3 years' duration that compared nifedipine with placebo in 1,632 subjects aged 60–79 years.[50] Patients with essential hypertension were allocated to either nifedipine or placebo treatment. The terminating events for the survival analysis were stroke, heart failure, uremia, myocardial infarction, angina pectoris, severe arrhythmia, death, and hospitalization for severe illness. An intention-to-treat analysis was used. Nifedipine produced a highly significant ($p < 0.001$) decrease in the total number of terminating events. The decrease in events among patients in the nifedipine group was significant whether or not concomitant treatment was included and whether the analysis included all events or was restricted to events related to hypertension.

Recent Developments: An Update

Within the past year, several important randomized control trials have been reported that strongly argue for the safety of calcium antagonists as initial antihypertensive therapy. The recent report of the Systolic Hypertension in Europe (Syst-Eur) study lends strong support to the safety of calcium antagonists and their efficacy in reducing cardiovascular events.[53] Syst-Eur investigated whether antihypertensive treatment could decrease the risk of cardiovascular complications in elderly patients with isolated systolic hypertension. Patients ≥ 60 years of age were randomly assigned to treatment with the DHP calcium antagonist nitrendipine ($n = 2{,}398$), with addition of enalapril and hydrochlorothiazide if needed, or to matching placebo ($n = 2{,}297$). Active treatment decreased the total incidence of stroke (the primary endpoint) by 42% ($p = 0.003$), of all cardiac endpoints by 26% ($p = 0.03$), and of all cardiovascular endpoints combined by 31% ($p < 0.001$). Per-protocol analysis largely confirmed the intent-to-treat results. In view of concerns about the use of calcium antagonists as first-line

TABLE 2. Mortality as a Function of Baseline Coronary Heart Disease Status and Calcium Antagonist Use

	No Coronary Heart Disease		Coronary Heart Disease Present	
	Calcium Antagonist Users (n = 404)	Calcium Antagonist Nonusers (n = 2,566)	Calcium Antagonist Users (n = 171)	Calcium Antagonist Nonusers (n = 397)
Deaths	67 (17%)	628 (24%)	62 (36%)	213 (54%)

Adapted from Abascal VM, Larson MG, Evans JC, et al: Calcium antagonists and mortality risk in men and women with hypertension in the Framingham Heart Study. Arch Intern Med 158:1882–1886, 1998.

antihypertensive therapy, the demonstration by Syst-Eur of an almost 50% reduction ($p \leq 0.004$) in several endpoints, including total and cardiovascular mortality, militates against such concerns.

A recent analysis of two of the largest databases of nifedipine GITS and amlodipine[54] indicated that there were no deaths, and the incidence of myocardial infarction (MI) was similar to the rate suggested for diuretics and beta blockers by Psaty et al.[12]

A recent assessment of the association of calcium antagonist use with mortality risk in hypertensive subjects from the Framingham Heart Study fails to support the allegations that calcium antagonists are harmful.[55] Separate analyses were performed for patients with a history of coronary heart disease (CHD) at baseline and those without CHD. A total of 3,547 subjects (52% female) were eligible for analysis. The mean follow-up was 7 years (maximum 12.7 years). Mortality as a function of baseline CHD status and calcium antagonist use is summarized in Table 2. Proportional hazard analyses, adjusting for age, sex, blood pressure and other covariables, revealed no evidence of increased mortality risk in association with calcium antagonist use. Results were similar for short-term mortality (4 years). In this substantive population-based sample the authors found no evidence of increased mortality risk in hypertensive patients being treated with calcium antagonists.[55]

Studies in the Diabetic Hypertensive Patient

In 1998, the calcium antagonist controversy has centered on studies in the diabetic hypertensive patient. Two major studies published in 1998 propose that calcium antagonists may increase cardiovascular risk in the diabetic patient—the ABCD trial[56] and the FACET study.[57] The following section briefly critiques the design of these two studies and reviews their results.

The Appropriate Blood Pressure Control in Diabetes (ABCD) trial. The ABCD trial was a prospective, randomized, double-blind study comparing the effects of moderate vs. intensive control of blood pressure.[56] The primary endpoint was 24-hour creatinine clearance; secondary endpoints included a variety of cardiovascular events, retinopathy, clinical neuropathy, urinary albumin excretion, and left ventricular hypertrophy.

The ABCD investigators reported a significantly higher incidence of fatal and nonfatal MI among obese patients with poorly controlled hypertension and type 2 diabetes assigned to receive therapy with the calcium antagonist nisoldipine (25 of 235) than among those assigned to receive enalapril (5 of 235). At baseline, 55% of the total patient population had cardiovascular disease and a surprisingly large proportion (33%) had electrocardiographic evidence of left ventricular hypertrophy. Of critical importance, during the 5-year study, 57% of the patients discontinued the study medication.

Because of the above-cited concerns, the conclusions and proposals of Estacio et al. should be interpreted with caution until additional information is available.[58,59] First, the decision to stop the arbitrarily defined "hypertensive" limb of the ABCD study is difficult to understand. This decision was based on the findings in a subgroup of one of the secondary endpoints of the trial, comprising only 470 patients, which was not designed or powered to evaluate the effect of the drugs used on these endpoints.

The primary endpoint was 24-hour creatinine clearance, and among the secondary endpoints—there were at least five—was a composite of cardiovascular events. There were at least 20 possible subtypes or combinations of cardiovascular events that included, for example, fatal MI, nonfatal MI, and fatal or nonfatal MI.[104] Thus, including the primary endpoint, the 20 possible types of cardiovascular events, and the 4 other secondary endpoints, 25 types of events were potentially open to comparison between the two drug groups in each of the two limbs of the trial—hypertensive and normotensive subjects with diabetes. Consequently, overall, the two drug groups could have been compared relative to at least 50 types or combinations of events. Chance would predict that any two of such comparisons could be significantly different.

In the only report of this trial published to date,[56] the rates of two of the possible 50 types or combinations of events (nonfatal MIs and the combination of nonfatal and fatal MIs) were reported as significantly different (fatal MIs alone were not significantly different) in the hypertensive limb. In the normotensive limb of the study, no such differences or trends regarding these two types of cardiovascular events were apparent. Thus we are obliged to conclude that the two significant differences between the event rates observed were compatible with chance.[58]

Failure to report comprehensive information on microalbuminuria is a crucial shortcoming of the ABCD study. Estacio et al.[56] report a prevalence of overt albuminuria (> 200 μg/min) of 18% in the nisoldipine group and 19% in the enalapril group, but failed to present data on the urinary albumin excretion rate of < 200 μg/min at baseline or during the 5-year follow-up. As detailed earlier, urinary albumin excretion rate is highly predictive of morbidity and mortality from cardiovascular disease in diabetes.[60–62] Consequently, this lack of information is problematic. Furthermore, a beneficial effect of primary and secondary prevention of cardiovascular disease with low-dose aspirin in diabetes is well established.[63] Unfortunately, no information on use and dosage of aspirin (or other concomitant medications except for lipid-modifying agents) was presented in the ABCD trial report. Aspirin is also known to decrease urinary albumin excretion rate, potentially leading to misclassification.[59]

Additional drawbacks of the study include failure to correct for multiple comparisons (numerous cardiovascular outcome events) and a high drop-out rate, raising the possibility of systematic bias. Finally, the rate of MI among patients treated with nisoldipine was similar to that seen in historic control groups, suggesting that differences may have been attributable to a beneficial effect of the ACE inhibitor rather than to an adverse effect of the calcium antagonist. In summary, various factors related to the experimental design and the conduct of the study confound the certainty of the conclusions.

Fosinopril vs. Amlodipine Cardiovascular Events Randomized Trial (FACET). The second major study that has amplified the controversy is the FACET study. The FACET study[57] compared the effects of fosinopril and amlodipine on serum lipid levels and diabetes control in patients with type 2 diabetes mellitus and hypertension. Prospectively defined cardiovascular events were assessed as secondary outcomes. A total of 380 subjects were randomly assigned to open-label fosinopril (20 mg/day) or amlodipine (10 mg/day) and followed for as long as 3.5 years. If blood pressure was not controlled, the other study drug was added.

This trial had as an original endpoint two cardiovascular disease surrogates: glucose control (to measure the effect on diabetes) and blood pressure reduction (to measure the effect on hypertension), adding lipid modification as an important metabolic parameter. Both antihypertensive drugs lowered blood pressure. At the end of follow-up, the two groups were not significantly different in total serum lipid levels or glycemic control. In the recent full-length paper that was reported in *Diabetes Care*,[57] subjects receiving fosinopril had a significantly lower risk of the combined outcome of acute MI, stroke, or hospitalization for angina than the patients receiving amlodipine (14 of 189 vs. 27 of 191, hazards ratio = 0.49).

Nevertheless, several factors related to the design and the conduct of the study confound the certainty of a definitive conclusion. First, several important aspects of the study have changed since the original abstract was published in 1996.[64] The original abstract concluded that the combination of amlodipine and fosinopril provided the best outcomes for patients with type 2 diabetes mellitus and hypertension. Patients assigned to this therapy had failed one of the randomized monotherapies and so might be presumed to have been at greater risk.[57] In addition, the number of patients assigned to each study group changed. In the first 6 months after randomization, 20 patients in the amlodipine group had dropped out and 50 had crossed over to amlodipine + fosinopril, whereas in the fosinopril group 13 had dropped out and 60 had crossed over. Thus, 70 of 190 (37%) and 73 of 190 (40%) were no longer in their original groups by June 1993, 6 months after recruitment ended (and the starting point for follow-up in the original trial).

More important was the misinterpretation that calcium antagonists were harmful for hypertensive patients with diabetes. The trial did not include a placebo arm; therefore, no such conclusion could be drawn. Further, despite the limited numbers involved and the nonrandomized design of part of this trial, patients who received both fosinopril and amlodipine had a lower rate of cardiovascular events than those receiving either agent alone.[58,59,65] Therefore, it is unlikely that the calcium antagonist had an adverse impact on cardiovascular events; rather, by itself, it was less beneficial than the ACE inhibitor.

Also, this was an open-label, single-center study, which exposes the results to sponsor and investigator bias. There were no significant differences between the study groups for any of the individual cardiovascular events, and all-cause mortality was the same in the amlodipine and fosinopril groups. Only when cardiovascular events were arbitrarily grouped together did any of the events reach statistical significance.[57]

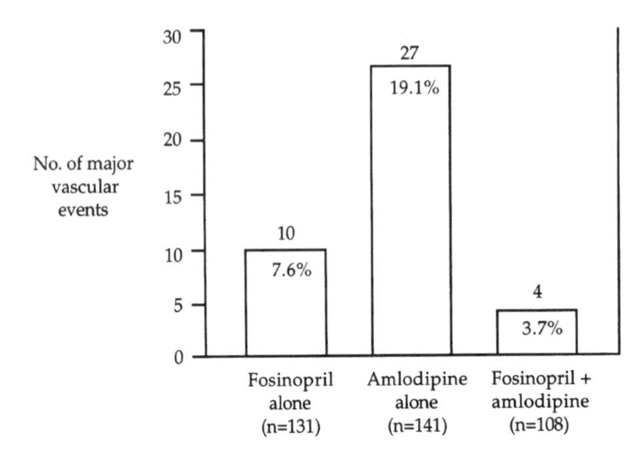

FIGURE 4. FACET: Major cardiovascular events according to treatment. (Adapted from Tatti P, Pahor M, Byington RP, et al: Outcome results of the Fosinopril versus Amlodipine Cardiovascular Events Trial (FACET) in patients with hypertension and NIDDM. Diabetes Care 21:597–602, 1998.)

Although the FACET investigators reported that the combined incidence of MI, stroke, and hospitalized angina was significantly lower in the group treated with the ACE inhibitor fosinopril, the incidence of major cardiovascular events was lowest (3.7%) in the 108 patients treated with both fosinopril and amlodipine (Fig. 4). This effect did not appear to reflect significantly lower blood pressure in the combined-treatment group. Although this observation was not extensively discussed in either the original article,[57] or the subsequent commentary,[66] it could clearly be interpreted as the most relevant finding in this report.[58,59] That the combined therapy resulted in a lower incidence of cardiovascular events than either treatment alone could be interpreted as evidence that combination therapy is the preferable strategy in this high-risk population.[67,68] These results further suggest that the combination of an ACE inhibitor and a calcium antagonist may be an excellent combination for treating hypertension in the patient with type 2 diabetes. At the very least, these results should lead to hypothesis-driven, double-blind, randomized controlled trials testing this notion.

Are these Criticisms Relevant to Clinical Practice?

One might argue that these criticisms of the published data are technical or purist, but they clearly demonstrate that although trials are considered the standard for investigating drug effects, they may be misinterpreted if the potential effects of power and the role of chance are not considered. Such concerns are particularly relevant when dealing with subgroup or post hoc analyses of secondary endpoints of small studies. It is likely that the controversy evoked by the reports of the ABCD and FACET studies is not an isolated event; we may witness additional studies alleging that calcium antagonists pose a health hazard. Consequently, the reader must bear these concerns in mind and cautiously interpret future studies to ensure that they are appropriately powered.

Inferences from Recent Randomized Controlled Trials

Rather than focus solely on the negative allegations of the ABCD and FACET studies, it is constructive to consider the results of several recent randomized trials conducted in diabetic hypertensive patients that contribute substantively to the interpretation and understanding of the controversy.

The Systolic Hypertension in Europe (Syst-Eur) trial. In Syst-Eur trial, Tuomilehto et al.[69] compared outcomes of treatment with nitrendipine vs. placebo in patients with isolated systolic hypertension with diabetes (n = 492) and without diabetes (n = 4,203). They found that the endpoint reduction achieved by active treatment of patients with diabetes exceeded that of patients without diabetes: total mortality: –55% vs. –6%; cardiovascular mortality: –76% vs. –13%; all cardiovascular endpoints: –69% vs. –26%; stroke: –73% vs. –38%; and all cardiac endpoints: –63% vs. –21%. Thus, in the Syst-Eur trial, the benefits of antihypertensive therapy with a calcium antagonist were much more pronounced in diabetic than in nondiabetic patients.[53] Data attesting to benefits of long-acting DHP calcium antagonists in hypertensive patients with diabetes do not necessarily conflict with those of the ABCD study,[56] in which in a similar population there was no difference between an ACE inhibitor and the long-acting DHP calcium antagonist in primary outcome (renal function), although admittedly more patients taking the calcium antagonists had MIs. However, since a placebo arm was lacking in that study, the absolute reduction in rates of MI could not be compared between the two treatment groups. From these two studies one may conclude that ACE inhibitors have a more favorable effect than calcium antagonists in the hypertensive patient with diabetes, not that calcium antagonists are harmful.

The Hypertension Optimal Treatment (HOT) study. Hansson et al.,[70] reporting the results of the HOT trial, noted that the lowest incidence of major cardiovascular events occurred at a mean achieved (with treatment) diastolic blood pressure of 82.6 mm Hg. The HOT data were based on 18,790 hypertensive patients treated with the long-acting DHP calcium antagonist felodipine for an average of 3.8 years. Aspirin (75 mg/day) decreased the aggregate risk of MI by 36%, with no effect on stroke. Finally, patients with diabetes mellitus (n = 1,501) in the group with a target diastolic blood pressure of 80 mm Hg had half the rate of major cardiovascular events compared with those in the target group with blood pressure of 90 mm Hg. The aggregate MI event rate per 1,000 patient years was 5.1 (unadjusted for the beneficial effect of aspirin) and 6.0 when adjusted for aspirin use. This event rate is comparable to the aggregate MI event rate per 1,000 patient years of 5.3 in the ACE inhibitor–treated diabetic patients with hypertension (n = 235) in the ABCD trial.[56] The finding of comparable benefit with felodipine in the HOT study was obtained despite a mean older age, higher arterial blood pressure, and higher serum cholesterol level compared with the ABCD trial.[71] Data from the HOT trial can probably be generalized, because it was the largest calcium antagonist trial in patients with diabetes and it was conducted in 26 countries.

Salutary Effects of Calcium Antagonists

To date, much of the controversy regarding calcium antagonists has revolved around allegations that they promote adverse cardiovascular events. What has not received sufficient attention is the possibility that they may confer beneficial effects independent of blood pressure lowering. Recently, Forette et al.[72] conducted a subanalysis on patients in the Syst-Eur study to assess the ability of the long-acting DHP calcium antagonist to lessen/attenuate vascular dementia.

They reported that in elderly patients with isolated systolic hypertension (ISH), active treatment starting with the calcium antagonist nitrendipine halved the rate of dementia from 7.7 to 3.8 cases per 1000 patient years. They proposed that the potential 50% reduction of the incidence of dementia by antihypertensive drug treatment, initiated with nitrendipine, "... may have important public health implications in view of the increasing longevity of populations worldwide."

DO CALCIUM ANTAGONISTS INCREASE THE RISK OF GASTROINTESTINAL HEMORRHAGE?

Pahor et al.[73] proposed that calcium antagonists increase the risk of gastrointestinal hemorrhage (GIH). Their first paper looked at GIH in 1,636 hypertensive patients aged 67 years or older (mean age: 75 years).[73] Of interest, 16.9% had cancer, 28.7% were taking aspirin, 13.1% were taking NSAIDs, 3.9% were taking warfarin, and 3.2% were taking steroids. The authors suggested that the calcium antagonists were associated with a higher rate of GIH than beta blockers and ACE inhibitors. Surprisingly, the risk was greater than for aspirin or NSAIDs. Furthermore, whereas verapamil and diltiazem were significantly associated with all and severe GIH events, nifedipine was not. As expected, GIH was related to gastrointestinal ulcer and gastritis, cancers, and diverticular disease. Of the three areas covered by the study (Boston, Iowa, and Connecticut), less difference was observed with calcium antagonists in the Boston area. This puzzling point highlights the difficulty of pooling data from geographically and socially diverse areas. The study provides no data about duration of therapy and no records of who was on what therapy at the time of GIH. As many investigators have learned, relying on the memory of elderly patients is fraught with difficulty.

As pointed out in several subsequent assessments, the Pahor study is not a randomized trial, nor was it a study of GIH per se. It is therefore open to many confounding factors. It is most odd that calcium antagonists conferred a greater relative risk than aspirin or NSAIDs and that the risk varied among the calcium antagonists. Verapamil, the only calcium antagonist shown to reduce mortality after infarction, was associated with a significantly increased risk of GIH. The verapamil infarction trials, however, reported no increase in GIH. The accumulated evidence makes it difficult to accept the view that calcium antagonists significantly increase GIH.

DO CALCIUM ANTAGONISTS INCREASE THE RISK OF CANCER?

The indictments against first-generation calcium antagonists seem to lengthen steadily. Recently the controversy has been extended beyond the relatively narrow limits of cardiovascular safety. In June 1996, the lead article in the *American Journal of Hypertension* was a report by Pahor and colleagues of an apparent increase in the risk of cancer in patients taking calcium antagonists for hypertension.[5] Subsequently, they extended the findings to a larger cohort of elderly patients.[6] The validity of the association with cancer necessitates careful scrutiny. Pahor et al.[5,6] hypothesized that calcium antagonists increase the risk of cancer by interfering with the physiologic mechanisms that regulate cancer cell growth. Evidence is emerging that calcium antagonists can block apoptosis,[74,75] an efficient mechanism for limiting cancer growth.[76–79] Calcium antagonists might affect cancer risk generally or be limited to specific sites at which calcium mechanisms predominate. For example, colon cancer has been related to reduced calcium ingestion.[80] Recently, in the lead article in the *Lancet*, Pahor et al. extended the initial analyses performed in patients receiving treatment for hypertension[5] to the older population in general.[6] The aim was to assess whether patients taking calcium antagonists for any indication were at higher risk of developing cancer than patients not taking such drugs. The authors observed that the hazard ratio for cancer associated with calcium antagonists (1,549 person-years; 47 events) compared with patients not taking calcium antagonists (17,225 person-years; 373 events) was 1.72 (95% CI = 1.27–2.34; p = 0.0005) after adjustment for confounding factors. A significant dose-response gradient was found. Hazard ratios associated with verapamil, diltiazem, and nifedipine did not differ significantly. The association between calcium antagonists and cancer was found with most of the common cancers.

As pointed out by Dargie in an accompanying editorial,[81] the validity of the association of calcium antagonists with cancer is questionable. The evidence is based entirely on observational data in elderly patients who will have first received calcium antagonists late in life. Their previous drug history is unknown. In their latest paper, Pahor et al. compared a group of patients treated with calcium antagonists for various reasons with a heterogeneous control group, thereby rendering adjustment for possible confounders highly difficult.

Exposure information. A major limitation of this analysis, as with the previous analyses carried out by Pahor et al.,[5,6] is the weak exposure information. Information about drug use was collected only at the baseline visit in 1988 using a 2-week window, and was not collected after that time. For the current analysis, patients were followed from the baseline 1988 time period for a mean of 3.7 years until a cancer event, death, or study completion in 1992. For patients receiving calcium antagonists, exposure was assumed to be continuous from the baseline period until the outcomes described above. For patients not exposed, it was assumed that calcium antagonists were not used during the entire time period. This assumption of continuous exposure or nonexposure is the basis of the proportional hazards model and the use of person-time as a denominator. The use of person-time is appropriate if the tested hypothesis is that a one-time exposure (i.e., atomic bomb) confers a long-term continuing risk of cancer. It is not appropriate when continual exposure or exposure close in time to the event is the basis of the hypothesis (i.e., apoptosis). Therefore, the use of person-time as the denominator for the incidence calculations on which the reported relative risks are based is inappropriate, and attributing subsequent follow time to calcium antagonist use is not valid. If person-time is eliminated and the crude rates are based on the number of actual persons known to be exposed or not exposed at baseline, the result is a smaller relative risk that is not statistically significant.

Pahor et al. suggested that inhibition of apoptosis is a possible explanation for the development of cancer on the basis that calcium is an intracellular messenger in apoptosis. Calcium, however, is such a ubiquitous component of cell signaling systems that almost any adverse drug reaction may be said to be explained on the basis of this mechanism. Such an explanation, therefore, must be regarded as highly speculative. Indeed, the converse may be possible; calcium antagonists have shown favorable results in the treatment of cancer.[82]

Lessons from History

We have encountered previous cancer scares with respect to antihypertensive agents, and history has taught us to be cautious. Two examples come readily to mind. In the atenolol study of more than 2,000 elderly hypertensive patients, the relative risk of cancer in men was increased two- or threefold.[83] A later retrospective study that linked data from the Glasgow Blood Pressure Clinic with data from the West of Scotland Cancer Registry refuted this claim.[84] In 6,528 patients, there was no overall association between cancer and use of atenolol. A likely explanation is that an increased risk of cancer (up to threefold) is often a nonspecific finding of questionable significance.[85]

Another cancer scare centered on reserpine. In 1974, the *Lancet* published three separate retrospective studies from Oxford, Boston, and Helsinki, linking reserpine to breast cancer. An accompanying editorial commented on the three- to fourfold increase of breast cancer in women exposed to reserpine compared with unexposed women.[86] The authors of the three papers included internationally recognized authorities such as Sir Richard Doll, who is credited with establishing the links between smoking and lung cancer. Yet further work disproved these claims. The same group that had provided evidence for the increased risk with reserpine in 1974[87] disproved the hypothesis a full 10 years later.[88] These discordant findings were attributable to differences in sample size. The 1974 study included 150 women with breast cancer, of whom only 11 had used reserpine, whereas the 1984 study included 1,881 women with breast cancer, of whom 65 had used reserpine. Thus, even an apparent increase in relative cancer risk of more than threefold, coming from three centers in three different countries and involving highly respected investigators, was subsequently discounted. These lessons from the past highlight the danger of focusing on small numbers of cases, such as the 11 fatal cancers associated with calcium antagonists in the recent study by Pahor.[5]

FUTURE PERSPECTIVES

What, if anything, should physicians make of the apparently disparate hazards associated with a group of drugs that they prescribe so widely? Various authorities still believe that many patients benefit from the judicious use of calcium antagonists and that current evidence offers no reason to suspend their use. The only evidence that would justify suspension of use is the randomized, controlled trial. Ironically, if the use of calcium antagonists was suspended, the large randomized, controlled trials that we have espoused and that are now underway might flounder through lack of interest or concern. In this case, the current debate over many of the adverse effects, particularly cancer, would not be resolved. As I noted in a recent editorial, the evidence for the safety of third-generation, slow-acting calcium antagonists, currently under investigation in large outcome studies, gives no cause for concern.[14]

Statement of the World Health Organization and International Society of Hypertension

Because of the recent controversy in the medical and lay press about the safety of calcium antagonists, the Liaison Committee of the World Health Organization and the International Society of Hypertension formed an ad hoc subcommittee whose primary objective was to review the relevant evidence about the effects of calcium antagonists on the risk of coronary heart disease (CHD) and cancer. A secondary objective was to review the available evidence about the effects of calcium antagonists on bleeding. The subcommittee's report, issued on February 10, 1997, reached the following conclusions:[89]

1. The available evidence does not prove the existence of either beneficial or harmful effects of calcium antagonists on the risks of major CHD events, defined as fatal or nonfatal myocardial infarction and other deaths (sudden or not) from CHD. This conclusion applies to the collective evidence for calcium antagonists as a class and to the evidence for specific subgroups.
2. The available observational studies do not provide good evidence of an adverse effect of calcium antagonists on the risk of cancer (fatal or nonfatal).
3. The available observational studies and randomized trials do not provide clear evidence of an adverse effect of calcium antagonists on bleeding risks.

Why Should We Care Whether Calcium Antagonists Continue to Serve as First-line Antihypertensive Agents?

Clearly, an extensive antihypertensive armamentarium with many efficacious agents is now available. If the safety of calcium antagonists has raised concerns, why not merely discard them? Such a simplistic approach is patently inappropriate; there are compelling reasons for retaining calcium antagonists as part of the antihypertensive armamentarium. Despite a massive publicity effort and promulgation of management guidelines by many national advisory bodies, attempts to control blood pressure have not achieved great success.[90,91] The recent National Health and Nutrition Examination Survey (NHANES)[90] clearly demonstrated that during the 20 years from 1971 to 1991, patient awareness of hypertension increased markedly in the United States from 51% to 69% of the population. Concomitantly, the percentage of patients treated for hypertension has increased from 36% to 53%.[90] Nevertheless, at present only 24% of patients achieve blood pressure control at a level below 140/90 mm Hg[90] (Table 3).

A recent survey indicates that the experience in Europe is similar. The extent of the problem is revealed in data from the Cardiomonitor study, an independent survey of approximately 1,500 physicians performed biannually in 10 European countries and the United States by an independent healthcare market research company.[92] Hypertensive patients were treated with diuretics, calcium antagonists, beta blockers, and ACE inhibitors (plain and combined). The survey, conducted between April and June 1992, demonstrated that of the 11,613 hypertensive patients receiving treatment, only 37% had reached the target diastolic blood pressure set by the physician.

Although the reasons for this striking failure are complex and multifactorial, a major factor is compliance and the need for patients to remain on prescribed therapy. The Treatment of Mild Hypertension Study (TOMHS)[93] disclosed that

TABLE 3. Hypertensive Patients in the United States Who Are Aware of Their Disease, Treat It, or Control It

	Aware (%)	Treated (%)	Controlled (%)*
Overall	69	53	24
Men	63	44	19
Women	76	61	28

*Systolic blood pressure < 140 mm Hg and diastolic blood pressure < 90 mm Hg.
Adapted from Burt VL, Whelton P, Roccella EJ, et al: Prevalence of hypertension in the US adult population. Results from the Third National Health and Nutrition Examination Survey, 1988–1991. Hypertension 25:305–313, 1995.

the percentage of patients remaining on initial therapy at 4 years was highest in the calcium antagonist group, who received amlodipine. Elliott[94] also analyzed drug discontinuation in a large group of hypertensive patients followed for 7 years. He observed that the relative likelihood of discontinuing hypertensive therapy was greatest with beta blockers and least with calcium antagonists.

CONCLUSIONS

The above-enumerated clinical trials in diabetic patients offer some compelling lessons. First, it is readily apparent that aggressive approaches to lowering blood pressure are appropriate for diabetic patients with hypertension. Second, the results of the diabetic cohorts in Syst-Eur,[105] the HOT study,[70] and the PRAISE study[106] collectively suggest that long-acting calcium antagonists do not enhance cardiovascular risk in the diabetic patient and therefore are safe.

On the basis of limited trial evidence,[95,96] the consensus among diabetologists and hypertension specialists, reinforced in recent management guidelines,[71,97] is that aggressive blood pressure lowering is particularly important in diabetic patients. Consequently, recently recommended antihypertensive treatment thresholds for diabetics are substantively lower than those for nondiabetic persons, with targets of < 130/85 mm Hg, and even lower in certain subgroups of patients. Single-drug therapy may not be sufficient to lower pressures to these levels; two or more drugs may be required. It is therefore crucial in this subgroup of hypertensive patients to assure that effective agents are used along with ACE inhibitors. The most recent analyses arising from the Syst-Eur trial,[69] in which the diabetic subgroup was treated with a long-acting calcium antagonist (usually combined with an ACE inhibitor), provides compelling evidence that this combination of medications is especially effective in preventing cardiovascular events. Additional comparative studies will be required to extend and validate this observation.

Despite its many drawbacks, the calcium antagonist controversy was helpful in alerting physicians to the fact that hypertension remains a surrogate endpoint and that not all drugs that reduce blood pressure will paribus passu reduce morbidity and mortality. A prime example of failure of the so-called surrogate endpoint concept is provided by the recent meta-analysis by Messerli et al. of the effects of antihypertensive therapy in the elderly.[98,99] Although beta blockers did lower blood pressure, they consistently failed to reduce the rate of

MI and cardiovascular and all-cause mortality.[99] The analysis suggested that numerous elderly hypertensive patients were exposed to adverse effects, inconvenience, and cost of beta blockers without harvesting any benefits.

It is clear that the excessive news media coverage of the calcium antagonist controversy was inappropriate and led to panic and confusion among patients and frustration among physicians. Perhaps we should remember the first rule of treatment set forth by Sir George Pickering: "Never frighten your patients."[100]

A compelling body of evidence indicates that reports of an increase in myocardial infarction in hypertensive patients treated with calcium antagonists may be restricted to the more rapidly absorbed and rapid-acting agents. Short-acting calcium antagonists can induce recurrent sympathetic neurohormonal activation and evoke a reactive cardioacceleration that may be detrimental to patients with underlying myocardial disease. Recent studies suggest that such changes are not encountered with slow-release formulations, which achieve a more gradual and sustained antihypertensive effect. Moreover, evidence suggests that the slower onset and longer duration of action of the formulations currently used in clinical practice are more efficacious and produce fewer side effects than the older formulations. Large prospective, randomized outcome studies using slow-release agents are urgently needed to validate use of these formulations in patients with chronic stable angina and hypertension. Hopefully, such studies will allay the apprehension engendered by the recent retrospective observational studies.

Fortunately, a number of such prospective studies have recently been initiated, including the Antihypertensive and Lipid-Lowering Treatment to Prevent Heart Attack (ALLHAT),[101] the International Nifedipine Study, Intervention as a Goal in Hypertension Treatment (INSIGHT),[102] and the Anglo-Scandinavian Cardiac Outcomes Trial (ASCOT).[107] ALLHAT, an $85-million, 9-year study supported by the National Heart, Lung and Blood Institute, will examine the ability of several antihypertensive agents, including the long-acting calcium antagonist amlodipine, to reduce coronary heart disease among hypertensive patients.[101] INSIGHT, a randomized, 3-year study of 6,600 high-risk hypertensive patients in Europe, will compare the effects of nifedipine GITS and diuretic therapy on cardiovascular and cerebrovascular morbidity and mortality.[102] Another pivotal ongoing study is ASCOT, which is comparing the effects of beta-blocker ± diuretic therapy vs. calcium antagonist ± ACE inhibitor therapy on nonfatal MI and fatal CHD in 18,000 high-risk hypertensive patients. The aim of the study is to achieve 1,150 primary endpoints.[107]

In addition, the second Swedish Trial in Old Patients (STOP) with Hypertension[103] is under way to compare beta blockers and/or diuretics with ACE inhibitors (enalapril and lisinopril) and calcium antagonists (isradipine and felodipine) in 6,600 elderly hypertensive patients. The primary aim is to assess the effects of the different interventions on cardiovascular mortality; the primary endpoints are stroke, myocardial infarction, and other cardiovascular events.

While we await the results of these trials, a prudent interim approach is to include long-acting calcium antagonists in the armamentarium for treating hypertensive patients. Furthermore, such formulations should promote patient compliance and thereby favorably influence hypertension-related morbidity and mortality.

Acknowledgments

The author thanks Elsa V. Reina for her secretarial assistance. This chapter was adapted from an earlier review by the author in the *American Journal of Hypertension* (9:110–121, 1996).

References

1. Epstein M: Calcium antagonists in the management of hypertension. In Epstein M (ed): Calcium Antagonists in Clinical Medicine. Philadelphia, Hanley & Belfus, 1992, pp 213–230.
2. Frishman WH: Current status of calcium channel blockers. Curr Probl Cardiol 19:637–688, 1994.
3. Epstein M: Calcium antagonists and renal protection: Current status and future perspectives. Arch Intern Med 152:1573–1584, 1992.
4. Messerli FH, McLoughlin M: Cardiac effects of calcium antagonists in hypertension. In Epstein M (ed): Calcium Antagonists in Clinical Medicine. Philadelphia, Hanley & Belfus, 1992, pp 137–150.
5. Pahor M, Guralnik JM, Salive ME, et al: Do calcium channel blockers increase the risk of cancer? Am J Hypertens 9:695–699, 1996.
6. Pahor M, Guralnik JM, Ferrucci L, et al: Calcium channel blockade and incidence of cancer in aged populations. Lancet 348:493–497, 1996.
7. Held PH, Yusuf S, Furberg CD: Calcium channel blockers in acute myocardial infarction and unstable angina: An overview of the result from the randomized trials. BMJ 299:1187–1192, 1989.
8. Furberg CD, Psaty BM: Calcium antagonists: Not appropriate as first line antihypertensive agents. Am J Hypertens 9:122–125, 1996.
9. Angell M: Science on Trial. New York, W.W. Norton, 1996.

9a. Stason WB, Schmid CH, Niedzwiecki D, et al: Safety of nifedipine in patiens with hypertension: A meta-analysis. Hypertension 30(Pt 1):7–14, 1997.

10. Buring JE, Glynn RJ, Hennekens CH: Calcium channel blockers and myocardial infarction: A hypothesis formulated but not yet tested. JAMA 274:654–655, 1995.
11. Opie LH, Messerli FH: Nifedipine and mortality. Grave defects in the dossier. Circulation 92:1068–1073, 1995.
12. Psaty BM, Heckbert SR, Koepsell TD, et al: The risk of myocardial infarction associated with antihypertensive drug therapies. JAMA 274:620–625, 1995.
13. Epstein M: Calcium antagonists should continue to be used for first-line treatment of hypertension. Arch Intern Med 155:2150–2156, 1995.
14. Epstein M: Calcium antagonists: Still appropriate as first-line antihypertensive agents. Am J Hypertens 9:110–121, 1996.
15. Morris AD, Meredith PA, Reid JL: Pharmacokinetics of calcium antagonists: Implications for therapy. In Epstein M (ed): Calcium Antagonists in Clinical Medicine. Philadelphia, Hanley & Belfus, 1992, pp 49–67.
16. Triggle DJ: Biochemical and pharmacologic differences among calcium channel antagonists: Clinical implications. In Epstein M (ed): Calcium Antagonists in Clinical Medicine. Philadelphia, Hanley & Belfus, 1992, pp 1–27.
17. Ruzicka M, Leenen FHH: Relevance of intermittent increases in sympathetic activity for adverse outcome on short-acting calcium antagonists. In Laragh JH, Brenner BM (eds): Hypertension: Pathophysiology, Diagnosis, and Management, 2nd ed. New York, Raven Press, 1995, pp 2815–2825.
18. Kleinbloesem CH, van Brummelen P, Danlöf M, et al: Rate of increase in the plasma concentration of nifedipine as a major determinant of its hemodynamic effects in humans. Clin Pharmacol Ther 41:26–30, 1987.
19. Myers MG, Raemsch KD: Comparative pharmacokinetics and antihypertensive effects of the nifedipine tablet and capsule. J Cardiovasc Pharmacol 10(Suppl 10):S76–S78, 1987.
20. Swanson DR, Barclay BL, Wong PSL, Theeuwes F: Nifedipine gastrointestinal therapeutic system. Am J Med 83(Suppl 6B):3–9, 1987.
21. Chung M, Reitberg DP, Gafney M, Singleton W: Clinical pharmacokinetics of nifedipine gastrointestinal therapeutic system. A controlled release formulation of nifedipine. Am J Med 83(Suppl 6B):10–14, 1987.
22. Stokes GS, Shenfield GM, Johnston HJ, et al: Timing of blood pressure measurements in determining anomalies in duration of effect of an antihypertensive drug: Assessment of isradipine. J Cardiovasc Pharmacol 15(Suppl 1):S65–S69, 1990.

23. Frohlich ED, McLoughlin MJ, Losem CJ, et al: Hemodynamic comparison of two nifedipine formulations in patients with essential hypertension. Am J Cardiol 68:1346–1350, 1991.
24. Rousseau MF, Melin J, Benedict CR, et al: Effects of nisoldipine therapy on myocardial perfusion and neuro-hormonal status in patients with severe ischaemic left ventricular dysfunction. Eur Heart J 15:957–964, 1994.
25. van Zwieten PA, Hansson L, Epstein M: Slowly acting calcium antagonists and their merits. Blood Pressure 6:78–80, 1997.
26. Abernethy DR: Amlodipine: Pharmacokinetic profile of a low-clearance calcium antagonist. J Cardiovasc Pharmacol 17(Suppl 1):S4–S7, 1991.
27. Elliot HL, Pasanisi F, Sumner DJ, Reid JL: The effect of calcium channel blockers on alpha 1- and alpha 2-adrenoreceptor-mediated vascular responsiveness in man. J Hypertens 3(Suppl 3):S235–S237, 1985.
28. Pasanisi F, Elliot HL, Meredith PA, et al: Effect of calcium channel blockers on adrenergic and nonadrenergic vascular responses in man. J Cardiovasc Pharmacol 7:1166–1170, 1985.
29. Ueda S, Meredith PA, Howie CA, Elliot HL: A comparative assessment of the duration of action of amlodipine and nifedipine GITS in normotensive subjects. Br J Clin Pharmacol 36:561–566, 1993.
30. Thadani U, Zellner S, Glasser S, et al: Double-blind, dose-response, placebo-controlled multicenter study of nisoldipine: A new second-generation calcium channel blocker in angina pectoris. Circulation 84:2398–2408, 1991.
31. Glasser S, Ripa S, MacCarthy EP, for the Nisoldipine CC Multicenter Study Group: Efficacy and safety of extended release nisoldipine as monotherapy for chronic stable angina pectoris [abstract]. J Am Coll Cardiol 23:267A, 1994.
32. Bond G, Del Palu C, Hansson L, et al: ELSA: European Lacidipine Study on Atherosclerosis [abstract]. J Hypertens 11(Suppl 5):S405, 1993.
33. Borhani NO, Mercuri M, Borhani PA, et al: Final outcome results of the multicenter isradipine diuretic atherosclerosis study (MIDAS). JAMA 276:785–791, 1996.
34. Parati G, Pomidossi G, Albini F, et al: Relationship of 24-hour blood pressure mean and variability to severity of target organ damage in hypertension. J Hypertens 5:93–98, 1987.
35. Frattola A, Parati G, Cuspidi C, et al: Prognostic value of 24-hour blood pressure variability. J Hypertens 11:1133–1137, 1993.
36. Billman GE: The antiarrhythmic effects of calcium antagonists. In Epstein M (ed): Calcium Antagonists in Clinical Medicine. Philadelphia, Hanley & Belfus, 1992, pp 183–212.
37. Roine RO, Kaste M, Kinnunen A, et al: Nimodipine after resuscitation from out of hospital ventricular fibrillation. JAMA 264:3171–3177, 1990.
38. Bassett AL, Chakko S, Epstein M: Are calcium antagonists proarrhythmic? J Hypertens 15:915–923, 1997.
39. Waters D: Proischemic complications of dihydropyridine calcium channel blockers. Circulation 84:2598–2600, 1991.
40. Egstrup K, Andersen PE: Transient myocardial ischemia during nifedipine therapy in stable angina pectoris, and its relation to coronary collateral flow and comparison with metoprolol. Am J Cardiol 71:177–183, 1993.
41. Schulz W, Jost S, Kober G, Kaltenbach M: Relation of antianginal efficacy of nifedipine to degree of coronary arterial narrowing and to presence of coronary collateral vessels. Am J Cardiol 55:26–32, 1985.
42. Kloner RA: Nifedipine in ischemic heart disease. Circulation 92:1074–1078, 1995.
43. Malacoff RF, Lorell BH, Mudge GH, et al: Beneficial effects of nifedipine on regional myocardial blood flow in patients with coronary artery disease. Circulation 65(Suppl I):I-32–I-37, 1982.
44. Kugiyama K, Yasue H, Horio Y, et al: Effects of propranolol and nifedipine on exercise thallium-201 myocardial scintigraphy with quantitative rotational tomography. Circulation 2:374–380, 1986.
45. Bonaduce D, Muto P, Morgano G, et al: Effect of nifedipine on dipyridamole thallium-201 myocardial scintigraphy. Clin Cardiol 9:285–288, 1986.
46. Hayward R, Hunter GJS, Swanton H: Effects of nifedipine on coronary perfusion; recent assessment by parametric digital subtraction. Eur Heart J 8(Suppl K):1–7, 1987.
47. Nesto RW, White HD, Ganz P, et al: Addition of nifedipine to maximal beta blocker nitrate therapy: Effects on exercise capacity and global left ventricular performance at rest and during exercise. Am J Cardiol 55:3E–8E, 1985.
48. Parmley WW, Nesto RW, Singh BN, et al, for the N-CAP Study Group: Attenuation of the circadian patterns of myocardial ischemia with nifedipine GITS in patients with chronic stable angina. J Am Coll Cardiol 19:1380–1389, 1992.

49. Packer M, O'Connor CM, Ghali JK, et al, for the Prospective Randomized Amlodipine Survival Evaluation Study Group (PRAISE). Effect of amlodipine on morbidity and mortality in severe chronic heart failure. N Engl J Med 335:1107–1114, 1996.
50. Gong L, Zhang W, Zhu Y, et al: Shanghai Trial of Nifedipine in the Elderly (STONE). J Hypertens 14:1237–1245, 1996.
51. Cohn JN, Ziesche SM, Loss LE, Anderson GF, and the VHeFT Study Group: Effect of felodipine on short term exercise and neurohormone and long term mortality in heart failure: Results of VHeFT VIII. Circulation 92:I-143, 1995.
52. Poole-Wilson PA and the DEFIANT-II Research Group: Doppler flow and echocardiography in functional cardiac insufficiency: Assessment of nisoldipine therapy. Eur Heart J 18:31–40, 1996.
53. Staessen JA, Fagard R, Thijs L, and the Systolic Hypertension in Europe (Syst-Eur) Trial Investigators: Randomised double-blind comparison of placebo and active treatment for older patients with isolated systolic hypertension. Lancet 350:757–764, 1997.
54. Kloner RA, Vetrovec GW, Materson BJ, Levenstein M: Safety of long-acting dihydropyridine calcium channel blockers in hypertensive patients. Am J Cardiol 81:163–69, 1998.
55. Abascal VM, Larson MG, Evans JC, et al: Calcium antagonists and mortality risk in men and women with hypertension in the Framingham Heart Study. Arch Intern Med 158:1882–1886, 1998.
56. Estacio RO, Jeffers BW, Hiatt WR, et al: The effect of nisoldipine as compared with enalapril on cardiovascular outcomes in patients with non–insulin-dependent diabetes and hypertension. N Engl J Med 338:645–652, 1998.
57. Tatti P, Pahor M, Byington RP, et al: Outcome results of the Fosinopril versus Amlodipine Cardiovascular Events Trial (FACET) in patients with hypertension and NIDDM. Diabetes Care 21:597–602, 1998.
58. Poulter NR: Calcium antagonists and the diabetic patient: A response to recent controversies. Am J Cardiol 82(9B):40R–41R, 1998.
59. Parving HH: Calcium antagonists and cardiovascular risk in diabetes. Am J Cardiol 82(9B):42R–44R, 1998.
60. MacLeod JM, Lutale J, Marshall SM: Albumin excretion and vascular deaths in NIDDM. Diabetologia 38:610–616, 1995.
61. Gall M-A, Borch-Johnsen K, Hougaard P, et al: Albuminuria and ROOF glycemic control predicts mortality in NIDDM. Diabetes 44:1303–1309, 1995.
62. Dinneen SF, Gerstein HG: The association of microalbuminuria and mortality in non–insulin-dependent diabetes mellitus. Arch Intern Med 157:1413–1418, 1997.
63. Colwell JA: Aspirin therapy in diabetes. Diabetes Care 20:1767–1773, 1997.
64. Tatti P, Guarisco R, De Mauro P, et al: Reduced risk of major cardiac events with the association of two antihypertensive therapies in non–insulin-dependent diabetic population [abstract]. Diabetes 45:A198, 1996.
65. Sowers JR: Comorbidity of hypertension and diabetes: The Fosinopril versus Amlodipine Cardiovascular Events Trial (FACET). Am J Cardiol 82 (9B):15R–19R, 1998.
66. Califf RM, Granger CB: Hypertension and diabetes and the Fosinopril versus Amlodipine Cardiovascular Events Trial (FACET): More ammunition against surrogate end points. Diabetes Care 21:655–657, 1998.
67. Lash JP, Bakris GL: Effects of ACE inhibitors and calcium antagonists alone or combined on progression of diabetic nephropathy. Nephrol Dialysis Transpl 10(Suppl 9):56–62, 1995.
68. Epstein M, Bakris GL: Newer approaches to antihypertensive therapy: Use of fixed-dose combination therapy. Arch Intern Med 156:1969–1978, 1996.
69. Tuomilehto J, Rastenyte D, Thisj L, Staessen JA: Reduction of mortality and cardiovascular events in older diabetic patients with isolated systolic hypertension in Europe treated with nitrendipine-based antihypertensive therapy (Syst-Eur Trial) [abstract]. Diabetes 47:A54, 1998.
70. Hansson L, Zanchetti A, Carruthers SG, et al: Effects of intensive blood-pressure lowering and low-dose aspirin in patients with hypertension: Principal results of the Hypertension Optimal Treatment (HOT) randomised trial. HOT Study Group. Lancet 351:1755–1762, 1998.
71. Parving HH: Calcium channel blockers and cardiovascular risk in diabetes. Lancet 352:574, 1998.
72. Forette F, Seux M-L, Staessen JA, et al: Prevention of dementia in randomised double-blind placebo-controlled Systolic Hypertension in Europe (Syst-Eur) trial. Lancet 352:1347–1351, 1998.
73. Pahor M, Guralnik JM, Furberg CD, et al: Risk of gastrointestinal haemorrhage with calcium antagonists in hypertensive persons over 67 years old. Lancet 347:1061–1065, 1996.

74. Connor J, Sawczuk IS, Benson MC, et al: Calcium channel antagonists delay regression of androgen dependent tissues and suppress gene activity associated with cell death. Prostate 13:119–130, 1988.
75. Ray SD, Kamendulis LM, Gurule MW, et al: Ca^{2+} antagonists inhibit DNA fragmentation and toxic cell death induced by acetaminophen. FASEB J 7:453–463, 1993.
76. Carson DA, Ribeiro JM: Apoptosis and disease. Lancet 341:1251–1254, 1993.
77. Trump BF, Berezesky IK: Calcium mediated cell injury and cell death. FASEB J 9:219–228, 1995.
78. Martin SJ, Green DR: Apoptosis and cancer: the failure of controls on cell death and cell survival. Crit Rev Oncol Hematol 18:137–153, 1995.
79. Whitfield JF: Calcium signals and cancer. Crit Rev Oncol 3:55–90, 1992.
80. Garland C, Shekelle RB, Barrett Connor E, et al: Dietary vitamin D and calcium and risk of colorectal cancer: A 19-year prospective study in men. Lancet i:307–309, 1985.
81. Dargie HJ: Calcium channel blockers and the clinician. Lancet 348:488–489, 1996.
82. Zacharski LR, Moritz TE, Haakenson CM, et al: Chronic calcium antagonist use in carcinoma of the lung and colon: A retrospective cohort observational study. Cancer Invest 8:451–459, 1990.
83. MRC Working Party: Medical Research Council trial of treatment of hypertension in older adults: Principal results. BMJ 304:405–412, 1992.
84. Hole DJ, Hawthorne VM, McGhee SM, et al: Incidence of mortality from cancer in hypertensive patients. BMJ 306:609–611, 1993.
85. Taubes G: Epidemiology faces its limits. Science 269:164–169, 1995.
86. Rauwolfia derivatives and cancer. Lancet ii:701–702, 1974.
87. Boston Group, Program Boston Collaborative Drug Surveillance: Reserpine and breast cancer. Lancet ii:669–671, 1974.
88. Shapiro S, Parsells JL, Rosenberg L, et al: Risk of breast cancer in relation to the use of rauwolfia alkaloids. Eur J Clin Pharmacol 26:143–146, 1984.
89. Ad Hoc Subcommittee of the Liaison Committee of the World Health Organization and the International Society of Hypertension: Effects of calcium antagonists on the risks of coronary heart disease, cancer and bleeding. J Hypertens 15:105–115, 1997.
90. Burt VL, Whelton P, Roccella EJ, et al: Prevalence of hypertension in the US adult population. Results from the Third National Health and Nutrition Examination Survey, 1988–1991. Hypertension 25:305–313, 1995.
91. Frohlich ED: There is good news and not so good news [editorial]. Hypertension 25:303–304, 1995.
92. Hosie J, Wiklund I: Managing hypertension in general practice: Can we do better? J Hum Hypertens 9:S15–S18, 1995.
93. Neaton JD, Grimm RH Jr, Prineas RJ, et al: Treatment of Mild Hypertension Study. JAMA 270:713–724, 1993.
94. Elliott WJ: Drug therapy and "dropouts" in a tertiary hypertension clinic [abstract]. Am J Hypertens 7:26A, 1994.
95. UK Prospective Diabetes Study Group: Tight blood pressure control and risk of macrovascular and microvascular complications in type 2 diabetes: UKPDS 38. BMJ 317:703–713, 1998.
96. UK Prospective Diabetes Study Group: Efficacy of atenolol and captopril in reducing risk of macrovascular and microvascular complications in type 2 diabetes: UKPDS 39. BMJ 317:713–720, 1998.
97. Epstein M: Introduction. In Epstein M (ed): Symposium: Calcium Antagonists and the Diabetic Patient: Recent Controversies and Future Perspectives. Am J Cardiol 82(9B):1R–3R, 1998.
98. Messerli FH, Grossman E: The calcium antagonist controversy: A Posthumous commentary. Am J Cardiol 82(9B):35R–39R, 1998.
99. Messerli FH, Grossman E, Goldbourt U: Are beta blockers efficacious as first-line therapy for hypertension in the elderly? A systematic review. JAMA 279:1903–1907, 1998.
100. Pickering GA: Part I: Hypertension. Definitions, natural histories, and consequences. In Laragh JH, Brenner BM (eds): Hypertension: Pathophysiology, Diagnosis, and Management. Vol. 1, 2nd. ed. New York, Raven Press, 1995, pp 3–21.
101. Davis BR, Cutler JA, Gordon DJ, et al: Rationale and design for the Antihypertensive and Lipid Lowering Treatment to Prevent Heart Attack Trial (ALLHAT). Am J Hypertens 9(4 Pt 1):342–360, 1996.
102. Brown MJ, Castaigne A, Ruilope LM, et al: INSIGHT: International nifedipine GITS study intervention as a goal in hypertension treatment. J Hum Hypertens 10(Suppl 3):S157–S160, 1996.
103. Dahlöf B, Hansson L, Lindholm LH, et al: STOP Hypertension 2: A prospective intervention trial of "newer" versus "older" treatment alternatives in old patients with hypertension. Blood Pressure 2:136–141, 1993.

104. Savage S, Johnson Nagel N, Estacio RO, et al: The ABCD (Appropriate Blood Pressure Control in Diabetes) Trial. Online J Curr Clin Trials Nov 24 1993 (Doc. No. 104).
105. Tuomilehto J, Rastenyte D, Birkenhäger WH, et al: Effects of calcium channel blockade in older patients with diabetes and systolic hypertension. N Engl J Med 1999 [in press].
106. Miller A, Cropp A, Sussex B, et al: Safety of amlodipine in diabetic patients with advanced congestive heart failure: Results of the PRAISE trial [abstract]. J Cardiac Failure 4(Suppl 1):36, 1998.
107. Dahlöf B, Sever PS, Pulter NR, et al: The Anglo-Scandinavian Cardiac Outcomes Trial (ASCOT) [abstract]. Am J Hypertens 11:9A–10A, 1998.

Index

Page numbers in boldface type indicate complete chapters.